CHEMISTRY IN CONTEXT

GRAHAM HILL AND JOHN HOLMAN

FIFTH EDITION

Nelson

Contents

Preface

Aims and intentions of the book

This book is about chemistry and chemists and the contributions they make towards science, industry and society. It provides a modern, relevant course text for Advanced Subsidiary (AS) and Advanced GCE Chemistry. In planning this entirely new and extended edition, we have been particularly influenced by:

- the introduction of new specifications (syllabuses) and courses for AS and Advanced GCE from September 2000,
- the increasing contextualisation of chemistry,
- the views of students, teachers and awarding bodies (examination boards) who have found the previous editions of *Chemistry in Context* so helpful.

The book provides an excellent course text for all AS and A2 specifications, irrespective of the awarding body, as the reduction in the number of specifications has enabled us to cover virtually their whole content.

In a single concise volume, *Chemistry in Context* presents chemistry as a unified subject using physical principles as a basis for the inorganic and organic sections which follow. Our continued aim has been to show chemistry in its wider context as a relevant and developing science, which makes a vital contribution towards society, industry and civilisation.

Our approach in the book

The text draws on experimental evidence to develop key ideas and establish laws and theories. The book is intended to complement a course of practical work and consequently space is not devoted to instructions or suggestions for experiments. However, the companion volume *Chemistry in Context Laboratory Manual* would be an ideal source for such work.

Experiments described in the text should not be undertaken without a careful assessment of the risks involved and the safety precautions required.

Evidence from earlier editions indicates that the clear text and informative artwork make *Chemistry in Context* suitable for students of all abilities. To help those who are new to post-16 studies, each chapter is divided into fairly short sections, using figures, tables, photographs, flow diagrams and graphs to reinforce the text.

In addition, there are frequent **in-text questions**, which encourage the student to read more carefully and critically.

Each chapter concludes with a **summary** of important facts and ideas to enable students to review the contents of each chapter with ease.

Review questions at the end of each chapter (with full answers at the back of the book) provide interesting material for tests, class discussion, revision and homework.

A large bank of **assessment (examination) questions** has been added at the end of the book, giving further opportunities for private study and test (exam) preparation. These assessment questions are taken from Advanced GCE past papers and are, in general, of A2 standard. Those associated with AS content can be used as whole questions to develop A2 skills, and appropriate subsections of these past papers can be used for AS practice and test preparation.

Selected answers for the assessment questions are provided at the back of the book.

Detailed **marking schemes** for the questions and other information related to Chemistry in Context are available on the associated website at www.chemistryincontext.co.uk

As far as nomenclature, units and abbreviations are concerned, we have been guided by *Chemical Nomenclature, Symbols and Terminology for Use in School Science* (ASE, 3rd edition, 1985) and we have used the International System of Units (SI). Modern nomenclature has been used throughout the book. In general, organic chemicals are named systematically following rules of the International Union of Pure and Applied Chemistry (IUPAC), though trivial names of common substances such as ethanoic acid and phenylamine are sometimes bracketed after the systematic name.

Structure of the book

The first four chapters begin by revising and extending key ideas concerning amount of substance, reactions, equations, redox and periodicity. These chapters have been revised to ensure a smooth transition from GCSE or its equivalent to post-16 studies.

The next ten chapters (5–14) develop the major themes of structure, bonding, energetics, periodicity and competition which are then used throughout the following five chapters of inorganic chemistry (15–19). Here, emphasis is placed on patterns and on the use of the periodic table and the activity series to summarise information. Five chapters (20–24) on equilibria, entropy and reaction kinetics complete the development of principles and illustrate the importance of these factors in industrial processes. The book concludes with ten chapters (25–34) of organic chemistry. This is developed systematically in terms of functional groups, and the patterns of behaviour are related to structure, kinetics and energetics. Mechanistic explanations are introduced where appropriate.

Acknowledgements

Many friends and colleagues have influenced this and earlier editions of *Chemistry in Context*. Countless students at different schools have influenced our styles of teaching and the way in which topics are introduced and developed in the book. This edition includes a number of improvements suggested by users of the first four editions. We are especially grateful to Professor Bruce Gilbert and his colleagues at the University of York for their valuable comments on the manuscript.

To all these people, to our publishers, particularly Sonia Clark, Mark Lawrence and Jonathan Crowe, and to our wives Elizabeth and Wendy, we owe our sincere thanks.

Graham Hill
John Holman

May 2000

Graham C. Hill, MA(Cantab), FRSA Headmaster, Dr Challoner's Grammar School, Amersham

John S. Holman, MA(Cantab), FRSC Professor of Chemical Education, University of York

How to get the best from *Chemistry in Context*

To the student

From Graham Hill and John Holman

Chemistry gives you a special window on the world. As a chemist, you have insight into the world at an atomic and molecular level that helps you understand why the materials around us behave as they do. We want *Chemistry in Context* to give you this insight – and at the same time, we want the book to help you get the best possible grades.

We know from the many students who talk and write to us that *Chemistry in Context* is used in different ways. You may be using the book under the direction of your teacher, you may be using it on your own, or you may be doing a bit of both. However you are using the book, make sure you know how to get the most from it.

How the book is structured

We realise that getting started on AS and Advanced GCE courses can be a challenge. We have designed the book so that the early chapters build a bridge from pre-16 courses such as GCSE.

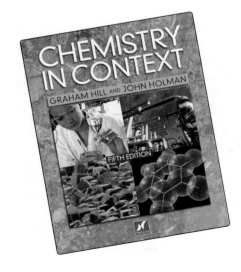

- Chapters 1 to 4 revise and extend key ideas concerning chemical calculations, equations, redox and the periodic table, to give you a smooth transition from GCSE.

- Chapters 5 to 14 develop the major principles of chemistry that are used throughout your chemistry course: structure, bonding, energetics, periodicity and competition.

- Chapters 15 to 19 apply these principles to inorganic chemistry, using patterns, the periodic table and the activity series to summarise information.

- Chapters 20 to 24 introduce further principles: equilibrium, entropy and reaction kinetics.

- Chapters 25 to 34 deal with organic chemistry, using the idea of functional groups and mechanisms to explain and summarise.

Finding your way around

The table of contents and the index are very detailed and will quickly help you to find the topic you are looking for.

Contents

Index

The text

The text provides clear explanations and information. We have divided it into short sections to make it manageable and we've used diagrams and tables to explain and summarise. Photographs show chemistry in its everyday context.

We use **bold text** to introduce key terms that are then defined and explained.

We have not included details of practical work in this book, but you will find our supporting *Laboratory Manual* very useful. See our website at www.chemistryincontext.co.uk for details.

---

31 Carbonyl Compounds

▲ Aldehydes and ketones contribute to the flavour of some fruit. The smell of ethanal is reminiscent of apples.

31.1 The carbonyl group

In chapter 27 we looked at compounds containing C═C double bonds and saw that their typical reactions involves electrophilic addition. In chapter 30 we looked at compounds containing the C—O single bond, which is polar and tends to bring about substitution reactions. We will now turn our attention to the **carbonyl** group, C═O, which is the functional group in aldehydes and ketones. We might expect the reactions of this group to show similarity to the reactions of both

The double bond between C and O in the carbonyl group, like the double bond in alkenes, can be considered to consist of a σ-bond and a π-bond. Unlike the C═C group, however, the carbonyl group does not have an even electron distribution between the two atoms. There is a greater electron density over the more electronegative oxygen atom, as shown in figure 31.1.

Figure 31.1
Electron distribution in the C═O bond.

This electron distribution makes the carbon atom attractive towards nucleophiles. Nucleophiles tend to attack and bond to this carbon, breaking the π-bond and resulting eventually in addition. With a general nucleophile X̄—Y, the overall reaction is

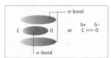

The exact mechanism depends on the nature of X and Y. In most cases, Y is hydrogen. This sort of reaction, which is typical of the carbonyl group, is called **nucleophilic addition**: compare it with the *electrophilic* addition which is typical of alkenes. The mechanism of the nucleophilic addition reaction involving HCN is described in section 31.3.

The reactions of the carbonyl group are important because the group is common in

---

The in-text questions

These questions help you to check your understanding as you read. Try to get in the habit of answering the in-text questions as you go along – you don't always need to write the answers down. You can often find the answers to these questions later on in the chapter.

lithium oxide. In figure 8.1, the nucleus of each atom is represented the electrons in each shell are represented by circled dots or crosses an Ions are shown in square brackets with the charge at the top right-han

1 What is the electron structure of: (i) the lithium ion (ii) the oxide i

2 Which noble gas has an electron structure like Li^+?

3 Which noble gas has an electron structure like O^{2-}?

4 Why is it that two lithium atoms react with only one oxygen atom?

Summaries

At the end of each chapter there's a summary of the main ideas and facts. You will find these particularly useful when you are revising.

...learn to use and conserve our fin power from the wind.

Summary

1 The transfer of energy to or from chemicals plays a crucial part in chemical processes in industry and in living things.

2 In an exothermic reaction, heat is lost from the reacting materials and $\Delta H_{reaction}$ is negative. In an endothermic reaction, heat is gained by the reacting materials and $\Delta H_{reaction}$ is positive.

3 If m g of a substance (specific heat capacity = c J g^{-1} K^{-1}) increase in temperature by ΔT K (ΔT °C), the enthalpy change, $\Delta H = + m \times c \times \Delta T$ J.

4 The standard enthalpy change of a reaction is the amount of heat absorbed or evolved when the molar quantities of reactants as stated in the equation react

10 A substanc decompose

11 The enthal provides a compound the compo

12 It is impor and kinetic are energeti But, they d pressures b slow. They

---

Review questions

Each chapter has a full set of questions at the end. You can use these to check and reinforce your understanding of the work. We have provided full answers at the back of the book – though we are sure you won't want to look at these until you have completed the questions!

Review questions

1 (a) How do each of the following properties of the elements in Group II change with increasing atomic number?
(i) Atomic radius
(ii) Ionisation energy
(iii) Strength as reducing agents
(iv) Vigour of reaction with chlorine
(v) Electropositivity
(b) In each case, explain why the five properties change in

potent
(e) It is e
magne
magne
(f) Group
the co
anhydr

3 The eleme
barium (B

Assessment questions

There is a large bank of assessment questions from recent past examination papers at the end of the book. You will find these questions especially useful when you are revising and practising for tests and the exams themselves. There are some selected answers at the end of the book and you can find full answers and marking schemes on our website at www.chemistryincontext.co.uk

25 (a) (i) Using the data provided, construct a Born-Haber cycle for magnesium chloride, $MgCl_2$, and from it determine the electron affinity of chlorine.

	ΔH/kJ mol^{-1}
Enthalpy of atomisation of chlorine	+122
Enthalpy of atomisation of magnesium	+148
First ionisation energy of magnesium	+738
Second ionisation energy of magnesium	+1451
Lattice enthalpy of magnesium chloride	−2526
Enthalpy of formation of magnesium chloride	−641

(5)

(ii) The theoretically calculated value for lattice

26 Potassium ch
the equation

$4KClO_3(s) \rightarrow$

The standard
the reaction i

KClO
KClO
KCl(s

Atoms, Atomic Masses and Moles

1

1.1 Atoms and molecules

The first chemist to use the name '**atom**' was John Dalton (1766–1844). Dalton used the word 'atom' to mean the smallest particle of an element. He then went on to explain how atoms could react together to form **molecules**, which he called 'compound atoms'.

▶ John Dalton collecting 'marsh gas' (mainly methane) from the rotting vegetation at the bottom of a pond. Dalton was born in 1766 in the remote village of Eaglesfield in Cumbria. He was the son of a handloom weaver. For most of his life, Dalton lived and worked in Manchester at what was then the Presbyterian College.

▶▶ Dalton's symbols for the elements. In 1803, Dalton published his Atomic Theory. He suggested that all matter was composed of small particles which he called 'atoms'. Later, Dalton went on to suggest symbols for the atoms of different elements and these are shown in the photograph. Azote, the second element in Dalton's list, is now called nitrogen. Which substances in Dalton's list are not elements, but compounds?

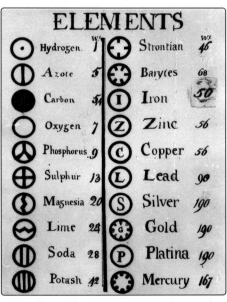

Don't confuse the terms atom and molecule.

> *An atom is the smallest part of an element which can ever exist.*
> *A molecule is the smallest part of an element or a compound which can exist alone under ordinary conditions.*

For example, chlorine consists of particles of Cl_2 under ordinary conditions, but at very high temperatures these split up to form particles of Cl. So, molecules of chlorine are written as Cl_2 and atoms of chlorine are written as Cl.

1 Name an element (other than chlorine) whose molecules consist of two atoms under normal conditions.
2 Name an element whose molecules consist of one atom under normal conditions.
3 How many atoms are there in one molecule of sugar (sucrose), $C_{12}H_{22}O_{11}$?
4 How many different atoms are there in one molecule of sugar?

Most atoms have a radius of about 10^{-10} m.

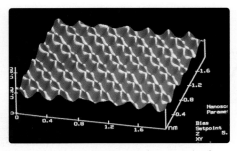

▲ An electron microscope photograph of the surface of graphite showing the regular pattern of carbon atoms.

Nowadays, the unit used in measuring atomic distances is usually the nanometre (nm)

$$1 \text{ m} = 10^9 \text{ nanometres}$$
$$10^{-9} \text{ m} = 1 \text{ nm}$$
$$\Rightarrow 10^{-10} \text{ m} = 0.1 \text{ nm}$$

so, the radii of atoms are about 0.1 nm.

Atoms, of course, are far too small to be seen even with the most powerful light microscope. However, chemists have discovered that it is possible to identify single atoms of certain elements using electron microscopes. In 1958, scientists in the USSR reported that they could pick out atoms of barium (which have a diameter of about 0.4 nm). They used an electron microscope with a magnification of 2 000 000. In 1967, Japanese scientists demonstrated their identification of particles as small as 0.1 nm.

1.2 Comparing the masses of atoms

Individual atoms are too small to be weighed. Nevertheless, early in the nineteenth century Dalton suggested that it should be possible to *compare* the masses of atoms even though the atoms themselves could not be weighed. During the nineteenth century various methods were devised for obtaining approximate values for the relative masses of different atoms.

Then, in 1919, Aston invented the mass spectrometer. This gave chemists a reliable and accurate method of comparing the relative masses of atoms. At one time, the relative masses of atoms were known as atomic weights, but nowadays we refer to them as **relative atomic masses.**

The basic idea of a mass spectrometer can be demonstrated using the apparatus in figure 1.1. Wooden balls of *different* sizes, but with *identical* iron cores, roll down a sloping plane. At the bottom of the slope, a powerful magnet attracts the iron cores and the moving balls are deflected.

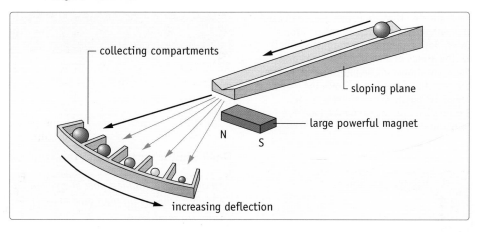

Figure 1.1
A simple model to illustrate the working of a mass spectrometer.

1 Why does the magnet have the same attraction for all the balls?
2 Which size of ball will be deflected the most? Why?

As the balls have identical iron cores, they are all attracted equally by the magnet. But, the smaller balls are lighter and therefore they are deflected the most. The balls collect in different compartments depending on their mass. All balls of the same mass collect in the same compartment. Using this simple apparatus, it is possible to separate the different-sized balls according to their mass and to find the relative numbers of each present.

A real mass spectrometer works in a similar fashion to this simple model. It separates atoms according to their mass and shows the relative numbers of the different atoms present. Before the atoms can be deflected and separated, they must be converted to positively charged ions.

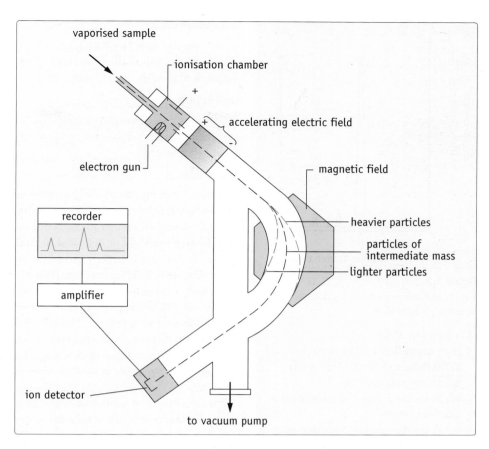

Figure 1.2
A diagram of a mass spectrometer.

Figure 1.2 shows a simple mass spectrometer. There are five main stages.

1 Vaporisation – the sample of element is vaporised.
2 Ionisation – positive ions are obtained from the vapour.
3 Acceleration – the positive ions are accelerated by an electric field.
4 Deflection – the positive ions are deflected by a magnetic field.
5 Detection – the ions are detected and a record is made.

Vaporisation

Gases, liquids and volatile solids are injected into the instrument just before the ionisation chamber. Less volatile solids must be pre-heated to vaporise them.

Ionisation

After vaporisation, the element passes into the ionisation chamber. Here, atoms of the element are bombarded with a stream of high-energy electrons. This causes ionisation. One or occasionally two electrons are knocked out of the atoms leaving positive ions. Thus, for an atom X we have:

$$e^- \quad + \quad X \quad \rightarrow \quad X^+ \quad + \quad e^- \quad + \quad e^-$$

| high-energy electron | atom | positive ion | electron knocked out of X | high-energy electron retreating |

or occasionally

$$e^- \quad + \quad X \quad \rightarrow \quad X^{2+} \quad + \quad e^- \quad + \quad e^- \quad + \quad e^-$$

2 electrons
knocked
out of X

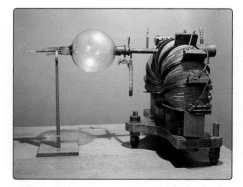

▲ Aston's mass spectrometer. A vaporised sample of the element was bombarded by electrons in the large glass bulb on the left. The ions produced were then accelerated by an electric field towards the magnetic field on the right (produced by the many coils of the electromagnet).

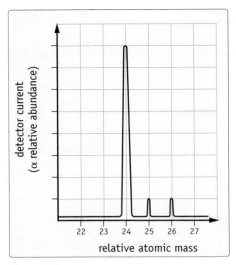

Figure 1.3
A mass spectrometer trace for naturally occurring magnesium.

Look closely at figure 1.3
1 How many different ions are detected in the mass spectrum of naturally occurring magnesium?
2 What are the relative masses of these different ions?
3 What are the relative proportions of these different ions?

Acceleration

These positive ions, such as X^+ and X^{2+}, now pass through holes in parallel plates to which an electric field is applied. The electric field accelerates the ions into the instrument towards the magnetic field.

Deflection

As the ions pass through the magnetic field, they are deflected according to their mass and their charge. Notice that particles can only pass through the instrument if they are positively charged.

Detection

If the accelerating electric field and the deflecting magnetic field stay constant, ions of only one particular mass/charge ratio will hit the ion detector at the end of the apparatus. Ions of smaller mass/charge ratio will be deflected too much. Ions of greater mass/charge ratio will be deflected too little.

The ion detector is usually linked through an amplifier to a recorder. As the strength of the magnetic field is slowly increased, ions of increasing mass will be detected and a mass spectrum similar to that shown in figure 1.3 is traced out by the recorder. The relative heights of the peaks in the mass spectrum give a measure of the relative amounts of the different ions present. (Strictly speaking, it is the areas under the peaks and not the peak heights which give the relative amounts or abundances.)

In practice, a reference peak using a known substance is first obtained on the mass spectrum. The relative masses of other particles can then be obtained by comparison with this.

The calculation of relative atomic masses and the interpretation of mass spectrometer traces are taken up again in sections 5.5 and 25.5.

1.3 Relative atomic masses – the relative atomic mass scale

Chemists use a **relative atomic mass scale** to compare the masses of different atoms. At first the element hydrogen was chosen as the standard against which the masses of other atoms were compared. Hydrogen was chosen initially because chemists realised that it had the smallest atoms and these could be assigned a relative atomic mass of one (H = 1). At a later date, when relative atomic masses could be obtained more accurately and when scientists realised that one element could contain atoms of different mass (**isotopes** – see section 5.4), it became necessary to choose a single isotope as the reference standard for relative atomic masses.

In 1961, carbon-12 (^{12}C) was chosen as the new standard. This is the isotope of carbon with a relative atomic mass of 12. An isotope of carbon was chosen rather than an isotope of hydrogen because carbon is a very common element. Being a solid, it is also easier to store and transport than hydrogen, which is a gas.

On the carbon-12 scale, atoms of the isotope ^{12}C are assigned a relative atomic mass of 12. The relative masses of all other atoms are obtained by comparison with the mass of a carbon-12 atom. A few relative atomic masses are listed in table 1.1. For example, the relative atomic mass of magnesium, on the carbon-12 scale, is 24.312. This means that the average mass of a magnesium atom and the mass of a carbon-12 atom are in the ratio 24.312 : 12.000. Notice in table 1.1 that the relative atomic mass of carbon is 12.011, i.e. the average mass of a carbon atom is 12.011, not 12.000. This is because naturally occurring carbon contains a few atoms of carbon-13 and carbon-14 mixed in with those of carbon-12.

Use table 1.1 to answer the following questions.
1 Roughly how many times heavier are magnesium atoms than ^{12}C atoms?
2 Approximately how many times heavier are carbon atoms than hydrogen atoms?
3 Which element has atoms approximately twice as heavy as sulphur atoms?

A complete list of relative atomic masses can be found on page 535.

Table 1.1
The relative atomic masses of some elements

Element	Symbol	Relative atomic mass
Carbon-12	^{12}C	12.000
Carbon	C	12.011
Chlorine	Cl	35.453
Copper	Cu	63.540
Hydrogen	H	1.008
Iron	Fe	55.847
Magnesium	Mg	24.312
Sulphur	S	32.064

1.4　The Avogadro constant, moles and molar masses

Since one atom of carbon is 12 times as heavy as one atom of hydrogen, it follows that 12 grams of carbon and one gram of hydrogen contain the same number of atoms. In the same way, since an atom of sulphur is 32 times as heavy as an atom of hydrogen, it follows that 32 grams of sulphur will also contain the same number of atoms as one gram of hydrogen. In fact, *the relative atomic mass in grams of all elements will contain the same number of atoms*. Experiments show that this number is 6.0×10^{23}. Written out in full, this is 600 000 000 000 000 000 000 000. It is called the **Avogadro constant**.

> The Avogadro constant, represented by the symbol L, *is defined as the number of atoms in exactly 12 grams of* ^{12}C.

Since 12 g of carbon contains 6.0×10^{23} atoms,

1 g of carbon contains $\dfrac{6.0 \times 10^{23}}{12}$ atoms

5 g of carbon contains $\dfrac{5 \times 6.0 \times 10^{23}}{12}$ atoms

In this way, it is easy for chemists to count the number of atoms in a sample of an element by weighing. Chemists are not the only people who 'count by weighing'. Bank clerks use the same idea when they count money by weighing it. Since 100 pennies weigh 356 grams, it is quicker to take one hundred 1p coins by weighing out 356 grams of them than by counting.

As the relative atomic mass in grams of all elements contains 6.0×10^{23} atoms, chemists refer to 6.0×10^{23} atoms of an element as **one mole** of atoms. The mole (symbol 'mol') is the basic unit for measuring amounts of substances in the same way that the metre (symbol 'm') is the basic unit for measuring distances and the kilogram (symbol 'kg') is the basic unit for measuring masses.

Strictly speaking, the mole is defined as the amount of substance which contains the same number of particles (atoms, ions or molecules) as there are atoms in exactly 12 grams of ^{12}C. The mass of one mole of a substance is called the **molar mass**. The symbol for the molar mass of an element is A and that for relative atomic mass is A_r. The unit for molar mass is g/mol or g mol^{-1}. Thus, one mole of magnesium is 24.312 g, the molar mass of magnesium is 24.312 g mol^{-1} ($A(\text{Mg}) = 24.312$ g mol^{-1}) and the relative atomic mass of magnesium is 24.312 ($A_r(\text{Mg}) = 24.312$). The relative atomic mass of magnesium is often simply written as Mg = 24.312. Notice that the term 'mol' is the symbol for mole. It is *not* an abbreviation for 'molecule' or 'molecular'.

In equations, the symbol for an element can be used to represent one mole of the element as well as one particle of the element. Thus, C represents one mole of carbon atoms (12 g of carbon) and $\frac{1}{10}$C represents one-tenth of a mole of carbon atoms (1.2 g of carbon). O represents one mole of oxygen atoms (16 g of oxygen) but O_2 represents one mole of oxygen molecules (32 g of oxygen).

Notice how important it is to specify precisely which particles you mean in discussing the number of moles of different substances. For example, the statement 'one mole of oxygen' is ambiguous. It could mean one mole of oxygen atoms (O), i.e. 16 g; or it could mean one mole of oxygen molecules (O_2), i.e. 32 g. Because of this ambiguity, it is important to state the formula of the substance involved.

In considering giant structures such as sodium chloride, Na^+Cl^-, it is again important to quote the formula in discussing amounts of substance. Thus one mole of sodium chloride, Na^+Cl^-, will have a mass equal to the formula mass in grams, i.e. 58.5 g. It contains one mole of sodium ions, Na^+ (23.0 g), and one mole of chloride ions, Cl^- (35.5 g). The molar mass of sodium chloride is 58.5 g mol^{-1}. The symbol for the molar mass of a compound is M and that for relative molar mass or relative formula mass is M_r. So, we can write, $M(\text{NaCl}) = 58.5$ g mol^{-1} or $M_r(\text{NaCl}) = 58.5$. On the other hand, one mole of methane (CH_4) is 16.043 g, containing one mole of carbon, C (12.011 g) and four moles of hydrogen, H (4×1.008 g).

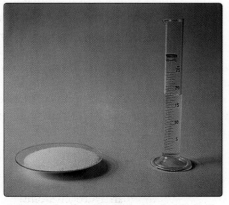

▲ One mole of sodium chloride (salt), NaCl (58.5 g) and one mole of water, H_2O (18.0 g).

Use the relative atomic masses in table 1.1 to answer the following questions.

1. What is the mass of 6.0×10^{23} atoms of copper?

2. What is the mass of 6.0×10^{23} atoms of magnesium?

3. How many atoms are there in 6 g of carbon?

4. How heavy is one atom of carbon?

5. What is the mass of 1 mole of hydrogen molecules, H_2?

6. What is the mass of 0.5 moles of SCl_2?

7. What is the molar mass of CCl_4?

8. How many atoms of S are there in 4.008 g of S?

The answers to the above questions are:

1. 63.54 g

2. 24.312 g

3. 3×10^{23}

4. $\dfrac{12}{6 \times 10^{23}} = 2 \times 10^{-23} g$

5. $2 \times 1.008 = 2.016 g$

6. $0.5 \times (32.064 + 2 \times 35.453) = 51.485 g$

7. $12.000 + 4 \times 35.453 = 153.812 \, g \, mol^{-1}$

8. $\dfrac{4.008}{32.064} = \frac{1}{8}$ mole, $\frac{1}{8} \times 6 \times 10^{23} = \frac{3}{4} \times 10^{23}$ atoms

Check that you have understood this section by answering the questions in the margin.

From the discussion above, you will see that it is more sensible for chemists to consider amounts which contain the same number of particles (atoms, ions, molecules or formal units) rather than amounts which contain the same number of grams. Thus, in making one mole of iron(II) sulphide (88 g of FeS) from iron and sulphur, it is necessary to weigh out 56 grams of Fe and 32 grams of S (and not 44 g of Fe and 44 g of S).

1.5 Empirical formulas and molecular formulas

The formula of a compound shows the relative number of atoms of each element in it.

The formula can be calculated once we know the actual masses and relative atomic masses of the different elements in a sample of the compound.

The following example shows how the formula of a compound can be obtained after its composition has been determined by experiment.

When ethene is analysed, it is found to contain 85.72% carbon and 14.28% hydrogen.

	C	H
masses combined	85.72 g	14.28 g
mass of 1 mole	12 g	1 g
$\therefore$ moles present	$\dfrac{85.72}{12}$	$\dfrac{14.28}{1}$
	= 7.14	= 14.28
ratio of moles	1	2
$\therefore$ ratio of atoms	1	2
$\Rightarrow$ formula	CH_2	

This formula for ethene shows only the simplest ratio of carbon atoms to hydrogen atoms. The actual formula showing the correct number of carbon atoms and hydrogen atoms in one molecule of ethene could be CH_2, C_2H_4, C_3H_6, C_4H_8, etc., since all these formulas give CH_2 as the simplest ratio of atoms.

The simplest formula for a substance, such as CH_2 for ethene, is called the **empirical formula.** This shows the simplest whole-number ratio for the atoms of different elements in the compound.

In the case of molecular compounds such as ethene, it is generally more helpful to use the **molecular formula.** This shows the actual number of atoms of the different elements in one molecule of the compound. The molecular formula for ethene is not CH_2 but C_2H_4.

In substances composed of giant lattices, whether these are giant covalent structures, such as silicon(IV) oxide, SiO_2, or giant ionic structures, such as sodium chloride, Na^+Cl^-, it is of course meaningless to talk of molecules and molecular formulas. In this case, the formulas we use are empirical formulas. Experimental methods of determining the structures and formulas of compounds are described in sections 10.2, 10.3, 11.5, 11.8 and 25.4.

1.6 Solutions and moles

As many reactions only take place in solution, it is useful for chemists to know the concentrations of these solutions. Chemists usually measure the concentration of solutions as the number of **moles of solute per cubic decimetre of solution**. The cubic decimetre (dm^3) is now the agreed international name for the unit of volume also known as the litre. In this book, we shall use cubic decimetres (not litres) and cubic centimetres, cm^3 (not millilitres, ml).

$$1 \text{ decimetre (dm)} = \tfrac{1}{10} \text{ metre} = 10 \text{ cm}$$

$$\therefore 1 \, dm^3 = (10 \text{ cm})^3 = 1000 \, cm^3 = 1 \text{ litre}$$

A solution of HCl(aq) containing 1.0 mole per cubic decimetre will have 36.5 g of HCl in 1 dm^3 of solution. A solution with 2.0 moles of HCl per cubic decimetre contains 73 g (2 × 36.5 g) in 1 dm^3 of solution.

Notice that these concentrations are expressed as the number of moles in 1 dm^3 of *solution*, not 1 dm^3 of *solvent* (figure 1.4)

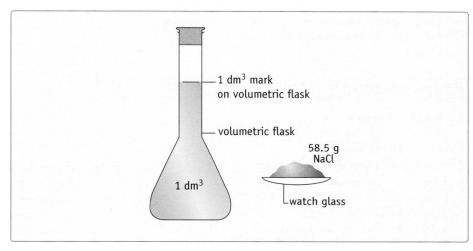

Figure 1.4
Preparing 1 dm^3 of 1.0 M sodium chloride.

Weigh out 1 mole of NaCl (58.5 g).
Dissolve this solid in about 500 cm^3 of distilled water in a beaker.
Add this solution to the 1 dm^3 volumetric flask plus washings from the beaker.
Add more distilled water to the solution up to the 1 dm^3 mark on the neck of the flask.
Finally, mix the solution thoroughly.
Can you see that **1 dm^3 of 1.0 M NaCl will contain less than 1 dm^3 of distilled water?**

A solution of HCl(aq) containing 1.0 mole per cubic decimetre is usually written as:

$$[HCl] = 1.0 \, mol \, dm^{-3}, \text{ or}$$
$$[HCl] = 1.0 \, M.$$

Notice that square brackets are used to indicate the substance whose concentration is given. The use of square brackets to represent the concentration of substances is taken up more fully in chapter 20.

Since HCl(aq) is fully dissociated into H^+(aq) and Cl^-(aq) ions, and if

$$[HCl] = 1.0 \, mol \, dm^{-3},$$

we can deduce that:

$$[H^+] = 1.0 \, mol \, dm^{-3} \text{ and } [Cl^-] = 1.0 \, mol \, dm^{-3}.$$

The amounts of solute needed to make solutions of different concentrations can be worked out by simple proportion. For example:

1 dm^3 of 1.0 M solution requires 1 mol of solute

250 cm^3 of 1.0 M solution requires $1 \times \dfrac{250}{1000}$ mol of solute

250 cm^3 of 0.1 M solution requires $1 \times \dfrac{250}{1000} \times 0.1$ mol of solute.

Summary

1 An atom is the smallest part of an element which can ever exist.

2 A molecule is the smallest part of an element or a compound which can exist alone under ordinary conditions.

3 Most atoms have a radius between 0.07 and 0.20 nanometres(nm).
   1 nanometre $\equiv 10^{-9}$ m.

4 Chemists use the relative atomic mass scale to compare the masses of different atoms. Atoms of the isotope ^{12}C are assigned a relative atomic mass of 12 and the relative masses of all other atoms are obtained by comparison with the mass of a carbon-12 atom.

5 The relative atomic mass in grams (molar mass) of any element contains 6×10^{23} atoms.

6 The Avogadro constant $(6 \times 10^{23} \text{ mol}^{-1})$ is defined as the number of atoms in exactly 12 g of ^{12}C.

7 The mole (symbol 'mol') is the basic unit for measuring amounts of substances.

8 The mole is the amount of substance which contains the same number of particles (atoms, ions, molecules or formula units) as there are atoms in exactly 12 g of ^{12}C.

9 An empirical formula shows the simplest whole-number ratio for the atoms of different elements in a compound.

10 A molecular formula shows the actual number of atoms of the different elements in one molecule of a compound.

11 The concentrations of solutes in solutions are usually expressed in moles per cubic decimetre (mol dm^{-3}). M is sometimes used as an abbreviation for mol dm^{-3}.

Review questions

1 What is the mass of
(a) 6×10^{23} atoms of O,
(b) 6×10^{23} atoms of P,
(c) 6×10^{23} molecules of O_2,
(d) 6×10^{23} molecules of P_4,
(e) one mole of carbon dioxide, CO_2,
(f) two moles of silver, Ag,
(g) 0.2 moles of sulphur dioxide, SO_2,
(h) NaOH in 2 dm^3 of 1.5 mol dm^{-3} solution?

2 How many moles of
(a) Cl_2 are there in 7.1 g of chlorine,
(b) $CaCO_3$ are there in 10.0 g of calcium carbonate,
(c) Ag are there in 10.8 g of silver,
(d) NH_3 are there in 3.4 g of ammonia,
(e) S are there in 32 g of sulphur,
(f) S_8 are there in 32 g of sulphur,
(g) NaOH are there in 1 dm^3 of 3.0 mol dm^{-3} solution,
(h) NaOH are there in 20 cm^3 of 0.1 mol dm^{-3} solution?

3 How many atoms are there in
(a) two moles of iron, Fe,
(b) 0.1 moles of sulphur, S,
(c) 18 g of water, H_2O,
(d) 0.44 g of carbon dioxide, CO_2?

4 Read section 1.2 again.
(a) Why must a mass spectrometer be evacuated to a very low pressure before being used?
(b) How would the accelerating field differ in its effect on X^+ and X^{2+}?
(c) How would the deflecting magnetic field differ in its effect on X^+ and X^{2+}?

(d) Relative atomic masses are not really masses. What are they?
(e) What units do relative atomic masses have?

5 15.3 g of element X ($A_r = 27$) will combine with 13.6 g of oxygen to form an oxide.
(a) Express this in moles.
(b) How many moles of oxygen, O, will combine with one mole of X?
(c) What is the simplest formula for the oxide?

6 5.34 g of a salt of formula M_2SO_4 (where M is a metal) were dissolved in water. The sulphate ion was precipitated by adding excess barium chloride solution when 4.66 g of barium sulphate ($BaSO_4$) were obtained.
(a) How many moles of sulphate ion were precipitated as barium sulphate?
(b) How many moles of M_2SO_4 were in the solution?
(c) What is the formula mass of M_2SO_4?
(d) What is the relative atomic mass of M?
(e) Use a table of relative atomic masses to identify M.

7 2.4 g of a compound of carbon, hydrogen and oxygen gave, on combustion, 3.52 g of CO_2 and 1.44g of H_2O. The relative molecular mass of the compound was found to be 60.
(a) What are the masses of carbon, hydrogen and oxygen in 2.4 g of the compound?
(b) What are the empirical and molecular formulas of the compound?

8 Figure 1.5(a) shows the mass spectrum of HCl.
The peak at mass 36 corresponds to the molecular ion
$(H^{35}Cl)^+$. Assume that chlorine has only two isotopes,
$^{35}_{17}Cl$ and $^{37}_{17}Cl$.

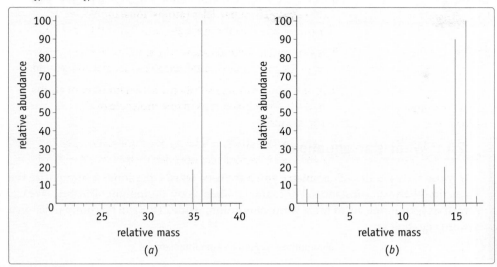

Figure 1.5
Mass spectra for (a) hydrogen chloride
and (b) methane.

(a) What particle is responsible for the prominent peak at
mass 38?

(b) What particles are responsible for the two peaks
showing lower abundance?

(c) How do you explain the relative heights of the peaks at
mass 36 and 38?

Figure 1.5(b) shows the mass spectrum of methane.
The peak at mass 16 corresponds to the molecular ion
$(CH_4)^+$.

(d) How do you explain the peaks of relative mass 1, 2,
12, 13, 14, 15 and 17?

2 Reacting Quantities and Equations

▲ Formulas and equations help us to calculate how much product we can get from a given amount of reactant. This is important in processes such as the manufacture of lime (calcium oxide) from limestone (calcium carbonate). This was once carried out in kilns like the one shown above.

▲ A drawing of the original balance with which John Dalton studied the reacting quantities of different substances.

2.1   What is an equation?

In your early studies of chemistry you have used word equations to summarise the reactions between substances. For example, when shiny aluminium tarnishes, it reacts with oxygen in the air to form white aluminium oxide. This can be summarised in a word equation as:

aluminium + oxygen → aluminium oxide

Although word equations are useful, it is more convenient to use the formula of a substance in equations. For example, in the word equation above we could write aluminium as Al, oxygen as O_2 and aluminium oxide as Al_2O_3. The formula Al_2O_3 can be used to represent 1 mole of aluminium oxide or 1 particle of Al_2O_3. In the same way, H_2O can be used to represent 1 mole of water or 1 molecule of water.

When iron (Fe) reacts with sulphur (S), experiments show that the product is iron(II) sulphide (FeS). This can be summarised in a word equation as

iron + sulphur → iron(II) sulphide

or more simply using symbols and formulas as

Fe + S → FeS

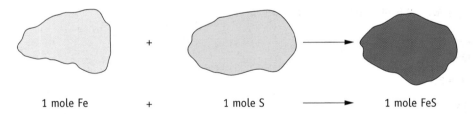

| 1 mole Fe | + | 1 mole S | ⟶ | 1 mole FeS |

2.2   Relative numbers of moles of reactants and products

It is helpful to summarise every reaction with a balanced equation. However, equations cannot be written simply by looking at the formulas of the reactants.

In order to write an equation, we must know:

• the formulas of the reactants and products;
• how many moles of each reactant are involved;
• how many moles of each product are formed.

Various methods can be used to find the number of moles of substances which react. In acid/base reactions, we can use indicators to decide how much acid reacts with a given amount of base.

In precipitation reactions, we can investigate the volumes of solutions which react by adding increasing volumes of one liquid to a fixed volume of the other liquid until no more precipitate forms.

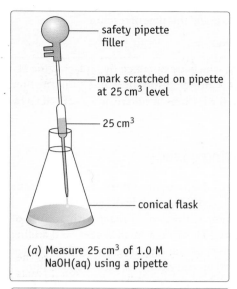

(a) Measure 25 cm³ of 1.0 M NaOH(aq) using a pipette

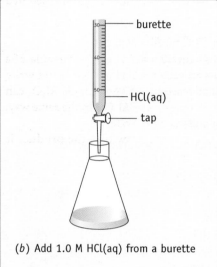

(b) Add 1.0 M HCl(aq) from a burette

Figure 2.1

Table 2.1

Volume of 1.0 M HCl(aq) added/cm³	Colour of indicator
0	Purple
5	Purple/blue
10	Blue
15	Blue
20	Blue
21	Blue
22	Blue
23	Blue
24	Blue/green
25	Yellow/green
26	Pink
27	Pink
28	Pink
29	Pink
30	Pink

For reactions in which heat is evolved (or absorbed) we can find how much of one substance reacts with another by measuring temperature changes.

For reactions involving gases we can measure pressure or volume changes.

2.3   Equations for reactions involving acids and bases

When acids react with bases, the pH of the solution changes. We can use indicators which show the pH change to find how much base (e.g. sodium hydroxide solution) reacts with the acid (e.g. hydrochloric acid).

25 cm³ of sodium hydroxide solution containing 1.0 mol dm⁻³ (i.e. 1.0 M NaOH(aq)) were pipetted into a conical flask (figure 2.1(a)). 5 drops of universal indicator were added and the colour was recorded.

5 cm³ of 1.0 M hydrochloric acid were then added from a burette (figure 2.1(b)), thoroughly mixed and the colour recorded again. The colour was also recorded when 10, 15 and 20 cm³ of the hydrochloric acid had been added. Separate 1 cm³ portions of the hydrochloric acid were then added, recording the colour after each 1 cm³ until a total of 30 cm³ 1.0 M HCl had been added.

The results of the experiment are shown in table 2.1.

Notice that the indicator shows a neutral pH colour (yellow/green) when 25 cm³ of hydrochloric acid have been added.

So, 25 cm³ 1.0 M HCl just react with (neutralise) 25 cm³ 1.0 M NaOH.

> This method of adding one solution from a burette to another solution in order to find out how much of the two solutions will just react with each other is called a **titration**. When the two solutions just react and neither is in excess, we have found the **neutral point** or **end point** of the titration.

From the results of this experiment:

25 cm³ 1.0 M HCl just react with 25 cm³ 1.0 M NaOH

Now 1000 cm³ of a 1.0 M solution contain 1 mole, so

25 cm³ of 1.0 M HCl contain $\frac{25}{1000} \times 1 = 0.025$ moles of HCl and

25 cm³ of 1.0 M NaOH contain $\frac{25}{1000} \times 1 = 0.025$ moles of NaOH

∴ 0.025 moles of HCl react with 0.025 moles of NaOH

⇒ 1 mole of HCl reacts with 1 mole of NaOH

We can, therefore, write the left hand side of the equation as:

$$HCl(aq) + NaOH(aq) \rightarrow$$

If all the water from the end-point solution is evaporated, then the only product which remains is sodium chloride, NaCl. The complete equation is therefore:

$$HCl(aq) + NaOH(aq) \rightarrow NaCl(aq) + H_2O(l)$$

In this reaction, the acid is **neutralised** by the base and this type of reaction –

acid + base → salt + water is called **neutralisation**.

Substances, such as sodium chloride (NaCl), which are obtained by replacing the H^+ ions in acids by metal ions, are called **salts**. Copper sulphate ($CuSO_4$) is another example of a salt. This is obtained by replacing the H^+ ions in sulphuric acid (H_2SO_4) by Cu^{2+} ions.

2.4 Equations for reactions involving gases

When a small piece of magnesium is added to dilute hydrochloric acid a vigorous effervescence occurs, showing that a gas is produced. This gas is hydrogen, H_2.

We can investigate how many moles of hydrochloric acid react with 1 mole of magnesium by pipetting 25 cm^3 of 1.0 mol dm^{-3} HCl into a small beaker, and adding 0.4 g of magnesium ribbon. A vigorous reaction occurs and hydrogen is evolved. When the effervescence stops, all the acid has been used up and excess unreacted magnesium is left. The unreacted magnesium is washed, dried and weighed. Here are some typical results:

$$\text{volume of 1.0 mol } dm^{-3} \text{ HCl taken} = 25.0 \text{ cm}^3$$
$$\text{mass of magnesium before reaction} = 0.41 \text{ g}$$
$$\text{mass of magnesium unreacted} = 0.11 \text{ g}$$
$$\text{fi mass of magnesium reacting} = 0.30 \text{ g}$$

∴ 25 cm^3 of 1.0 mol dm^{-3} HCl reacts with 0.30 g of Mg

25 cm^3 of 1.0 M HCl contains $\dfrac{25}{1000} \times 1.0$ moles HCl $\qquad$ = 0.025 moles HCl

0.30 g of Mg = $\dfrac{0.30}{24}$ moles Mg $\qquad$ = 0.0125 moles Mg

∴ 0.0125 moles of Mg react with 0.025 moles of HCl
⇒ 1 mole of Mg reacts with 2 moles of HCl
So, the left-hand side of the equation for the reaction is

$$Mg(s) + 2HCl(aq) \rightarrow$$

The next question to ask is 'how many moles of hydrogen are produced from 1 mole of magnesium?' We can find the answer to this question by adding 0.03 g of magnesium to about 5 cm^3 (an excess) of 2 mol dm^{-3} HCl. The volume of hydrogen produced can be measured by connecting the tube in which the reaction takes place to a syringe (figure 2.2). Here are some typical results:

$$\text{mass of magnesium taken} = 0.030 \text{ g}$$

$$\text{volume of hydrogen produced} = 30 \text{ cm}^3$$

1. How many moles of magnesium are there in 0.030 g?
2. What is the mass of 30 cm^3 of hydrogen? (1 dm^3 (1000 cm^3) at room temperature has a mass of 0.084 g.)
3. How many moles of $H_2(g)$ are there in this mass of hydrogen?
4. How many moles of $H_2(g)$ are produced from 1 mole of magnesium?

The results of this experiment show that 1 mole of magnesium reacts with hydrochloric acid to produce 1 mole of hydrogen (H_2). If the remaining solution is evaporated to

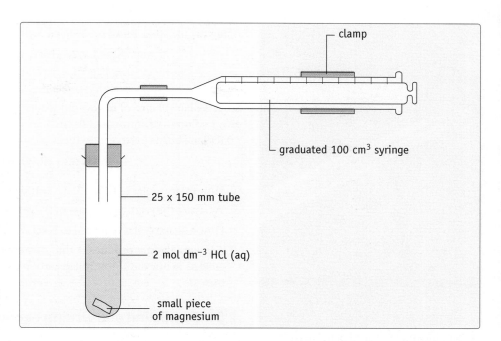

Figure 2.2
Measuring the volume of hydrogen produced when magnesium reacts with hydrochloric acid.

Diagram labels:
- clamp
- graduated 100 cm³ syringe
- 25 x 150 mm tube
- 2 mol dm⁻³ HCl (aq)
- small piece of magnesium

dryness, pure white solid magnesium chloride, $MgCl_2$, is left. Thus, the overall equation for this reaction is

$$Mg(s) + 2HCl(aq) \rightarrow H_2(g) + MgCl_2(aq)$$

Both hydrochloric acid and magnesium chloride are ionic substances. They consist of ions which are free to move apart in the aqueous solutions. The hydrochloric acid, $HCl(aq)$, contains $H^+(aq)$ and $Cl^-(aq)$ ions, and the aqueous magnesium chloride, $MgCl_2(aq)$, contains $Mg^{2+}(aq)$ and $Cl^-(aq)$ ions. An ionic equation for the reaction is

$$Mg(s) + 2H^+(aq) + 2Cl^-(aq) \rightarrow H_2(g) + Mg^{2+}(aq) + 2Cl^-(aq)$$

Notice that neither magnesium nor hydrogen are written as ions. *The only ionic compounds are salts, bases and aqueous solutions of acids. Metals, such as magnesium, consist of atoms. Non-metals, such as hydrogen, and non-metal/non-metal compounds, such as water, consist of molecules.* Both atoms and molecules are uncharged particles. For example, magnesium consists of uncharged magnesium atoms (Mg) and hydrogen gas consists of uncharged hydrogen molecules (H_2).

2.5 Reacting volumes of gases

The volume occupied by 1 mole of any gas at s.t.p. is 22.4 dm³. This volume is usually called the **molar volume**.

Experiments which lead to this result for the molar volume are described in section 11.5. Remember that s.t.p. means standard temperature and pressure. This is 0°C (273 K) and 1 atmosphere pressure. At room temperature (say 25°C) and pressure (1 atmosphere), the volume of 1 mole of any gas is about 24.4 dm³.

Using the idea of molar volumes, it is possible to calculate the volumes of gases which react. For example, consider the manufacture of ammonia from nitrogen and hydrogen:

$$nitrogen + hydrogen \rightarrow ammonia$$
$$N_2(g) + 3H_2(g) \rightarrow 2NH_3(g)$$

From the equation:

$$1 \text{ mole of } N_2 + 3 \text{ moles } H_2 \rightarrow 2 \text{ moles } NH_3$$

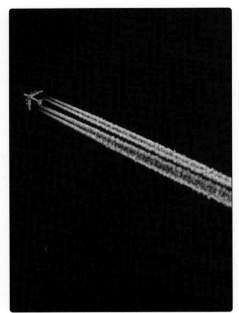

▲ The white trail from a high flying jet is actually a trail of tiny ice crystals.
A typical jumbo jet burns fuel at a rate of 200 kg per minute. Assuming jet fuel is $C_{11}H_{24}$, what mass of water does a jumbo jet produce per minute?

$$1 \text{ mole } C_{11}H_{24} \quad \rightarrow \quad 12 \text{ moles } H_2O$$
$$\Rightarrow 156 \text{ g } C_{11}H_{24} \quad \rightarrow \quad 216 \text{ g } H_2O$$
$$\therefore 1 \text{ g } C_{11}H_{24} \quad \rightarrow \quad \frac{216}{156} \text{ g } H_2O$$
$$\therefore 200 \text{ kg } C_{11}H_{24} \rightarrow \quad 200 \times \frac{216}{156} \text{ kg } H_2O$$

Mass of water produced per minute
$= 200 \times \frac{216}{156} \text{ kg} = 277 \text{ kg}$

Using the idea that 1 mole of each gas will occupy 22.4 dm³ at s.t.p., we can write:

$$22.4 \text{ dm}^3 \text{ N}_2 + (3 \times 22.4) \text{ dm}^3 \text{ H}_2 \rightarrow 2 \times 22.4 \text{ dm}^3 \text{ NH}_3$$
$$\Rightarrow 1 \text{ dm}^3 \text{ N}_2 + 3 \text{ dm}^3 \text{ H}_2 \rightarrow 2 \text{ dm}^3 \text{ NH}_3$$
$$\Rightarrow 1 \text{ volume N}_2 + 3 \text{ volumes H}_2 \rightarrow 2 \text{ volumes NH}_3$$

So, if we had 10 cm³ of nitrogen it would react with 3 volumes of hydrogen (30 cm³) to give 2 volumes of ammonia (20 cm³).

Methane burns in oxygen to form carbon dioxide and water:

$$CH_4(g) + 2O_2(g) \rightarrow CO_2(g) + 2H_2O(g)$$

1 How many moles of CH_4, O_2, CO_2 and H_2O does the equation show?

2 What are the relative volumes of the reactants and products in the equation?

3 How much oxygen is needed to react with 10 dm³ of methane?

4 A biogas digester produces 1 dm³ of methane every hour. How much methane collects in one day? How much carbon dioxide is produced when this methane burns?

2.6　Predicting and writing equations

You will realise by now that it is impossible to write an equation unless you know:

- the formulas of the reactants and products;
- the relative numbers of moles of reactants.

However, if you know the names of all the reactants and products, it is possible to predict their formulas and a possible equation. It is very convenient to predict equations in this way, instead of carrying out an experiment to establish the equation for every reaction.

There are three stages in writing equations.

1 Write out the equation in words. For example

$$\text{copper} + \text{oxygen} \rightarrow \text{copper(II) oxide}$$

2 Write symbols for elements and formulas for compounds in the equation.
This gives

$$\text{Cu} + O_2 \rightarrow \text{CuO}$$

Remember that the elements hydrogen, oxygen, nitrogen and the halogens exist as molecules containing two atoms. They are called **diatomic molecules** and represented in equations as H_2, O_2, N_2, Cl_2, Br_2 and I_2.

All other elements are regarded as single atoms and represented by a monatomic symbol in equations.

3 Balance the equation by making the number of atoms of each element the same on both sides. In the last equation, there are two oxygen atoms on the left and only one on the right. Therefore CuO on the right must be doubled (2CuO). There are now two copper atoms on the right, so two copper atoms must be shown on the left (2Cu). The balanced equation is

$$2\text{Cu} + O_2 \rightarrow 2\text{CuO}$$

Remember that **formulas must *never* be altered in balancing an equation**. Atoms can only be balanced by putting a number in front of a formula, thus doubling or trebling, etc., the whole formula.

Normally state symbols are also included, so the final equation is:

$$2\text{Cu}(s) + O_2(g) \rightarrow 2\text{CuO}(s)$$

Here are two more examples to show how equations are written:

1 The reaction between hydrogen and chlorine:

(i) Write the equation in words:

$$\text{hydrogen} + \text{chlorine} \rightarrow \text{hydrogen chloride}$$

(ii) Write symbols for elements and formulas for compounds:

$$\Rightarrow H_2 + Cl_2 \rightarrow HCl$$

Notice that hydrogen and chlorine are written as H_2 and Cl_2 since they exist as diatomic molecules.

(iii) Balance the equation. There are two hydrogen atoms on the left and only one on the right. There are also two chlorine atoms on the left and only one on the right. By doubling the formula of HCl (i.e. 2HCl), both hydrogen and chlorine are balanced:

$$\Rightarrow H_2 + Cl_2 \rightarrow 2HCl$$

2 The reaction between magnesium and nitric acid:

(i) Write the equation in words:

$$\text{magnesium} + \text{nitric acid} \rightarrow \text{magnesium nitrate} + \text{hydrogen}$$

(ii) Write symbols for elements and formulas for compounds:

$$Mg + HNO_3 \rightarrow Mg(NO_3)_2 + H_2$$

(iii) Balance the equation. Groups of atoms that form a single ion, such as nitrate (NO_3^-), are regarded as a single unit in balancing equations. There are two nitrate ions on the right and only one on the left. There are two hydrogen atoms on the right and only one on the left. By writing $2HNO_3$ on the left, the whole equation is balanced:

$$Mg + 2HNO_3 \rightarrow Mg(NO_3)_2 + H_2$$

These equations in which all substances are represented by their full chemical formulas are called **formal** or **balanced chemical equations**.

> Write balanced chemical equations for the reactions of
> 1 zinc with oxygen
> 2 copper oxide with hydrochloric acid
> 3 sodium with water.

2.7 What do equations tell us?

A balanced chemical equation tells us

- the reactants and products;
- the formulas of reactants and products;
- the relative numbers of moles of the reactants and products.

For example, the equation

$$2Cu(s) + O_2(g) \rightarrow 2CuO(s)$$

tells us that

- copper reacts with oxygen to form copper oxide;
- the formula of copper(II) oxide is CuO and that of oxygen is O_2;
- 2 moles of copper atoms react with 1 mole of oxygen molecules (O_2) to produce 2 moles of copper oxide (CuO).

Some of the most useful data we can obtain from balanced equations are the relative masses of reactants and products.

The mass of 1 mole of any substance can be calculated from the relative atomic masses of its constituent elements. So, from the last equation, we can calculate that

$$(2 \times 63.5) \text{ g Cu} + 32 \text{ g O}_2 \rightarrow (2 \times 79.5) \text{ g CuO}$$

i.e. 127 g copper react with 32 g oxygen to form 159 g of copper(II) oxide (see figure 2.3).

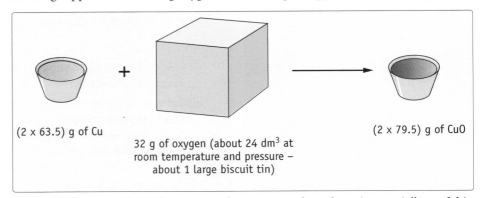

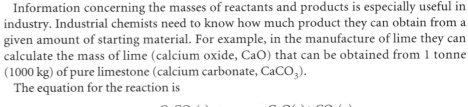

(2 x 63.5) g of Cu

32 g of oxygen (about 24 dm³ at room temperature and pressure – about 1 large biscuit tin)

(2 x 79.5) g of CuO

Figure 2.3
The masses of reactants and products when copper reacts with oxygen to form copper oxide.

▼ The photographs below show three stages in the manufacture of copper from copper ore (copper pyrites). The left-hand photo shows copper ore being mined at Brigham Canyon near Salt Lake City, USA. The ore is crushed and then reduced in furnaces to produce molten copper (centre). This can be made into solid bars or rolls of copper cable (right).

Information concerning the masses of reactants and products is especially useful in industry. Industrial chemists need to know how much product they can obtain from a given amount of starting material. For example, in the manufacture of lime they can calculate the mass of lime (calcium oxide, CaO) that can be obtained from 1 tonne (1000 kg) of pure limestone (calcium carbonate, $CaCO_3$).

The equation for the reaction is

$$CaCO_3(s) \rightarrow \qquad CaO(s) + CO_2(g)$$
$$\Rightarrow 1 \text{ mole of } CaCO_3 \rightarrow 1 \text{ mole of } CaO + 1 \text{ mole of } CO_2$$
$$(40 + 12 + (3 \times 16)) \text{g} \rightarrow \qquad (40 + 16) \text{ g} + (12 + (2 \times 16)) \text{ g}$$
$$\Rightarrow 100 \text{ g } CaCO_3 \rightarrow \qquad 56 \text{ g } CaO + 44 \text{ g } CO_2$$

∴ 1000 kg of limestone produces 560 kg of lime.

1. What is the relative formula mass of $CuFeS_2$? (Cu = 64, Fe = 56, S = 32)
2. How much copper can be obtained from 1 mole of $CuFeS_2$?
3. How much copper can be obtained from 1 tonne (1000 kg) of $CuFeS_2$?

Equations are very useful, but we must remember their limitations. An equation will not tell us:

- the rate of the reaction – whether it is fast or slow, whether it will explode or whether it will take years to react;
- how or why a reaction takes place. The equation for the reaction of magnesium with oxygen is

$$2Mg(s) + O_2(g) \rightarrow 2MgO(s)$$

2 Reacting Quantities and Equations

The equation does not show that the magnesium must be heated before a reaction will take place. Nor does it show that once the reaction has started it will continue vigorously of its own accord. The equation does not tell us why the magnesium reacts with oxygen or how the atoms and molecules behave in the reaction. Do oxygen molecules (O_2) each combine with two magnesium atoms, or do oxygen molecules split to form separate oxygen atoms which then combine with separate magnesium atoms? The equation cannot answer these questions.

Equations can only tell us about the amounts and states of the reactants and products. They cannot tell us anything about what happens on the route between the reactants and the products. Equations can only tell us what overall change occurs.

2.8 Ionic equations

Many reactions involve ionic compounds. The part played by the separate ions of these compounds in a reaction can be shown in an ionic equation. Here are some examples of ionic equations. Remember that the only substances which contain ions are salts, bases and aqueous acids.

1 When hydrochloric acid reacts with sodium hydroxide solution, the equation for the reaction is

$$HCl(aq) + NaOH(aq) \rightarrow NaCl(aq) + H_2O(l)$$

$HCl(aq)$, $NaOH(aq)$ and $NaCl(aq)$ consist of ions, so we can write an ionic equation for the reaction as:

$$\underbrace{H^+(aq) + Cl^-(aq)}_{\text{hydrochloric acid}} + \underbrace{Na^+(aq) + OH^-(aq)}_{\text{sodium hydroxide solution}} \rightarrow \underbrace{Na^+(aq) + Cl^-(aq)}_{\text{sodium chloride solution}} + \underbrace{H_2O(l)}_{\text{water}}$$

2 The combination of iron and sulphur can also be represented by an ionic equation:

$$Fe(s) + S(s) \rightarrow Fe^{2+}S^{2-}(s)$$

3 When liquids are electrolysed, the reactions which take place at the electrodes involve ions. The reactions which occur during electrolysis are discussed in sections 3.5, 3.6 and 3.7.

4 We can use an ionic equation to represent the precipitation of lead iodide during the reaction between lead nitrate and potassium iodide:

$$Pb^{2+}(aq) + 2I^-(aq) \rightarrow Pb^{2+}(I^-)_2(s)$$

5 Earlier in this chapter we used an ionic equation to summarise the reaction between magnesium and hydrochloric acid:

$$Mg(s) + 2H^+(aq) + 2Cl^-(aq) \rightarrow Mg^{2+}(aq) + 2Cl^-(aq) + H_2(g)$$

These five examples illustrate the commonest types of reaction for which ionic equations can be used:

- The reactions between acids and bases (neutralisation).
- The reactions of metals with non-metals.
- The reactions during electrolysis.
- The precipitations of insoluble ionic solids.
- The reactions of metals with acids.

True or false? In chemical reactions:

1 atoms may be converted to ions of the same element

2 atoms may be converted to atoms of a different element

3 molecules may be converted to ions

4 ions may be converted to molecules

5 there is sometimes a smaller number of molecules after a reaction than before.

2.9 Writing ionic equations

Ionic equations are often simpler and easier to write than formal equations. In order to obtain an ionic equation for a reaction:

1 Write ionic formulas for the ionic substances that take part in the reaction. The three main types of ionic substance are salts, bases and aqueous solutions of acids. For example

$$\underbrace{2H^+(aq) + SO_4^{2-}(aq)}_{\text{sulphuric acid}} + \underbrace{2Na^+(aq) + 2OH^-(aq)}_{\text{sodium hydroxide}} \rightarrow \underbrace{2Na^+(aq) + SO_4^{2-}(aq)}_{\text{sodium sulphate}} + \underbrace{2H_2O(l)}_{\text{water}}$$

2 Cancel spectator ions which appear unchanged on both sides of the equation and take no part in the reaction:

$$2H^+(aq) + \cancel{SO_4^{2-}(aq)} + \cancel{2Na^+(aq)} + 2OH^-(aq) \rightarrow \cancel{2Na^+(aq)} + \cancel{SO_4^{2-}(aq)} + 2H_2O(l)$$

$$\Rightarrow 2H^+(aq) + 2OH^-(aq) \rightarrow 2H_2O(l)$$

3 Check that ionic substances are written in the correct state.

Aqueous solutions of ionic substances are written with a + sign between the different ions because the ions are free to move apart in the aqueous solution (e.g. $2Na^+(aq) + SO_4^{2-}(aq)$).

The ions in molten (liquid) ionic substances are also free to move apart. Thus, liquid sodium sulphate is $2Na^+(l) + SO_4^{2-}(l)$.

In solid ionic substances, the ions are held together by the attraction of their oppositely charged neighbours. The ions are *not* free to move apart and so they are *not* separated by a + sign (e.g. $(Na^+)_2SO_4^{2-}(s)$ for solid sodium sulphate).

4 Check that the equation is balanced with respect to atoms and with respect to positive and negative charges. Some ionic equations (in particular those for electrolyses) involve electrons which carry one unit of negative charge.

The main advantage of ionic equations is that they provide a clearer picture of what is happening in chemical reactions. It is possible to see which ions are taking part in the reaction and which are 'spectators'.

Summary

1 An equation is a summary of the reactants and the products in a chemical reaction.

2 In order to write an equation we must know:
   (i) the formulas of reactants and products, and
   (ii) the relative numbers of moles of reactants and products.

3 The rules for writing equations are
   (i) write a word equation,
   (ii) write symbols for elements and formulas for compounds,
   (iii) balance the equation.

4 Equations can only tell us about the amounts and states of the reactants and products. They cannot tell us how or why the reaction occurs.

5 Equations in which all substances are represented by their full chemical formulas are called formal or balanced chemical equations.

6 The part played by separate ions in a reaction can be shown in an ionic equation.

7 Ionic equations are useful in summarising:
   (i) the reactions between acids and bases,
   (ii) the reactions between acids and metals,
   (iii) the reactions between metals and non-metals,
   (iv) electrolyses,
   (v) the precipitation of insoluble ionic solids.

1 0.27 g of aluminium reacts with 2.4 g of bromine.
   (a) How many moles of Al are there in 0.27 g of aluminium?
   (b) How many moles of Br are there in 2.4 g of bromine?
   (c) Write a possible formula for the compound of aluminium and bromine which is produced.
   (d) What is the relative molecular mass of a compound with this formula?
   (e) The *actual* relative molecular mass of the compound of aluminium with bromine is 534. What is the actual molecular formula of the compound?
   (f) Write an equation for the reaction of aluminium with bromine.

2 A compound of nitrogen and oxygen was found to contain 28 g of nitrogen to every 64 g of oxygen ($N = 14$, $O = 16$). Which of the following formulas could represent the compound?
   A N_2O B N_2O_2 C NO_2 D N_2O_4 E N_4O_2

3 (a) Explain what is meant by 'hydrogen peroxide has the molecular formula H_2O_2'.
   (b) $2H_2O_2 \rightarrow 2H_2O + O_2$
       What mass of oxygen can be formed from 17 g of H_2O_2?
   (c) What mass of oxygen can be formed from 17 g of H_2O_2 if 1 g of manganese dioxide catalyst is used?

4 When lithium perchlorate, $LiClO_4$, is heated, a gas is given off and a white solid is left. The gas is colourless and re-lights a glowing splint. The solid residue dissolves completely in water. When silver nitrate is added to this solution, a white precipitate is formed.
   (a) What is the name and formula of the gas produced?
   (b) What is the name and formula of the solid left on heating?
   (c) Write an equation for the reaction which takes place when lithium perchlorate is heated.
   (d) Write an equation for the reaction of the aqueous solution of the solid residue with silver nitrate.
   (e) 2.13 g of lithium perchlorate were heated to constant mass. What mass of residue was produced?

5 By calculating relative molecular masses, find whether it is cheaper to buy washing soda (sodium carbonate) as the decahydrate ($Na_2CO_3.10H_2O$) at 4p per kg, or as the anhydrous salt (Na_2CO_3) at 8p per kg.

6 Write balanced equations to summarise the following:
   (a) When aqueous solutions of sodium chloride and silver nitrate are mixed, a white precipitate of silver chloride forms.
   (b) The electrolysis of molten sodium chloride produces sodium and chlorine.
   (c) Iron reacts with chlorine to form iron(III) chloride.
   (d) Methane (CH_4) burns in oxygen to produce carbon dioxide and water.
   (e) Copper(II) oxide reacts with hot dilute sulphuric acid to produce copper(II) sulphate and water.
   Rewrite (a), (b) and (c) as ionic equations.

7 True or false? A balanced equation shows:
   A the formulas of the products
   B the molar proportions in which the products are formed
   C that a reaction can occur
   D the relative numbers of atoms, ions and molecules which react
   E that a reaction is exothermic.

8 Sulphuric acid reacts with 'thio' (sodium thiosulphate) producing a precipitate of sulphur. Samples of 10 cm^3 of 0.1 mol dm^{-3} 'thio' were taken in similar tubes and varying quantities of 0.2 mol dm^{-3} acid were added. The sulphur precipitate was allowed to settle and its depth measured. The results are given in figure 2.4.

Volume of 0.2 mol dm^{-3} H_2SO_4 added/cm^3	1	3	5	7	9	11
Depth of precipitate/mm			1	3	5	4.5	5	5

Figure 2.4

   (a) Draw a careful sketch graph (it is unnecessary to use graph paper) of the depth of precipitate against the volume of acid added.
   (b) Why does the depth of the sulphur precipitate increase as more acid is added, but then reach a constant value?
   (c) In what molar proportions do 'thio' and acid react?

9 A hydrate of potassium carbonate has the formula $K_2CO_3.xH_2O$. Ten grams of the hydrate leave 7.93 g of anhydrous salt on heating.
   (a) What is the mass of anhydrous salt in 10 g of hydrate?
   (b) What is the mass of water in 10 g of hydrate?
   (c) How many moles of anhydrous salt are there in 10 g of hydrate?
   (d) How many moles of water are there in 10 g of hydrate?
   (e) How many moles of water are present in the hydrate for every 1 mole of K_2CO_3?
   (f) What is the formula of the hydrate?

10 25 cm^3 of a solution of NaOH required 28 cm^3 of 1.0 mol dm^{-3} H_2SO_4 to neutralise it.
   (a) Write the equation for the reaction.
   (b) How many moles of H_2SO_4 were needed?
   (c) How many moles of NaOH were thus neutralised?
   (d) How many moles of NaOH are there in 25 cm^3 of solution?
   (e) What is the concentration of the NaOH in mol dm^{-3}?

3 Redox

3.1 Introduction

The term **redox** is used by chemists as an abbreviation for the processes of **red**uction and **oxi**dation. These two processes usually occur simultaneously. Redox reactions include processes such as burning, rusting and respiration. Originally, chemists had a very limited view of redox, using it to account for only the reactions of oxygen and hydrogen. Nowadays, our ideas of redox have been extended to include all electron-transfer processes. An important feature of some electron-transfer redox reactions is that the energy of the chemical reaction may be released in the form of electrical energy and harnessed to provide electricity. This is what happens in the dry cell of a small torch, in the button cell of a hearing aid and in the battery of a motor car. Commercial cells and batteries are discussed in some detail in section 14.7.

▶ Antoine Lavoisier (1743–1794) was the first chemist to explain the redox reactions which occur during burning. In 1775, Lavoisier was appointed to a post at the French government munitions factory. Here he carried out experiments in combustion and respiration. His wife, Marie, assisted in the experiments and illustrated their many publications.

3.2 Redox processes in terms of electron transfer

When metals react with oxygen they form oxides. For example:

$$2Mg + O_2 \rightarrow 2Mg^{2+}O^{2-}$$

$$4Na + O_2 \rightarrow 2(Na^+)_2O^{2-}$$

The metal is oxidised and the oxygen is reduced. During this process the metal atoms lose electrons to form positive ions and oxygen gains electrons to form negative oxide ions, O^{2-}. The oxygen takes the electrons given up by the metal.

$$2Mg \rightarrow 2Mg^{2+} + \boxed{4e^-}$$

$$O_2 + \boxed{4e^-} \rightarrow 2O^{2-}$$

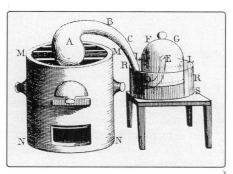

▲ Lavoisier's apparatus for preparing oxygen. Red mercury(II) oxide was heated in the retort (A) on the left. The oxygen which formed was collected in the bell jar (E) above mercury.

Electron-transfer reactions such as this are called **redox reactions**. The separate equations showing which substance gains electrons and which substance loses electrons are known as **half-equations**.

In the half-equations above, Mg loses electrons and is oxidised to Mg^{2+}; O_2 gains electrons and is reduced to O^{2-}.

This leads to the following definitions.
- **Oxidation** *is the loss of electrons.*
- **Reduction** *is the gain of electrons.*
- **Oxidising agents,** such as oxygen, *accept electrons.*
- **Reducing agents** *donate electrons.*

When metals react they lose electrons and form their ions. Thus, metals are oxidised and act as reducing agents in their reactions. Notice that the oxidised substance (magnesium in the above example) acts as the reducing agent and the reduced substance (oxygen in the above example) acts as the oxidising agent.

When redox is viewed in terms of electron transfer it is easy to see why oxidation and reduction always take place together. One substance cannot lose electrons and be oxidised unless another substance gains electrons and is reduced.

Do the following processes involve oxidation, reduction, both oxidation and reduction or none of these?

1. $2H^+ + 2e^- \rightarrow H_2$
2. $Cu^+ \rightarrow Cu^{2+} + e^-$
3. $Mg \rightarrow Mg^{2+} + 2e^-$
4. $Ag^+ + Cl^- \rightarrow AgCl$
5. $NH_3 + H^+ \rightarrow NH_4^+$
6. $2Cu^+ \rightarrow Cu^{2+} + Cu$

3.3 Electron transfer in redox reactions

When powdered zinc is added to copper(II) sulphate solution an exothermic reaction occurs. Copper ions are reduced to a deposit of red-brown copper and zinc goes into solution as zinc ions.

$$Zn(s) + Cu^{2+}(aq) \rightarrow Zn^{2+}(aq) + Cu(s)$$

The overall reaction can be separated into two simpler processes involving electron transfer.

and

$$Zn(s) \rightarrow Zn^{2+}(aq) + 2e^-$$
$$Cu^{2+}(aq) + 2e^- \rightarrow Cu(s)$$

The apparatus in figure 3.1 shows how these two half-reactions can be separated to demonstrate that electron transfer is occurring. As soon as the circuit is complete, the bulb lights, showing that electrons are flowing through the wire. An ammeter used in place of the bulb will measure the electric current.

Zinc dissolves from the zinc rod, which loses weight, whilst copper is deposited on the copper rod. The overall reaction is

$$Zn(s) + Cu^{2+}(aq) \rightarrow Zn^{2+}(aq) + Cu(s)$$

This is identical to the reaction which occurs when zinc is added to copper(II) sulphate solution. In this experiment, however, the energy of the reaction is released as electrical energy (electron transfer), whereas on direct mixing the energy is liberated as heat.

▲ This cartoon (published in the eighteenth century) pokes fun at Joseph Priestley (1733–1804). Priestley was a nonconformist English clergyman who carried out experiments on combustion. Priestley's discovery of oxygen provided the Lavoisiers with the vital information they needed for their redox theory of combustion.

Using the apparatus in figure 3.1 the overall process has been separated into two distinct half-reactions:

$$Zn(s) \rightarrow Zn^{2+}(aq) + 2e^- \text{ at the zinc rod}$$

and

$$Cu^{2+}(aq) + 2e^- \rightarrow Cu(s) \text{ at the copper rod}$$

At the zinc rod, zinc atoms give up electrons and form zinc ions which go into solution as $Zn^{2+}(aq)$. The electrons flow from the zinc rod through the external circuit to the copper rod where they combine with Cu^{2+} ions to form copper atoms.

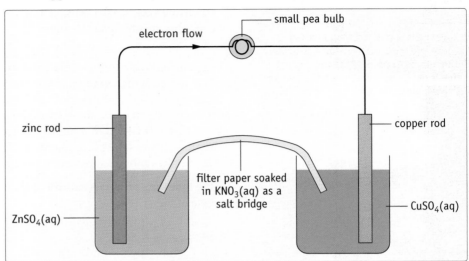

Figure 3.1
Electron transfer between zinc and copper(II) sulphate solution.

1. Will electrons flow from positive to negative or from negative to positive?
2. Will the zinc rod be positive or negative?
3. What is the function of the salt bridge?
4. What happens when the salt bridge is removed?

When the circuit in figure 3.1 is complete, the zinc rod dissolves away and the concentration of $Zn^{2+}(aq)$ in the left-hand beaker rises. If this net increase in positive charge is not 'neutralised', the reaction will soon stop because the excess positive charge in the solution will prevent any more Zn^{2+} ions from entering it. The excess positive

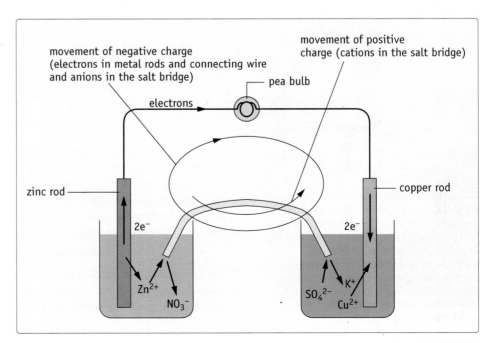

Figure 3.2
Movement of charge around a circuit.

▲ Photochromic spectacles in low light intensity (above) and in strong light (below). Photochromic glass darkens on exposure to strong light and becomes clearer when the light is not so strong. The glass contains small particles of silver chloride and copper(I) chloride. When the light is strong, Ag^+ ions oxidise Cu^+ to Cu^{2+}.

$$Ag^+ + Cu^+ \rightarrow Ag + Cu^{2+}$$

The silver produced reflects light from the glass and cuts out transmission (glare) to the eye. In low light intensity, copper(II) ions oxidise Ag back to Ag^+ ions and the glass becomes transparent again.

▲ An unusual redox reaction!

charge from Zn^{2+} ions in the solution is reduced by nitrate ions, NO_3^-, moving out of the salt bridge into the solution or by zinc ions moving into the salt bridge.

In the right-hand beaker, copper ions are converted to copper atoms leaving an excess of sulphate ions, SO_4^{2-}, in solution. The excess negative charge in the right-hand half-cell is reduced by SO_4^{2-} ions moving into the salt bridge or potassium ions, K^+, diffusing out of the salt bridge into the solution.

Notice how the salt bridge serves two main functions:

(a) It completes the circuit by allowing ions carrying charge to move towards one half-cell from the other. When the salt bridge is removed, the current ceases, since charge (whether it is ions or electrons) can no longer flow around the circuit.

(b) It provides cations and anions to replace those consumed at the electrodes and balance the charges on any ions formed from the electrodes. The movement of charge around the circuit is shown in figure 3.2.

These arrangements which generate an electric current from chemicals are called **electrochemical cells.** They are discussed in more detail in chapter 14.

3.4 Balancing redox reactions

Many substances can act as oxidising agents (e.g. MnO_2, Cl_2, H_2O_2, Fe^{3+}); and there is a similar number of reducing agents (e.g. I^-, Fe^{2+}, Zn). This leads to a variety of possible redox reactions from various combinations of oxidising agents and reducing agents. The following simple rules can be used to obtain a balanced equation from these redox processes.

1 Write down the oxidising and reducing agents and their products.

2 Write separate half-equations for the oxidation and reduction processes and balance these with respect to atoms and charge.

3 Combine the half-equations to obtain the overall equation.

We can use these simple rules to write an equation for the reaction between bromine and iron(II) ions.

1 The oxidising agent is Br_2, which reacts to form Br^-. The reducing agent is Fe^{2+}, which forms Fe^{3+}.

2 The balanced Br_2/Br^- half-equation is

$$Br_2 + 2e^- \rightarrow 2Br^-$$

One Br_2 molecule gains two electrons and is reduced to two Br^- ions. The balanced Fe^{2+}/Fe^{3+} half-equation is

$$Fe^{2+} \rightarrow Fe^{3+} + e^-$$

3 During this reaction, Br_2 molecules take electrons from Fe^{2+} ions. In order to obtain a balanced overall equation, it is necessary to double the Fe^{2+}/Fe^{3+} equation so that $2Fe^{2+}$ gives up two electrons which are taken by one Br_2:

$$2Fe^{2+} \rightarrow 2Fe^{3+} + \boxed{2e^-}$$
$$Br_2 + \boxed{2e^-} \rightarrow 2Br^-$$

Adding these together gives the overall equation:

$$2Fe^{2+} + Br_2 \rightarrow 2Fe^{3+} + 2Br^-$$

Many other oxyanions, including $Cr_2O_7^{2-}$, BrO_3^-, IO_3^-, ClO^- and ClO_3^- also act as oxidising agents in acid solution. Oxygen from the oxyanions combines with H^+ ions from the acid to form water, and atoms such as Cr and Cl from the oxyanion are reduced to stable ions such as Cr^{3+} and Cl^-. As a second example of redox reaction equations, consider the reaction between manganate(VII), MnO_4^-, and iodide, I^-, in acid solution.

Try to write balanced half-equations and an overall equation for the following:

1. The reaction between I_2 and $S_2O_3^{2-}$. ($S_2O_3^{2-}$ is oxidised to $S_4O_6^{2-}$ during the reaction and the I_2 forms I^- ions.)
2. The reaction between H_2O_2 and I^- in acid solution. (The H_2O_2 reacts with H^+ ions to form water.)

1 The oxidising agent is MnO_4^-. In acid solution this is reduced to Mn^{2+} and H_2O.

2 The unbalanced MnO_4^-/Mn^{2+} half-equation is therefore

$$MnO_4^- + H^+ \rightarrow Mn^{2+} + H_2O$$

Since the MnO_4^- ion contains four oxygen atoms it can produce four H_2O molecules. This means that eight H^+ ions are required to balance the hydrogen atoms in four H_2O molecules. The equation, balanced with respect to atoms, is:

$$MnO_4^- + 8H^+ \rightarrow Mn^{2+} + 4H_2O$$

We must now balance the equation with respect to charge.
On the left-hand side, charge $= -1 + 8 = +7$.
On the right-hand side, charge $= +2$.
This means that $5e^-$ must be added to the left-hand side to obtain a balanced half-equation:

$$MnO_4^- + 8H^+ + 5e^- \rightarrow Mn^{2+} + 4H_2O$$

The balanced I^-/I_2 half-equation is

$$2I^- \rightarrow I_2 + 2e^-$$

3 In order to obtain the overall equation, the MnO_4^-/Mn^{2+} half-equation must be multiplied by 2 and the I^-/I_2 half-equation must be multiplied by 5 so that 10 electrons are transferred.

$$2MnO_4^- + 16H^+ + 10e^- \rightarrow 2Mn^{2+} + 8H_2O$$
$$10I^- \rightarrow 5I_2 + 10e^-$$
$$\overline{2MnO_4^- + 16H^+ + 10I^- \rightarrow 2Mn^{2+} + 8H_2O + 5I_2}$$

3.5   Electrolysis

Electrolysis *involves the decomposition of a molten or aqueous compound by electricity. The process takes place in an* **electrolytic cell**. *The compound decomposed during electrolysis is called an* **electrolyte**.

Electrolysis is important in industry. It is used:
- to extract sodium (section 15.8), aluminium (section 19.9) and other reactive metals (section 19.14),
- to manufacture chlorine and sodium hydroxide (section 16.3),
- to purify 'blister copper' (section 19.7).

The energy which causes the chemical changes during electrolysis is provided by an electric current.

An electric current is simply a flow of electrons.

We could measure the quantity of electricity (electric charge) that has flowed along a wire by counting the number of electrons which pass a certain point. The charge on one electron is, however, much too small to be used as a practical unit in measuring the quantity of electricity.

The practical unit normally used is the **coulomb** (C). This is approximately six million million million (6×10^{18}) electron charges.

If one coulomb of charge passes along a wire in one second, then the rate of charge flow (i.e. the electric current) is 1 coulomb per second or 1 **ampere.**
If Q coulombs flow along a wire in t seconds, the electric current (I) is given by:

$$I = \frac{Q}{t}$$

This equation can be rearranged to give

$$Q \qquad = \qquad I \qquad \times \qquad t$$

charge in coulombs   current in amps   time in seconds

1 The current in a small torch bulb is 0.25 A. How much electric charge flows if the torch is used for 20 minutes?

3.6 Explaining electrolysis

When an electric current passes through molten sodium chloride, sodium is produced at the **cathode** (the negative electrode) and chlorine forms at the **anode** (the positive electrode) (figure 3.3).

During electrolysis, Na^+ ions in the electrolyte are attracted to the cathode and Cl^- ions in the electrolyte are attracted to the anode.

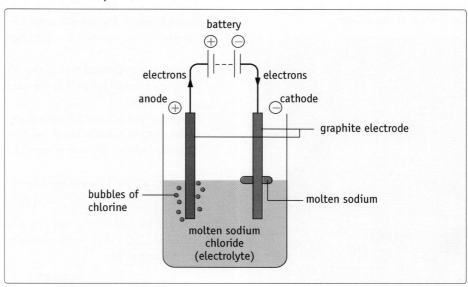

Figure 3.3
Electrolysis of molten sodium chloride.

When Na^+ ions reach the cathode, they combine with negative electrons from the battery forming neutral sodium atoms.

$$Na^+ \qquad + \qquad e^- \qquad \rightarrow \qquad Na$$

sodium ion   electron on   sodium atom
in sodium chloride cathode from   in metal
electrolyte   battery

At the anode, Cl^- ions lose electrons to the positive anode forming neutral chlorine atoms.

$$Cl^- \qquad \rightarrow \qquad e^- \qquad + \qquad Cl$$

chloride ion   electron given   chlorine atom
in electrolyte   to anode

The Cl atoms immediately join up in pairs to form molecules of chlorine gas, Cl_2.

$$Cl + Cl \rightarrow Cl_2$$

During electrolysis:
* *a metal or hydrogen forms at the negative cathode. This confirms that metals and hydrogen have positive ions. These ions are called* **cations** *because they are attracted to the cathode.*
* *a non-metal (except hydrogen) forms at the positive anode. This confirms that non-metals (except hydrogen) have negative ions. These ions are called* **anions** *because they are attracted to the anode.*

K	K$^+$
Na	Na$^+$
Mg	Mg^{2+}
Al	Al^{3+}
Zn	Zn^{2+}
Fe	Fe^{2+}
Pb	Pb^{2+}
H	H$^+$
Cu	Cu^{2+}
Ag	Ag$^+$

metals become less reactive (i.e. less likely to form ions) →

ions become more likely to form atoms →

Figure 3.4
The electrochemical series, showing the reactivity of atoms and ions.

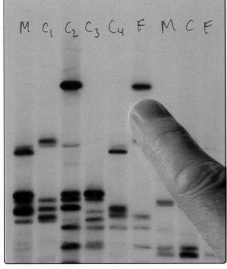

▲ DNA fingerprinting uses a process called electrophoresis. Electrophoresis of DNA involves the migration of charged particles from a person's DNA. The DNA is absorbed onto damp paper like chromatography paper or onto a gel. The charged particles move over the paper or through the gel in an electric field of several thousand volts. The 'fingerprinting' of DNA in body fluids is used in crime detection and to prove paternity. This photo shows bands of DNA from different people after electrophoresis. The pattern of DNA bands is unique to each individual, but some bands are shared by related people. The bands in these DNA fingerprints suggest that F is the father of child C_2.

The equations for the electrolysis of molten sodium chloride show that Na$^+$ ions remove electrons from the cathode and Cl$^-$ ions give up electrons to the anode. In this way, the electric current is being carried through the electrolyte by ions moving to the electrode of opposite charge.

The electrolysis of other molten and aqueous substances can also be explained in terms of ions.

Electrolysis of mixtures

Electrolysis is more complicated if there is more than one cation or more than one anion or water present in the electrolyte. In most cases, however, there is only one product at each electrode.

- When there is more than one cation in the electrolyte, the ion discharged at the cathode can be predicted from the **reactivity (electrochemical) series** (figure 3.4).

It is not surprising that metals like potassium and sodium which are keen to form ions do not readily re-form the metal during electrolysis. *So, the ions of metals lower down in the electrochemical series are discharged in preference to those higher up.*
This simple rule is, however, complicated by two other factors.

(i) If a cation is present in very high concentration, it may be discharged in preference to one below it in the electrochemical series at much lower concentration.

(ii) Hydrogen is discharged from aqueous solutions of the salts of metals above it in the electrochemical series. So, hydrogen is produced at the cathode when aqueous sodium chloride is electrolysed.

In pure water and in aqueous solutions like NaCl(aq), water is slightly ionised.

$$H_2O(l) \rightleftharpoons H^+(aq) + OH^-(aq)$$

Even though the concentration of H$^+$ ions in aqueous solution is only about 10^{-7} mol dm^{-3}, H$^+$ ions are still discharged in preference to cations like Na$^+$ at much higher concentrations.

$$\textit{Cathode (–)} \quad 2H^+(aq) + 2e^- \rightarrow H_2(g)$$

- When there is more than one anion present in the electrolyte, experiments show that the order of discharge is

$$\underset{\text{ions discharge more easily}}{\underrightarrow{SO_4^{2-}, NO_3^-, Cl^-, OH^-, Br^-, I^-}}$$

This simple rule, like discharge at the cathode, is complicated by two other factors.

(i) If an anion is present in very high concentration, it may discharge in preference to one that would be discharged more readily if their concentrations were equal.

(ii) Oxygen is usually discharged from the aqueous solutions of nitrates and sulphates, even though the concentration of OH$^-$ ions from the water may be as low as 10^{-7} mol dm^{-3}.

$$\textit{Anode (+)} \quad 4OH^-(aq) \rightarrow 2H_2O(l) + O_2 + 4e^-$$

Consider the electrolysis of aqueous sodium sulphate.

	Cations	Anions
Ions present from sodium sulphate	Na$^+$(aq)	SO$_4^{2-}$(aq)
Ions present from water	H$^+$(aq)	OH$^-$(aq)

Cathode   H$^+$(aq) ions are discharged in preference to Na$^+$(aq)
$$2H^+(aq) + 2e^- \rightarrow H_2(g)$$

Anode   OH$^-$(aq) ions are discharged in preference to SO$_4^{2-}$(aq)
$$4OH^-(aq) \rightarrow 2H_2O(l) + O_2(g) + 4e^-$$

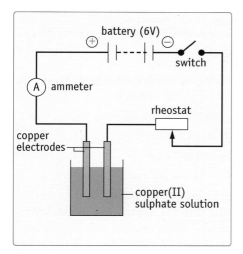

Figure 3.5
Finding the amount of charge required to deposit 1 mole of copper.

Table 3.1
The charge required to produce 1 mole of seven different elements

Element	Charge required to produce one mole/C
copper	193 000
silver	96 500
sodium	96 500
aluminium	289 500
lead	193 000
hydrogen	96 500
oxygen	193 000

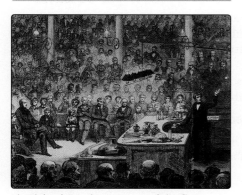

▲ Michael Faraday was one of the first scientists to study electrolysis. Here he is giving the Christmas Lecture at the Royal Institution in 1855. Prince Albert, Consort to Queen Victoria, is in the audience.

3.7 The Faraday constant: How much charge is needed to deposit one mole of copper during electrolysis?

Using the apparatus in figure 3.5, we can find the amount of charge required to deposit one mole of copper (63.5 g) on the cathode during electrolysis. The rheostat (variable resistor) is used to keep the current constant and quite low. If the current is too large, the copper deposits too fast and flakes off the cathode.

The copper cathode is cleaned, dried and weighed. The circuit is connected up and a current of about 0.15 A is passed for at least 45 minutes.

After electrolysis, the cathode is removed and washed, first with distilled water and then with propanone. When it is completely dry, it is reweighed.

1 Why must the cathode be clean and dry when it is weighed before electrolysis?

2 Why is the cathode washed in distilled water and then propanone after electrolysis?
Here are the results of one experiment.

Mass of copper cathode before electrolysis	= 54.07 g
Mass of copper cathode after electrolysis	= 54.25 g
Mass of copper deposited	= 0.18 g

Time of electrolysis = 60 min $\quad$ = 3600 s
Current $\qquad\qquad\qquad\qquad$ = 0.15 A
Quantity of electric charge used $\quad = I \times t$
$\qquad\qquad\qquad\qquad\qquad\qquad = 0.15\,\text{A} \times 3600\,\text{s}$
$\qquad\qquad\qquad\qquad\qquad\qquad = 540$ coulombs (C)

From these results:
0.18 g of copper is produced by 540 C
∴ 1 mole of copper (63.5 g) is produced by $\dfrac{540}{0.18} \times 63.5\,\text{C} = 190\,500\,\text{C}$

Accurate experiments show that 1 mole of copper is deposited by 193 000 coulombs. This quantity of electricity would operate a 2-bar electric fire for about 6 hours.

3 A similar experiment to the last one was carried out using silver electrodes in silver nitrate solution.
0.60 g of silver was deposited in 60 minutes using a current of 0.15 A.
Calculate the quantity of charge needed to deposit 1 mole (108.0 g) of silver.

Table 3.1 shows the charge required to produce 1 mole of seven different elements.

When molten liquids and aqueous solutions are electrolysed, the quantity of electricity needed to produce one mole of an element is always a multiple of 96 500 coulombs (i.e. 96 500 C or 193 000 C ($2 \times 96\,500$ C) or 289 500 C ($3 \times 96\,500$ C)).

Because of this, 96 500 coulombs is called the **Faraday constant** (F), in honour of Michael Faraday. Faraday was one of the first scientists to measure the masses of elements produced during electrolysis.

During electrolysis, 1 mole of sodium ions requires 1 mole of electrons in order to form one mole of sodium atoms.

$$Na^+ + e^- \rightarrow Na$$

These electrons carry the 96 500 C of charge which are required to deposit one mole of sodium. So, 96 500 C is the charge on 1 mole of electrons.

Copper ions (Cu^{2+}) carry twice as much charge as sodium ions (Na^+), so it is not surprising that twice as much charge ($2 \times 96\,500$ C = 193 000 C) is required to deposit one mole of copper.

3.8 Important types of redox reaction

1 The reactions of metals with non-metals

In this case, metals give up electrons to form positive ions and non-metals (O_2, Cl_2, S) take these electrons to form negative ions (O^{2-}, Cl^-, S^{2-}). For example:

$$Fe + S \rightarrow Fe^{2+}S^{2-} \quad \begin{cases} Fe \rightarrow Fe^{2+} + 2e^- \\ S + 2e^- \rightarrow S^{2-} \end{cases}$$

$$2Fe + 3Cl_2 \rightarrow 2FeCl_3 \quad \begin{cases} 2Fe \rightarrow 2Fe^{3+} + 6e^- \\ 3Cl_2 + 6e^- \rightarrow 6Cl^- \end{cases}$$

The metal is oxidised and the non-metal is reduced. Metals higher in the activity (electrochemical) series lose electrons more easily than the less reactive metals lower down. Thus, moving down the activity series, metals become weaker reducing agents. The reactions of metals and the activity series are discussed more fully in section 14.2 and in chapter 19.

2 The reactions of metals with water

Metals at the top of the activity series (K, Na, Ca and Mg) are sufficiently reactive to form hydrogen with water.

$$Ca + 2H_2O \rightarrow Ca^{2+}(OH^-)_2 + H_2$$

The next few metals in the activity series (e.g. Al, Zn and Fe) do not react noticeably with water, but they will react with steam to form hydrogen.

In these reactions the metal atoms are oxidised to form positive ions. The electrons which they release are accepted by water molecules, which are reduced to hydroxide ions (OH^-) and hydrogen (H_2).

$$Ca \rightarrow Ca^{2+} + \boxed{2e^-}$$
$$2H_2O + \boxed{2e^-} \rightarrow 2OH^- + H_2$$

3 The reactions of metals with acids

In this case, the metal atoms lose electrons which are taken by aqueous H^+ ions in the acid to form H_2. For example:

$$Zn(s) + 2H^+(aq) \rightarrow Zn^{2+}(aq) + H_2(g)$$

The half-equations are:

$$Zn \rightarrow Zn^{2+} + \boxed{2e^-}$$
$$2H^+ + \boxed{2e^-} \rightarrow H_2$$

The reactivity of metals with acids also depends on the ease with which the metal loses electrons to form its aqueous ions. Metals which form ions more easily than hydrogen will therefore react with acids releasing electrons, which are taken by H^+ ions in the acid.

Metals below hydrogen in the activity series, such as copper and silver, form ions less readily than hydrogen and so do not react with acids to form hydrogen.

4 Reactions at the electrodes during electrolysis

During electrolysis, cations are attracted to the negatively charged cathode where they gain electrons and are reduced. For example, during the electrolysis of molten lead(II) bromide:

at the cathode (−)

$$Pb^{2+} + 2e^- \rightarrow Pb$$

At the same time, Br^- anions are attracted to the positively charged anode and oxidised by the loss of electrons.

at the anode (+)

$$2Br^- \rightarrow Br_2 + 2e^-$$

▲ Tutankhamun's gold funerary mask: gold is such an unreactive metal that it remains unoxidised and untarnished after centuries.

▲ Redox reactions drive these Patriot missiles as they intercept enemy missiles.

▲ Electrical gear and batteries under the bonnet of an experimental Peugeot 106 car powered by fuel cells. In a fuel cell, electrical energy is produced by the controlled oxidation of a fuel. The batteries store enough charge for about 75 km of driving before they need to be recharged.

▲ A coil of copper wire has been suspended in silver nitrate solution. A reaction has taken place forming solid silver and copper nitrate solution. What redox processes have occurred?

3.9 Oxidation number

Many redox reactions involve a *complete* transfer of electrons from one substance to another. These redox processes usually have ions as either the reactants or the products. In some cases, the reactants and the products are both ions.

There are, however, some reactions which are regarded as redox processes in spite of the fact that they involve only molecules and do not have a *complete* transfer of electrons from one substance to another. For example, the reactions

$$2H_2 + O_2 \rightarrow 2H_2O$$

and

$$C + O_2 \rightarrow CO_2$$

clearly involve redox, yet there is no complete transfer of electrons from one substance to another.

In order to overcome this problem, the concept of **oxidation number** (or oxidation state) was introduced. This provided a similar, but alternative, definition of redox to that involving electron transfer. An oxidation number is a number assigned to an atom or an ion to describe its relative state of oxidation or reduction.

Using oxidation numbers it is possible to decide whether redox has occurred in chemical reactions. For example:

- Atoms in uncombined elements are given an oxidation number of zero. Thus, the oxidation number of Mg is 0, as is the oxidation number of chlorine atoms in Cl_2.
- For simple ions, the oxidation number is simply the charge on the ion. Thus, the oxidation numbers of Cl^-, Fe^{2+} and Fe^{3+} are -1, $+2$ and $+3$, respectively.
- For compounds and complex ions, the oxidation numbers of the atoms within them are obtained by assuming the compounds and complex ions are *wholly ionic* and then working out the charge associated with each atom. For example:

$H_2O((H^+)_2O^{2-})$	Ox. No. of H in H_2O	$= +1$
	Ox. No. of O in H_2O	$= -2$
$HCl(H^+Cl^-)$	Ox. No. of H in HCl	$= +1$
	Ox. No. of Cl in HCl	$= -1$

In assigning oxidation numbers in this way it is necessary to assume that the electrons in each bond of the molecule or ion belong to the *more electronegative atom* (i.e. the atom with the greater attraction for electrons).

Since hydrogen is the least electronegative non-metal, the oxidation number of H in its compounds is usually $+1$. For example:

$PH_3(P^{3-}(H^+)_3)$	Ox. No. of H in PH_3	$= +1$
	Ox. No. of P in PH_3	$= -3$
$H_2S((H^+)_2S^{2-})$	Ox. No. of H in H_2S	$= +1$
	Ox. No. of S in H_2S	$= -2$

In metal hydrides, however, the metal is more electropositive than hydrogen and in this case the oxidation number of H is -1. For example:

Na^+H^-	Ox. No. of H in NaH	$= -1$
	Ox. No. of Na in NaH	$= +1$

Apart from fluorine, oxygen is the most electronegative element. This means that the oxidation number of oxygen in its compounds is usually -2.

$Na_2O((Na^+)_2O^{2-})$	Ox. No. of O in Na_2O	$= -2$
	Ox. No. of Na in Na_2O	$= +1$
$CO_2(C^{4+}(O^{2-})_2)$	Ox. No. of O in CO_2	$= -2$
	Ox. No. of C in CO_2	$= +4$

However, in OF_2 the oxidation number of oxygen is +2 and in peroxides the oxidation number of oxygen is −1.

$OF_2(O^{2+}(F^-)_2)$	Ox. No. of O in OF_2	= +2
	Ox. No. of F in OF_2	= −1
$Na_2O_2((Na^+)_2(O_2)^{2-})$	Ox. No. of O in Na_2O_2	= −1
	Ox. No. of Na in Na_2O_2	= +1

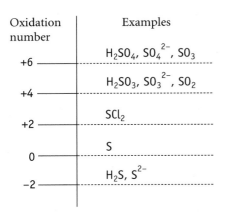

Oxidation number	Examples
+6	H_2SO_4, SO_4^{2-}, SO_3
+4	H_2SO_3, SO_3^{2-}, SO_2
+2	SCl_2
0	S
−2	H_2S, S^{2-}

Figure 3.6
An oxidation number chart for sulphur.

Rules for assigning oxidation numbers

The points we have just made about oxidation numbers can be summarised in six simple rules.

1 The oxidation number of atoms in uncombined elements is 0.

2 In neutral molecules, the algebraic sum of the oxidation numbers is 0.

3 In ions, the algebraic sum of the oxidation numbers equals the charge on the ion.

4 In any substance, the more electronegative atom has the negative oxidation number, and the less electronegative atom has the positive oxidation number.

5 The oxidation number of hydrogen in all its compounds, except metal hydrides, is +1.

6 The oxidation number of oxygen in all its compounds, except in peroxides and in OF_2, is −2.

---

1 What are the oxidation numbers of each element in the following?

$$MgCl_2, \ SO_2, \ CO, \ NaOH, \ PCl_3, \ SO_4^{2-}, \ MnO_4^-$$

2 What are the oxidation numbers of sulphur in the following compounds?

$$NaHSO_4, \ CS_2, \ SO_2Cl_2, \ Na_2S, \ S_2Cl_2$$

Some elements have five or more possible oxidation states. The principal oxidation states of sulphur are shown in an oxidation number chart in figure 3.6.

Oxidation numbers and nomenclature

The oxidation number concept is often important in naming compounds.

Simple molecular compounds containing only two elements, such as CO_2, $SiCl_4$ and S_2Cl_2, are usually named by reference to the numbers of their different atoms without stating oxidation numbers. Hence, CO_2 – carbon *di*oxide, $SiCl_4$ – silicon *tetra*chloride and S_2Cl_2 – *di*sulphur *di*chloride.

The systematic names of more complex molecular compounds, such as HClO and H_2CrO_4, are obtained by reference to the oxidation numbers of constituent elements which can have variable oxidation numbers.

For example:

	Recommended name	Common (trivial) name
HClO	chloric(I) acid	hypochlorous acid
H_2CrO_4	chromic(VI) acid	chromic acid

However, the names sulphuric, sulphurous, sulphate, sulphite, nitric, nitrous, nitrate and nitrite are recommended by IUPAC (The International Union of Pure and Applied Chemistry) in preference to sulphuric(VI), sulphuric(IV), sulphate(VI), sulphate(IV), nitric(V), nitric(III), nitrate(V) and nitrate(III), respectively.

The systematic names of giant ionic compounds and giant molecular compounds usually include the oxidation numbers of those constituent elements which have a variable oxidation state.

For example:

$FeSO_4$	iron(II) sulphate
$KMnO_4$	potassium manganate(VII)
$FeCl_3$	iron(III) chloride

NaClO   sodium chlorate(I)

$Cu(NO_3)_2$ copper(II) nitrate

Remember that Roman numerals are used in writing the oxidation number of an element within a compound to prevent any confusion with the real charge on an ion. Thus, CuO is named copper(II) oxide *not* copper(2) oxide and MnO_4^- is named manganate(VII) not manganate(7).

3.10 Explaining redox in terms of oxidation number

We are now in a position to define oxidation and reduction in terms of oxidation number. This definition of redox is an alternative to that involving electron transfer, although the two definitions are quite closely related.

> *An atom is said to be oxidised when its oxidation number increases and reduced when its oxidation number decreases.*

Consider the following reactions.

(a)
$$\overset{0}{2Na} + \overset{0}{Cl_2} \rightarrow \overset{+1 \; -1}{2NaCl}$$

The oxidation number of sodium has increased from 0 to +1: it has been oxidised. The oxidation number of chlorine has decreased from 0 to −1: it has been reduced.

(b)
$$\overset{+1}{H^+} + \overset{-2 \; +1}{OH^-} \rightarrow \overset{+1 \; -2}{H_2O}$$

This ionic equation summarises the neutralisation of an acid with an alkali. Notice that the oxidation number of each element remains unchanged and so the reaction does not involve redox.

(c) Another important ionic reaction which does not involve redox is precipitation. Here again the oxidation number of each element remains unaltered during the reaction.

$$\overset{+1}{Ag^+}(aq) + \overset{-1}{Cl^-}(aq) \rightarrow \overset{+1 \; -1}{AgCl}(s)$$

$$\overset{+2}{Ba^{2+}}(aq) + \overset{+6 \; -2}{SO_4^{2-}}(aq) \rightarrow \overset{+2 \; +6 \; -2}{BaSO_4}(s)$$

(d)
$$\overset{+2 \; -2}{CO} + \overset{0}{\tfrac{1}{2}O_2} \rightarrow \overset{+4 \; -2}{CO_2}$$

In this case, the oxidation number of carbon rises from +2 to +4: it has been oxidised. The oxidation number of the oxygen atom in CO remains unchanged, but the elemental oxygen is reduced: its oxidation number falls from 0 to −2.

Which of the following reactions involve redox?

1. $Cl_2 + 2OH^- \rightarrow Cl^- + ClO^- + H_2O$
2. $Cu^{2+} + 2OH^- \rightarrow Cu(OH)_2$
3. $H_2O + SO_3 \rightarrow H_2SO_4$
4. $2CrO_4^{2-} + 2H^+ \rightarrow Cr_2O_7^{2-} + H_2O$

3.11 The advantages and disadvantages of the oxidation number concept

Oxidation numbers can help us decide whether or not redox is involved in a particular process. Oxidation numbers show that neutralisation and precipitation are not redox reactions, even though they involve ions. This point highlights the importance of oxidation numbers as an electron book-keeping device that allows us to recognise redox processes.

The second advantage in using oxidation numbers is that they allow us to see exactly which part of a molecule or a complex ion is reduced or oxidised. For example, the half-equation

$$MnO_4^- + 8H^+ + 5e^- \rightarrow Mn^{2+} + 4H_2O$$

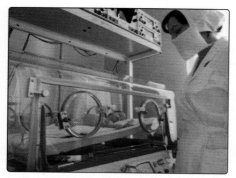

▲ This premature baby is being supplied with oxygen in an incubator. The oxygen is needed for the baby's respiration, which is an important redox reaction.

shows that MnO_4^- and H^+ ions are reduced to Mn^{2+} and $4H_2O$. But, which element or elements in MnO_4^- and H^+ are reduced? If oxidation numbers are assigned to the atoms in the half-equation

$$\overset{+7 \; -2}{MnO_4^-} + \overset{+1}{8H^+} + 5e^- \rightarrow \overset{+2}{Mn^{2+}} + \overset{+1 \; -2}{4H_2O}$$

it is clear that manganese is the reduced element because its oxidation number changes from +7 to +2.

The main disadvantage of the oxidation number concept is that it can lead to a misunderstanding about the structure of molecular substances. No physical or structural significance can be attached to oxidation numbers of atoms in molecular substances. The oxidation number of carbon in CO_2 is +4, but it must *not* be supposed that there is a charge of +4 on the carbon atom.

In a few cases ambiguities can arise with oxidation numbers. For example, the rules for assigning oxidation numbers suggest that each sulphur atom in the thiosulphate ion, $S_2O_3^{2-}$, has an oxidation number of +2. However, the structure of the $S_2O_3^{2-}$ ion shows that the two sulphur atoms in it are quite different. One sulphur atom is at the centre of a tetrahedron bonded to the other four atoms (one S and three O atoms) similar to the S atom in the SO_4^{2-} ion. With this in mind, we could assign an oxidation number of +6 to the central S atom in $S_2O_3^{2-}$ (similar to the central S atom in SO_4^{2-}) and an oxidation number of -2 to each of the surrounding atoms, including the second sulphur atom.

Two further problems with oxidation numbers concern their use with organic compounds.

> 1   What is the oxidation number of carbon in
> (i) CH_4; (ii) C_2H_6; (iii) C_3H_8?

The carbon atoms in CH_4, C_2H_6 and C_3H_8 all have four covalent bonds. In spite of this similarity, they have different oxidation numbers i.e -4, -3 and $-2\frac{2}{3}$ respectively.

The other problem (which C_3H_8 highlights) is that in some compounds, atoms can have oxidation numbers which are not whole numbers.

In spite of these disadvantages the concept of oxidation number is still very useful.

Summary

1   There are two related definitions of redox, one involving electron transfer, the other involving the concept of oxidation number.

2   In terms of electron transfer, oxidation is defined as a loss of electrons, and reduction is defined as a gain of electrons.

3   Electron-transfer reactions are called redox reactions.

4   The energy evolved by many electron-transfer reactions can be released in the form of electrical energy and harnessed to provide electricity.

5   An electrochemical cell converts chemical energy into electrical energy by producing an electric current. An electrolytic cell converts electrical energy into chemical energy by producing new substances at the electrodes.

6   Electrolysis is the decomposition of a molten or aqueous compound by electricity. The compound decomposed during electrolysis is called an electrolyte.

7   When an electric current flows:

$$\underset{\text{charge}}{Q} \quad = \quad \underset{\text{current}}{I} \quad \times \quad \underset{\text{time}}{t}$$

8   During electrolysis, one mole of an element is always produced by a multiple of 96 500 coulombs. Because of this, 96 500 coulombs is called the Faraday constant, F.

9   An oxidation number is a number assigned to an atom or ion to describe its relative state of oxidation or reduction. Using oxidation numbers it is possible to decide whether redox has occurred.

10   In terms of oxidation numbers, oxidation is defined as an increase in oxidation number and reduction is defined as a decrease in oxidation number.

1 What are the oxidation numbers of
(a) chlorine in HCl, $HClO$, ClO_3^-, PCl_3, Na_3AlCl_6, $POCl_3$;
(b) nitrogen in N_2O, NO, NO_2, NO_3^-, N_2H_4, HCN?

2 (a) Write the formulas for substances containing sulphur in which it shows the following oxidation states: $-2, -1, 0, +1, +2, +4, +6$.
(b) The following equations summarise redox reactions involving sulphur compounds. Deduce the oxidation number of all the atoms and ions in these equations and hence determine precisely which species is oxidised and which is reduced.
(i) $2MnO_4^- + 6H^+ + 5SO_3^{2-} \rightarrow 2Mn^{2+} + 3H_2O + 5SO_4^{2-}$
(ii) $2NaI + 3H_2SO_4 \rightarrow 2NaHSO_4 + 2H_2O + I_2 + SO_2$
(iii) $S_2O_3^{2-} + 2H^+ \rightarrow S + SO_2 + H_2O$
(iv) $SO_3^{2-} + H_2O + 2Ce^{4+} \rightarrow SO_4^{2-} + 2H^+ + 2Ce^{3+}$
(v) $2S_2O_3^{2-} + I_2 \rightarrow S_4O_6^{2-} + 2I^-$

3 Which of the following may be regarded as redox reactions? Explain your answers.
(a) $Cu^{2+} + 4NH_3 \rightarrow Cu(NH_3)_4^{2+}$
(b) $Cl_2 + 2OH^- \rightarrow Cl^- + ClO^- + H_2O$
(c) $Ca^{2+} + 2F^- \rightarrow CaF_2$
(d) $Ca + F_2 \rightarrow CaF_2$
(e) $2CCl_4 + CrO_4^{2-} \rightarrow 2COCl_2 + CrO_2Cl_2 + 2Cl^-$
(f) $NH_3 + H^+ \rightarrow NH_4^+$

4 Write redox half-equations for the following reactions:
(a) When copper is added to concentrated nitric acid the solution becomes pale blue, and brown fumes of nitrogen dioxide are produced.
(b) When potassium iodide is added to acidified hydrogen peroxide a brown colour appears.
(c) Sodium sulphite reduces an acidified solution of orange dichromate(VI) ions, $Cr_2O_7^{2-}$, to a green solution containing Cr^{3+} ions.
(d) Manganese(IV) oxide will oxidise concentrated hydrochloric acid to chlorine.
(e) When zinc is added to silver nitrate solution, crystals of silver form on the zinc surface.

5 Look at the apparatus in figure 3.7.

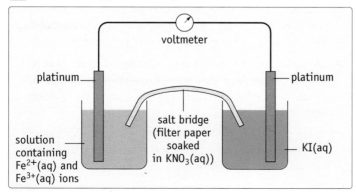

Figure 3.7
Electron transfer between KI(aq) and a solution containing Fe^{2+} and Fe^{3+} ions.

When the circuit is complete a brown colour appears around the platinum in the right-hand beaker.
(a) Write a half-equation to summarise the reactions at each of the platinum terminals.
(b) In which direction do electrons flow?
(c) Explain the function of the salt bridge, stating which ions are moving into it and out of it in each beaker.
(d) Do you think the voltage of the cell will increase, decrease or remain the same if the concentration of KI in the right-hand beaker is increased? Explain your answer.

6 What do you understand by the terms 'oxidation' and 'reduction'? In each of the following reactions say what (if anything) has been reduced and what has been oxidised. Write electron-transfer equations to explain your answers:
(a) $2FeCl_2 + Cl_2 \rightarrow 2FeCl_3$
(b) $CuO + H_2 \rightarrow Cu + H_2O$
(c) $3Cu + 8HNO_3 \rightarrow 3Cu(NO_3)_2 + 4H_2O + 2NO$
(d) $2Na + H_2 \rightarrow 2NaH$
(e) $AgNO_3 + NaCl \rightarrow AgCl + NaNO_3$

7 (a) What is the oxidation number of chromium in each of the substances A–F in the reaction scheme in figure 3.8?
(b) Which steps in the scheme involve redox?
(c) Write equations or half-equations to describe each step in the scheme.

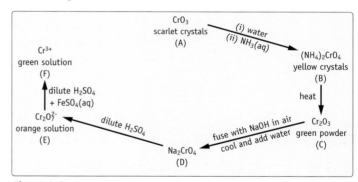

Figure 3.8
A reaction scheme involving compounds of chromium.

8 Certain features of the element vanadium, V, are presented in figure 3.9.
Consider the diagram carefully.
(a) What is the oxidation number of vanadium in the compounds A–I?
(b) What can you deduce about the oxidising power of the halogens towards vanadium?
(c) How does the action of chlorine on vanadium compare with the action of hydrogen chloride?

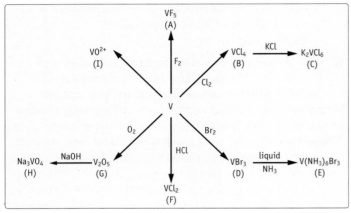

Figure 3.9
Products from vanadium.

9 0.50 g of hydrated iron(II) sulphate ($FeSO_4.7H_2O$) was dissolved in dilute sulphuric acid and titrated with 0.1 mol dm^{-3} potassium manganate(VII) (potassium permanganate). What volume of the potassium manganate(VII) is required to complete the titration? What assumptions have you made in the calculations?
(Fe = 55.8; S = 32; O = 16; H = 1)

10 10 g of an impure iron(II) salt were dissolved in water and made up to 200cm^3 of solution. 20 cm^3 of this solution, acidified with dilute sulphuric acid, required 25 cm^3 of 0.04 mol dm^{-3} $KMnO_4$(aq) before a faint pink colour appeared.
(a) Write a balanced ionic equation (or half-equations) for the reaction of acidified manganate(VII) (permanganate) ions with iron(II) ions.
(b) How many moles of iron(II) ions react with one mole of MnO_4^- ions?
(c) How many moles of Fe^{2+} react with 25 cm^3 of 0.04 mol dm^{-3} $KMnO_4$(aq)?
(d) How many grams of Fe^{2+} are there in the 200 cm^3 of original solution?
(Fe = 56)
(e) Calculate the percentage by mass of iron in the impure iron(II) salt.

Patterns and Periodicity

4

4.1 Introduction

Chemists have always searched for patterns and similarities in the properties and reactions of substances. From your own studies of chemistry you will already appreciate the classification of elements as metals and non-metals and you will have looked at the patterns in reactivity of different metals with oxygen, water and acids. Unfortunately, there are some elements that cannot be classified easily and unambiguously as either a metal or a non-metal. Non-metals are usually volatile and non-conductors of electricity, but graphite (carbon) and silicon (section 4.8), usually classed as non-metals, have very high melting points and very high boiling points, and both conduct electricity. Can you think of other elements which do not fit neatly into the classification of elements as metals or non-metals?

4.2 Families of similar elements

Limitations in the overall classification of elements as metals or non-metals led chemists to search for trends and similarities in the properties of much smaller groups of elements. Pairs of similar elements, such as sodium and potassium, calcium and magnesium, and chlorine and bromine, will already be familiar to you.

To what extent do sodium and potassium resemble each other?

1. Do they have similar properties (e.g. melting point, boiling point, hardness, density, lustre)?

2. Do they react in the same way with oxygen in the air?

3. What colour are their oxides, chlorides, sulphates, etc?

4. How do sodium and potassium react with water? Write equations for their reactions with water. Are their reactions and products similar?

5. Are the formulas of their compounds similar? Do sodium and potassium have the same oxidation number in their compounds?

In this chapter, we shall follow the search for patterns amongst elements leading to the development of the periodic table. Then, we shall study the physical and chemical periodicity of the elements in periods 2 and 3. In chapter 13, these periodic trends are explained in terms of atomic properties such as electronic structure, bonding and electronegativity.

Early in the nineteenth century, the German chemist Döbereiner pointed out that many of the known elements could be arranged in groups of three similar elements. He called these families of three elements '**triads**'. Two of Döbereiner's triads were lithium, sodium and potassium (alkali metals – figure 4.1) and chlorine, bromine and iodine (halogens – figure 4.2). Döbereiner showed that when the three elements in each triad were written in order of relative atomic mass, the middle element had properties in

▲ Johann Wolfgang Döbereiner was born in 1780 at Hof and died in 1849 at Jena. Döbereiner studied as a pharmacist at Münchberg and then studied chemistry at Strasbourg. In 1810 he became professor of chemistry and pharmacy at Jena. Döbereiner made the first observations on platinum as a catalyst and on families of similar elements. Döbereiner and the poet Goethe were close friends and colleagues.

Li
6.9
Na
23.0
K
39.1

Figure 4.1
Döbereiner's triad of alkali metals.

Cl
35.5
Br
79.9
I
126.9

Figure 4.2
Döbereiner's triad of halogens.

between those of the other two. What is more, the relative atomic mass of the middle element was very close to the average of the relative atomic masses of the other two elements. In the triad of chlorine, bromine and iodine, the relative atomic mass of bromine (79.9) is close to the average of the relative atomic masses of chlorine and iodine $((35.5 + 126.9)/2 = (162.4)/2 = 81.2)$.

> **1** How close to the relative atomic mass of sodium is the average of the relative atomic masses of lithium and potassium?

4.3 Patterns and periodicity

The relationships that Döbereiner had discovered encouraged other chemists to search for patterns relating the properties of elements to their relative atomic masses.

Lothar Meyer's curves

In 1869, the German chemist Julius Lothar Meyer and the Russian chemist Dmitri Mendeléev separately published results which showed that when the elements were arranged in order of their relative atomic masses, any one element had properties similar to those of elements in front of it and behind it on the list. This periodic repetition of similar elements led to the idea of periodicity and then to a **periodic table** of the elements.

Lothar Meyer plotted various physical properties (melting point, boiling point, density) of the known elements against their relative atomic masses. One of Lothar Meyer's curves is shown in figure 4.3. It reveals the periodic variation in melting point with relative atomic mass. Notice that elements in the same chemical family occur at similar points on the curves.

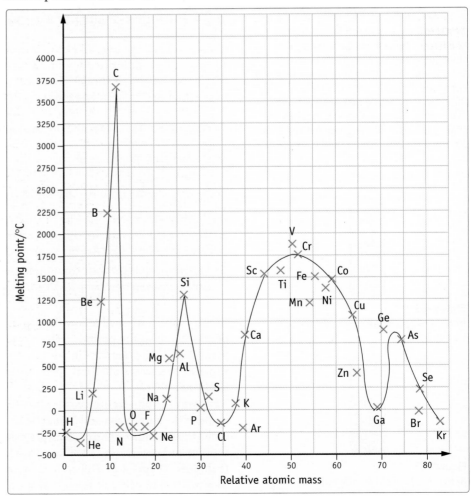

Figure 4.3
Lothar Meyer's curve of melting point against relative atomic mass.

1. Where do the alkali metals appear on the curves in figure 4.3? Do they occupy similar positions?

2. Where do the halogens appear on the curves in figure 4.3? Do they occupy similar positions?

3. Which elements appear on the peaks in figure 4.3? Are these elements alike in their properties?

Mendeléev's periodic table

Although Lothar Meyer's curves showed a periodic repetition of properties with relative atomic mass, most of the credit for arranging the elements in a **periodic table** is given to Mendeléev.

Mendeléev arranged all the elements known to him in order of increasing relative atomic mass and showed that elements with similar properties recurred at regular intervals. Figure 4.4 shows part of Mendeléev's periodic table published in 1869. Elements with similar properties fall in the same vertical column. These vertical columns of similar elements are called **groups** and the horizontal rows of elements are called **periods**.

Figure 4.4
Mendeléev's periodic table.

	Group I	Group II	Group III	Group IV	Group V	Group VI	Group VII	Group VIII
Period 1	H							
Period 2	Li	Be	B	C	N	O	F	
Period 3	Na	Mg	Al	Si	P	S	Cl	
Period 4	K	Ca	*	Ti	V	Cr	Mn	Fe Co Ni
	Cu	Zn	*	*	As	Se	Br	
Period 5	Rb	Sr	Y	Zr	Nb	Mo	*	Ru Rh Pd
	Ag	Cd	In	Sn	Sb	Te	I	

Mendeléev's periodic table was far more successful than earlier periodic tables proposed by other chemists. There were four important reasons for his success.

In the first place, Mendeléev left gaps in his table in order that similar elements would fall in the same vertical column.

Secondly, he suggested that, in due course, elements would be discovered to fill these gaps.

Thirdly, he predicted properties for some of the missing elements.

Fourthly, and most importantly, elements with the properties he had predicted were discovered in Mendeléev's own lifetime to fill the gaps in his table.

There are four gaps (indicated by *) in the portion of Mendeléev's periodic table reproduced in figure 4.4. Which elements have since been discovered to fill these gaps?

Mendeléev also proposed that periods 4, 5, 6 and 7 should contain more than just seven elements as in periods 2 and 3. In order to fit these longer periods into his pattern, he divided them into halves and placed the first half of the elements in the top left-hand corner of their space (e.g. K, Ca, etc., in period 4), and the second half in the bottom right-hand corners (e.g. Cu, Zn, etc.).

The accuracy of Mendeléev's predictions convinced scientists that his ideas were correct and his periodic table was accepted as a valuable overall summary of the properties of elements. Indeed, Mendeléev's concept of a periodic table has absorbed more and more new knowledge and the importance and usefulness of his original idea has been demonstrated many times.

4.4 Modern forms of the periodic table

Mendeléev's periodic table arranged the elements in order of relative atomic mass. His periodic law stated that *the properties of the elements are a periodic function of their relative atomic masses*. This law fulfilled two important functions.

1 It summarised the properties of elements, putting them into groups with similar properties.
2 It enabled predictions to be made about the properties of known and unknown elements and led to considerable research activity.

Although Mendeléev used the order of relative atomic masses as a basis for his periodic table, he wrote the elements tellurium (Te = 127.6) and iodine (I = 126.9) in the reverse order. Mendeléev found that this was necessary if Te and I were to fall in their correct vertical groups. He argued that iodine must occupy the same group as chlorine and bromine.

Mendeléev's periodic table was proposed long before chemists could understand its fundamental relationship to electronic structure. Although the reverse order of Te and I worried scientists at the time of Mendeléev's proposals, it is now explained by modern forms of the periodic law. This states that *the properties of elements are a periodic function of their atomic numbers*. (The atomic number of Te is 52 and that of I is 53.)

If the elements are numbered along each period from left to right starting at period 1, then period 2, etc., the number given to each element is its **atomic number**. The real significance of atomic number is discussed in sections 5.2 and 5.4.

In modern periodic tables, all the elements are in strict order of their atomic numbers.

Although there are various forms of the periodic table suitable for one purpose or another, the 'wide form' shown in figure 4.5 is probably the most useful. In this figure the atomic numbers are shown below the symbols for each element.

Those groups with elements in periods 2 and 3 are numbered from I to VII followed by Group O. Some of these groups have names besides numbers. The most common names used for particular groups are:

Group number	Group name
I	alkali metals
II	alkaline-earth metals
VII	halogens
0	noble (inert) gases

The most obvious difference between the modern periodic table (figure 4.5) and that proposed by Mendeléev is the removal of the **transition elements** from the simple groups. In period 4, for example, 10 transition elements (Sc, Ti, V, Cr, Mn, Fe, Co, Ni, Cu, Zn) have been taken out of the simple groupings suggested by Mendeléev and placed between Ca in Group II and Ga in Group III.

At the present time there are 106 known elements. The most recently discovered elements are kurchatovium (Ku, atomic number 104) which was first synthesised by Russian chemists in 1964, hahnium (atomic number 105) synthesised at Berkeley, USA, in 1970, and meitnerium (Mt, atomic number 109) synthesised by West German scientists in 1983.

Besides dividing the periodic table into vertical groups of similar elements, it is also useful to split it into five blocks of elements with similar properties. These five blocks are coloured differently in figure 4.5.

▲ Dmitri Ivanovich Mendeléev (1839–1907). In 1869, Mendeléev published his periodic table. This was the forerunner of modern periodic tables.

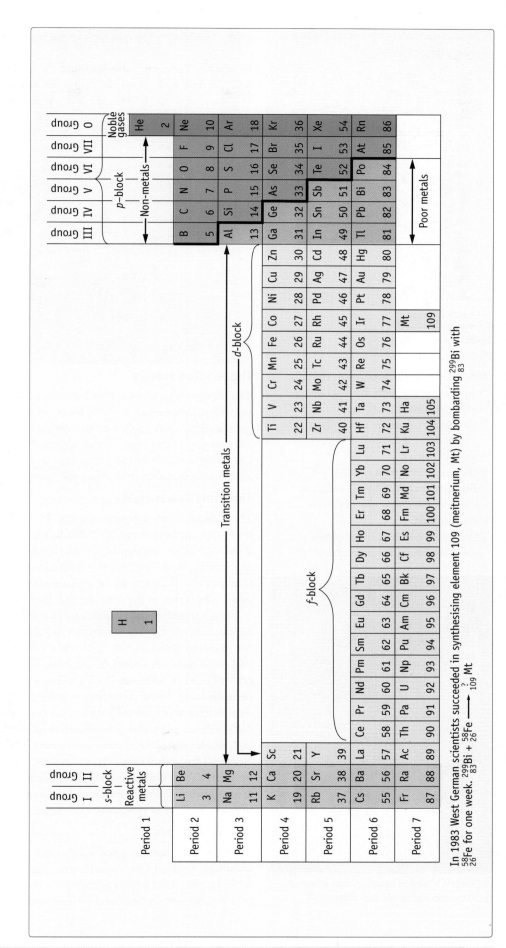

Figure 4.5
The modern periodic table (wide form).

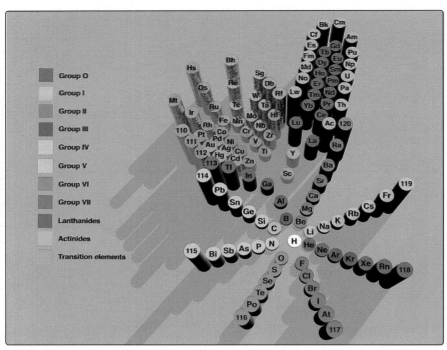

How do the transition metals iron and copper compare with the reactive metals sodium and calcium in:

1. their melting points and boiling points,
2. their densities,
3. their reactions with water,
4. the number of different oxidation states which they show in their compounds,
5. the colours of their compounds as solids and in aqueous solution?

▲ Crystals of xenon fluoride, first isolated at the Argonne National Laboratory, USA.

The reactive metals

The elements in Group I and Group II form a block of reactive metals. They are sometimes referred to as the '*s*-block' elements, since the outermost electrons in these metals are in *s* sub-shells (see section 6.5). These metals (including potassium, sodium, calcium and magnesium) are all high in the activity (electrochemical) series. They have lower densities, lower melting points and lower boiling points than most other metals and they form stable, involatile ionic compounds.

The transition metals

These elements occupy two rectangles between Group II and Group III. Some of the transition metals are called '*d*-block' elements, because electrons are being added to *d* sub-shells across this block of elements (section 6.5). These metals (including chromium, iron, copper, zinc and silver) are much less reactive than the metals in Groups I and II. In this block of elements there is also a marked horizontal similarity in properties as well as the usual vertical likeness.

The lanthanides and actinides form a second block of elements within the transition metals. Indeed, they are sometimes called the **inner transition elements**. Another name for these elements is the '*f*-block' elements, since electrons are being added to *f* sub-shells across this block of elements. The lanthanides consist of the 14 elements from cerium (Ce) to lutetium (Lu) which come immediately after lanthanum (La) in the periodic table. These elements resemble each other very closely. In fact, the horizontal similarities across this block are so great that chemists experienced considerable difficulty in separating the lanthanides from one another. Lanthanum and the lanthanides are sometimes known as the rare earth elements. The actinides are the 14 elements from thorium (Th) to lawrencium (Lr) which follow actinium (Ac) in the periodic table. Only the first three elements in the actinide series (thorium (Th), protactinium (Pa) and uranium (U)) are naturally occurring. All the elements beyond uranium have been synthesised by chemists since 1940.

The poor metals

The term 'poor metals' is not widely used, but it is a useful description for several metals including tin, lead and bismuth. These metals fall in a triangular block of the periodic table to the right of the transition metals. They are usually low in the activity (electrochemical) series and they have some resemblances to non-metals.

The non-metals

These elements also form a triangular block in the periodic table. The elements in this block and the previous one are sometimes called the '*p*-block' elements since the outermost electrons in these elements are going into *p* sub-shells.

The noble gases

The atoms in these elements have completely filled *s* and *p* sub-shells of electrons. They are very unreactive and the first noble gas compound was not obtained until 1962. Because of their chemical unreactivity these elements were originally called 'inert gases'. Nowadays, several compounds of these elements (mainly oxides and fluorides of xenon and krypton) are known and the name 'inert' has been replaced by 'noble'.

4.5 Metals, non-metals and metalloids

Although the periodic table does not classify elements as metals and non-metals, there is a fairly obvious division between the two. '*Fairly obvious*', you will note, not '*clear cut*'. Separating the elements into either metals or non-metals is rather like trying to separate all the shades of grey into either black or white. It cannot be done without ambiguity. The fairly obvious division between metals and non-metals is shown by a thick stepped line in figure 4.5. The 20 or so non-metals are packed into the top right-hand corner above the thick stepped line. Some of the elements next to the thick steps, such as germanium, arsenic and antimony, have similarities to both metals and non-metals and it is difficult to place these in one class or the other. Chemists sometimes use the name **metalloid** (or **semi-metal**) for these elements which are difficult to classify one way or the other.

Figure 4.6 shows a classification of elements as metals, metalloids and non-metals on the basis of their electrical conductance. In this classification:

1 *Metals are good conductors of electricity* with atomic conductance* (atomic electrical conductivity) greater than 10^{-3} ohm^{-1} cm^{-4}. Their conductivity slowly falls as the temperature rises.

2 *Metalloids are poor conductors of electricity* with atomic conductance usually less than 10^{-3} but greater than 10^{-5} ohm^{-1} cm^{-4}. The conductivity of a metalloid increases as the temperature rises and is also considerably affected by the presence of impurity.

3 *Non-metals are virtually non-conductors* (insulators). Their atomic conductance is usually less than 10^{-10} ohm^{-1} cm^{-4}.

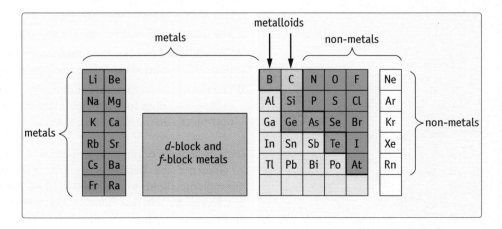

Figure 4.6
Classifying the elements as metals, metalloids and non-metals on the basis of their electrical conductance. In this figure, metalloids are coloured grey.

*The atomic conductance (atomic electrical conductivity) is the conductivity of a block of the substance 1 cm^2 in cross-section but long enough to contain one mole of atoms of the element. It is a measure of the conductivity of one mole of atoms of the element. Notice that the units of atomic conductance (atomic electrical conductivity) are ohm^{-1} cm^{-4}. The electrical conductivity of a substance is usually defined as the reciprocal of the electrical resistivity. This is the resistance of a section of the substance 1 cm^2 in cross-section and one centimetre long. The units of electrical resistivity are ohm cm so the units of electrical conductivity are therefore ohm^{-1} cm^{-1}. This compares equal volumes of substances. More appropriately, the atomic conductance (atomic electrical conductivity) compares the conductivity of equal numbers (i.e. 1 mole) of atoms and its value is obtained by dividing the electrical conductivity of an element by its molar volume. Hence, the units of atomic conductance are ohm^{-1} cm^{-4}.

Notice in figure 4.6 that the cell for carbon has been shaded less heavily than those for the other metalloids. This is because carbon exists in two different solid forms called **allotropes**. One of these allotropes, graphite, is a poor conductor of electricity and would be classed as a metalloid; the other allotrope, diamond, is an insulator and therefore classed as a non-metal. In spite of problems such as this, the classification of elements into metals, metalloids and non-metals is useful and convenient.

One other criterion which has been used to classify elements as metals, metalloids and non-metals is the acidic/basic nature of their oxides.

> *A metal can be defined as an element whose oxide (with the group oxidation number) is basic. A metalloid is an element whose oxide is amphoteric and a non-metal is an element whose oxide is acidic.*

1. Which elements in periods 2 and 3 are classed as metalloids on the criteria of electrical conductivity described above?
2. Which elements in periods 2 and 3 are classed as metalloids on the criterion of amphoteric oxides? (Section 13.7 will help you with this question.)

4.6 Periodic properties

In the following sections of this chapter we shall look at the variation in properties of the elements across the periodic table. Some of these features and other properties of the elements will be taken up again and extended in chapter 13. In chapters 15, 16 and 17, we shall look at the vertical trends in properties down the periodic table.

In this chapter we shall pay particular attention to elements in the second and third periods, Li, Be, B, C, N, O, F, Ne, and Na, Mg, Al, Si, P, S, Cl, Ar, respectively.

4.7 The periodicity of physical properties

Tables 4.1 and 4.2 show the values of various physical properties for the elements in the second and third rows, respectively, of the periodic table. Notice the following trends across the second and third periods.

1 The melting point and molar heat of fusion ($\Delta H_{fus.}$) rise to the element in Group IV and then fall to low values (figure 4.7).

2 The boiling point and molar heat of vaporisation (ΔH_{vap}) rise to the element in Group IV and then fall to low values.

3 The density rises to the elements in Groups III and IV and then falls (figure 4.8).

4 The molar volume (i.e. the volume occupied by 1 mol of the element) falls to the centre of the periods and then rises again (figure 4.8).

5 The electrical and thermal conductivity is relatively high for the metals on the left of each period, lower for metalloids in the centre of the periods and almost negligible for non-metals on the right of each period.

This periodicity of physical properties is, of course, related to a periodicity in the types of elements and their structures. Across a period, the elements change from metals through metalloids to non-metals. In period 3, sodium, the left-hand element, is a very reactive metal, whereas chlorine, next to the extreme right, is a very reactive non-metal. In between, the elements show a gradual transition from metals to non-metals. These periodic changes in the physical properties of the elements across the table are reflected in a periodic change in structure. The structure of the elements varies from metallic, through giant molecular in the metalloids to simple molecular structures in the non-metals.

Table 4.1
The values of various physical properties for the elements in the second row of the periodic table

	Li	Be	B	C (graphite)	C (diamond)	N	O	F	Ne
Melting point/°C	180	1280	2030	3700	3550	−210	−219	−220	−250
Heat of fusion/kJ mol^{-1}	+3.0	+11.7	+22.2	–	–	+0.36	+0.22	+0.26	+0.33
Boiling point/°C	1330	2480	3930	sublimes	4830	−200	−180	−190	−245
Heat of vaporisation /kJ mol^{-1}	+135	+295	+539	+717 (sub)	–	+2.8	+3.4	+3.3	+1.8
Density/g cm^{-3} (at 25°C)*	0.53	1.85	2.55	2.25	3.53	0.81	1.14	1.11	1.21
Molar volume/cm^3 mol^{-1} (conditions as for density)	13.1	4.9	4.6	5.3	3.4	17.3	14.0	17.1	16.7
Atomic conductance × 1000 /ohm^{-1} cm^{-4}	8	51	–	0.14	–	–	–	–	–
Thermal conductivity /J cm^{-1} s^{-1} K^{-1} (at 25°C)	0.71	1.6	0.01	0.24	–	0.00025	0.00025	–	0.00042
Type of element	Metals		Metalloids			Non-metals			
	Li	Be	B	C(graphite)	C(diamond)	N	O	F	Ne
Type of structure	Giant metallic		Giant molecular			Simple molecular			
						N$_2$	O$_2$	F$_2$	Ne

*For those elements which are gaseous at 25°C the density quoted is that of the liquid at its boiling point.

Table 4.2
The values of various physical properties for the elements in the third row of the periodic table

	Na	Mg	Al	Si	P (white)	S (rhombic)	Cl	Ar
Melting point/°C	98	650	660	1410	44	119	−101	−189
Heat of fusion/kJ mol^{-1}	+2.60	+8.95	+10.75	+ 46.4	+0.63	+1.41	+3.20	+1.18
Boiling point/°C	890	1120	2450	2680	280	445	−34	−186
Heat of vaporisation/kJ mol^{-1}	+89.0	+128.7	+293.7	+376.7	+12.4	+9.6	+10.2	+6.5
Density/g cm^{-3} (at 25°C)*	0.97	1.74	2.70	2.33	1.82	2.07	1.57	1.40
Molar volume/cm^3 mol^{-1} (conditions as for density)	23.7	14.6	10.0	12.1	16.9	15.6	22.8	28.5
Atomic conductance × 1000 /ohm^{-1} cm^{-4}	10	16	38	4	10^{-16}	10^{-22}	–	–
Thermal conductivity /J cm^{-1} s^{-1} K^{-1} (at 25°C)	1.34	1.6	2.1	0.84	–	0.00029	0.00008	0.00017
Type of element	Metals			Metalloid	Non-metals			
	Na	Mg	Al	Si	P	S	Cl	Ar
Type of structure	Giant metallic			Giant molecular	Simple molecular			
					P$_4$	S$_8$	Cl$_2$	Ar

*For those elements which are gaseous at 25°C the density quoted is that of the liquid at its boiling point.

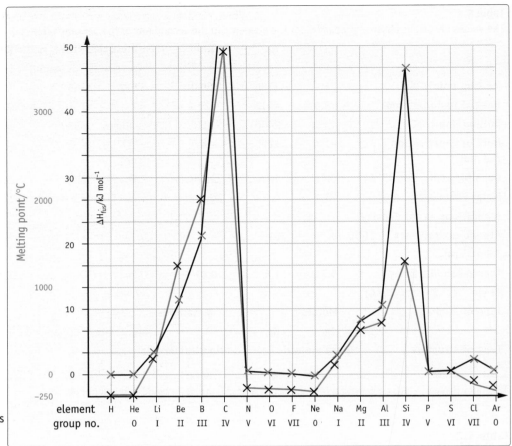

Figure 4.7
Variation of melting point (red line) and ΔH_{fus} (black line) for the elements hydrogen to argon.

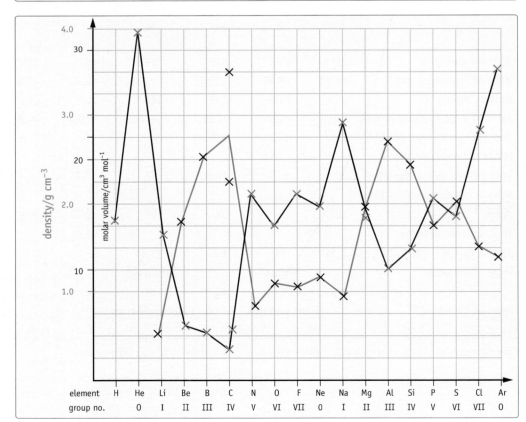

Figure 4.8
Variation of density (red line) and molar volume (black line) for the elements hydrogen to argon.

1 Which elements occur at or near the peaks on the graph in figure 4.7? What does this tell you about the strength of forces between their particles? In which groups of the periodic table are these elements found? What type of structure do these elements possess?

2 Which elements occur at or near the troughs on the molar volume graph in figure 4.8? What does this tell you about the average distances between their atoms?

3 Why are the heat of fusion, heat of vaporisation and molar volume values in tables 4.1 and 4.2 given for one mole of the element and not one gram?

Notice that elements on the left-hand side of the periodic table exist as giant structures (metallic or giant molecular), whereas those on the right consist of small molecules (simple molecular). The graphs in figure 4.7 reflect this change in structure across periods 2 and 3 by the sharp drops from C to N and from Si to P. In period 2, carbon is the last element with a giant structure and nitrogen is the first element with simple molecules (N_2). In period 3, silicon is the last element with a giant structure and phosphorus is the first element with simple molecules (P_4).

4.8   Relating the properties of elements to their structure

Giant metallic structures

Metals usually have high melting points, high boiling points and high heats of fusion and vaporisation. These high values suggest that there are strong forces between the separate atoms in the metal. How does the structure of a metal explain these and other typical metal properties? The physical properties of a metal can be explained using a model in which the outer shell electrons of the metal move randomly throughout a l attice of regularly spaced positive ions (figure 4.9). The moving electrons are sometimes described as a 'sea' or 'cloud' of fluctuating negative charge.

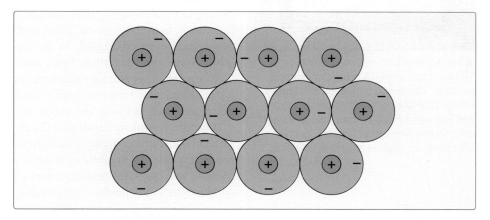

Figure 4.9
A model of metallic structure.

In the metal lattice, each positively charged ion is attracted to the 'cloud' of negative electrons and vice versa. These electrostatic attractions bind the entire crystal together as a single unit. In this model, one particular electron does not belong to one particular metal ion but is attracted to all the positive ions in the lattice and vice versa. Hence the name **giant metallic structure**.

Since the mobile outer shell electrons are responsible for the bonding in metals, it is not surprising that moving from sodium (one outer shell electron) through magnesium (two outer shell electrons) to aluminium (three outer shell electrons) the bonding gets gradually stronger. Thus, the melting point, the boiling point and the heats of fusion and vaporisation rise from Na to Mg to Al. A similar trend is evident in the second period from Li → Be → B (see figure 4.7). The bonding in these elements is discussed more fully in sections 13.3 and 13.4.

The stronger bonding from Na → Mg → Al means that the atoms are pulled closer in Mg than Na and even closer in Al. This explains the increasing density (mass per unit volume) and the decreasing molar volume (volume of one mole of atoms) from Na → Mg → Al. Similar trends are observed in the second period from Li → Be → B (see figure 4.8).

The mobile outer electrons in metals account for their high electrical and thermal conductivity. Under normal conditions these mobile electrons will move fairly randomly throughout the lattice of positive ions in the metal crystal. When the metal is connected across a difference of potential there is an overall movement of electrons (superimposed on their random motion) away from the repelling negative electrode towards the attracting positive electrode. Thus, metals have good electrical conductivity.

The high thermal conductivity of metals can also be explained in terms of their mobile electrons. Electrons in the regions of high temperature (i.e. electrons with high kinetic energy) move rapidly and randomly towards the cooler regions of the metal, transferring their energy to other electrons throughout the metal.

Look at the atomic conductance (atomic electrical conductivity) and the thermal conductivity of elements in tables 4.1 and 4.2.

1 Why do you think the electrical conductivity rises from Na → Mg → Al?

2 Why do you think the thermal conductivity rises from Na → Mg → Al and from Li → Be?

3 Why does B have a lower thermal conductivity than Be? What type of structure does B have?

Giant molecular structures

The metalloids (boron, graphite and silicon) and the non-metal diamond have giant molecular structures. The structures of these elements are also described as giant covalent and giant atomic. Figure 4.10 shows the arrangement of atoms in the structure of silicon and diamond. Each atom can be imagined to be situated at the centre of a regular tetrahedron strongly bound by covalent bonds to four other atoms. The covalent linking in these elements extends from one atom to the next through the whole lattice forming a three-dimensional **giant molecule** (**macromolecule**). The strong covalent bonds hold each atom tightly in the crystal and it is extremely difficult to break one atom away from the lattice. Thus, these elements have very high melting points and boiling points. They also have very high heats of fusion and vaporisation and they are very hard. Indeed, they have even higher values than metals for their melting points, boiling points and heats of fusion and vaporisation (figure 4.7).

The strong covalent bonds pull the atoms together and result in high densities and small molar volumes (figure 4.8). Furthermore, the electrons within the covalent bonds are held much more tightly in these elements than in metals and consequently their thermal and electrical conductivities are lower. Compare the thermal and electrical conductivity of Si with the corresponding values for Na, Mg and Al in table 4.2.

▲ The element silicon is used to make 'silicon chips' on which complex circuits can be made. This Inmos transputer combines 2 kilobytes of RAM, a 32-bit microprocessor and four serial input /output links in a single unit. Shown here actual size, this unit is effectively a complete computer on a chip.

Figure 4.10
Models showing the arrangement of atoms in the structure of silicon and diamond. (*a*) The tetrahedral arrangement of atoms around a central atom in silicon and diamond. (*b*) An extended section of the structure of silicon and diamond.

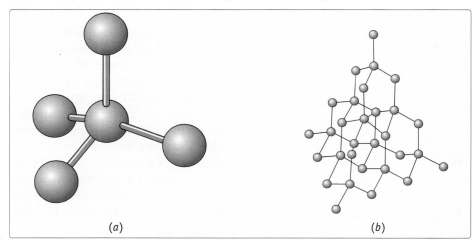

(a)   (b)

Simple molecular structures

All the non-metals in periods 2 and 3 (except diamond) form **simple molecular structures**. Each of these elements consists of separate, small molecules, i.e. N_2, O_2, F_2, Ne, P_4, S_8, Cl_2, Ar. There are strong covalent bonds between the atoms within these molecules (i.e. between the two O atoms in an O_2 molecule), but only very weak molecular (Van der Waals) forces (sections 9.3 and 9.4) between the separate molecules (i.e. between one O_2 molecule and other O_2 molecules). Consequently, the small distinct molecules can be separated easily and these non-metals have low melting points, low boiling points and low heats of fusion and vaporisation (figure 4.7). The weak Van der Waals forces between molecules such as N_2, O_2, P_4 and Cl_2 mean that their crystals are not packed tightly. So these non-metals have relatively low densities and high molar volumes (figure 4.8). There are no mobile electrons in the crystal structures of these elements and so they have very low conductivities.

4.9   The periodicity of chemical properties

Table 4.3 describes some reactions of the elements in period 3. Notice how the reactivities of the elements change across the period. Notice also that the trends in reactivity vary with different reagents.

Table 4.3
Some reactions of the elements in period 3

Element	Heat element in dry chlorine	Heat element in dry oxygen	Heat element in dry hydrogen	Add cold dilute acid (e.g. $H_2SO_4(aq)$) to the element	Add conc. HNO_3 as oxidising agent
Na	very vigorous reaction forming Na^+Cl^-	very vigorous reaction forming $(Na^+)_2O^{2-}$ + $(Na^+)_2(O_2^{2-})$	very vigorous reaction forming Na^+H^-	violent reaction $\rightarrow Na_2SO_4(aq)$ + $H_2(g)$	explosive reaction $\rightarrow Na^+$ salt + H_2O + oxides of nitrogen
Mg	vigorous reaction forming $Mg^{2+}(Cl^-)_2$	very vigorous reaction forming $Mg^{2+}O^{2-}$	vigorous reaction forming $Mg^{2+}(H^-)_2$	very vigorous reaction $\rightarrow$ $MgSO_4(aq)$ + $H_2(g)$	violent reaction $\rightarrow Mg^{2+}$ salt + H_2O + oxides of nitrogen
Al	vigorous reaction forming Al_2Cl_6	vigorous reaction at first forming $(Al^{3+})_2(O^{2-})_3$, then the oxide layer prevents further attack	no reaction	vigorous reaction (after oxide layer is removed)	oxide layer on Al reduces vigour of reaction
Si	slow reaction forming $SiCl_4$	slow reaction forming SiO_2	no reaction	no reaction with dilute acids	no reaction
P	slow reaction forming PCl_3 + PCl_5	vigorous reaction forming P_4O_6 and P_4O_{10}	no reaction	no reaction with dilute acids	vigorous reaction on heating forming P_4O_6 + P_4O_{10} + oxides of nitrogen
S	slow reaction forming SCl_2 + S_2Cl_2	slow reaction forming SO_2	very slow reaction forming H_2S	no reaction with dilute acids	slow reaction on heating forming a little SO_2
Cl	no reaction	no reaction	vigorous reaction in sunlight forming HCl	no reaction with dilute acids	no reaction
Ar	no reaction	no reaction	no reaction	no reaction with dilute acids	no reaction

For example, the reactivity of the elements with chlorine gradually falls across the period from Na to Ar. However, the reactivity with hydrogen falls at first to elements in the centre of the period and then rises for S and Cl_2. In spite of these differences, all the trends in table 4.3 can be related with considerable accuracy to the action of the elements as oxidising or reducing agents.

The metals (sodium, magnesium and aluminium) are strong reducing agents. They readily give up electrons to form their corresponding ions.

$$Mg \rightarrow Mg^{2+} + 2e^-$$

Of course, these three metals never exist freely in nature, but only in compounds as Na^+, Mg^{2+} or Al^{3+} ions. Thus, these three metals react vigorously with non-metals and acids which are oxidising agents.

$$Cl_2 + 2e^- \rightarrow 2Cl^-$$
$$O_2 + 4e^- \rightarrow 2O^{2-}$$
$$H_2 + 2e^- \rightarrow 2H^-$$
$$2H^+ + 2e^- \rightarrow H_2$$

Silicon in Group IV is a very weak reducing agent. It will react slowly with Cl_2 and O_2, which are strong oxidising agents, but not with H_2 and acids which are weaker oxidising agents.

$$Si + 2Cl_2 \rightarrow SiCl_4$$
$$Si + O_2 \rightarrow SiO_2$$

The next two elements, phosphorus and sulphur, are weak reducing agents and weak oxidising agents. They react slowly with oxygen and chlorine (strong oxidising agents), moderately with concentrated HNO_3 on heating, but they have no reaction with dilute acids (weaker oxidising agents).

$$S + Cl_2 \rightarrow SCl_2$$
$$2S + Cl_2 \rightarrow S_2Cl_2$$
$$S + 2HNO_3 \rightarrow SO_2 + H_2O + NO + NO_2$$

Phosphorus does not react directly with hydrogen, but molten sulphur will react slowly with hydrogen to form hydrogen sulphide.

$$H_2 + S \rightarrow H_2S$$

Notice in this case that the sulphur is now acting as an oxidising agent and the hydrogen as a reducing agent.

Chlorine, at the other extreme to the metals in chemical reactivity, is a strong oxidising agent. Thus, it has no reaction with O_2, dilute acids or concentrated acids which are themselves oxidising agents. However, in sunlight, it will react violently with hydrogen (which can act as a reducing agent) to form hydrogen chloride.

$$H_2 + Cl_2 \rightarrow 2HCl$$

On the extreme right, argon (a noble gas) shows no reactivity with any of these reagents.

Other periods show a similar trend in reactivity to period 3. Excluding the noble gases, elements change from strong reducing agents on the extreme left to elements which are weak reducing agents and/or weak oxidising agents in the centre and finally to strong oxidising agents.

4.10 Periodicity in the structure and properties of chlorides

Look closely at table 4.4. It shows the formulas, states and boiling points of chlorides of the elements in the first three periods.

1 Work out the oxidation numbers of each element in the chlorides shown. How do the oxidation numbers of the elements vary across period 2 and period 3?

2 How does the volatility of the chlorides (as indicated by their boiling points) vary across period 2 and period 3?

3 The structures of the chlorides change from giant structures on the left of each period to simple molecular structures on the right. Which chlorides in periods 2 and 3 have giant structures and which have simple molecular structures?

We shall look at the trends in properties across a period more fully in chapter 13.

Table 4.4
The formulas, states and boiling points of chlorides of the elements in the first three periods

Period 1			H					He
Formula of chloride			HCl					No chloride
State of chloride (at 20°C)			g					–
B.pt. of chloride/°C			–85					–
Period 2	**Li**	**Be**	**B**	**C**	**N**	**O**	**F**	**Ne**
Formula of chloride	LiCl	$BeCl_2$	BCl_3	CCl_4	NCl_3	OCl_2 (O_7Cl_2)	FCl	No chloride
State of chloride (at 20°C)	s	s	g	l	l	g (g)	g	–
B.pt. of chloride/°C	1350	487	12	77	71	2 (decomposes)	–101	–
Period 3	**Na**	**Mg**	**Al**	**Si**	**P**	**S**	**Cl**	**Ar**
Formula of chloride	NaCl	$MgCl_2$	Al_2Cl_6	$SiCl_4$	PCl_3 (PCl_5)	SCl_2 (S_2Cl_2)	Cl_2	No chloride
State of chloride (at 20°C)	s	s	s	l	l (s)	l (l)	g	–
B.pt. of chloride/°C	1465	1418	423	57	74 (164)	59 (138)	–35	–

Summary

1 Most of the credit for arranging the elements in a periodic table is given to Mendeléev. Mendeléev arranged the elements in order of increasing relative atomic mass and showed that elements with similar properties recurred at regular intervals.

2 The vertical columns of similar elements are called groups and the horizontal rows of elements are called periods.

3 In modern periodic tables, all the elements are in strict order of atomic number. The modern periodic law states that the properties of elements are a periodic function of their atomic numbers.

4 Besides a division into vertical groups of similar elements, it is also useful to split the periodic table into five blocks of elements with similar properties:
> The reactive metals (elements in Groups I and II)
> The transition metals
> The poor metals (metals in Groups III, IV, V and VI)
> The non-metals in Groups III, IV, V, VI and VII
> The noble gases

5 In spite of its limitations, the classification of elements as metals, metalloids and non-metals is very useful. The commonest classification is based on electrical conductivity.

Metals are good conductors of electricity.
Metalloids are poor conductors of electricity.
Non-metals are non-conductors.

6 Metals form giant metallic structures. They have high melting points, high boiling points, high density and good conductivity.

7 Metalloids and diamond form giant molecular structures. They have very high melting points, very high boiling points, high density but poor conductivity.

8 Non-metals (except diamond) form simple molecular structures. They have low melting points, low boiling points, low density and they are non-conductors.

9 From left to right across the periodic table, elements change from being reactive metals, through less reactive metals, metalloids, less reactive non-metals to reactive non-metals. (On the extreme right are the noble gases.)

10 These trends in metal/non-metal character are reflected in the changing redox behaviour across a period. Excluding the noble gases, the elements change from strong reducing agents (e.g. Na, Mg) through elements which are weak reducing agents (e.g. Si), then weak oxidising agents (e.g. S) to strong oxidising agents (e.g. Cl_2).

1 Consider the elements (Li, Be, B, C, N, O, F, Ne) in the second period.
(a) Which elements
  (i) form cations,
  (ii) form a chloride of empirical formula, XCl_3,
  (iii) react together to form a compound of formula, XY,
  (iv) exist as diatomic molecules at room temperature?
(b) How do each of the following properties vary for this sequence of elements?
  (i) The boiling points of the elements.
  (ii) The oxidation numbers of the elements in the chlorides.
  (iii) The boiling points of the chlorides of the elements.
(c) Give a brief explanation of the trends in oxidation number in terms of the electronic structures of the atoms of these elements.
(d) How do you account for
  (i) the trend in boiling points of the elements,
  (ii) the trend in boiling points of the chlorides of the elements?

2 Use a data book to plot (on the same graph) the melting points and boiling points of the elements from H to Ca against atomic number.
(a) Which elements occur at or near the peaks on
  (i) the melting point curves,
  (ii) the boiling point curves?
(b) Which elements occur in the troughs on
  (i) the melting point curves,
  (ii) the boiling point curves?
(c) What type of structure is found in those elements which occur
  (i) at or near the peaks,
  (ii) in the troughs?
(d) Make a note of elements which are liquid over
  (i) unusually large temperature ranges,
  (ii) unusually small temperature ranges.
(e) What explanation can you give for the fact that some elements have
  (i) unusually large liquid ranges,
  (ii) unusually small liquid ranges?

3 Table 4.5 shows the specific heat capacity for most of the elements from Li to Ca in the periodic table.

Li	Be	B	C	N	O	F	Ne
3.36	1.66	1.30	0.50				

Na	Mg	Al	Si	P	S	Cl	Ar
1.26	1.05	0.88	0.71	0.82	0.76	0.97	–

Li	Be
0.76	0.71

Table 4.5 The specific heat capacities for elements Li to Ca inclusive (all values are in $J\ g^{-1}\ K^{-1}$)

(a) Plot the values of specific heat capacity against atomic number for the elements listed. Does the graph show a periodic variation?
(b) Calculate the specific heat capacities per mole of atoms for those elements for which values are listed in table 4.5. Now plot these values of specific heat capacity per mole against atomic number. Does the graph show a periodic variation?

4 An element A reacts with another element B to form a compound of formula AB_2. The element B exists as molecules of formula B_2. Some properties of A, B_2 and AB_2 are tabulated below.

	A	B_2	AB_2
Melting point	High (in the range 700–1200°C)	Very low (less than –50°C)	Moderately high (in the range 400–700°C)
Electrical conductivity of the solid	High	Very low	Very low
Electrical conductivity of molten material	High	Very low	High
Electrical conductivity of aqueous solution of the material			High

(a) Which particles will move when a potential difference is applied across a sample of (i) solid A, (ii) molten AB_2?
(b) Explain why the electrical conductivity of molten AB_2 is high, whereas that of the solid is very low.
(c) Electrolysis of an aqueous solution of AB_2 with Pt electrodes gave A at the cathode and B_2 at the anode. Suggest possible names for the elements A and B consistent with all the above information.

5 The following table shows the melting point and conductivity of five substances

	Melting point/K	Electrical conductivity in solid state	Electrical conductivity in molten state
Magnesium oxide, MgO	3173	poor	good
Sodium chloride, NaCl	1081	poor	good
Magnesium, Mg	923	good	good
Carbon dioxide, CO_2	217	poor	poor
Silicon(IV) oxide, SiO_2	1883	poor	poor

(a) Why is the electrical conductivity of MgO(l) good, but that of MgO(s) poor?
(b) Why is the melting point of MgO considerably higher than that of NaCl?
(c) Why is the electrical conductivity of magnesium good in *both* solid and liquid states?
(d) Why is the melting point of SiO_2 so much higher than that of CO_2 in spite of the fact that C and Si are both in Group IV of the periodic table?

6 Look closely at the boiling points of the elements Na to Ar in table 4.2.

(a) Why do the boiling points rise from Na → Si? Explain the trend in terms of the structures of the elements.

(b) Why are the boiling points of the second four elements (P to Ar) much lower than those of the first four?

(c) Why is there no clear trend in the boiling points of P, S, Cl and Ar? Plot a graph of the boiling points against the relative molecular masses of the simple molecules (P_4, S_8, Cl_2, Ar) of these elements to help you answer this question.

7 An element, X, reacts with hydrogen at 300°C to form a black compound, Y. When a sample of Y was heated to 1000°C, it decomposed forming 1.00 g of X and 141 cm^3 of hydrogen at s.t.p. If Y is heated with chlorine at 250°C, it forms a green chloride, Z, containing 37.3% by mass of chlorine and 62.7% by mass of X. Deduce

(a) the relative atomic mass of X,

(b) the element X,

(c) the formulas of Y and Z.

(The molar volume of a gas at s.t.p. is 22.4 dm^3.)

5 Atomic Structure

5.1 Introduction

Just over a century ago, scientists believed that atoms were solid indestructible particles like minute snooker balls. Since then, they have obtained a great deal of evidence about the detailed structure of atoms.

Experiments involving electrolysis suggested that certain compounds contained charged particles called ions. The formation of these ions from atoms could be explained by the loss or gain of negatively charged particles (electrons). This led to the idea that atoms consisted of a small positive nucleus surrounded by negative electrons. Later, scientists discovered that the nuclei of atoms contained two kinds of particle – protons and neutrons. But, what is the evidence for these particles? In this chapter we shall consider those experiments which have been important in shaping our ideas about atomic structure.

5.2 Evidence for atomic structure

1897 – Thomson experiments with electrons

In 1874, G.J. Stoney suggested the name 'electron' for the tiny negative particles which made up an electric current. Stoney realised that the experiments involving electrolysis which Faraday had carried out earlier in the nineteenth century could be explained in terms of electrons.

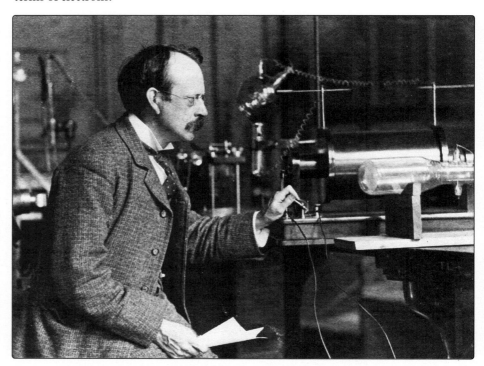

▶ J. J. Thomson (1856–1940) investigating the conductivity of electricity by gases. Using discharge tubes (similar to that on the right of the photograph), Thomson discovered the electron.

However, firm evidence for the existence of electrons was not obtained until 1897. In that year, J.J. Thomson was investigating the conduction of electricity by gases at *very low pressure*. At ordinary pressures gases are electrical insulators, but when they are subjected to very high voltages at very low pressures (below 0.01 atm) they 'break down' and conduct electricity. When the gas pressure is lowered to 0.0001 atm and a potential difference of 5000 volts is applied across the tube containing the gas, the glass container begins to glow.

When Thomson applied 15 000 volts across the electrodes of a tube containing a trace of gas, a bright green glow appeared on the glass. The green glow results from the bombardment of the glass by rays travelling in straight lines from the cathode until they strike the anode or the glass walls of the tube. Thomson called these rays **cathode rays** (figure 5.1). Thomson also showed that when cathode rays were deflected onto an electrode of an electrometer, the instrument became negatively charged and the rays could be deflected by a magnetic or by an electric field (figure 5.2).

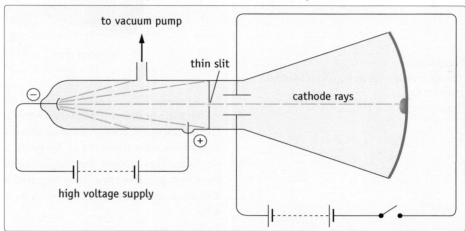

Figure 5.1
Cathode rays, deflecting plates uncharged.

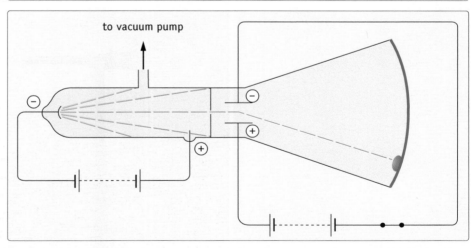

Figure 5.2
Cathode rays, deflecting plates charged.

When the rays were deflected by an electric field across a pair of charged plates, the rays moved away from the negative plate towards the positive plate. This suggested that the cathode rays were negative. Thomson studied the bending of a thin beam of cathode rays by magnetic and electric fields and concluded that they consisted of **electrons** – tiny negatively charged particles. Later, electrons were shown to be 1840 times lighter than hydrogen atoms.

Now, 6×10^{23} H atoms weigh 1 g

$$\therefore \text{1 H atom weighs} \quad \frac{1}{6 \times 10^{23}} \quad \text{g}$$

$$\therefore \text{1 electron weighs} \quad \frac{1}{6 \times 10^{23}} \quad \times \quad \frac{1}{1840} \text{ g} = 9 \times 10^{-28} \text{ g}$$

1 Where do you think the electrons in the cathode rays have come from?
2 Why do the cathode rays cause the glass to fluoresce?
3 Which piece of Thomson's evidence showed that cathode rays could not be rays of light?

The same cathode rays were obtained using different gases in the tube and with tubes and electrodes of different materials. This suggested that electrons were present in the atoms of all substances.

Thomson discovers positive rays

In his experiments with cathode rays, Thomson had noticed a red glow near the cathode. This red glow appeared on the opposite side of the cathode to the anode. To investigate this red glow, Thomson designed a discharge tube with a central cathode which had a hole in it (figure 5.3).

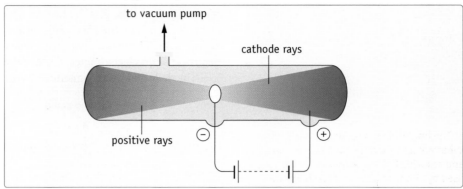

Figure 5.3
The apparatus used by Thomson to obtain positive rays.

This time he saw a red fluorescence in the tube as well as a green fluorescence. The green fluorescence was caused by cathode rays – electrons. The red glow was caused by rays which were deflected by magnetic and electric fields in the opposite direction to electrons. This showed that they contained positively charged particles. Thomson called these rays **positive rays**. They required much larger fields than electrons to cause their deflection, and the mass of the particles in them depended on the nature of the gas in the tube.

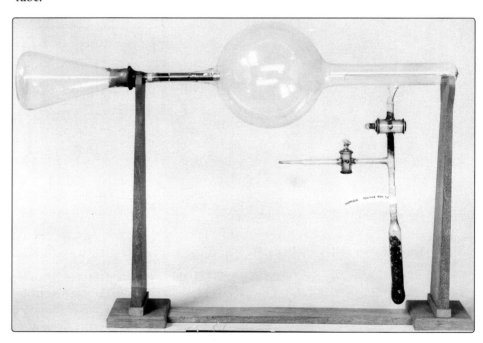

▶ The apparatus with which Thomson investigated positive rays. Compare this with the diagram in figure 5.3.

For example, when the gas in the tube is hydrogen, the mass of the positive particles is almost identical to that of a hydrogen atom. If the gas in the tube is oxygen, the positive particles have a mass almost identical to that of oxygen atoms. Thomson concluded that the positive rays were produced when electrons in the cathode rays collided with molecules of gas in the tube. When a fast-moving electron, streaming away from the cathode, collides with a gas molecule it splits the molecule into atoms and dislodges one or more electrons. The atoms are converted into positive ions by this removal of electrons.

5 Atomic Structure

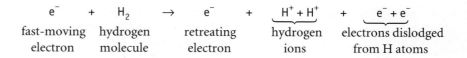

$$e^- \quad + \quad H_2 \quad \rightarrow \quad e^- \quad + \quad \underbrace{H^+ + H^+}_{} \quad + \quad \underbrace{e^- + e^-}_{}$$

| fast-moving electron | hydrogen molecule | | retreating electron | | hydrogen ions | electrons dislodged from H atoms |

The positive ions are attracted towards the cathode and pass through the hole in it to the space behind.

The lightest positive ions are produced when the gas in the tube is hydrogen. Just as the unit of negative charge is the electron, so the unit of positive charge is that associated with a hydrogen ion, H^+. This unit of positive charge is called a **proton**.

Since Thomson's day, the bright glows produced by positive rays have been applied in neon advertising signs and sodium vapour street lamps.

1899 – Thomson's model of atomic structure

As a result of his experiments, Thomson suggested that atoms consisted of negative electrons embedded in a sphere of positive charge. The negative and positive charges balanced, making the atom neutral. Thomson believed, wrongly of course, that the mass of the atom was due only to the electrons and, since one electron was $\frac{1}{1840}$ th of the mass of a hydrogen atom, Thomson concluded that each hydrogen atom must contain 1840 electrons.

Thomson's model for the atom was compared to a Christmas pudding. The electrons in a sphere of positive charge were likened to currants in a spherical Christmas pudding.

1909 – Geiger and Marsden explore the atom

Thomson's ideas of atomic structure led scientists to investigate how electrons were arranged within the atom. Ernest Rutherford, who had been one of Thomson's research students at Cambridge University, had the idea of probing inside the atom using alpha-particles. Rutherford used alpha-particles from radioactive substances as 'nuclear bullets'. Alpha-particles are helium ions, He^{2+}. They travel a few centimetres in air and produce a tiny pinpoint of light on striking a fluorescent screen.

Most of the experimental work was carried out by Geiger and Marsden, two of Rutherford's colleagues at Manchester University. Geiger and Marsden built the apparatus shown in figure 5.4. The source of alpha-particles was a piece of radium inside a protective lead shield. A narrow hole through the lead allowed the alpha-particles to travel in only one direction through the evacuated vessel towards a fluorescent zinc sulphide screen.

▲ The bright glows produced in discharge tubes are now used for advertising signs and roadside lighting. Discharge tubes were used by Thomson to obtain evidence for electrons.

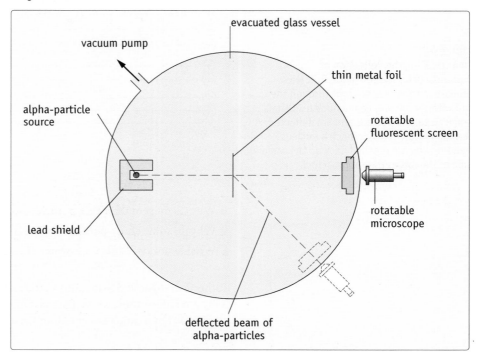

Figure 5.4
The apparatus used by Geiger and Marsden to investigate the deflection of alpha-particles by thin metal foil.

These two might look like Laurel and Hardy, but nothing could be further from the truth. Thomson (left) and Rutherford are unquestionably two of the most brilliant scientists of the twentieth century.

When an alpha-particle hits the fluorescent screen a tiny pinpoint of light can be observed through the microscope. Even when a thin metal foil was placed in the path of the alpha-particles, flashes were still seen on the screen.

Using Thomson's model of atomic structure, Rutherford expected that most of the very fast-moving alpha-particles would pass straight through the foil or be deviated a little. Most of the alpha-particles did pass straight through the foil, but when the detecting screen and microscope were rotated well away from the straight-on position flashes could still be seen. Clearly, some of the alpha-particles were deflected by the foil, but to everyone's surprise, one particle in every 10 000 appeared to rebound from the foil.

Here is Rutherford's own account of the startling discovery.

'I remember Geiger coming to me in great excitement and saying, "We have been able to get some of the alpha-particles coming backwards . . .".

It was quite the most incredible event that has ever happened to me in my life. It was almost as incredible as if you fired a 15-inch shell at a piece of tissue paper and it came back and hit you. On consideration, I realised that this scattering backwards must be the result of a single collision . . . with a system in which the greater part of the mass of the atom was concentrated in a minute nucleus.'

1911 – Rutherford explains the structure of the atom

Alpha-particles are positively charged. This led Rutherford to suggest that deflections and reflections could only be caused by the particles coming close to a concentrated region of positive charge. Only a very small fraction of the alpha-particles were deflected, so he concluded that the region of positive charge which caused the scattering must occupy only a small part of the atom (figure 5.5).

Figure 5.5
Investigating the deflection of alpha-particles by thin metal foil. Most alpha-particles pass straight through the foil without coming near a nucleus. Others come close to a nucleus and are deflected. A few alpha-particles approach a nucleus head-on and are repelled back on the same side of the foil as they approached.

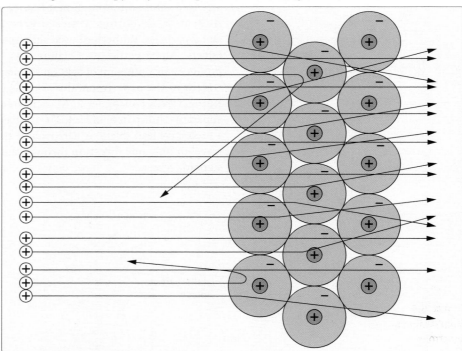

1. Why do most of the alpha-particles pass straight through the metal foil?
2. Why do some alpha-particles appear to rebound from the metal foil?
3. What factors will affect the extent of scattering of the alpha-particles?

Rutherford suggested that atoms in the metal foil consisted of a central positive nucleus composed of protons, where the mass of the atom was concentrated. This nucleus was surrounded by a much larger volume in which the electrons moved.

5 Atomic Structure

▲ If we magnified an atom one million million times to the size of Wembley Stadium, the nucleus would be the size of a pea at the centre of the pitch.

From the angles through which alpha-particles were deflected, Rutherford calculated that the nucleus of an atom would have a radius of about 10^{-14} m. This is about one ten-thousandth of the size of the whole atom which has a radius of about 10^{-10} m. Thus, if we magnified an atom to the size of a football stadium (about 100 m across), the nucleus would be represented by a pea at the centre of the pitch.

Notice the difference between Rutherford's atomic model and that proposed by Thomson a few years earlier. Rutherford's atomic model has been compared to the solar system. Rutherford pictured each atom as a miniature solar system with electrons orbiting the nucleus, like planets orbiting the sun. He suggested that hydrogen, which had the smallest atoms, would have *one* proton in its nucleus, balanced by *one* orbiting electron. Atoms of helium which were next in size would have *two* protons in their nucleus, balanced by *two* orbiting electrons and so on.

Although Rutherford's idea of planetary electrons has now been discarded, his idea of a small positive nucleus has been supported by many experiments.

1913 – Moseley explores the nucleus

Calculations based on the results obtained by Geiger and Marsden showed that the number of positive charges (protons) in the nucleus of an atom was *about* half its relative atomic mass. At about the same time, it was noticed that the number of positive charges on the nucleus was equal to the atom's numbered position in the periodic table, which had been given the symbol Z.

Initially, Z was thought to be simply a number corresponding to the order of the element in the periodic table. Later, experiments by Charles Moseley, a young research student at Oxford University, showed that Z could be obtained from properties of the element. This led to an appreciation of its fundamental significance as the number of protons in the nucleus.

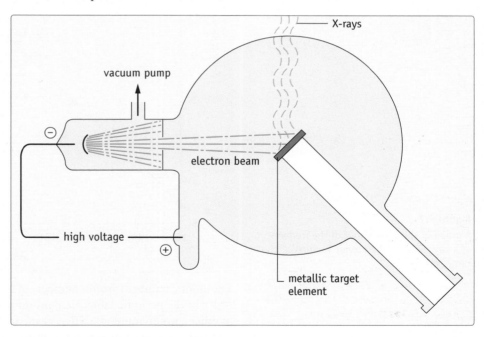

Figure 5.6
Bombarding metals with high speed electrons.

In 1913, Moseley bombarded various metallic elements with high-speed electrons and showed that X-rays were produced (figure 5.6). X-rays are electromagnetic waves like light, but with a very short wavelength and very high frequency. Moseley found that the wavelength and frequency of the X-rays depended on the element he used. Whatever the conditions of his experiment, a particular target element always formed X-rays of the same frequency. What is more, the frequency of the X-rays, v, was found to be given by a formula which involved Z:

$$\sqrt{v} = a(Z - b)$$

Some of Moseley's results are shown in figure 5.7. The equation describing his results is of the form $y = mx + c$, in which $\sqrt{v}$ corresponds to y, and Z corresponds to x. a and b are constants where $a = m$, the gradient of the line, and $-ab = c$, the intercept on the vertical axis.

Z is now known as the **atomic number** or the **proton number** of an element. It is the most important feature of an element's individuality because it represents
(i) the number of protons in the nucleus,
(ii) the number of electrons in the neutral atom,
(iii) the position in which the element appears in the periodic table.

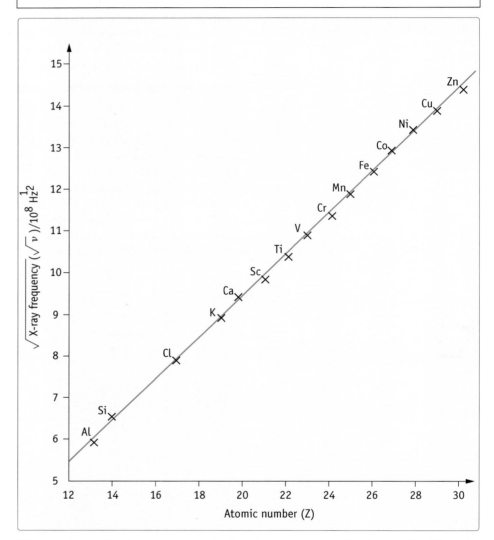

Figure 5.7
A graph of the square root of the frequency of X-rays against the atomic number of the target element.

The atomic number (proton number) of an element can therefore be predicted from its order in the periodic table; element number five in the periodic table has an atomic number (proton number) of five, element 25 has an atomic number of 25, and so on.

Moseley's equation was also important in showing the values of Z for three undiscovered elements. Blank spaces occurred in Moseley's results between molybdenum (Mo) and ruthenium (Ru), between neodymium (Nd) and samarium (Sm), and between tungsten (W) and osmium (Os). Moseley's experiments sparked off a search for these missing elements which were found to occupy the blank spaces.

▌ What are the names of the three missing elements? You may need the periodic table in section 4.4.

Another valuable outcome of Moseley's work was that the frequency of the X-rays from a particular element provided an easy method of identifying the element.

1932 – Chadwick discovers the neutron

In spite of the success of Rutherford and Moseley in explaining atomic structure, one major problem remained unsolved.

If the hydrogen atom contains one proton and the helium atom contains two protons, then the relative atomic mass of helium should be twice that of hydrogen. Unfortunately, the relative atomic mass of helium is four and not two.

James Chadwick, one of Rutherford's collaborators, was able to show where the extra mass in helium atoms came from. Chadwick bombarded a thin sheet of beryllium with alpha-particles. The alpha-particles can be traced by a counter (figure 5.8(a)) which detects charged particles. When the beryllium is in place, the counter registers nothing (figure 5.8(b)), showing that the alpha-particles are being stopped by the beryllium. However, if a piece of paraffin wax is placed between the beryllium and the counter, charged particles are detected again (figure 5.8(c)).

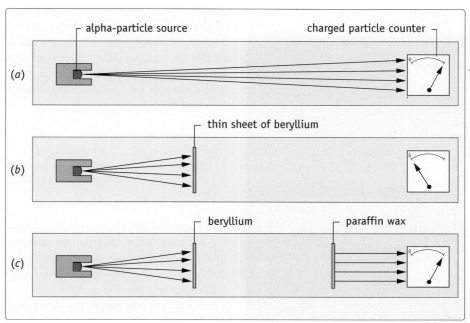

Figure 5.8
The experiment in which Chadwick detected neutrons.

It seems that alpha-particles are stopped by the beryllium foil, yet charged particles are shooting out from the paraffin wax. How can this happen? What has caused charged particles to be ejected from the paraffin wax? Chadwick provided an explanation. He suggested that the alpha-particles striking the beryllium foil displaced *uncharged* particles called **neutrons** from the nuclei of beryllium atoms. These uncharged neutrons could not affect the charged-particle counter, but they could displace positively charged protons from the paraffin wax which would affect the counter. A summary of Chadwick's explanation is shown in figure 5.9. Further experiments showed that neutrons had almost the same mass as protons and Chadwick was able to explain the difficulty concerning the relative atomic masses of hydrogen and helium.

Figure 5.9
Chadwick's explanation of his experiment to detect neutrons.

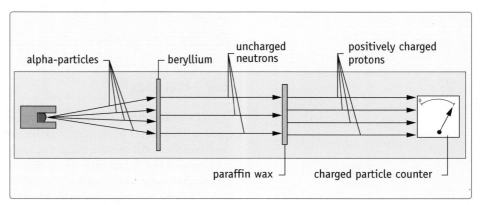

Lithium is the third element in the periodic table. Its relative mass is very nearly seven.

1. What is the atomic number of lithium?
2. How many electrons does each lithium atom contain?
3. How many protons does one lithium atom contain?
4. How many neutrons does one lithium atom contain?
5. How many protons, neutrons and electrons does one Li$^+$ ion contain?

▲ A bubble chamber photograph showing the tracks of sub-atomic particles. As the particles pass through the liquid, they lose energy and cause the liquid to vaporise, forming bubbles. Each track is therefore a line of tiny bubbles.

Hydrogen atoms have one proton, no neutrons and one electron. Since the mass of the electron is negligible compared to the masses of the proton and neutron, a hydrogen atom has a relative mass of one unit. Helium atoms have two protons, two neutrons and two electrons, so the relative mass of a helium atom is four units. This means that a helium atom is four times as heavy as a hydrogen atom, i.e. the relative atomic mass of helium = 4.

5.3 Sub-atomic particles

As a result of the experiments described in section 5.2, scientists believe that all atoms are composed of three important particles – protons, neutrons and electrons.

The nucleus of an atom is composed of protons and neutrons. Because protons and neutrons occupy the nucleus, they are sometimes collectively called **nucleons**. Both the proton and the neutron have a mass almost equal to that of a hydrogen atom. Virtually all the mass of the atom is concentrated in the nucleus, which occupies only a small fraction of the total volume of the atom. The neutron has no charge, whereas the proton carries one positive charge. Electrons with one negative charge occupy the space outside the nucleus. The mass of an electron is 1840 times less than that of a proton.

The nucleus of a fluorine atom has a diameter of about 10^{-12} cm and a mass of 3.1×10^{-23} g

$$\therefore \text{The density of the fluorine nucleus} = \frac{\text{mass}}{\text{volume}}$$

$$= \frac{3.1 \times 10^{-23}}{\frac{4}{3}\pi(5 \times 10^{-13})^3} \text{ g cm}^{-3}$$

$$= 6 \times 10^{13} \text{ g cm}^{-3}$$

The incredibly high density within the fluorine nucleus suggests that the particles in it are drawn very close together by extremely powerful forces. These forces are capable of overcoming the repulsion of the protons for each other. But these powerful forces within the nucleus are effective over only a very short range because they do not pull the outer electrons into the nucleus.

Table 5.1 summarises the relative masses, the relative charges and the position within the atom of these sub-atomic particles. The atoms of all elements are built up from these three particles, different atoms having different numbers of the three. Other particles have been detected during atom-splitting experiments, but these other particles are less stable than protons, neutrons and electrons and only exist under extreme conditions.

Table 5.1
The relative masses, relative charges and positions within the atom of protons, neutrons and electrons

Particle	Mass (relative to that of a proton)	Charge (relative to that on a proton)	Position within the atom
Proton	1	+1	nucleus
Neutron	1	0	nucleus
Electron	$\frac{1}{1840}$	−1	shells

5.4 Isotopes

In 1913 Thomson discovered that some elements could have atoms with different masses. Thomson's equipment was developed by Aston into a mass spectrometer which could compare the relative masses of atoms (section 1.2).

Thomson and Aston discovered that some naturally occurring elements contained atoms that were not exactly alike. When atoms of these elements were ionised and deflected in a mass spectrometer, the beam of ions separated into two or more paths.

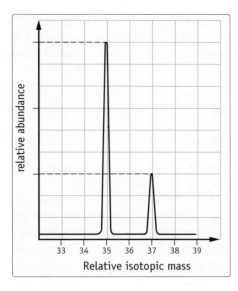

Since all ions in the beam have the same electrical charge, those in the separated paths must have different masses. Consequently, the atoms from which the ions were formed must also have different masses.

> These atoms of the same element with different masses are called **isotopes**. All the isotopes of one particular element have the *same atomic number* because they have the same number of protons. but they have *different mass numbers* because they have different numbers of neutrons.

Figure 5.10 shows a mass spectrometer trace for atoms of chlorine. This shows that chlorine consists of two isotopes: each has an atomic number of 17, but they have different isotopic masses, 35 and 37, respectively.

Atoms of chlorine-35 have 17 protons and 18 neutrons, whereas atoms of chlorine-37 have 17 protons and 20 neutrons.

> The number of protons + the number of neutrons in an atom is called the **mass number** or the **nucleon number**.

Thus, chlorine-35 has a mass number (nucleon number) of 35 (17 + 18) and chlorine-37 has a mass number (nucleon number) of 37 (17 + 20).

We can write the symbol $^{35}_{17}Cl$ in order to specify the mass number and the atomic number of the chlorine-35 atom. Figure 5.11 shows both the chlorine isotopes represented in this way. The mass number is written at the top and to the left of the symbol. The atomic number is written at the bottom and to the left of the symbol.

Isotopes have the same number of electrons and hence the same chemical properties, because chemical properties depend upon the transfer and redistribution of electrons. But isotopes have different numbers of neutrons, so they will have different masses and hence different physical properties. For example, pure $^{37}_{17}Cl_2$ has a higher density, higher melting point and higher boiling point than pure $^{35}_{17}Cl_2$.

5.5 Mass numbers, relative isotopic masses and relative atomic masses

The relative mass of a single isotope is called its **relative isotopic mass**. Relative isotopic masses relate to the same scale as relative atomic masses on which the isotopic $^{12}_{6}C$ is given a relative mass of 12.0000 units. (Relative atomic masses were introduced in section 1.3.) On this scale, the mass of the proton (1.0074 units) is almost the same as that of the neutron (1.0089 units), and the mass of the electron is very small in comparison (0.0005 units). Now since the *relative* masses of the proton and neutron are both very close to one and the electron has a negligible mass, it follows that all relative isotopic masses will be very close to whole numbers. In fact, the relative isotopic mass of an isotope will be very close to its mass number (number of protons + number of neutrons) and the two are assumed to be almost identical in all but the most accurate work.

However, naturally occurring elements often consist of a mixture of isotopes and this results in relative atomic masses which are not close to whole numbers. The relative atomic mass of an element represents the *average* mass of one atom, taking into account the different isotopes and their relative proportions. For example, chlorine consists of two isotopes – $^{35}_{17}Cl$ and $^{37}_{17}Cl$, with relative isotopic masses of 35 and 37, respectively. If chlorine consisted of pure $^{35}_{17}Cl$, its relative atomic mass would be 35. If it consisted of only $^{37}_{17}Cl$, its relative atomic mass would be 37. A 50 : 50 mixture of $^{35}_{17}Cl$ and $^{37}_{17}Cl$ would have a relative atomic mass of 36.

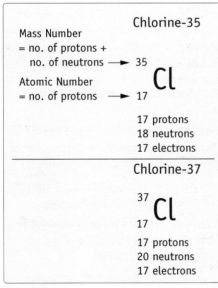

Figure 5.11
Specifying the mass number and atomic number with the symbol of an isotope.

Table 5.2
Relating the relative atomic mass of chlorine to the composition of its isotopic mixture

%$^{35}_{17}$Cl	100	75	50	25	0
%$^{37}_{17}$Cl	0	25	50	75	100
Relative Atomic Mass	35.0	35.5	36.0	36.5	37.0

Look at figure 5.10. The relative heights of the peaks corresponding to $^{35}_{17}Cl$ and $^{37}_{17}Cl$ indicate that the isotopes occur in the ratio of about 3 : 1 (i.e. 75% chlorine-35 and 25% chlorine-37). This results in a relative atomic mass for chlorine of 35.5. Table 5.2 shows this more clearly.

> Can you see that *the relative atomic mass is a weighted mean of the relative isotopic masses* of the different isotopes; weighted, that is, in the proportions in which they occur?

$\therefore$ The relative atomic mass of chlorine

	35	$\times$	$\frac{75}{100}$	+	37	$\times$	$\frac{25}{100}$
	$\uparrow$		$\uparrow$		$\uparrow$		$\uparrow$
	relative isotopic mass of chlorine-35		proportion of Cl-35		relative isotopic mass of chlorine-37		proportion of Cl-37

$= 26.25 + 9.25$

$= 35.50$

Figure 5.12 shows a mass spectrometer trace for the isotopes of neon. The atomic number of neon is 10.

1 What are the relative isotopic masses of the two neon isotopes?
2 What are the mass numbers of the two neon isotopes?
3 How many protons, neutrons and electrons has each of the neon isotopes?
4 What are the percentage abundancies of the two isotopes?
5 Calculate the relative atomic mass of neon.

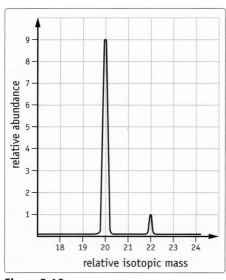

Figure 5.12
A mass spectrometer trace for neon.

Summary

1 Cathode rays consist of tiny negatively charged particles called electrons. Electrons have a mass of $\frac{1}{1840}$th the mass of a hydrogen atom.

2 Rutherford pictured the atom as a miniature solar system in which electrons orbited the positively charged nucleus like planets orbiting the sun.

3 The nucleus is very small compared to the size of the atom as a whole. The volume of the nucleus is only about one million millionth (i.e. one 10^{12}th) of the total volume of the atom.

4 The nucleus is composed of two types of particle:

protons with a mass of one relative to the hydrogen atom and a relative charge of +1, and

neutrons with a mass of one relative to the hydrogen atom but with no charge.

5 Atomic number (proton number) = number of protons

= order of element in the periodic table.

6 Mass number (nucleon number) = number of protons + number of neutrons.

7 Isotopes are atoms with the same atomic number, but different mass numbers. Isotopes have the same number of electrons and therefore the same chemical properties.

Isotopes have different numbers of neutrons (i.e. different masses) and therefore different physical properties.

8 The relative isotopic mass of a single isotope is the relative mass of this isotope compared with an isotope of $^{12}_{6}C$ which is assigned a relative mass of 12.0000.

$$\text{Relative isotopic mass} = \frac{\text{mass of 1 isotope of the element}}{\frac{1}{12} \times \text{mass of 1 atom of } ^{12}_{6}C}$$

9 The relative atomic mass of an element is the average of the relative isotopic masses of the different isotopes weighted in the proportions in which they occur.

1 (a) What are the atomic numbers of tellurium (Te) and iodine (I)?

(b) What are the relative atomic masses of Te and I?

(c) What are the numbers of protons, neutrons and electrons in the commonest isotopes of tellurium (^{128}Te) and iodine (^{127}I)?

(d) Why does Te come before I in the periodic table?

(e) Te comes before I in the periodic table, but Te has a larger relative atomic mass than I. Explain.

(f) Look closely at the periodic table and write down the names of two other *pairs* of elements which are placed in the periodic table in the reverse order to their relative atomic masses.

2 The five main proposals in Dalton's Atomic Theory of matter were:

1 All matter is composed of tiny indestructible particles called atoms.

2 Atoms cannot be created or destroyed.

3 Atoms of the same element are alike in every way.

4 Atoms of different elements are different.

5 Atoms can combine together in small numbers to form molecules.

(a) In the light of modern knowledge about atoms, isotopes, molecules and atomic structure comment on the truth of each of these proposals.

(b) Why is Dalton's Atomic Theory still useful in spite of these limitations?

3 Assume that the fluorine atom ($^{19}_{9}$F) is a sphere of diameter 10^{-10} m and that its nucleus is a sphere of diameter 10^{-14} m.

(a) What is (i) the atomic number, (ii) the mass number of $^{19}_{9}$F?

(b) What is the *actual* mass of the nucleus in *one* $^{19}_{9}$F atom? (6×10^{23} $^{1}_{1}$H atoms have a mass of 1 g.)

(c) What is the density of the nucleus in a $^{19}_{9}$F atom?

(d) What does the value in (c) suggest about the forces within the $^{19}_{9}$F nucleus?

(e) What is the ratio of the volume of the atom to the volume of the nucleus in a $^{19}_{9}$F atom?

4 Natural silicon in silicon-containing ores contains 92% silicon-28, 5% silicon-29 and 3% silicon-30.

(a) What is the atomic number of silicon?

(b) What are the relative isotopic masses of the three silicon isotopes?

(c) What is the relative atomic mass of silicon?

(d) Samples of pure silicon obtained from natural silicon ores, mined in different parts of the world, have slightly different relative atomic masses. Explain.

5 Natural hydrogen contains two isotopes: $^{1}_{1}$H and $^{2}_{1}$H (sometimes called deuterium).

(a) Write down the possible formulas of a hydrogen molecule.

(b) What are the possible relative molecular masses for a hydrogen molecule?

(c) 1 dm^3 (1 litre) of a deuterium-enriched sample of hydrogen gas was found to weigh 0.10 g at s.t.p. What is the isotopic composition of this gas?

6 The accurate relative isotopic masses of five isotopes are shown below.

$^{1}_{1}$H	$^{2}_{1}$H(D)	$^{12}_{6}$C	$^{14}_{7}$N	$^{16}_{8}$O
1.0078	2.0141	12.0000	14.0031	15.9949

(a) Calculate the accurate relative molecular masses (molecular weights) for

(i) N$_2$ (ii) DCN (iii) CO (iv) C$_2$H$_4$ (v) C$_2$D$_2$

(b) The relative molecular mass of a certain gas in a high-resolution mass spectrometer was 28.0171. What gas is probably under observation?

7 A compound containing only carbon, hydrogen and oxygen has peaks in its mass spectrum which correspond to relative masses of 60 (the molecular ion), 43, 31, 29 and 17. These masses are relative to the standard $^{12}_{6}$C = 12.

(a) What fragmented particles could be responsible for relative masses of 17, 29 and 31? Assume that the only isotopes involved are $^{1}_{1}$H, $^{12}_{6}$C and $^{16}_{8}$O.

(b) Deduce the *two* possible structures for the compound.

8 The mass spectrum of dichloromethane shows peaks corresponding to 84, 86 and 88 atomic mass units (a.m.u.) with relative intensities of 9 : 6 : 1, respectively.

(a) What particles cause the three peaks?

(b) Why are their relative intensities 9 : 6 : 1?

You can assume that the only isotopes involved are $^{1}_{1}$H, $^{12}_{6}$C, $^{35}_{17}$Cl and $^{37}_{17}$Cl and that the relative abundances of $^{35}_{17}$Cl to $^{37}_{17}$Cl are 3 : 1.

9 Suppose that you are a research chemist and that you have just discovered a new element.

(a) How would you identify its different isotopes and determine its relative atomic mass?

(b) How would you attempt to determine its atomic number?

Electronic Structure

6.1 Evidence for the electronic structure of atoms

When atoms react, electrons are redistributed. These electrons may be transferred from one atom to another or they may be shared between the reacting atoms in a different way.

▲ Chemical reactions involve electron transfer. The outcome can be explosive.

- Remember that it is *only* electrons which are involved in chemical reactions: protons and neutrons take no part.

As chemical reactions involve electrons, the similarities in chemical properties of certain elements (e.g. Na, K and Li; F, Cl and Br) suggest that these elements may have similar electronic structures.

It is possible to obtain information about the arrangement of electrons in atoms by studying the ease with which atoms lose electrons. The energy needed to remove one electron from an atom is known as the **ionisation energy** (ionisation enthalpy) and is given the symbol ΔH_i.

> *The first ionisation energy of an element* is the energy required to remove one electron from each atom in a mole of gaseous atoms producing one mole of gaseous ions with one positive charge. Thus, the first ionisation energy of sodium is the energy required for the process
>
> $$Na(g) \rightarrow Na^+(g) + e^- \qquad \Delta H = \Delta H_{i_1}$$

(The subscript number one in the symbol ΔH_{i_1} indicates the *first* ionisation energy.) The most useful method of determining the ionisation energies of elements involves a study of their emission spectra.

6.2 Obtaining ionisation energies from emission spectra

When sodium chloride is heated strongly in a bunsen, it gives off a brilliant yellow light. Other sodium compounds, such as sodium nitrate and sodium sulphate also give a bright yellow light when heated strongly in a bunsen flame. The compounds of some other elements also emit light when they are heated in this way. For example, the flame colour of lithium compounds is red and that of potassium compounds is lilac (section 15.9).

What colour of light is emitted by:
1. copper compounds,
2. calcium compounds
   when they are heated strongly on a nichrome wire in the hottest bunsen flame?

Some gaseous elements will also emit light when they are subjected to large potential differences in electric discharge tubes. Neon advertising signs work in this way and the yellow sodium street lamps are discharge tubes containing sodium vapour. Thermal energy from the bunsen flame and electrical energy from the discharge tube have caused the elements to emit coloured light.

▲ Copper sulphate being heated strongly in a bunsen flame.

Figure 6.1
Photographs of the line emission spectra of hydrogen, helium and mercury in the visible region. Hydrogen's prominent lines have wavelengths (from right to left) of 656.3 nm (red), 486.1 nm (cyan) and 434.0 nm (blue). Helium's lines have wavelengths (from right to left) of 667.8 nm (red), 587.5 nm (yellow), 501.5 nm and 492.1 nm (cyan) and 471.3 nm, 447.1 nm and 438.8 nm (blue). Those for mercury (right to left) are 579.1 and 577.0 nm, together in this photo (yellow), 546.1 nm (green), 491.6 nm (cyan) and 435.8 nm (blue).

▲ Neon lights in Las Vegas, USA. Certain gaseous materials such as neon emit light when they are subjected to high voltages in discharge tubes.

Figure 6.2

The colours and frequencies of the more prominent lines in the visible region of the atomic hydrogen spectrum.

If the light emitted by these substances is examined using a spectroscope, it does *not* consist of a continuous range of colours like the spectrum of white light or the colours in a rainbow. Instead, the light emitted by these substances is composed of separate lines of different colour. This kind of spectrum is called a **line emission spectrum**.

The photographs in figure 6.1 show the line emission spectra of hydrogen, helium and mercury in the visible region. Notice that each element has its own characteristic set of lines different from those of any other element. Consequently, elements can be identified from their line emission spectra.

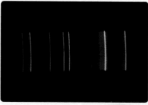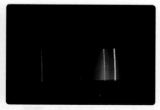

Each line in an emission spectrum corresponds to light of a particular frequency. The frequencies and colours of the prominent lines in the visible region of the spectrum of hydrogen are shown in figure 6.2. This particular series of lines is known as the **Balmer Series** for hydrogen.

What do you notice about the spacing of the lines? How do you explain the spacing of the lines? Why does an element only emit light of certain frequencies?

In order to answer these questions scientists assume that the electrons in an atom can exist only at certain energy levels. Under normal conditions, the electrons in an atom or ion fill the lowest energy levels first. When sufficient energy is supplied to the atom, it is sometimes possible to promote (excite) an electron from a lower energy level to a higher one. This process is called **excitation**. The electron is unstable in the higher energy level, so it will emit the excess energy as radiation and drop back into a lower energy level. Now, the energy difference between the higher and lower energy levels can have only certain fixed values because the energy levels themselves are fixed. This means that the radiation emitted when an electron falls from a higher to a lower energy level can have only certain fixed frequencies (i.e. certain specific colours) because the frequency of any radiation is determined by its energy.

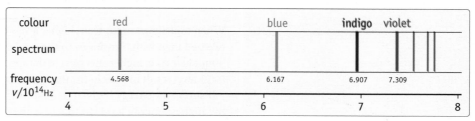

The small amount of radiation emitted by an electron when it falls from a higher to a lower energy level is referred to as a **quantum** of radiation.

The relationship between the energy (E) of a quantum of radiation and its frequency (v) is

$$E = h \times v$$

h is a constant, called Planck's constant. The value of h is 4×10^{-13} kJ s mol^{-1} or $4 \times 10^{-13}/6 \times 10^{23} = 6.66 \times 10^{-37}$ kJ s molecule^{-1}.

Figure 6.3 shows the energy levels in a hydrogen atom and the electron transitions which produce the lines in the visible region of the atomic hydrogen spectrum. Notice the following points.

1 The electronic energy levels are numbered ($n = 1$, $n = 2$, $n = 3$, etc.). These numbers are sometimes referred to as the **principal quantum numbers** for the energy levels which correspond to the shells of electrons. The level of lowest energy is given the principal quantum number 1, the next lowest 2, and so on.

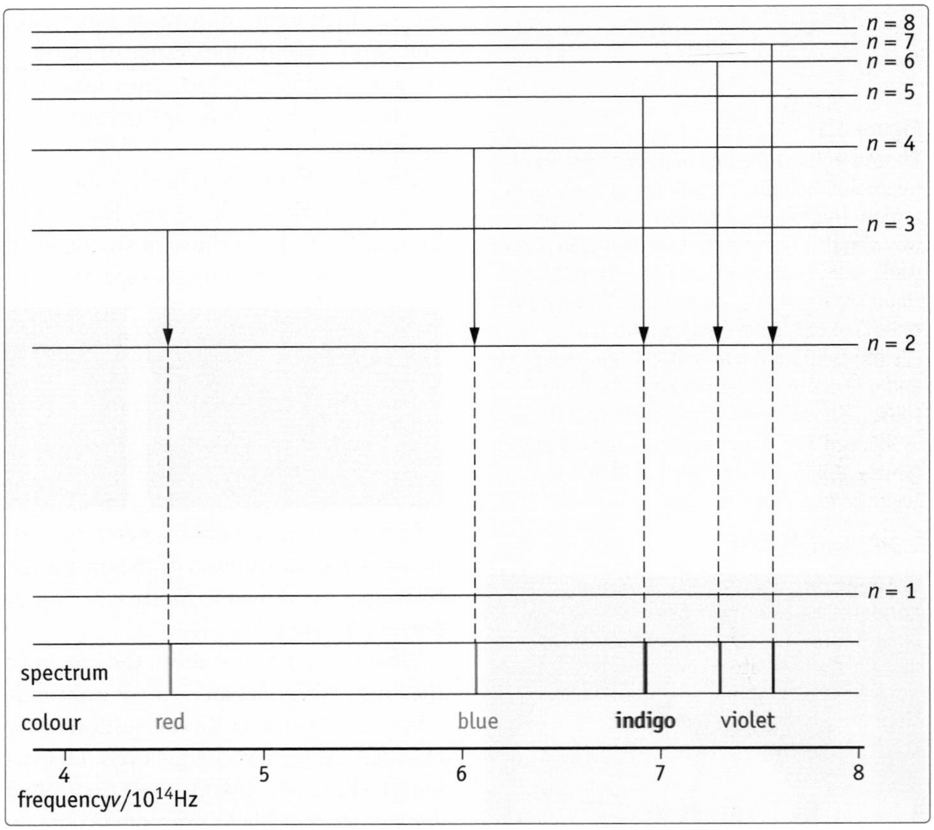

Figure 6.3

The energy levels in a hydrogen atom and the electron transitions which produce the lines in the visible region of the atomic hydrogen spectrum. Notice that the lines in the visible region are caused by transitions to the $n = 2$ level. Lines which result from transitions to the $n = 2$ level in an atom are known as the Balmer series of lines.

2 The colours in the visible region of the hydrogen spectrum are caused by electron transitions from higher levels to the level $n = 2$ and not to the lowest energy level ($n = 1$). For example, transitions from $n = 3$ to $n = 2$ result in a red line (at frequency 4.568×10^{14} Hz) in the hydrogen spectrum, while transitions from $n = 4$ to $n = 2$ produce a blue line.

3 As the energy levels get closer and eventually come together, it follows that the spectral lines also get closer and eventually come together.

When transitions occur from higher levels to the lowest energy level ($n = 1$), more energy is released than with transitions to the $n = 2$ level. Consequently, these lines appear at higher frequencies (i.e. higher energies) in the spectrum. In the case of hydrogen, transitions to the $n = 1$ level result in lines in the ultraviolet region of the spectrum.

If sufficient energy is given to an atom, it is possible to excite an electron just beyond the highest energy level. In this case the electron will escape and the atom becomes an ion. **Ionisation** has taken place.

In an atom, the highest possible energy level corresponds to the frequency at which the lines in the spectrum come together. So, by determining the frequency at which the converging spectral lines come together, we can find the ionisation energy of an element. This particular frequency is called the **convergence limit**.

The frequencies in table 6.1 relate to electron transitions to the $n = 1$ level in hydrogen. Lines corresponding to these frequencies occur in the ultraviolet region of the hydrogen spectrum. These lines which result from transitions to the lowest energy level in an atom are known as the **Lyman Series** of lines.

Table 6.1

Frequencies of the Lyman Series for hydrogen

Frequency $v/10^{14}$ Hz	Transition to which frequency corresponds		
24.66	$n = 2$	to	$n = 1$
29.23	$n = 3$	to	$n = 1$
30.83	$n = 4$	to	$n = 1$
31.57	$n = 5$	to	$n = 1$
31.97	$n = 6$	to	$n = 1$
32.21	$n = 7$	to	$n = 1$
32.37	$n = 8$	to	$n = 1$

1 Work out the difference in frequency (Δv) between successive lines in the Lyman Series for hydrogen.

2 Plot a graph of Δv (vertically) against frequency, v. (Use the value of the lower frequency for plotting v.)

3 Use your graph to estimate the frequency when Δv becomes 0.

▲ Fireworks over the River Thames in London. The colours of fireworks are caused by electrons moving from higher to lower energy levels in atoms or ions. When they do this, radiation of characteristic frequency is emitted, which sometimes corresponds to light of a particular colour.

4 Δv becomes 0 when the difference in energy between the electronic energy levels becomes 0. Use the relationship $E = hv$ to find the energy which corresponds to the frequency when Δv becomes 0. This energy is the ionisation energy (ionisation enthalpy) for hydrogen.

An accurate value for the frequency at the convergence limit for hydrogen is 32.7×10^{14} Hz. Using $E = hv$, the energy of radiation with this frequency

$$= 4 \times 10^{-13} \times 32.7 \times 10^{14} \, \text{kJ mol}^{-1}$$
$$= 1308 \, \text{kJ mol}^{-1}$$

∴ The ionisation energy of hydrogen = 1308 kJ mol^{-1}.

6.3 Using ionisation energies to predict electronic structures – evidence for shells

If an atom containing several electrons is provided with sufficient energy it will lose one electron. Additional supplies of energy will result in the removal of a second electron, then a third, then a fourth, and so on. A succession of ionisations is therefore possible, each of which has its associated ionisation energy. For example, the first ionisation energy of sodium corresponds to

$$Na(g) \rightarrow Na^+(g) + e^- \qquad \Delta H_{i1} = +494 \, \text{kJ mol}^{-1}$$

whereas the second ionisation energy of sodium corresponds to

$$Na^+(g) \rightarrow Na^{2+}(g) + e^- \qquad \Delta H_{i2} = +4564 \, \text{kJ mol}^{-1}$$

The successive ionisation energies of an element are usually obtained from spectroscopic experiments.

Table 6.2
The successive ionisation energies of beryllium

Element	Ionisation energy/kJ mol^{-1}			
	First	**Second**	**Third**	**Fourth**
Beryllium	900	1758	14905	21060
	2 electrons relatively easy to remove		2 electrons very difficult to remove	

Table 6.2 shows the successive ionisation energies of beryllium. Twice as much energy is needed to remove the second electron from beryllium as to remove the first, and eight times as much energy is required to remove the third electron as the second. The values for the different ionisation energies in table 6.2 suggest that beryllium has two electrons which are relatively easy to remove and two electrons which are very difficult to remove. Chemists have deduced that the beryllium atom has two electrons in a low energy level ($n = 1$) and therefore very difficult to remove; and two other electrons in a higher energy level ($n = 2$) and therefore easy to remove. This is represented on an energy level diagram in figure 6.4.

The electron arrangement in beryllium is written as 2, 2. Lithium, with three electrons, has an electron arrangement of 2, 1 which shows two electrons in the $n = 1$ energy level and one electron in the $n = 2$ energy level.

Figure 6.4
Energy levels and quantum shells for electrons in a beryllium atom.

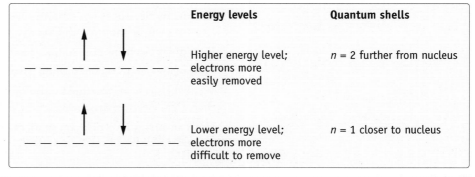

	Energy levels	Quantum shells
↑ ↓ — — — — — — —	Higher energy level; electrons more easily removed	$n = 2$ further from nucleus
↑ ↓ — — — — — — —	Lower energy level; electrons more difficult to remove	$n = 1$ closer to nucleus

In a beryllium atom, the two electrons in the lowest energy level ($n = 1$) spend most of their time closer to the nucleus than do the two electrons in the $n = 2$ level. These two electrons in the $n = 1$ level are said to occupy the first **quantum shell**. The two electrons in the second energy level ($n = 2$) spend most of their time further from the nucleus and are said to occupy the second quantum shell.

In figure 6.4 the electrons have been represented by arrows. When an energy level is filled the electrons have paired up with each other and in each of these pairs the electrons are spinning in opposite directions. Evidence that electrons pair up and spin in opposite directions is provided by spectroscopic and magnetic measurements. Chemists believe that paired electrons can only be stable when they spin in opposite directions so that the magnetic attraction which results from their opposite spins can counterbalance the electrical repulsion which results from their identical negative charges. In figure 6.4, the opposite spins of the paired electrons are shown by drawing the arrows in opposite directions.

6.4 How are the electrons arranged in larger atoms?

Figure 6.5 shows a graph of the logarithm to base ten (lg) for the successive ionisation energies of sodium.

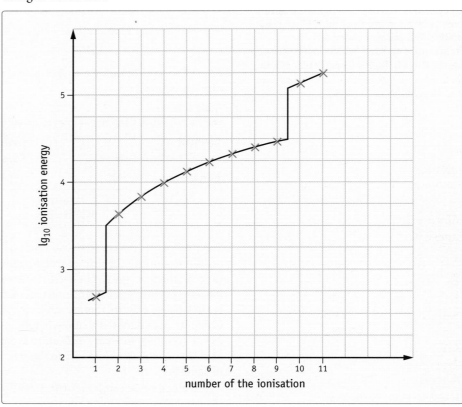

Figure 6.5
A graph of the logarithm to base ten of the successive ionisation energies for sodium.

Look closely at figure 6.5.

1 How many electrons does sodium have in the first shell close to the nucleus? These electrons will be the most difficult to remove.

2 How many electrons does sodium have in the second shell?

3 How many electrons does sodium have in the third shell far away from the nucleus?

4 Write the electron structure for sodium showing the number of electrons in each shell. (The electron structure for beryllium is 2, 2.)

5 Draw an energy level diagram similar to figure 6.4 for the electrons in a sodium atom.

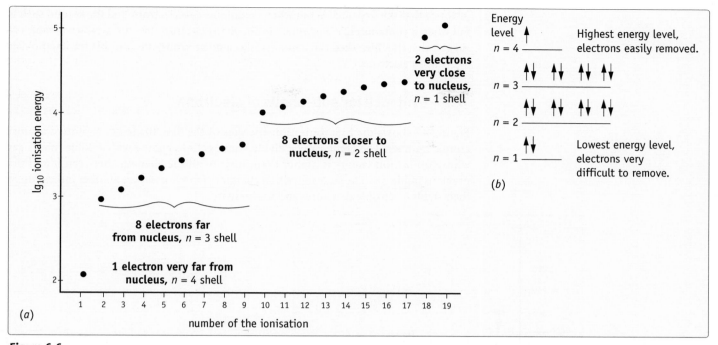

Figure 6.6
(a) A sketch graph of the logarithm to base ten of the successive ionisation energies for potassium. (b) An energy level diagram for the electrons in a potassium atom.

Figure 6.6(a) shows a sketch graph of the logarithm to base ten for the successive ionisation energies of potassium. This graph has been used to construct an energy level diagram (figure 6.6(b)) for the electrons in a potassium atom.

The electron structure of potassium is written as 2, 8, 8, 1, which shows the number of electrons in each of the quantum shells as we move out from the nucleus. Using ionisation energy data, it is possible to predict the electron patterns in other elements, and table 6.3 shows the electron structures of the first 20 elements in the periodic table.

Table 6.3
Electron structures of the first 20 elements in the periodic table

Atomic Number	Element	Symbol	Number of electrons in the first shell	Number of electrons in the second shell	Number of electrons in the third shell	Number of electrons in the fourth shell
1	Hydrogen	H	1	–	–	–
2	Helium	He	2	–	–	–
3	Lithium	Li	2	1	–	–
4	Beryllium	Be	2	2	–	–
5	Boron	B	2	3	–	–
6	Carbon	C	2	4	–	–
7	Nitrogen	N	2	5	–	–
8	Oxygen	O	2	6	–	–
9	Fluorine	F	2	7	–	–
10	Neon	Ne	2	8	–	–
11	Sodium	Na	2	8	1	–
12	Magnesium	Mg	2	8	2	–
13	Aluminium	Al	2	8	3	–
14	Silicon	Si	2	8	4	–
15	Phosphorus	P	2	8	5	–
16	Sulphur	S	2	8	6	–
17	Chlorine	Cl	2	8	7	–
18	Argon	Ar	2	8	8	–
19	Potassium	K	2	8	8	1
20	Calcium	Ca	2	8	8	2

Notice that the first shell is full when it contains two electrons and the second shell is full when it contains eight electrons. Although the electron structures shown in table 6.3 suggest that the third shell can contain only eight electrons, it is possible for it to hold as many as 18 electrons.

6.5 Evidence for sub-shells of electrons

Figure 6.7 shows the first ionisation energies of the first 40 elements plotted against atomic number. The graph can be divided into sections ending with a noble (inert) gas whose ionisation energy is higher than those of all the elements between it and the previous noble gas. The large ionisation energies of noble gases are another indication of their stable electronic structures and unreactivity.

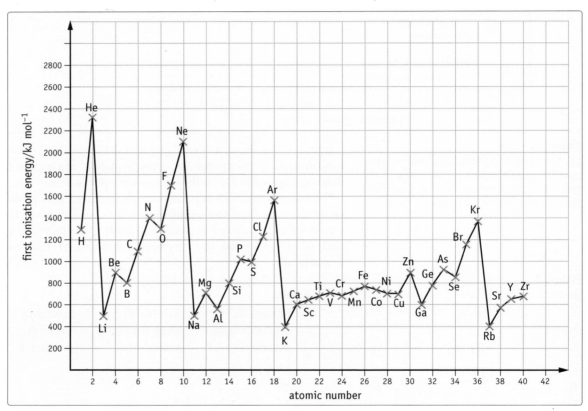

Figure 6.7
A graph of the first ionisation energies of the elements plotted against atomic number.

The graph between one noble gas and the next can also be divided into sub-sections. These sub-sections contain either two, six or ten points on the graph. For example, after both He and Ne there are deep troughs in the graph followed by a small intermediate peak (Li to Be and Na to Mg), i.e. sub-sections of two points. Immediately after Be and Mg, there are groups of six points (B to Ne and Al to Ar). From Sc to Zn there is a sub-section containing 10 points.

In fact, the points on the graph in figure 6.7 between one noble gas and the next correspond to the filling of one shell with electrons. This means that the sub-sections of points correspond to **sub-shells** of electrons. By studying ionisation energies and atomic spectra in this way, scientists have concluded that:

the $n = 1$ shell can have 2 electrons in the same sub-level (sub-shell).

the $n = 2$ shell can have 2 electrons in one sub-level, and
                  6 electrons in a slightly higher sub-level.

the $n = 3$ shell can have 2 electrons in one sub-level,
                  6 electrons in a slightly higher sub-level, and
                  10 electrons in a still slightly higher sub-level.

6 Electronic Structure

the $n = 4$ shell can have   2 electrons in one sub-level,

                                 6 electrons in a slightly higher sub-level,

                        10 electrons in a still slightly higher sub-level, and

                        14 electrons in a still slightly higher sub-level.

The sub-shells (or sub-levels) containing 2 electrons are called s sub-shells (s electrons).
The sub-shells (or sub-levels) containing 6 electrons are called p sub-shells (p electrons).
The sub-shells (or sub-levels) containing 10 electrons are called d sub-shells (d electrons).
The sub-shells (or sub-levels) containing 14 electrons are called f sub-shells (f electrons).

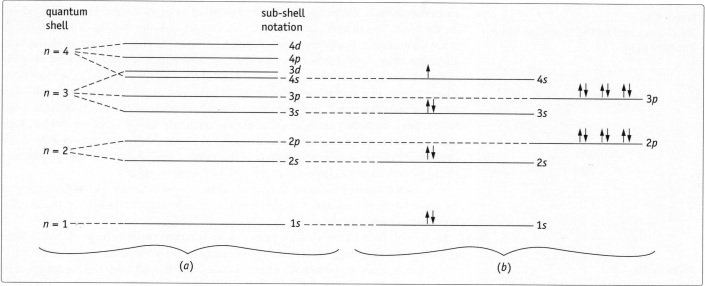

Figure 6.8
(a) The positions of energy sub-levels in a potassium atom. (b) The distribution of electrons amongst the sub-levels in a potassium atom.

The relative positions of these various sub-shells are shown on the left of figure 6.8 and the distribution of electrons amongst sub-levels in a potassium atom is shown on the right. Compare figure 6.8 with figure 6.6(b).

When energy sub-levels are being filled, *electrons always occupy the lowest available energy sub-level first* and *the electrons 'pair-up' as soon as each sub-level is half-filled.*

The electron structure of an atom can therefore be described in terms of its sub-shells occupied by electrons. A number (1, 2, 3, etc.) is used to denote the quantum shell, a letter (s, p, d or f) to denote the sub-shell and a superscript to indicate the number of electrons in the sub-shell.

Thus, potassium has 2 electrons in the $1s$ sub-shell (i.e. $1s^2$),

                       2 electrons in the $2s$ sub-shell (i.e. $2s^2$),

                       6 electrons in the $2p$ sub-shell (i.e. $2p^6$),

                       2 electrons in the $3s$ sub-shell (i.e. $3s^2$),

                       6 electrons in the $3p$ sub-shell (i.e. $3p^6$), and

                       1 electron in the $4s$ sub-shell (i.e. $4s^1$)

This means that the electronic structure for potassium can be written in terms of energy levels as 2, 8, 8, 1; and more precisely in terms of energy sub-shells as

$$1s\overset{2}{\underset{2}{\downarrow}}, 2s\overset{2}{\downarrow}2p\overset{6}{\underset{8}{\downarrow}}, 3s\overset{2}{\downarrow}3p\overset{6}{\underset{8}{\downarrow}}, 4s\overset{1}{\underset{1}{\downarrow}}$$

Notice in figure 6.8 that the $3d$ sub-shell is just above the $4s$ sub-shell, but just below the $4p$ sub-shell. This means that once the $4s$ level is filled (at element Ca in the periodic table) further electrons enter the $3d$ level, not the $4p$ level. Thus, scandium (the element after Ca in the periodic table) has the electronic structure $1s^2, 2s^2 2p^6, 3s^2 3p^6 3d^1, 4s^2$ and not $1s^2, 2s^2 2p^6, 3s^2 3p^6, 4s^2 4p^1$. The filling of the $3d$ sub-shell is very important in the chemistry of the elements from Sc to Zn. These are known as the **transition elements** and we shall deal with them in some detail in chapters 18 and 19.

An element has the electronic structure $1s^2, 2s^2 2p^3$.
1 What is the name of this element?
2 Write the electronic structure of this element in terms of energy levels.
The electron structure of silicon is 2, 8, 4.
3 Write the electronic structure of silicon in terms of energy sub-levels.
4 Draw an energy sub-level diagram for the electrons in a silicon atom.

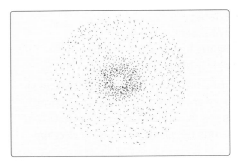

Figure 6.9
The charge cloud for the 1s electron in a hydrogen atom.

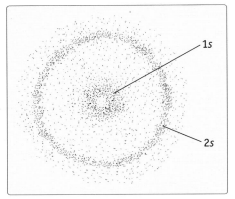

Figure 6.10
The charge clouds for electrons in a lithium atom.

6.6 Electrons and orbitals

So far, we have described the electrons in atoms as occupying different quantum shells at increasing distances from the nucleus. But, suppose we could photograph the electron in a hydrogen atom at any given moment. The electron is moving in an unknown path at high speed, and if we took a second photograph an instant later, the electron would occupy a different position. If we superimposed millions of such photographs, the composite picture would resemble a cloud composed of an enormous number of dots, each dot representing one position of the electron (figure 6.9). Thus, in the hydrogen atom the electron can be imagined to be a cloud of negative charge. The diffused cloud for the 1s electron in hydrogen is spherical in shape, but notice that the density of charge is not uniform throughout the cloud. Figure 6.10 shows the charge clouds for electrons in a lithium atom. The charge cloud for the two 1s electrons is again spherical as is the charge cloud for the 2s electron. Figure 6.10 also shows that the electron in the 2s level has a larger mean radius than electrons in the 1s level. The 2s electron can be imagined to occupy a 'spherical band'. In theory, there is no sharp boundary to the charge cloud, but in practice a boundary surface can be drawn to enclose almost all of the charge. Such regions which enclose almost all the charge cloud within them are called **orbitals**. Thus, orbitals are the regions in which there is the greatest probability of finding particular electrons, although the electrons are not confined to these regions.

Your own movements can be compared to those of an electron. At any one time you are most likely to be found either at home or in school, but fortunately you are not confined to these two places.

Each orbital can hold either one or a maximum of two electrons. If the orbital contains two 'paired-up' electrons they will be spinning in opposite directions.

Using complex calculations scientists have deduced that all s orbitals are spherical in shape; p orbitals are not spherical, but approximately 'dumb-bell' shaped with the nucleus located between the two halves of the 'dumb-bell'. In fact, each p sub-shell has three separate p orbitals, each of which can hold a maximum of two electrons, making a total of six electrons in a filled p sub-shell. Figure 6.11 shows the shapes of the three orbitals in a p sub-shell. They are identical except for their axes of symmetry which, like the axes of a cartesian co-ordinate system, are mutually at right angles. Thus, it is convenient to distinguish between the p orbitals by labelling them p_x, p_y and p_z. Electrons will, of course, occupy the three p orbitals singly at first because of the repulsion between them. When the fourth electron is added to a p sub-shell, one orbital will contain a pair of electrons.

The charge clouds for d electrons are even more complex than those of p electrons and we need not discuss them at this stage.

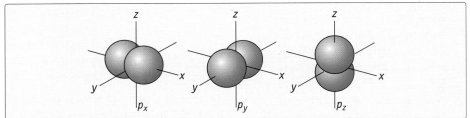

Figure 6.11
The shapes and relative positions of three p orbitals in a p sub-shell.

Figure 6.12 shows how the electronic structures of beryllium, carbon, nitrogen and oxygen can be represented to show the number of electrons in each orbital. Each box represents an orbital.

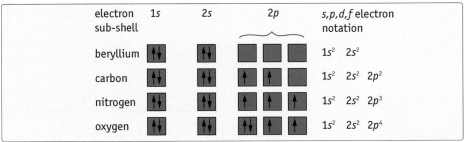

Figure 6.12
An orbital (electrons-in-boxes) representation of the electronic structures of beryllium, carbon, nitrogen and oxygen.

6 Electronic Structure

6.7   Electronic structures and the periodic table

In the periodic table, elements are arranged in groups with similar chemical properties. The chemical similarities of elements in the same group arise because of their similar electron structures. Table 6.4 shows the electron structures of the first 20 elements in the periodic table. Note that elements in the same group have similar electron structures.

Table 6.4
The electron structures of the first 20 elements in the periodic table

Period 1			H					He
			1					2
Period 2	Li	Be	B	C	N	O	F	Ne
	2, 1	2, 2	2, 3	2, 4	2, 5	2, 6	2, 7	2, 8
Period 3	Na	Mg	Al	Si	P	S	Cl	Ar
	2, 8, 1	2, 8, 2	2, 8, 3	2, 8, 4	2, 8, 5	2, 8, 6	2, 8, 7	2, 8, 8
Period 4	K	Ca						
	2, 8, 8, 1	2, 8, 8, 2						

▲ Crystals of xenon fluoride (XeF_2) on the inside surface of a glass vessel. The crystals form when Xe and F_2 are mixed and exposed to sunlight.

Group 0: The noble (inert) gases

All of the noble gases have completely filled s and p orbitals in their outer shell: helium ($1s^2$), neon ($1s^2, 2s^2 2p^6$), argon ($1s^2, 2s^2 2p^6, 3s^2 3p^6$), etc. Because of their complete and stable electronic structures, noble gases have large first ionisation energies. The atoms of these elements are very stable and exist as monatomic molecules. They do not combine with each other under any conditions to form diatomic molecules nor do they combine readily with the atoms of other elements.

The first noble gas compound was not prepared until 1962 and chemists have still not succeeded in preparing a single stable compound of helium or neon. The inert character of these elements is attributed to their stable electronic structures with filled s and p sub-shells.

As a result of their stable electronic structures, the atoms of noble gases have little interaction with each other and the forces of attraction between separate atoms are very small. Thus, noble gases have very low melting points and very low boiling points. Helium has the lowest melting point and the lowest boiling point of any known substance (3 K and 4 K, respectively). Radon, the heaviest noble gas, boils at −62°C (211 K).

Helium, the simplest noble gas, is used in meteorological balloons because of its low density and because it is non-flammable. Helium mixed with oxygen is also used as the gas breathed by divers.

Neon is used in brightly lit advertising signs and other forms of illumination because of the beautiful orange-red glow which it produces in discharge tubes.

Argon and krypton are used as the gas in electric light bulbs. If there was a vacuum inside the bulbs, metal atoms would evaporate from the very hot tungsten filament. To reduce this evaporation and to prolong the life of the filament, the bulb is filled with these unreactive noble gases which will not react with the hot tungsten.

Argon is also used during some welding operations. The argon provides an inert atmosphere and prevents a reaction between the metals being welded and oxygen in the air.

Group I: The alkali metals

Each alkali metal follows a noble gas in the periodic table. Thus, alkali metals have one electron in their outermost shell (figure 6.13) which is fairly easily removed (low first ionisation energy). By losing their outermost electron, atoms of alkali metals form singly charged positive ions with stable electron structures like the previous noble gas (Li^+, Na^+, K^+, etc.). This means that all the alkali metals have an oxidation number of +1 in their compounds and they are very reactive because of the ease with which they lose the single electron in their outermost shell. The chemistry of the alkali metals is taken further in chapter 15.

Li 2, 1
Na 2, 8, 1
K 2, 8, 8, 1
Rb 2, 8, 18, 8, 8, 1
Cs
Fr

Figure 6.13
Electron structures of the first four alkali metals.

▲ A diver explores the sea bed. Strapped to her back are cylinders containing a mixture of oxygen/helium rather than oxygen/nitrogen. Helium is much less soluble in the blood than nitrogen. The diver is therefore less likely to suffer from 'bends' (pains caused by gases coming out of the blood) as she rises to the surface and the pressure falls.

▼ Up she goes! An airship filled with helium gently floats upwards, though still held to the ground by ropes.

1 How many electrons do the atoms of Group VII elements have in the outermost shell?

2 Write down the symbols with the correct charge for the ions of three elements in Group VII.

3 Why do elements in Group VII form ions with the same charge?

4 How many electrons do the atoms of Group II elements have in the outermost shell?

5 Why do Group II elements have an oxidation number of +2 in all their compounds?

Summary

1   Each element has a characteristic line emission spectrum which can be used to identify it.

2   The electrons in an atom can exist only at certain energy levels. If an atom is given sufficient energy, electrons can be promoted to higher energy levels. When such a promoted electron falls from a higher to a lower energy level, a quantum of radiation is emitted. This radiation can be detected as a line in the emission spectrum of the element.

3   The ionisation energy of an element can be obtained from the convergence limit of the Lyman Series of lines in its emission spectrum.

4   The principal quantum numbers ($n = 1, 2, 3$, etc.) are used to denote the shells of electrons in atoms.

5   In a stable atom, the electrons occupy the lowest available energy sub-levels. When each sub-level is half-filled, the addition of further electrons causes them to 'pair up'.

6   s sub-levels can hold a maximum of 2 electrons, p sub-levels can hold a maximum of 6 electrons, d sub-levels can hold a maximum of 10 electrons, and f sub-levels can hold a maximum of 14 electrons

7   The volume in space in which there is a high probability of finding an electron is called an orbital. Each orbital can hold a maximum of two electrons. s sub-shells contain one orbital; p sub-shells contain three orbitals.

Review questions

1 (a) Draw a sketch graph of the logarithm to base 10 for the successive ionisation energies of phosphorus.

(b) What conclusions could you draw from the graph concerning the electronic structure of phosphorus?

(c) Write the electronic structure of phosphorus using s, p, d, f notation.

2 The following table shows the ionisation energies (in kJ mol^{-1}) of five elements lettered A, B, C, D and E.

Element	1st ionisation energy	2nd ionisation energy	3rd ionisation energy	4th ionisation energy
A	500	4600	6900	9500
B	740	1500	7700	10500
C	630	1600	3000	4800
D	900	1800	14800	21000
E	580	1800	2700	11600

(a) Which of these elements is most likely to form an ion with a charge of 1+? Give reasons for your answer.

(b) Which two of the elements are in the same group of the periodic table? Which group do they belong to?

(c) In which group of the periodic table is element E likely to occur? Give reasons for your answer.

(d) Which element would require the least energy to convert one mole of gaseous atoms into ions carrying two positive charges?

3 The electron energy levels of a certain element can be represented as $1s^2$, $2s^2 2p^6$, $3s^2 3p^4$.

(a) Sketch a graph for the first seven ionisation energies of the element against the number of the ionisation.

(b) What is the atomic number of the element?

(c) What is the name of the element?

(d) Draw an energy level diagram for the electrons in an atom of the element.

4 (a) List three factors which influence the size of the first ionisation energy of an element.

(b) The first ionisation energies of the elements in Group I are shown below.

Element	First ionisation energy /kJ mol^{-1}
Li	520
Na	500
K	420
Rb	400
Cs	380

Explain the variation in the first ionisation energy with atomic number.

(c) Explain, with examples, how ionisation energies can provide evidence for the arrangement of electrons in shells.

5 Figure 6.14 shows a simplified diagram of the visible line emission spectrum of the hydrogen atom.

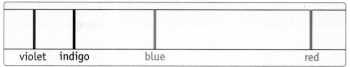

violet   indigo   blue   red

Figure 6.14
A simplified diagram of the visible line spectrum of the hydrogen atom.

(a) What quantity is represented along the horizontal axis in figure 6.14?

(b) Is this quantity increasing or decreasing from left to right?

(c) Explain why the atomic spectrum consists of a series of lines.

(d) To which energy level do the transitions corresponding to the visible lines in the spectrum of hydrogen relate?

(e) Why is the line spectrum of an element sometimes compared to the 'fingerprint of a criminal'?

6 The graph in figure 6.15 shows the first and second ionisation energies of the elements nitrogen to calcium.

(a) Why is there a large decrease in the first ionisation energy after neon and after argon?

(b) Why is the first ionisation energy of magnesium greater than that for aluminium?

(c) Why is the second ionisation energy of each element greater than the corresponding first ionisation energy?

(d) Why do the maxima for the two graphs occur at different atomic numbers?

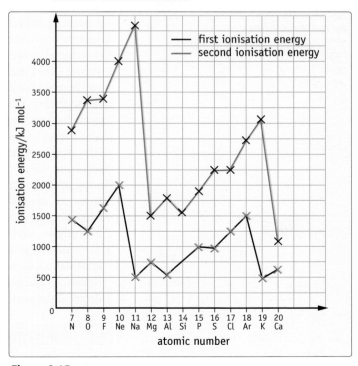

Figure 6.15
The first and second ionisation energies of the elements nitrogen to calcium.

7 Nuclear Structure and Radioactivity

7.1 Introduction

In 1896, the Frenchman Henri Becquerel was investigating the possible production of X-rays from uranium salts exposed to bright sunlight. By chance, he left some crystals of a uranium salt on a photographic plate wrapped in black paper. Poor sunshine caused him to abandon his intended investigation, but, for interest's sake, he decided to develop the photographic plate. To his surprise, the plate was considerably darkened in the region where the crystals had been. Further experiments showed that the uranium salt was emitting a radiation which had quite different properties from X-rays, and Becquerel called this phenomenon **radioactivity**.

Intrigued by Becquerel's discovery, Marie and Pierre Curie examined the radioactivity of uranium salts in some detail. In particular they looked at a uranium ore called pitchblende. They found that pitchblende was much more radioactive than they had expected from its uranium content. In 1898, after a long and tedious extraction process, they isolated two other radioactive elements from pitchblende. They called these new radioactive elements polonium and radium. Radium was found to be about two million times more radioactive than uranium. By 1900, the elements thorium and actinium were also known to be radioactive.

The fogging of a photographic plate by radioactive materials still has practical applications. For example, the metabolism of CO_2 by a plant leaf can be studied by exposing the leaf to an atmosphere of radioactive $^{14}CO_2$ and then holding the leaf against a photographic plate. When the plate is developed, the darkest parts indicate the presence of ^{14}C.

Before very long, radioactive substances were shown to cause another effect which could be used to detect their radiations. This was called **scintillation**. When a screen coated with zinc sulphide is placed near a radioactive radium salt and examined in the dark with a magnifying lens, tiny flashes of light appear on the surface of the zinc sulphide.

The flashes of light on the zinc sulphide result from alpha-particles hitting the surface of the crystal. When alpha-particles hit the crystal, the energy they have is passed on to particles in the crystal and eventually emitted as light. The alpha-particles are produced during the radioactive decay of radium nuclei in the radioactive radium salt. From time to time, a radium nucleus 'explodes' and ejects an alpha-particle at great speed.

- This spontaneous disintegration of atoms is responsible for radioactivity. Those isotopes which decay (break up) in this way are said to be **radioactive**.
- Nowadays, crystals of zinc sulphide are combined with photomultipliers to form scintillation counters. These can detect even the faintest of flashes and measure the radiations falling on the crystal. Scintillation counters are used by research scientists to detect alpha-particles.

7.2 Alpha-particles, beta-particles and gamma-rays

When radium-226 decays it emits alpha-particles, but other radioactive substances may behave differently when they disintegrate.

For example, potassium-43 emits much lighter beta-particles during decay and other materials emit gamma-rays.

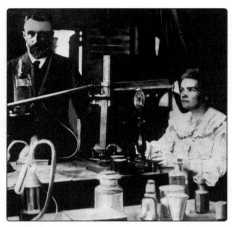

▲ Marie (1867–1934) and Pierre Curie (1859–1906) spent nearly four years isolating the radioactive elements polonium and radium from pitchblende. In 1903, they shared the Nobel Physics Prize with Becquerel. Then in 1911, Marie was awarded the Chemistry Prize – the first person ever to win two Nobel Prizes.

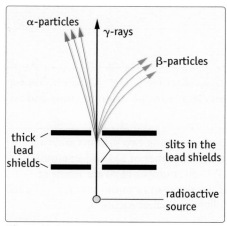

Figure 7.1
The effect of a magnetic field, perpendicular to the paper, on alpha-particles, beta-particles and gamma-rays.

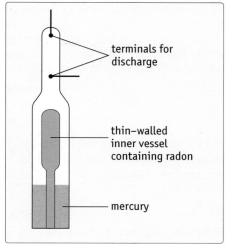

Figure 7.2
The experiment designed by Rutherford and Royds to investigate the nature of alpha-particles.

Evidence for the nature of alpha-particles and beta-particles

Alpha-particles, beta-particles and gamma-rays can be separated easily using magnetic or electric fields (figure 7.1).

Alpha-particles are deflected in a direction indicating that they are positively charged. Beta-particles are deflected in the opposite direction and must be negatively charged. Gamma-rays are unaffected by magnetic or electric fields which indicates that they have no charge.

Rutherford determined the charge/mass ratio for alpha-particles by measuring their deflection in magnetic and electric fields. The value of charge/mass was half the value for $_1^1H^+$ ions (protons), showing that alpha-particles were singly charged particles of atomic mass two, or doubly charged particles of atomic mass four, or triply charged particles of atomic mass six, etc. But, which of these suggestions was correct? The matter was settled by Rutherford and Royds using an ingenious yet simple experiment (figure 7.2).

Radioactive gaseous radon, known to emit alpha-particles, was allowed to decay for several days in a very thin-walled glass tube inside an evacuated thick-walled container. Alpha-particles from the decaying radon passed through the thin walls of the inner tube into the evacuated space outside. After a few days, the gas that had accumulated in the outer space was compressed into the top of the apparatus by raising the mercury level. It was then analysed by passing an electric discharge through it. When the radiation was analysed using a spectrometer it became clear that the gas collecting in the outer space was helium. This confirmed that alpha-particles had a mass of four units and thus a charge of +2.

- Alpha-particles are therefore doubly charged helium ions, $_2^4He^{2+}$.

Beta-particles were found to have similar properties to cathode rays.

- Measurement of charge/mass confirmed that beta-particles were electrons.

The symbol $_{-1}^0e$ can be used to represent an electron since its mass number (number of protons (n_p) + number of neutrons (n_n) is zero, and it carries a unit charge of -1.

Evidence for the nature of gamma-rays

What conclusion can you make from each of the following properties of gamma rays?

1. Gamma-rays are unaffected by an electric field or by a magnetic field.
2. Gamma-rays can penetrate several centimetres of lead.
3. Gamma-rays can be diffracted by the lattice of a crystal.
4. When an atom emits gamma-rays there is no change in atomic number or mass.

This evidence shows that gamma-rays are an uncharged, highly penetrating form of radiation, and have no mass. Unlike alpha- and beta-particles, which cause superficial damage, gamma-rays can penetrate deep inside our bodies, damaging and even killing cells and tissues.

In fact, gamma rays are electromagnetic waves similar to light rays and X-rays. They have a wavelength of only 10^{-12} metres, which is about one-tenth of the wavelength of X-rays and about one hundred-thousandth of the wavelength of visible light. The emission of gamma-rays enables a nucleus to lose surplus energy.

The nature and properties of alpha-particles, beta-particles and gamma-rays are summarised in table 7.1.

Table 7.1
The nature and properties of alpha-particles, beta-particles and gamma-rays

Emission	Nature	Relative effect of electric and magnetic fields	Distance travelled through air	Relative penetrating power
α-particles	helium nuclei ($_2^4He^{2+}$)	small deflection	a few cm	1
β-particles	electrons ($_{-1}^0e$)	large deflection	a few m	100
γ-rays	electromagnetic waves	no deflection	a few km	10 000

7.3   Nuclear equations

Alpha-decay

When an atom disintegrates by losing an alpha-particle ($_2^4\text{He}^{2+}$), the remaining fragment will have a mass number four units less than the original atom and an atomic number two units lower.

For example, when $_{92}^{238}\text{U}$ loses an alpha-particle (containing two protons and two neutrons), the breakdown product will have a mass number of 234 and an atomic number of 90. All isotopes (section 5.4) with an atomic number of 90 are atoms of thorium, Th. Thus, the products from the nuclear decay of uranium-238 are $_{90}^{234}\text{Th}$ and $_2^4\text{He}$ and the process can be summarised by the following nuclear equation.

$$_{92}^{238}\text{U} \rightarrow _{90}^{234}\text{Th} + _2^4\text{He}$$

Remember that both mass (indicated by the superscripts) and charge (indicated by the subscripts) must balance in nuclear equations.

   Most radioactive isotopes with atomic numbers greater than 83 (bismuth) undergo alpha-decay. Isotopes which decay by loss of alpha-particles include radium-226, plutonium-238, polonium-218 and radon-220.

> 1   Write nuclear equations for the alpha-decay of radium-226 ($_{88}^{226}\text{Ra}$) and plutonium-238 ($_{94}^{238}\text{Pu}$), using the periodic table in section 4.4 to find the symbols of the breakdown products.

Beta-decay

Elements with an atomic number less than 83 do not usually exhibit alpha-decay. Instead, various isotopes of these elements emit beta-particles. For example, carbon-14 ($_6^{14}\text{C}$) decays by beta-particle emission forming $_7^{14}\text{N}$. The decay process can be represented by the equation

$$_6^{14}\text{C} \rightarrow _7^{14}\text{N} + _{-1}^{0}\text{e}$$

Notice that both mass and charge are conserved in beta-decay as in alpha-decay.

> 2   What happens to the nucleus of the carbon-14 atoms during the beta-emission?
> 3   Where does the ejected electron come from?

> During beta-decay, mass number remains constant, so the number of protons + neutrons stays constant. However, the atomic number increases by one .

This means that a neutron in the $_6^{14}\text{C}$ nucleus is converted to a proton + an electron:

$$_0^1\text{n} \rightarrow _1^1\text{p} + _{-1}^{0}\text{e}$$

The proton remains in the nucleus (now $_7^{14}\text{N}$), but the electron is ejected as a beta-particle. Notice that the ejected electron comes from the breakdown of a neutron *in the nucleus*. It is *not* one of the electrons in the shells of the isotope.

• Thus, the overall effect of beta-decay is an increase of one proton in an isotope and a corresponding decrease of one neutron.

> 4   Write nuclear equations for the beta-decay of $_{38}^{90}\text{Sr}$ and $_{53}^{131}\text{I}$.
> 5   How do chemical reactions differ from nuclear reactions such as alpha-decay and beta-decay?

   During chemical reactions electrons are redistributed either by transfer from one atom to another or by sharing between atoms. Chemical reactions involve only the outer parts of atoms – the electrons.

During nuclear reactions one element may be converted to another either by radioactive decay or by atomic fission or fusion. Nuclear reactions involve the nucleus – the protons and neutrons.

Another important difference between chemical reactions and nuclear reactions is that the energy changes in nuclear reactions are usually much greater than those in chemical reactions.

7.4 Stable and unstable isotopes

At the beginning of the twentieth century detailed investigations were made of the disintegration of naturally occurring radioactive isotopes. Scientists were interested in:

1 the types of radioactive change (alpha-decay or beta-decay),

2 the products of decay (alpha-particles, beta-particles or gamma-rays),

3 the rate of decay, and

4 the relative stability of different radioactive isotopes.

Experiments showed that the natural decay of a radioactive isotope could not be accelerated or retarded by physical or chemical means. For example, extremes of temperature and pressure had virtually no effect on the rate of decay.

> Scientists found that the most convenient method of expressing the rate of disintegration of a radioactive element was its **half-life**. The half-life of a radioactive isotope is the time taken for the amount or concentration of the isotope to fall to half of its original value. Since the radioactivity of the isotope will also halve in this time, the half-life can also be defined as the time for the radioactivity of the isotope to be reduced to half its initial value.

The half-life of a radioactive isotope remains constant. For example, the half-life for $^{60}_{27}$Co is 5.2 years. This means that starting with one gram of cobalt-60, only half a gram would remain after 5.2 years. After another 5.2 years, the half gram would have decayed to one-quarter of a gram and 5.2 years after that (ie. 15.6 years from the start of the experiment) only one-eighth of a gram would remain. Table 7.2 gives the half-lives of a few radioactive isotopes.

Table 7.2
The half-lives of some radioactive isotopes

Radioactive isotope	Half-life
Uranium-238, $^{238}_{92}$U	4.5×10^9 years
Carbon-14, $^{14}_{6}$C	5.7×10^3 years
Radium-226, $^{226}_{88}$Ra	1.6×10^3 years
Strontium-90 $^{90}_{38}$Sr	28 years
Iodine-131, $^{131}_{53}$I	8.1 days
Bismuth-214, $^{214}_{83}$Bi	19.7 minutes
Polonium-214, $^{214}_{84}$Po	1.5×10^{-4} seconds

The shorter the half-life, the faster the isotope decays and the more unstable it is. The longer the half-life, the slower the decay process and the more stable the isotope. Thus, the half-life of a radioactive isotope provides a quantitative measure of its relative stability.

Uranium-238 with a half-life of 4.5×10^9 years could be described as 'almost stable', but polonium-214 is quite the reverse.

Suppose you had one gram of $^{214}_{84}$Po.

1 How much of it would be left after 1.5×10^{-4} seconds?

2 How much of it would be left after 3×10^{-4} seconds?

3 How much of it would be left after 1.5×10^{-3} seconds?

The answer to the last question is slightly less than one thousandth of a gram.

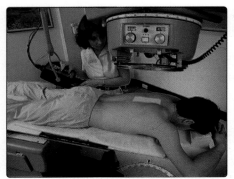

▲ Radioactive cobalt-60 being used in the treatment of cancer. Cobalt-60 is the most common source of gamma-radiation.

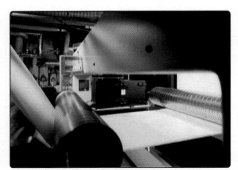

▲ A radiation thickness gauge being used to control the thickness of plastic sheets. The radiation source is krypton-85.

Treatment of cancer

The structure of DNA within the genes of animals and plants can be changed by radiation. For reasons which are not yet clear, cancer cells are more susceptible than normal cells to destruction by radiation. Because of this, gamma-rays can be used in the treatment of cancer. Penetrating gamma-radiation from cobalt-60 ($^{60}_{27}$Co) is used in treating inaccessible growths. Superficial cancers, such as skin cancer, can be treated by less penetrating radiation from $^{32}_{15}$P or $^{90}_{38}$Sr in plastic sheets strapped on the affected area.

Studying metabolic pathways

Radioactive isotopes can be used to trace the uptake and metabolism of various elements by animals and plants. For example, the uptake of phosphate and the metabolism of phosphorus by plants can be studied using a phosphate fertiliser containing $^{32}_{15}$P. Radioactive tracer studies using $^{14}_{6}$C have helped in the understanding of photosynthesis and protein synthesis. $^{131}_{54}$I has been used in the diagnosis and treatment of thyroid diseases and in research into the working of the thyroid gland.

Thickness gauges and empty-packet detectors

The radiation passing through a material decreases as the material gets thicker. Thus, the amount of penetrating beta- or gamma-radiation can be used to estimate the thickness of various materials such as paper, metal or plastic. The advantage of using radiation for measuring the thickness of materials is that there need be no touching, marking or tearing of the article concerned. Thus, radiation thickness gauges can be used to control the thickness of sheet emerging from a high-speed rolling mill. Beta-particles can be used for steel thicknesses up to about 0.2 cm. More penetrating gamma-rays can be used with steel up to 10 cm thick.

A similar use of radioactive isotopes is made in level gauges and empty-packet detectors. The level of liquid in a closed vessel can be found by placing a source on one side of the container and a detector on the other. A sudden decrease in the detected radiation, when source and detector are moved down the vessel simultaneously, shows the level of liquid within the vessel. Level gauges of this type are used to measure the amount of liquid in fire extinguishers and gas cylinders. In a similar fashion, empty-packet detectors can be set to reject empty or insufficiently filled packets of biscuits or cigarettes.

Archaeological and geological uses

How old is the earth?

If we know the rate at which an isotope decays, we can work out the time it takes for a certain amount of it to disappear. Thus, the decay of uranium-238 to lead-206 provides a method of dating rocks in the earth's crust. We must begin by assuming that the uranium-bearing rocks in the earth's crust originally contained uranium-238, but no lead-206. The present ratio of U-238 to Pb-206 in the rocks can then be used to calculate the time which has elapsed since the rocks formed. Using this method, the ages of rocks are found to vary from forty million to four thousand million years. This second value is often taken by geologists as the age of the earth.

Dating ancient remains from carbon-14

The basis of carbon dating is the simultaneous production and disintegration of carbon-14 ($^{14}_{6}$C). $^{14}_{6}$C is continuously formed in the upper atmosphere by the collision of neutrons with nitrogen atoms:

$$^{14}_{7}\text{N} + ^{1}_{0}\text{n} \rightarrow ^{14}_{6}\text{C} + ^{1}_{1}\text{H}$$

The colliding neutrons are produced when cosmic rays strike atoms and split them into fragments.

At the same time as $^{14}_{6}$C is being formed, the carbon-14 already present is decaying to $^{14}_{7}$N by the emission of electrons (beta-decay):

$$^{14}_{6}\text{C} \rightarrow ^{14}_{7}\text{N} + ^{0}_{-1}\text{e}$$

As a result of this simultaneous formation and decay of $^{14}_{6}C$, the atmosphere contains a constant concentration of $^{14}_{6}C$ as $^{14}_{6}CO_2$. This $^{14}_{6}CO_2$ gets into plants via photosynthesis and from plants it is passed into animals. Thus, all living things have a constant proportion of their carbon in the form of carbon-14.

Now, when the animal or plant dies, replacement of $^{14}_{6}C$ atoms ceases, but decay of $^{14}_{6}C$ atoms continues. Just suppose that $^{14}_{6}C$ makes up x% of all carbon in living things. About 5700 years after the plant or animal has died it will contain only half the percentage of $^{14}_{6}C$ atoms (i.e. $\frac{x}{2}$%). After another 5700 years, the concentration of $^{14}_{6}C$ will have halved again from $\frac{x}{2}$ to $\frac{x}{4}$%. By comparing the concentration of $^{14}_{6}C$ in archaeological specimens with the concentration of $^{14}_{6}C$ in similar materials living at the present time, it is possible to calculate the age of the specimen. In this way, carbon dating has been used to establish Egyptian chronology and to check the authenticity of ancient remains such as the Dead Sea Scrolls.

The most infamous forgery to be uncovered by radioactive dating is probably Piltdown Man. In 1912, the archaeologist Charles Dawson claimed he had discovered the fossilised skull of a primitive apeman on Piltdown Common, near Lewes in England. In 1953, the remains were re-examined more closely. Results showed that the cranium and the jaw contained different percentages of fluorine, nitrogen, uranium and carbon. This made people very suspicious. In 1959, the age of the bones was determined from the radioactivity of carbon-14 in them. The remains were shown to be skilfully disguised fragments of a modern human cranium and an orang-utan jaw. Chemical tests revealed that Dawson had stained the fragments with chromium and iron(II) salts. Neither of these salts occurred in the region where the remains were found.

▶ Reconstructed skulls of Piltdown Man, a forgery uncovered by radioactive dating.

Summary

1 Radioactivity results from the spontaneous disintegration of nuclei.

2 Naturally occurring elements which undergo radioactive decay emit alpha-particles ($_2^4He^{2+}$ ions), beta-particles (electrons) and gamma-rays (electromagnetic radiation).

3 Gamma-radiation is far more penetrating than alpha- and beta-radiation. It can penetrate thick metal sheets and is responsible for most of the damage caused by the disintegration of radioactive substances.

4 The half-life of a radioactive isotope is the time taken for the mass or concentration of the isotope (or for its rate of decay) to fall to half its initial value. The half-life of a radioactive isotope is virtually unaffected by external conditions. It can be used as a measure of the relative stability of an isotope.

5 During alpha-decay, the mass number falls by four units and the atomic number falls by two units.

6 During beta-decay, the mass number remains constant, but the atomic number rises by one unit.

7 Chemical reactions involve electrons in the outer parts of atoms, whereas nuclear reactions involve protons and neutrons in the nucleus.

Review questions

1 (a) What are the relative masses and charges of alpha- and beta-particles?
(b) What is the nature of gamma-radiation?
(c) What evidence can you provide to support your statements concerning the nature of alpha-, beta- and gamma-radiation?

2 A radioactive source with a very long half-life emits alpha-particles of energy 8×10^{-12} J at the rate of 4.2×10^{10} per second. The source is embedded in material sufficient to absorb the alpha-particles. The material loses heat at the rate of 8.4 J per minute for every 10°C its temperature is above that of its surroundings and eventually a steady temperature is attained.
(a) What conditions apply when the steady temperature is attained?
(b) How much has the temperature of the material risen above its surroundings when the steady state is achieved?

3 (a) How do the atomic number and the mass number of an isotope change when its nucleus loses
  (i) an alpha-particle,   (ii) a beta-particle,
  (iii) gamma-rays,   (iv) a neutron?
(b) $_{90}^{232}Th$ emits a total of six alpha-particles and four beta-particles in its natural decay sequence. What is the atomic number, mass number and symbol of the final product?
(c) Write nuclear equations to summarise the following changes:
  (i) when a $_3^7Li$ nucleus absorbs a colliding proton the product disintegrates into two exactly similar fragments,
  (ii) $_{90}^{232}Th$ atoms undergo alpha-decay to form radium atoms.

(d) Scintillation counting shows that 1 mg of polonium emits 3×10^{18} alpha-particles in the course of complete decay. What is the relative atomic mass of polonium? (State the assumptions in your calculation.)

4 $_{83}^{214}Bi$ has a half-life of 20 minutes.
(a) Plot a graph of the percentage of the $_{83}^{214}Bi$ remaining (vertical) against time.
What percentage of the $_{83}^{214}Bi$ remains after 70 minutes?
(b) Rewrite the following nuclear equations substituting symbols and numbers for the question marks.
  (i) $_?^{14}? + _?^?He \rightarrow _?^?O + _1^1?$
  (ii) $_?^?H + _?^{56}Fe \rightarrow _?^{57}Co + _0^1?$

5 Outline the importance or uses of two of the following isotopes.

$$_{27}^{60}Co, _{15}^{32}P, _{53}^{131}I.$$

6 (a) How is carbon-14 produced in nature?
(b) Write an equation for the beta-decay of $_6^{14}C$.
(c) How is $_6^{14}C$ used in 'dating' archaeological remains?
(d) How was radioactive 'dating' used to confirm either the validity of the Dead Sea Scrolls or the forgery of Piltdown Man?

The Electronic Theory and Chemical Bonding

8

8.1 Introduction

Look at the information in table 8.1 showing the electron structures of the atoms and ions of the elements in period 3.

Atoms of the first three elements in the period (Na, Mg and Al) *lose* electrons from their outermost shell to form positively charged ions (Na^+, Mg^{2+} and Al^{3+}). These ions have an electron structure like neon, the previous noble gas.

Elements at the end of the period in Groups VI and VII (S and Cl) *gain* electrons to form negatively charged ions (S^{2-} and Cl^-). These ions have the same electron structure as the next noble gas, argon.

But what about elements in the middle of the period such as silicon and phosphorus? These elements do not usually form ions in their compounds, but they do obtain an electron structure similar to that of a noble gas.

Table 8.1
Electron structures of the atoms and ions of the elements in period 3

Element	Na	Mg	Al	Si	P	S	Cl	Ar
Electron structure of the atom	2, 8, 1	2, 8, 2	2, 8, 3	2, 8, 4	2, 8, 5	2, 8, 6	2, 8, 7	2, 8, 8
No. of electrons in outermost shell of atom	1	2	3	4	5	6	7	8
Ion formed	Na^+	Mg^{2+}	Al^{3+}	–	–	S^{2-}	Cl^-	–
Electron structure of the ion	2, 8,	2, 8	2, 8	–	–	2, 8, 8	2, 8, 8	–

Like other elements in Groups IV and V, silicon and phosphorus achieve electron structures similar to noble gases by sharing electrons with other atoms and not by electron transfer.

In 1916, Kossel and Lewis realised that all the noble gases, except helium, had an outer shell containing eight electrons. They suggested that this arrangement was responsible for the stability and inertness of the noble gases.

Thus, *when elements form compounds, they either lose, gain or share electrons so as to achieve stable (low-energy) electron configurations similar to the next higher or lower noble gas in the periodic table.* This simple idea forms the basis of the electronic theory of bonding.

Since Kossel and Lewis put forward their ideas in 1916, the noble gases have been shown to be more reactive than expected and we now know of many compounds in which the elements do not have a noble gas structure. For example, most of the ions of transition metals (e.g. Fe^{2+}, Fe^{3+}, Cu^{2+}) do not have an electron structure like a noble gas, nor does the sulphur atom in SO_2, SO_3 or SF_6. Nevertheless, the ideas suggested by Kossel and Lewis still form the basis of modern theories of bonding. These theories explain the formulas and structures of most compounds and the forces holding atoms and ions together.

8.2 Transfer of electrons – electrovalent (ionic) bonding

Typical ionic compounds are formed when metals in Group I or Group II react with non-metals in Group VI or Group VII. When the reaction occurs, electrons are transferred from the metal to the non-metal until the outer electron shells of the resulting ions are identical to those of a noble gas.

Figure 8.1 shows how the transfer of electrons from lithium to oxygen forms ions in lithium oxide. In figure 8.1, the nucleus of each atom is represented by its symbol and the electrons in each shell are represented by circled dots or crosses around the symbol. Ions are shown in square brackets with the charge at the top right-hand corner.

1 What is the electron structure of: (i) the lithium ion (ii) the oxide ion?

2 Which noble gas has an electron structure like Li^+?

3 Which noble gas has an electron structure like O^{2-}?

4 Why is it that two lithium atoms react with only one oxygen atom?

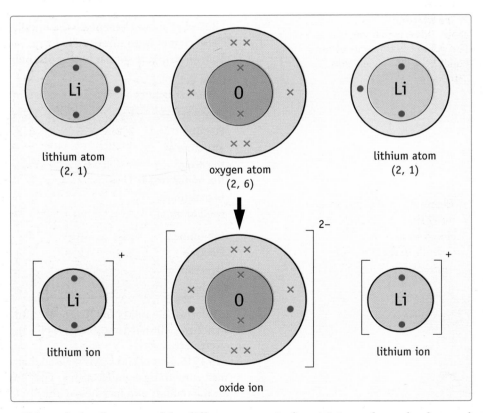

Figure 8.1
Transfer of electrons from lithium to oxygen in the formation of lithium oxide.

1 Draw similar 'dot/cross' diagrams to represent the electron transfers which take place in the formation of magnesium sulphide and potassium oxide.

Although the electrons of the different atoms in figure 8.1 are shown by dots and crosses, you must not think that electrons of lithium are any different from those of oxygen. All electrons are identical. They are shown differently in the diagram so that you can follow their transfer more easily.

Figure 8.2 shows the electron transfers which take place during the formation of sodium chloride and magnesium fluoride. In this figure, only those electrons in the outer shell of each atom are represented by dots or crosses round their symbol. The full electron structures are shown in brackets below the symbols.

Figure 8.2
Electron transfers during the formation of sodium chloride and magnesium fluoride.

Na •   +   × Cl × ⟶ [Na]⁺   [× Cl ×]⁻
(2, 8, 1)   (2, 8, 7)   (2, 8)   (2, 8, 8)
Sodium atom   Chlorine atom   Sodium ion Chloride ion

Mg ⦂   +   × F ×
(2, 8, 2)   (2, 7)   [Mg]²⁺   [× F ×]⁻
             × F ×   (2, 8)   (2, 8)
             (2, 7)   [× F ×]⁻
Magnesium   2 Fluorine   Magnesium   (2, 8)
atom   atoms   ion   2 Fluoride
                                         ions

▲ The White Cliffs of Dover are composed of chalk. This is mainly calcium carbonate ($CaCO_3$) which is held together by ionic bonds between calcium ions (Ca^{2+}) and carbonate ions (CO_3^{2-}).

Remember when writing 'dot/cross' diagrams that the dots and crosses are simply a means of counting electrons. They cannot show the precise location of electrons within the atom since electrons are distributed in space as diffuse negative charge clouds.

> The formation of ions in such compounds as lithium oxide, sodium chloride and magnesium fluoride involves a *complete transfer* of electrons. In the crystals of these substances ions are held together by **ionic (electrovalent) bonds**. These bonds involve an electrostatic attraction between oppositely charged ions.

The structure, properties and bonding in ionic compounds are discussed in sections 10.7 and 10.8.

8.3   Factors influencing the formation of ions

What factors determine whether an atom will form an ion?

(A) Cations – ionisation energies

In the case of metals, ionisation involves losing electrons. Thus, the ionisation energy of the metal provides a quantitative measure of the ease with which the metal atoms will form ions. In the case of doubly charged cations we must take the sum of the first and second ionisation energies into consideration. In section 6.2, we discussed the measurement of ionisation energies. The influence of nuclear charge and atomic radius on the sizes of ionisation energies is discussed more fully in section 13.3.

The lower the ionisation energy of an atom or ion the more easily it will lose an electron. The further an electron is from the nucleus, the less firmly it is held and the more easily it can be lost. Thus, in Group I, caesium, which has the largest atoms, forms ions most easily in spite of having the largest positive nuclear charge. In fact, caesium loses electrons so easily that it is used in photoelectric cells (section 15.9). In general, it becomes easier to form positive ions on passing down a group from atoms of smaller to those of larger relative atomic mass.

Once an electron has been lost from an atom, the overall positive charge can hold the remaining electrons more firmly. So, although many metals form doubly charged cations, those with three units of charge are less common and ions with four positive charges are very rare. In fact, it requires four times as much energy to form $Mg^{2+}(g)$ from $Mg(g)$ than to form $Na^+(g)$ from $Na(g)$, and the process $Al(g) \rightarrow Al^{3+}(g)$ requires ten times as much energy as $Na(g) \rightarrow Na^+(g)$.

(B) Anions – electron affinity

Ionisation energies provide an indication of the ease with which atoms or ions can lose electrons. They are therefore relevant to the formation of cations, but are of no use in considering the formation of anions which involves the gain of an electron or electrons.

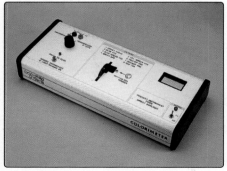

1 Element X has two electrons in its outer shell and element Y has six electrons in its outer shell. What is the formula of the most likely compound formed from X and Y?

▲ A photoelectric colorimeter which can be used to determine the concentration of a coloured solution. There is an explanation of the working of the instrument in section 24.4. This often involves the ionisation of caesium atoms.

A more appropriate measurement in the case of anions is the electron affinity, which gives the energy change for the process:

$$X(g) + e^- \rightarrow X^-(g)$$

Notice that the electron affinity for $X(g)$ is simply the reverse of the first ionisation energy for $X^-(g)$. Electron affinities can be obtained by spectroscopic methods similar to those used in the determination of ionisation energies (section 6.2). Table 8.2 shows some values of electron affinities. The more negative (i.e. more exothermic) the electron affinity, the more stable is the anion formed. Thus, Cl^- is more stable than Br^- with respect to their corresponding atoms. Likewise, Br^- is more stable than I^-.

Table 8.2
Electron affinities for some atoms and ions (values are in kJ mol^{-1})

$H \rightarrow H^-$			
-72			
$F \rightarrow F^-$	$Cl \rightarrow Cl^-$	$Br \rightarrow Br^-$	$I \rightarrow I^-$
-333	-364	-342	-295
$O \rightarrow O^-$	$O^- \rightarrow O^{2-}$	$S \rightarrow S^-$	$S^- \rightarrow S^{2-}$
-141	$+791$	-200	$+649$

1 Why are the electron affinities of halogen atoms more exothermic than those of the oxygen atom or the sulphur atom?

2 Why are the electron affinities of O^- and S^- endothermic?

In general, electron affinities become more exothermic as a period is crossed from left to right because the incoming electron is attracted more strongly by the progressively smaller atoms with an increasing positive charge in their nucleus.
The electron affinities of negative ions are always endothermic because energy has to be used in forcing another electron onto an already negative ion.

(C) Lattice enthalpies

Although ionisation energies and electron affinities provide information about the ease with which atoms form ions, these processes are often endothermic. For example, the conversion of solid sodium to gaseous sodium ions $(Na(s) \rightarrow Na^+(g) + e^-)$ is endothermic and this means that $Na^+(g)$ is less stable than $Na(s)$. However, the gaseous sodium ion (Na^+) is stabilised when it comes into contact with Cl^- ions forming solid NaCl. The process

$$Na^+(g) + Cl^-(g) \rightarrow Na^+Cl^-(s)$$

is highly exothermic ($\Delta H = -781$ kJ mol^{-1}). The energy change for this reaction is known as the **lattice enthalpy (lattice energy)** for sodium chloride.
   The more exothermic the lattice enthalpy the more stable the ionic compound formed. Lattice enthalpies are discussed in more detail in section 12.12.

8.4   Sharing electrons – covalent bonding

Look closely at figure 8.3. This shows an electron density map for the hydrogen molecule. Lines on this map join points with the same electron density in the same way that contours on a geographical map join points at the same height above sea-level. Notice that, although the highest concentration of electrons is near each nucleus, there is also a high concentration of electrons between the two nuclei. This suggests that in molecules such as H_2, electrons are shared by the two hydrogen atoms. The two atoms are held together by the strong attractions of their nuclei for the electrons in between.

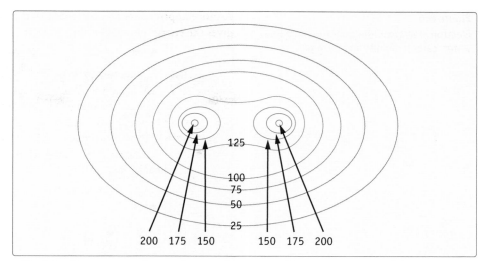

Figure 8.3

An electron density map for the hydrogen molecule (units for the contours are electrons per nm^3). After C.A. Coulson, *Proc. Cam. Phil. Soc.*, **34**, 210.

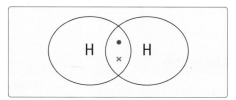

Figure 8.4

A 'dot/cross' representation of the electron structure of a hydrogen molecule.

Each hydrogen atom has only one electron and is very reactive. If, however, two hydrogen atoms come close together their electron orbitals can overlap. The pair of electrons is then attracted to each nucleus and shared by each atom. A schematic 'dot/cross' representation of this is shown in figure 8.4. Each hydrogen atom now has two electrons, which is the same electron structure as helium. Thus, the molecule H$_2$ is much more stable than the H atom.

The sharing of a pair of electrons between two atoms makes a covalent bond.

In a normal covalent bond, each atom contributes one electron to the shared pair. The electrons of the shared pair help to fill the outermost shell of both atoms.

Covalent bonding can be used to explain the structures and formulas of non-metals (e.g. Cl$_2$, P$_4$, S$_8$) and also of non-metal/non-metal compounds (e.g. HCl, CH$_4$, CO$_2$). In these substances, *each atom usually gains a noble gas electron structure* as a result of electron sharing. Figure 8.5 shows how this happens in the case of chlorine, Cl$_2$. Each chlorine atom has the electronic structure 2, 8, 7 with seven electrons in its outermost shell. By sharing one pair of electrons they acquire an electron structure similar to argon (2, 8, 8). Figure 8.6 shows the electron 'dot/cross' diagrams for ammonia, NH$_3$; water, H$_2$O; carbon dioxide, CO$_2$; and nitrogen N$_2$. In figures 8.5 and 8.6 only the electrons in the outer shell of each atom are shown.

Figure 8.5

The formation of a covalent bond between two chlorine atoms.

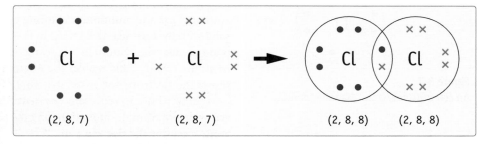

▲ Quartz (silicon dioxide) is a very hard substance. The silicon and oxygen atoms in it are held together by covalent bonds.

1 What is the electron shell structure for a nitrogen atom?
2 Which noble gas has an electron structure similar to N in NH$_3$?
3 How many double covalent bonds are there in one CO$_2$ molecule?
4 How many electrons are shared in each double covalent bond?
5 How would you describe the bond between nitrogen atoms in N$_2$?

In carbon dioxide, double bonds are formed by the sharing of four electrons, two contributed by each atom. Similarly, triple bonds can be formed by the sharing of six electrons. In this case, three electrons are contributed by each atom (see nitrogen in figure 8.6). Remember that 'dot/cross' diagrams are merely a convenient method for counting electrons. They do not represent the positions of electrons, nor do they give any indication of the shape of the molecules.

Figure 8.6
Electron 'dot/cross' diagrams for ammonia, water, carbon dioxide and nitrogen.

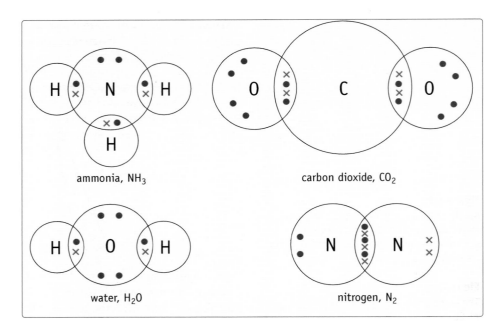

ammonia, NH₃

carbon dioxide, CO₂

water, H₂O

nitrogen, N₂

8.5 Co-ordinate (dative covalent) bonding

In a normal covalent link, each atom provides one electron for the bond.

> However, in a few compounds a bond is formed by the sharing of a pair of electrons both of which are provided by one atom. Since both electrons in the bond are donated by one atom, the bonding is known as **dative covalent** or **co-ordinate**.

Dative covalent bonding plays an important part in the formation of the addition compound, $AlCl_3.NH_3$.

In the vapour phase at high temperatures, aluminium chloride consists of simple molecules of $AlCl_3$ with covalent bonds between the aluminium and chlorine atoms (figure 8.7). This covalent bonding is very unexpected because metal/non-metal compounds like $AlCl_3$ are normally ionic. Notice that aluminium has only six electrons in its outer shell – two short of the noble gas structure for argon. Now look at the 'dot/cross' structure of ammonia in figure 8.6. The nitrogen atom has eight electrons in its outer shell, but two of these electrons are not shared with any other atom. When ammonia gas and aluminium chloride vapour are mixed, they react rapidly to form a solid with the formula $AlCl_3.NH_3$. In this compound, the NH_3 and $AlCl_3$ have formed a bond because the N atom in NH_3 can donate its unshared pair of electrons towards the Al atom in $AlCl_3$. This enables the Al atom to achieve a noble gas structure. Figure 8.8 shows the electron 'dot/cross' structure of the product and also the simpler notation using single lines to represent covalent bonds. The dative bond is represented by an arrow drawn from the atom donating the pair of electrons (in this case nitrogen) to the atom accepting the electron pair.

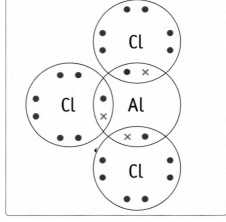

Figure 8.7
An electron 'dot/cross' diagram for $AlCl_3$.

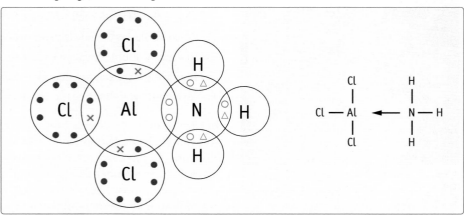

Figure 8.8
The electronic structure and bonding in $AlCl_3.NH_3$.

8 The Electronic Theory and Chemical Bonding

When gaseous aluminium chloride is cooled, $AlCl_3$ molecules dimerise to form molecules of Al_2Cl_6. Monomers of $AlCl_3$ are held together in the Al_2Cl_6 dimer by dative covalent bonding (figure 8.9).

Compounds containing unshared pairs of electrons readily form dative covalent bonds. For example, when ammonia is mixed with gaseous hydrogen chloride, white clouds of ammonium chloride form. The reaction can be thought to result from the formation of a dative covalent bond between the unshared pair of electrons on the N atom in NH_3 and an H^+ ion from the HCl (figure 8.10).

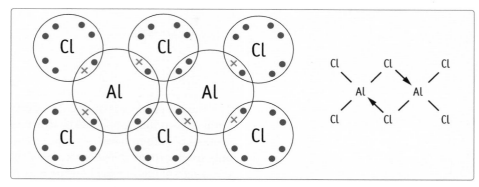

Figure 8.9
The electronic structure and bonding in Al_2Cl_6.

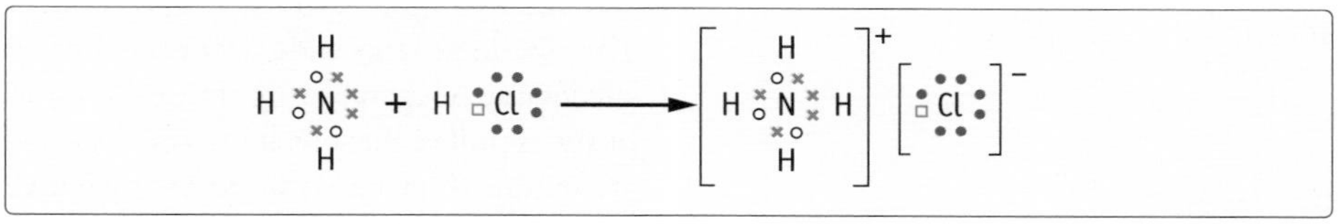

Figure 8.10
An electron 'dot/cross' representation for the formation of NH_4Cl from NH_3 and HCl.

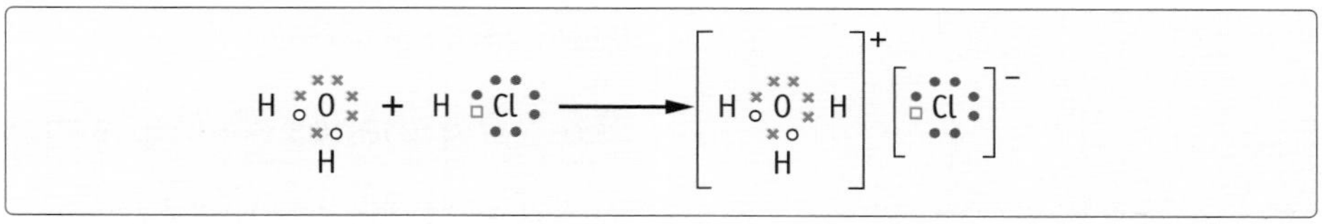

Figure 8.11
An electron 'dot/cross' representation for the reaction between water and hydrogen chloride.

A similar reaction to this occurs when hydrogen chloride dissolves in water (figure 8.11).

Notice that both the ammonium ion and the oxonium ion (H_3O^+) have an overall charge of +1 which originates from the H^+ ions. In these ions, this charge is distributed over the whole structure. It is not localised on any one atom.

Once a dative bond has formed, it is indistinguishable from a covalent bond. The only difference between the two lies in the way we pictured the bond forming. Thus, although the NH_4^+ and H_3O^+ ions are sometimes represented as

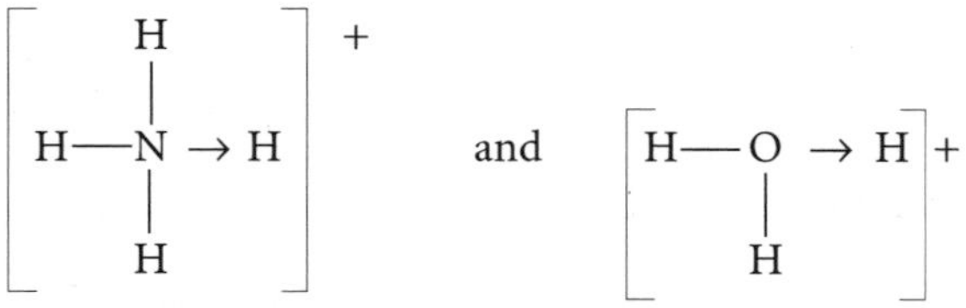

$$\left[\begin{array}{c} H \\ | \\ H-N \rightarrow H \\ | \\ H \end{array} \right]^+ \qquad \text{and} \qquad \left[\begin{array}{c} H-O \rightarrow H \\ | \\ H \end{array} \right]+$$

showing the co-ordinate bond as an arrow, it is important to remember that all four N—H bonds in NH_4^+ are identical as are the three O—H bonds in H_3O^+. There is no difference between any of the four hydrogen atoms in NH_4^+ and all four N—H bonds have the same length and the same strength.

The idea of dative covalent bonding has also been used to explain the existence of more than one oxide of the same non-metal. The structure of carbon dioxide can be explained purely in terms of covalent bonds. But the structure of carbon monoxide must involve a dative covalent bond if both atoms are to attain a noble gas structure.

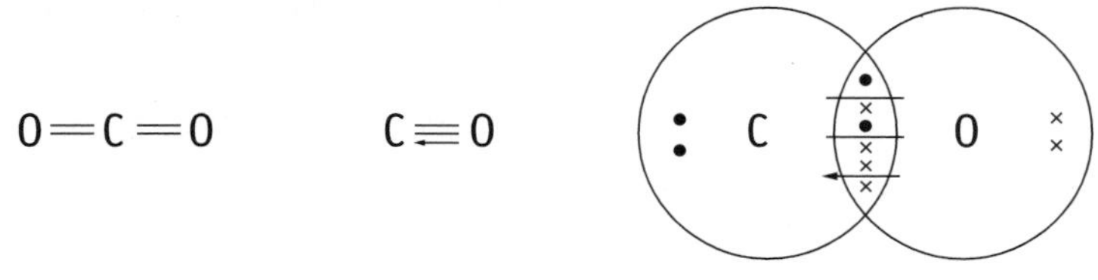

$$O=C=O \qquad\qquad C\equiv O$$

▲ Ordinary silicones are excellent sealants for bathrooms, kitchens and windows. They are water repellent and flexible over a wide temperature range. In hot weather, they do not become runny, and in cold weather, they do not become brittle. Silicones are long chain molecules in which silicon, oxygen, carbon and hydrogen are held together by covalent bonds.

Figure 8.12
Co-ordinate bonding in 'potty putty'.

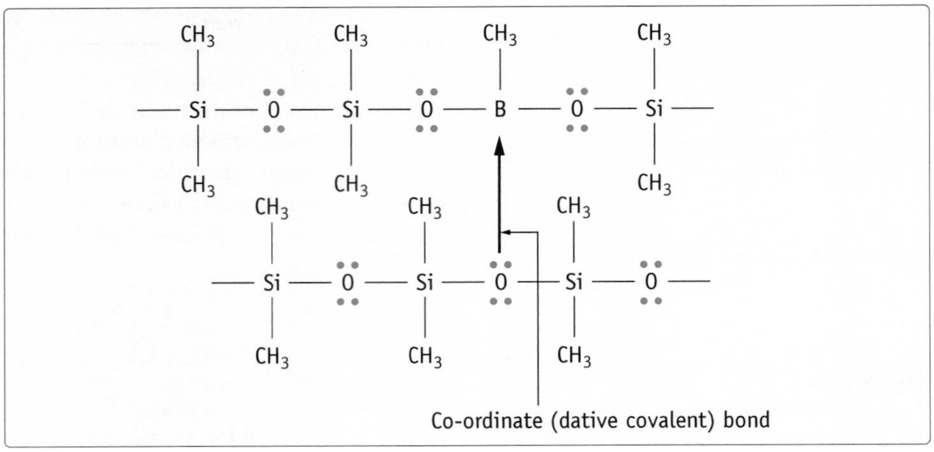

Co-ordinate (dative covalent) bond

The strange properties of silicone bouncing putty ('potty putty') can be explained in terms of co-ordinate bonds. The structure of 'potty putty' resembles a silicone in which some silicon atoms have been replaced by boron (figure 8.12). These boron atoms have only six electrons in their outermost shell and readily form dative covalent bonds with the oxygen atoms in neighbouring silicone chains.

When a sample of 'potty putty' is pulled steadily, it extends like a piece of plasticine. The silicone chains slide over each other as the boron atoms form co-ordinate bonds to successive oxygen atoms along the same neighbouring chain. If, however, the 'potty putty' is pulled sharply, it breaks like a piece of crumbly cheese because the boron atoms are unable to progress smoothly by dative bonding from one oxygen to the next. Thus, 'potty putty' can be both plastic and brittle. Not surprisingly, it has been described as a schizophrenic material.

8.6 The shapes of simple molecules

Look closely at the 'dot/cross' diagrams and structures of the simple molecules in figure 8.13.

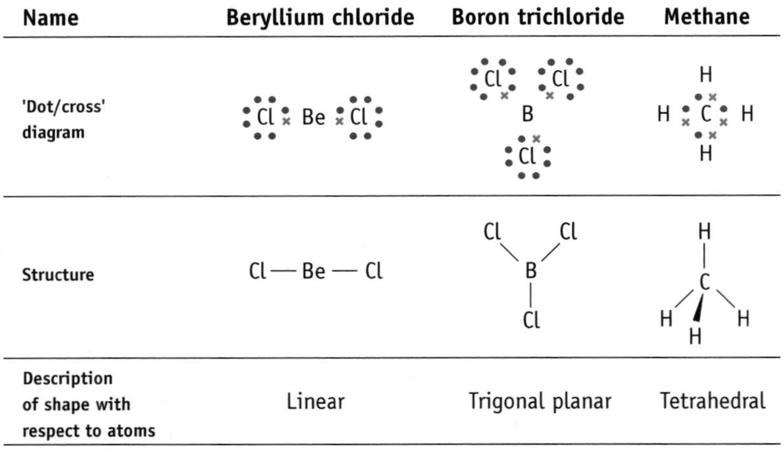

Name	Beryllium chloride	Boron trichloride	Methane
'Dot/cross' diagram			
Structure	Cl — Be — Cl		
Description of shape with respect to atoms	Linear	Trigonal planar	Tetrahedral

Figure 8.13
Electron 'dot/cross' diagrams and structures for $BeCl_2$, BCl_3 and CH_4.

Notice that the bonds in $BeCl_2$, BCl_3 and CH_4 spread out so as to be as far apart as possible. The three atoms in $BeCl_2$ are in a line. The shape is described as **linear**. The four atoms in BCl_3 are in the same plane with the chlorines at the corners of a triangle. The shape is described as **trigonal planar**. In CH_4, the four H atoms lie at the apices of a tetrahedron with the C atom at its centre. This shape is described as **tetrahedral**.

1 Why do the bonds get as far apart as possible in $BeCl_2$, BCl_3 and CH_4?

2 What is unusual about the electron structures of Be and B in $BeCl_2$ and BCl_3, respectively?

Now look at the 'dot/cross' diagrams and structures of ammonia and water in figure 8.14. NH_3, in which the three hydrogen atoms form the base of a pyramid, is described as

8 The Electronic Theory and Chemical Bonding

Name	Ammonia	Water
'Dot/cross' diagram	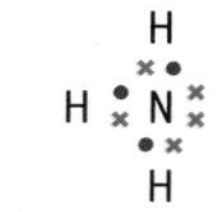 H ⨯•N⨯ (with H above, H below)	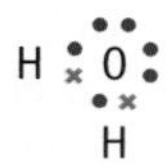 H ⨯O (with H below)
Structure	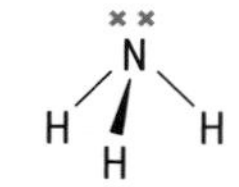 N with H, H, H (pyramidal)	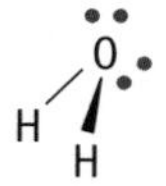 O with H, H (bent)
Description of shape with respect to atoms	Pyramidal	V-shaped or non-linear (bent)

Figure 8.14
Electron 'dot/cross' diagrams and structures for ammonia and water.

pyramidal. But why is ammonia pyramidal? Why is it not planar like BCl_3 which also has three atoms attached to the central atom? The answer lies in the non-bonded **lone pair of electrons** on the nitrogen atom. This lone pair on the nitrogen atom occupies the fourth tetrahedral position around the nitrogen atom in the NH_3 molecule. Each of the N—H bonds in ammonia is composed of a region of negative charge similar to the lone pair. The nitrogen atom is therefore surrounded by four regions of negative charge. These four negative charge clouds repel each other as far apart as possible. Consequently, the shape of ammonia, though pyramidal with respect to atoms, can be described as tetrahedral with respect to negative centres around the central nitrogen atom (figure 8.15).

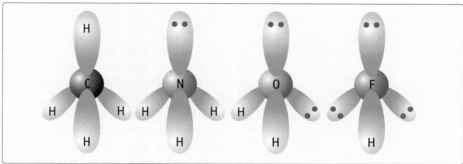

Figure 8.15
Electron charge cloud models of CH_4, NH_3, H_2O and HF.

What about the water molecule? Can its **V-shaped** or **non-linear** structure be explained in a similar manner to that of ammonia?

1 How many lone pairs does the O atom in water possess?
2 How many covalent bonds are there to the O atom in a water molecule?
3 How many centres of negative charge are there around the O atom in water?
4 What is the shape of water with respect to negative centres around the central O atom?

Figure 8.15 shows the electron charge cloud models of CH_4, NH_3, H_2O and HF. You will realise from this discussion that it is the number of regions of negative charge (not the number of bonds) around the central atom which dictates the shape of a molecule. In predicting the shape of a molecule you must, therefore, count the number of negative centres around the central atom. Thus, CO_2 (figure 8.6) and HCN (figure 8.16) with two negative centres around their central C atoms are both linear with respect to negative centres and linear with respect to atoms. In CO_2, each double covalent bond counts as a single negative centre as does the triple covalent bond in HCN. In contrast to CO_2, SO_2 has three centres of negative charge around the S atom, and is V-shaped with respect to atoms (figure 8.16). Our discussions concerning the shapes of molecules are summarised in table 8.3. These predictions about the shapes of molecules and bond angles are usually described as the **electron-pair repulsion theory**.

The electron 'dot/cross' diagram and the **octahedral** structure of sulphur hexafluoride are shown in figure 8.17. There are six regions of negative charge around the central S atom

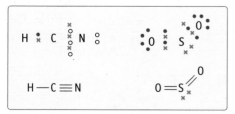

Figure 8.16
The electronic structure and bonding in HCN and SO_2.

Table 8.3
The number of negative centres, the number of lone pairs and the shapes of some simple molecules

No. of negative centres around central atom	Angular separation of negative centres	Shape w.r.t. negative centres	Example	Shape w.r.t. atoms	No. of lone pairs
2	180°	linear	$BeCl_2$, CO_2	linear	0
3	120°	trigonal planar	BCl_3, BF_3	trigonal planar	0
3	120°	trigonal planar	SO_2	V-shaped or non-linear (bent)	1
4	109°	tetrahedral	CH_4, NH_4^+, BH_4^-, CCl_4	tetrahedral	0
4	109°	tetrahedral	NH_3, H_3O^+	pyramidal	1
4	109°	tetrahedral	H_2O, F_2O	V-shaped or non-linear (bent)	2
6	90°	octahedral	SF_6	octahedral	0

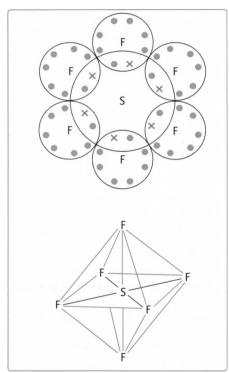

Figure 8.17
The electron 'dot/cross' diagram and the octahedral structure of SF_6.

in SF_6. These push each other as far apart as possible. This leaves four F atoms arranged in a square around the S atom and the other two F atoms above and below the square. If lines are drawn from each F atom to its nearest neighbours, the result is an octahedron (a solid with eight sides).

The fine structure of methane, ammonia and water

Look closely at table 8.4. Molecules of CH_4 are perfectly symmetrical. Consequently, the bond angle for H—C—H is that of a regular tetrahedron, 109°.

1 The bond angle for H—N—H in NH_3 is 107°, i.e. 2° less than the bond angle in CH_4. Does a bonded electron pair or a lone electron pair exert the greater repelling force?

2 The bond angle for H—O—H in water is 105°, i.e. 2° less than the bond angle in NH_3. Write the following repulsions in order of decreasing strength: bond pair/bond pair repulsion, bond pair/lone pair repulsion, lone pair/lone pair repulsion.

The region in space occupied by a lone pair of electrons is smaller and closer to the nucleus of an atom than a bonded pair. Bonded pairs of electrons are drawn out between the nuclei of the two atoms which they bind together. This means that a lone pair can exert a greater repelling effect than a bonded pair and this results in a decreasing bond angle from CH_4 to NH_3 to H_2O.

Table 8.4
Bond angles in methane, ammonia and water

	Methane CH_4	Ammonia NH_3	Water H_2O
HXH bond angle (X = C, N or O)	109°	107°	105°

8.7 Delocalisation of electrons

Look closely at the structural formula and the 'dot/cross' diagram for methanoic acid (HCOOH) below.

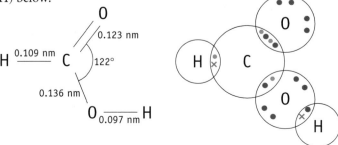

The bond lengths and bond angles in molecules of methanoic acid vapour are also shown on the structural formula. Notice, as you would expect, that the stronger $C == O$ bond is shorter than the weaker $C — O$ bond. In addition, there are three negative centres around the carbon atom giving the molecule a trigonal planar structure with bond angles of approximately 120° at the carbon atom.

Now look at the structural formula of the methanoate ion and its 'dot/cross' structure below. Although the diagram shows a carbon–oxygen double bond ($C == O$) and a carbon–oxygen single bond ($C — O$) as in methanoic acid, both of these carbon–oxygen bonds are the same length (0.127 nm) in the methanoate ion. Why is this?

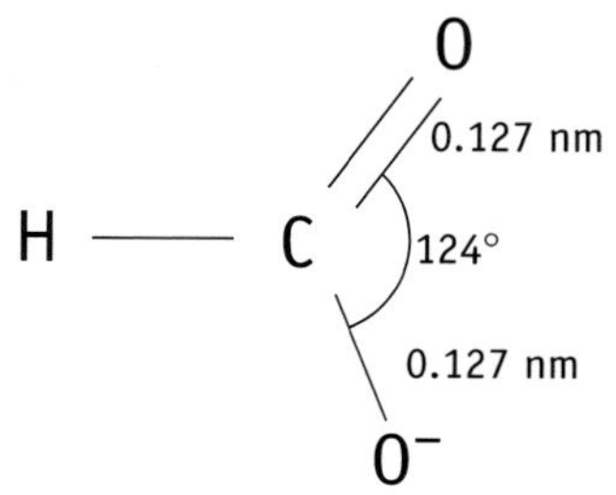

$$
H — C \overset{\displaystyle O}{\underset{\displaystyle O^-}{\Big\langle}} \quad
\begin{array}{l} 0.127\ \text{nm} \\ 124° \\ 0.127\ \text{nm} \end{array}
$$

The answer lies in the alternative 'dot/cross' diagrams below.

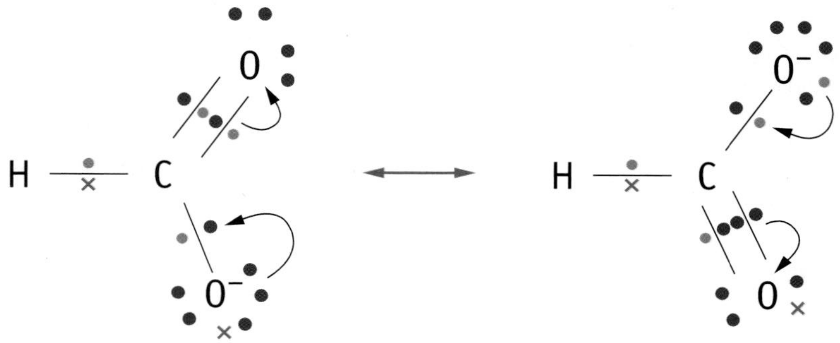

By shifting pairs of electrons, as indicated by the arrows, it is possible to switch the negative charge and the double bond to different oxygen atoms. This means that the negative charge is spread over both the carbon–oxygen bonds and the bonds are somewhere between single and double bond in character. The bond lengths in methanoate of 0.127 nm are between the $C — O$ bond length of 0.136 nm and the $C == O$ bond length of 0.123 nm in methanoic acid.

A similar situation to the methanoate ion arises with carbonic acid and carbonate ions (figure 8.18).

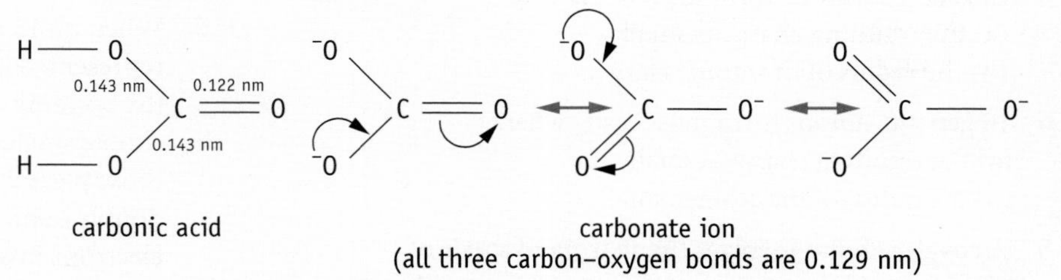

Figure 8.18
The structural formulas of carbonic acid and carbonate ions with alternative structures.

carbonic acid

carbonate ion
(all three carbon–oxygen bonds are 0.129 nm)

Another similar situation arises with sulphuric acid and sulphates (figure 8.19).

These examples show that when molecules or ions containing double bonds can be represented by two or more alternative and interchangeable structures, the bonding cannot be described accurately in terms of simple single and double bonds. In these structures, the bond lengths are in between those for single and double bonds and the bond angles are also affected.

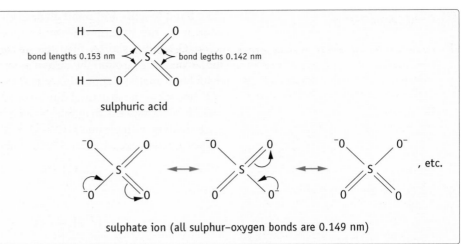

bond lengths 0.153 nm — bond legths 0.142 nm

sulphuric acid

sulphate ion (all sulphur–oxygen bonds are 0.149 nm)

Figure 8.19
The structural formulas of sulphuric acid and sulphate ions with alternative structures.

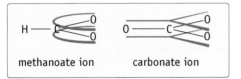

methanoate ion   carbonate ion

Figure 8.20
Delocalised charged-clouds of electrons in the methanoate and carbonate ions.

When equivalent alternative structures, such as those in figures 8.18 and 8.19 can be drawn for a molecule or an ion without changing the positions of any atoms, the actual structure is not like any of the alternatives but somewhere in between, with the available electrons distributed among the atoms involved. In those parts of the structure which can be shown with either a single or a double bond, each pair of atoms is bonded by two electrons between their nuclei forming a single covalent bond, whilst the remaining electrons are distributed as a charged cloud above and below the atoms involved (figure 8.20). The electrons in the charged clouds are not fixed on any one atom but are mobile over the three or more atoms involved. They are therefore described as **delocalised electrons** and the phenomenon is called **delocalisation**.

This idea of delocalised electrons will arise again in the context of metal structures (section 10.5) and benzene (section 28.2).

Summary

1   In many cases, when elements form compounds they lose, gain or share electrons so as to achieve stable electron configurations similar to the next higher or lower noble gas in the periodic table.

2   An ionic (electrovalent) bond involves the complete transfer of electrons from one atom to another forming oppositely charged ions. The ions are held together in the solid crystal by the electrostatic attraction between their opposite charges.

3   In general, cations form most easily when:
    (a) the resulting charge is small
    (b) the radius of the atom is large.

4   In general, anions form most easily when:
    (a) the resulting charge is small
    (b) the radius of the atom is small.

5   A covalent bond involves the sharing of a pair of electrons between two atoms, each atom contributing one electron to the shared pair.
    A double covalent bond involves the sharing of four electrons, two contributed by each atom.

6   A co-ordinate (dative covalent) bond involves the sharing of a pair of electrons between two atoms, both electrons in the bond being donated by one atom.

7   The shape of a simple molecule is dictated by the number of regions of negative charge around the central atom and the fact that the regions of negative charge will try to get as far away from each other as possible. The regions of negative charge could be a non-bonded (lone) pair of electrons, a shared pair of electrons, four shared electrons in a double covalent bond or even six electrons in a triple bond.

8   A lone pair of electrons exerts a greater repelling effect than a bonded pair.

9   When molecules or ions containg double bonds can be represented by two or more interchangeable structures, the bonding cannot be described accurately in terms of simple single and double bonds. In these structures, those parts which can be shown by either a single or a double bond have each pair of atoms bonded by two electrons in a *single* covalent bond, with the remaining electrons distributed as a charged cloud above and below the atoms involved. These electrons are described as delocalised electrons and the phenomenon is called delocalisation.

1 Draw 'dot/cross' electron structures showing electrons in the outermost shell of each atom in the following compounds. Show the overall charge on each ion in the ionic compounds. (None of the compounds involves co-ordinate bonding except those in part (d).)
(a) Cl_2, H_2O_2, O_2, C_2H_4, C_2H_2, HCN
(b) LiF, $CaCl_2$, Na_3P, Al_2S_3
(c) PCl_3, Cl_2O, HNO_2, H_2CO_3, C_2H_5OH
(d) SO_3, undissociated HNO_3, undissociated H_2SO_4.

2 (a) How is the type of bonding in the chlorides of the elements Na, Mg, Al, Si, P and S related to
(i) their position in the periodic table,
(ii) the number of electrons in the outermost shells of these elements?
(b) Explain the terms ionic, covalent and co-ordinate as applied to bonds in compounds. Illustrate your answer by clear 'dot/cross' diagrams to show the positions of all electrons in the outer shells of each atom or ion in four of the following compounds: hydrogen chloride, ammonia, ammonium chloride, aluminium chloride, sodium hydride.

3 Draw 'dot/cross' diagrams and predict the shapes with respect to atoms for molecules of the following compounds:
SF_6, $POCl_3$, $SOCl_2$, H_3O^+, BF_3, C_2H_2

4 (a) State the electronic configurations of the following atoms (e.g. Be would be 2, 2):
C, N, O, F
(b) Draw a series of 'dot/cross' diagrams to show the structures of the simplest hydrides formed by carbon, nitrogen, oxygen and fluorine.
(c) Sketch and describe the shapes of the molecules in part (b) and discuss the influence of any lone pairs of electrons on these shapes.
(d) What shape would you predict for the following:
NH_4^+, NH_3, NH_2^-?

5 X, Y and Z represent elements of atomic numbers 9, 19 and 34.
(a) Write the electronic structures for X, Y and Z (e.g. Be would be 2, 2).
(b) Predict the type of bonding which you would expect to occur between
(i) X and Y, (ii) X and Z, (iii) Y and Z.
(c) Draw 'dot/cross' diagrams for the compounds formed, showing only the electrons in the outermost shell for each atom.
(d) Predict, giving reasons, the relative
(i) volatility,
(ii) electrical conductance and
(iii) solubility in water
of the compound formed between X and Y compared with that formed between X and Z.

6 Explain the following.
(a) Tin and lead form the ions Sn^{4+} and Pb^{4+}, respectively, but carbon and silicon, elements in the same group do not form C^{4+} and Si^{4+} ions.
(b) The carbon–oxygen bond length in methoxymethane (dimethyl ether, CH_3OCH_3) is 0.14 nm, while that in carbon dioxide is 0.12 nm and that in carbon monoxide is only 0.11 nm.
(c) The C—O bond lengths in CO_3^{2-} are all identical.
(d) Aluminium fluoride has a much higher melting point than aluminium chloride.

7 The first electron affinities of the elements Na to Cl in period 3 are given below.

Element	Na	Mg	Al	Si	P	S	Cl
First electron affinity/kJ mol^{-1}	−20	+67	−30	−135	−60	−200	−364

(a) What is the general trend in the electron affinities from Na to Cl?
(b) Explain the general trend noticed in part (a).
(c) Why is the first electron affinity of Mg more positive than one might expect from the general trend in the values above?
(d) Why is the first electron affinity of silicon more exothermic than that of phosphorus?

Intermolecular Forces

9.1 Polar and non-polar molecules

Fill a burette with water. Open the tap and bring a charged ebonite rod close to the stream of water running from the jet. The water is deflected from its vertical path towards the charged rod (figure 9.1). Why is this?

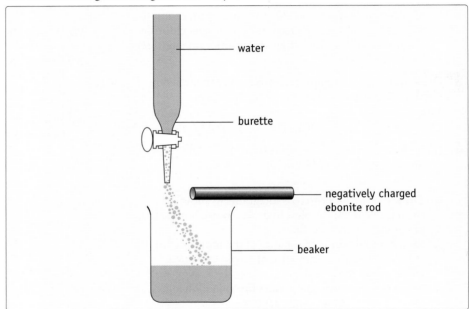

Figure 9.1
The effect of a charged rod on a thin stream of water.

The ebonite rubbed with fur has a negative charge. When the ebonite rod is replaced by a positively charged rod, the water is again deflected *towards* the rod. Why is this?

The results in table 9.1 show what happens when water is replaced by other liquids. Those liquids which are affected are always deflected *towards* the charged rod.

Table 9.1
Testing the deflection of a jet of liquid using a charged rod

Liquids showing a marked deflection	Liquids showing no deflection
Water	
Trichloromethane	Tetrachloromethane
Propanone (acetone)	
Ethoxyethane (ether)	
Nitrobenzene	Benzene
Cyclohexene	Cyclohexane
Ethanol	

1 Why is trichloromethane deflected, but not tetrachloromethane?

2 Why is nitrobenzene deflected, but not benzene?

3 Why is cyclohexene deflected, but not cyclohexane?

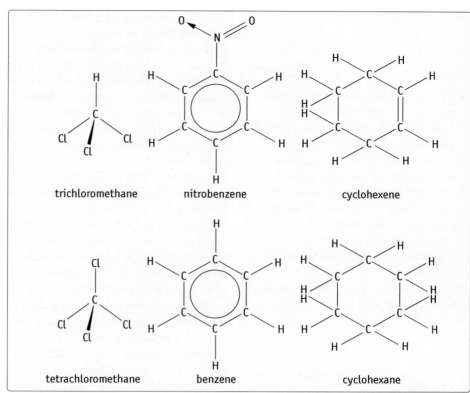

Figure 9.2
The structures of some simple molecules.

trichloromethane   nitrobenzene   cyclohexene

tetrachloromethane   benzene   cyclohexane

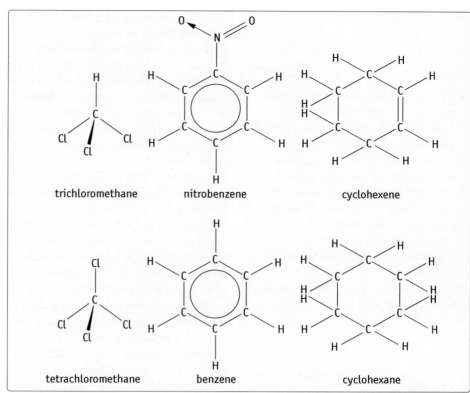

Figure 9.3
The centres of positive and negative charge in a molecule of HCl.

Centre of positive charge

Centre of negative charge

Look at the structures of these six molecules in figure 9.2. Notice that those molecules which are unaffected by a charged rod are symmetrical. Those which are deflected are not symmetrical. This lack of symmetry, in $CHCl_3$ for example, means that its centre of positive charge does not coincide exactly with its centre of negative charge. This means that the molecule is affected by an electrostatic field from the charged rod.

When two different atoms are joined by a covalent bond, their attractions for the bonding electrons will not be the same. For example, in a molecule of HCl the bonding electrons will not be shared equally by the hydrogen and chlorine atoms. In fact, the chlorine atom has a greater attraction for the electrons in the covalent bond. In chemical language, we say that chlorine is more **electronegative** than hydrogen. Consequently, the centre of negative charge in the HCl molecule is drawn towards the chlorine atom and it is closer than the centre of positive charge to the nucleus of the chlorine atom (figure 9.3).

The overall distortion of charge in molecules such as HCl, which results from unequal sharing of electrons, is known as **polarisation**. The molecules are said to be **polar** and the separation of charge in the molecule is referred to as a **dipole**.

Molecules, such as tetrachloromethane, benzene and cyclohexane, with a symmetrical distribution of similar atoms and in which equal dipoles cancel each other exactly are **non-polar**. In these molecules, the centres of positive and negative charge are at the same point.

Molecules, such as water, trichloromethane, nitrobenzene and cyclohexene, in which dipoles do not cancel each other, are polar. When these liquids stream from a burette past a charged rod, molecules are attracted towards the charged rod and the jet is deflected. When a positive rod is used, it is the negative ends of the dipoles in the polar molecules that are attracted towards the rod. With a negative rod, positive ends of the dipoles are attracted.

9.2   Electronegativity, polarity and bonding

In non-symmetrical molecules such as HCl and H_2O, the centre of positive charge and the centre of negative charge do not coincide. The polarity in these molecules results from the presence of highly electronegative elements such as chlorine and oxygen.

1 Which element will have the highest electronegativity value?

2 Which element will have the lowest electronegativity value?

3 How will the electronegativities of the elements vary across a period?

4 How will the electronegativities of the elements vary down a group as relative atomic mass increases?

5 Will the trend in electronegativity values down Group I be the same as the trend down Group VII?

Check your answers to these questions by referring to table 13.3.

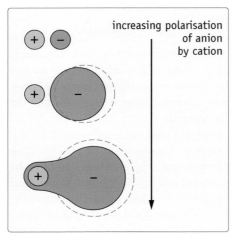

increasing polarisation of anion by cation

Figure 9.4
Polarisation of the negative charge cloud in anions of different sizes. (The dashed circles represent the shapes of the isolated anions before polarisation.)

Various attempts have been made to quantify the electronegativity or electron-attracting power of an atom within a molecule. The electron-attracting power of an atom will remain the same in different compounds. But, the *effects* of a particular atom's electron-attracting power will vary from one compound to another because it will depend on the other atoms to which the atom under consideration is attached. For this reason, it is difficult to estimate values of electronegativity with real accuracy. In spite of this, the concept of electronegativity is very useful.

One of the most widely used scales of electronegativity is that devised by the American chemist Linus Pauling (section 13.3). Pauling defined the electronegativity of an atom *as the power of that atom in a molecule to attract electrons*. He obtained values of electronegativity by considering the strengths of the bonds between atoms in different molecules. Non-metals with a strong desire to gain electrons have the highest values of electronegativity. Metals have low values.

Notice that the concept of electronegativity is concerned with the attraction for electrons of *atoms in molecules*. Don't confuse electronegativity with electron affinity which is concerned with the attraction for electrons of *single gaseous atoms*. Unlike electronegativities, electron affinities can be measured directly.

Electronegativity values can be used to estimate the polarity of different bonds. The bonds between elements of widely different electronegativities (i.e. between a metal and non-metal) will be ionic. The bonds between elements of similar electronegativity will be non-polar or slightly polar. If the two elements are non-metals the bonding will be covalent. If the elements are metals the bonding will be metallic. Thus, the bonding between sodium and fluorine, like that between potassium and oxygen, will be ionic. The bonding between fluorine and oxygen will be covalent and that between potassium and sodium will be metallic. Of course, two metals would not react with each other, but two liquid metals could mix intimately and there would be metallic bonding between different metal atoms in the mixture.

The existence of a dipole confers partial ionic character on a polar molecule. As the polarity of the molecule increases so does the extent of its ionic character.

As a result of differing electronegativities there is a distortion of the equal sharing which one would expect in a pure covalent bond. Similarly, chemists believe that there is also distortion of the charge in ionic substances. Cations attract the negative charge cloud of the anions with which they are associated (figure 9.4). This polarisation of the ions confers partial covalent character on the ionic bonding. Although it is convenient to regard bonds as either ionic or covalent, it is as well to remember that the wholly ionic and wholly covalent bonds described in chapter 8 are extreme types. All ionic bonds have some covalent character and most covalent bonds have some ionic character.

Polarisation in aluminium compounds

The chemistry of aluminium compounds is strongly influenced by the high charge and small radius of the Al^{3+} ion. These result in an unusually large charge density. The high charge density of Al^{3+} is reflected in its large charge/radius ratio which is compared with the values for some other cations in table 9.2.

Table 9.2
Values of charge, ionic radius and charge/ radius ratio for some common cations

Cation	Unit charge	Ionic radius /nm	Charge/radius ratio charge nm^{-1}
Na^+	+1	0.098	10
Mg^{2+}	+2	0.065	31
Al^{3+}	+3	0.048	63
Zn^{2+}	+2	0.074	27
Cu^{2+}	+2	0.069	29

The high charge density of the Al^{3+} ion will distort the electron cloud around any ion in contact with it causing polarisation. If the polarisation is significant, there is an effective electron density in the region between the aluminium ion and its neighbouring anion. This comprises partial covalent bonding (figure 9.4).

Aluminium oxide is normally regarded as ionic containing Al^{3+} and O^{2-} ions. However, the oxide ions are polarised by the Al^{3+} ions and the bonds have a marked degree of covalent character. This feature and the small size of both the Al^{3+} and O^{2-} ions combine to make the bonding in aluminium oxide very strong indeed. The strong bonding in Al_2O_3 results in its insolubility in water and very high melting point (2050°C).

The influence of polarisation on bonding, structure and properties is well illustrated by the aluminium halides (figure 9.5).

	Aluminium fluoride	Aluminium chloride	Aluminium bromide
Bonding:	Ionic	Intermediate between ionic and covalent	Covalent
	F^- ion is too small to be polarised	Cl^- ion is intermediate in size and partly polarised	Br^- ion easily polarised forming covalent bond to Al
Structure:	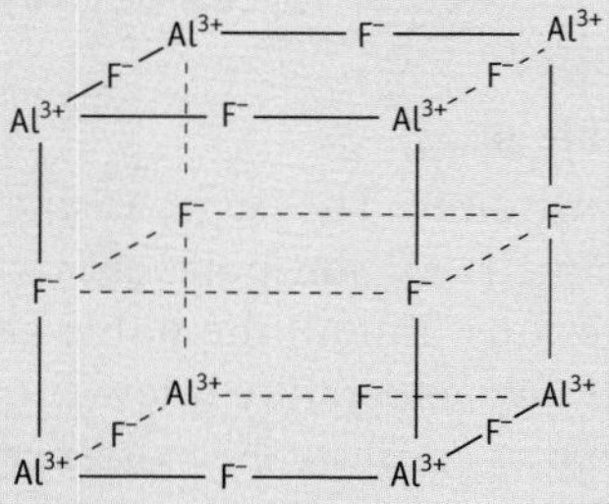	Solid shows intermediate bonding, but this sublimes to yield Al_2Cl_6 dimers with the same structure as Al_2Br_6. At high temps $AlCl_3$ molecules form. (see section 8.5 and figure 8.9)	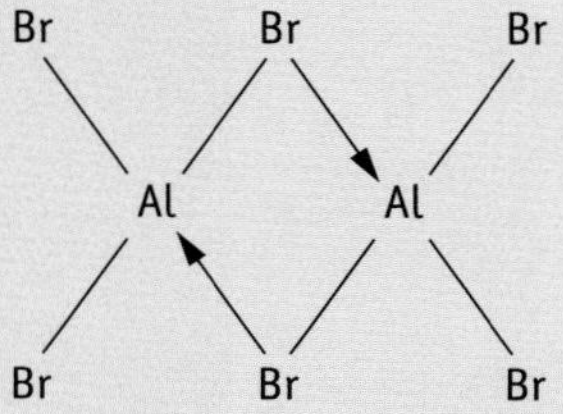 (The outer shell of each Al atom is completed by a co-ordinate bond from a Br atom in the associated $AlBr_3$ molecule.)
	Symmetrical ionic crystal	Simple molecular dimer in the vapour phase	Simple molecular (dimer) structure
Melting point:	1265°C	Sublimes at 180°C	97°C

Figure 9.5
The influence of polarisation on the bonding, structure and properties of aluminium halides.

1. Why are large anions polarised more easily than small anions?

2. What bonding and structure would you predict for aluminium iodide?

3. Gallium is in Group III, immediately below aluminium. Will Ga^{3+} have a greater or smaller polarising effect than Al^{3+} on anions? Explain.

9.3 Intermolecular forces

In the last chapter, we considered the bonding between ions in ionic compounds and that between atoms in covalent compounds. We must now consider the forces between molecules. We know, for example, that within a molecule of $CHCl_3$ the three Cl atoms and the H atom are joined to the central carbon atom by strong covalent bonds. But what kind of forces hold the separate $CHCl_3$ molecules together? What are the forces like between one $CHCl_3$ molecule and its neighbours?

(A) Dipole–dipole attractions

In section 9.1 we noted the deflection of a jet of $CHCl_3$ by a charged ebonite rod. Our observations led us to conclude that $CHCl_3$ molecules were polar as a result of the non-symmetrical distribution of charge within each molecule. The interactions between permanent dipoles explain the attractions between neighbouring $CHCl_3$ molecules. These attractions are called **permanent dipole–permanent dipole attractions** (figure 9.6). The existence of dipole–dipole attractions will explain the forces holding together *polar* molecules in liquids such as trichloromethane ($CHCl_3$), propanone (acetone, CH_3COCH_3) and nitrobenzene ($C_6H_5NO_2$). But, what about *non-polar* molecules in liquids such as tetrachloromethane (CCl_4) and benzene (C_6H_6)? How can we explain the forces between non-polar molecules in these substances which have no permanent dipole? For example, what holds CCl_4 molecules together in liquid CCl_4? We must now look for evidence of intermolecular forces between non-polar molecules.

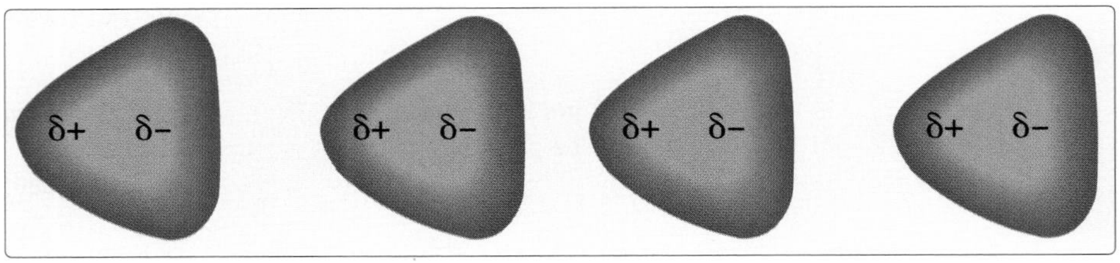

Figure 9.6
Permanent dipole–permanent dipole attractions in polar molecules.

(B) Evidence for intermolecular forces between non-polar molecules – Van der Waals forces

(i) The properties of noble gases

The noble gases are monatomic. They exist as single atoms in the gas phase at room temperature. These symmetrical, non-polar atoms have no permanent dipole and do not form any normal bonds. But all the noble gases will condense to liquids and ultimately form solids if the temperature is low enough. The possible liquefaction and solidification of noble gases suggests the existence of intermolecular forces in these non-polar substances. These intermolecular forces hold the molecules together in the solid and liquid states.

Furthermore, energy is required both to melt the solid and to boil the liquid noble gases, showing that cohesive forces are operating between molecules. For example, the energy of sublimation for solid xenon is $+14.9 \text{ kJ mol}^{-1}$.

(ii) The non-ideal behaviour of gases

> According to the kinetic theory, molecules of an ideal gas
> • occupy negligible volume, and
> • exert no forces on one another.

Using this model, the properties of an ideal gas can be summarised in the **Ideal Gas Equation**.

$$pV = nRT$$

where n is the number of moles of gas (see section 11.6).
For one mole of gas $pV = RT$.
This equation represents the behaviour of gases accurately at low pressures (i.e. no more than atmospheric pressure) and high temperatures. (These conditions are well above the condensation point of a gas, when it is most like a gas and least like a liquid.) There are, however, increasing deviations at low temperatures and high pressures.

> 1 What is the value of pV/RT for one mole of an ideal gas?
>
> 2 Do you think pV/RT will increase, remain constant or decrease as p increases for an ideal gas?

Look closely at figure 9.7.

> 3 How does pV/RT vary as p increases for N_2 (a real gas) at 673 K?
>
> 4 How does pV/RT deviate from ideal behaviour as temperature is reduced?

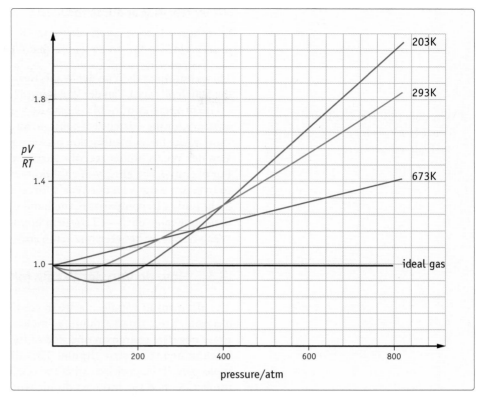

Figure 9.7
Variation of pV/RT with increasing pressure for N_2 at 673 K, 293 K and 203 K.

Figure 9.7 shows that the properties of nitrogen deviate from ideal behaviour at
• low temperatures and
• high pressures.

Other real gases show similar deviations which can be understood by reconsidering the two fundamental assumptions of the kinetic theory of gases stated at the beginning of this sub-section.

1 *In deriving the Ideal Gas Equation, it is assumed that the molecules occupy a negligible volume.* Strictly speaking, the molecules do not have a negligible volume. Each one has a finite size which excludes a certain volume of the container from all the others. If we call this 'excluded' volume b, then the 'true' volume in which the molecules move is $(V - b)$.

Thus, the equation $pV = nRT$ should be written as $p(V - b) = nRT$.

The constant b has a value of about 30 cm^3 mol^{-1} for many gases. Since one mole of many gases occupies approximately 30 cm^3 on liquefaction, this provides further evidence that b is roughly the same as the volume of the molecules.

2 *In deriving the Ideal Gas Equation, it is assumed that the molecules exert no forces on each other.* Unfortunately, intermolecular forces cannot be neglected. This is particularly so at high pressure. The pressure of a gas results from particles of the gas bombarding the walls of their container. Within the bulk of the gas, intermolecular forces cancel each other out, but those molecules near the walls experience an overall net force tending to pull them back into the bulk. This results in a measured pressure lower than the 'true' pressure. The magnitude of the 'pressure reduction' will be proportional to both the concentration of molecules near the wall ($\propto \frac{1}{V}$) and the concentration of molecules within the bulk (also $\propto \frac{1}{V}$). Thus, for one mole of gas the 'pressure reduction' can be written as a/V^2 where a is a constant. By adding the term a/V^2 to the measured pressure, we obtain the corrected pressure term $[p + (a/V^2)]$.

If we replace p and V in the Ideal Gas Equation by their corrected expressions we obtain

$$[p + (\frac{a}{V^2})].(V - b) = nRT$$

This is known as the **Van der Waals Real Gas Equation**. This equation was first used by the Dutch physicist Van der Waals in 1873.

The properties of noble gases and the non-ideal behaviour of real gases provide evidence for the existence of cohesive forces between non-polar molecules. These weak, short-range forces of attraction between molecules are known as **Van der Waals forces**. Van der Waals bonds are, of course, much weaker than covalent and ionic bonds. For example, the energy of sublimation for solid chlorine (i.e. the energy required to overcome the Van der Waals forces between one mole of Cl_2 molecules) is only 25 kJ mol^{-1}. In comparison, the bond energy of chlorine (i.e. the energy required to break one mole of Cl—Cl covalent bonds) is 244 kJ mol^{-1}. Roughly speaking, Van der Waals bonds are between one-tenth and one-hundredth the strength of covalent bonds.

1 Why does this Real Gas Equation reduce to $pV = nRT$ at low pressure and high temperature?

9.4 How do Van der Waals forces arise?

The electrons in a molecule are in continual motion. At any particular moment, the electron charge cloud around the molecule will not be perfectly symmetrical. There is more negative charge on one side of the molecule than on the other. It possesses an instantaneous electric dipole. This dipole will induce dipoles in neighbouring molecules. If the positive end of the original dipole is pointing towards a neighbouring molecule, then the induced dipole will have its negative end pointing towards the positive of the original dipole. In this way, weak **induced dipole–induced dipole attractions** exist between molecules.

These induced dipoles will act first one way, then another way, continually forming and then disappearing as a result of electron movement. Notice that the force between the original dipole and the induced dipole is always an attraction. Consequently, even though the average dipole on every molecule over a period of time is zero, the resultant forces between molecules at any instant are not zero.

As the size of a molecule increases, the number of constituent electrons increases. As a result the induced dipole–induced dipole attractions become stronger.

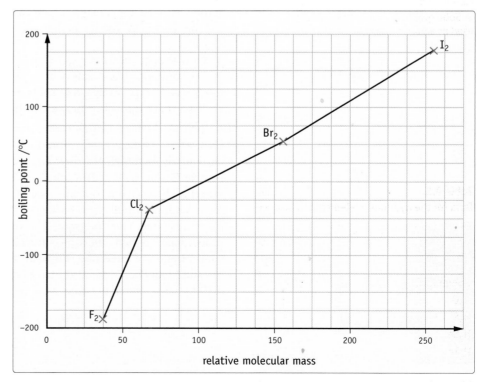

Figure 9.8
Boiling points for the elements in Group VII plotted against relative molecular mass.

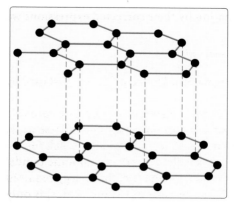

Figure 9.9
The structure of graphite.

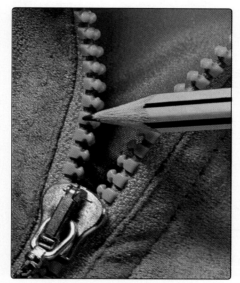

▲ Solid graphite in a soft pencil can be used as a lubricant in this zip fastener.

The increase in boiling point for the elements in Group VII (F_2, Cl_2, Br_2 and I_2 – figure 9.8) and the increase in boiling point for the homologous series of alkanes (CH_4, C_2H_6, C_3H_8, etc., in figure 26.4) result from stronger Van der Waals attractions with increasing relative molecular mass.

Although the Van der Waals forces between simple molecules such as CCl_4, Cl_2 and CH_4 are very small, the total Van der Waals forces between the molecules of a large polymer with many contacts can be very large. The strength of non-polar polymers such as poly(ethene) and poly(propene) is dependent on significant Van der Waals bonding between parallel molecules. Furthermore, experiments have shown that the tensile strength of high-density poly(ethene), which has tightly packed parallel molecules is three times as large as that of low-density poly(ethene), which is packed less tightly and therefore has weaker Van der Waals attractions.

Van der Waals forces also account for the properties of graphite. Crystals of graphite are composed of parallel layers of hexagonally arranged carbon atoms (figure 9.9). Within each layer, carbon atoms are linked by strong covalent bonds. In comparison, the parallel layers are held together by Van der Waals forces. The C—C distance within each layer is 0.14 nm but the distance between adjacent layers is 0.34 nm. The Van der Waals bonding between the layers is strong enough to hold the layers together, but weak enough to allow them to slide over each other. Because of this, graphite is soft and acts as a solid lubricant.

9.5 Hydrogen bonding

Look closely at the graphs in figure 9.10. These show the boiling points of hydrides in Group IV, Group V, Group VI and Group VII. Notice that the boiling points of the Group IV hydrides decrease with decreasing relative molecular mass from SnH_4 to CH_4. Is there a similar decrease for the hydrides of Groups V, VI and VII?

Figure 9.10 shows the expected decrease in boiling point from H_2Te through H_2Se to H_2S, but H_2O has a much higher boiling point than one would expect. A similar pattern appears with the hydrides in Group V and those in Group VII. Here we find much higher boiling points for NH_3 and HF than extrapolation of the graphs would suggest.

How can we account for these unusually high boiling points for H_2O, NH_3 and HF?

In water, liquid ammonia and liquid hydrogen fluoride there must be unusually strong intermolecular forces. Why is this?

H_2O, NH_3 and HF are all very polar because they contain the three most electronegative elements (oxygen, nitrogen and fluorine) linked directly to hydrogen, which is weakly electronegative. This results in exceptionally polar molecules with stronger intermolecular forces than usual. These particularly strong intermolecular forces are known as **hydrogen bonds**.

> 1 Look closely at the data in table 9.3. Do you think the melting point, the molar enthalpy change of fusion and the molar enthalpy change of vaporisation of water are influenced by hydrogen bonding?
>
> 2 Which other physical properties of a substance, besides those shown in table 9.3, will be influenced by the existence of hydrogen bonds?

Table 9.3
Data for the hydrides of elements in Group VI

Compound	Melting point /K	Molar enthalpy change of fusion, $\Delta H_{fus}^{\ominus}$ /kJ mol^{-1}	Molar enthalpy change of vaporisation, $\Delta H_{vap}^{\ominus}$ /kJ mol^{-1}
H_2O	273	+6.02	+40.7
H_2S	188	+2.39	+18.7
H_2Se	207	+2.51	+19.3
H_2Te	225	+4.18	+23.2

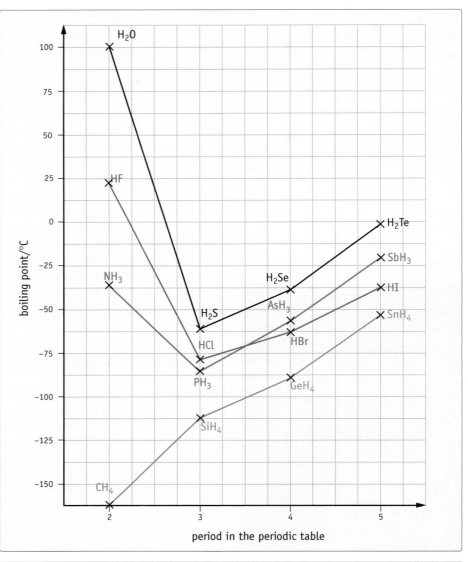

Figure 9.10
Variation of the boiling points of the hydrides in Groups IV, V, VI and VII.

9.6 What is a hydrogen bond?

Nitrogen, oxygen and fluorine are the three most electronegative elements. When they are bonded to a hydrogen atom, the electrons in the covalent bond are drawn towards the electronegative atom. Remember also that the H atom has no electrons other than its share of those in this covalent bond and these are being pulled away from it by the more electronegative N, O or F.

Since the H atom has no inner shell of electrons, the single proton in its nucleus is unusually 'bare' and readily available for any form of dipole–dipole attraction. Thus, H atoms attached to N, O or F are able to interpose themselves between two of these atoms exerting an attractive force on them and bonding them together (figure 9.11). The two larger atoms are drawn closer with an H atom effectively buried in their electron clouds.

H-bonds are therefore extra-strong intermolecular, permanent dipole–permanent dipole attractions.
  The essential requirements for an H-bond are:
  • a hydrogen atom attached to a highly electronegative atom and
  • an unshared pair of electrons on the electronegative atom.

In practice, this usually means that hydrogen bonding will occur in any substance containing a hydrogen atom attached to an oxygen a nitrogen or a fluorine atom.
  In the NH_3 molecule, there are three N—H bonds, but only one non-bonded electron pair.

Figure 9.11
Hydrogen bonding in water.

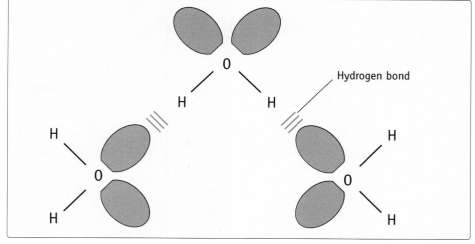

1 How many H—F bonds are there in one HF molecule?

2 How many unshared electron pairs are there in each HF molecule?

3 What is the maximum number of H-bonds per HF molecule?

This means that there can be an average of only one H-bond per molecule.

In water, however, there are two O—H bonds and two unshared electron pairs per molecule. This means that each H_2O molecule can form two hydrogen bonds and this helps to explain the three-dimensional lattice structure in ice (section 9.8(A)).

9.7 Estimating the strength of hydrogen bonds in water

Which one of the following values could be used to estimate the strength of an H-bond in water? Remember that H-bonds are intermolecular forces.

A The strength of an O–H bond in H_2O.

B The heat evolved when one mole of H_2O forms from its elements.

C The enthalpy change of vaporisation of water.

D The melting point of ice.

H-bonds in water are intermolecular forces. They are the forces which hold the molecules together. When water is vaporised, the H_2O molecules, which are relatively close to each other in the liquid, are pulled much further apart. The energy which vaporises the water is needed to overcome the forces holding the water molecules together in the liquid state. Thus, the enthalpy change of vaporisation of water will give a reasonable estimate of the strength of H-bonds between H_2O molecules. But, the enthalpy change of vaporisation overcomes Van der Waals bonds between H_2O molecules as well as the H-bonds. How can we estimate the strength of H-bonding alone?

Figure 9.12 shows a graph of the molar enthalpy change of vaporisation for the hydrides of elements in Group VI plotted against relative molecular mass. Now, if we assume that H_2S, H_2Se and H_2Te have intermolecular forces due only to Van der Waals bonds (negligible H-bonding), we can estimate a value for the strength of Van der Waals forces in water. We do this by extrapolating the curve through the values of ΔH_{vap} for H_2S, H_2Se and H_2Te. The value we want is indicated by the circled cross in figure 9.12. This gives a predicted enthalpy change of vaporisation for water of $+18.5$ kJ mol^{-1} assuming that water has only Van der Waals bonds.

Total strength of intermolecular forces (Van der Waals bonds + H-bonds) in
water = 40.7 kJ mol^{-1}
Estimated strength of Van der Waals forces in water = 18.5 kJ mol^{-1}
∴ Approximate strength of H-bonds in water = 40.7 − 18.5 = 22.2 kJ mol^{-1}

Usually, the strengths of H-bonds are in the range 5–40 kJ mol^{-1}. Thus H-bonds are about one-tenth as strong as covalent bonds and roughly the same strength as Van der Waals forces. Remember, however, that molecules which are H-bonded will also be attracted by Van der Waals forces. In general, we can say that H-bonds are weak forces. They are strong enough to influence physical properties but they are not strong enough to change the chemical reactions of a substance.

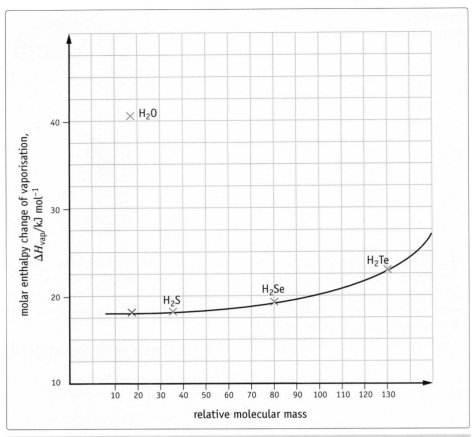

Figure 9.12
Variation of the molar enthalpy change of vaporisation for the hydrides of Group VI elements.

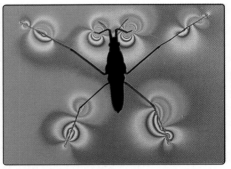

▲ Walking on water! A pond skater standing on the surface of pond water – it's all done by hydrogen bonding and surface tension.

9.8   The influence and importance of hydrogen bonding

(A) The structure and properties of water and ice

Water has unexpectedly high values for its melting point, boiling point, molar heat of fusion and molar heat of vaporisation. These unexpectedly high values result from extra attractions between water molecules due to the formation of H-bonds. This extra bonding between H_2O molecules also causes high surface tension and high viscosity. The high surface tension of water provides a sort of 'skin effect' and it is possible for small but relatively dense articles, such as razor blades and beetles, to float on an undisturbed water surface.

The presence of *two* hydrogen atoms and *two* lone electron pairs in each water molecule results in a three-dimensional tetrahedral structure in ice. Each oxygen atom in ice is surrounded tetrahedrally by four others. Hydrogen bonds link each pair of oxygen atoms (figure 9.13).

The distance between adjacent oxygen atoms in ice is 0.276 nm, more than twice the O–H distance of 0.096 nm in gaseous water molecules. This suggests that the hydrogen atom linking the two oxygen atoms in ice is not midway between them.

The arrangement of water molecules in ice creates a very open structure. This accounts for the fact that ice is less dense than water at 0°C. When ice melts, the regular lattice breaks up. The water molecules can then pack more closely, so the liquid has a higher density. In liquid water, the strong hydrogen bonding does however result in some ordered packing of water molecules over a short range. But there is no long-range order as the H-bonds are continually being broken and formed. Also the regions of short-range order continually change.

The anomalous physical properties of ice and water which result from H-bonding have a great influence biologically and environmentally.

The fact that ice is less dense than water at 0°C means that ponds and lakes freeze from the surface downwards. The layer of ice insulates the water below, preventing complete solidification. If ice were the more dense, water would freeze from the bottom upwards.

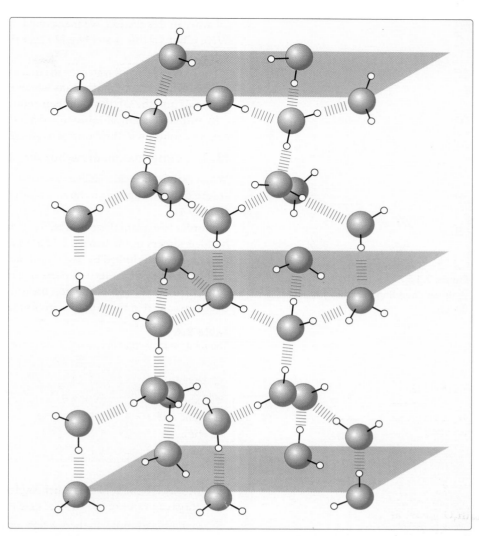

Figure 9.13
The structure of ice.

▶ Ice floats on water. The arrangement of water molecules in ice creates a very open structure, making it less dense than water.

In this case, ponds would freeze completely, killing fish, aquatic plants and other water-living creatures.

Water also has an unusually high boiling point owing to hydrogen bonds. Without these H-bonds, water would probably be a gas under normal atmospheric conditions. Oceans, lakes and rivers would never exist and it would never rain!

Earlier in this section, we mentioned the exceptionally high surface tension of water. Were it not for this, water would never rise through the capillary tubes in the roots and stems of plants.

The high polarity of water means that polar and ionic substances are usually soluble in water. Plants can obtain the salts which they require for growth by absorption of these materials into the sap through their roots.

In animals, essential ionic substances are absorbed into the bloodstream from the aqueous solution in their stomachs and intestines.

(B) The dimerisation of carboxylic acids

When carboxylic acids, such as ethanoic acid (acetic acid), are dissolved in benzene, the solute particles appear to have a relative molecular mass twice as large as expected (table 9.4).

In order to explain these results, it is believed that the carboxylic acid forms hydrogen-bonded dimers in the benzene. The dimers are particularly stable because each pair of acid molecules is linked by two H-bonds, not just one (figure 9.14).

The presence of dimerised pairs of carboxylic acid molecules in benzene has been confirmed by electron diffraction measurements. Studies of the relative molecular mass of ethanoic acid in the vapour state also suggest the presence of dimers.

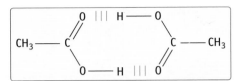

Figure 9.14
Hydrogen bonding in carboxylic acid dimers.

Table 9.4
The relative molecular masses of ethanoic acid and benzoic acid

Carboxylic acid	Formula	Relative molecular mass, M_r	Apparent relative molecular mass, M_r, in benzene
Ethanoic acid (acetic acid)	CH_3COOH	60	120
Benzoic acid	C_6H_5COOH	122	244

1 One measurement of the relative molecular mass of ethanoic acid in the vapour state gave a value of 90. To what extent is the acid dimerised?

2 Ethanoic acid could, in theory, form trimers, tetramers, etc., in benzene. Why, do you think, are dimers formed in preference?

3 When ethanoic acid is dissolved in water, dimers do not form. Why is this?

4 Name one other solvent, besides benzene, in which you would expect ethanoic acid to form H-bonded dimers.

When ethanoic acid molecules are dissolved in water, they are more likely to form H-bonds with water molecules than with themselves, so dimers do not form. However, in solvents which cannot H-bond, the ethanoic acid molecules will dimerise.

The effect of dimerisation on the boiling points of carboxylic acids is discussed in section 32.1.

(C) The hardness of ionic crystals

Crystals of anhydrous calcium sulphate (anhydrite, $CaSO_4$) are very hard and very difficult to cleave. On the other hand, crystals of hydrated calcium sulphate (gypsum, $CaSO_4.2H_2O$) are soft and easily cleaved. Why is there a big difference in hardness between such similar materials?

The structure of gypsum consists of layers containing both Ca^{2+} and SO_4^{2-} ions separated by layers of water molecules. Within the Ca^{2+}/SO_4^{2-} layers the ions are held together by strong ionic bonds. But these Ca^{2+}/SO_4^{2-} layers are linked by relatively weaker H-bonds. The weaker H-bonds link SO_4^{2-} ions in alternate layers to water molecules in the intermediate region. Consequently, the gypsum can be readily cleaved and scratched along the layers of water molecules.

In contrast, anhydrite has a completely ionic structure involving only Ca^{2+} and SO_4^{2-} ions. It is therefore much harder than gypsum and cannot be easily cleaved.

(D) The structures and properties of biological molecules

H-bonds are present in the structures of proteins, carbohydrates and nucleic acids. The properties and functions of these biological molecules are dependent on their H-bonding.

Proteins

Proteins are composed of long sequences of amino acids joined by **peptide links**. The general formula of an amino acid is:

The carboxylic acid group, $-C\overset{O}{\underset{OH}{\big\langle}}$, of one amino acid molecule can react with the amino group, $-N\overset{H}{\underset{H}{\big\langle}}$, of a second amino acid to eliminate water and form a peptide bond:

Proteins contain anything from 10 to 10 000 amino acids linked by peptide bonds. Each is a long-chain macromolecule which may be represented as:

The sequence of amino acids in the protein chain is referred to as the **primary structure** of the protein. The detailed configuration of the chain is called the **secondary structure**.

One of the commonest secondary structures in proteins is a spiral arrangement known as the α-helix. The amino acid units are arranged in a coiled helix. This is stabilised by H-bonds between the $>$N—H group of one amino acid unit and the $>$C=O group of the fourth unit further along the chain (figure 9.15).

H-bonding plays an important part in maintaining the structure of proteins. The precise position and sequence of amino acids in the protein structure means that proteins can carry coded information. This enables them to control growth and metabolism in plants and animals. In fact, most biological catalysts (which are usually known as **enzymes**) are proteins. So, life itself is very dependent on H-bonding.

The presence of polar $>$C=O and $>$N—H groups in proteins which can H-bond with water means that many proteins are water soluble. These soluble proteins are therefore present in the blood and in cellular fluids. They can take part in metabolism in plants and animals.

The properties of amino acids and the peptide link are discussed further in section 33.8.

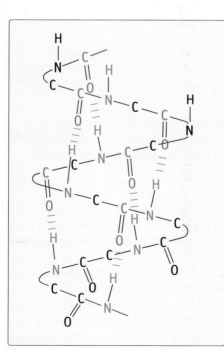

Figure 9.15
Hydrogen bonding in the coiled helical structure of proteins.

Carbohydrates (polysaccharides)

Most common polysaccharides (e.g. starch, cellulose, glycogen) are built up from glucose. Glucose is a monosaccharide with the two possible structures below. α-glucose and β-glucose are stereoisomers with different arrangements of the atoms attatched to the carbon atom number one ($_1$C).

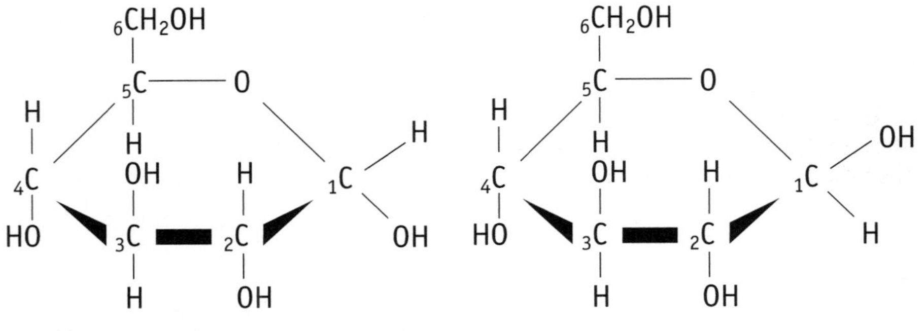

α-glucose

β-glucose

In polysaccharides, the glucose units are linked by C—O—C bonds between the carbon atoms numbered 1, 4 and 6 in the structure above. Figure 9.16 shows the structure of cellulose in which the carbon atom labelled $_1$C in one β-glucose unit is linked to $_4$C in the next. The relative molecular masses of natural polysaccharides range from 10^5 to 10^7.

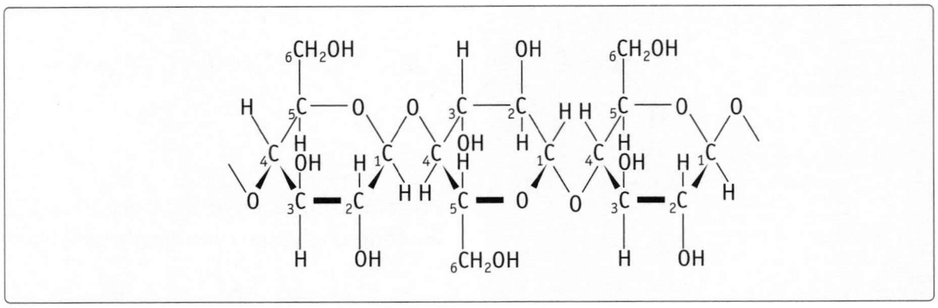

Figure 9.16
The structure of cellulose.

1. Use figure 9.16 to calculate the relative molecular mass of one glucose unit in a polysaccharide.

2. Approximately how many glucose units are there in a polysaccharide of relative molecular mass 10^6?

3. Cellulose is an important structural polysaccharide in the cell walls of plants. It is the best known example of a polysaccharide with considerable mechanical strength. What forces might operate between parallel chains of cellulose in order to hold them together?

Carbohydrates are non-ionic, but they contain a large number of —OH groups. These give rise to the possibility of extensive hydrogen bonding with themselves or with water. In cellulose, the long chains of glucose units are packed very closely. They form a strong bundle of parallel strands linked together by H-bonds.

Small carbohydrates, such as glucose, are very water soluble owing to their ability to hydrogen bond with H_2O molecules. Carbohydrates also release a large amount of energy on reaction with oxygen. Because of this both animals and plants make use of glucose as an energy provider and polysaccharides as energy storage materials.

In plants, the principal energy storage carbohydrate is starch, whereas that in animals is glycogen. When energy is required, the polysaccharides (starch or glycogen) break down releasing glucose. This dissolves easily in the sap or blood for rapid transport to those parts of the organism where energy supplies are needed. The chemistry of carbohydrates is discussed further in section 31.7.

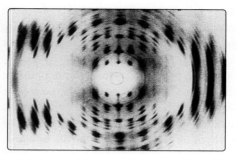

▲ An X-ray diffraction photograph of DNA. The repeated pattern provides evidence for the helical (spiral) structure of DNA.

Nucleic acids

The monomer units in nucleic acids are nucleotides. These consist of a complex organic base, B, a sugar, S, and a phosphate group as shown below:

$$B - S - O - P(=O)(O^-) - O^-$$

In the nucleic acid molecule, nucleotides are linked together to form a chain through their sugar/phosphate groups (figure 9.17).

Essentially, there are two types of nucleic acids, deoxyribose nucleic acids (DNA) in which the sugar is deoxyribose, and ribose nucleic acids (RNA) in which the sugar is ribose. Both ribose and deoxyribose are monosaccharides containing five carbon atoms.

DNA is composed of two parallel helices. This double helix is held together by hydrogen bonds between bases in the parallel helices (figure 9.18).

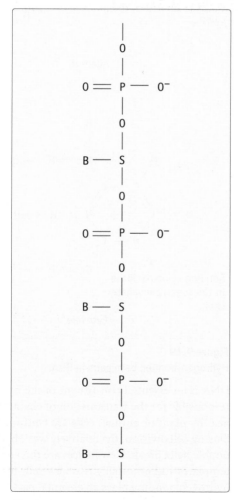

Figure 9.17
The simplified structure of three nucleotide monomers in a single-stranded nucleic acid molecule.

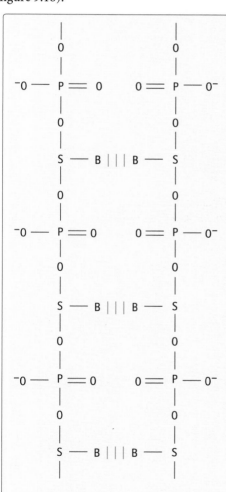

Figure 9.18
The simplified structure of DNA showing the hydrogen bonds between bases in the parallel chains. For simplicity, the parallel chains are drawn, without showing the helical structure.

No attempt has been made to show the coiling of the helix in figure 9.18. The structure of DNA can be compared to a spiral staircase in which the sides of the staircase are composed of alternate sugar and phosphate groups and the steps of the staircase are composed of hydrogen-bonded base pairs.

There are four different bases in DNA; adenine, thymine, cytosine and guanine. Adenine can only form hydrogen bonds with thymine, and vice versa. In the same way, cytosine can only form hydrogen bonds with guanine and vice versa (figure 9.19). Thus, the sequence of bases in each strand of the double helix must be complementary. Adenine must be opposite thymine and cytosine opposite guanine.

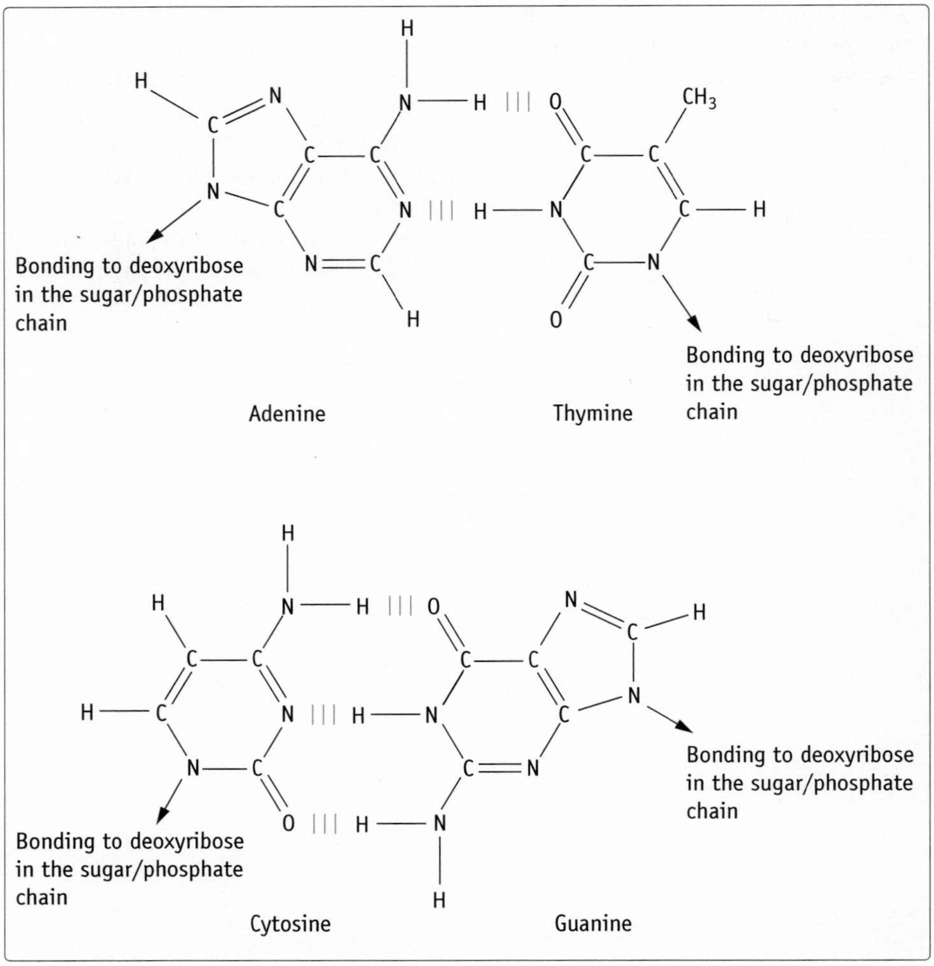

Figure 9.19
Hydrogen-bonded base pairs in DNA.

DNA is an essential constituent of the nuclei of cells. It makes up the genes which are responsible for the transmission of characteristics from one generation to the next. In a healthy plant or animal, cells are continually dividing and replacing those which die. During cell division, the relatively weak hydrogen bonds between the two strands in the double helix break. New helices are then synthesised using the two 'daughter' strands as templates and eventually two new cells form – yet another biological process in which hydrogen bonding plays an essential part.

Summary

1 When two different atoms are joined by a covalent bond, their attractions for the bonding electrons will not be the same. The distortion of charge which results from the unequal sharing of electrons is known as polarisation. Molecules with an overall distortion of charge are said to be polar.

2 The electronegativity of an atom can be defined as the power of that atom in a molecule to attract electrons.

3 Do not confuse electronegativity with electron affinity. Electronegativity is concerned with the attraction of atoms for electrons in covalent bonds. Electron affinity is concerned with the attraction for electrons of single gaseous atoms.

4 Van der Waals forces are weak, short-range forces of attraction between molecules. Essentially, they are induced dipole–induced dipole attractions.

5 Hydrogen bonds are extra-strong intermolecular forces. They are permanent dipole–permanent dipole attractions

6 Hydrogen bonding can occur when
(i) a hydrogen atom is attached to a highly electronegative atom (usually N, O or F) and
(ii) this highly electronegative atom has an unshared pair of electrons.

7 Hydrogen bonding in water is responsible for the unexpectedly high values of its melting point, boiling point, enthalpy change of vaporisation, surface tension and viscosity.

8 Hydrogen bonds are present in the structures of proteins, carbohydrates and nucleic acids. The biological properties and functions of these molecules in living things are very dependent on their H-bonding.

Review questions

1 (a) Which of the following molecules would you expect to have a permanent dipole?
   (i) GeH_4 (ii) ICl (iii) SiF_4 (iv) CH_2Cl_2
   (v) CO_2
 (b) The following molecules have no permanent dipole. What is their shape?
   (i) BCl_3 (ii) CS_2 (iii) C_2Cl_2 (iv) CBr_4
 (c) In which of the following compounds will hydrogen bonding occur?
   (i) $C_2H_5NH_2$ (ii) CH_3OH (iii) CH_3I
   (iv) CF_4 (v) H_2SO_3 (vi) CH_3OCH_3
 (d) Which of the following molecules have a structure with a bond angle greater than 109° 28′?
   (i) SCl_2 (ii) CO_2 (iii) BF_3 (iv) NF_3
   (v) CH_4

2 The solubility of iodine in tetrachloromethane at 10°C is 3 g per 100 g of solvent. The solubility of iodine in water at 10°C is 0.02 g per 100 g of solvent. The solution of iodine in tetrachloromethane is violet; the solution of iodine in water is yellow.
 (a) State whether the following are polar or non-polar:
   (i) iodine
   (ii) tetrachloromethane
   (iii) water.
 (b) Why is iodine very soluble in CCl_4?
 (c) Why is iodine only slightly soluble in water?
 (d) Would you expect CCl_4 and water to mix? Explain your answer.
 (e) A yellow solution of iodine in water is shaken with an equal volume of tetrachloromethane. Describe and explain what happens.

3 (a) The boiling point of cis-dichloroethene is 333 K, whereas that of trans-dichloroethene is 321 K. Draw the structural formulas of these two isomers and explain the difference in boiling point.

 (b) The structural formulas, boiling points and densities of the isomers pentane and 2,2-dimethylpropane are shown below.

	pentane	2,2-dimethylpropane
Structural formula	$CH_3{-}CH_2{-}CH_2{-}CH_2{-}CH_3$	$CH_3{-}\overset{\displaystyle CH_3}{\underset{\displaystyle CH_3}{C}}{-}CH_3$
Boiling point/°C	36	9
Density/g cm^{-3}	0.626	0.591

   (i) Why does pentane have a higher boiling point than 2,2-dimethylpropane?
   (ii) Why does pentane have a higher density than 2,2-dimethylpropane?
   (iii) 2-Methylbutane is an isomer of pentane and 2,2-dimethylpropane. How do you think its boiling point and density will compare with these two substances? Explain your answer.

4 The relative molecular masses and molar enthalpy changes of vaporisation for three of the hydrides of elements in Group V are given below.

Compound	Relative molecular mass	$\Delta H_{vap}^{\ominus}$/kJ mol^{-1}
NH_3	17	+23.4
PH_3	34	+14.6
AsH_3	78	+17.5

 (a) Plot a graph of ΔH_{vap} against relative molecular mass for the three hydrides.
 (b) Why is the value of ΔH_{vap} for NH_3 unexpectedly high?
 (c) Use your graph to estimate a value for the ΔH_{vap} of NH_3 assuming that it has only Van der Waals bonds.

(d) Predict a value for the strength of H-bonds in NH_3.

(e) The strength of H-bonding in water is approximately 22 kJ mol^{-1}. How do you explain the fact that H-bonding in NH_3 is only about half the strength of that in H_2O?

5 (a) The strength of the H-bonds in ice is 22 kJ mol^{-1}. How much energy is required to break one mole of hydrogen bonds in ice?

(b) What percentage of the hydrogen bonds are broken when ice melts, assuming that all the energy involved in the heat of fusion is used to break hydrogen bonds? (The enthalpy change of fusion of ice is +6 kJ mol^{-1}.)

(c) What is the shape of the H_2O molecule? Explain why the H_2O molecule is polar.

(d) What shape would you predict for the hypothetical molecule H_2X if it were non-polar? Explain your prediction.

6 Experiments were carried out with the three liquids A, B and C (each with the empirical formula CHF), in order to find their polarity and relative molecular masses. The results of these experiments are given in the table below.

Compound	Empirical formula	Effect of a charged rod on a thin stream of the liquid issuing from a burette	Approx. mass of 1 dm^3 of gas at s.t.p./g
A	CHF	nil	3
B	CHF	liquid is attracted to charged rod	3
C	CHF	nil	6

(a) Which of the three liquids is (are) polar?

(b) What is the relative molecular mass of:
  (i) A   (ii) B   (iii) C?
  (The molar mass of a gas at s.t.p. occupies 22.4 dm^3)

(c) Draw *one* possible structural formula for A.
  (H = 1, C = 12, F = 19)

(d) Draw *two* possible structural formulas for B.

(e) Draw *three* possible structural formulas for C.

7 Proteins are polypeptides with relative molecular masses in the range 10^3 to 10^6. They are composed of a sequence of amino acids joined by the peptide link.

(a) Write a general formula for the monomer amino acid units which make up proteins.

(b) Suppose that the approximate relative molecular mass for *one* amino acid unit *in* proteins is 100. About how many amino acid units will there be in a protein of relative molecular mass 10^5?

(c) Draw a section of a protein structure.

(d) How does hydrogen bonding explain:
  (i)   the water solubility of certain proteins,
  (ii) the precise configurations of protein enzymes,
  (iii) the elasticity of natural protein fibres such as hair, wool and silk?

8 Suggest reasons for the following:

(a) The boiling points of water, ethanol and ethoxyethane ($CH_3CH_2OCH_2CH_3$, diethyl ether) are in the reverse order of their relative molecular masses, unlike those of their analogous sulphur compounds H_2S, C_2H_5SH and $C_2H_5SC_2H_5$.

(b) BF_3 is non-polar, but NF_3 is polar.

Structure, Bonding and Properties – The Solid State

▲ A small cylindrical magnet floating freely above a cylindrical superconducting ceramic made from yttrium, barium and copper oxide. The glowing vapour is from liquid nitrogen which keeps the ceramic in its superconducting temperature range. Superconducting ceramics are leading to new advances in technology.

▶This photograph shows strands of glass fibres. Glass fibres have been developed in recent years for use in fibre optics and for strengthening other materials.

10.1 Introduction

One of the major achievements of chemistry has been the synthesis of new materials, such as ceramics, plastics and alloys. Many of these new materials can be designed to have specific properties. In order to design these new substances, it is necessary to know how the structure of materials can affect their properties. Thus, it is not surprising that one of the most important aspects of chemistry is the investigation of the structure of materials and the bonds holding the particles together.

In this chapter, we shall begin by looking at the methods used to investigate the structures of solid materials. Then we shall consider the properties of these materials and the way in which properties are dictated by structure and bonding.

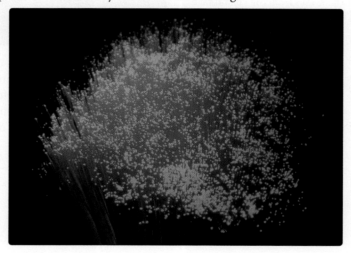

10.2 Evidence for the structure of materials

The physical properties of a material can often provide evidence for its structure. For example, the diffraction of X-rays by a solid provides evidence for the arrangement of particles within it. The amount of heat required to melt the solid gives us information about the forces of attraction between these particles, whilst the effect of an electric current on the molten material can tell us something about the nature of the constituent particles.

Each of these phenomena gives different information about the material, but together they provide a detailed picture of its structure and bonding.

Evidence from X-ray diffraction

Look through a piece of thin stretched cloth (possibly your handkerchief) at a bright light. The pattern you see is caused by the deflection of light as it passes through the regularly spaced threads of the fabric. This deflection of light is called **diffraction** and the patterns produced are **diffraction patterns**.

▲ Professor Dorothy Hodgkin, who won a Nobel Prize for her work on X-ray crystallography.

▶ An early version of the X-ray spectrometer designed by W.L. Bragg. X-rays generated at A pass through the slit at B and on to the inclined face of the crystal at C. The reflection of the X-rays from the crystal face is measured in the ionisation chamber D.

Figure 10.1
(*a*) Reinforcement of waves in phase.
(*b*) Interference of waves out of phase.

When the cloth is rotated in front of the light, the diffraction pattern rotates also. If the cloth is stretched so that the strands of the fabric get closer, then the pattern spreads out further. If the strands are arranged differently the pattern changes. From the diffraction pattern which we can observe, it is possible to deduce how the strands are arranged in the fabric. The same idea is used to determine the arrangement of particles in a crystal.

In 1912, the German scientist von Laue obtained a photograph of the diffraction pattern produced by reflecting X-rays off a crystal. X-rays behave in a similar way to ordinary light rays, but they have a much smaller wavelength. Using a similar technique W.L. Bragg was able to determine the simple cubic structure of sodium chloride. Nowadays, the structures of extremely complex substances such as proteins and nucleic acids are investigated using X-ray diffraction.

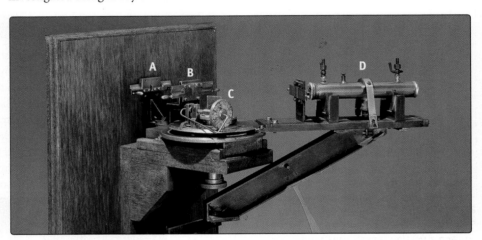

When a beam of X-rays falls on a crystal composed of regularly spaced atoms or ions the X-rays are reflected. In most instances, waves from the reflected X-rays interfere with and destroy each other. However, it is possible for the X-rays to be reflected by the particles so that their waves coincide and reinforce each other (figure 10.1). Under what conditions will reinforcement occur?

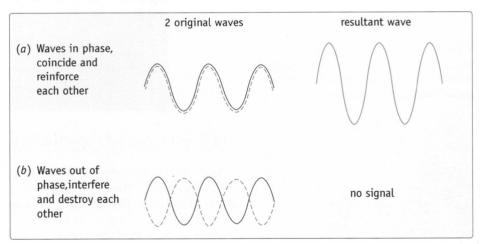

Figure 10.2 shows a beam of X-rays being directed onto a crystal such as sodium chloride. Only three layers of the crystal are represented.

1. Figure 10.2 shows two waves in phase approaching the crystal. After reflection, the ray emerging along XV is ahead of the ray emerging along YZ. Why is this?

2. How much further does the ray reflected at the second layer travel?

10 Structure, Bonding and Properties – The Solid State

Figure 10.2
The diffraction of two waves of X-rays by the regularly spaced particles in a crystal.

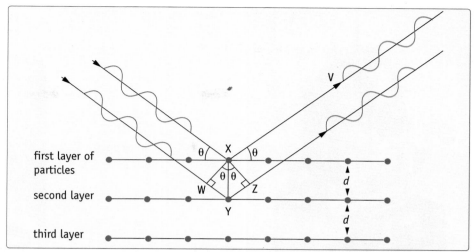

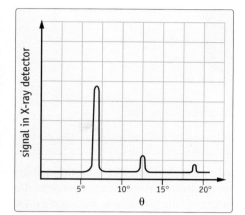

Figure 10.3
Signals from a crystal of sodium chloride as the angle (θ) between the direction of the X-ray beam and the crystal face increases.

If the waves are in phase after reflection, the difference in distance travelled by the two rays (i.e. the **path difference**) must equal a whole number of wavelengths, $n\lambda$.

$$\text{Path difference} = WY + YZ$$
$$= XY \sin \theta + XY \sin \theta$$
$$= 2XY \sin \theta = 2d \sin \theta$$
where d = distance between layers.

Thus, for reinforcement of the reflected X-rays

$$2d \sin \theta = n\lambda$$

This is known as the **Bragg equation**. For X-rays of a particular wavelength, λ, and for a given value of d, maximum reinforcement will occur only at certain values of θ. If θ is measured and λ is known, then the distance, d, between the layers in the crystal can be obtained.

By slowly increasing θ from 0° to 90°, an experimenter will obtain a series of reinforced, strong signals interspersed with regions of cancellation. Figure 10.3 shows some results for sodium chloride.

> 1 Why are there several values of θ at which reinforcement occurs?
>
> 2 What is the smallest value of θ for reinforcement?
>
> 3 What is the distance between layers in the sodium chloride crystal? (Assume the wavelength of the X-rays is 5.8×10^{-2} nm.)

The X-ray goniometer

Crystal structures can be determined by the methods described above using X-ray goniometers similar to the one shown in figure 10.4.

X-rays are directed through the narrow tube labelled X in figure 10.4 towards a single crystal. The crystal is mounted on a fine point of the goniometer head at the centre of the instrument. The goniometer head (figure 10.5) can be rotated in two arcs at right angles so that the crystal axis can be set in the correct orientation relative to the direction of the X-ray beam. In some instruments, X-ray film can be placed around the crystal in order to detect the diffracted X-rays. In figure 10.4, the diffracted X-rays are detected on a flat sensitive fibre optics screen. This is shown behind the crystal. The 'messages' received on the screen are intensified and converted for display on a TV monitor.

Figure 10.6 shows how a pattern of spots is produced during X-ray diffraction. As the crystal is rotated on the goniometer head, different planes of the crystal come into the correct angle for reinforcement. X-ray beams are then diffracted towards the photographic film where spots appear. If a flat film were used, the spots would occur along curves. Usually the film is bent into a cylinder with the crystal at its centre, so that the spots appear along straight lines. From the position of the spots and their relative intensities it is possible to obtain information about the structure of a crystal. Even non-crystalline materials such as raw cotton fibres (figure 10.7) produce diffraction patterns which can sometimes help in the elucidation of their structures.

Figure 10.4
An X-ray goniometer.

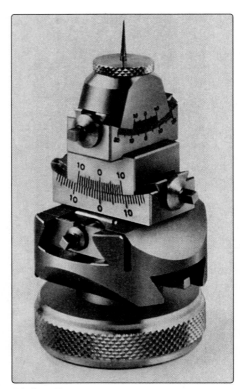

Figure 10.5
The goniometer head.

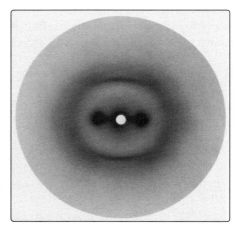

Figure 10.7
An X-ray diffraction photograph of raw cotton fibres.

Figure 10.8
(*a*) An electron density map of naphthalene (contours in electrons per nm^3). (*b*) The structural formula of naphthalene.

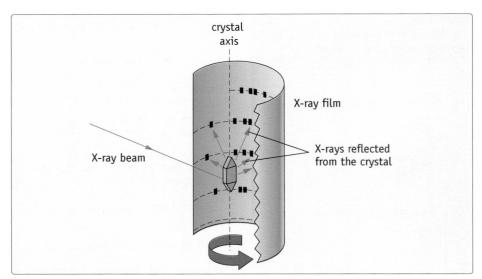

Figure 10.6
X-rays deflected from a crystal produce a pattern of spots on X-ray film.

10.3 Electron density maps

When X-rays strike a crystal, they are diffracted by the electrons in the atoms or ions. Consequently, the larger the atom and the more electrons it possesses, the brighter the spot will be on the diffraction pattern. By analysing both the positions and the intensities of the spots on a diffraction pattern, it is possible to determine the charge density of electrons within the crystal. The charge density is measured in terms of electrons per cubic nanometre. Points of equal density in the crystal are joined by contours giving an electron density 'map' similar to the one in figure 10.8.

Look closely at figure 10.8.

1 How many atoms can be located from the electron density map?
2 Which atoms are they?
3 Why are the other atoms not evident from the electron density map?

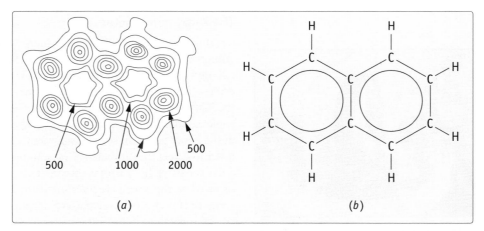

By mounting electron density maps on clear plastic or perspex it has been possible to assemble accurate three-dimensional structures of complex substances such as proteins and nucleic acids. It is then possible to see the positions of different atoms within them. There are many different arrangements in which atoms can be packed in repeating units to form crystals. A few of the arrangements that occur in natural crystals are considered in the following sections.

10 Structure, Bonding and Properties – The Solid State

Fruit and vegetables can be stacked in the same way that metal atoms pack in close-packed structures.

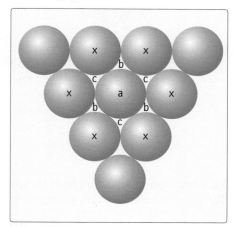

Figure 10.9
A layer of close-packed atoms.

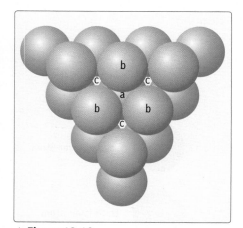

▲ **Figure 10.10**
First and second layers of close-packed atoms.

▶ **Figure 10.11**
First, second and third layers of close-packed atoms (ababab arrangement).

Close-packed structures

The crystals of some substances can be imagined to consist of identical atoms packed together as closely as possible. Most metals and many molecular substances have a close-packed structure. X-ray analysis shows that there are two possible close-packed arrangements: **hexagonal close-packed (h.c.p.)** and **cubic close-packed (c.c.p.)**.

In a close-packed layer, each atom is in contact with six others (figure 10.9). The atom marked 'a' has six other atoms (marked 'x') in the same layer in contact with it. Figure 10.9 also emphasises the hexagonal arrangement of atoms within each layer.

In the second layer, atoms pack as closely as possible to those in the first layer by 'sitting' in the depressions between atoms in the first layer. Around each first-layer atom there are six depressions. These depressions are marked alternately 'b' and 'c' around the atom 'a'. If the depressions marked 'b' are used, those marked 'c' cannot be. Figure 10.10 shows the second layer of atoms in place. Have the 'b' or 'c' depressions been used? Notice that three atoms in the second layer touch each atom in the first layer.

This means that any one atom touches 12 others in these close-packed arrangements – six in its own layer, three in the layer above and three in the layer below. This is summarised by saying that the **co-ordination number** is 12.

The third layer of atoms can now be added to figure 10.10 in two quite distinct ways. There are two types of depression available for atoms in the third layer. These two types of depression are not exactly the same. One type of depression, labelled 'a' in figure 10.10, lies directly above the centre of atom 'a' in the first layer. The other type of depression, denoted by 'c', lies directly over a hole in the first layer.

In figure 10.11, the brown sphere is part of a third layer whose atoms are directly above the atoms of the first layer. In this case, the brown sphere occupies an 'a' type depression. Try to construct this arrangement using marbles, ping-pong balls or polystyrene spheres. Use books to hold the bottom layer in place. This type of arrangement in which alternate layers have a similar positioning of atoms is denoted as 'ababab', etc., because alternate layers of atoms occupy either the 'a' depressions or the 'b' depressions. This arrangement is known as **hexagonal close-packing (h.c.p.)**.

The diagrams in figure 10.12 show the arrangement of atoms in the h.c.p. structure. (The spheres are coloured differently to show the packing of consecutive layers more clearly. There is, of course, no difference in the atoms in the different layers of a metal.) Zinc and magnesium are examples of metals with hexagonal close-packed structures.

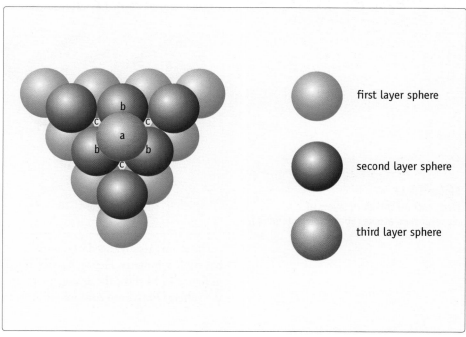

first layer sphere

second layer sphere

third layer sphere

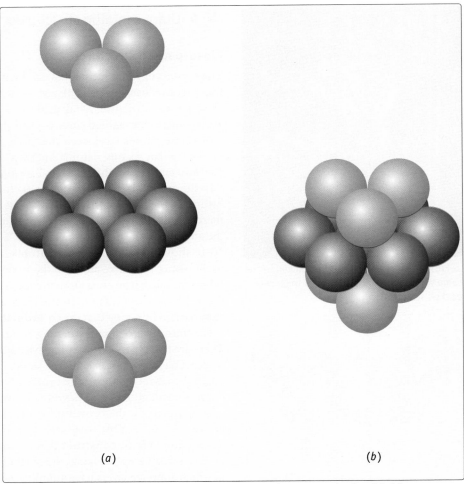

Figure 10.12
Hexagonal close-packed spheres:
(*a*) exploded view; (*b*) normal view.

(*a*)

(*b*)

An alternative close-packed arrangement occurs when atoms in the third layer occupy 'c' depressions, above neither the first layer nor the second layer atoms. This arrangement is shown in figure 10.13.

This type of arrangement in which the first layer of atoms occupy 'a' depressions, the second layer of atoms occupy 'b' depressions and the third layer of atoms occupy 'c' depressions can be repeated to give a sequence of layers denoted as 'abcabc', etc. This arrangement is known as **cubic close-packing** (**c.c.p.**).

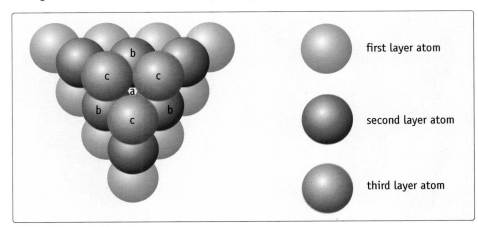

first layer atom

second layer atom

third layer atom

Figure 10.13
First, second and third layers of
close-packed atoms (abcabc arrangement).

The atoms are arranged hexagonally within each layer, but this sequence of layers also has cubic symmetry. Hence, it is described as a *cubic* close-packed structure. The diagrams in figure 10.14 show the arrangement of three layers with respect to each other. Some of the spheres have been coloured differently so that the cubic symmetry is more obvious.

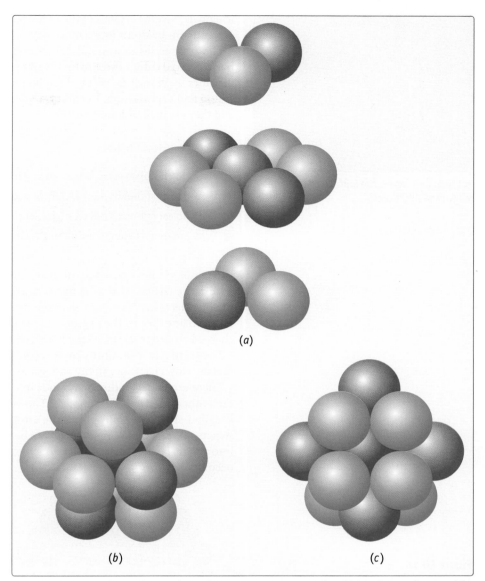

Figure 10.14
Cubic close-packed spheres:
(*a*) exploded view; (*b*) normal view;
(*c*) rotation of the normal view to show cubic symmetry more clearly.

(*a*)

(*b*)   (*c*)

Figure 10.14(*c*) which shows the cubic symmetry most clearly has been obtained by rotating figure 10.14(*b*) through 60°.

Aluminium, copper, lead, silver, gold and platinum are examples of metals with a cubic close-packed structure.

The body-centred cubic structure

Some metals do not have their atoms arranged in a close-packed structure. These metals have a **body-centred cubic** (**b.c.c.**) structure.

This is shown in figure 10.15. As its name implies, this structure is basically cubic with an atom at the centre of each cube. In this case, each atom is surrounded by eight others. Thus, the co-ordination number is eight and the packing is more open than in the close-packed structures.

All the alkali metals, iron and manganese have a b.c.c. structure. There is no clear relationship between the structure of a metal and its position in the periodic table.

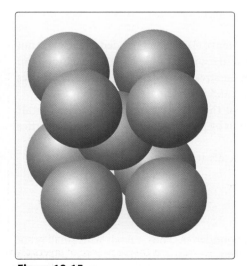

Figure 10.15
A model of the body-centred cubic structure of metal atoms.

10.5 The properties of metals

If the surface of galvanised iron or unpolished tin plate is examined closely, it is possible to see small irregularly shaped areas clearly separated from each other by distinct boundaries. These irregularly shaped areas are known as **grains**. X-ray analysis shows that the metal atoms are packed regularly within the grains. Thus, these grains are small

▲ Crystal grains on the surface of cast iron magnified 200 times.

irregularly shaped crystals of the metal pushed tightly together and containing millions and millions of atoms in a giant structure.

The properties of a metal depend on its crystal structure (i.e. close-packed or body-centred cubic) and also on the size of its crystals (grains). Metals usually have high densities, high melting points, high boiling points and high molar enthalpy changes of fusion and vaporisation. In addition, they are good conductors of heat and electricity, and they are malleable and ductile.

Look back at section 4.8.

1 How do chemists visualise the strong forces between metal atoms?
2 Why do most metals have a high density?
3 Why do most metals have a high melting point and a high boiling point?
4 Why are metals good conductors of electricity?

Chemists believe that the outermost electrons in metal atoms move about freely within the lattice. Thus, the metal consists of positive ions surrounded by a 'sea' of moving electrons. The outer electrons are said to be **delocalised** as they move from one place to another in the crystal, rather than staying in one locality. The negatively charged electrons attract *all* the positively charged ions and bind the nuclei together.

The relatively low affinity for electrons results in two other properties common to metals – low electronegativity and low ionisation energy.

There are no rigid, directed bonds in a metal, so that layers of atoms can slide over each other when a force is applied. This relative movement of layers in the metal lattice is called **slip**. After 'slipping', the atoms settle into new positions and the crystal structure is restored. Thus, a metal can be hammered into different shapes or drawn out into a wire, i.e. it is malleable and ductile. Figure 10.16 shows what happens when 'slip' occurs.

Figure 10.16
The arrangement of metal atoms (*a*) before, and (*b*) after 'slip'.

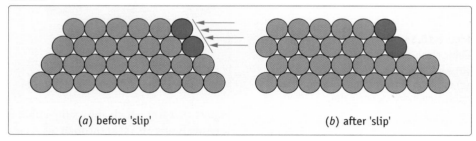

(*a*) before 'slip'   (*b*) after 'slip'

When a metal is placed under stress, it will behave elastically provided the stress is not too great. This is the way an engineer would like a metal to behave in machines and structures. However, a point is reached when the stress is so great that the metal behaves plastically and the changes in it are permanent and irreversible. This is 'slip'.

Can 'slip' be prevented so as to improve the elastic behaviour of the metal? Metallurgists can increase the strength of metals by inserting barriers in the metal lattice which prevent 'slip' occurring.

Two important methods of strengthening metals are:

(a) reducing the size of the crystal grains
(b) alloying.

▲ When metals are twisted or beaten into different shapes, slip is occurring.

Slip does not readily take place across grain boundaries. Thus, metals with small grain size and more grain boundaries per unit volume are harder to deform, less malleable and less ductile than metals with larger grains. By careful choice of metallurgical techniques and casting processes the grain size can be reduced.

Metals will readily form alloys since the metallic bond is non-specific. The presence of small quantities of a second element in the metal frequently increases its strength.

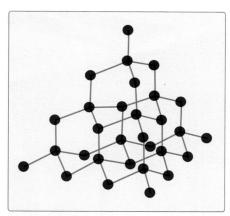

Figure 10.17
The structure of diamond.

▲ A glass engraver using a diamond-tipped wheel to produce a design on a glass goblet.

For example, brass – an alloy of zinc and copper – is much stronger than either pure copper or pure zinc. Atoms of the second metal are different in size to those of the original metal. These differently sized atoms interrupt the orderly arrangement of atoms in the lattice and prevent them sliding over each other.

10.6 Giant molecular (giant covalent) structures

Elements, such as carbon and silicon, which have the ability to form four covalent bonds form giant molecular structures of covalently bonded atoms.

The structure of diamond

In diamond, every carbon atom can be imagined to be at the centre of a regular tetrahedron surrounded by four other carbon atoms whose centres are at the corners of the tetrahedron (figure 10.17). Within the structure, every carbon atom forms four covalent bonds by sharing electrons with each of its four nearest neighbours.

1 What is the co-ordination number of carbon atoms in diamond?
2 Will the carbon atoms on the outside of the diamond form four covalent bonds?

Silicon (section 4.8) and silicon carbide (SiC, used in the abrasive 'carborundum') exist in a similar crystal structure to diamond. Silicon(IV) oxide (SiO_2) is another example of a compound with a giant molecular structure (section 10.10). In each of these structures, atoms are linked by localised electrons in strong covalent bonds throughout the whole three-dimensional lattice.

It is therefore very difficult to distort a covalently bonded crystal, because this would involve breaking covalent bonds. Consequently, diamond, silicon carbide and silicon(IV) oxide are hard and brittle with very high melting points and very high boiling points. Furthermore, the localisation of the electrons within covalent bonds prevents their moving freely in an applied electric field and thus these materials do not conduct electricity.

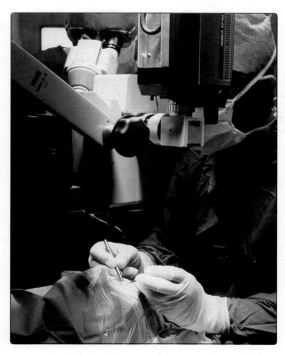

▶ A special diamond-tipped cutting tool being used for eye surgery.

Almost all the industrial uses of diamond and silicon carbide depend on their hardness. Diamond is one of the hardest natural substances. Diamonds unsuitable for gemstones are used in glass cutters and in diamond-studded saws. Powdered diamond and carborundum (SiC) are used as abrasives for smoothing very hard materials.

▶ A giant diamond-tipped circular saw being used to cut through granite.

The structure of graphite

In graphite, the carbon atoms are arranged in flat, parallel layers. Each layer contains millions of hexagonally arranged carbon atoms (figure 10.18). Each carbon atom is covalently bonded to three other atoms in its layer. Each layer can be viewed as a two-dimensional sheet polymer or layer lattice. The carbon–carbon bond length within a layer is 0.142 nm. This suggests some multiple bond character, because the value is intermediate in length between a single and a double carbon–carbon bond. The strong covalent bonds within the layers account for the very high melting point (3730°C) of graphite. Owing to its high melting point, graphite is used to make crucibles for molten metals. A special form of heat-resistant graphite, pyrographite, is used for the exhaust cones of rockets.

The distance between the layers in graphite is 0.335 nm which is much longer than a single carbon–carbon bond. Thus, the bonding between the layers is restricted to relatively weak forces and the layers can slide over each other easily.

This accounts for the softness of graphite and its use as a lubricant.

1. Why is it better to lubricate a car lock or a zip-fastener with a pencil rather than with oil?
2. Why is graphite sometimes more suitable than oil for lubricating the moving parts of machinery?
3. How do you think the proportion of graphite to clay affects the hardness of a pencil?

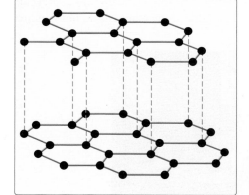

Figure 10.18
The structure of graphite.

The bonding in graphite can be pictured as three trigonally arranged covalent bonds. These are formed by three of the four valence electrons of carbon, whilst the fourth electron is delocalised over the whole layer. This delocalisation of electrons similar to that in metals results in graphite conducting electricity and in its shiny appearance. The electrical conductivity of graphite enables it to be used as electrodes in electric furnaces and during electrolyses.

10.7 Giant ionic structures

Ionic structures are formed when atoms with large differences in electronegativity form compounds. Electrons are transferred from atoms of low electronegativity to those of high electronegativity. The oppositely charged ions which result are held together by strong electrostatic forces of attraction. The electrical force binding the ions together is described as an ionic or electrovalent bond.

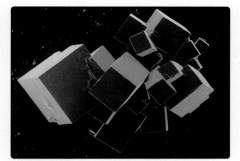

These crystals of sodium chloride show their cubic shape quite clearly.

The sodium chloride lattice

X-ray analysis shows that the particles in different ionic structures can be arranged in different patterns. One of the simplest structures for ionic compounds is the cubic arrangement of ions as in sodium chloride. Figure 10.19 shows a space-filling model of the structure of sodium chloride. The sodium and chloride ions are shown as solid spheres. The sizes of the ions are in the correct ratio. Notice that the ions are arranged in a cubic pattern. Although figure 10.19 shows only a few Na^+ and Cl^- ions, there will be millions and millions of ions in even the smallest visible crystal of sodium chloride.

The positions of Na^+ and Cl^- ions in the cubic lattice of sodium chloride are emphasised in figure 10.20. The small blue and orange circles in figure 10.20 represent the centres of Na^+ and Cl^- ions, respectively. The solid lines in the diagram show the cubic geometry of the lattice.

Notice in figure 10.20 that each positive sodium ion is surrounded by six Cl^- ions and each negative chloride ion is surrounded by six Na^+ ions. The six Cl^- ions around the central Na^+ ion in figure 10.20 have been numbered.

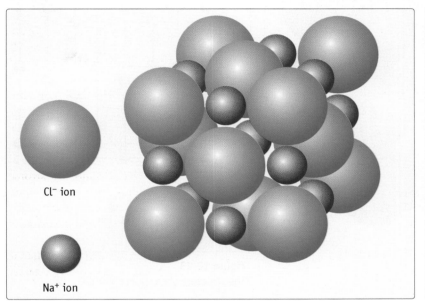

Figure 10.19
A space-filling model of the structure of sodium chloride.

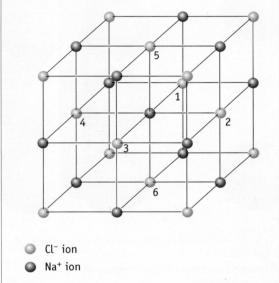

Figure 10.20
The structure of sodium chloride.

Four of the Cl^- ions (numbered 1, 2, 3 and 4) are in the same horizontal layer of the crystal as the central Na^+ ion. One Cl^- ion (number 5) is in the layer above. The final Cl^- ion (number 6) is in the layer below.

The structure of sodium chloride is said to have 6:6 co-ordination because the Na+ ions have a co-ordination number of 6 and the Cl^- ions also have a co-ordination number of 6.

Measuring the size of ions

X-ray measurements on ionic solids can be presented in the form of electron density maps in order to determine the size of different ions. Figure 10.21 shows an electron density map for sodium chloride. The circular contours suggest that the electron distribution in these ions is spherical. The spacing of the contour lines enables us to distinguish particles with different numbers of electrons.

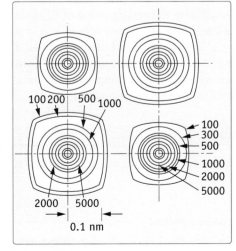

Figure 10.21
An electron density map for sodium chloride. (Electron densities are expressed as electrons per nm^3.)

Look closely at figure 10.21.

1. Which are the sodium ions?
2. What is the interionic distance between neighbouring Na^+ and Cl^- ions?
3. What is the ionic radius of: (i) Na^+, (ii) Cl^-?

Using electron density maps of this kind, it is possible to compile tables of ionic radii. But remember, the size of a particular ion can vary slightly depending on the size and charge of other ions in the crystal. So, values of ionic radii for the same ion do not always agree. The trends in the size of ionic radii in relation to the periodic table are discussed in section 13.3.

The caesium chloride lattice

Another relatively simple structure for ionic compounds is that shown by caesium chloride. Figure 10.22 shows a space-filling model of the structure of caesium chloride. The caesium and chloride ions are shown as solid spheres. As in figure 10.19, the sizes of the ions are in the correct ratio. Unlike Na^+, the Cs^+ ion is similar in size to the Cl^- ion. The general shape of the lattice is again cubic, but it is different in detail from the sodium chloride lattice.

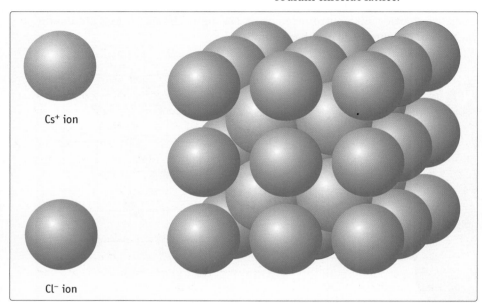

Figure 10.22
A space-filling model of the structure of caesium chloride.

Figure 10.23
The structure of caesium chloride.

The positions of Cs^+ and Cl^- ions in the caesium chloride lattice are shown in figure 10.23. The blue and green dots in figure 10.23 represent the centres of Cs^+ and Cl^- ions, respectively. The solid lines in the diagram show the overall cubic shape of the crystal.

The dashed diagonal lines in figure 10.23 emphasise the arrangement of Cl^- ions relative to Cs^+ ions and vice versa. This shows a single cube of Cl^- ions with a Cs^+ ion at its centre. There are eight Cl^- ions around each Cs^+ ion, so the co-ordination number of Cs^+ ions in the structure is eight. In the same way, the co-ordination number of Cl^- ions is also eight. In fact, the lattice could be redrawn with Cl^- ions exchanged for Cs^+ ions. The structure of caesium chloride is therefore said to have 8:8 co-ordination.

10.8 The properties of ionic crystals

▲ Slate has a layered structure of atoms and ions. It can be cleaved neatly between the layers of ions to produce thin sheets.

Crystals of ionic solids are:

- Hard and brittle.

- Involatile with high melting points and high boiling points.

- Good conductors of electricity when molten, but non-conductors when solid.

- Soluble in polar solvents such as water, but insoluble in non-polar solvents such as tetrachloromethane.

10 Structure, Bonding and Properties – The Solid State

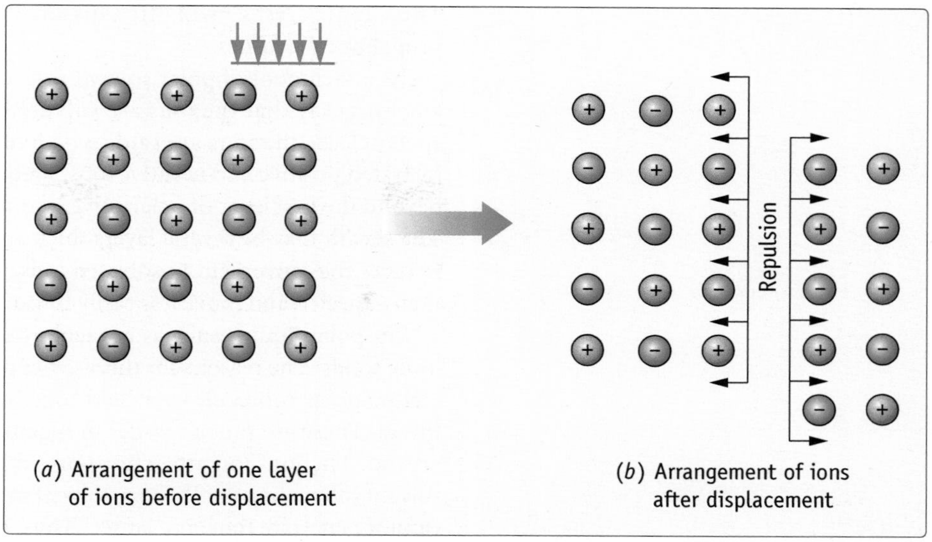

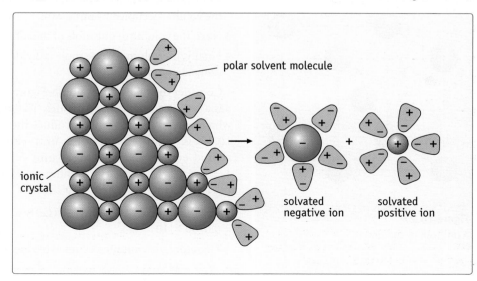

Figure 10.24
Crystal cleavage in ionic compounds.

(a) Arrangement of one layer of ions before displacement

(b) Arrangement of ions after displacement

1 Why is it easier to cleave ionic solids in some directions rather than in others?
2 Why do molten ionic compounds conduct electricity?
3 Why are solid ionic compounds non-conductors?
4 Why are ionic compounds such as sodium chloride soluble in water, but insoluble in tetrachloromethane?

In an ionic solid, every ion is held in the crystal by strong attractions from the oppositely charged ions around it. Consequently, ionic solids like sodium chloride are hard and difficult to cut. However, they are also very brittle and may be split cleanly (**cleaved**) using a sharp-edged razor. When the crystal is tapped sharply along a particular plane it is possible to displace one layer of ions relative to the next. As a result of this displacement, ions of similar charge come together. Repulsion then occurs, forcing apart the two portions of the crystal (figure 10.24).

The forces of attraction between oppositely charged ions in an ionic lattice are very strong. Large quantities of energy must be supplied to the crystal before the ions vibrate vigorously enough to overcome the forces of attraction between one another. The ions then move away from their relative positions. Thus, ionic solids have high melting points and high molar enthalpy changes of fusion. Even in the molten ionic liquid there will still be strong forces between the oppositely charged ions. Consequently, ionic solids have high boiling points and high molar enthalpy changes of vaporisation.

What happens when an ionic solid dissolves in water?

When sodium chloride dissolves in water, the crystal lattice is broken up forming separate Na^+ and Cl^- ions in aqueous solution. Where does the energy required to separate the oppositely charged ions come from? We have seen already that water contains highly polar molecules. The positive ends of polar water molecules are attracted to negative ions in the crystal, and negative ends of the water molecules are attracted to positive ions. The formation of ion–solvent bonds results in a release of energy. This is sufficient to cause the detachment of ions from the lattice (figure 10.25).

polar solvent molecule

ionic crystal

solvated negative ion

solvated positive ion

Figure 10.25
Polar solvent molecules dissolving an ionic solid.

Thus, ionic crystals will often dissolve in polar solvents such as water, ethanol and propanone (acetone).

The attachment of polar solvent molecules to ions is known generally as **solvation**. Chemists say that the ions are **solvated**. Very often the solvent is water and in this specific case the ions are said to be **hydrated**. Figure 10.26 shows the structures of hydrated positive and negative ions. Frequently the attraction of the central ion extends beyond the first layer of polar solvent molecules to a whole sheath of solvent molecules. The sheath may be several layers thick and the envelope of attracted solvent molecules reduces the movement of solvated ions. Thus solvated ions diffuse much more slowly than expected and move less easily towards electrodes during electrolysis.

Non-polar liquids such as tetrachloromethane, benzene and hexane will not solvate ionic solids. The reasons for this are not difficult to see.

Non-polar molecules are held together by weak intermolecular (Van der Waals) forces. These are much smaller in magnitude than the forces between ions in an ionic crystal. The ion–ion attractions are in fact much stronger than either the solvent–solvent interactions or the ion–solvent interactions. So the non-polar solvent molecules cannot penetrate the ionic lattice. Thus, sodium chloride is virtually insoluble in tetra-chloromethane and benzene.

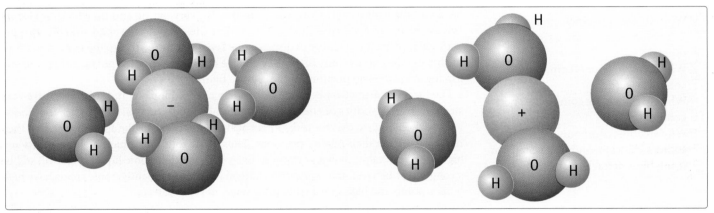

Figure 10.26
The structures of hydrated negative and positive ions. Only those water molecules in the horizontal plane are shown.

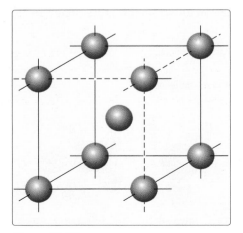

Figure 10.27
A unit cell of sodium.

10.9 Determining the Avogadro constant from X-ray studies

One of the most valuable results of X-ray diffraction studies has been the accurate determination of the Avogadro constant, L. Using this method, L can be found to an accuracy of one in ten thousand.

1 The spacing of particles in a crystal is first determined.

2 Knowing the distance between atoms (or ions) in the crystal, it is then possible to find the volume occupied by one atom.

3 Next, the volume of one mole of the substance is determined.

4 Finally, the volume of one mole is divided by the volume of one atom to obtain the Avogadro constant.

Consider the following example. Figure 10.27 shows a **unit cell** of sodium metal. This has a body-centred cubic structure. The unit cell is the simplest arrangement of atoms which when repeated will reproduce the whole structure. The central atom in figure 10.27 is wholly inside the unit cell. The eight atoms at the corners are shared equally between eight unit cells. Thus, the unit cell contains a total of $1 + 8 \times \frac{1}{8}$ (=2) atoms. X-ray diffraction measurements show that the width of the unit cell is 0.429 nm (i.e. 0.429×10^{-7} cm).

Thus, the volume of the unit cell (i.e. two atoms) = $(0.429 \times 10^{-7})^3$ cm^3
= 0.0790×10^{-21} cm^3

Therefore, the volume occupied by one sodium atom = 0.0395×10^{-21} cm^3

The relative atomic mass of sodium = 22.99

The density of sodium = 0.97 g cm^{-3}

Using the equation, volume = $\dfrac{\text{mass}}{\text{density}}$,

the volume of one mole (L atoms) of sodium = $\dfrac{22.99}{0.97}$ cm^3

$$= 23.70 \text{ cm}^3$$

The Avogadro constant, $L = \dfrac{\text{volume of one mole of atoms}}{\text{volume of one atom}}$

$\therefore$ The Avogadro constant, $L = \dfrac{23.70}{0.0395 \times 10^{-21}} = 6.0 \times 10^{23}$ mol^{-1}.

10.10 Sand and silicates – more giant structures

After water, sand is probably the most common material on Earth. Sand is mainly silicon(IV) oxide, SiO_2, sometimes known as silica. There are several forms of SiO_2 with different crystal structures. These different solid forms of the same compound are called **polymorphs**. Polymorphism in compounds can be compared to **allotropy** for elements (section 4.5). Allotropes are different forms of the same element in the same state.

The most common form of silicon(IV) oxide is quartz. Sand is an impure form of quartz. Its brown colour is due to impurities of iron(III) compounds.

The structure of sand is based on tetrahedra of silicon atoms covalently bonded to four oxygen atoms (figure 10.28).

In silicon(IV) oxide, each SiO_4 tetrahedron shares its corners with four other SiO_4 tetrahedra. Therefore, each silicon atom has a half-share in four oxygen atoms. So, its formula is SiO_2 and silicon(IV) oxide is a pure giant molecular (giant covalent) compound. The strong covalent bonds linking the three-dimensional network of atoms (average bond energy, $E(Si—O) = 464$ kJ mol^{-1}, compared with $E(C—C) = 346$ kJ mol^{-1}) account for the hardness, very high melting point, electrical insulating and thermal insulating properties of silica. Its structure and physical properties are similar to those of diamond.

Anions derived from silica are called **silicates**. Some of them are very important compounds. All silicates, like SiO_2, have structures based on SiO_4 tetrahedra.

In silica itself, each SiO_4 tetrahedron shares its corners with four others. If *none* of the oxygen atoms are shared, then the silicate ion, SiO_4^{4-}, results. Between these extremes, various ring, chain and sheet structures are possible for silicates (figure 10.29).

In each of these structures, negative charges on the silicate chains and sheets are balanced by cations, such as Na^+, Ca^{2+}, Fe^{2+}, Fe^{3+}, Mg^{2+} and Al^{3+} held within the giant structure.

In chain silicates, the charge on each tetrahedron is 2–, because two of the four oxygen atoms are not bonded to other tetrahedra. Notice also that silicates have bonding characteristics of both giant covalent and giant ionic structures.

Probably the most important products from sand and silicates are the various kinds of glass. The main constituents in glass are sodium and calcium silicates.

Look carefully at figure 10.29.

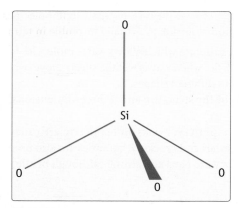

Figure 10.28
The SiO_4 tetrahedron is the basic unit for the structure of sand and all silicates.

1 How many oxygen atoms in each SiO_4 tetrahedron are bonded to other tetrahedra in sheet silicates?

2 What is the charge on each tetrahedron in sheet silicates?

3 Why do you think talcum powder feels smooth when rubbed between your fingers?

4 Thin sheets can be stripped off flat pieces of mica. Why is this?

Another set of compounds closely related to the silicates is the **aluminosilicates**. In these compounds, some silicon atoms are replaced by aluminium atoms. Clay is probably the

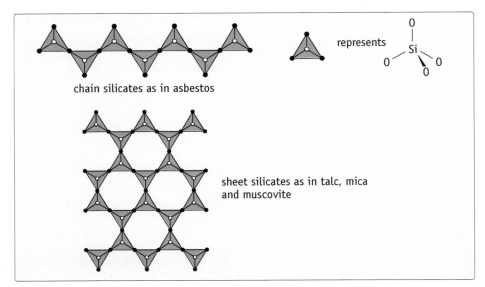

chain silicates as in asbestos

represents

sheet silicates as in talc, mica
and muscovite

Figure 10.29
Chain and sheet silicates are based on SiO_4 tetrahedra.

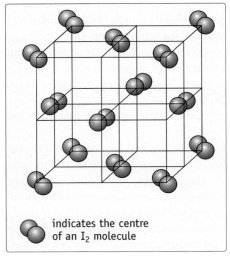

indicates the centre
of an I_2 molecule

Figure 10.30
The crystal structure of iodine.

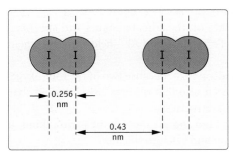

Figure 10.31
The distance between iodine atoms in a
molecule and the distance between iodine
atoms in adjacent molecules.

most important aluminosilicate. It has a sheet structure which allows water molecules to get between the sheets when the two are mixed. So, when clay is wet the sheets move over each other and the material can be moulded into different shapes.

When the clay is fired, water is driven out of the structure and a three-dimensional network of bonds is formed.

The products from fired clay such as china, pottery, bricks and concrete are called **ceramics**. Besides being used as building materials and crockery, ceramics are also used as furnace linings and electrical insulators. They are hard and strong (although brittle), heat-resistant and chemically unreactive.

10.11 Simple molecular structures

Non-metal elements and compounds of non-metals such as iodine, carbon dioxide, water and naphthalene ($C_{10}H_8$) are usually composed of simple molecules. In these molecular substances, the atoms are joined together *within* the molecule by strong covalent bonds. But the separate molecules are attracted to each other by much weaker Van der Waals bonds.

X-ray diffraction measurements on crystals of molecular substances have been used to determine both the distance between atoms within each molecule and the distance between separate molecules. Figure 10.30 shows the arrangement of I_2 molecules in solid iodine. The arrangement of molecules in the crystal lattice is described as **face-centred cubic.** The molecules are arranged in a *cube* with a molecule at each corner and a molecule at the *centre* of each *face*. Notice that the relative positions of molecules with 12 nearest neighbours in the face-centred cubic structure are the same as the relative positions of metal atoms in the cubic close-packed structure (figure 10.14(*c*)).

Analysis of the X-ray diffraction pattern for solid iodine suggests that the distance between two iodine atoms in a molecule is 0.256 nm. Thus, the I—I covalent bond length is 0.256 nm. The distance between the centres of iodine atoms in adjacent molecules is 0.43 nm, which gives a value of 0.215 nm for the Van der Waals radius for iodine (figure 10.31).

The packing of molecules in crystals is often more complicated than the packing of simple atoms or ions. Even so, each substance is found to have a characteristic and uniform lattice arrangement.

The properties of simple molecular substances

Molecular substances such as iodine, naphthalene ($C_{10}H_8$), sucrose (sugar, $C_{12}H_{22}O_{11}$) and ice
- are usually soft,
- have low melting points and boiling points, and
- do not conduct electricity when molten or aqueous.

Non-polar molecular compounds such as iodine and naphthalene are almost insoluble in polar solvents such as water, though they are usually very soluble in non-polar solvents such as benzene and tetrachloromethane.

1 How does the bonding in ice and sugar differ from that in non-polar simple molecules like naphthalene and iodine?

2 Why are ice and sugar harder than iodine and naphthalene?

3 Why is sugar soluble in water unlike iodine and naphthalene?

The properties of molecular compounds can be explained in terms of their structures, which consist of simple molecules. Although the atoms within these molecules are joined by strong covalent forces, the separate molecules are held together by very weak intermolecular forces. These weak intermolecular bonds permit the molecules to be separated easily. Hence the crystals of simple molecular substances are usually soft with low melting points and low boiling points.

Molecular compounds contain neither delocalised electrons (like metals) nor ions (like ionic compounds). Thus, they cannot conduct electricity when molten or when dissolved in water.

Why are non-polar substances such as iodine insoluble in water, but soluble in benzene?

In liquids of high polarity such as water, there are strong water–water attractions. These are considerably stronger than either iodine–iodine attractions or iodine–water attractions. Consequently, iodine molecules cannot penetrate the water structure and there is little tendency for water molecules to solvate uncharged iodine molecules. Iodine is therefore almost insoluble in water.

In non-polar liquids, such as benzene and tetrachloromethane, there are weak intermolecular forces. The benzene–benzene attractions are similar in strength to iodine–benzene and iodine–iodine attractions. Thus, it is easy for benzene molecules to penetrate into the iodine crystal and solvate the iodine molecules. Consequently, iodine dissolves easily in benzene.

10.12 Comparing typical solid structures

The structure, bonding and properties of the four common solid structures (giant metallic, giant molecular, giant ionic and simple molecular) are summarised and compared in table 10.1.

Use table 10.1 to consider each of the following in the solid state: Cu, Si, NH_3, NaI, Xe. Which solid

1 is a good electrical conductor?

2 is a poor electrical conductor, but conducts on melting?

3 is hard and brittle and insoluble in water?

4 is easy to cleave?

5 has strong hydrogen bonds between molecules?

6 has the lowest melting point?

Table 10.1
Comparing typical solid structures

	Giant metallic	Giant molecular (Giant covalent)	Giant ionic	Simple molecular
1 Structure				
(i) Examples	Na, Fe, Cu	Diamond, SiC, SiO_2	Na^+Cl^-, $Ca^{2+}O^{2-}$, $(K^+)_2SO_4^{2-}$	I_2, S_8, $C_{10}H_8$, HCl, CH_4
(ii) Constituent particles	Atoms	Atoms	Ions	Molecules
(iii) Type of substance compound	Metal element with low electronegativity with a large difference in electronegativity)	Non-metal element in Group IV or its compound (elements with high electronegativity)	Metal/non-metal compound (a compound of elements	Non-metal element or non-metal/non-metal
2 Bonding in the solid	Attraction of outer mobile electrons for positive nuclei binds atoms together by strong metallic bonds	Atoms are linked through the whole structure by very strong covalent bonds from one atom to the next	Attraction of positive ions for negative ions results in *strong* ionic bonds	Strong covalent bonds hold atoms together within the separate molecules; separate molecules are held together by weak intermolecular forces
3 Properties				
(i) Volatility	Non-volatile High m.pt., high b.pt.	Non-volatile Very high m.pt., very high b.pt.	Non-volatile High m.pt., high b.pt.	Volatile Low m.pt., low b.pt.
State at room temp.	Usually solid	Solid	Solid	Usually gases or volatile liquids
(ii) Hardness/ malleability	Hard, yet malleable	Very hard and brittle	Hard and brittle	Soft
(iii) Conductivity	Good conductors when solid or liquid	Non-conductors (graphite is an exception)	Non-conductors when solid. Good conductors when molten or in aqueous solution – *electrolytes*	Non-conductors when solid, liquid and in aqueous solution. (A few (e.g. HCl) react with water to form electrolytes.)
(iv) Solubility	Insoluble in polar and non-polar solvents, but soluble in liquid metals	Insoluble in all solvents	Soluble in polar solvents (e.g.H_2O), insoluble in non-polar solvents (e.g. CCl_4)	Polar molecules (e.g. HCl) are soluble in polar solvents such as H_2O, but insoluble in non-polar solvents and vice-versa

Summary

1 When X-rays are directed at a crystalline solid they are diffracted by particles in the lattice. The diffracted X-rays can be detected using an X-ray film. The diffraction pattern which is obtained can be used to determine the arrangement of particles in the crystal.

2 For reinforcement of reflected X-rays, $2d \sin \theta = n\lambda$

This is known as the Bragg equation (d = distance between layers in the crystal, θ = angle between direction of X-rays and the crystal face, $n = 1, 2, 3$, etc., λ = wavelength of X-rays).

3 The structure and bonding of a substance determine its properties. Table 10.1 summarises the structure, bonding and properties of the four typical solid structures (giant metallic, giant molecular, giant ionic and simple molecular).

4 The co-ordination number of an atom or ion is the number of its nearest neighbours.

5 Metals form either close-packed structures (co-ordination number = 12) or body-centred structures (co-ordination number = 8).

6 Electrons which are not held tightly between atoms in directed bonds, but which move freely from one atom to another, are called delocalised electrons.

7 In diamond, each carbon atom is joined by tetrahedrally spaced covalent bonds to four other atoms (co-ordination number = 4).

8 In graphite, carbon atoms are arranged hexagonally in flat, parallel layers. Graphite is described as a two-dimensional sheet polymer or layer lattice.

9 Sodium chloride forms a cubic structure in which the co-ordination numbers of both Na^+ and Cl^- ions are six. Caesium chloride also has a cubic arrangement of ions, but in this case the co-ordination numbers of both Cs^+ and Cl^- ions are eight.

10 The attachment of solvent molecules to the particles of a solute during the dissolving process is known as solvation. Very often the solvent is water and in this particular case the process is called hydration.

Review questions

1 (a) What do you understand by the terms:
   (i) atom, (ii) molecule, (iii) ion?
   (b) Use one or more of these terms to describe the following structures.
   (i) copper, (ii) solid carbon dioxide, (iii) graphite.
   (c) How are the properties of copper and graphite related to their structure and bonding?

2 Consider the following five types of crystalline solids:
   A Metallic,
   B Ionic,
   C Giant molecular (macromolecular),
   D Composed of monatomic molecules, and
   E Composed of molecules containing a small number of atoms.
   Select the letter (A–E) for the structure most likely to show the following properties.
   (a) An element which conducts electricity and boils at 1600°C to form gaseous monatomic atoms.
   (b) A solid which melts at –250°C.
   (c) A solid with a very high molar enthalpy change of vaporisation which does not conduct when liquid.
   (d) A hard, brittle solid which easily cleaves.
   (e) A substance which boils at –50°C and decomposes at high temperatures.

3 Consider the following five substances in the solid state:
   A sodium   D argon
   B silicon   E potassium bromide
   C tetrachloromethane
   Select the letter (A–E) for the substance most likely to show the structure or property described in each of the following:

   (a) A monatomic substance held together by Van der Waals forces.
   (b) A compound of low melting point composed of small molecules.
   (c) A network solid of covalently bonded atoms.
   (d) A non-conducting solid which melts to form a liquid which conducts electricity.
   (e) A substance which exists as a liquid over a temperature range of only 3 K.
   (f) A substance which is decomposed by an electric current in the liquid state.

4 Diamond is one of the hardest natural substances. Before the production of carborundum, powdered diamond was the most widely used abrasive. Carborundum (silicon carbide) is made by heating coke with sand (silicon(IV) oxide) at 2500°C.
   (a) What is an abrasive?
   (b) Write an equation for the manufacture of carborundum from coke and sand.
   (c) Why do you think carborundum has superseded diamond as the most widely used abrasive?
   (d) 'Diamonds are a girl's best friend'! Why? Carborundum has not superseded diamonds in this context. Why not?
   (e) It has been said that the discovery of carborundum enabled the industrial revolution to occur swiftly during the late nineteenth and early twentieth centuries. Why was this?

5 Solid iodine has a dark purple lustrous appearance. In addition, solid iodine displays a small electrical conductivity. When liquid iodine chloride (ICl) is electrolysed, iodine is liberated at the cathode. Discuss and explain each of these observations, explaining whether they are consistent with the expected structure of the substances involved and the position of iodine in the periodic table.

6 Discuss and explain the following:
(a) The ionic nature of $MgCl_2$ is greater than that of $AlCl_3$, which in turn is greater than that of $SiCl_4$.
(b) Silicon(IV) oxide is a solid at room temperature which does not melt until 1973 K, whereas carbon dioxide (m.pt. 217 K) is a gas at room temperature.
(c) Both calcium oxide and sodium chloride have a simple cubic structure and similar interionic distances yet calcium oxide melts at 2973 K whereas sodium chloride melts at 1074 K.
(d) Glucose ($C_6H_{12}O_6$) is much more soluble in water than in benzene, but cyclohexane (C_6H_{12}) is much more soluble in benzene than in water.

7 When X-rays of wavelength 0.1537 nm are directed towards a crystal of KCl, reinforcement of the X-rays first occurs when the angle between the direction of the rays and the crystal face is 14°.
(a) Using the Bragg equation, calculate the distance between layers of ions in the crystal.
(b) At what angle would the second reinforcement of X-rays occur with the crystal in the same orientation?
(c) Figure 10.32 shows a unit cube for potassium chloride.

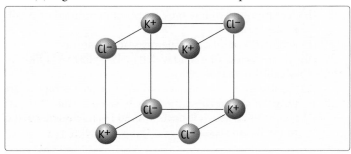

Figure 10.32
A unit cube of potassium chloride.

    (i) Use your result in part (a) to calculate the volume of the unit cube.
    (ii) How many K^+Cl^- 'ion pairs' does the unit cube contain?
    (Remember that ions at the corners of a cube are shared equally amongst eight cubes.)
    (iii) What is the volume occupied by one K^+Cl^- 'ion pair'?
(d) The density of KCl is 2.0 g cm^{-3} and its relative formula mass is 74.6. Using these values, calculate the total volume of one mole of K^+Cl^-.
(e) Using the volume of one mole of K^+Cl^- 'ion pairs' (obtained in part (d)) and the volume of one K^+Cl^- 'ion pair' (obtained in part (c)(iii)), calculate a value for the Avogadro constant, L.

8 (a) What is meant by the term co-ordination number?
(b) Name two metals with a close-packed structure and two metals with a body-centred cubic structure.
(c) What is the co-ordination number of atoms.
    (i) in a close-packed structure,
    (ii) in a body-centred cubic structure?
(d) Suppose one particular metal can have either a close-packed or a body-centred cubic structure. In which of these two forms would you expect it to have:
    (i) the higher density,
    (ii) the higher melting point,
    (iii) the greater malleability?
    Explain your choice in each case.

9 The following data apply to the compounds XCl_x and YCl_y.

	Melting point /°C	Boiling point /°C	Solubility in water /g per 100 g	Solubility in benzene /g per 100 g
XCl_x	801	1443	37	0.063
YCl_y	−22.6	76.8	0.08	miscible with benzene in all proportions.

(a) What types of bond(s) are present in these two chlorides?
(b) Explain clearly how the bonding in each chloride leads to such great differences in volatility and solubility.

10 Figure 10.33 shows the arrangement of ions in sodium chloride in which the smaller orange Na^+ ions are in contact with the larger green Cl^- ions.

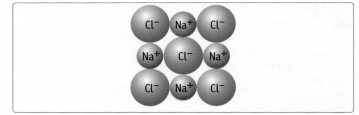

Figure 10.33
The arrangement of ions in sodium chloride.

(a) Draw a diagram for an alkali metal halide in which the metal ions are in contact with larger halide ions and the halide ions also touch each other.
(b) Show that the ratio between the radii of metal ions and halide ions in the structure you have drawn in part (a) is $\sqrt{2} - 1$.
(c) Use the values in table 10.2 to find the alkali metal halide in which the situation described in part (a) is most likely to occur.

Ion	Ionic radius/nm	Ion	Ionic radius/nm
F^-	0.133	Li^+	0.068
Cl^-	0.181	Na^+	0.098
Br^-	0.196	K^+	0.133
I^-	0.219	Rb^+	0.148
		Cs^+	0.167

Table 10.2 The ionic radii of some ions

The Gaseous State

11

▲ The Roman poet Lucretius. In 60 B.C. Lucretius, in his poem 'The Nature of the Universe', suggested that matter was made up of invisible particles.

▲ A male emperor moth can smell the scent from a female moth as far as 8 km away. How would you explain this?

11.1 Evidence for moving particles

The Greeks were probably the first people to believe that matter was made up from particles. In 60 B.C., Lucretius suggested that matter existed in the form of invisible particles. He wrote about this idea in his poem, 'The Nature of the Universe.'

Unfortunately, the Greeks rarely performed experiments. Their theories remained nothing more than good ideas. They had, however, a vast range of everyday experience which supported their beliefs that matter was particulate. They knew, for instance, that a small amount of flavouring such as pepper or ginger could give a whole dish a strong, distinctive taste. This suggested that tiny particles in the pepper could spread throughout the stew to give it a particular flavour.

The ancient Greeks also knew that when dyes such as Tyrian purple were dissolved in water, a tiny amount of the dye could colour an enormous volume of solution. This supported the idea that there must be many particles of the purple pigment in only a little solid and that the particles must be very small.

Use the idea of particles to explain the following:

1. It is possible to smell the perfume a person is wearing from some distance.
2. Solid blocks of air freshener provide a pleasant smell in bathrooms. They disappear after a time without leaving any liquid.

Particles of vapour from the perfume and the air freshener which have a distinctive smell *mix* with air particles and *move* to other parts of the room. This mixing and movement of matter which results from the kinetic energy of moving particles is called **diffusion**.

The idea that particles in gases and liquids are moving has been confirmed by many other experiments involving diffusion.

One other phenomenon which provides strong evidence for moving particles is **Brownian Motion**.

During the 1820s, the botanist, Robert Brown, was carrying out a study of pollen grains. At first, Brown believed that he would be able to observe the pollen grains more effectively through his microscope if they were suspended in water. To his annoyance he found that the pollen continually jittered around in the water in a random manner. As a result of Brown's observation, the random movement of solid particles suspended in a liquid or in a gas is called Brownian Motion.

3. What particles are present in water? How are these particles moving?
4. What caused the pollen grains to jitter about in Brown's experiment? Why do they move so randomly and haphazardly?

11.2 Examining the Brownian Motion of smoke particles in air

Brown's observation of pollen grains was the first recorded example of Brownian

Motion. It is possible to watch Brownian Motion more conveniently using smoke particles in air. Figure 11.1 shows a suitable arrangement which you could use for this experiment.

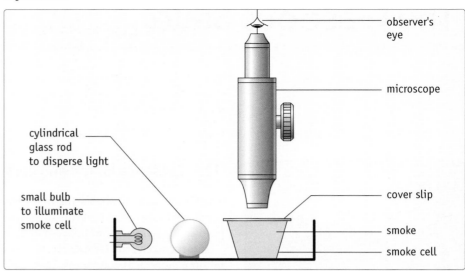

Figure 11.1
An arrangement for examining the Brownian Motion of smoke particles in air.

Use a teat pipette to inject smoke from a smouldering piece of string into the small glass smoke cell. Close the cell with a cover slip. When the illuminated cell is viewed through the microscope the smoke particles look like tiny jittering pin points of light. Why is this?

The smoke particles which you see through the microscope are under constant bombardment from even smaller and invisible molecules in the air. Each individual smoke particle is bombarded first on one side and then on another side. So, it moves first this way and then that way in a random jittery motion. The movement of the smoke particles which you can see provides strong evidence for the movement of air molecules which, of course, you cannot see.

1 What do you think will happen if the heat from the small illuminating lamp causes the temperature of the air and smoke in the cell to rise significantly?

11.3 The Kinetic Theory of Matter

Our ideas about the movement of particles in solids, liquids and gases are summarised in the **Kinetic Theory of Matter**. The word kinetic is derived from the Greek word *kineo* which means 'I move'. The main points in the Kinetic Theory can be summarised as follows:

1 All matter is composed of tiny, invisible particles.

2 Particles of different substances are different in size.

3 In solids, the particles are relatively close together. They have smaller amounts of energy than the same particles in the liquid and gaseous states at higher temperatures. Consequently, solid particles cannot overcome the strong forces of attraction holding them together. They can only vibrate about fixed positions in the solid crystal. So, particles in a solid have vibrational and rotational motion, but no translational motion (figure 11.2).

4 In liquids, the particles are slightly further apart than in solids and have larger amounts of energy. Thus, they are able to overcome the forces between each other to some extent. They can move freely around each other whilst in close proximity. The liquid particles have vibrational, rotational and translational motion.

5 In gases, the particles are much more widely separated than those in solids and liquids and have much larger amounts of energy. The gas particles have sufficient energy to overcome the forces of attraction between each other almost completely. They move rapidly, randomly and haphazardly into any space available. According to the Kinetic Theory, an ideal gas can be imagined to consist of a collection of point mass particles in random motion.

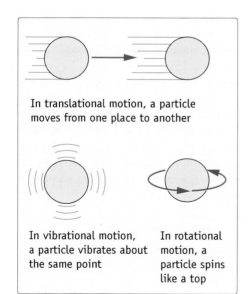

In translational motion, a particle moves from one place to another

In vibrational motion, a particle vibrates about the same point

In rotational motion, a particle spins like a top

Figure 11.2
The translational, vibrational and rotational motion of a particle.

6 An increase in temperature causes an increase in the average kinetic energy of particles. The kinetic energy is manifested as vibrational, rotational and translational energy in gases and liquids and as vibrational and rotational energy in solids.

11.4 Comparing solids, liquids and gases

In the last chapter, we looked at the structure, bonding and properties of different types of solids. This chapter concentrates on the properties and behaviour of gases, but before doing so it is instructive to compare solids with liquids and gases. Look closely at the information in table 11.1. This compares the bulk (macroscopic) characteristics of solids, liquids and gases.

Table 11.1
Comparing solids, liquids and gases

	Solids	Liquids	Gases
Volume	definite (fixed)	definite (fixed)	take the volume of their container
Shape	definite (fixed)	take the shape of their container, but do not necessarily occupy all of it	occupy the whole of their container
Relative compressibility	nil	almost nil	large
Relative density	large	large	small

Notice the similarities between solids and liquids, but the considerable differences between solids and gases. Solids and liquids have fixed volumes, they have similar densities and are almost non-compressible.

The particles in solids and liquids are very close, but in gases they are more widely spaced with much empty space in between. The closeness of the particles in solids and liquids means that these materials have relatively large densities and negligible compressibilities. There are strong forces of attraction between the particles and so the material is held together as a fixed volume.

In contrast, the widely spaced particles in a gas result in much lower densities and much larger compressibilities compared with solids and liquids. The forces between widely spaced gas molecules are very weak. So, the particles readily move away from each other and occupy the whole volume of their container.

In terms of their bulk characteristics, solids and liquids have certain features in common yet they are quite distinct. Liquids, unlike solids, are fluid, a property which they share with gases. However, there are strong forces of attraction between the liquid particles. The particles are held tightly together in a fixed volume even though they have sufficient energy to move freely around each other. This freedom of movement of the particles enables a liquid to flow freely and smoothly on pouring and to take the shape of its container.

In a solid, particles are confined to fixed places in the lattice and it is possible to predict the positions of particles in the structure with some accuracy. The solid is said to have **long-range order**. In a liquid, particles are not confined to fixed positions. They can move freely around each other and it is difficult to predict the positions of particles around an initial reference molecule with any accuracy. The liquid is said to have **short-range order**.

11.5 Molar volumes of gases

Table 11.2 shows the results of an experiment to determine the volume of 1 mole of four common gases. Notice that 1 mole of all the gases occupies about 24.4 dm^3.

▲ The particles in a gas are much more widely spaced than those in a liquid or solid. The forces between such widely spaced particles are weak. So the gas particles readily move away from each other and occupy the whole volume of their container such as this massive hot air balloon.

Table 11.2
Volumes occupied by 1 mole of various gases at 25°C and 1 atmosphere pressure

Gas	Mass of empty 1 dm³ flask/g	Mass of flask + gas/g	Mass of 1 dm³ of gas/g	Relative molecular mass of gas	Volume of 1 mole of gas/dm³
O_2	161.45	162.76	1.31	32	24.5
N_2	161.45	162.60	1.15	28	24.3
CO	161.45	162.60	1.15	28	24.3
CO_2	161.45	163.26	1.81	44	24.3

In fact, the results of a large number of experiments show that one mole of any gas at 25°C and one atmosphere pressure occupies 24.4 ± 0.1 dm³ or (22.4 ± 0.1 dm³ at s.t.p.).

Because of this, the volume of 1 mole of a gas is known as the **molar volume.**

Using this result, it is possible to obtain the formulas of gaseous compounds and to determine the volumes of gases which react. The following example illustrates how the formula of a hydrocarbon can be obtained.

Determining the formula of a hydrocarbon

10 cm³ of the gaseous hydrocarbon were mixed with 33 cm³ of oxygen which was in excess. The mixture was exploded and, after cooling to room temperature, the residual volume of gas occupied 28 cm³. On adding concentrated potassium hydroxide the volume decreased to 8 cm³. This remaining gas is excess oxygen.

Volume of hydrocarbon reacting = 10 cm³
Volume of oxygen reacting = 33 − 8 = 25 cm³
Volume of carbon dioxide (absorbed by potassium hydroxide)
= 28 − 8 = 20 cm³

If we give the hydrocarbon a hypothetical formula of C_xH_y, we can write a general equation for combustion as:

$$C_xH_y(g) + (x + \tfrac{y}{4})\, O_2(g) \rightarrow xCO_2(g) + \tfrac{y}{2}H_2O(l)$$

1 mole $\quad (x + \tfrac{y}{4})$ moles $\quad x$ moles $\quad \tfrac{y}{2}$ moles

This assumes the volumes are all measured at room temperature and the water produced is a liquid.

Using the idea of molar volumes, we can write:

$\Rightarrow$ 1 volume $C_xH_y + (x + \tfrac{y}{4})$ volumes $O_2 \longrightarrow x$ volumes CO_2 + negligible
vol. H_2O $\quad$ (1)
(liquid at room temp.)

10 cm³ $C_xH_y \quad$ 25 cm³ $\quad O_2 \quad\quad$ 20 cm³ $\quad CO_2$
$\Rightarrow$ 1 cm³ $C_xH_y \quad$ 2.5 cm³ $\quad O_2 \quad\quad$ 2 cm³ $\quad CO_2$
$\Rightarrow$ 1 volume $C_xH_y +$ 2.5 volumes $O_2 \longrightarrow$ 2 volumes CO_2 $\quad$ (2)

By comparing equations (1) and (2)

$x = 2$ and $x + \tfrac{y}{4} = 2.5 \quad \therefore y = 2$

$\therefore$ The molecular formula of the hydrocarbon is C_2H_2

Using the idea of molar volumes it also becomes possible to analyse gas mixtures.

A sample of air was analysed by explosion with excess hydrogen in order to determine its percentage by volume of oxygen. 25 cm³ of air was mixed with 100 cm³ of hydrogen and exploded. After cooling to room temperature, the final volume was 110 cm³.

1 Write an equation to represent the reaction of hydrogen with oxygen. Put in state symbols and remember that the measurements are all made at room temperature.

2 What are the relative volumes of hydrogen and oxygen which react?

3 What is the decrease in volume on reaction?

4 What proportion of the decrease in volume is due to oxygen?

⑤ How many cm^3 of oxygen are present in the original air sample?

⑥ What is the percentage by volume of oxygen in the air sample?

11.6 Important gas laws and the Ideal Gas Equation

During the seventeenth, eighteenth and nineteenth centuries, a great deal of scientific research involved the physical and chemical properties of gases. Careful investigations were carried out in order to discover how the volume of a gas varied with changes in temperature and pressure.

As early as 1662, Robert Boyle had discovered that *the volume of a fixed mass of gas is inversely proportional to its pressure, provided the temperature remains constant.*

This is known as **Boyle's Law** which can be expressed mathematically as

$$V \propto \frac{1}{p}$$

About a century later in 1787, Jacques Charles showed that *the volume of a fixed mass of gas is directly proportional to its absolute temperature, provided the pressure remains constant.*

This is known as **Charles's Law** which can be expressed mathematically as

$$V \propto T$$

By combining Boyle's Law and Charles's Law we get

$$V \propto \frac{T}{p}$$

$$\therefore pV \propto T$$

$$\text{and} \quad \frac{pV}{T} \text{ constant}$$

$$\text{so} \quad \frac{p_1 V_1}{T_1} = \frac{p_2 V_2}{T_2}$$

This last equation enables the volume of a gas to be obtained under any conditions of temperature and pressure (say T_1 and p_1), provided its volume (V_2) is known under some other conditions of temperature (T_2) and pressure (p_2).

What is the value of the constant in the expression pV/T = constant? Its value will, of course, depend on the amount of gas taken. Let us calculate the constant for one mole of gas. For one mole of gas, the constant is given the symbol R and is called **the gas constant.**

We know that one mole of gas occupies a volume of 22.4 dm^3 at one atmosphere pressure and 273 K.

$$\text{Using } R = \frac{pV}{T} = \frac{1 \times 22.4}{273}$$

$$\Rightarrow R = \textbf{0.082 atm dm}^3\,\textbf{K}^{-1}\,\textbf{mol}^{-1}$$

Operating in SI units, the volume of one mole of gas is 0.0224 m^3 at a pressure of 101 325 N m^{-2} (i.e. 101 325 Pa* ≃ 101 kPa) and 273 K.

$$\therefore \text{In SI units, } R = \frac{pV}{T} = \frac{101\,325 \times 0.0224}{273}$$

$$= 8.31 \text{ N m K}^{-1}\,\text{mol}^{-1}$$

$$\Rightarrow R = \textbf{8.31 J K}^{-1}\,\textbf{mol}^{-1}$$

So for one mole of gas, $pV = RT$
and for n moles of gas, $\boldsymbol{pV = nRT}$, which is known as the **Ideal Gas Equation.**

A gas which obeys this equation exactly is called a **'Perfect' gas** or an **'Ideal' gas.** In practice, real gases obey the equation very closely at low pressure and high temperature. Under these conditions a gas is most like a gas and least like a liquid (section 9.3(B)(ii)).

THE HON$^\text{ble}$ ROBERT BOYLE

▲ Robert Boyle (1627–1691), one of the foremost scientists of the seventeenth century. Boyle carried out investigations into many areas of science including combustion, besides his research into the volume and pressure of gases.

* The internationally accepted unit of pressure is the pascal, Pa. One pascal is defined as a pressure of one newton per square metre, 1.0 N m^{-2}.

Suppose the air in a room at one atmosphere pressure contains exactly one-fifth by volume of oxygen and four-fifths by volume of nitrogen.

> 1 What pressure is exerted by the oxygen?
>
> 2 What pressure is exerted by the nitrogen?

Imagine that you have two identical cylinders, joined by a tap, one containing pure oxygen at one atmosphere pressure, the other containing pure nitrogen at one atmosphere pressure. If all the oxygen in the first cylinder is forced through the tap into the second cylinder:

> 3 What is the total pressure in the second cylinder?
>
> 4 What is the pressure of nitrogen in the second cylinder?
>
> 5 What is the pressure of oxygen in the second cylinder?

In order to answer the last five questions you have assumed **Dalton's law of partial pressures**. This says that *the total pressure of a mixture of gases is equal to the sum of the pressures that each gas would exert if it occupied the space alone.* This can be expressed mathematically as:

$$p_{total} = p_1 + p_2 + p_3 \ldots p_n$$

p_{total} is the total pressure of the mixture, whilst p_1 is the partial pressure of gas 1, p_2 is the partial pressure of gas 2, etc.

11.7 Investigating the distribution of molecular speeds in gases – the Zartmann experiment

All the molecules in a sample of gas will not have the same speed. At any particular moment, some will be almost stationary whilst others will be moving very rapidly. They will also be moving in very different directions as a result of their haphazard motion. During the early 1930s, Zartmann and Ko developed an ingenious experiment for determining the distribution of molecular speeds. Their apparatus is shown in figure 11.3.

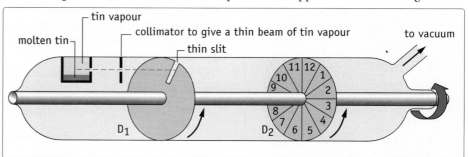

Figure 11.3
The Zartmann experiment.

D_1 and D_2 are discs which rotate rapidly and at the same speed. The discs are on a common axle, inside an evacuated container. In front of D_1 is an oven containing molten tin at a controlled temperature.

Tin vapour streams from a small hole in the oven and strikes D_1. Tin atoms pass through D_1 in small bursts each time the slit in D_1 is in line with the opening in the oven and the collimator. So, D_1 acts as a 'starting gate' for repeated 'molecular races' between D_1 and D_2.

No more molecules can get past D_1 until the disc has completed a full rotation and the slit is opposite the opening in the oven again. Those molecules passing D_1 travel towards the second rotating disc, D_2, and spread throughout its pie slices.

D_1 lets through burst after burst of tin atoms and a layer of tin builds up on D_2. The faster molecules hit the plate before the slower ones, since they all started from the slit simultaneously. Figure 11.4 shows how the deposited tin spreads over the pie slices if the rotating speed of D_1 and D_2 is carefully controlled. The distribution of tin on D_2

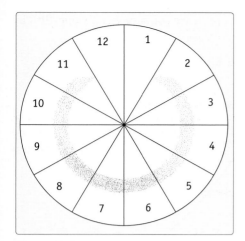

Figure 11.4
The distribution of tin over the pie slices of the target disc in the Zartmann experiment.

depends on the distribution of the speeds of the tin atoms. In order to find the distribution of tin on D_2, it is cut up into pie slices which are weighed to determine the amount of solid deposited on each slice.

1. Why is no tin deposited on pie slice 1?
2. Why is no tin deposited on pie slice 12?
3. What happens to the distribution of tin on the pie slices if D_1 and D_2 rotate faster?
4. Why must the space between D_1 and D_2 be continuously evacuated?

Figure 11.5 shows a distribution curve for the mass of tin deposited on each pie slice.

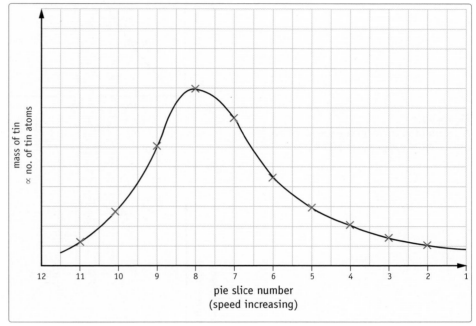

Figure 11.5
A distribution curve for the mass of tin deposited on the different pie slices in the Zartmann experiment.

If we know the distance between D_1 and D_2 and the rate of rotation of the discs, we can calculate the velocity of those molecules hitting a particular point on D_2.

For example, suppose the distance from D_1 and D_2 is 1 metre and the discs rotate at 50 revolutions per second.

1 revolution of the discs takes $\frac{1}{50}$th second.

Now the pie slice 1–pie slice 2 boundary $= \frac{1}{12}$th of the disc.

∴ Atoms on the pie slice 1–pie slice 2 boundary travel 1 metre (from D_1 to D_2) in $\frac{1}{12}$th of a revolution (i.e. $\frac{1}{600}$th second).

∴ Speed of molecules on pie slice 1–pie slice 2 boundary $= 600$ m s^{-1}.

5. What is the speed of those molecules on the pie slice 2–pie slice 3 boundary?

Proceeding in this way, it is possible to obtain a distribution curve for the molecular speeds.

Figure 11.6 shows a distribution curve for the speeds of tin atoms in the gas phase at 500 K. The most probable speed corresponds to the maximum of the curve. More molecules possess this speed than any other.

6. What is the effect of increased temperature on the distribution of molecular and atomic speeds?

Figure 11.7 shows the distribution curves for the speeds of tin atoms at 500 K and 1000 K. There are some important points to notice about the effect that increased temperature has had on the distribution and sizes of the atomic speeds.

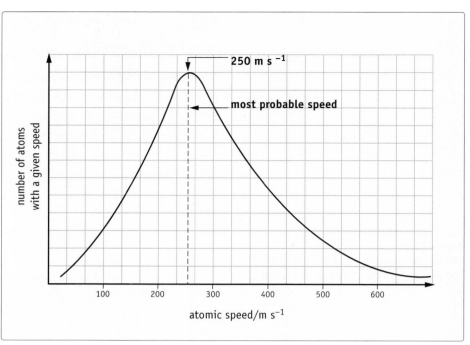

Figure 11.6
A distribution curve for the speeds of tin atoms in the gas phase at 500 K.

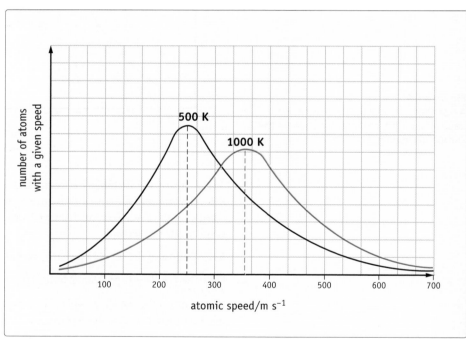

Figure 11.7
Distribution curves for the speeds of tin atoms at 500 K and 1000 K.

1 The most probable speed increases as the temperature increases. At 500 K, the most probable speed for tin atoms is 250 m s^{-1}. At 1000 K the most probable speed has increased to 360 m s^{-1}.

2 The distribution curve becomes flatter as the temperature rises. This means that there are fewer molecules with the most probable speed, but there is a greater proportion of high-speed molecules.

3 It is important to remember that the graph in figure 11.7 is really a histogram. Thus, the area beneath each curve is proportional to the total number of molecules involved. As we are dealing with the *same total* number of molecules at 500 K and 1000 K, the areas beneath the two curves must be identical.

11.8 Determining the relative molar masses of gases and volatile liquids

Nowadays, the most accurate and convenient method of measuring relative molar masses is by mass spectrometry. Using this method, relative molar masses can be obtained in a few minutes with an accuracy of one part in a million.

Originally, the relative molar masses of gases and volatile substances were obtained by methods dependent on the gas laws discussed earlier in this chapter. Two of the more common methods employed were those designed by Regnault and Victor Meyer.

Finding the relative molar mass of a gas by direct weighing – Regnault's method.

Using the Ideal Gas Equation, $pV = nRT$, and $n = \dfrac{m}{M}$

where m is the mass of gas whose molar mass is M.

We can write
$$pV = \frac{m}{M} \cdot RT$$

$$\Rightarrow M = \frac{m}{pV} \cdot RT$$

By measuring m, p, V and T for a sample of gas and assuming R, it is possible to calculate M.

A container of known volume (V) is weighed full of gas at pressure p and temperature T. The same container is then weighed after evacuation in order to obtain the mass of gas inside (m).

This method is very accurate provided the following precautions are taken.

1 The container must be large, so as to give an appreciable weight of gas. Gases are so light that any slight error in weighing produces a large percentage error if the mass of the substance weighed is very small.

2 The gas used must be perfectly pure and perfectly dry.

3 The container should be filled and emptied several times to make sure that all the air and any previous gas has been removed.

4 The container must be evacuated as completely as possible.

5 All weighings must be carried out at the same temperature and pressure so that the upthrust on the container remains constant.

Regnault's method has been used to determine relative molar masses and relative atomic masses to an accuracy of more than one part in a thousand. For example, the relative molar mass of H_2S ($M_r(H_2S)$) could be obtained accurately by this technique and then used to determine the relative atomic mass of sulphur ($A_r(S)$):

$$A_r(S) = M_r(H_2S) - 2 \times A_r(H)$$

Finding the relative molar mass of a volatile liquid

This method can be used with volatile liquids. It is a development of an earlier technique used by Victor Meyer. Although it gives only an approximate value for the relative molar mass, it enables an accurate value to be obtained if the empirical formula of the substance can be determined.

The principle of the experiment is the same as that described for gases earlier in this section using the equation:

$$\Rightarrow M = \frac{m}{pV} \cdot RT$$

Thus, the molar mass (M) of the liquid can be calculated by finding the volume of vapour (V) formed from a known mass of liquid (m) at a measured temperature (T) and pressure (p), provided we assume the value of the gas constant (R).

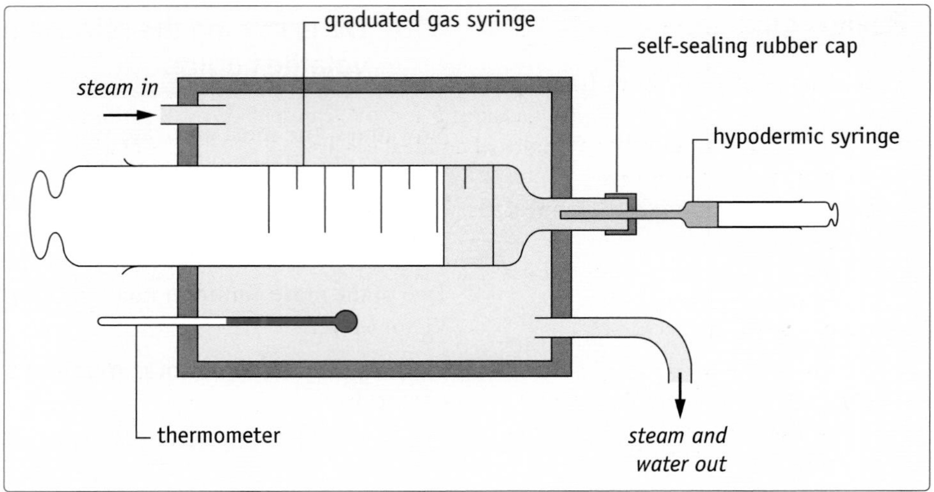

Figure 11.8
Determination of the relative molar mass of a volatile liquid.

1. Why is this method unsuitable for liquids which boil above 100°C?

2. Why should the hypodermic syringe be handled as little as possible between weighings?

3. Suggest two sources of error in this experiment.

4. The substance X, for which the typical results were given, has the percentage composition by mass of 22.0% C, 4.6% H and 73.4% Br.

   (i) Calculate the empirical formula of X.
   (ii) What is the molecular formula of X?
   (iii) What is the accurate relative molecular mass of X?
   Look at section 1.5 if you are uncertain about answering these last questions.

Figure 11.8 shows a suitable apparatus to use. Draw a few cubic centimetres of air into the graduated gas syringe and fit the self-sealing rubber cap over the nozzle. Pass steam through the outer jacket until the thermometer reading and the volume of air in the syringe become steady. Continue to pass steam through the jacket and record the temperature (T) and the volume of air in the syringe.

Now fill the hypodermic syringe with about 1 cm^3 of the liquid under investigation, ensuring that all air is expelled from the needle. Weigh the hypodermic syringe and its contents and then push the needle through the self-sealing cap of the graduated gas syringe so that it is well clear of the nozzle. Inject about 0.2 cm^3 of liquid into the large syringe and withdraw the hypodermic syringe. Immediately, re-weigh the hypodermic syringe and its contents.

The liquid injected into the graduated gas syringe will evaporate. The final volume of air plus vapour in the graduated syringe should be recorded when the volume becomes steady. Finally, record the atmospheric pressure. Here are some typical results:

Mass of liquid, X, vaporised, m = 0.16 g
Initial volume of air in graduated gas syringe = 10 cm^3
Final volume of air + vapour in gas syringe = 56 cm^3
∴ Volume of X vaporised = 46 cm^3
= $46 \times 10^{-6} m^3$
Temperature of X = 100°C = 373 K
Pressure of X = 1 atm = 101 325 N m^{-2}
Gas constant, R = 8.31 J K^{-1} mol^{-1}
Using $M = \dfrac{m}{pV} . RT$

$$M = \frac{0.16}{101\,325 \times 46 \times 10^{-6}} \times 8.31 \times 373 = 106.4 \text{ g mol}^{-1}$$

⇒ relative molar mass of X, $M_r \approx 106$

Summary

1. Diffusion is the mixing and movement of matter which results from the kinetic energy of moving particles.

2. Brownian Motion is the random movement of particles of solid suspended in a liquid or in a gas.

3. All matter is composed of tiny, invisible particles.

4. In solids, the particles are relatively close and can only vibrate about fixed positions.

5. In liquids, the particles are further apart than in solids. They have sufficient energy to overcome the forces of attraction between each other and they move freely around each other.

6. In gases, the particles are much more widely separated. They have sufficient energy to overcome the forces of attraction between each other almost completely and move rapidly and randomly into any space available.

7. Solids are said to have long-range order. Liquids have short-range order.

8. At s.t.p., one mole of any gas occupies 22.4 dm^3.

9. The Ideal Gas Equation is $pV = nRT$. Under ordinary conditions (and at low pressure and high temperature) real gases obey the Ideal Gas Equation very closely.

1 (a) Draw sketch graphs of (i) p against V, (ii) p against $1/V$
(iii) pV against p, (iv) pV against V.
for a constant number of moles of an ideal gas at
constant temperature.

(b) Plot on separate sketch graphs, (i) V against T (°C)
(ii) V against T (K)
for a constant number of moles of an ideal gas at
constant pressure.

(c) Plot pV against T and pV/T against p for a constant
number of moles of ideal gas.

(d) Draw a sketch graph to show how the pressure of a
constant mass of ideal gas will vary as the temperature
rises from absolute zero in a container of constant
volume. How will the graph change if the gas tends to
dissociate ($X_2 \rightarrow 2X$) as temperature increases? (For
each graph in sections (a), (b), (c) and (d), label the axes
and show the zero for each axis.)

2 A balloon can hold $1000\ cm^3$ of air before bursting. The
balloon contains $975\ cm^3$ of air at 5°C. Will it burst when it
is taken into a house at 25°C? Assume that the pressure of the
gas in the balloon remains constant.

3 What would you expect to be the effect of a change in
(i) pressure, (ii) volume, (iii) temperature,
on the rate of diffusion of a particular gas? Explain your
answer in terms of the kinetic theory.

4 (a) A mixture of two gases in a container exerts a pressure of
800 mm Hg and occupies a volume of $400\ cm^3$. If one of
these gases (A) occupies a volume of $300\ cm^3$ at the same
temperature and pressure, what pressure does the other
gas (B) exert in the mixture?

(b) When gaseous argon is allowed to diffuse, it separates
into a lighter and a heavier fraction. What does this
suggest about the nature of argon?

5 0.50 g of a volatile liquid was introduced into a globe of
$1000\ cm^3$ capacity. The globe was heated to 91°C so that all
the liquid vaporised. Under these conditions the vapour
exerted a pressure of 0.25 atm. What is the relative molecular
mass of the liquid?

6 An industrial chemist who works for a firm which
manufactures ethyne (acetylene, C_2H_2) for use in oxyacety-
lene welding discovers a very cheap way of producing ethyne.
Unfortunately, his ethyne is contaminated with ethene
(ethylene, C_2H_4). Unless the mixture contains at least 50% by
volume of ethyne, it is useless. In order to determine the relative
proportions of the two in the mixture, he exploded $10\ cm^3$ of the
mixture with $30\ cm^3$ of oxygen. After absorbing the residual CO_2
in KOH, the uncombined O_2 occupied $2\ cm^3$. What was the
composition of the mixture by volume?
(*Hint*: let the volume of one gas be $x\ cm^3$. Write an equation for
the combustion of that gas and hence work out, in terms of x, how
many cm^3 of O_2 it reacts with. Repeat this for the other gas, $(10 -
x)\ cm^3$. Find the total volume of O_2 used in terms of x. Find the
actual volume of O_2 used from the data. Hence find x.)

7 The extent of combustion of petrol vapour in an internal
combustion engine can be shown in terms of the percentage
of CO_2 in the exhaust gases. $60\ cm^3$ of exhaust gas (assumed
to contain only CO, CO_2 and N_2) were mixed with $20\ cm^3$
of O_2 (excess) and exploded. On cooling, there had been a
decrease to $70\ cm^3$. On adding KOH, the volume decreased
to $35\ cm^3$.

(a) What is the composition by volume of the mixture?

(b) What is the percentage by volume of CO_2 in the
exhaust fumes?

8 On decomposition with electric sparks, two volumes of a
gaseous compound containing only phosphorus and
hydrogen, gave a deposit of phosphorus (solid) and three
volumes of hydrogen at the same temperature and pressure.

(a) Deduce the number of atoms of hydrogen in one molecule
of the phosphorus hydride. (Assume hydrogen is
diatomic.)

(b) Write an empirical formula for the phosphorus hydride
in as much detail as you are able.

(c) What do you think its molecular formula is?

(d) What measurement would you make to try to confirm
this?

(e) Predict two physical and two chemical properties of this
compound using your knowledge of the periodic table.

9 It is required to find the composition by volume of a sample
of gas which contains only hydrogen, carbon monoxide and
nitrogen. $40\ cm^3$ of the gas were carefully exploded with
$40\ cm^3$ of oxygen (known to be in excess). Under the
conditions used, only hydrogen and carbon monoxide
would react with oxygen. On cooling to room temperature,
the volume was $51\ cm^3$. On adding concentrated KOH, the
volume decreased further to $41\ cm^3$.

(a) Write equations (including state symbols) for the
reactions which occur on explosion. (Remember that
water will be liquid, as the measurements are made at
room temperature.)

(b) What are the relative volumes of gaseous reactants and
products in these reactions?

(c) What volume of CO_2 is produced?

(d) What is the volume of CO in the original $40\ cm^3$ of gas?

(e) What is the total decrease in volume as a result of the
explosion?

(f) What decrease in volume is caused by
(i) CO (ii) H_2 on explosion?

(g) What are the volumes of
(i) H_2 (ii) N_2 in the original $40\ cm^3$ of gas?

10 When $20\ cm^3$ of a gaseous hydrocarbon, A, are exploded
with $150\ cm^3$ of oxygen, the residual gases occupied 110
cm^3. After shaking the residual gases with aqueous sodium
hydroxide, the final volume was $30\ cm^3$. (All volumes were
measured at the same temperature and pressure.)

(a) What is the molecular formula of A?

(b) Write six possible structural formulas for A and name
each of these structures.

Energy Changes and Bonding

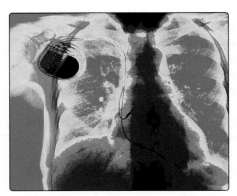

▲ This heart pacemaker, fitted over the ribs, is powered by a miniature lithium battery. Wires from the pacemaker battery supply electrical impulses to maintain a regular heartbeat.

12.1 Energy and energy changes

Energy is the most precious commodity we have. Without it there could be no life, no warmth, no movement. Energy gives us the power to do work. In every country, people's living standards are closely related to the availability of energy.

From the earliest times, people worshipped the sun. This is not surprising. Through the process of photosynthesis, the sun provides us with most of our food and, over millions of years, it has created our supplies of fossil fuels, coal, oil and natural gas. More than ninety per cent of our present energy needs are supplied by these fossil fuels – valuable reserves of energy that cannot last forever.

Until the late 1970s, Western Europe was entirely dependent on the Middle East for its increasing demands for oil. After the outbreak of the Arab/Israeli war in 1973, the price of oil rose alarmingly and created serious inflation in the oil-importing countries such as Britain. Since 1975, however, the prospects for Europe have brightened with the discovery of first natural gas, and then vast oilfields, below the North Sea.

The transfer of energy to or from chemicals plays a crucial part in the chemical processes in industry and in living things. Consequently, the study of these energy changes are as important as the study of the changes in the materials themselves.

Our present-day living conditions rely heavily on the availability of energy in its various forms. Chemical energy is converted to heat energy when fuels such as gas, oil and coal are burnt in our homes and in industry. Just think of the vast quantities of chemical energy converted to mechanical energy each day from the petrol burnt in our vehicles!

Within our own bodies, energy changes are vital. Foods such as fats and carbohydrates are important biological fuels. During metabolism, the chemical energy in these foods is converted into heat energy to keep us warm, to mechanical energy in our muscles and to electrical energy in the signals within our nerve fibres. Chemical energy is also converted to electrical energy when the materials in cells and batteries are used to generate electricity.

All these important processes involve energy changes. This in itself would be a sufficient answer to the question 'Why study energy changes?' Apart from their importance to industry and society, the study of energy changes can lead to a better understanding of the fundamental chemistry involved.

12.2 The ideas and language of thermochemistry

Very often chemical changes are accompanied by changes in the **heat content** of the materials which are reacting. The correct term for heat content is **enthalpy**, H. Usually, this change in the heat content or enthalpy is shown by a change in temperature. Indeed, the change in temperature when substances react often provides evidence that a chemical change has taken place.

When an exothermic reaction occurs at room temperature, heat is given out and the temperature of the products rises. Eventually, the temperature of the products falls to

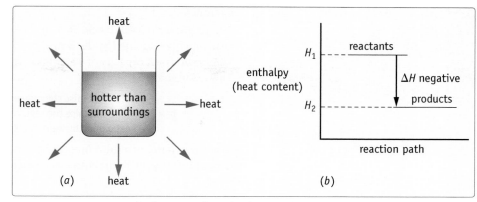

Figure 12.1
In an exothermic reaction, heat is lost to the surroundings and ΔH is negative.

room temperature as the heat produced is lost to the surroundings (figure 12.1(a)). Thus, the heat content (enthalpy) of the products (H_2) is less than that of the reactants (H_1). Since the materials have *lost* heat, we can see that the **enthalpy change for the reaction, ΔH** (sometimes called the **heat of reaction**) is negative (figure 12.1(b)).

For example, when magnesium reacts with oxygen, heat is evolved:

$$2Mg(s) + O_2(g) \rightarrow 2MgO(s) \qquad \Delta H = -1204 \text{ kJ}$$

Chemical energy in the magnesium and in the oxygen is partly transferred to chemical energy in the magnesium oxide and partly evolved as heat. Thus, the magnesium oxide has less energy than the starting materials, magnesium and oxygen (figure 12.2). The value of ΔH, the enthalpy change of the reaction, relates to the amounts shown in the equation – 2 moles of Mg atoms, 1 mole of O_2 molecules and 2 moles of MgO.

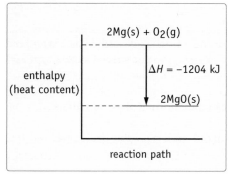

Figure 12.2
An energy level diagram for the exothermic reaction of magnesium with oxygen.

When an endothermic reaction occurs, the heat required for the reaction is taken from the reacting materials themselves. At first, the temperature of the products falls below the initial temperature, but eventually the temperature of the products rises to room temperature again as heat is *absorbed* from the surroundings. In this case, the heat content of the products is greater than that of the reactants and the enthalpy change (heat of reaction), ΔH, is positive (figure 12.3).

We can summarise these ideas as:

Enthalpy change = heat of reaction
$$\Delta H = H_2 - H_1$$

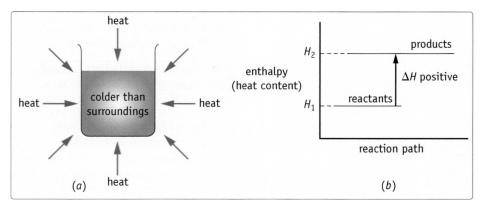

Figure 12.3
In an endothermic reaction, heat is gained from the surroundings and ΔH is positive.

Since the enthalpy change manifests itself as heat, the term 'heat of reaction' is often used in place of 'enthalpy change of reaction'.

Remember that ΔH refers only to the energy change *for the reacting materials*. The surroundings will obviously gain whatever heat the reacting materials lose, and vice versa. Thus, the *total* energy is unchanged during a chemical reaction. *Energy may be exchanged between the materials and the surroundings but the total energy of the materials and the surroundings remains constant.* This important concept is known as the **Law of Conservation of Energy** and also as the **First Law of Thermodynamics**.

Example

A GAZ burner is used to heat 500 cm^3 (500 g) of water from 20°C to 100°C. How much heat (energy) is gained by the water?

The specific heat capacity of water is 4.2 J g^{-1} K^{-1}.

This means that:

4.2 J will raise the temperature of 1 g of water by 1 K (1°C)

∴ $m \times 4.2 \times \Delta T$ J will raise the temperature of m g of water by ΔT K (ΔT°C)

So, heat gained by water = $500 \times 4.2 \times 80 = 168\,000$ J = 168 kJ

In general, if m g of a substance (specific heat capacity = c J g^{-1} K^{-1}) increase in temperature by ΔT K, the enthalpy change $\Delta H = +m.c.\Delta T$ J.

12.3 The standard conditions for thermochemical measurements

In order to compare energy changes, it is important to state the conditions under which a reaction is performed and ensure that the conditions of the system are the same before and after the reaction. In particular, the temperature and pressure should be stated.

> *The standard conditions of temperature and pressure for thermochemical measurements are 298 K (25°C) and 1 atmosphere, respectively.*

Any enthalpy change measured under these conditions is described as a **standard enthalpy change** and given the symbol $\Delta H^{\ominus}$ or $\Delta H^{\ominus}_{298}$ with a special superscript and subscript. The symbol $\Delta H^{\ominus}_{298}$ also implies that

- all the substances involved in the reaction are in their normal physical states at 298 K and 1 atmosphere and
- any solutions involved have a concentration of one mole per cubic decimetre.

Thus, $\Delta H^{\ominus}_{298}$ for the reaction

$$2H_2(g) + O_2(g) \rightarrow 2H_2O(l)$$

must relate to gaseous hydrogen, gaseous oxygen and liquid water (not steam).

In the case of elements which exist as different allotropes, and compounds which exist as different polymorphs, the most stable form at 298 K and 1 atm is chosen as the standard. Consequently, $\Delta H^{\ominus}_{298}$ values for reactions involving carbon should relate to the allotrope graphite rather than diamond.

> We can therefore define the **standard enthalpy change of a reaction** as *the amount of heat absorbed or evolved when the molar quantities of reactants as stated in the equation react together under standard conditions.*
>
> i.e. at a pressure of 1 atmosphere,
> at a temperature of 298 K,
> with substances in their normal physical states under these conditions and solutions having a concentration of 1.0 mol dm^{-3}.

12.4 Standard enthalpy changes of formation and combustion

> The **standard enthalpy change of formation** of a substance is the heat evolved or absorbed when one mole of the substance is formed from its elements under standard conditions.

The standard enthalpy change of formation of a substance is given the symbol $\Delta H^{\ominus}_f$. The superscript $^{\ominus}$ indicates standard conditions and the subscript $_f$ refers to the formation reaction.

Consider the reaction

$$2H_2(g) + O_2(g) \rightarrow 2H_2O(l) \quad \Delta H^{\ominus}_{298} = -575 \text{ kJ}$$

1 Is the reaction endothermic or exothermic?

2 Which have the greater enthalpy, the products or the reactants?

3 What is the value of $\Delta H^{\ominus}_{298}$ for
(i) $2H_2O(l) \rightarrow 2H_2(g) + O_2(g)$
(ii) $H_2(g) + \frac{1}{2}O_2(g) \rightarrow H_2O(l)$

4 Given also that

$$H_2O(l) \rightarrow H_2O(g) \quad \Delta H^{\ominus}_{298} = +44 \text{ kJ},$$

calculate the value of $\Delta H^{\ominus}_{298}$ for

$$2H_2(g) + O_2(g) \rightarrow 2H_2O(g)$$

Thus, the statement $\Delta H_f^{\ominus}(\text{MgO(s)}) = -602 \text{ kJ mol}^{-1}$ relates to the formation of 1 mole of magnesium oxide from 1 mole of Mg atoms and $\frac{1}{2}$ mole of O_2 molecules, i.e.

$$\text{Mg(s)} + \tfrac{1}{2}O_2(g) \rightarrow \text{MgO(s)}$$

One important consequence of the definition of standard enthalpy change of formation is that the enthalpy change of formation of an element in its normal physical state under standard conditions is zero, since no heat change is involved when an element is formed from itself.

$$\text{Cu(s)} \rightarrow \text{Cu(s)} \quad \Delta H_f^{\ominus}(\text{Cu(s)}) = 0$$
$$\text{H}_2(g) \rightarrow \text{H}_2(g) \quad \Delta H_f^{\ominus}(\text{H}_2(g)) = 0$$

Obviously, $\Delta H_{298}^{\ominus}$ for the process $\text{H}_2(g) \rightarrow \text{H}_2(g)$ is zero, but this is not so for the process $\text{H}_2(g) \rightarrow 2\text{H}(g)$. This involves atomisation, i.e. the conversion of H_2 molecules to single H atoms.

> The **standard enthalpy change of atomisation of an element** is defined as *the enthalpy change when one mole of gaseous atoms is formed from the element under standard conditions.*

Thus, the standard enthalpy change of atomisation of hydrogen ($\Delta H_{at}^{\ominus}(\text{H}(g))$) refers to the process

$$\tfrac{1}{2}\text{H}_2(g) \rightarrow \text{H}(g) \quad \Delta H_{at}^{\ominus}(\text{H}(g)) = +218 \text{ kJ mol}^{-1}$$

A wide range of experimental techniques is available for determining the atomisation energies of elements. One technique widely used with gaseous diatomic elements is to find the minimum frequency of radiation (v) required to dissociate the gaseous molecules into atoms. For example:

$$\text{I}_2(g) \xrightarrow{E = hv} \text{I}(g) + \text{I}(g) \quad \Delta H = hv$$

Since the energy (E) of the radiation is given by hv, where h is Planck's Constant,

$$\Delta H_{at}^{\ominus}(\text{I}(g)) = \frac{\Delta H}{2} = \frac{hv}{2}$$

The enthalpy changes of atomisation of solid and liquid elements are more difficult to measure. For many elements, they can be obtained using specific heat capacities and molar enthalpy changes of fusion and vaporisation.

In discussing the heat changes in chemical reactions, it is useful to give each compound a definite heat content or enthalpy. For convenience, all elements are assigned a heat content of zero under standard conditions. The standard enthalpy change of formation of a compound then provides a measure of the heat content of the compound relative to its constituent elements. Remember, however, that these enthalpy values are *only* relative values. They give no information about the absolute energy content of a substance. This will depend on the potential energy in the electrical and nuclear interactions of the constituent particles plus the kinetic energy possessed by the atoms and sub-atomic particles.

The standard enthalpy change of formation of water ($\Delta H_f^{\ominus}(\text{H}_2\text{O(l)}) = -286 \text{ kJ mol}^{-1}$) tells us that, under standard conditions, water has a lower energy content than the hydrogen and oxygen from which it is formed (figure 12.4). On an atomic scale, we can imagine that some of the potential energy in the molecules of H_2 and O_2 has been converted to kinetic energy in the molecules of H_2O and this has resulted in a temperature rise. We can compare this with the conversion of potential energy to kinetic energy and heat when a stone rolls down a hill.

Enthalpy changes of combustion

One other enthalpy change of considerable importance is the standard enthalpy change of combustion. This is very relevant to the study of fuels and foods.

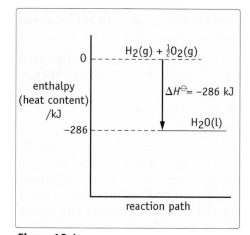

Figure 12.4
An energy level diagram for the standard enthalpy change of formation of water.

1 Why is it necessary to say 'completely burnt in oxygen' in the definition of $\Delta H_c^{\ominus}$?

2 Draw the enthalpy changes of combustion of both graphite and diamond on the same energy level diagram.

$$\Delta H_c^{\ominus}(\text{C(graphite)}) = -393 \text{ kJ mol}^{-1}$$
$$\Delta H_c^{\ominus}(\text{C(diamond)}) = -395 \text{ kJ mol}^{-1}$$

3 Which allotrope has the larger enthalpy (energy content)?

4 Which allotrope is the more stable?

5 What is the enthalpy change for the process

$$\text{C (graphite)} \rightarrow \text{C (diamond)}?$$

Figure 12.5
A bomb calorimeter.

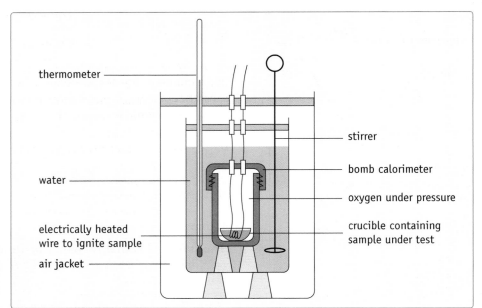

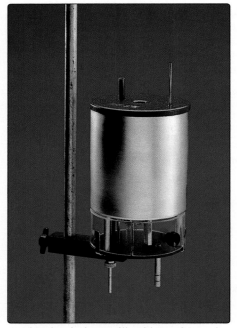

▲ A bomb calorimeter like this can be used to determine the energy values of food and fuels.

The **standard enthalpy change of combustion** of a substance, $\Delta H_c^{\ominus}$, is defined as *the enthalpy change when one mole of the substance is completely burnt in oxygen under standard conditions.*

For example, the standard enthalpy change of combustion of ethanol ($\Delta H_c^{\ominus}$ ($C_2H_5OH(l)$)) is -1368 kJ mol^{-1}, i.e.

$$C_2H_5OH(l) + 3O_2(g) \rightarrow 2CO_2(g) + 3H_2O(l) \qquad \Delta H^{\ominus} = -1368 \text{ kJ}$$

Review question 1 at the end of this chapter shows a simple apparatus in which the enthalpy change of combustion of a liquid fuel could be determined. The apparatus in figure 12.5 shows an instrument, known as a bomb calorimeter, used for accurate thermochemical determinations. Scientists use equipment such as this to determine the energy values of foods and fuels. The apparatus is specially designed to avoid heat losses by completely surrounding the 'bomb' with water. Heat losses can be eliminated altogether if the thermochemical investigation is coupled with an electrical calibration. First of all, the chemical reaction is carried out in the calorimeter and the temperature is plotted against time before, during and after the reaction (figure 12.6). The experiment is then repeated, but this time an electrical heating coil replaces the reactants. The current in the coil is carefully adjusted so as to give a temperature/ time curve identical to that obtained in the chemical reaction. By recording the current during the time of this electrical calibration, it is possible to calculate the electrical energy supplied with great accuracy. This electrical energy is exactly the same as the energy change in the chemical reaction. As it includes both the heat absorbed by the system and the heat lost from the system, it eliminates the need for a heat-loss correction.

The practical significance of enthalpy changes of combustion is clear enough. Fuel engineers and dieticians call them energy values (calorific values). The prices of fuels are closely related to their energy values. Energy-providing foods, such as sugar, are classified by the amount of heat they liberate when they are metabolised in the body.

12.5 Measuring standard enthalpy changes of formation

The standard enthalpy changes of formation of carbon dioxide, magnesium oxide and many other oxides can be measured directly using a bomb calorimeter similar to that discussed in the last section.

However, there are many compounds for which $\Delta H_f^{\ominus}$ cannot be measured directly. Consider, for example, tetrachloromethane (CCl_4). Graphite and chlorine do not combine readily, nor does CCl_4 decompose easily into its constituent elements. In other words, neither the reaction

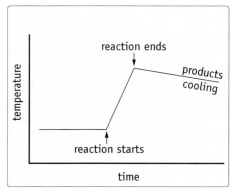

Figure 12.6
A graph of temperature against time for a reaction in a bomb calorimeter.

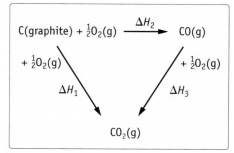

Figure 12.7
An energy cycle incorporating the enthalpy change of formation of carbon monoxide.

1. Write an equation for the standard enthalpy change of formation of carbon monoxide.

2. Why is it impossible to obtain $\Delta H_f^{\ominus}$ (CO(g)) directly?

$$C(graphite) + 2Cl_2(g) \rightarrow CCl_4(l)$$

nor the reaction

$$CCl_4(l) \rightarrow C(graphite) + 2Cl_2(g)$$

can be carried out in a calorimeter.

Boron(III) oxide (B_2O_3) and aluminium oxide (Al_2O_3) provide a different problem in attempting to measure their standard enthalpy changes of formation. In their case, it is difficult to burn the elements, boron and aluminium, completely in oxygen because a protective layer of oxide coats the unreacted element.

As a result of the problems involved in these measurements, chemists have had to obtain standard enthalpy changes of formation indirectly. One indirect method involves measuring the enthalpy change of combustion of the compound and the enthalpy changes of combustion of its constituent elements. These values can be linked in an energy cycle or an energy level diagram with the enthalpy change of formation of the compound. The energy cycle in figure 12.7 shows how this can be done for carbon monoxide. Notice that figure 12.7 shows two routes for converting graphite and oxygen to CO_2. One of these is the direct route straight from graphite and oxygen to CO_2. The alternative route goes via CO. It would seem reasonable that the overall enthalpy change for the conversion of graphite to carbon dioxide is independent of the route taken, so we can write

$$\Delta H_1 = \Delta H_2 + \Delta H_3 \qquad \text{Equation (1)}$$

$$\text{now } \Delta H_1 = \Delta H_c^{\ominus} \text{ (C(graphite))} \qquad = -393 \text{ kJ mol}^{-1},$$
$$\text{and } \Delta H_3 = \Delta H_c^{\ominus} \text{ (CO(g))} \qquad = -283 \text{ kJ mol}^{-1},$$
$$\text{so } \Delta H_2 = \Delta H_f^{\ominus} \text{ (CO(g))} \qquad = \Delta H_1 - \Delta H_3$$
$$= -393 - (-283)$$
$$= -110 \text{ kJ mol}^{-1}$$

The chemical processes and the different routes involved are represented in an energy level diagram in figure 12.8.

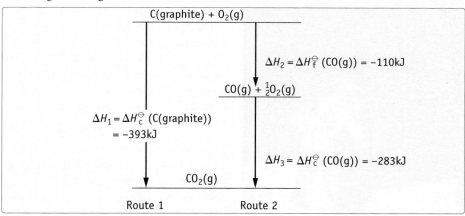

Figure 12.8
An energy level diagram incorporating the enthalpy change of formation of carbon monoxide.

The argument we used to obtain equation (1) above is a specific example of

Hess's Law of Constant Heat Summation. This says that *the energy change in converting reactants, A + B, to products, X + Y, is the same, regardless of the route by which the chemical change occurs, provided the initial and final conditions are the same.*

Of course, Hess's Law is simply an application of the more fundamental law of conservation of energy. If Hess's Law were not true, then we could create or destroy energy by going one way or the other round an energy cycle, such as that in figure 12.7.

We have used Hess's Law to find the standard enthalpy change of a reaction which cannot be measured directly. Before moving on, let us look at a more complex example involving butane (C_4H_{10}). Carbon and hydrogen will not react directly to produce butane. So, the formation of butane, represented by the equation.

$$4C(graphite) + 5H_2(g) \rightarrow C_4H_{10}(g)$$

cannot be carried out directly in a calorimeter. However, the standard enthalpy change of formation of butane can be obtained by measuring the standard enthalpy changes of combustion of butane, carbon and hydrogen. An energy level diagram for the reactions concerned is shown in figure 12.9.

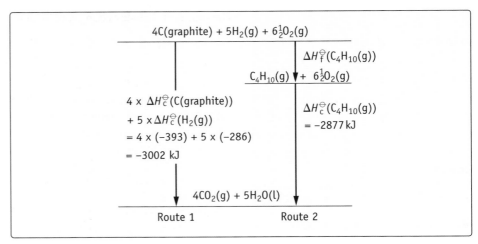

Figure 12.9
An energy level diagram used to obtain the enthalpy change of formation of butane indirectly.

Notice that route 1 involves four times the standard enthalpy change of combustion of graphite and five times the standard enthalpy change of combustion of hydrogen.

By Hess's Law,

$$\text{enthalpy change for route 1} = \text{enthalpy change for route 2}$$
$$\therefore -3002 = \Delta H_f^{\ominus}(C_4H_{10}(g)) - 2877$$
$$\Rightarrow \text{standard heat of formation of butane, } \Delta H_f^{\ominus}(C_4H_{10}(g)) = -125 \text{ kJ mol}^{-1}$$

12.6 Using standard enthalpy changes of formation to calculate the energy changes in reactions

One of the most important uses of $\Delta H_f^{\ominus}$ values is in calculating the enthalpy changes in chemical reactions.

The *Apollo 11* project landed the first man on the moon on 21 July 1969. During this project, engines of the lunar module used methylhydrazine (CH_3NHNH_2) and dinitrogen tetraoxide (N_2O_4). These liquids were carefully chosen since they ignite spontaneously and very exothermically on contact. How can we calculate the enthalpy change for the reaction?

First we write the equation for the reaction:

$$4CH_3NHNH_2(l) + 5N_2O_4(l) \rightarrow 4CO_2(g) + 12H_2O(l) + 9N_2(g)$$

Now draw an energy cycle by adding the formation equations from the same elements to both sides of the equation (figure 12.10).

By Hess's Law, the total enthalpy change for the formation of carbon dioxide, water and nitrogen will be the same whether they are formed directly from their elements or via the intermediates, CH_3NHNH_2 and N_2O_4.

$$\therefore 4\Delta H_f^{\ominus}(CH_3NHNH_2(l)) + 5\Delta H_f^{\ominus}(N_2O_4(l)) + \Delta H_{reaction}^{\ominus}$$
$$= 4\Delta H_f^{\ominus}(CO_2(g)) + 12\Delta H_f^{\ominus}(H_2O(l))$$

▲ Edwin Aldrin, the second man on the moon, steps from the ladder of the Apollo 11 Lunar Module.

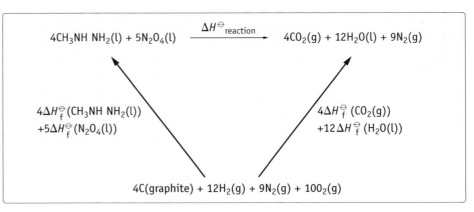

Figure 12.10
An energy cycle incorporating the reaction of methylhydrazine with dinitrogen tetraoxide.

12 Energy Changes and Bonding

The standard enthalpy changes of formation are:

$$\Delta H_f^{\ominus}(CH_3NHNH_2(l)) = +53 \text{ kJ mol}^{-1}$$
$$\Delta H_f^{\ominus}(N_2O_4(l)) = -20 \text{ kJ mol}^{-1}$$
$$\Delta H_f^{\ominus}(CO_2(g)) = -393 \text{ kJ mol}^{-1}$$
$$\Delta H_f^{\ominus}(H_2O(l)) = -286 \text{ kJ mol}^{-1}$$
$$\Rightarrow 4(+53) + 5(-20) + \Delta H_{reaction}^{\ominus} = 4(-393) + 12(-286)$$
$$\text{so, } \Delta H_{reaction}^{\ominus} = -5116 \text{ kJ}$$

Thus

$$4CH_3NHNH_2(l) + 5N_2O_4(l) \rightarrow 4CO_2(g) + 12H_2O(l) + 9N_2(g)$$
$$\Delta H^{\ominus} = -5116 \text{ kJ}$$

The enthalpy changes in other reactions can be calculated in a similar fashion to this by drawing an energy cycle for the reaction involved and for the reactions of the elements in forming both the reactants and products (figure 12.11).

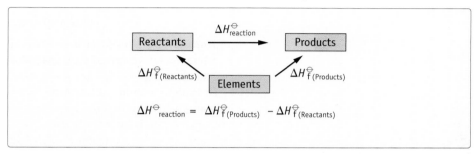

Figure 12.11
A general energy cycle to calculate the enthalpy change in any reaction.

12.7 Using standard enthalpy changes of formation to predict the relative stabilities of compounds

▲ The launch of Apollo 11 from the Kennedy Space Centre, Florida, on 16 July, 1969. Two of the crew members, Neil Armstrong and Edwin Aldrin, were the first people to step onto the Moon on 20 July, 1969. Apollo 11 was launched using a Saturn V rocket which burns 15 tonnes of kerosene/oxygen mixture every second for the first $2\frac{1}{2}$ minutes of the flight.

Most compounds are formed exothermically from their elements. Thus, the standard enthalpy changes of formation of water, carbon dioxide, aluminium oxide and many other compounds are negative. These compounds are therefore at a lower energy level than their constituent elements. This means that the compounds are energetically more stable than the elements from which they are formed. But, consider the following problem.

The standard enthalpy change of formation of hydrogen peroxide is -188 kJ mol^{-1}. From this, we would expect H_2O_2 to be stable. But, H_2O_2 decomposes fairly readily into water and oxygen. How can this be explained?

The answer lies in the fact that $\Delta H_f^{\ominus}(H_2O_2(l))$ only describes the stability of hydrogen peroxide *relative to its elements*:

$$H_2(g) + O_2(g) \rightarrow H_2O_2(l) \qquad \Delta H = -188 \text{ kJ}$$

H_2O_2 is obviously more stable than its elements, but on decomposition it produces not $H_2(g) + O_2(g)$ for which $\Delta H^{\ominus}$ is $+188$ kJ, but $H_2O(l) + \frac{1}{2}O_2(g)$ for which $\Delta H^{\ominus}$ is -98 kJ, i.e.

$$H_2O_2(l) \rightarrow H_2O(l) + \frac{1}{2}O_2(g) \qquad \Delta H = -98 \text{ kJ}$$

Thus, hydrogen peroxide is energetically stable with respect to its elements, but unstable with respect to water and oxygen. *This example shows how important it is to specify with respect to what substances a compound is stable or unstable.*

A few compounds, such as ethyne (C_2H_2), carbon disulphide (CS_2), and nitrogen oxide (NO), are formed endothermically from their elements:

$$2C(\text{graphite}) + H_2(g) \rightarrow C_2H_2(g) \qquad \Delta H^{\ominus} = +227 \text{ kJ}$$
$$\frac{1}{2}N_2(g) + \frac{1}{2}O_2(g) \rightarrow NO(g) \qquad \Delta H^{\ominus} = +90 \text{ kJ}$$

These compounds have positive standard enthalpy changes of formation. They are therefore energetically unstable with respect to their elements. So, why don't these compounds decompose instantaneously into their constituent elements?

Ethyne (acetylene), carbon disulphide and nitrogen oxide can all be stored for long periods at room temperature and pressure in the absence of a catalyst. They do, however, begin to decompose at high temperatures or in the presence of a catalyst. In order to explain the unexpected stability of these compounds, we must distinguish between *energetic* stability and *kinetic* stability.

Thus ethyne and nitrogen oxide are energetically unstable with respect to their elements. But at low temperatures and pressures the decomposition reactions are so slow that both ethyne and nitrogen oxide are kinetically stable. The kinetic stability of these energetically unstable compounds can be compared to the situation of a stone resting on a hillside. The stone is energetically unstable. Given the opportunity, it would roll to the bottom of the hill where it would come to rest in a position of lower energy. Resting on the hillside, stuck behind a tuft of grass, the stone's movement has been prevented. It is kinetically stable in spite of its energetic instability.

Diamond provides another example of an energetically unstable, yet kinetically stable, substance. At normal temperatures and pressures, diamond is unstable with respect to its allotrope, graphite:

$$C(\text{diamond}) \rightarrow C(\text{graphite}) \qquad \Delta H^\ominus = -2 \text{ kJ}$$

Fortunately, the rate of transformation of diamond to graphite is immeasurably slow at room temperature and so the diamond is kinetically stable. 'Diamonds are for ever', or are they?

The kinetic stability of nitrogen oxide with respect to its elements shows, yet again, how important it is to state clearly with respect to what a substance is stable. Obviously, nitrogen oxide is kinetically stable with respect to nitrogen and oxygen, but in the presence of air or oxygen it is energetically and kinetically unstable with respect to nitrogen dioxide. Hence, nitrogen oxide reacts rapidly with oxygen to form brown fumes of nitrogen dioxide:

$$2NO(g) + O_2(g) \rightarrow 2NO_2(g)$$

12.8 Predicting whether reactions will occur

The enthalpy change of a reaction is sometimes used as a rough guide to the likelihood that the reaction will occur. If $\Delta H^\ominus$ for a reaction is negative, energy is lost when the reaction occurs. The products are more stable than the reactants.

Thus, exothermic reactions are more likely to occur than endothermic reactions. Reactions which occur spontaneously are often very exothermic.

Although the value of $\Delta H^\ominus$ can provide some indication of the likelihood of a reaction, there are limitations to its use and it is important to bear these in mind.

1 $\Delta H^\ominus$ shows the relative energetic stabilities of the reactants and products for a reaction. It says nothing about the *kinetic* stability of the products relative to the reactants. Thus, $\Delta H^\ominus$ *is no guide to the rate of a reaction*. It cannot tell us whether the reaction is fast or slow. A reaction may be enormously exothermic, yet nothing happens. This is because the reaction rate is immeasurably slow and the reactants are kinetically stable with respect to the products (for example, a mixture of hydrogen and oxygen at room temperature). Reaction rates are studied in more detail in chapter 24.

2 In order to make accurate predictions concerning the relative energy levels of reactants and products, it is necessary to consider the energy lost or gained by the reacting system, and also any energy changes inside that system. For example, when a gas is produced in a reaction or a solid dissolves in a liquid, there is a marked increase in the disorder of the system itself and an increase in the number of ways in which the energy is distributed in the system. Consequently, in predicting the likelihood of any reaction we should take into account the change in order (or disorder) introduced into the system. At normal temperatures, this additional disorder is usually

▲ The reaction between paper (carbohydrate) and oxygen in the air is very exothermic. Paper and oxygen are energetically very unstable with respect to their possible products (CO_2 and water). Fortunately, however, the reactants are kinetically stable and it is safe to read the paper at leisure.

unimportant. Even so, it explains why certain endothermic reactions happen so spontaneously. For example, many solids dissolve in water readily and easily in spite of the fact that the process of dissolving is endothermic. The reason for this is that, although the enthalpy change is positive, there is an enormous increase in disorder (or entropy) as the solid dissolves, causing the reaction to occur spontaneously. Entropy is discussed further in chapter 23.

3 Predictions from $\Delta H_f^{\ominus}$ values relate only to standard conditions, i.e. 298 K and atmospheric pressure. The situation may be very different under different conditions or in the presence of a catalyst.

> ▯ The enthalpy change of combustion of octane (in petrol) is -5513 kJ mol^{-1}. Would you expect petrol to react with oxygen?
>
> ▯ Why are petrol/oxygen mixtures stable at room temperature before a spark is applied?
>
> ▯ How does the spark initiate a reaction between petrol and oxygen?

12.9 Enthalpy changes of combustion and molecular structure

Look closely at the information in table 12.1. Notice the similarity in the enthalpy changes of combustion for butane and methylpropane. Perhaps the similarity is not surprising. After all, they contain the same atoms and the same bonds.

Table 12.1
The formulas, bonds and enthalpy changes of combustion of butane and methylpropane

	Butane	Methylpropane
Molecular formula	C_4H_{10}	C_4H_{10}
Structural formula	H H H H | | | | H—C—C—C—C—H | | | | H H H H	H H H | | | H—C—C—C—H | | | H | H H—C—H | H
Bonds	3 C—C 10 C—H	3 C—C 10 C—H
$\Delta H_c^{\ominus}$/kJ mol^{-1}	-2877	-2869

This suggests that each kind of bond makes a fixed contribution to the total enthalpy change. This idea is confirmed when we inspect the enthalpy changes of combustion of the simpler alkanes (table 12.2). Two important points emerge from the data in table 12.2.

First, there is a regular difference in structure of one CH_2 group as we pass from one alkane to the next. Secondly, there is a regular difference of about 650 kJ in the value of $\Delta H_c^{\ominus}$ from one alkane to the next. This suggests that

- each additional CH_2 group is responsible for a fixed increment in the enthalpy change of combustion,
- each bond makes a characteristic contribution to the overall energy content of a substance.

What happens to the different bonds when an alkane burns? During combustion, the bonds between the atoms in the alkane molecule are broken. New bonds form between carbon and oxygen in CO_2 and between hydrogen and oxygen in H_2O. Notice from the structural formulas in table 12.2 that each alkane has one C—C bond and two C—H bonds more than the previous alkane. Check this for yourself.

Table 12.2
The formulas and enthalpy changes of combustion of the simpler alkanes

Alkane	Molecular formula	Structural formula	$\Delta H_c^{\ominus}$ /kJ mol^{-1}	Difference in $\Delta H_c^{\ominus}$ /kJ
Methane	CH_4	H | H— C —H | H	−890	
				670
Ethane	C_2H_6	H H | | H—C— C—H | | H H	−1560	
				660
Propane	C_3H_8	H H H | | | H—C—C—C—H | | | H H H	−2220	
				657
Butane	C_4H_{10}	H H H H | | | | H—C—C—C—C—H | | | | H H H H	−2877	
				643
Pentane	C_5H_{12}	H H H H H | | | | | H—C—C—C—C—C—H | | | | | H H H H H	−3520	

Obviously, energy must be supplied in order to break a bond between two atoms. Consequently, energy is released when the reverse occurs and a bond forms.

So, *bond-breaking is an endothermic process whereas bond-making is exothermic.*

Chemical reactions involve bond-breaking followed by bond-making. This means that the enthalpy change of a reaction is the energy difference between bond-breaking and bond-making processes (figure 12.12). In section 12.11, we shall see how information concerning the strengths of bonds can be used to calculate the enthalpy changes of reactions.

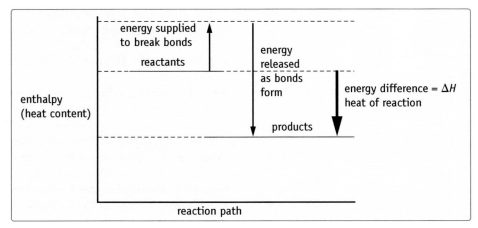

Figure 12.12
The enthalpy change of a reaction is the energy difference between bond-breaking and bond-making processes.

As the alkane molecules in table 12.2 get larger, more energy is required to break the bonds between their carbon and hydrogen atoms. But, even larger amounts of energy are released as these atoms form carbon dioxide and water. On burning, each alkane molecule forms one CO_2 molecule and one H_2O molecule more than the previous alkane.

$$CH_4 + 2O_2 \rightarrow CO_2 + 2H_2O$$
$$C_2H_6 + 3\tfrac{1}{2}O_2 \rightarrow 2CO_2 + 3H_2O$$
$$C_3H_8 + 5O_2 \rightarrow 3CO_2 + 4H_2O$$

Thus, the enthalpy change of combustion gets larger by a constant amount as we progress along the homologous series of alkanes. The exothermic reactions show that the energy released on making the bonds in CO_2 and H_2O is greater than the energy required to break the bonds in the alkanes and O_2. An exothermic change is therefore evidence for the formation of stronger bonds. Indeed, the energy we get from burning such fuels as coal, oil and natural gas results from the formation of strong $C{=}O$ and $H{-}O$ bonds in CO_2 and H_2O, respectively.

12.10 Finding the strength of the C—H bond and the C—C bond

From our discussions in the last section it would seem that a definite quantity of energy, known as the **bond energy**, may be associated with each type of bond. This energy is absorbed when the bond is broken and evolved when the bond is formed. In order to find the strength of the C—H bond, it would seem appropriate to study methane. This contains C—H bonds only. Consider the following equation which relates to the enthalpy change of atomisation of methane:

$$\overset{\displaystyle H}{\underset{\displaystyle H}{H{-}\overset{|}{\underset{|}{C}}{-}H}}(g) \rightarrow C(g) + 4H(g) \qquad\qquad \text{Equation (2)}$$

1 How many C—H bonds are broken during the atomisation of 1 mole of methane?
2 Suppose ΔH for this process in Equation (2) is x kJ mol^{-1}. What is the bond energy of one C—H bond in kJ mol^{-1}?

The energy change in equation (2), $\Delta H^{\ominus}_{at}(CH_4(g))$, is simply the breakage of four C—H bonds. The C—H bond energy is usually written as $E(C{-}H)$, so

$$\Delta H^{\ominus}_{at}(CH_4(g)) = 4\,E(C{-}H).$$

This process can be related to the elements carbon and hydrogen using the energy cycle in figure 12.13.

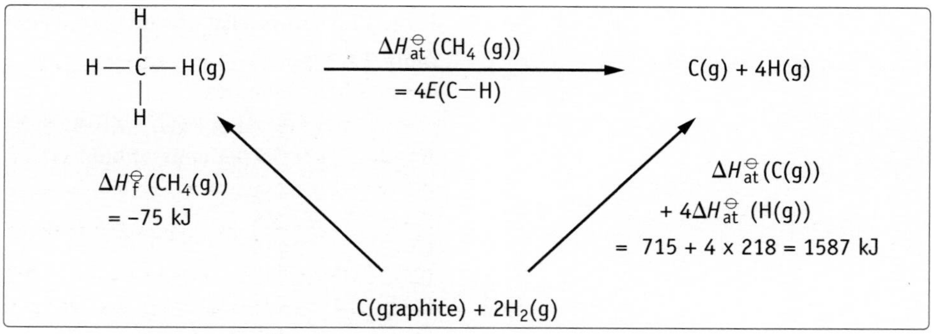

Figure 12.13
An energy cycle which can be used to find the C—H bond energy.

From figure 12.13, $\Delta H^{\ominus}_{at}(CH_4(g)) = 4E(C{-}H)$
$$= -(-75) + 1587 \text{ kJ}$$
$$= +1662 \text{ kJ}$$
$$\Rightarrow 4E(C{-}H) = 1662 \text{ kJ}$$
∴ the average C—H bond energy in methane, $E(C{-}H) = 415.5$ kJ mol^{-1}

If we assume that the C—H bonds in ethane have the same strength as those in methane, we can now calculate the carbon–carbon bond energy, $E(C—C)$. Figure 12.14 shows the relevant enthalpy changes in an energy cycle.

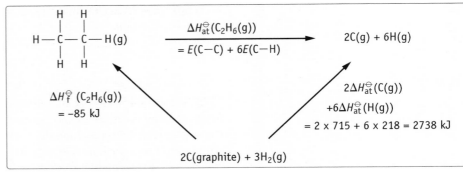

Figure 12.14
An energy cycle which can be used to find the C—C bond energy.

Equating the two routes from $C_2H_6(g)$ to $2C(g) + 6H(g)$, we can write:

$$E(C—C) + 6E(C—H) = -(-85) + 2738 \text{ kJ}$$
$$\Rightarrow E(C—C) + 6E(C—H) = 2823 \text{ kJ}$$

Assuming that $E(C—H) = 415.5 \text{ kJ mol}^{-1}$ as in methane,

$$E(C—C) + 6 \times 415.5 = 2823 \text{ kJ}$$
$$\Rightarrow E(C—C) = 2823 - 2493 = 330 \text{ kJ}$$

∴ **the C—C bond energy in ethane, $E(C—C) = 330 \text{ kJ mol}^{-1}$**

Proceeding in this way, we could calculate the strengths of other bonds. The idea that bond energies are additive and transferable from one molecule to another holds very well in most cases. However, results show small deviations in the strength of one particular type of bond in different molecules.

For example, the bond energy for C—C has been determined using several different compounds. The results range from 330 to 346 kJ mol^{-1}. This shows that a bond energy value depends to some extent on the compound from which it was determined and upon the environment of the bond within the molecule. Thus, the strength of a C—C bond varies slightly with the nature of the different atoms that are attached to the two carbon atoms.

12.11 What are the uses of bond energies?

The concept of bond energies is rather artificial because it does not relate to any everyday processes. It is nevertheless very useful. Four of its important uses are:

(a) comparing the strengths of bonds,
(b) understanding structure and bonding,
(c) estimating the enthalpy changes in reactions, and
(d) understanding the mechanisms of chemical reactions.

Table 12.3 shows a list of some bond energies.

Table 12.3
Average bond energies

Bond	Bond energy, $E(X—Y)$ /kJ per mole of bonds	Bond	Bond energy, $E(X—Y)$ /kJ per mole of bonds
C—H	413	O—O	146
C—C	346	O=O	497
C=C	610	N—N	163
C≡C	835	N≡N	945
C—F	495	N—H	390
C—Cl	339	O—H	463
C—Br	280	H—F	565
C—I	230	H—Cl	431
F—F	158	H—Br	365
Cl—Cl	242	C—O	360
Br—Br	193	C=O	740
I—I	151	Si—O	464
Si—Si	226		

The most important application of bond energies is probably their use in estimating the enthalpy changes in chemical reactions. These estimates are particularly useful when calorimetric or other experimental measurements cannot be made.

The following example illustrates the usefulness of bond energies very well.

Hydrazine is often used as a rocket fuel because it can be stored conveniently as a liquid and it reacts very exothermically with oxygen, forming purely gaseous products.

$$N_2H_4(g) + O_2(g) \rightarrow N_2(g) + 2H_2O(g) \qquad \Delta H^{\ominus} = -622 \text{ kJ}$$

It has been suggested that hydrazine/fluorine mixtures might be more exothermic than hydrazine/oxygen mixtures. Using bond energies from table 12.3, we can estimate ΔH for the reaction of hydrazine with fluorine:

$$N_2H_4(g) + 2F_2(g) \rightarrow N_2(g) + 4HF(g)$$

The following equation shows clearly those bonds which are broken and those which form.

$$
\begin{array}{c}
\text{H} \qquad \text{H} \\
\backslash \qquad / \\
\text{N---N} \quad + 2\text{F---F} \rightarrow \text{N} \equiv \text{N} + 4\text{H---F} \\
/ \qquad \backslash \\
\text{H} \qquad \text{H}
\end{array}
$$

Bonds broken		Bonds made	
one	N—N	one	N≡N
four	N—H	four	H—F
two	F—F		

1. Calculate the energy required to break the bonds in this reaction.
2. Calculate the energy evolved when the bonds form in the products.
3. What is the value of $\Delta H^{\ominus}$ for this reaction per mole of hydrazine?
4. Is the reaction with fluorine more or less exothermic than the reaction with oxygen?
5. If you had to decide whether to use oxygen or fluorine for the rocket flight, what other factors besides the exothermicity of the reaction would influence your final decision?

Your estimate of $\Delta H^{\ominus}$ for the hydrazine/fluorine reaction should be −1166 kJ. For most reactions, the values of ΔH estimated from bond energies agree closely with experimentally determined values. This has further established the usefulness of bond energies.

Sometimes there are small differences in enthalpy changes calculated from average bond energies compared with those determined by experiment. These small differences result from variations in the strength of one particular bond in different molecules.

Occasionally, the estimated values for ΔH disagree strongly with the experimental results. This has resulted in a re-examination of the structure of some compounds and an improvement in our understanding of bonding. A striking example of this is provided by benzene and other aromatic compounds. At one time, molecules of benzene were thought to contain hexagons of carbon atoms linked by alternate double and single bonds (figure 12.15).

When cyclohexene, containing one C=C bond, undergoes an addition reaction with hydrogen, the enthalpy change is −120 kJ mol^{-1}.

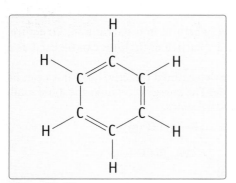

Figure 12.15
The Kekulé structure for benzene.

$$+ H_2 \longrightarrow \qquad \Delta H = -120 \text{ kJ}$$

This would suggest that the addition of hydrogen to Kekulé's benzene, containing three C=C bonds, should have an enthalpy change of −360 kJ. When experiments are performed, the heat of hydrogenation of benzene is found to be −208 kJ mol^{-1} and not −360 kJ mol^{-1}.

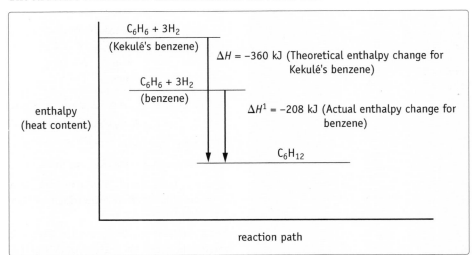

It would seem that one mole of benzene is 152 kJ more stable than the Kekulé structure would suggest. This increased stability is emphasised in figure 12.16.

As a result of this discrepancy, chemists were forced to re-examine the structure of benzene. It is now known that all six C—C bonds in its molecule are equal in length and in strength. The molecule does not contain alternate double and single bonds as Kekulé had proposed. Furthermore, this new structure for benzene has improved our understanding of its properties and reactions and those of other aromatic compounds. The structure of benzene is discussed further in section 28.2

Figure 12.16
An energy level diagram to emphasise the unexpected stability of benzene.

12.12   Energy changes in forming ionic substances

In the last section, we found that bond energies were extremely useful in helping us to understand the structure and properties of molecular substances. Bond energies for molecular substances can be compared with **lattice enthalpies** (lattice energies) for ionic compounds.

> **The lattice enthalpy of an ionic crystal** is *the enthalpy change of formation for one mole of the ionic compound from gaseous ions under standard conditions.* Thus, the lattice enthalpy of sodium chloride corresponds to the process:
>
> $$Na^+(g) + Cl^-(g) \rightarrow Na^+Cl^-(s) \quad \Delta H_{latt}(Na^+Cl^-(s))$$

As one might expect, lattice enthalpies are very helpful in discussing the structure, bonding and properties of ionic compounds. Lattice enthalpies can be used as a measure of ionic bond strength.

Unfortunately, lattice enthalpies cannot be determined directly, but values can be obtained indirectly by means of an energy cycle. The energy cycle links the ionic solid, the gaseous ions and the elements in their standard states.

$$Na(s) + \tfrac{1}{2}Cl_2(g) \xrightarrow{\Delta H_x} Na^+(g) + Cl^-(g)$$
$$\Delta H_f(Na^+Cl^-(s)) \searrow \qquad \swarrow \Delta H_{latt}(Na^+Cl^-(s))$$
$$Na^+Cl^-(s)$$

Since ΔH_f, the enthalpy change of formation of sodium chloride, can be measured conveniently in a calorimeter, ΔH_{latt} can be obtained if ΔH_x can be found.

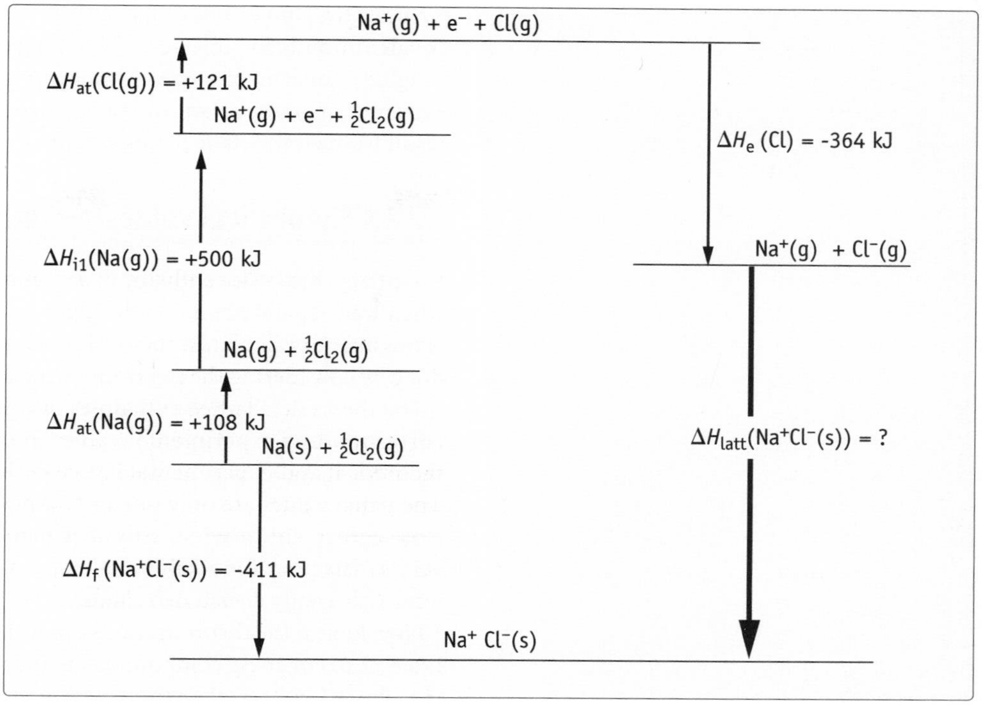

Figure 12.17
The Born–Haber cycle for sodium chloride.

In figure 12.17, the previous energy triangle has been extended to show the various stages involved in finding ΔH_x. The complete energy cycle, as shown in figure 12.17, is usually called a **Born–Haber Cycle**.

Notice that

$$\Delta H_x = \Delta H_{at}(Na(g)) + \Delta H_i(Na(g)) + \Delta H_{at}(Cl(g)) + \Delta H_e(Cl)$$

The first two stages in this process involve atomising and then ionising sodium. The enthalpy change of atomisation of sodium

$$Na(s) \xrightarrow{\Delta H_{at}(Na(g))} Na(g)$$

can be obtained from its enthalpy change of fusion, enthalpy change of vaporisation and specific heat capacity.

The first ionisation energy of sodium,

$$Na(g) \xrightarrow{\Delta H_{i1}(Na(g))} Na^+(g) + e^-$$

can be determined from spectra measurements (section 6.2). The third and fourth stages, in the expression ΔH_x, involve the atomisation of chlorine followed by the conversion of gaseous chlorine atoms to chloride ions. The latter process is, of course, called the electron affinity for chlorine (section 8.3).

The enthalpy change of atomisation of chlorine,

$$\tfrac{1}{2}Cl_2(g) \xrightarrow{\Delta H_{at}(Cl(g))} Cl(g)$$

can be obtained from spectroscopic studies (section 12.4), whilst the electron affinity for chlorine,

$$Cl(g) + e^- \xrightarrow{\Delta H_e(Cl)} Cl^-(g)$$

can be found by valve measurements.

The lattice enthalpy for sodium chloride can now be obtained since all other values in the Born–Haber cycle can be determined experimentally.

Using the values in figure 12.17,

$$\Delta H_{latt}(Na^+Cl^-(s)) = -(-364) - 121 - 500 - 108 - 411 \text{ kJ mol}^{-1}$$
$$= -776 \text{ kJ mol}^{-1}$$

This lattice enthalpy gives us some idea of the force of attraction between Na^+ and Cl^- ions in crystalline sodium chloride. The lattice enthalpies of various other ionic crystals are listed in table 12.4.

Table 12.4
Lattice enthalpies of some ionic solids

Compound	Lattice enthalpy/ kJ mol^{-1}
NaF	−915
NaCl	−776
NaBr	−742
NaI	−699
MgCl$_2$	−2489
MgO	−3933

1 Why does the lattice enthalpy become more exothermic along the series NaI, NaBr, NaCl, NaF?

2 The interionic distance between Na^+ and F^- ions in NaF is very similar to that between Mg^{2+} and O^{2-} ions in MgO. Why, then, is the lattice enthalpy of MgO about four times more exothermic than that of NaF? (The high lattice enthalpy of MgO results in its use as a refractory lining.)

Notice the similar size of the lattice enthalpies in table 12.4 with the bond energies of covalent bonds in table 12.3. This suggests that ionic bonds and covalent bonds are roughly similar in strength. Although ionic and covalent bonds are often regarded as extremes, it is important to realise their underlying similarity. They both, of course, result from electrostatic forces, but involve different distributions of electrons.

12.13 Theoretical values for lattice enthalpies – the ionic model

Essentially, the lattice enthalpy of an ionic compound is the energy change which occurs when well-separated ions are brought together in forming the crystal. Consequently, it is possible to calculate a theoretical value for the lattice enthalpy of a crystal. This is done by considering the interionic attractions and repulsions within the lattice.

The theoretical lattice enthalpies of some ionic substances are compared with their corresponding experimental values in table 12.5. Notice the similarity between the theoretical and experimental lattice enthalpies for the three alkali metal halides. The calculated values are only one or two per cent less than the experimental values. This close agreement provides strong evidence that the simple model of an ionic lattice is very satisfactory for alkali metal halides. This model is pictured as discrete spherical ions with evenly distributed charge.

Now look at the theoretical and experimental lattice enthalpies for the silver halides in table 12.5. For these compounds the theoretical values are approximately 130 kJ mol^{-1} (i.e. about 15%) less than the experimental values. In this case, the simple ionic model is not completely satisfactory and requires some modification.

Table 12.5
Theoretical and experimental lattice enthalpies

Compound	Theoretical lattice enthalpy/kJ mol^{-1}	Experimental lattice enthalpy (obtained via Born–Haber cycle) /kJ mol^{-1}
NaCl	−766	−776
NaBr	−731	−742
NaI	−686	−699
AgCl	−768	−890
AgBr	−759	−877
AgI	−736	−867

When the difference in electronegativity between the ions in the crystal is large, as in the case of alkali metal halides, the ionic model is satisfactory. However, when the difference in electronegativity is smaller, as in the case of silver halides, there is a significant disagreement between experimental results and those calculated in terms of a simple ionic model. The experimental values suggest that the bonding in silver halides is stronger than the ionic model predicts. The explanation of this anomaly is that the bonding is not purely ionic but is intermediate in character between ionic and covalent. The partly covalent nature of the bonds can be interpreted by saying that the ionic bonds have been polarised (section 9.2) or by suggesting that the electrons are incompletely transferred in forming the ions. This further emphasises the fact that ionic and covalent bonds are simply extreme types. The bonding in most substances is intermediate in character between purely ionic and purely covalent.

12.14 Solution, hydration and lattice enthalpies

When ionic solids dissolve in water, heat is usually evolved or absorbed. Why is this? Can we explain the enthalpy changes in terms of the processes taking place at a molecular and ionic level?

When sodium chloride is dissolved in water, the overall change can be represented as:

$$NaCl(s) + (aq) \rightarrow Na^+(aq) + Cl^-(aq)$$

For one mole of the solute and the formation of an infinitely dilute solution, this process is described as the **enthalpy change of solution** of sodium chloride, $\Delta H_{soln}(NaCl(s))$.

$$\Delta H_{soln}(NaCl(s)) = +5 \text{ kJ}$$

In order to understand this enthalpy change, the overall process of solution can be divided into two distinct stages which are shown in figure 12.18. First, the solid ionic crystal is separated into gaseous ions. This is the reverse of the lattice enthalpy process,

$$NaCl(s) \rightarrow Na^+(g) + Cl^-(g) \qquad -\Delta H_{latt} = +776 \text{ kJ}$$

Secondly these gaseous ions are solvated by water molecules which involves the **hydration enthalpy**, ΔH_{hyd}.

$$Na^+(g) + Cl^-(g) + (aq) \rightarrow Na^+(aq) + Cl^-(aq)$$

The first of these stages is, of course, always endothermic since it involves separating the ions of the solute. The second stage, on the other hand, is always exothermic since it involves the attraction of ions in the solute for water molecules. The overall enthalpy change on solution will depend on whether the endothermic or the exothermic stage has the larger enthalpy change.

In the case of sodium chloride, the enthalpy change of the endothermic process is marginally greater than the enthalpy change of the exothermic process. So the enthalpy change of solution has a small positive value. The lattice enthalpy and the hydration enthalpy for ionic substances are nearly always large values. The enthalpy change of solution, which is the difference between these two values, is positive in some cases and negative in others. The relationship between these three quantities is:

$$\Delta H_{soln} = -\Delta H_{latt} + \Delta H_{hyd}$$

Check this for yourself in figure 12.18.

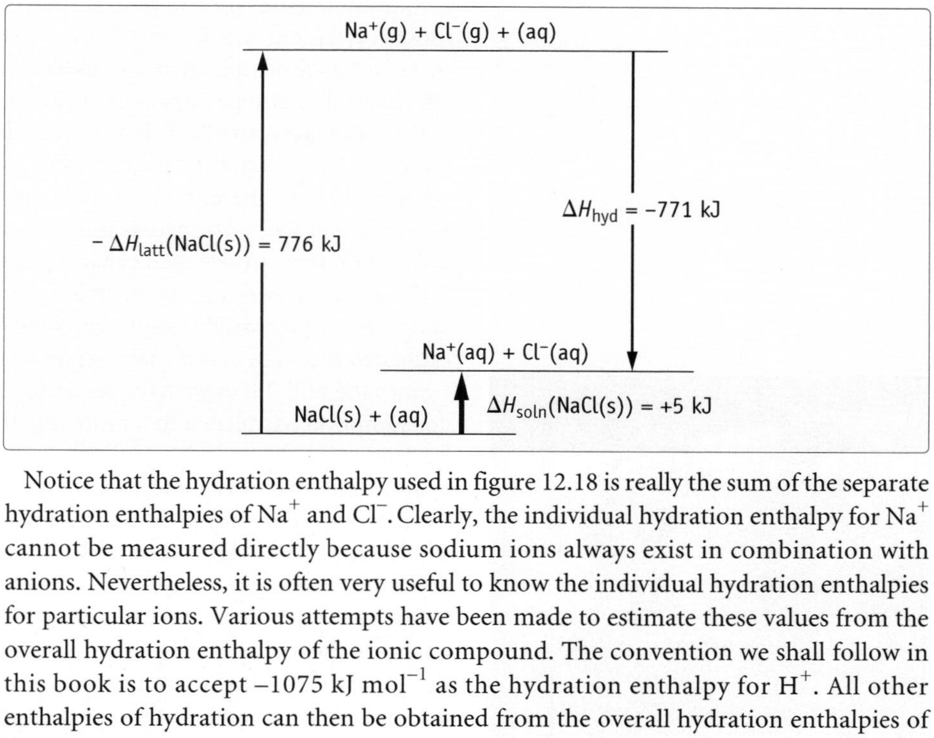

Figure 12.18
The relationship between lattice enthalpy, hydration enthalpy and enthalpy change of solution.

Look closely at table 12.6.
1 Why are the hydration enthalpies of *both* anions and cations negative?
2 What is the hydration enthalpy of $MgCl_2$?
3 Why does the hydration enthalpy get less exothermic along the series F^-, Cl^-, Br^-, I^-?
4 Why does the hydration enthalpy get more exothermic along the series Na^+ Mg^{2+}, Al^{3+}?

Table 12.6
Enthalpies of hydration of some ions

Ion	ΔH_{hyd} /kJ mol^{-1}	Ion	ΔH_{hyd} /kJ mol^{-1}
H^+	−1075	F^-	−457
Li^+	−499	Cl^-	−381
Na^+	−390	Br^-	−351
K^+	−305	I^-	−307
Mg^{2+}	−1891		
Ca^{2+}	−1562		
Al^{3+}	−4613		

Notice that the hydration enthalpy used in figure 12.18 is really the sum of the separate hydration enthalpies of Na^+ and Cl^-. Clearly, the individual hydration enthalpy for Na^+ cannot be measured directly because sodium ions always exist in combination with anions. Nevertheless, it is often very useful to know the individual hydration enthalpies for particular ions. Various attempts have been made to estimate these values from the overall hydration enthalpy of the ionic compound. The convention we shall follow in this book is to accept −1075 kJ mol^{-1} as the hydration enthalpy for H^+. All other enthalpies of hydration can then be obtained from the overall hydration enthalpies of the compounds.

A few individual hydration enthalpies are given in table 12.6. It does, of course, follow from what we have just said that the hydration enthalpies of compounds can be obtained by addition of the hydration enthalpies of their constituent ions.

12.15 Energy sources for the future

Until the Industrial Revolution, civilisations depended for their energy on the labour of people and animals or on the harnessing of wind and water.

Then, in the early nineteenth century, society began to use coal in large quantities as a source of heat and power in their homes and industries. Civilisation progressed beyond the level of a simple agrarian society. At first, the use of coal was very limited, but as the Industrial Revolution gained momentum towards the middle of the century, the consumption of coal began to double every fifteen years.

▶ Before the Industrial Revolution, societies had to depend on the labour of people as their supply of energy. This engraving shows the construction of the pyramids.

Early in the twentieth century, oil and natural gas began to make an impact. Discoveries multiplied fast and oil became plentiful, cheap and a source of many other products. Without oil, the internal combustion engine would have been impossible and the revolution in land, sea and air transport could never have taken place. Oil became the most convenient fuel for many industrial and domestic purposes and the basic raw material for the organic chemicals industry. Its consumption rose by leaps and bounds. By 1960, oil had outstripped coal as the major source of energy in most industrial countries.

A decade later, the first danger signals appeared as the rate of oil consumption increased faster than the discovery of new reserves.

Unfortunately, there is only a finite amount of oil on the Earth. In one decade from 1970 to 1980 we consumed as much oil as in the previous one hundred years. If we continue at this rate, we shall exhaust our supplies of oil well before the year 2050.

The situation with coal is much less alarming, since its rate of consumption is lower and reserves are possibly twenty times greater than those of oil. Nevertheless, it has been estimated that 80% of our coal reserves will have disappeared by the year 2500.

Since the mid-1980s we have begun to conserve our resources. Our thoughtless over-consumption has turned to a more intelligent use of fuel. We have reduced excessive heating and lighting in homes, offices, schools and factories. It has been shown that more than 20% of the energy used in heating buildings could be saved by better insulation of roofs and walls, and by double glazing windows. We have begun to accept slightly lower, but still comfortable, indoor temperatures. It is vital, however, that we continue to conserve our resources of fuels.

One of the most wasteful users of fuel is the private car in which only 10–20% of the energy in the fuel ultimately goes towards moving the car. Hopefully, batteries will one day supersede the inefficient internal combustion engine. Vast quantities of precious oil and gas are also being squandered on industrial and domestic heating and on the generation of electricity. Oil and gas are much too valuable for this purpose. They should be conserved as feedstock for essential chemicals such as plastics, paints, pesticides and pharmaceuticals. Natural gas is needed for cooking. A more suitable fuel for power generation is coal. In spite of these factors, it is likely that any new power stations built in the UK in the next decade or so will be gas-fired or oil-fired rather than coal-burning or nuclear. Indeed UK government predictions indicate that, if no nuclear stations are built, the electricity generated from nuclear reactors will fall from a level of 18% in the late 1980s to only 1% by the year 2020.

▲ Cutting peat, another form of fossil fuel, for domestic heating. Peat is still used as a source of fuel in some areas. It consists of plant fibres and mud and is probably the first stage in the formation of coal.

12 Energy Changes and Bonding

Faced with these problems of dwindling energy resources and limited capital resources, it is essential to examine all possibilities for future energy provision. At the present time, hydroelectric schemes supply less than 1% of the world's electricity, yet there are possibilities for expansion in many areas. The production of hydroelectricity is essentially limited in its location. It has only been developed so far in those areas which are both mountainous and industrially advanced. It has been estimated that the full development of all suitable areas could increase the production of hydroelectricity by 20 times.

► Huge turbines in a nuclear power station. Heat produced from nuclear reactions generates steam to drive the turbines.

▲ Water flooding from the turbine house below the dam at the hydroelectric power station at the Warsak Dam in Pakistan.

A less obvious source of water power lies in the rise and fall of the tides. The enormous movement of water between high and low tides results from the moon's gravitational pull on the seas. By damming tidal basins, it has been possible to generate electricity using low-speed turbines. One such installation is across the estuary of the Rance in Northern Brittany. There are many other places in the world, including the Bristol Channel, with potential for such development.

Another way in which we can conserve fossil fuels is to develop the use of renewable fuels from plants. These projects are usually on a small scale, but the use of biofuels does not deplete our finite energy sources. Biogas generators are increasingly used in less-developed, rural areas, whilst the production of biodiesel from rape seeds is becoming significant.

By far the biggest source of our energy is radiation from the sun. Scientists have calculated that one-tenth of the solar energy falling on the Arizona Desert could provide enough energy for the whole of the USA. Unfortunately, we make very little direct use of solar energy. This is because technology has not yet provided an economic method of transforming solar radiation into useable energy. Solar cells are still very expensive. The most promising possibility for solar energy is the heating of buildings. Although sunlight is intermittent and unreliable in many areas, developments in the storage of solar energy may help to solve this problem. Alternatively, solar energy can be used in conjuction with conventional heating systems. This could save considerable quantities of our fossil fuels.

The sun's inexhaustible source of energy comes from nuclear fusion. The origin of this energy is similar to that of the hydrogen bomb. Matter, below the sun's surface, is crushed together under a pressure millions upon millions of times greater than atmospheric pressure. Under this immense pressure hydrogen atoms (protons) collide and fuse forming deuterium and releasing a positron (like a positive electron) and lots of energy.

$$^1_1H + ^1_1H \rightarrow ^2_1H + ^0_1e$$

Deuterium atoms then fuse with hydrogen atoms to form helium-3 atoms and release more energy.

$$^2_1H + ^1_1H \rightarrow ^3_2He$$

Finally helium-3 atoms fuse to form helium-4 with the release of two protons and yet more energy.

$$^3_2He + ^3_2He \rightarrow ^4_2He + ^1_1H + ^1_1H$$

As a result of this fusion process, the energy radiated from the sun's surface in one year is equivalent to 250 thousand million million million tonnes of oil. If only we could harness the energy from such a fusion process on earth, we would have solved our energy problems for ever. It is impossible to predict when, or indeed if ever, controlled fusion will be achieved. The major problem is that it requires temperatures of about 100 000 000°C before atomic collisions are vigorous enough to cause fusion. Although such elevated temperatures have not yet been reached, experiments have achieved temperatures close to those required.

In the meantime, however, a world of limitless energy is still only a dream and we must learn to use and conserve our finite resources more wisely and more efficiently.

▲ A yacht uses its spinnaker to obtain power from the wind.

Summary

1 The transfer of energy to or from chemicals plays a crucial part in chemical processes in industry and in living things.

2 In an exothermic reaction, heat is lost from the reacting materials and $\Delta H_{reaction}$ is negative. In an endothermic reaction, heat is gained by the reacting materials and $\Delta H_{reaction}$ is positive.

3 If m g of a substance (specific heat capacity = c J g^{-1} K^{-1}) increase in temperature by ΔT K (ΔT °C), the enthalpy change, $\Delta H = + m \times c \times \Delta T$ J.

4 The standard enthalpy change of a reaction is the amount of heat absorbed or evolved when the molar quantities of reactants as stated in the equation react together under standard conditions.

5 Standard conditions for thermochemical measurements are:

a pressure of one atmosphere,
a temperature of 298 K,
substances in their normal physical states for these conditions,
concentrations of 1 mol dm^{-3}.

6 The standard enthalpy change of formation of a substance, $\Delta H^{\ominus}_f$, is the heat evolved or absorbed when one mole of the substance is formed from its elements under standard conditions.

7 The enthalpy change of atomisation of an element, $\Delta H^{\ominus}_{at}$, is the enthalpy change when one mole of gaseous atoms is formed from the element under standard conditions.

8 The standard enthalpy change of combustion of a substance, $\Delta H^{\ominus}_c$, is the enthalpy change when one mole of the substance is completely burnt in oxygen under standard conditions.

9 Hess's Law of Constant Heat Summation says that the energy change in converting reactants, A + B, to products, X + Y, is the same regardless of the route by which the chemical change occurs.

10 A substance is said to be stable if it tends neither to decompose nor to react spontaneously.

11 The enthalpy change of formation of a compound provides a measure of the energetic stability of that compound relative to its elements. If $\Delta H^{\ominus}_f$ is negative, the compound is more stable than its elements.

12 It is important to distinguish between energetic stability and kinetic stability. Compounds such as ethyne (C_2H_2) are energetically unstable with respect to their elements. But, they do not change at ordinary temperatures and pressures because their reaction rate is immeasurably slow. They are said to be kinetically stable.

13 Bond-breaking is an endothermic process whereas bond-making is exothermic.

14 A definite quantity of energy can be associated with each type of covalent bond. This energy is absorbed when the bond is broken and evolved when the bond is formed. Thus, the bond energy for the C—Cl bond, E(C—Cl), is defined as the energy required to break one mole of C—Cl bonds forming uncharged products under standard conditions.

15 Bond energies for molecular substances can be compared with lattice enthalpies for ionic compounds. Lattice enthalpies give the enthalpy change of formation of one mole of the ionic crystal from gaseous ions under standard conditions.

16 The close agreement between calculated and experimental lattice enthalpies for many ionic compounds provides strong evidence that the simple model of an ionic lattice composed of discrete spherical ions with evenly distributed charge is generally very satisfactory.

12 Energy Changes and Bonding

1 Figure 12.19 shows a diagram of the apparatus used by a student to determine the enthalpy change of combustion of ethanol. The heat produced by the burning fuel warms a known mass of water. By measuring the mass of fuel burnt and the temperature rise of the water, it is possible to obtain an approximate value for the enthalpy change of combustion of the fuel.

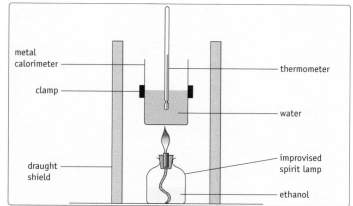

Figure 12.19
A simple apparatus to determine the heat of combustion of a liquid fuel.

Volume of water in calorimeter	$= 400 \text{ cm}^3$
Initial temperature of water	$= 12°C$
Final temperature of water	$= 22°C$
Mass of ethanol burnt	$= 0.92 \text{ g}$
(Specific heat capacity of water	$= 4.2 \text{ J g}^{-1} \text{ K}^{-1}$)

(a) How much heat is required to raise the temperature of the water from 12°C to 22°C?
(This is the amount of heat produced when 0.92 g of ethanol burn.)

(b) How much heat would be produced when 1 mole of ethanol burns?
(Call this the heat of combustion of ethanol.)

(c) Why is the answer to (b) not described as the *standard* enthalpy change of combustion of ethanol?

(d) An accurate value for $\Delta H_c^{\ominus} (C_2H_5OH(l))$ is −1368 kJ mol^{-1}. Mention four serious errors in the simple experiment which could be responsible for the poor result.

2 Ethanol (C_2H_5OH) cannot be prepared directly from its elements so the standard enthalpy change of formation of ethanol must be obtained indirectly.

(a) Define the terms: $\Delta H_f^{\ominus}$ and $\Delta H_c^{\ominus}$.

(b) Write an equation for the formation of ethanol from its elements in their standard states.

(c) Draw an energy cycle linking the enthalpy change of formation of ethanol with its enthalpy change of combustion and the enthalpy changes of combustion of its constituent elements.

(d) Calculate $\Delta H_f^{\ominus} (C_2H_5OH(l))$.

$\Delta H_c^{\ominus} (C(\text{graphite}))$	$= −393 \text{ kJ mol}^{-1}$
$\Delta H_c^{\ominus} (H_2(g))$	$= −286 \text{ kJ mol}^{-1}$
$\Delta H_c^{\ominus} (C_2H_5OH(l))$	$= −1368 \text{ kJ mol}^{-1}$

3 Two campers are desperately short of GAZ, yet they badly need a hot drink. They estimate that they have the equivalent of 1.12 dm^3 of GAZ (measured at 0°C and 1 atm) in their 'gas bottle'.

(a) What is the maximum volume of water (at 20°C) which they could boil in order to make some hot coffee?
(Assume that GAZ is pure butane, $\Delta H_c(C_4H_{10}(g)) = −3000$ kJ mol^{-1} and that 75% of the heat evolved in burning the GAZ is absorbed by the water.)

(b) State any other assumptions you make.

4 (a) Write equations, including state symbols, for the complete combustion of glucose ($C_6H_{12}O_6(s)$) and ethanol ($C_2H_5OH(l)$).

(b) Calculate the enthalpy change during the fermentation of glucose.

$$C_6H_{12}O_6(s) \rightarrow 2C_2H_5OH(l) + 2CO_2(g)$$

Assume that the enthalpy changes of combustion of glucose and ethanol are −2820 and −1368 kJ mol^{-1}, respectively.

(c) In breweries, the vats in which fermentation is carried out are sometimes fitted with copper pipes to promote cooling during fermentation. Why is this?

(d) With your answer to part (c) in mind, why is it that people who brew beer at home usually put their fermentation vessel in a warm place (such as an airing cupboard)?

5 A possible mechanism for the reaction of fluorine with methane is

$$CH_4 + F_2 \xrightarrow{\text{slow}} CH_3\cdot + HF + F\cdot$$

$$CH_3\cdot + F\cdot \xrightarrow{\text{fast}} CH_3F$$

(a) Use the bond energies in table 12.3 to calculate the enthalpy change in each step of the reaction mechanism.

(b) Is the suggested mechanism viable? Explain your answer.

(c) Write the equations for a reaction between CH_4 and Cl_2 assuming a similar mechanism to that for CH_4 and F_2 and calculate the enthalpy change in each step.

(d) Is such a mechanism viable for the CH_4/Cl_2 reaction? Explain your answer.

(e) Why is it that fluorine will react with methane in the dark, whereas chlorine only reacts appreciably with methane in sunlight?

6 (a) A simple rule regarding the solubility of solutes in solvents is 'Like dissolves like'. In general, non-polar solutes dissolve readily in non-polar solvents but not in polar ones. On the other hand, polar solutes dissolve readily in polar solvents but are insoluble in non-polar solvents. Discuss and explain this pattern in solubility, mentioning specific examples and referring to solute–solute, solvent–solvent and solute–solvent attractions.

(b) Explain the terms lattice enthalpy, hydration energy and enthalpy change of solution with reference to the hypothetical substance, X^+Y^-.

(c) Draw an energy diagram relating the three terms in part (b).

(d) Calculate the hydration energy of potassium iodide, assuming that its enthalpy change of solution is $+21$ kJ mol^{-1} and its lattice enthalpy is -642 kJ mol^{-1}.

7 (a) Explain what is meant by the terms
   (i)   ionisation energy,
   (ii) atomisation energy,
   (iii) lattice enthalpy.

(b) Draw a complete, fully labelled Born–Haber Cycle for the formation of potassium bromide.

(c) Using the information in the table below, calculate the lattice enthalpy of potassium bromide.

Reaction	ΔH/kJ mol^{-1}
$K(s) + \frac{1}{2}Br_2(l) \rightarrow K^+Br^-(s)$	-392
$K(s) \rightarrow K(g)$	$+90$
$K(g) \rightarrow K^+(g) + e^-$	$+420$
$\frac{1}{2}Br_2(l) \rightarrow Br(g)$	$+112$
$Br(g) + e^- \rightarrow Br^-(g)$	-342

(d) The values of the lattice enthalpies of the other potassium halides are:

Compound	KF	KCl	KI
Lattice enthalpy/kJ mol^{-1}	-813	-710	-643

What explanation can you give for the trend in these values?

8 The enthalpy changes involved in the synthesis of calcium oxide are represented in a Born–Haber cycle below. (The numerical values printed beside the cycle are in kJ mol^{-1}.)

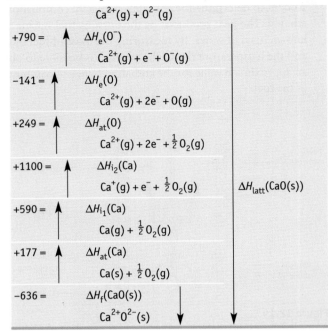

(a) Calculate the lattice enthalpy for calcium oxide, $\Delta H_{latt}(CaO(s))$.

(b) Why is $\Delta H_{i_2}(Ca)$ greater than $\Delta H_{i_1}(Ca)$?

(c) Why is $\Delta H_e(O)$ negative whereas $\Delta H_e(O^-)$ is positive?

(d) State and explain how the value of the first ionisation energy of magnesium would compare with the corresponding value for calcium.

Patterns Across the Periodic Table

13

13.1 Periodic properties

In chapter 4 we noticed that a large number of chemical and physical properties of the elements vary periodically with atomic number. *This idea of periodicity embodied in the periodic table is one of the major unifying themes in chemistry.* It provides a structure for our knowledge and understanding of inorganic chemistry. It also reduces the need for memorising isolated pieces of factual information.

In this chapter, we shall look more closely at the trends and gradations in properties of the elements in periods 2 and 3. We shall also look at their more important compounds. Several of the bulk physical properties of the elements, such as melting point and density, depend upon their structure and bonding. We discussed these relationships in chapter 4. However, the structure and bonding of the elements are in turn related to atomic properties such as electron structures, ionisation energies and atomic radii. It is therefore profitable to discuss the trends in these atomic properties and then relate them to physical and chemical properties. Table 13.1 provides a summary of the structure, bonding and properties of the elements in periods 2 and 3.

Table 13.1
A summary of the structure, bonding and properties of the elements in periods 2 and 3

Period 2	Li	Be		B		C		N	O	F	Ne
Period 3	Na	Mg	Al		Si			P	S	Cl	Ar
Type of element		metal			metalloid				non-metal		
Structure	Metallic			Giant molecular				Simple molecular			
	Close-packed or body-centred cubic arrangement of atoms			Infinite lattice structure				Discrete small molecules (e.g. P_4, Cl_2, O_2, Ne)			
Bonding	Metallic: strong forces of attraction of positive ions for mobile outer electrons			Covalent: very strong forces of attraction between atoms due to the attraction of nuclei for shared electrons				Molecular (Van der Waals): weak forces between molecules (strong covalent forces hold atoms together within the molecule)			
Co-ordination number	8 or 12			4 or less				–			
Properties m.pt./b.pt.	High			Very high				Low			
$\Delta H_{fus}/\Delta H_{vap}$	High			Very high				Low			
Molar volume	Moderate			Low				High			
Conductivity	Good			Poor				Nil			

13.2 Atomic structure and the periodic table

Chemists now realise that the atomic number of an element dictates the number and arrangement of its electrons.

Table 13.2 shows the electron (shell and sub-shell) structures for the elements in periods 2 and 3.

Table 13.2
The electron structures of the elements in periods 2 and 3

Period 2	Li	Be	B	C	N	O	F	Ne
Electron shell structure	2, 1	2, 2	2, 3	2, 4	2, 5	2, 6	2, 7	2, 8
Electron sub-shell structure	$1s^2 2s^1$	$1s^2 2s^2$	$1s^2 2s^2 2p^1$	$1s^2 2s^2 2p^2$	$1s^2 2s^2 2p^3$	$1s^2 2s^2 2p^4$	$1s^2 2s^2 2p^5$	$1s^2 2s^2 2p^6$

Period 3	Na	Mg	Al	Si	P	S	Cl	Ar
Electron shell structure	2, 8, 1	2, 8, 2	2, 8, 3	2, 8, 4	2, 8, 5	2, 8, 6	2, 8, 7	2, 8, 8
Electron sub-shell structure	$1s^2 2s^2 2p^6$ $3s^1$	$1s^2 2s^2 2p^6$ $3s^2$	$1s^2 2s^2 2p^6$ $3s^2 3p^1$	$1s^2 2s^2 2p^6$ $3s^2 3p^2$	$1s^2 2s^2 2p^6$ $3s^2 3p^3$	$1s^2 2s^2 2p^6$ $3s^2 3p^4$	$1s^2 2s^2 2p^6$ $3s^2 3p^5$	$1s^2 2s^2 2p^6$ $3s^2 3p^6$

Elements in the same group of the periodic table have similar electron configurations. For example, Group I elements (Li to Fr) have one electron in their outer shells (ns^1). Group II elements (Be to Ra) have two electrons in their outer shell (ns^2) and Group VII elements (F to At) have seven outer shell electrons ($ns^2 np^5$).

Since electron configurations recur in a periodic manner, it is not surprising that properties show a similar periodic recurrence. We shall investigate the similarities in the properties of elements within a *group* more thoroughly in chapters 15, 16 and 17. Meanwhile, we can investigate the variation of atomic properties across a period.

13.3 Atomic properties and the periodic table

Atomic radii

The electron cloud of an atom has no definite limit. Because of this, the size of an atom cannot be defined simply and easily. However, *the radius of an atom is often defined as the distance of closest approach to another identical atom.* This means that the size of an atom is determined by the effective radius of the outer electrons. But, what do we mean when we say the atomic radius of iodine in the solid is 0.133 nm? Is it half the distance between the two iodine atoms in an I_2 molecule or half the distance between two iodine atoms in adjacent molecules which are not chemically bonded (figure 13.1)? The following terms clarify this situation. *Half the distance*

Figure 13.1
Covalent (atomic) and Van der Waals radii for iodine.

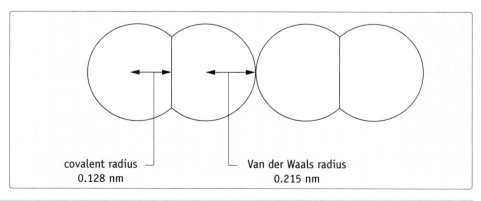

covalent radius
0.128 nm

Van der Waals radius
0.215 nm

between two covalently bonded iodine atoms is defined as the **covalent radius** *for iodine. Half the distance between two iodine atoms which are not chemically bonded is defined as the* **Van der Waals radius**. The former gives a measure of the length of a covalent bond. The latter gives a measure of the length of a Van der Waals bond. The covalent radius for iodine is 0.128 nm and the Van der Waals radius is 0.215 nm. Clearly, the type of bonding greatly influences the size of an atom. When chemists refer to *atomic radii* they are usually thinking of *covalent radii*.

The size of atoms in metallic elements can be obtained by determining the inter-nuclear distance in the metallic crystal using X-ray diffraction and then dividing by two to give an atomic radius.

Figure 13.2 shows the atomic radii (covalent radii) of the elements Li—F and Na—Cl in the periodic table. Notice that along each period there is a gradual decrease in atomic size as the outer electron shell is being filled.

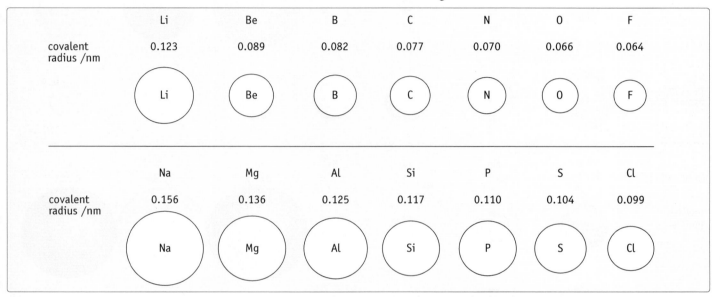

Figure 13.2
Covalent (atomic) radii of the elements Li to F and Na to Cl.

1. What causes the general decrease in atomic size across a period?
2. How is the attraction of the positive nucleus for the outer electrons changing across the period?

Moving from one element to the next across a period, electrons are being added to the same shell at about the same distance from the nucleus. At the same time, protons are being added to the nucleus. Therefore, the electrons are attracted and pulled towards the nucleus by an increasing positive charge, so the radius of the atom decreases. However, the rate of decrease in the radius becomes smaller as the atoms get heavier. The addition of one more proton to the 11 already present in sodium causes a greater proportional change in nuclear attractive power than the addition of one more proton to the 16 already present in sulphur. Thus, the atomic radius falls by 0.020 nm from Na to Mg, but by only 0.005 nm from S to Cl.

Ionic radii

So far we have considered only atomic sizes. Ionic sizes are also useful in explaining and understanding chemical properties. They provide a measure of the space occupied by an ion in the crystal lattice. Figure 13.3 shows the radii of the most stable ions for the first 20 elements in the periodic table.

Look for the following general patterns in figure 13.3.

(a) *The ionic radii of positive ions are smaller than their corresponding atomic radii.*

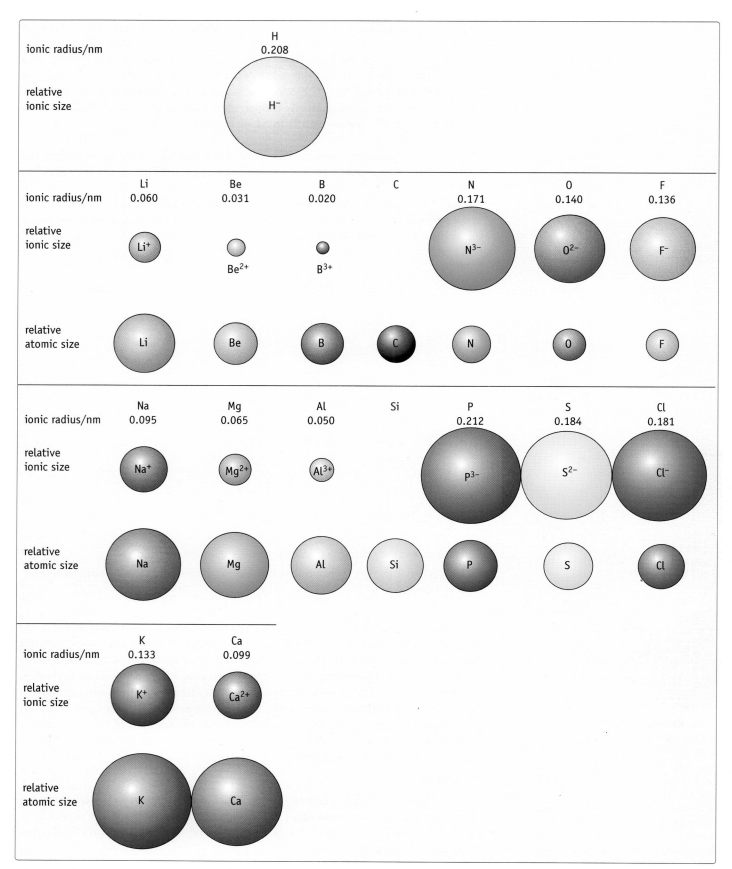

Figure 13.3
Radii of the most stable ions for the first 20 elements in the periodic table and the relative sizes of their atoms and ions.

1 Which electrons are usually removed from metals in forming their positive ions?

2 How does the electronic structure of a positive ion (e.g. Na^+ or Mg^{2+}) compare with that of the corresponding metal atom?

3 Why is the radius of a positive ion less than that of the corresponding atom?

(b) *The ionic radii of negative ions are greater than their corresponding atomic radii.* When one or more electrons are added to the outer shell of an atom forming a negative ion, there is an increase in the repulsion between negative charge clouds. This results in an overall increase in size.

(c) *In a series of ions with the same number of electrons (an isoelectronic series), the ionic radius decreases as the atomic number increases.* In figure 13.4 the solid lines link those ions which are isoelectronic.

Figure 13.4
Variation in ionic size with atomic number.

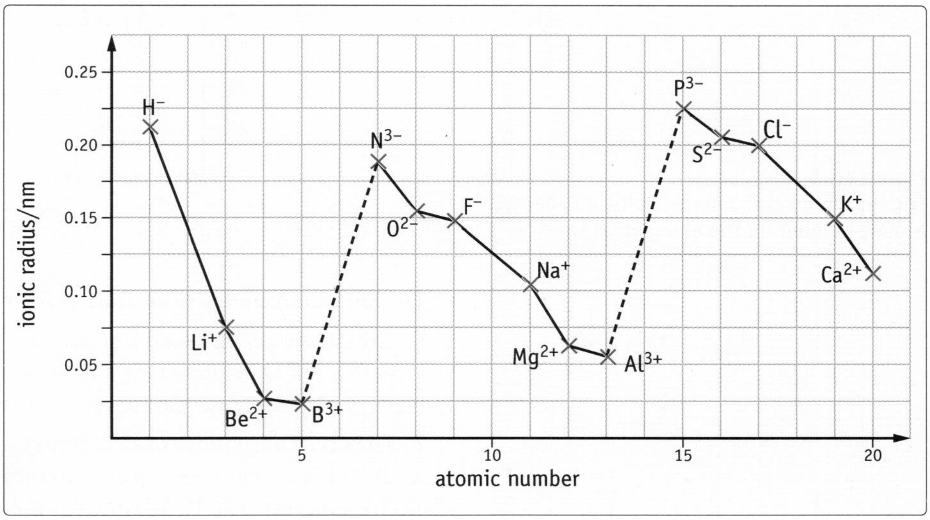

Why does ionic size decrease along these isoelectronic sequences? The nuclear charge increases along the isoelectronic series (e.g. from N^{3-} to Al^{3+}). This makes the electron cloud contract, because it is pulled in more effectively by an increasing positive charge.

Ionic radii provide a measure of ionic size in a *solid* crystal, but they give no accurate indication of the size of *aquated* ions in solution. Here the situation can be very different. The increasing ionic size from Li^+ through Na^+ to K^+ might lead us to expect the electrical mobility to decrease in the order Li^+, Na^+, K^+. In practice, K^+(aq) is the most mobile and Li^+(aq) is the least mobile of the three aqueous ions. This is due to solvation by water molecules. The Li^+ ion has the largest charge density per unit of surface area and therefore attracts polar solvent (water) molecules around it more strongly than the Na^+ ion. Similarly, Na^+ exerts more attraction on polar solvent molecules than K^+. Thus, the effective size of the aqueous ions is Li^+(aq) > Na^+(aq) > K^+(aq) and this makes Li^+(aq) least mobile.

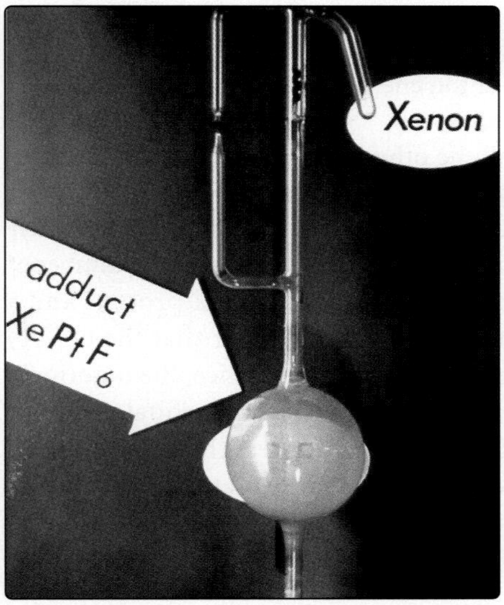

▲ Neil Bartlett produced the first noble gas compound in 1962. Whilst carrying out research he allowed oxygen in the air to react with PtF_6 producing $O_2^+Pt\,F_6^-$. He quickly realised that PtF_6 had oxidised O_2 and, as the first ionisation energy of xenon (+1170 kJ mol^{-1}) was similar to that of molecular oxygen (+1175 kJ mol^{-1}), it should be possible to make $Xe^+\,PtF_6^-$. Within days he had produced a yellow/orange solid which was $XePtF_6$.

Ionisation energy (Ionisation enthalpy)

The striking periodic variation in the first ionisation energies of the elements was discussed in section 6.5. Figure 13.5 shows a graph of the first ionisation energy against atomic number for the first 20 elements. Notice that within a period the first ionisation energy tends to rise as atomic number increases. This increase in ionisation energy is associated to some extent with a decrease in metallic character from left to right in the periodic table. Metals with mobile outer-shell electrons have low ionisation energies. The ionisation energy gradually rises across the period as metallic character disappears and electrons are more tightly held in covalent bonds. But, how are ionisation energies related to atomic structure?

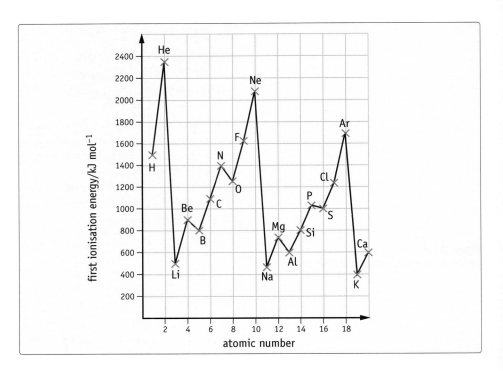

Figure 13.5
The first ionisation energies plotted against atomic number for the elements H to Ca.

The ionisation energy of an atom is strongly influenced by three atomic parameters.

1 *The distance of the outermost electron from the nucleus*
   As this distance increases, the attraction of the positive nucleus for the negative electron will decrease. Consequently the ionisation energy will decrease.

2 *The size of the positive nuclear charge*
As the nuclear charge gets more and more positive, its attraction for the outermost electron increases and consequently the ionisation energy increases.

3 *The screening (shielding) effect of inner electrons*
   The outermost electrons are repelled by all the other electrons in the atom besides being attracted by the positive nucleus. Chemists say that the outermost electron is **screened** (shielded) from the attraction of the nucleus by the repelling effect of inner electrons. For example, the outermost electron in an atom with an atomic number of Z (i.e. Z protons in the nucleus) experiences an attraction from Z protons and a repulsion from $(Z - 1)$ electrons. At first sight, it might appear that the $(Z - 1)$ electrons could cancel all but one of the nuclear charges. In practice, the outermost electron occupies an orbital which is not completely outside the orbitals of other inner electrons. So the screening (shielding) effect is much less than perfect. This means that the electron experiences an overall effective nuclear attraction which is much greater than that from one proton. In general, however, *the screening effect by inner electrons is more effective the closer these inner electrons are to the nucleus.* Thus,

(a) electrons in shells of lower principal quantum number are more effective
     shields than electrons in shells of higher quantum number.
(b) electrons in the same shell exert a negligible shielding effect on each other.

This means that we need only consider inner shells of electrons when we discuss the screening effect on an outermost electron.

Moving from left to right across any period, there is a general increase in the first ionisation energy because:

• the nuclear charge is increasing,
• the atomic radius is decreasing, and
• the screening effect remains almost the same for elements in the same period.

Look closely at elements Li to Ne in figure 13.5. Notice that the ionisation energy of beryllium is higher than that of boron.

1 Which other element in period 2 has a higher first ionisation energy than the element immediately after it?
2 The electron structure of beryllium is $1s^2 2s^2$. Write out the electron structure of boron in the same notation.
3 Which of the two elements, beryllium or boron, has the more stable electron structure?

The electron configuration of beryllium is $1s^2 2s^2$, whereas that of boron is $1s^2 2s^2 2p^1$. All the sub-shells in beryllium are filled, but the outer sub-shell of boron contains only one electron. From our studies in chapter 6, we know that filled electron shells are associated with extra stability. *There is also some extra stability associated with filled sub-shells.* This means that the electron structure in beryllium is rather more stable than we might have expected and its first ionisation energy is greater than that of boron.

A similar situation arises with nitrogen, which has a higher first ionisation energy than oxygen. The electron structures of nitrogen and oxygen are $1s^2 2s^2 2p^3$ and $1s^2 2s^2 2p^4$, respectively. The half-filled $2p$ sub-shell in nitrogen with one electron in each of the three $2p$ orbitals and its evenly distributed charge is more stable than the $2p$ sub-shell in oxygen which contains four electrons. This results in a higher first ionisation energy for nitrogen than oxygen.

Electronegativity

Elements that tend to acquire electrons in their chemical interactions are said to be **electronegative**. The electronegativity of an atom provides a numerical measure of the power of that atom in a molecule to attract electrons (section 9.2). The electronegativity values for main-group elements (based on Pauling's scale) are given in table 13.3. These values for the first 20 elements in the periodic table are plotted graphically against atomic number in figure 13.6.

H							He
2.1							—

Li	Be		B	C	N	O	F	Ne
1.0	1.5		2.0	2.5	3.0	3.5	4.0	—
Na	Mg		Al	Si	P	S	Cl	Ar
0.9	1.2		1.5	1.8	2.1	2.5	3.0	—
K	Ca		Ga	Ge	As	Se	Br	Kr
0.8	1.0		1.6	1.8	2.0	2.4	2.8	—
Rb	Sr		In	Sn	Sb	Te	I	Xe
0.8	1.0		1.7	1.8	1.9	2.1	2.5	—
Cs	Ba		Tl	Pb	Bi	Po	At	Rn
0.7	0.9		1.8	1.8	1.9	2.0	2.2	—
Fr	Ra							
0.7	0.9							

Table 13.3
Electronegativities for the main-group elements.

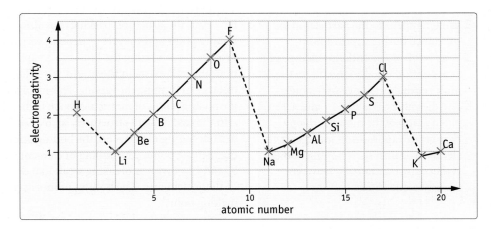

Figure 13.6
Variation in electronegativity with atomic number.

Notice that electronegativities decrease down a group, but increase across a period. As expected, the most electronegative elements are the reactive non-metals in the top right-hand corner of the periodic table. In contrast, the least electronegative elements are the reactive metals in the bottom left-hand corner. Why does electronegativity increase from left to right across the periodic table? Moving from one element to the next across a period, the nuclear charge increases by one unit and one electron is added to the outer shell. As the positive charge on the nucleus rises, the atom has an increasing electron-attracting power and therefore an increasing electronegativity.

13.4 Physical properties and the periodic table

Melting point

The melting point of a substance is the temperature at which pure solid is in equilibrium with pure liquid at atmospheric pressure. Moving across a period from left to right, the melting point rises through the metals and metalloids and then drops abruptly to low values for the non-metals (see figure 4.7). Although there is some evidence of periodicity in the melting points of elements, it is not so obvious as the periodicity of other properties. This is partly because there are several factors which influence a melting point. Melting points depend on both the structure (packing) and the bonding in a substance. Since there are abrupt changes in bonding and structure across a period, there are also abrupt changes in the melting point. For example, in the period Na→Ar, the melting point rises sharply from the last metallic element, aluminium, to the giant molecule, silicon. It then falls even more abruptly from silicon to the simple molecule phosphorus. In the period Li→Ne, the melting point rises steeply through the metals lithium and beryllium and on through the giant molecules boron and carbon. It then falls sharply to the simple molecules starting at nitrogen.

Molar enthalpy change of fusion

The molar enthalpy change of fusion is the amount of energy required to melt one mole of the solid at its melting point. As expected, there is a close similarity in the periodicity of the enthalpy changes of fusion and the periodicity of the melting points (figure 4.7). When a marked change of structure or bonding occurs, there is a sharp change in the ΔH_{fus} in the same way that we noticed sharp changes in the melting point in the last sub-section.

Consider the changes in ΔH_{fus} across the third period from Na to Ar. In metals (Na, Mg, Al) a good deal of the bonding remains when the metal liquefies. Only restricted bond-breaking has occurred, so the enthalpy change of fusion is only moderate. In giant molecules such as silicon, nearly all the bonds must be broken before the solid melts. Therefore the enthalpy change of fusion is very large. In simple molecules (P_4, S_8, Cl_2 and Ar), the enthalpy changes of fusion are very small because melting in these elements involves only the breaking of non-polar (Van der Waals) forces.

Boiling point and molar enthalpy changes of vaporisation

The periodic trends in this case can be related to those in the last two sub-sections. However, the structure (packing) has already been disrupted in forming the liquid, so, the boiling point and the ΔH_{vap} depend very much on the strength of the bonds which are broken in converting the liquid to a vapour. In the case of metals, most of the metallic bonding still exists in the liquid state. The atoms must be separated to a considerable distance in forming the vapour. This involves complete breakage of the metal bonds. Thus, the boiling points of metals are much higher than their melting points.

In the case of giant structures, such as carbon and silicon, most of the bonds have been broken during melting. So, the boiling point, though very high, is not much higher than the melting point. The boiling of elements with simple molecular structures, such as P_4, S_8 and Cl_2, involves breaking weak Van der Waals bonds. This is relatively easy for most of these elements. The boiling points occur at low temperatures, not much above the melting point.

1 What happens to the particles of a substance during vaporisation?

2 Why is the molar enthalpy change of vaporisation usually larger than the molar enthalpy change of fusion?

3 Use the values in tables 4.1 and 4.2 to plot a graph of the molar enthalpy changes of vaporisation for at least 10 elements against their boiling points. Is there any relationship between these two quantities?

Density

The density of a substance shows its mass per unit volume. Gaseous elements present a problem in comparing densities. With these elements, it is usual to take the density of the element as a liquid. What factors will influence the density of an element? The factors to consider are atomic mass, atomic radius (which determines the volume of each atom) and crystal structure (which determines how close to each other the atoms are packed). Although each of these factors varies from one element to the next across a period, there is a periodic trend in density (see figure 4.8).

Across a period there is a general increase in density until a maximum is reached at Group III or IV for periods 2 and 3 and at the end of the transition metals for the later periods. This increase in density can be related to the increasing atomic mass and decreasing atomic radius. A maximum is reached when the atomic radius reaches a low value and the strength of the metallic or giant covalent bonding is at a maximum. In spite of having higher atomic masses and lower atomic radii, elements at the right-hand side of each period have relatively low densities. This is because they form simple molecular structures in which molecules are only weakly bonded by Van der Waals forces.

Molar volume

The molar volume of an element is the volume occupied by 1 mole (6×10^{23} atoms) of the element. It is very dependent on atomic radius and structure (packing). Suppose the relative atomic mass of the element is A_r and its density is ρ g cm^{-3}.

6×10^{23} atoms of the element have a mass of A_r g

$\therefore 6 \times 10^{23}$ atoms of the element have a volume of $\dfrac{A_r}{\rho}$ cm^3

$$\therefore \text{molar volume} = \frac{A_r}{\rho} = \frac{\text{relative atomic mass in grams}}{\text{density}}$$

Since molar volume is *inversely related to density*, it is not surprising that trends in molar volume are the *reverse* of trends in density. Thus molar volumes gradually fall across a period to a minimum value and then rise again (figure 4.8). However, the major factor affecting volume is structure rather than atomic radius. So, there are abrupt changes in molar volume across a period where there are obvious changes in structure. For example, the atomic radius of phosphorus is less than that of silicon as expected, but phosphorus has the larger molar volume because it contains weakly bonded P_4 molecules whereas silicon is a strongly bonded giant molecule.

13.5 Patterns in the formulas of compounds

Table 13.4 shows the formulas of the chlorides, oxides and hydrides of the elements in periods 2 and 3.

Table 13.4
Formulas of the chlorides, oxides and hydrides of the elements in periods 2 and 3

Period 2	Li	Be	B	C	N	O	F	Ne
Formula of chloride	LiCl	$BeCl_2$	BCl_3	CCl_4	NCl_3	Cl_2O	ClF	—
Formula of oxide	Li_2O	BeO	B_2O_3	CO	N_2O	O_2	OF_2	—
				CO_2	NO			
					NO_2			
					N_2O_3			
					N_2O_4			
					N_2O_5			
Formula of simplest hydride	LiH	BeH_2	BH_3	CH_4	NH_3	H_2O	HF	—
Period 3	Na	Mg	Al	Si	P	S	Cl	Ar
Formula of chloride	NaCl	$MgCl_2$	Al_2Cl_6	$SiCl_4$	PCl_3	SCl_2	Cl_2	—
					PCl_5	S_2Cl_2		
Formula of oxide	Na_2O	MgO	Al_2O_3	SiO_2	P_4O_6	SO_2	Cl_2O	—
	Na_2O_2				P_4O_{10}	SO_3	Cl_2O_7	
Formula of hydride	NaH	MgH_2	AlH_3	SiH_4	PH_3	H_2S	HCl	—

Look at the formulas of the chlorides.

1 What is the oxidation number of each element in its chloride?
2 How do these oxidation numbers vary across periods 2 and 3?
3 Why do the oxidation numbers vary in this way?

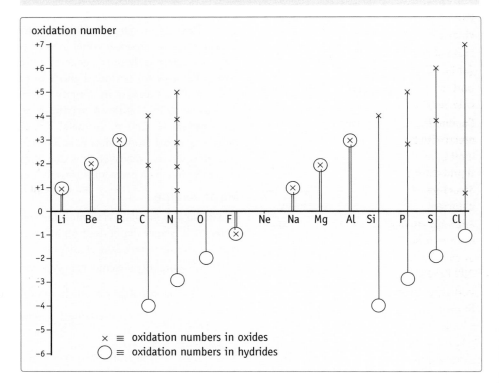

Figure 13.7
Oxidation numbers of the elements from Li to Cl.

The oxidation numbers of elements in their oxides are always positive (except for fluorine) because oxygen is the most electronegative element apart from fluorine. The oxidation numbers of the elements from Li to Cl are represented graphically in figure 13.7. Oxidation numbers of the elements in compounds with oxygen are indicated by a '×'. The pattern in these oxidation numbers provides further evidence of periodic properties. The maximum oxidation number of each element (apart from fluorine) is the same as its group number. For example, lithium in Group I has an oxidation number of +1, boron in Group III has an oxidation number of +3 and nitrogen in Group V has a maximum oxidation number of +5. Thus, the maximum oxidation number of each element corresponds to the number of electrons in the outermost shell of its atoms.

The oxidation numbers of the elements in their hydrides (indicated by a 'O') are also shown in figure 13.7. These have been obtained by assuming that the oxidation number of hydrogen in non-metal compounds is +1, whilst its oxidation number in metal hydrides is −1. Another interesting periodic pattern emerges this time. The oxidation numbers of the elements with hydrogen rise from +1 to +3 for the metals in Groups I, II, and III, plunge to −4 for the elements in Group IV and then rise through −3, −2 and −1 for the elements in Groups V, VI and VII.

Many of the oxidation numbers correspond to the loss or gain of enough electrons for the atoms to obtain a completely filled outer shell of the type ns^2np^6. This tendency is certainly the case for elements in Groups I, II and III which contain 1, 2 and 3 electrons, respectively, in their outer shell. In forming compounds, the elements in these three groups lose 1, 2 or 3 electrons, respectively, forming ions such as Na^+, Mg^{2+} and Al^{3+} with oxidation numbers of +1, +2 and +3. In contrast, the elements in Groups V, VI and VII gain 3, 2 and 1 electrons, respectively, to achieve stable electron structures in ions such as P^{3-}, S^{2-} and Cl^-. Thus, the elements in these groups have oxidation numbers of −3, −2 and −1, respectively.

13.6 Patterns in the properties of chlorides

Table 13.5 shows various properties of the chlorides of the elements in the third period.

Table 13.5
Properties of the chlorides of elements in period 3

Formula of chloride	NaCl	MgCl$_2$	Al$_2$Cl$_6$	SiCl$_4$	PCl$_3$ (PCl$_5$)	S$_2$Cl$_2$	Cl$_2$
State of chloride (at 20°C)	s	s	s	l	l (s)	l	g
b.pt. of chloride/°C	1465	1418	423	57	74 (164)	136	−35
Conduction of electricity by molten or liquid chloride	good	good	v. poor	nil	nil	nil	nil
Structure of chloride	Giant structures		Simple molecular structures				
Enthalpy change of formation of chloride at 298 K/kJ mol^{-1}	−411	−642	−1408	−640	−320	−60	0
Enthalpy change of formation of chloride at 298 K per mole of Cl atoms/kJ	−411	−321	−235	−160	−107	−30	0
Effect of adding chloride to water	solid dissolves readily		chloride reacts with water, HCl fumes are produced				some Cl$_2$ reacts with water
pH of aqueous solution of chloride	7	6.5	3	2	2	2	2

1. How do the states of the chlorides at 20°C vary across the third period?
2. How do the boiling points of the chlorides vary across the period?
3. How do the conductivities of the molten chlorides vary across the period?

Structure and bonding

The variations to which the above three questions point can be explained in terms of the structure and bonding in the chlorides.

NaCl and MgCl$_2$ are giant structures composed of oppositely charged ions attracted to each other by strong electrostatic forces in ionic bonds. This means that the melting points and boiling points of these compounds are high. But the molten substances will conduct electricity because the ions which they contain can move towards the electrode of opposite charge.

All the other chlorides are simple molecular structures composed of small discrete molecules attracted to each other by relatively weak intermolecular forces. So, the melting points and boiling points of these compounds are low and as liquids they will not conduct electricity.

Enthalpy changes of formation

Elements on the left of the periodic table are highly electropositive metals keen to give up electrons. Elements on the right are highly electronegative non-metals (excluding the noble gases). Between these extremes, the electronegativity slowly increases from left to right. Consequently, there is a gradual decrease in the heat evolved when the elements react with 1 mole of chlorine (Cl). Sodium and chlorine react violently liberating 411 kJ per mole of Cl.

$$Na(s) + \tfrac{1}{2}Cl_2(g) \rightarrow Na^+Cl^-(s) \quad \Delta H = -411 \text{ kJ}$$

The strong ionic bonding in Na^+Cl^- results in a highly stable product and a reaction which is very exothermic. On the other hand, the reaction between sulphur and chlorine (two electronegative elements) is relatively feeble and much less exothermic.

$$S(s) + \tfrac{1}{2}Cl_2(g) \rightarrow \tfrac{1}{2}S_2Cl_2(l) \quad \Delta H = -30 \text{ kJ}$$

Reaction with water

When ionic chlorides are added to water, there is an immediate attraction of polar water molecules for ions in the chloride. The solid dissolves forming single aquated ions such as $Na^+(aq)$ and $Cl^-(aq)$. These are separate metal and non-metal ions surrounded by polar water molecules.

The solution of sodium chloride is neutral (pH = 7).

In contrast to this, the solution of aluminium chloride in water is very acidic (pH = 3). The acidic nature of the solution can be explained in terms of the relatively small but highly charged Al^{3+} ion. This draws electrons away from its surrounding water molecules and causes them to give up H^+ ions.

$$[Al(H_2O)_6]^{3+}(aq) \rightarrow [Al(H_2O)_5(OH)]^{2+}(aq) + H^+(aq)$$

The non-metal chlorides in period 3 all react with water forming acidic solutions.

$$SiCl_4(l) + 2H_2O(l) \longrightarrow SiO_2(s) + \underbrace{4H^+(aq) + 4Cl^-(aq)}_{\text{hydrochloric acid}}$$

$$PCl_3(l) + 3H_2O(l) \longrightarrow \underset{\substack{\text{phosphonic} \\ \text{(phosphorous)} \\ \text{acid}}}{H_3PO_3(aq)} + \underbrace{3H^+(aq) + 3Cl^-(aq)}_{\text{hydrochloric acid}}$$

$$2S_2Cl_2(l) + 2H_2O(l) \longrightarrow 3S(s) + SO_2(aq) + \underbrace{4H^+(aq) + 4Cl^-(aq)}_{\text{hydrochloric acid}}$$

A similar pattern is found in the chlorides of period 2 (study question 2).

13.7 Patterns in the properties of oxides

Tables 13.6 and 13.7 show various properties of the oxides of the elements in period 2 and period 3, respectively. There are marked periodic patterns in the structure, bonding and properties of these oxides.

Structure and bonding

In each period, the oxides of metals and metalloids have giant structures, whereas the oxides of non-metals are composed of simple molecules. Thus, lithium oxide, Li_2O; beryllium oxide, BeO; sodium oxide, Na_2O; magnesium oxide, MgO; and aluminium oxide, Al_2O_3, may be regarded as ionic structures. Consequently, they are solids at room temperature, with high melting points and boiling points. These ionic oxides will conduct electricity in the molten state.

The metalloids, boron and silicon, form oxides with giant molecular structures. In boron oxide, B_2O_3, boron and oxygen atoms are arranged in layers. Strong covalent bonds link one atom to the next in a giant sheet structure. The giant structures are therefore

Table 13.6
Properties of the oxides of elements in period 2

Formula of oxide	Li_2O	BeO	B_2O_3	CO_2 (CO)	N_2O NO NO_2 N_2O_4 N_2O_5	O_2	OF_2
State of oxide (at 20°C)	solid	solid	solid	gases	gases (except N_2O_5, a solid)	gas	gas
Conduction of electricity by molten or liquid oxide	good	moderate	v. poor	nil	nil	nil	nil
Structure of oxide	Giant structures			Simple molecular structures			
Enthalpy change of formation of oxide at 298 K/kJ mol^{-1}	−596	−611	−1273	−394 (CO_2)	+33 (NO_2)	−	+22
Enthalpy change of formation of oxide at 298 K per mole of O/kJ	−596	−611	−424	−197	+17	−	+22
Effect of adding oxide to water	Reacts to form LiOH(aq) alkaline solution	BeO does not react with water but it is amphoteric	Reacts to form H_3BO_3, a very weak acid	CO_2 reacts to form H_2CO_3, a weak acid	NO_2 reacts to form an acid solution of HNO_3 and HNO_2	−	OF_2 reacts slowly forming O_2 and an acidic solution of HF
Nature of oxide	Basic (alkaline)	Amphoteric	Acidic	Acidic	Acidic	−	Acidic

solids with high melting points and boiling points. The melting point of B_2O_3 is 577°C and that of SiO_2 is 1700°C. Unlike ionic solids, giant molecular solids do not conduct electricity in the molten state.

Carbon dioxide (CO_2), nitrogen dioxide (NO_2), oxygen difluoride (OF_2), phosphorus(V) oxide (P_4O_{10}), sulphur(VI) oxide (SO_3) and dichlorine heptoxide (Cl_2O_7) consist of discrete small molecules. These simple molecular oxides are much more volatile than the ionic metal oxides and the giant molecular metalloid oxides. They have low melting points and low boiling points and they do not conduct electricity in the liquid state.

Notice the gradations in structure and bond type across each period from ionic oxides and chlorides to simple molecular oxides and chlorides. The gradations can be correlated with changes in electronegativity across the period from low values on the left to high values on the right. Thus, atoms of low electronegativity, such as Na, Mg and Li, form compounds in which they have given up electrons to either chlorine atoms or oxygen atoms. In contrast to this, compounds formed between the more electronegative atoms (such as Si, P, S and F) and either chlorine or oxygen exist as discrete molecules (e.g. $SiCl_4$, ClF) or as giant covalent structures (e.g. SiO_2).

Thus, bonding in the oxides and chlorides becomes less ionic and more covalent as we move across a period from atoms of low to atoms of high electronegativity.

Enthalpy changes of formation

Tables 13.6 and 13.7 show the standard enthalpy changes of formation of some of the oxides in periods 2 and 3. Notice that most of the oxides have a negative enthalpy change of formation. This means that oxides are usually very stable. Indeed, only fluorides are generally more stable than oxides.

The standard enthalpy changes of formation of the oxides vary from very large negative values to relatively small positive values. At first sight, there appears to be no

Table 13.7

Properties of the oxides of elements in period 3

Formula of oxide	Na_2O	MgO	Al_2O_3	SiO_2	P_4O_{10} (P_4O_6)	SO_3 (SO_2)	Cl_2O_7 (Cl_2O)
State of oxide (at 20°C)	solid	solid	solid	solid	solid (solid)	liquid (gas)	liquid (gas)
Conduction of electricity by molten or liquid oxide	good	good	good	v. poor	nil	nil	nil
Structure of oxide		Giant structures			Simple molecular structures		
Enthalpy change of formation of oxide at 298 K/kJ mol^{-1}	−416	−602	−1676	−910	−2984 (P_4O_{10})	−395 (SO_3)	+80 (Cl_2O)
Enthalpy change of formation of oxide at 298 K per mole of O/kJ	−416	−602	−559	−455	−298	−132	+80
Effect of adding oxide to water	reacts to form NaOH(aq) alkaline solution	reacts to form $Mg(OH)_2$ weakly alkaline solution	does not react with water but it is amphoteric	does not react with water but it is acidic	P_4O_{10} reacts to form H_3PO_4 acid solution	SO_3 reacts to form H_2SO_4 acid solution	Cl_2O_7 reacts to form $HClO_4$ acid solution
Nature of oxide	Basic (alkaline)	Basic (weakly alkaline)	Amphoteric	Acidic	Acidic	Acidic	Acidic

distinct pattern in the standard enthalpy changes of formation of the oxides. However, when we compare the enthalpy changes of formation *per mole of oxygen atoms*, an obvious pattern appears (tables 13.6 and 13.7 and figure 13.8). This value gives the stability of bonds to one mole of oxygen atoms in the oxide. The enthalpy change of formation per mole of oxygen atoms becomes less negative as the atomic number increases across a period. This means that, in general, oxygen forms its most stable compounds with elements such as Li, Na, Mg and Al which are furthest removed from it in the periodic table. This is generally the case with other pairs of elements also.

Reaction with water: acid–base character of oxides

1 Write equations for the reaction of each of the following oxides with water: Na_2O, MgO, P_4O_{10}, SO_3.

2 How does the acid–base character of the oxides vary across periods 2 and 3?

As we pass across a period from left to right there is a steady change in the structure of the oxide from ionic, through giant molecular to simple molecular. This change in structure leads to a profound difference in the way in which the oxides react with water, acids and alkalis.

The ionic oxides contain O^{2-} ions in the crystal lattice. These O^{2-} ions in Li_2O and Na_2O, the oxides of Group I metals, react vigorously with water to form alkaline solutions.

$$O^{2-} + H_2O \rightarrow 2OH^-$$
$$Li_2O(s) + H_2O(l) \rightarrow 2Li^+(aq) + 2OH^-(aq)$$
$$Na_2O(s) + H_2O(l) \rightarrow 2Na^+(aq) + 2OH^-(aq)$$

These oxides would react even more vigorously with acids forming a solution containing cations of the metal.

$$Na_2O(s) + 2H^+(aq) \rightarrow 2Na^+(aq) + H_2O(l)$$

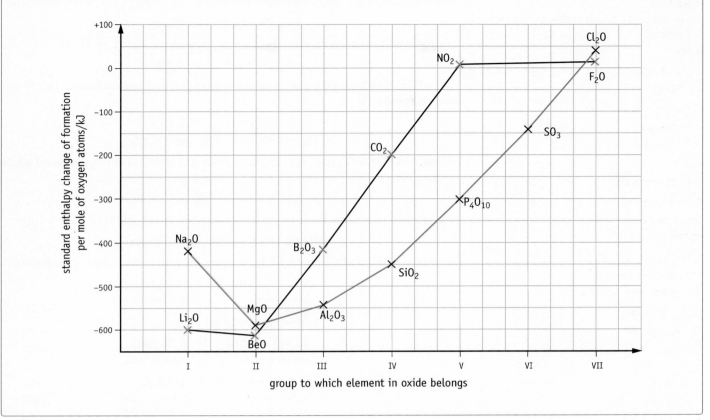

Figure 13.8
Variation in the enthalpy changes of formation of oxides in periods 2 and 3 per mole of oxygen atoms.

The oxides of Group II metals do not react so readily with water or acids since the large charge density on the Group II cation holds the O^{2-} ions more firmly.

Thus, MgO is only slightly soluble in water, although it reacts readily with acids to form a solution of magnesium ions.

$$MgO(s) + H_2O(l) \rightleftharpoons Mg^{2+}(aq) + 2OH^-(aq)$$
$$MgO(s) + 2H^+(aq) \rightarrow Mg^{2+}(aq) + H_2O(l)$$

BeO is insoluble in water, but it shows basic properties by dissolving in acids to form Be^{2+} salts.

$$BeO(s) + 2H^+(aq) \rightarrow Be^{2+}(aq) + H_2O(l)$$

However, BeO also resembles acidic oxides by reacting with alkalis to form salts called beryllates.

$$BeO(s) + 2OH^-(aq) + H_2O(l) \rightarrow Be(OH)_4^{2-}(aq)$$
tetrahydroxoberyllate(II)
(beryllate)

Aluminium oxide has similar properties to beryllium oxide. It does not react with water, but it will react with both dilute acids (H^+ ions) and dilute alkalis (OH^- ions).

$$Al_2O_3(s) + 6H^+(aq) \rightarrow 2Al^{3+}(aq) + 3H_2O(l)$$
$$Al_2O_3(s) + 2OH^-(aq) + 3H_2O(l) \rightarrow 2[Al(OH)_4]^-(aq)$$
tetrahydroxoaluminate(III)
(aluminate)

Oxides such as BeO and Al_2O_3, which show *both basic and acidic properties*, are called **amphoteric oxides**.

The remaining oxides of the elements in periods 2 and 3 are all acidic except CO, N_2O, NO and ClO_2, which are described as **neutral**. *Neutral oxides show neither acidic nor basic character.*

Boron(III) oxide, B_2O_3, reacts with water to form boric(III) acid, H_3BO_3, a very weak acid.

$$B_2O_3(s) + 3H_2O(l) \rightarrow 2H_3BO_3(aq)$$

CO_2 dissolves in water and reacts slightly to form weak carbonic acid, H_2CO_3.

$$CO_2(g) + H_2O(l) \rightleftharpoons H_2CO_3(aq)$$

CO_2 will react with the OH^- ions in alkalis to form first hydrogencarbonate, HCO_3^- and then carbonate, CO_3^{2-}.

$$CO_2(g) + OH^-(aq) \rightarrow HCO_3^-(aq)$$
$$CO_2(g) + 2OH^-(aq) \rightarrow CO_3^{2-}(aq) + H_2O(l)$$

Silicon(IV) oxide, SiO_2, does not react with water, but it reacts with concentrated alkalis forming silicate, SiO_3^{2-}.

$$SiO_2(s) + 2OH^-(aq) \rightarrow SiO_3^{2-}(aq) + H_2O(l)$$

NO_2 reacts with water to form a mixture of two acids, HNO_2 and HNO_3.

$$2NO_2(g) + H_2O(l) \rightarrow HNO_2(aq) + HNO_3(aq)$$

The oxides of P, S and Cl (except ClO_2) react readily with water to form strong acids.

$$\underset{\substack{\text{phosphorus(v)} \\ \text{oxide} \\ \text{(phosphorus pentoxide)}}}{P_4O_{10}(s)} + 6H_2O(l) \rightarrow \underset{\text{phosphoric(v) acid}}{4H_3PO_4(aq)}$$

$$\underset{\text{sulphur trioxide}}{SO_3(g)} + H_2O(l) \rightarrow \underset{\text{sulphuric acid}}{H_2SO_4(aq)}$$

$$\underset{\substack{\text{dichlorine} \\ \text{heptoxide}}}{Cl_2O_7(l)} + H_2O(l) \rightarrow \underset{\substack{\text{chloric(vii) acid} \\ \text{(perchloric acid)}}}{2HClO_4(aq)}$$

$$\underset{\substack{\text{dichlorine} \\ \text{oxide}}}{Cl_2O(g)} + H_2O(l) \rightarrow \underset{\substack{\text{chloric(i) acid} \\ \text{(hypochlorous acid)}}}{2HClO(aq)}$$

Notice that *as we cross the periodic table, we move from the ionic oxides of metals, which are basic, to the oxides of metalloids with giant molecular structure, which are weakly basic, weakly acidic or amphoteric, and finally to the simple molecular oxides of non-metals, which are acidic.*

Summary

1 Periodicity as summarised in the periodic table is one of the major unifying themes in chemistry.

2 Elements in the same group of the periodic table have similar electron configurations. This results in similar chemical properties.

3 Moving from left to right across a period there is a decrease in atomic radius since the increasing positive charge in the nucleus pulls the electrons closer.

4 Moving from left to right across a period, there is a general increase in the first ionisation energy. This results from an increasing nuclear charge and a decreasing atomic radius across the period causing the outermost electron to be held more tightly.

5 The electronegativity of an atom provides a measure of the power of that atom in a molecule to attract electrons.

6 Moving from left to right across a period there is an increase in the electronegativities of elements.

7 In forming compounds, the elements in Groups I, II and III form positive ions with charges of +1, +2 and +3, respectively. In contrast, the elements in Groups V, VI and VII gain electrons forming negative ions with charges of −3, −2 and −1, respectively.

8 Moving from left to right across a period:

(i) The chlorides of the elements change from ionic, involatile metal chlorides to simple molecular, volatile non-metal chlorides.

(ii) The hydrides of the elements change from ionic, involatile metal hydrides to simple molecular, non-metal hydrides. The hydrides also become more acidic across the period.

(iii) The oxides of the elements change from being ionic, involatile and basic in metal oxides, through giant molecular, involatile, amphoteric, metalloid oxides to simple molecular, volatile, and acidic non-metal oxides.

1 Use the information in the following table to explain the statements below.

	Na	Mg	Al	Si	P	S	Cl
Atomic radius/nm	0.156	0.136	0.125	0.117	0.110	0.104	0.099
Ionic radius/nm	0.095	0.065	0.050			0.184	0.181
First ionisation energy/kJ mol^{-1}	+492	+743	+579	+791	+1060	+1003	+1254

(a) The atomic radius decreases across the period from Na to Cl.

(b) The ionic radii of Na^+, Mg^{2+} and Al^{3+} are less than their respective atomic radii, whereas the ionic radii of Cl^- and S^{2-} are greater than their respective atomic radii.

(c) The first ionisation energies show a general increase from Na to Cl.

(d) The first ionisation energy of Al is less than that for Mg.

2 Table 13.8 shows various properties of the chlorides of the elements in period 2.

(a) Explain the variation in state and boiling point of the chlorides across period 2.

(b) LiCl has a giant structure and LiCl(l) is a good conductor of electricity.
What type of structure does LiCl have?
Why does LiCl(l) conduct electricity?

(c) How would you expect LiCl and BCl_3 to behave when added to water?

(d) How is the type of bonding in the chlorides of the elements in period 2 related to their electronegativity.

Formula of chloride	LiCl	$BeCl_2$	BCl_3	CCl_4	NCl_3	Cl_2O	ClF
State of chloride at 20°C	s	s	g	l	l	g	g
B.pt. of chloride/°C	1350	487	12	77	71	2	−101
Conduction of electricity by liquid chloride	good	very poor	nil	nil	nil	nil	nil
Structure	giant structures		simple molecular structures				

Table 13.8 Properties of the chlorides of elements in period 2

3 The behaviour of the hydrides of the elements Na–Cl with water is summarised below.

NaH MgH$_2$ AlH$_3$ SiH$_4$ PH$_3$ H$_2$S   HCl

react forming H$_2$(g) and an alkaline soln. | no reaction | reacts to form a slightly alkaline solution | reacts to form a slightly acidic solution | to form an acidic solution

(a) Write equations to summarise the reactions of NaH and MgH_2 with water.

(b) Suggest a reason why SiH_4 has no reaction with water.

(c) Write an equation to account for the formation of a slightly alkaline solution when PH_3 reacts with water.

(d) Write an equation to account for the formation of an acidic solution when HCl reacts with water.

(e) Explain the trends and differences in the reactions of these hydrides with water in terms of their structure and bonding.

4 Consider the elements Li, Be, B, N, F, Ne.

(a) Which exist as diatomic molecules in the gaseous state at room temperature?

(b) Which has the highest boiling point?

(c) Which form a chloride of formula XCl_3?

(d) Which has the largest first ionisation energy?

(e) Which has the smallest second ionisation energy?

(f) Which form hydrides that dissolve in water to give an alkaline solution?

5 (a) Draw 'dot/cross' diagrams to show the electronic structures of
(i) LiF, (ii) CF_4, (iii) NF_3.

(b) Predict the shapes with respect to atoms of molecules of
(i) BF_3, (ii) CF_4, (iii) NF_3, (iv) OF_2.

(c) Why is it not meaningful to talk about the shape of LiF?

6 (a) What is the nature of the bonds in the oxides formed when Na, Mg, Al and S react with excess oxygen?

(b) How do these oxides react with
(i) water, (ii) dilute acids, (iii) alkali?

(c) Magnesium chloride is a high melting point solid, aluminium chloride is a solid which sublimes readily at about 180°C and silicon tetrachloride is a volatile liquid. Explain the nature of the chemical bonding in these chlorides and show how this accounts for the above differences in volatility.

7 (a) Write down the symbols of the elements in the third period of the periodic table (ending with the noble gas, argon) in order of increasing atomic number.

(b) Which of these elements are
(i) 's-block' elements,
(ii) 'd-block' elements?

(c) (i) Write the empirical formula of the chloride formed by the element of atomic number 13.
(ii) Describe briefly how you could prepare a sample of this chloride.

8 Some of the properties of elements P and Q are given below.

Element P
(i) is soft and malleable,
(ii) floats on water,
(iii) has a melting point less than 100°C,
(iv) forms an ionic hydride of formula PH.

Element Q
(i) has a density greater than 7 g cm^{-3},
(ii) has a melting point greater than 1500°C,
(iii) forms oxides of formulas QO, Q_2O_3 and QO_3,
(iv) does not react with concentrated nitric acid,
(v) forms some simple molecular compounds in its highest oxidation state.

(a) State the deductions you can make about P and Q from each piece of evidence.

(b) To which blocks of the periodic table do P and Q belong?

(c) Identify P and Q as precisely as possible.

14.1 Fundamental reactions in inorganic chemistry

The reactions of inorganic substances can be divided into four major classes. All these reactions involve competition in one way or another.

1 Redox

2 Acid–base

3 Precipitation

4 Complexing

Other processes such as decomposition and synthesis usually fit into one of these four categories.

Redox reactions

Redox processes have already been discussed in Chapter 3. It is worth recalling the following important types of redox reaction.

(a) The reactions of metals with non-metals

$$\text{e.g. } 2Na(s) + Cl_2(g) \rightarrow 2Na^+Cl^-(s)$$
$$2Mg(s) + O_2(g) \rightarrow 2Mg^{2+}O^{2-}(s)$$

(b) The reactions of metals with water and steam

$$\text{e.g. } Ca(s) + 2H_2O(l) \rightarrow Ca^{2+}(OH^-)_2(s) + H_2(g)$$

(c) The reactions of metals with acids

$$\text{e.g. } Mg(s) + 2H^+(aq) \rightarrow Mg^{2+}(aq) + H_2(g)$$

(d) The reactions at the electrodes during electrolysis

e.g. during the purification of impure copper

$$\text{Impure copper anode (+)} \quad Cu(s) \rightarrow Cu^{2+}(aq) + 2e^-$$
$$\text{Pure copper cathode (−)} \quad Cu^{2+}(aq) + 2e^- \rightarrow Cu(s)$$

▶ Rows of copper cells at a copper refinery. This photo shows one row of anodes withdrawn. In the purification process, copper from the impure anodes goes into solution as copper ions. Pure copper deposits on the cathode.

1 Which **one** of the following processes involves redox? (Remember that the best check for redox is to consider the oxidation numbers of the elements involved in the reaction.)

A $Ag^+ + 2NH_3 \rightarrow [Ag(NH_3)_2]^+$
B $H^+ + NH_3 \rightarrow NH_4^+$
C $Ba^{2+} + CrO_4^{2-} \rightarrow BaCrO_4$
D $2Al + 3Cl_2 \rightarrow Al_2Cl_6$
E $CuCO_3 \rightarrow CuO + CO_2$

▲ 'Scum' is a white precipitate which forms when calcium ions in hard water react with complex palmitate and stearate ions in soap.

(e) Disproportionation reactions (see section 16.8)

Acid–base reactions

You will have met many examples of acid–base reactions already. These include reactions of acids with insoluble bases (metal oxides and hydroxides)

$$CuO(s) + 2H^+(aq) \rightarrow Cu^{2+}(aq) + H_2O(l)$$
$$Cu(OH)_2(s) + 2H^+(aq) \rightarrow Cu^{2+}(aq) + 2H_2O(l)$$

reactions of acids with alkalis

$$Na^+(aq) + OH^-(aq) + H^+(aq) \rightarrow Na^+(aq) + H_2O(l)$$

and less obvious acid–base reactions such as that between aqueous ammonia and dilute acid.

$$NH_3(aq) + H^+(aq) \rightarrow NH_4^+(aq)$$

In this chapter we shall extend our ideas of acid–base processes to include such reactions as acid + carbonate.

Precipitation reactions

Precipitation occurs when aqueous solutions are mixed and an insoluble substance forms. For example, a white precipitate of silver chloride forms when aqueous solutions of silver nitrate and sodium chloride are mixed.

$$Ag^+(aq) + NO_3^-(aq) + Na^+(aq) + Cl^-(aq) \rightarrow Ag^+Cl^-(s) + Na^+(aq) + NO_3^-(aq)$$

'Scum' is a precipitate which forms when calcium ions in hard water react with complex palmitate and stearate ions in soap (section 14.17).

$$Ca^{2+}(aq) + 2X^-(aq) \rightarrow CaX_2(s)$$

| in hard water | complex ions in soap | 'scum' precipitate |

Reactions involving complexes

The formation of complexes is of great importance in the chemistry of transition metals (section 18.6), and in the chemistry of most aqueous metal ions. In aqueous solution, most cations (M^{2+}) are surrounded by water molecules to form hydrated ions of the form $[M(H_2O)_n]^{2+}$. The hydrated ions can be regarded as complex ions. Other important complexes form when cations are surrounded by anions or molecules other than water. For example, when aluminium hydroxide dissolves in excess NaOH(aq):

$$Al(OH)_3(s) + OH^-(aq) \rightarrow [Al(OH)_4]^-(aq)$$

tetrahydroxoaluminate(III)
(aluminate)

The complex ion, $[Al(OH)_4]^-$, consists of an Al^{3+} ion surrounded by four OH^- ions.

14.2 Redox: competition for electrons

Most redox reactions involve a transfer of electrons from one reactant to another. When metals react with non-metals, electrons are transferred from the metal to the non-metal with the formation of ions.

$$2Na \rightarrow 2Na^+ + 2e^-$$
$$Cl_2 + 2e^- \rightarrow 2Cl^-$$

The metal, which loses electrons, is oxidised whilst the non-metal, which gains electrons is reduced. The oxidising agent (oxidant) is the electron acceptor and the reducing agent (reductant) is the electron donor. In this case, the reactions occur because most non-metals are eager to gain electrons whilst metals will part with their electrons quite readily. In the competition for electrons, the non-metal wins 'hands down'. However, the competition for electrons in redox reactions is not always so clear cut.

Consequently, chemists have developed the idea of electrode potentials. These can be used to check whether (and in which direction) a redox reaction might occur between two systems.

In chapter 3, we discovered that electron transfer occurred when a $Zn(s)/Zn^{2+}(aq)$ half-cell was connected by a salt bridge to a $Cu(s)/Cu^{2+}(aq)$ half-cell (figure 14.1). The bulb lights, showing that electrons are flowing through the wire.

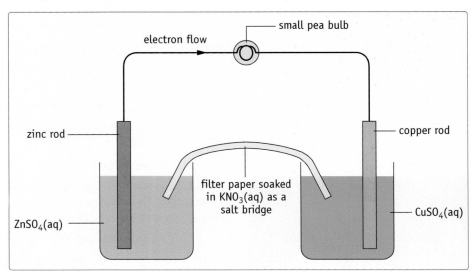

Figure 14.1
Electron transfer between two half-cells connected by a salt bridge.

Zinc dissolves from the zinc rod which loses weight and copper is deposited on the copper rod. The reactions at the terminals can be represented by the following half-equations.

$$Zn(s) \rightarrow Zn^{2+}(aq) + 2e^{-}$$
$$Cu^{2+}(aq) + 2e^{-} \rightarrow Cu(s)$$

Electrons flow from the negative zinc terminal through the external circuit to the positive copper terminal.

If the pea bulb is replaced by a high resistance voltmeter (e.g. a valve voltmeter) it is possible to measure the potential difference between the two half-cells.

By using a high-resistance voltmeter, the current in the external circuit is virtually zero and the cell registers its maximum potential difference. This maximum p.d. is called the **electromotive force (e.m.f.)**. This e.m.f. is the difference in electrical potential between the two metals. It gives a quantitative measure of the likelihood of the redox reaction taking place in the cell.

When the pea bulb in figure 14.1 is replaced by a valve voltmeter, a voltage of 1.10 volts is recorded.

This combination of half-cells and the resulting cell e.m.f. can be summarised in a diagram as

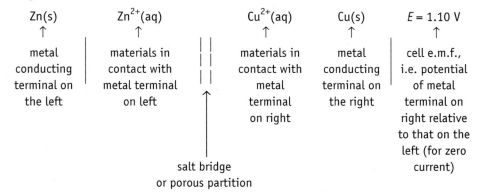

The voltage shown in the summary above relates to changes indicated by the order of material in the diagram, i.e.

$$Zn(s) \rightarrow Zn^{2+}(aq) + 2e^-$$

and

$$Cu^{2+}(aq) + 2e^- \rightarrow Cu(s)$$

So, for the change,

$$Zn(s) + Cu^{2+}(aq) \rightarrow Zn^{2+}(aq) + Cu(s) \qquad E = 1.10\,V$$

and for the reverse process,

$$Zn^{2+}(aq) + Cu(s) \rightarrow Zn(s) + Cu^{2+}(aq) \qquad E = -1.10\,V$$

Notice that this convention leads to a positive value of E if the reaction actually takes place in that direction when the cell is short-circuited. A negative value of E indicates that the cell reaction as written cannot take place.

When the $Zn(s)/Zn^{2+}(aq)$ half-cell in figure 14.1 was replaced by other metal/metal ion half-cells, the results shown in table 14.1 were obtained.

Table 14.1
The overall cell e.m.f. developed between various metal/metal ion half-cells and the $Cu(s)/Cu^{2+}(aq)$ half-cell

Metal/metal ion half-cell	Overall cell e.m.f./volts
$Zn(s)/Zn^{2+}(aq)$	+1.10
$Fe(s)/Fe^{2+}(aq)$	+0.78
$Pb(s)/Pb^{2+}(aq)$	+0.47
$Cu(s)/Cu^{2+}(aq)$	0.00
$Ag(s)/Ag^+(aq)$	−0.46

The first three metals (zinc, iron and lead) all form the negative terminal with respect to copper. These metals are stronger reducing agents than copper and go into solution as their ions when the cells are short-circuited. On the other hand, the $Ag(s)/Ag^+(aq)$ half-cell produces a negative e.m.f. relative to the copper half-cell. This means that the silver electrode is positive with respect to copper and in this case copper is acting as the reducing agent. The cell reaction can be summarised as

$$Ag(s)|Ag^+(aq) \;\vdots\; Cu^{2+}(aq)|Cu(s) \qquad E = -0.46\,V$$

As we might expect, the reaction

$$2Ag(s) + Cu^{2+}(aq) \rightarrow 2Ag^+(aq) + Cu(s)$$

has a negative e.m.f. of −0.46 volts. It is the reverse reaction which actually occurs when the cell is short circuited. Figure 14.2 shows diagrammatically the relative potentials for the copper, zinc and silver half-cells.

1. What other factors besides the particular metals and ions used will affect the size of the cell e.m.f. measured using apparatus like that in figure 14.1?
2. Look at figure 14.2. What would be the e.m.f. of a $Zn(s)/Zn^{2+}(aq) \;\vdots\; Ag^+(aq)/Ag(s)$ cell? Write half-equations for the reactions which take place in the two half-cells of this combination when it is short-circuited.

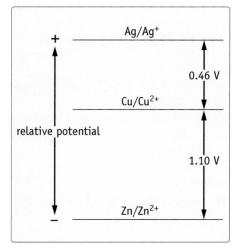

Figure 14.2
Relative cell potentials.

Notice that the e.m.f. values in table 14.1 can be used to compare the relative tendencies of the metals to release electrons and form ions. Thus zinc, at the head of the table, is a strong reducing agent, releasing electrons readily, whilst silver at the bottom of the table is a very poor reducing agent. The e.m.f. values also give a quantitative record of the position of the metal in the electrochemical series (ECS).

So far, we have seen that cell e.m.f.s enable metals to be placed in order of their relative ability as reducing agents. Conversely, we can also draw up a list of metal ions showing their ability as oxidising agents. Thus, $Ag^+(aq)$ is a stronger oxidising agent than $Cu^{2+}(aq)$, i.e. it is a better competitor for electrons.

You will appreciate that it is impossible to obtain the electrical potential or the e.m.f. for a single half-cell. E.m.f.s can only be measured for a complete circuit with two electrodes. In other words, *only differences in potential are measurable*. Nevertheless, it would be extremely useful if we could summarise all e.m.f. data by giving each half-cell a characteristic value. This can be done by arbitrarily assigning an **electrode potential** of zero to one particular half-cell. The electrode potentials of other half-cells can then be compared with this standard.

The standard chosen for electrode potentials is not a metal but hydrogen. The so-called **standard hydrogen half-cell** (sometimes loosely referred to as the standard hydrogen electrode) is shown in figure 14.3.

This consists of H_2 gas at one atmosphere pressure and 25°C bubbling around a platinised platinum electrode. The electrode is immersed in a 1.0 mol dm^{-3} solution of H^+ ions. It is alternately bathed in first $H^+(aq)$ and then $H_2(g)$. Hydrogen is adsorbed on the platinum and an equilibrium is established between the adsorbed layer of H_2 gas and aqueous H^+ ions in the solution.

$$\tfrac{1}{2}H_2(g) \rightleftharpoons H^+(aq) + e^-$$

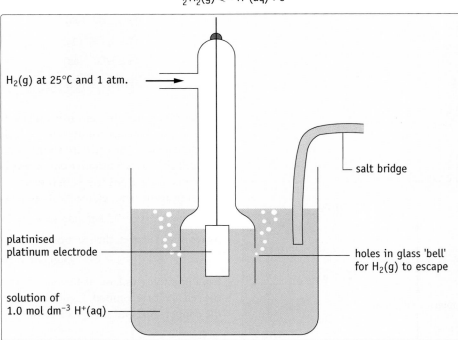

$H_2(g)$ at 25°C and 1 atm.

salt bridge

platinised platinum electrode

holes in glass 'bell' for $H_2(g)$ to escape

solution of 1.0 mol dm^{-3} $H^+(aq)$

Figure 14.3
The standard hydrogen half-cell.

The platinised platinum electrode has three functions.

1 It acts as an *inert* metal connection to the H_2/H^+ system. (There is no tendency for Pt to form ions itself.)

2 It allows H_2 gas to be adsorbed onto its surface.

3 It is covered by a loosely deposited layer of finely divided platinum (i.e. it is *platinised*). This increases its surface area so that it can establish an equilibrium between $H_2(g)$ and $H^+(aq)$ as rapidly as possible.

The standard electrode potential ($E^\ominus$) for this reference half-cell is taken as zero, i.e.

$$\tfrac{1}{2}H_2(g) \rightarrow H^+(aq) + e^- \qquad E^\ominus = 0.00\,V$$

14.4 Measuring standard electrode potentials

The potential of the platinum electrode in a standard hydrogen half-cell will depend on the temperature, the concentration of H^+ ions and the pressure of the $H_2(g)$.

Consequently, *in measuring and comparing electrode potentials we must choose the same standard conditions for all measurements*. The standard conditions chosen are similar to those for thermochemical measurements.

- All solutions have a concentration of 1 mol dm^{-3}.
- Any gases involved have a pressure of 1 atmosphere.
- The temperature is 25°C (298 K).
- Platinum is used as the electrode when the half-cell system does not include a metal.

Figure 14.4 shows a standard $Cu^{2+}(aq)/Cu(s)$ half-cell and a standard $Fe^{3+}(aq)/Fe^{2+}(aq)$ half-cell. The latter system does not involve a metal so a platinum electrode must be used as the electrical connection. Notice also that the solution is 1.0 mol dm^{-3} with respect to both $Fe^{3+}(aq)$ and $Fe^{2+}(aq)$.

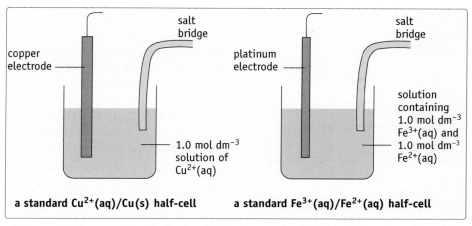

Figure 14.4
Two standard half-cells.

a standard $Cu^{2+}(aq)/Cu(s)$ half-cell   **a standard $Fe^{3+}(aq)/Fe^{2+}(aq)$ half-cell**

The standard potentials of all other half-cells are obtained by reference to that of the standard hydrogen half-cell. The standard electrode potential of the $Cu^{2+}(aq)/Cu(s)$ half-cell is thus the potential difference between the electrodes of a cell consisting of the standard hydrogen half-cell and the standard $Cu^{2+}(aq)/Cu(s)$ half-cell (figure 14.5). When this cell is set up the potential is 0.34 volts and copper is the positive terminal. The reactions occurring at the electrodes are

$$H_2 \rightarrow 2H^+ + 2e^-$$

and

$$Cu^{2+} + 2e^- \rightarrow Cu.$$

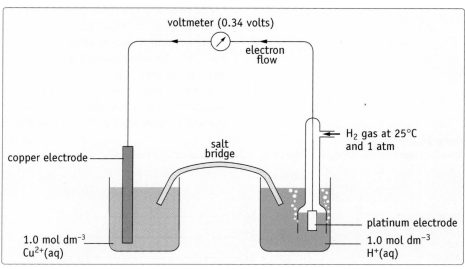

Figure 14.5
Measuring the standard electrode potential of the $Cu^{2+}(aq)/Cu(s)$ half-cell.

Now, since the standard electrode potential for the system $H_2 \rightarrow 2H^+ + 2e^-$ is arbitrarily taken as zero, the standard electrode potential for the system $Cu^{2+} + 2e^- \rightarrow Cu$ is $+0.34$ volts. The plus sign is used because the copper electrode is positively charged with respect to the standard hydrogen electrode,

$$Cu^{2+} + 2e^- \rightarrow Cu \qquad E^{\ominus} = +0.34\ V$$

$$Pt(s)|H_2(g), H^+(aq) \,\|\, Cu^{2+}(aq)|Cu(s) \qquad\qquad E^{\ominus} = +0.34\ V$$

By convention, the oxidised form is written first when a particular redox half-equation and its standard electrode potential are being referred to. Thus, $Cu^{2+}(aq)/Cu(s)$, $E^{\ominus} = +0.34$ volts means that the half-cell reaction

$$Cu^{2+}(aq) + 2e^- \rightarrow Cu(s)$$

has a standard electrode potential of $+0.34$ volts. **Standard electrode potentials** are sometimes called **standard reduction potentials** because they relate to the *reduction of the more oxidised species.*

When inert electrodes are present (as in the hydrogen half-cell), the reduced form of the components of the system should be written next to the electrode. Hence, the standard hydrogen electrode is written as $H^+(aq), H_2(g)/Pt(s)$ or as $Pt(s)/H_2(g)$, $H^+(aq)$, but not as $Pt(s)/H^+(aq), H_2(g)$.

We are now in a position to define the *standard electrode (reduction) potential,* $E^{\ominus}$, *of a standard half-cell. This is the potential of that half-cell relative to a standard hydrogen half-cell under standard conditions.*

When a standard zinc half-cell is connected to a standard hydrogen half-cell, the e.m.f. produced is 0.76 volts. In this case the zinc electrode is negative. Thus, the standard electrode potential for the $Zn^{2+}(aq)/Zn(s)$ half-cell is -0.76 volts, i.e.

$$Zn^{2+}(aq) + 2e^- \rightarrow Zn(s) \qquad E^{\ominus} = -0.76\ V$$

The reactions occurring at the electrodes are

$$Zn(s) \rightarrow Zn^{2+}(aq) + 2e^-$$

and

$$2H^+(aq) + 2e^- \rightarrow H_2(g)$$

Notice that the reaction at the hydrogen electrode is now in the opposite direction to that which occurred when a $Cu^{2+}(aq)/Cu(s)$ half-cell was connected to a standard hydrogen half-cell.

14.5 Relative strengths of oxidising agents and reducing agents

Standard electrode potentials provide a direct measure of the relative oxidising and reducing power of different species.

The standard electrode potentials of the zinc, hydrogen and copper half-cells have been tabulated in table 14.2 in order to emphasise the relative strengths of the different oxidising and reducing agents. Cu^{2+} is a stronger oxidising agent than either H^+ or Zn^{2+}. On the other hand, Zn is a more powerful reducing agent than either H_2 or Cu since it will reduce H^+ to H_2 and Cu^{2+} to Cu. Notice that *the relative size of $E^{\ominus}$ gives a measure of the strengths of both oxidants and reductants.* Cu^{2+}, the most powerful oxidising agent in this table, has the largest (most positive) value for $E^{\ominus}$; Zn^{2+}, the least powerful oxidising agent, has the smallest (most negative) value for $E^{\ominus}$.

True or false: The statement $Cu^{2+}(aq)/Cu(s)$, $E^{\ominus} = +0.34$ volts means that:

1 copper is more positive than Cu^{2+} ions.

2 a solution of $Cu^{2+}(aq)$ is 0.34 volts more positive than a Cu electrode immersed in it.

3 the reaction $Cu^{2+}(aq) + 2e^- \rightarrow Cu(s)$ has a standard electrode potential of $+0.34$ volts.

4 the copper electrode of a standard $Cu^{2+}(aq)/Cu(s)$ half-cell is 0.34 volts more positive than the platinum electrode in a standard hydrogen half-cell to which it is connected.

Table 14.2
Standard electrode potentials of the zinc, hydrogen and copper half-cells

	Oxidising agent		Reducing agent	$E^{\ominus}$ /volts
increasing strength of oxidising agent ↑	$Cu^{2+}(aq) + 2e^-$	$\longrightarrow Cu(s)$	increasing strength of reducing agent	+0.34
	$H^+(aq) + e^-$	$\longrightarrow \tfrac{1}{2}H_2(g)$		0.00
	$Zn^{2+}(aq) + 2e^-$	$\longrightarrow Zn(s)$	↓	−0.76

On the other hand, the value of $E^\ominus$ for the reaction

$$Zn(s) \rightarrow Zn^{2+}(aq) + 2e^-$$

is +0.76 volts and Zn is the most powerful reducing agent. It is important to remember that *the more positive the value of $E^\ominus$, the more energetically favourable is the reaction.*

The standard electrode potentials for a large number of redox half-reactions are shown in table 14.3. The table is arranged so that the strongest oxidising agent, F_2, which has the most positive value for $E^\ominus$, is at the top of the list. The weakest oxidising agent, K^+, with the most negative value for $E^\ominus$, is at the bottom. Thus, fluorine is a better competitor for electrons than any other oxidising agent in the list. Potassium ions have the weakest tendency to accept electrons.

Conversely, F^- is the weakest reducing agent with the least tendency to donate electrons, whilst K is the most powerful reducing agent in the list. This illustrates the reciprocal or conjugate character of an oxidising agent and its corresponding reducing agent. *The stronger the oxidising agent, the weaker is its corresponding (conjugate) reducing agent and vice versa.*

Notice also in table 14.3 that the order of metals as reducing agents according to their $E^\ominus$ values is exactly the same as the order of metals in the reactivity (electrochemical) series.

Table 14.3
Standard electrode potentials

Strongest oxidising agent			Weakest reducing agent	$E^\ominus$/volts
	$F_2(g) + 2e^-$	$\rightarrow 2F^-(aq)$		+2.87
	$H_2O_2(aq) + 2H^+(aq) + 2e^-$	$\rightarrow 2H_2O(l)$		+1.77
	$MnO_4^-(aq) + 4H^+(aq) + 3e^-$	$\rightarrow MnO_2(s) + 2H_2O(l)$		+1.70
	$2HClO(aq) + 2H^+(aq) + 2e^-$	$\rightarrow Cl_2(aq) + 2H_2O(l)$		+1.59
	$MnO_4^-(aq) + 8H^+(aq) + 5e^-$	$\rightarrow Mn^{2+}(aq) + 4H_2O(l)$		+1.51
	$Cl_2(aq) + 2e^-$	$\rightarrow 2Cl^-(aq)$		+1.36
	$MnO_2(s) + 4H^+(aq) + 2e^-$	$\rightarrow Mn^{2+}(aq) + 2H_2O(l)$		+1.23
	$Br_2(aq) + 2e^-$	$\rightarrow 2Br^-(aq)$		+1.09
	$NO_3^-(aq) + 2H^+(aq) + e^-$	$\rightarrow NO_2(g) + H_2O(l)$		+0.80
	$Ag^+(aq) + e^-$	$\rightarrow Ag(s)$		+0.80
	$Fe^{3+}(aq) + e^-$	$\rightarrow Fe^{2+}(aq)$		+0.77
	$2H^+(aq) + O_2(g) + 2e^-$	$\rightarrow H_2O_2(aq)$		+0.68
	$I_2(aq) + 2e^-$	$\rightarrow 2I^-(aq)$		+0.54
	$Cu^{2+}(aq) + 2e^-$	$\rightarrow Cu(s)$		+0.34
	$2H^+(aq) + 2e^-$	$\rightarrow H_2(g)$		0.00
	$Pb^{2+}(aq) + 2e^-$	$\rightarrow Pb(s)$		−0.13
	$Fe^{2+}(aq) + 2e^-$	$\rightarrow Fe(s)$		−0.44
	$Zn^{2+}(aq) + 2e^-$	$\rightarrow Zn(s)$		−0.76
	$Al^{3+}(aq) + 3e^-$	$\rightarrow Al(s)$		−1.66
	$Mg^{2+}(aq) + 2e^-$	$\rightarrow Mg(s)$		−2.37
	$Na^+(aq) + e^-$	$\rightarrow Na(s)$		−2.71
Weakest oxidising agent	$K^+(aq) + e^-$	$\rightarrow K(s)$	Strongest reducing agent	−2.92

increasing strength of oxidising agent (left axis, pointing up)
increasing strength of reducing agent (right axis, pointing down)

14.6 The uses of standard electrode potentials

Using standard electrode potentials to predict the possibility of reactions

Why can $Cu^{2+}(aq)$ oxidise $Zn(s)$, but $Zn^{2+}(aq)$ cannot oxidise $Cu(s)$?

As we have seen already, $E^\ominus$ values provide an indication of the relative strengths of oxidising agents and reducing agents.

The value of $E^\ominus$ for the reaction

$$Cu^{2+}(aq) + 2e^- \rightarrow Cu(s) \qquad \text{is +0.34 volts.}$$

That for the reaction

$$Zn(s) \rightarrow Zn^{2+}(aq) + 2e^- \quad \text{is +0.76 volts.}$$

Consequently, the overall potential or **standard cell potential** for the reaction

$$Cu^{2+}(aq) + Zn(s) \rightarrow Cu(s) + Zn^{2+}(aq) \quad \text{is 1.10 volts.}$$

i.e.

$Cu^{2+}(aq) + 2e^- \rightarrow Cu(s)$	+0.34 volts
$Zn(s) \rightarrow Zn^{2+}(aq) + 2e^-$	+0.76 volts
Add: $\overline{Cu^{2+}(aq) + Zn(s) \rightarrow Cu(s) + Zn^{2+}(aq)}$	+1.10 volts

The overall positive value for the reaction potential suggests that the process is energetically feasible.

Conversely, the overall potential for the reverse reaction

$$Cu(s) + Zn^{2+}(aq) \rightarrow Cu^{2+}(aq) + Zn(s) \quad \text{is -1.10 volts.}$$

The negative value indicates that this reaction is unlikely to occur. In general, *reactions with an overall positive potential are energetically feasible whereas those with an overall negative value are not so.* Relating this to table 14.3, it means that any oxidising agent on the left will oxidise any reducing agent below it on the right. In other words,

stronger oxidising agent	+	stronger reducing agent	$\longrightarrow$	weaker reducing agent	+	weaker oxidising agent

Use table 14.3 to predict whether a reaction will occur between

1. F_2 and Na
2. F^- and Na^+
3. F_2 and Na^+
4. Cl_2 and Fe^{2+}
5. I_2 and Br^-

In predicting reactions, it is important to remember that $E^{\ominus}$ *relates to the probability of a reaction occurring, not to the quantity of materials reacting.* The reaction is just as likely (or unlikely) to occur whether we have a milligram or a kilogram, a thimbleful or a bucketful, one mole or two moles. The electrode e.m.f. is independent of the number of electrons being transferred.

Thus, if $E^{\ominus}$ for the reaction $Fe^{3+}(aq) + e^- \rightarrow Fe^{2+}(aq)$ is +0.77 volts, $E^{\ominus}$ for the reaction $2Fe^{3+}(aq) + 2e^- \rightarrow 2Fe^{2+}(aq)$ is also +0.77 volts, and *not* +1.54 volts.

Although an overall positive value of $E^{\ominus}$ for a redox pair suggests that the reaction should take place, in practice the reaction may be too slow to occur. The important point is that $E^{\ominus}$ relates to the relative stabilities of reactants and products. Therefore, it indicates the feasibility of the reaction from an energetic standpoint. However, $E^{\ominus}$ gives no information whatsoever about *the rate of a reaction or its kinetic feasibility* (sections 12.7 and 12.8).

Thus, $E^{\ominus}$ values predict that $Cu^{2+}(aq)$ should oxidise $H_2(g)$ to $H^+(aq)$, i.e.

$Cu^{2+}(aq) + 2e^- \rightarrow Cu(s)$	+0.34
$H_2(g) \rightarrow 2H^+(aq) + 2e^-$	0.00
Add: $\overline{Cu^{2+}(aq) + H_2(g) \rightarrow Cu(s) + 2H^+(aq)}$	+0.34

But, nothing happens when hydrogen is bubbled into copper(II) sulphate solution because the reaction rate is effectively zero.

Another important point to remember in using $E^{\ominus}$ values is that they *relate only to standard conditions.* Changes in temperature, pressure and concentration will affect the values of electrode potentials. In particular, all electrode potentials for metal ion/metal systems become more positive when the concentration of the metal ion is increased. This is because the reduction of these ions is then more likely to occur. Conversely, electrode potentials become less positive when the concentration of reactant ions is reduced.

For example, the standard electrode potential of copper is +0.34 volts, but the electrode potential of copper in contact with 0.1 mol dm^{-3} $Cu^{2+}(aq)$ is only +0.31 volts. Similarly, the standard electrode potential of zinc is -0.76 volts, but the electrode potential of zinc in contact with 0.1 mol dm^{-3} $Zn^{2+}(aq)$ is reduced to -0.79 volts.

The standard electrode potential for the half-reaction

$$Fe^{3+}(aq) + e^- \rightarrow Fe^{2+}(aq) \quad \text{is +0.77 volts}$$

What happens to the electrode potential when the concentration of

1 $Fe^{3+}(aq)$ rises
2 $Fe^{3+}(aq)$ falls
3 $Fe^{2+}(aq)$ rises
4 $Fe^{2+}(aq)$ falls?

The effect of non-standard conditions on electrode potentials is nicely illustrated by the reaction between $MnO_2(s)$ and $HCl(aq)$. Under standard conditions, MnO_2 will not oxidise 1.0 mol dm^{-3} $HCl(aq)$ to Cl_2.

$$MnO_2(s) + 4H^+(aq) + 2e^- \rightarrow Mn^{2+}(aq) + 2H_2O(l) \qquad \text{+1.23 volts}$$
$$2Cl^-(aq) \rightarrow Cl_2(g) + 2e^- \qquad \text{−1.36 volts}$$
$$\overline{MnO_2(s) + 4H^+(aq) + 2Cl^-(aq) \rightarrow Mn^{2+}(aq) + 2H_2O(l) + Cl_2(g) \qquad \text{−0.13 volts}}$$

However, when manganese(IV) oxide is warmed with *concentrated* hydrochloric acid the electrode potentials for each half-equation become more positive, and the overall electrode potential becomes positive. Hence, chlorine is produced.

Using standard electrode potentials to calculate the e.m.f.s of cells (standard cell potentials)

Electrode potentials can also be used to predict the e.m.f.s of cells. One of the first practical cells to be used was the **Daniell cell**, invented in 1836 (figure 14.6). Notice how closely the Daniell cell resembles the arrangement in figure 14.1 in using $Zn^{2+}(aq)/Zn(s)$ and $Cu^{2+}(aq)/Cu(s)$ half-cells. The only difference is that the Daniell cell uses a rigid porous pot to allow movement of ions between the two solutions whereas the arrangement in figure 14.1 uses a soggy filter paper soaked in potassium nitrate solution.

When the Daniell cell is operating, zinc goes into solution as zinc ions, and copper deposits on the copper electrode. Under standard conditions, the e.m.f. of the cell is 1.10 volts, i.e. $E^{\ominus}_{cell} = 1.10$ volts

$$Zn(s) \rightarrow Zn^{2+}(aq) + 2e^- \qquad \text{+0.76 volts}$$
$$Cu^{2+}(aq) + 2e^- \rightarrow Cu(s) \qquad \text{+0.34 volts}$$

Add: $\overline{Cu^{2+}(aq) + Zn(s) \rightarrow Zn^{2+}(aq) + Cu(s) \qquad \text{+1.10 volts}}$

The arrangement in the Daniell cell can be summarised as:

$$Zn(s)|Zn^{2+}(aq) \; \vdots \; Cu^{2+}(aq)|Cu(s) \qquad E^{\ominus}_{cell} = 1.10 \text{ volts}$$

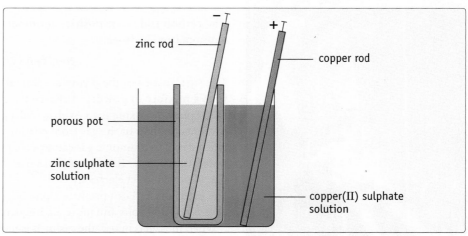

Figure 14.6
The Daniell cell.

Using standard electrode potentials to show the relative oxidising and reducing power of different species

The relative oxidising and reducing power of different species and the relative reactivity of metals and non-metals can be shown using standard electrode potentials (see section 14.5).

Cells and batteries provide a useful and economic way of obtaining energy from chemical reactions. The Daniell cell, discussed in the last section, has now been replaced by cheaper and more convenient cells.

The Leclanché dry cell

One of the commonest, cheapest and most convenient cells in use at the present time is the Leclanché dry cell (figure 14.7). It is used in a wide range of small electrical appliances such as torches, bicycle lamps, radios and electric bells.

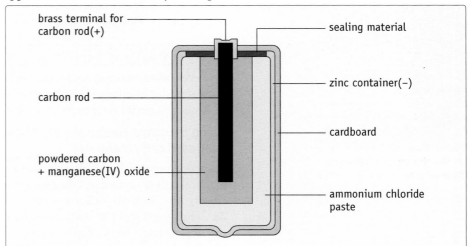

Figure 14.7
The Leclanché dry cell.

1 Why is the Leclanché dry cell cheaper than the Daniell cell?

2 Why is the Leclanché dry cell more convenient and more portable than the Daniell cell?

3 Why is the ammonium chloride used as a paste rather than as a dry solid?

The Leclanché cell can be summarised as:

$$Zn(s)|Zn^{2+}(aq) \;\vdots\; NH_4^+(aq), [2NH_3(g) + H_2(g)]|C \text{ (graphite)} \qquad E = 1.5 \text{ volts}$$

The negative terminal in the cell is zinc, as in the Daniell cell.

$$Zn(s) \rightarrow Zn^{2+}(aq) + 2e^-$$

The carbon rod is the positive terminal at which ammonium ions are converted to ammonia and hydrogen.

$$2NH_4^+(aq) + 2e^- \rightarrow 2NH_3(g) + H_2(g)$$

The surface area of the positive terminal is increased by surrounding the carbon rod with a mixture of powdered charcoal and manganese(IV) oxide. The purpose of the manganese(IV) oxide is to oxidise hydrogen produced at the electrode to water. This prevents bubbles of the gas from coating the carbon terminal, which would reduce its efficiency. The ammonia gas causes no such problems because it dissolves rapidly in the water of the paste. It is necessary to use a paste because dry solid ammonium chloride would not conduct electricity.

A single dry cell can produce a potential difference of 1.5 volts, but batteries of these cells giving 100 volts and more are frequently used.

When the cell is in use, the casing is eaten away as the zinc is converted to Zn^{2+} ions. In time, the zinc case disintegrates and the paste oozes through the outer cardboard.

A number of variants have been developed from the Leclanché dry cell in which the electrolyte is different. For example, the heavy duty type, which can deliver a higher current, has an electrolyte of zinc chloride in place of ammonium chloride and the 'long life' alkaline type has a concentrated potassium hydroxide electrolyte.

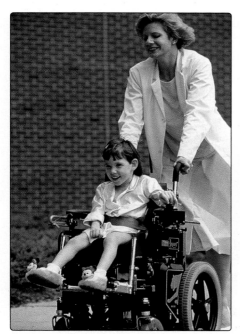

▲ A battery-operated wheelchair.

▲ The chassis of a battery-operated milk float with the batteries visible.

Accumulators (secondary cells)

Obviously, the dry cell cannot be used to provide continuous supplies of electrical energy indefinitely. Nor is there any means of restoring or *recharging* the cell so that it can be used again. Cells such as this which can be used once only and cannot be recharged are called **primary cells**. In contrast, **secondary cells (accumulators)** can be recharged and used again.

When an accumulator is recharged, an electric current is passed through it in the opposite direction to that which the cell generates. Chemical reactions occur at the terminals as the original substances reform. This recharging is an example of electrolysis and is the reverse of discharge.

The lead–acid accumulator

This is certainly one of the most commonly used types of accumulator. Virtually all car batteries are composed of either three or six lead–acid cells in series. This gives a total potential difference of approximately 6 or 12 volts, respectively.

Most milk floats are powered by lead–acid accumulators. They have only a limited mileage and can be conveniently recharged for each day's work. In recent years, the rising cost of petrol has made the electrically powered motor car an increasingly viable proposition. Unfortunately, attempts to develop and market an economic vehicle of this type are still unsuccessful. Problems are caused by high battery costs, the smaller power/weight ratio of the 'battery' engine compared to a petrol engine and the limited mileage available before recharging is needed. Battery cars will not become a reality until a cheap, low-weight, rechargeable battery is developed.

The negative terminal in the lead–acid accumulator is a lead plate. This gives up electrons, forming lead(II) ions during discharge.

$$Pb(s) \rightarrow Pb^{2+}(aq) + 2e^-$$

The positive terminal is lead(IV) oxide. This reacts with H^+ ions in the sulphuric acid electrolyte, also forming lead(II) ions and water.

$$PbO_2(s) + 4H^+(aq) + 2e^- \rightarrow Pb^{2+}(aq) + 2H_2O(l)$$

The Pb^{2+} ions formed during discharge react with SO_4^{2-} ions in the electrolyte forming insoluble lead(II) sulphate.

$$Pb^{2+}(aq) + SO_4^{2-}(aq) \rightarrow PbSO_4(s)$$

It is important that discharge does not continue for too long, otherwise the precipitate of fine lead sulphate changes to a coarser, inactive and non-reversible form. If this happens the accumulator becomes much less efficient.

Fuel cells

Fuel cells are electrochemical cells which convert the chemical energy of a fuel directly to electrical energy. Fuel cells differ from normal electrochemical cells (e.g. dry cells and lead–acid accumulators) in having a steady supply of reactants from which to produce an electric current. Fuel cells use a wide variety of fuels including hydrogen, hydrocarbons and alcohols.

One of the most important fuel cells is the **hydrogen/oxygen fuel cell** (figure 14.8).

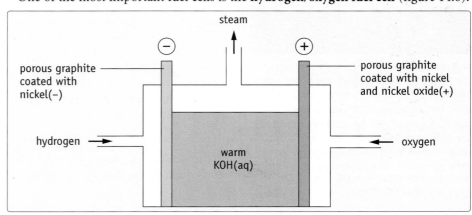

Figure 14.8
A hydrogen/oxygen fuel cell.

▲ A view of the hydrogen/oxygen fuel cell in Europe's first hydrogen-powered taxi, unveiled in London in 1998. The waste product from the fuel cell is water. This helps to reduce pollution from the emissions of city vehicles.

In this cell, the negative terminal is porous graphite coated with nickel and the positive terminal is porous graphite coated with nickel and nickel(II) oxide. The nickel and nickel oxide act as catalysts on the surface of the terminals.

Hydrogen is passed into the negative compartment of the cell and oxygen into the positive compartment. Under pressure, the gases diffuse through the porous graphite terminals into the warm potassium hydroxide solution between them.

At the negative terminal, hydrogen reacts with hydroxide ions from the warm KOH(aq) and electrons are released.

$$2H_2(g) + 4OH^-(aq) \rightarrow 4H_2O(l) + 4e^-$$

At the positive terminal, oxygen and water take electrons to form hydroxide ions.

$$O_2(g) + 2H_2O(l) + 4e^- \rightarrow 4OH^-(aq)$$

The overall reaction in the hydrogen/oxygen fuel cell (obtained by adding the last two equations together) is

$$2H_2(g) + O_2(g) \rightarrow 2H_2O(l)$$

Hydrocarbons and alcohols have also been used in fuel cells in place of hydrogen.

When methane is used, the following half-reactions occur.

$$CH_4(g) + 2H_2O(l) \rightarrow CO_2(g) + 8H^+(aq) + 8e^-$$
$$2O_2(g) + 8H^+(aq) + 8e^- \rightarrow 4H_2O(l)$$

Fuel cells are very efficient, with 70% or more of the chemical energy being converted into electricity. By comparison, the most modern power plants can only achieve a conversion of chemical energy to electricity of about 45%. Fuel cells are also pollution-free. Unfortunately, they are not yet in use on a large scale because impurities in the fuel and oxygen poison the electrocatalysts and this greatly reduces the cells' efficiency.

14.8 Acids and bases

During the nineteenth century, the Swedish chemist Arrhenius put forward ideas about acids and bases. He suggested that *acids were substances which dissociated in water to produce hydrogen ions, H$^+$*,

e.g. $$HCl \rightarrow H^+ + Cl^-$$

Arrhenius suggested that strong acids were completely dissociated in aqueous solution. Weak acids were only partially dissociated with a high proportion of the acid remaining in the undissociated form. Thus, a solution of ethanoic (acetic) acid could be represented by the following equilibrium.

e.g. $$CH_3COOH \rightleftharpoons CH_3COO^- + H^+$$

With the development of knowledge concerning atomic structure, chemists realised that the H$^+$ ion was simply a proton and it was highly improbable that such a small ion would exist independently in aqueous solution.

Thus, H$^+$ ions with their high charge density were believed to associate with polar water molecules in aqueous solution as H$_3$O$^+$ ions. The correct name for H$_3$O$^+$ is the hydroxonium ion. Experimental evidence for H$_3$O$^+$ ions was obtained by observing the movement of H$^+$ ions during electrolysis.

Consequently, the dissociation of acids in water can be represented more accurately as

$$HCl + H_2O \rightarrow H_3O^+ + Cl^-$$
$$CH_3COOH + H_2O \rightleftharpoons H_3O^+ + CH_3COO^-$$

Two important advantages of this refinement to Arrhenius's theory are immediately apparent.

1 It recognised the essential role of 'solvent' water molecules in the dissociation of acids.

2 It explains why substances such as hydrogen chloride and ethanoic acid only show their acidic properties in the presence of water.

▲ A life-saving acid–base reaction. During the Apollo 13 space project, the astronauts discovered that carbon dioxide was building up in their crippled spacecraft as they travelled towards the Earth. By an ingenious use of lithium hydroxide, they were able to repair the air purification equipment in their spacecraft. The photograph shows test pilot Scott MacLeod holding one of the lithium hydroxide containers.

1. Write an equation for the reaction of sodium oxide with dilute hydrochloric acid.
2. Write an equation for the reaction of sodium sulphide with dilute hydrochloric acid.
3. Compare the two equations you have just written. Do you regard the second equation as an acid–base reaction? Explain your answer.

Pure dry ethanoic acid and solutions of hydrogen chloride or ethanoic acid in organic solvents such as toluene contain no H^+ or H_3O^+ ions. Thus, they are non-electrolytes. They do not affect dry litmus paper and they do not react with carbonates to produce CO_2.

Although H^+ ions exist in aqueous solutions as H_3O^+, for the sake of simplicity we usually write H^+ rather than H_3O^+ in equations.

According to the Arrhenius theory, *bases were substances which reacted with H^+ ions to form water*. For example,

$$CuO + 2H^+ \rightarrow Cu^{2+} + H_2O$$
$$Cu(OH)_2 + 2H^+ \rightarrow Cu^{2+} + 2H_2O$$

Alkalis, soluble bases, were regarded as substances which dissociated in water producing OH^- ions.

$$NaOH(s) \xrightarrow{water} Na^+(aq) + OH^-(aq)$$

14.9 The Brønsted–Lowry theory of acids and bases

The definitions of acids and bases in the last section are unsatisfactory because they only apply to aqueous solutions.

Furthermore, the Arrhenius definition of a base is restricted to those substances which react with H^+ ions *to form water*.

In order to widen the scope of acid–base reactions and include non-aqueous systems, Brønsted and Lowry independently suggested the following definitions in 1923.

> *An acid is a proton donor.*
> *A base is a proton acceptor.*

Thus, the relationship between an acid and its corresponding base is

$$\underset{\substack{\text{acid} \\ \text{(proton donor)}}}{HB} \rightleftharpoons \underset{\text{proton}}{H^+} + \underset{\substack{\text{base} \\ \text{(proton acceptor)}}}{B^-}$$

HB and B^- are said to be *conjugate* and to form a **conjugate acid–base pair**. HB is the conjugate acid of B^- and B^- is the conjugate base of HB.

1. What is the conjugate acid of
   (i) CH_3COO^-, (ii) HSO_4^-, (iii) NH_3, (iv) OH^-?
2. What is the conjugate base of
   (i) HCl, (ii) H_3O^+, (iii) HSO_4^-, (iv) NH_3?

According to the Brønsted–Lowry theory, acid salts (such as $NaHSO_4$) and ammonium ions are recognised as acids. What is more, the definition of a base now includes all anions, water and ammonia as well as oxide and hydroxide ions.

Acid		Base		Base		Acid
HSO_4^-	+	OH^-	$\rightarrow$	SO_4^{2-}	+	H_2O
NH_4^+	+	OH^-	$\rightarrow$	NH_3	+	H_2O
$2H_3O^+$	+	S^{2-}	$\rightarrow$	$2H_2O$	+	H_2S
H_3O^+	+	NH_3	$\rightarrow$	H_2O	+	NH_4^+

Notice in the equations above that *water can act as a base and as an acid*. What is more, it does this simultaneously during its dissociation.

$$\underset{\text{acid}}{H_2O} + \underset{\text{base}}{H_2O} \rightleftharpoons \underset{\text{base}}{OH^-} + \underset{\text{acid}}{H_3O^+}$$

14.10 Relative strengths of acids and bases

When the acid HB dissociates in water the following equilibrium is established.

$$HB + H_2O \rightleftharpoons B^- + H_3O^+$$

Acids differ in the extent to which they donate protons to water in aqueous solution. Those which donate all their protons to water molecules are known as strong acids. The equilibrium in the equation above is well over to the right. Thus, the conjugate base of a strong acid is weak, and vice versa.

If HB is a weak acid, relatively few molecules will donate H^+ ions to water. The equilibrium will be well over to the left.

Using the idea of an acid as a proton donor and a base as a proton acceptor, we can arrange all acids and bases in a 'league table'. This shows their order of relative strengths (table 14.4).

Table 14.4
Relative strengths of acids and bases

	Name of acid	Acid		H⁺ + Base	Name of base	
acid strength increases	ethanol	C_2H_5OH	$\rightleftharpoons$	$H^+ + C_2H_5O^-$	ethoxide	base strength decreases
	water	H_2O	$\rightleftharpoons$	$H^+ + OH^-$	hydroxide	
	ammonium	NH_4^+	$\rightleftharpoons$	$H^+ + NH_3$	ammonia	
	hydrogen sulphide	H_2S	$\rightleftharpoons$	$2H^+ + S^{2-}$	sulphide	
	ethanoic acid	CH_3COOH	$\rightleftharpoons$	$H^+ + CH_3COO^-$	ethanoate	
	sulphurous acid	H_2SO_3	$\rightleftharpoons$	$2H^+ + SO_3^{2-}$	sulphite	
	hydroxonium	H_3O^+	$\rightleftharpoons$	$H^+ + H_2O$	water	
	sulphuric acid	H_2SO_4	$\rightleftharpoons$	$2H^+ + SO_4^{2-}$	sulphate	
	hydrochloric acid	HCl	$\rightleftharpoons$	$H^+ + Cl^-$	chloride	
	chloric(VII) acid	$HClO_4$	$\rightleftharpoons$	$H^+ + ClO_4^-$	chlorate(VII)	

If an acid is weak (i.e. has little tendency to donate protons) it follows that its conjugate base is strong. The base will have a strong affinity for protons. For example, hydrogen sulphide is a weak acid, but the sulphide ion is a strong base. Thus, in table 14.4 the acids increase in strength down the page, but their conjugate bases gradually become weaker. The relative strengths of acids are discussed further in sections 22.5 and 22.9.

14.11 Acid–base reactions: competition for protons

When dilute hydrochloric acid is added to a solution of sodium sulphide, hydrogen sulphide is produced.

$$\underbrace{S^{2-} + 2H^+ + 2Cl^-}_{\text{hydrochloric acid}} \rightarrow H_2S + 2Cl^-$$

This reaction emphasises that hydrochloric acid is a stronger acid than hydrogen sulphide. It donates protons to sulphide ions in forming H_2S. Alternatively, we might say that sulphide is a stronger base than chloride. In competition for protons the sulphide wins convincingly.

A similar reaction occurs when dilute hydrochloric acid is added to carbonates. In this case, the HCl(aq) donates H+ ions to carbonate ions forming carbonic acid (H_2CO_3), because hydrochloric acid is a stronger acid than carbonic acid. The carbonic acid is unstable and decomposes to form carbon dioxide and water.

$$2HCl + \underset{\text{carbonate}}{CO_3^{2-}} \rightarrow 2Cl^- + H_2CO_3 \rightarrow 2Cl^- + H_2O + CO_2$$

Thus, *acid–base reactions involve competitions between bases for protons. In this respect, they can be compared to redox reactions which involve competitions between oxidising agents for electrons.*

When dilute hydrochloric acid or sulphurous acid is added to a solution of sodium benzoate, a white precipitate of benzoic acid appears.

14 Competition Processes

1 When Universal Indicator solution
was added to separate portions of
(i) ammonium chloride solution,
(ii) a dilute solution of phenol
(C_6H_5OH) in water and
(iii) water,
solutions (i) and (ii) produced an
orange colour, whereas (iii) was
yellow–green. What can you conclude
from this about the relative acidities of
the three solutions?
2 When a few magnesium turnings were
added to the three solutions mentioned
in the last paragraph, solution (i)
evolved hydrogen most vigorously and
(iii) least vigorously. What can you
conclude from this about the relative
acidities of (i), (ii) and (iii)?
3 Where would you place phenol
(C_6H_5OH) in the 'league' of relative
acidity in table 14.4?

Figure 14.9
Co-ordinate bonding in an aqueous H^+ ion
(hydroxonium ion).

$$C_6H_5COO^-(aq) + \underline{H^+(aq) + Cl^-(aq)} \rightarrow C_6H_5COOH(s) + Cl^-(aq)$$

benzoate ion   hydrochloric acid   benzoic acid

$$C_6H_5COO^-(aq) + H_2SO_3(aq) \rightarrow C_6H_5COOH(s) + HSO_3^-(aq)$$

benzoate ion   sulphurous acid   benzoic acid

Hence, both hydrochloric acid and sulphurous acid are stronger than benzoic acid.

However, when ethanoic acid is added to a solution of sodium benzoate ions, no apparent reaction occurs. Ethanoic acid does not protonate benzoate ions because ethanoic acid is a weaker acid than benzoic acid. Thus, we could place benzoic acid between ethanoic acid and sulphurous acid in the 'league' showing relative strengths of acids in table 14.4.

14.12 Complex ions

In the last few sections, we have been discussing the reactions of acids with bases. Acids donate protons (H^+ ions). In aqueous solution these H^+ ions are strongly attracted to polar water molecules by co-ordinate bonds forming H_3O^+ ions (figure 14.9).

In the same way as H^+, other cations (e.g. M^{2+}) will also exist in aqueous solution as hydrated ions (e.g. $[M(H_2O)_n]^{2+}$). Since the H^+ ion (a bare proton) is more than 50 000 times smaller than any other cation, the bonding between H_2O and H^+ is much stronger than that between H_2O molecules and other cations. However, the larger size of other cations enables them to associate with two, four or even six water molecules. H^+ associates with only one.

The number of water molecules associated with a given cation has been firmly established in some cases. In other cases it has been established more tentatively through indirect evidence. Thus, $Mg^{2+}(aq)$ is represented as $[Mg(H_2O)_6]^{2+}$, $Cu^{2+}(aq)$ is represented as $[Cu(H_2O)_4]^{2+}$ and $Ag^+(aq)$ is represented as $[Ag(H_2O)_2]^+$.

In these ions, the water molecules are bound to the central cation by co-ordinate bonds (figure 14.10).

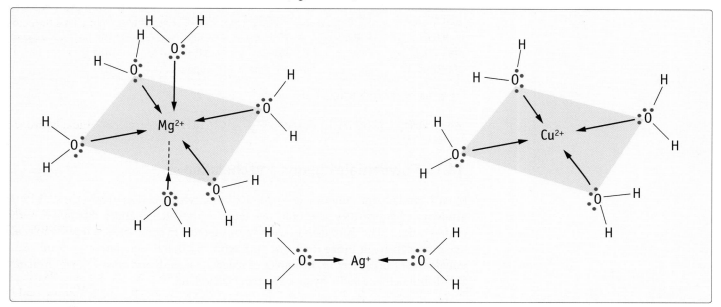

Figure 14.10
Co-ordinate bonding in hydrated cations.

Other polar molecules, besides water, can also co-ordinate to metal cations. For example, in ammonia solution Cu^{2+} exists principally as $[Cu(NH_3)_4]^{2+}$ whilst Ag^+ exists as $[Ag(NH_3)_2]^+$.

It is not only polar molecules but also anions that can associate with cations in this way. Thus, when anhydrous copper(II) sulphate is added to concentrated hydrochloric acid, the solution becomes yellow–green. This is due to the formation of $[CuCl_4]^{2-}$ ions.

Ions such as $[Cu(H_2O)_4]^{2+}$, $[Cu(NH_3)_4]^{2+}$ and $[CuCl_4]^{2-}$, in which a metal ion is associated with a number of anions or neutral molecules, are known as **complex ions**.

Table 14.5
Hydration enthalpies of some cations

Ion	Ionic radius /nm	Hydration enthalpy /kJ mol^{-1}
Li$^+$	0.068	−499
Na$^+$	0.098	−390
K$^+$	0.133	−305
Rb$^+$	0.148	−281
Cs$^+$	0.167	−248
Mg^{2+}	0.065	−1891
Ca^{2+}	0.094	−1562
Sr^{2+}	0.110	−1413
Ba^{2+}	0.134	−1273
Al^{3+}	0.045	−4613
Ga^{3+}	0.062	−4650

Look closely at table 14.5.

1 How does the hydration enthalpy change as ionic radius increases?

2 How does the hydration enthalpy change as the charge on an ion increases?

3 In general, how will the strengths of co-ordinate bonds from ligands to a central cation vary with the size of the cation and with the charge on the cation?

The anions and molecules firmly bonded to the central cation are called **ligands**. Each ligand contains at least one atom with a lone pair of electrons. These can be donated to the central cation forming a co-ordinate (dative) bond. The ligand is said to be co-ordinated to the central ion.

Water is by far the most common ligand. The strengths of the co-ordinate bonds between different cations and water molecules can be compared by determining their **hydration enthalpies** (section 12.14). These relate to the process

$$M^{x+}(g) + nH_2O(l) \rightarrow [M(H_2O)_n]^{x+}(aq)$$

The hydration enthalpies of some cations are listed in Table 14.5.

14.13 Naming complex ions

The systematic naming of complex ions is based on four simple rules.

1 *State the number of ligands around the central cation* using Greek prefixes: *mono-, di-, tri-,* etc.

2 *Identify the ligands* using names ending in *-o* for anions, e.g. F$^-$ -fluoro, CN$^-$ -cyano, Cl$^-$ -chloro, OH$^-$ -hydroxo, H$_2$O aqua, NH$_3$ -ammine.

3 *Name the cation* using the English name for the cation in a positively charged complex, but the Latinised name ending with the suffix *-ate* for the cation in a negatively charged complex, i.e. aluminate for aluminium, cuprate for copper, ferrate for iron, plumbate for lead, zincate for zinc, etc.

4 *Indicate the oxidation number of the central cation* using Roman numerals (I, II, III, etc.). The following examples show you how to apply these rules.

Formula of complex ion	1 State the number of ligands	2 Name the ligand	3 Name the central cation (suffix *-ate* for anions)	4 Indicate oxidation number of central cation
[Cu(NH$_3$)$_4$]$^{2+}$	tetra	ammine	copper	(II)
[CuCl$_4$]$^{2-}$	tetra	chloro	cuprate	(II)
[Fe(CN)$_6$]$^{3-}$	hexa	cyano	ferrate	(III)
[Cu(NH$_3$)$_2$]$^+$	di	ammine	copper	(I)

Try to name the following ions:

1 [Al(OH)$_4$]$^-$, [Zn(NH$_3$)$_4$]$^{2+}$, [Fe(CN)$_6$]$^{4-}$, [CrCl$_2$(H$_2$O)$_4$]$^+$.

14.14 Polydentate ligands and chelation

Most ligands form only one co-ordinate bond with a cation. They are said to be **unidentate** because they have only 'one tooth' with which to attach themselves to the central cation. (The word *dens* in Latin means tooth.) In some cases, a ligand molecule or anion can form more than one link with the metal ion. These are said to be **polydentate** ('many teeth'). Examples of bidentate ligands are: ethanedioate (oxalate), 1, 2-diaminoethane and 2-hydroxybenzoate (salicylate).

ethanedioate   1,2–diaminoethane   2–hydroxybenzoate

Some ligands such as edta (ethylenediaminetetraacetate) can form as many as six dative bonds with the central ion. They are known as hexadentate ligands.

The complex ions which form between polydentate ligands and cations are known as **chelates** or **chelated complexes**. These names come from the Greek word *chelos* meaning 'a crab's claw'. In these complexes the ligand forms a clawing pincer-like grip on the metal ion. Figure 14.11 shows the chelate formed between iron(III) and ethanedioate (oxalate) ions.

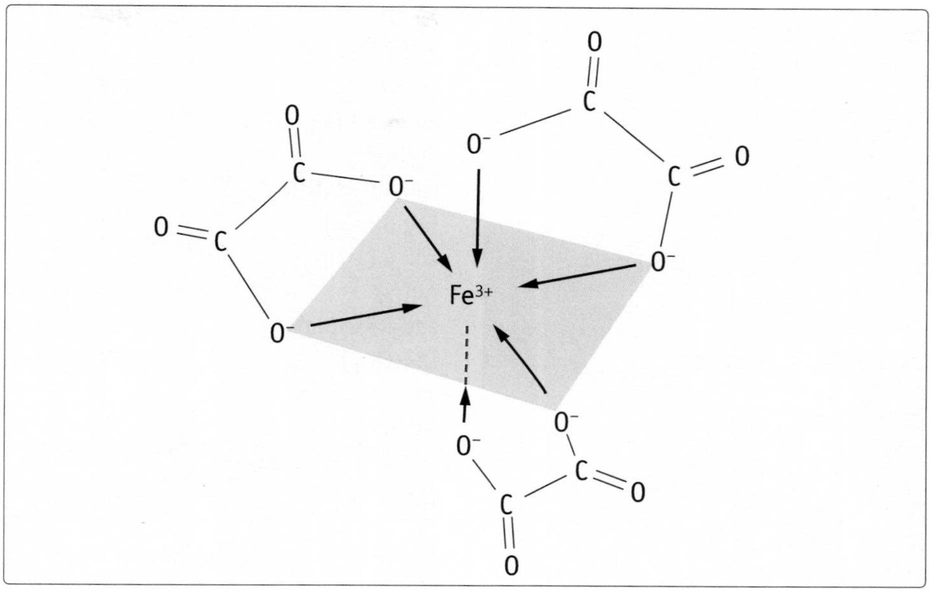

Figure 14.11
The structure of the iron(III)/ethanedioate chelate, $[Fe(C_2O_4)_3]^{3-}$.

In general, polydentate ligands are more powerful and more versatile than simple unidentate ligands. The stability of a complex is much enhanced by chelation as the pincer-like grip of the polydentate ligand can hold the central cation more securely.

14.15 Complexing: competition for cations

In section 14.11 we noticed that *acid–base reactions involve competitions between different bases for H$^+$ ions*. In a similar fashion, *complexing reactions involve competitions between different ligands for metal cations*. Thus, when excess ammonia solution is added to aqueous copper(II) sulphate, ammonia molecules displace water molecules from hydrated copper(II) ions forming $[Cu(NH_3)_4]^{2+}$. The colour changes from pale blue to a much deeper blue.

$$[Cu(H_2O)_4]^{2+}(aq) + 4NH_3(aq) \rightarrow [Cu(NH_3)_4]^{2+}(aq) + 4H_2O(l)$$

In this case, ammonia acts as a stronger ligand than water.

In general, it can be shown that a more powerful ligand will displace a less powerful ligand from a cation complex. Thus, when 2-hydroxybenzoate ions (salicylate, $HOC_6H_4COO^-$) are added to aqueous iron(III) chloride the colour changes from yellow to a deep purple. 2-Hydroxybenzoate ions displace water molecules from hydrated iron(III) ions forming purple $[Fe(HOC_6H_4COO)_3]$.

$$[Fe(H_2O)_6]^{3+}(aq) + 3HOC_6H_4COO^-(aq) \rightarrow [Fe(HOC_6H_4COO)_3](s) + 6H_2O(l)$$
$$\text{yellow} \qquad\qquad\qquad\qquad\qquad\qquad\qquad \text{purple}$$

When edta (an even stronger ligand than 2-hydroxybenzoate) is added to the purple suspension, the edta displaces 2-hydroxybenzoate ions. The colour changes from purple to pale yellow, which is the colour of the iron(III)/edta complex.

One of the most fascinating demonstrations of the relative strengths of different ligands for a particular cation is outlined in the sequence of reactions shown in table 14.6. Going down the table, the ligands are increasing in strength and becoming stronger competitors for Ag$^+$ ions, i.e.

$$edta^{2-} > S^{2-} > CN^- > I^- > S_2O_3{}^{2-} > Br^- > NH_3 > Cl^- > H_2O$$

Table 14.6
The relative strengths of different ligands for Ag^+

	Ligand	Colour and state of complex
$[Ag(H_2O)_2]^+$ add ↓ NaCl(aq)	H_2O	clear solution
AgCl(s) add ↓ NH_3(aq)	Cl^-	white precipitate
$[Ag(NH_3)_2]^+$ add ↓ KBr(aq)	NH_3	clear solution
AgBr(s) add ↓ $Na_2S_2O_3$(aq)	Br^-	cream precipitate
$[Ag(S_2O_3)_2]^{3-}$ add ↓ KI(aq)	$S_2O_3^{2-}$	clear solution
AgI(s) add ↓ KCN(aq)	I^-	yellow precipitate
$[Ag(CN)_2]^-$ add ↓ Na_2S(aq)	CN^-	clear solution
Ag_2S(s) add ↓ sodium edta(aq)	S^{2-}	black precipitate
$[Ag(edta)]^-$	$edta^{2-}$	clear solution

Notice that some reactions in the sequence result in the precipitation of insoluble solids. These precipitates can be regarded in some ways as neutral complexes. Being uncharged, they are less readily hydrated by polar water molecules than the charged complexes and so they are less likely to dissolve in water.

14.16 Investigating the formulas of complex ions

Determining the formula of a complex ion involves measuring the number of ligands complexing with one metal cation. This can be investigated by various methods. The most important methods are:

1 titration methods involving competitive complexing (described below), and

2 colorimetric methods which measure the colour intensity of the mixture as the proportion of metal cation to ligand is varied.

Determining the formula of the iron(III)/edta complex by competitive complexing

When sodium 2-hydroxybenzoate (sodium salicylate) is added to a solution of Fe^{3+} ions a deep purple complex ion is formed. As edta is added to this purple solution, the colour slowly fades to a pale yellow iron(III)/edta complex. Of course, the purple colour remains as long as there are more than enough Fe^{3+} ions for the edta. Any excess Fe^{3+} ions will form a purple complex with the 2-hydroxybenzoate. When the purple colour just disappears, the quantities of Fe^{3+} and edta just react and neither is in excess. Thus, 2-hydroxybenzoate ions can act as indicators. They will show when just sufficient edta has been added to react with all the Fe^{3+}.

In a particular experiment, 20 cm^3 of 0.1 mol dm^{-3} iron(III) chloride was pipetted into a flask. A few drops of sodium 2-hydroxybenzoate were added. 0.1 mol dm^{-3} edta was then added from a burette and the solution became a clear yellow colour after the addition of 20.1 cm^3.

1 How many moles of Fe^{3+} were taken?

2 How many moles of edta reacted with the Fe^{3+} taken?

3 How many moles of edta react with one mole of Fe^{3+}?

4 What is the formula of the Fe^{3+}/edta complex?

Complex ions are important in biology and in industry.

Complex ions of biological importance

Two essential biological macromolecules composed of complex ions are **chlorophyll** and **haemoglobin**. Chlorophyll is the green pigment in plant cells. It is responsible for absorbing the radiant energy of sunlight. It converts sunlight into chemical energy in the bonds of carbohydrate molecules synthesised by the plant.

The chlorophyll molecule is composed of a complicated cyclic tetradentate ligand. This contains four nitrogen atoms surrounding magnesium in a square planar arrangement (figure 14.12).

▶ An electron microscope photograph of a chloroplast. The white and yellow circles are the sites of carbohydrate (starch) produced during photosynthesis.

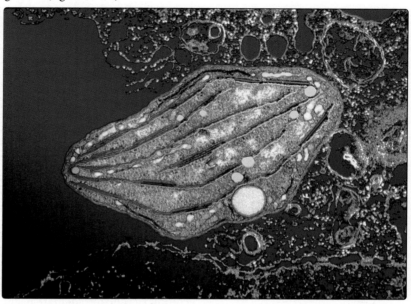

Chlorophyll a
(In chlorophyll b, the methyl group marked by an asterisk is replaced by a ——CHO group)

Haemoglobin

Figure 14.12
Complex ions of biological importance.

Sunlight reaching the Earth's surface has a maximum intensity in the blue–green region of the spectrum in the wavelength range 450–550 nm. Curiously, the chlorophyll molecule has its weakest absorption in this portion of the visible spectrum.

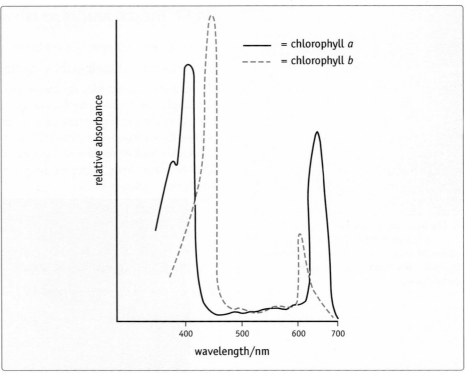

Figure 14.13
The absorption spectrum of chlorophyll.

The maximum absorption peaks for chlorophyll are, in fact, at 680 nm in the red region of the spectrum and at 440 nm in the violet (figure 14.13). In spite of this inefficiency in absorbing the radiation of greatest intensity, chlorophyll has other properties that make it particularly suitable as a photosynthetic pigment. It can receive energy both directly from light and indirectly from other pigments in plants such as carotenoids.

Figure 14.12 also shows the structure of haemoglobin, the red pigment present in red blood cells. Haemoglobin acts as the transport of oxygen in the blood. It is composed of the complex cyclic tetradentate haem ligand and the protein globin. Notice the striking similarity of the haem portion of the complex to the chlorophyll molecule. This suggests that the two may have been adapted from the same original substance during the course of evolution. In haemoglobin, the central metal ion is iron(II), not magnesium. The active part of the haemoglobin complex is the Fe^{2+} ion. This is co-ordinated to four nitrogen atoms in the haem ligand and also to two nitrogen atoms in the globin. One of the latter two positions can be occupied weakly and reversibly by oxygen. It is this property which enables haemoglobin to transport oxygen in the bloodstream from the lungs to other parts of the body. Unfortunately, much stronger ligands than oxygen, such as cyanide and carbon monoxide, can also occupy this position. Unlike oxygen, they attach themselves to the Fe^{2+} ion in the haem group irreversibly. So, they act as acute poisons by eliminating the oxygen-transporting ability of haemoglobin.

Complex ions of industrial importance

Complex ions play an important part in the methods used to soften hard water and in the extraction of metals from their ores.

Soap contains the sodium salts of long-chain carboxylic acids such as sodium hexadecanoate (sodium palmitate) and sodium octadecanoate (sodium stearate) (section 32.3). When hard water, containing calcium and magnesium ions, is mixed with soap, it reacts with the anions in these salts. The products are insoluble compounds which we see as 'scum' on the surface of the water.

$$Ca^{2+}(aq) \quad + \quad 2CH_3(CH_2)_{16}COO^- \quad \rightarrow \quad Ca(CH_3(CH_2)_{16}COO)_2(s)$$

| ions in hard water | octadecanoate (stearate) ions in soap | insoluble ppte of calcium octadecanoate (stearate) in 'scum' |

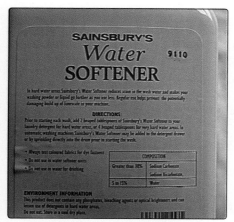

▲ The label from a packet of Sainsbury's water softener containing sodium carbonate and sodium hydrogencarbonate (sodium bicarbonate) as the main constituents.

Various ligands will react with calcium and magnesium ions to form soluble and stable complex ions. As a result of this, the calcium and magnesium ions are unable to react with the anions in soap and the water is softened. Thus, sodium polyphosphate ($Na_6P_6O_{18}$), which contains the powerful polyphosphate ligand ($P_6O_{18}^{6-}$), is sold for domestic and industrial water softening under the trade name Calgon. The name Calgon is derived from the expression 'calcium gone'. Edta and its sodium salt are also used in water softening. However, these two substances are poisonous so they cannot be used to soften drinking water. Consequently, their use is restricted to the softening of water required for industrial processes such as the dyeing of textiles.

Complex formation frequently plays an essential part in the purification of metal ores and the subsequent extraction of the metal. An excellent example of this is provided by the purification of aluminium oxide from bauxite (section 19.9). The extraction of gold and silver involves the formation of complex cyanides. Impure silver ores such as argentite (impure Ag_2S) and horn silver (impure $AgCl$) are first mixed with a solution of sodium cyanide. This forms a soluble complex cyanide $[Ag(CN)_2]^-$.

$$Ag_2S(s) + 4CN^-(aq) \rightarrow 2[Ag(CN)_2]^-(aq) + S^{2-}(aq)$$
$$AgCl(s) + 2CN^-(aq) \rightarrow [Ag(CN)_2]^-(aq) + Cl^-(aq)$$

Silver is then precipitated by adding zinc dust to the solution. Any excess zinc is removed by adding dilute acid.

$$2[Ag(CN)_2]^-(aq) + Zn(s) \rightarrow [Zn(CN)_4]^{2-}(aq) + 2Ag(s)$$

14.18 Titrations – determining amounts and concentrations

Titrations involve the addition of one solution to another in order to find out how much of the two solutions just react with each other (section 2.3). From the results of the titration, it is possible to calculate the amount or concentration of one substance if the values for the other substance are known.

For **acid–base reactions**, we can find the amount or concentration of an acid by titration with a suitable alkali (usually sodium hydroxide solution) of known concentration. An indicator such as methyl orange (red in acid, yellow in alkali) or phenolphthalein (colourless in acid, red in alkali) can be used to judge the *end point* of the titration when the two solutions just react and neither is in excess (section 2.3).

Similarly, we can determine the amount or concentration of an alkali by titration with a suitable acid of known concentration (normally hydrochloric acid or sulphuric acid) using an indicator.

In redox reactions, oxidising agents are reacting with reducing agents. So, we need a suitable oxidising agent to determine the amounts of reducing agents and vice versa. The oxidising agent usually chosen to estimate reducing agents is potassium manganate(VII), $KMnO_4$, because it can be obtained as a pure solid which is stable and easily stored.

Another advantage of potassium manganate(VII) is that it provides its own indication of the end point of the titration. In order to determine the concentration of a reducing agent (e.g. $Fe^{2+}(aq)$, $I^-(aq)$), a solution of purple potassium manganate(VII) of suitable concentration is added from a burette to the reducing agent in a titration flask until there is a slightly pink tinge. This pink tinge indicates a tiny excess of $KMnO_4$ and therefore the end point of the titration.

During titrations with $KMnO_4$, excess dilute sulphuric acid should be added to the reaction mixture because of the need for $H^+(aq)$ ions in the MnO_4^- half-equation.

$$MnO_4^-(aq) + 8H^+(aq) + 5e^- \rightarrow Mn^{2+}(aq) + 4H_2O(l)$$

Example

An iron tablet containing iron(II) sulphate was analysed using 0.005 M $KMnO_4$. The iron tablet (mass 0.65g) was dissolved in 100 cm^3 of dilute sulphuric acid. 10 cm^3 of this solution required 6.00 cm^3 of 0.005 M $KMnO_4$ to produce a faint pink colour. (A_r(Fe) = 56). What is the percentage of iron in the iron tablets?

The equations for the reaction are:

$$MnO_4^- + 8H^+ + 5e^- \rightarrow Mn^{2+} + 4H_2O$$
$$(Fe^{2+} \rightarrow Fe^{3+} + e^-) \times 5$$
$$\therefore 5 \text{ moles } Fe^{2+} \text{ react with 1 mole } MnO_4^-$$

Number of moles of MnO_4^- used in titration $= \dfrac{6}{1000} \times 0.005$

$\therefore$ number of moles of Fe^{2+} reacting in the titration $= \dfrac{6}{1000} \times 0.005 \times 5$

$\Rightarrow$ number of moles of Fe^{2+} in 100 cm^3 of dilute acid $= \dfrac{6}{1000} \times 0.005 \times 5 \times 10$
and in one tablet

$$= 0.0015$$

$\therefore$ mass of iron in one tablet $= 0.0015 \times 56 = 0.084$g

$\Rightarrow$ % of iron in the tablet $= \dfrac{0.084}{0.65} \times 100$

$$= 12.9\%$$

The determination of oxidising agents uses a combination of potassium iodide and sodium thiosulphate as the reducing agent.

The oxidising agent to be estimated (e.g. Cl_2(aq), Fe^{3+}(aq)) is first added to excess potassium iodide. This liberates iodine:

$$Cl_2(aq) + 2e^- \rightarrow 2Cl^-(aq)$$

$$2I^- \rightarrow I_2(aq) + 2e-$$

The amount of iodine liberated is then determined by titration against a solution of sodium thiosulphate of known concentration.

$$I_2(aq) + 2e^- \rightarrow 2I^-(aq)$$

$$2S_2O_3^{2-}(aq) \rightarrow S_4O_6^{2-}(aq) + 2e^-$$
$$\text{thiosulphate} \qquad \text{tetrathionate}$$

So, 1 mole Cl_2 ≡ 1 mole I_2 ≡ 2 moles $S_2O_3^{2-}$.

The end point of the titration can be detected by adding starch solution. The starch gives a dark blue colour as long as any iodine is present. By calculating the amount of thiosulphate used in the titration, we can determine the amount of iodine liberated and hence the amount of oxidising agent taken.

Substances, such as potassium manganate(VII) and sodium thiosulphate, which can be prepared to a high degree of purity and made up into solutions of precise concentration, are described as **primary volumetric standards**. They are used as the starting point in determining the concentration of (i.e. standardising) other solutions.

Summary

1 The major classes of reaction involving inorganic substances are redox, acid–base, precipitation and complexing.

2 Redox reactions involve competition for electrons. Oxidising agents (oxidants) accept electrons. The stronger the oxidising agent, the greater is its competitive desire for electrons. Reducing agents (reductants) are electron donors.

3 The electromotive force of a cell is the maximum potential difference which the cell can generate.

4 The standard conditions for electrochemical ($E^{\ominus}$) measurements are similar to those for thermochemical measurements: a temperature of 25°C, gases at a pressure of 1 atm, aqueous solutions with a concentration of 1.0 mol dm^{-3}, and platinum as the electrode when the half-cell system does not include a metal.

5 The standard electrode (reduction) potential, $E^{\ominus}$, of a standard half-cell is the potential of that half-cell relative to a standard hydrogen half-cell under standard conditions. $E^{\ominus}$, which relates to the reduction half-equation, has the same sign as the particular half-cell relative to the standard hydrogen half-cell.

6 The size and sign of $E^{\ominus}$ values provide a measure of the relative strength of oxidising agents and their conjugate reducing agents.

7 An overall positive $E^{\ominus}$ for a redox pair suggests that a reaction should take place. In practice:

(i) some likely reactions may be too slow to occur,

(ii) reactions which are unlikely under standard conditions (to which $E^{\ominus}$ values refer) may take place readily under different conditions of temperature, pressure and concentration.

8 Arrangements which generate electric currents from chemical reactions are called cells. A battery is composed of several cells.

9 Cells which can be used only once and which cannot be recharged are called primary cells. Cells which can be recharged and used again and again are called secondary cells or accumulators.

10 According to the Brønsted–Lowry theory of acids and bases, an acid is a proton donor, whereas a base is a proton acceptor.

11 Acid–base reactions involve competitions between bases for protons (H$^+$ ions). The stronger the base the greater is its competitive desire for protons.

12 Water can act as a base and as an acid. It does so simultaneously during its dissociation.

$$\underset{\text{acid}}{H_2O} + \underset{\text{base}}{H_2O} \rightleftharpoons \underset{\text{base}}{OH^-} + \underset{\text{acid}}{H_3O^+}$$

13 A complex ion is an ion in which a number of anions or neutral molecules are bound to a central metal cation by co-ordinate (dative) bonds.

14 The anions and molecules bound to the central cation in a complex ion are called ligands.

15 Chelates or chelated complexes are complex ions in which each ligand forms more than one link with the central cation.

16 Complexing reactions involve competitions between ligands for metal cations. The stronger the ligand the greater is its competitive desire for cations.

17 Titrations involve the addition of one solution to another in order to find out how much of the two solutions just react with each other.

18 In acid–base titrations, an acid of known concentration is used to determine the amount of base and vice versa. In redox titrations, an oxidising agent of known concentration is used to determine the amount of reducing agent and vice versa.

Review questions

1 (a) How did Arrhenius define (i) an acid, (ii) a base?

(b) How did Brønsted and Lowry define (i) an acid, (ii) a base?

(c) Show, by writing appropriate equations how:

(i) the HSO_4^- ion can act as an Arrhenius acid,

(ii) the HSO_4^- ion can act as a Brønsted–Lowry base.

(d) Explain, using examples, how the Brønsted–Lowry definition allows a much wider range of substances to be classified as bases.

2 2.5 g of a hydrate of sodium carbonate, $Na_2CO_3.xH_2O$, were dissolved in 250 cm^3 of water. 25 cm^3 of the resulting solution required 17.5 cm^3 of 0.1 mol dm^{-3} hydrochloric acid to neutralise it completely.

(a) How many moles of sodium carbonate react with one mole of hydrochloric acid?

(b) How many moles of hydrochloric acid are there in 17.5 cm^3 of a 0.1 mol dm^{-3} solution?

(c) How many moles of sodium carbonate react with 17.5 cm^3 of 0.1 mol dm^{-3} hydrochloric acid?

(d) How many moles of sodium carbonate were there in the original 2.5 g?

(e) How many moles of water of crystallisation must there have been in the original 2.5 g?

(f) What is the value of x?

(Na = 23, C = 12, O = 16, H = 1)

3 (a) Explain what you understand by the terms oxidation and reduction.

(b) Illustrate your answer to part (a) by writing separate half-equations involving electrons for the oxidation and reduction processes when the following substances react:
  (i)   Mg with Cl_2,
  (ii) I_2 with $S_2O_3^{2-}$,
  (iii) MnO_2 with conc. HCl,
  (iv) H_2O_2 with I^-,
  (v) Fe^{3+} with I^-.

(c) Suppose you were required to determine the standard electrode potential of a $Zn^{2+}(aq)/Zn(s)$ half-cell.
  (i) List all the essential materials and equipment required for this determination.
  (ii) Draw a diagram showing how the apparatus would be assembled for the determination.

4 Use oxidation numbers to decide which elements are oxidised or reduced in the following conversions.

(a) $Cr^{3+} \rightarrow Cr_2O_7^{2-}$
(b) $CrO_4^{2-} \rightarrow Cr_2O_7^{2-}$
(c) $S_2O_3^{2-} \rightarrow S_4O_6^{2-}$
(d) $C_2O_4^{2-} \rightarrow CO_2$
(e) $IO_3^- \rightarrow I_2$

5 (a) Draw a fully labelled diagram of a Daniell cell and show the direction in which electrons flow in the external circuit when it is used to generate electricity.

(b) Write equations for the reactions at each electrode when the cell produces an electric current.

(c) Deduce a value for the e.m.f. of the cell operating under standard conditions.
($Cu^{2+}(aq)/Cu(s) = +0.34$ V, $Zn^{2+}(aq)/Zn(s) = -0.76$ V.)

(d) How would the e.m.f. of the cell be affected if:
  (i) the concentration of Cu^{2+} ions was increased,
  (ii) the concentration of Zn^{2+} ions was increased?
Give an explanation of your reasoning in each case.

6 The rusting of iron is an oxidation process.

(a) Explain, with appropriate equations, what happens when iron rusts.

(b) Iron may be protected from rusting by coating with zinc or tin.

By reference to the following data, explain why zinc protects iron more effectively than tin, once the protective metal coating has been scratched to expose the iron below.

					$E^\ominus$ /volts
Zn	$\rightarrow$	Zn^{2+}	+	$2e^-$	+0.76
Fe	$\rightarrow$	Fe^{2+}	+	$2e^-$	+0.44
Sn	$\rightarrow$	Sn^{2+}	+	$2e^-$	+0.14

(c) Why do you think tin is used rather than zinc to coat and protect the inside of 'tin' cans, whereas zinc is used rather than tin to galvanise and protect buckets?

7 The following table gives the standard electrode potentials for a number of half-reactions.

						$E^\ominus$ /volts
Zn^{2+}	+	$2e^-$	$\rightarrow$		Zn	-0.76
Fe^{2+}	+	$2e^-$	$\rightarrow$		Fe	-0.44
I_2	+	$2e^-$	$\rightarrow$		$2I^-$	$+0.54$
Fe^{3+}	+	e^-	$\rightarrow$		Fe^{2+}	$+0.77$
Ce^{4+}	+	e^-	$\rightarrow$		Ce^{3+}	$+1.61$

(a) Which half-equation is used as the standard for these electrode potentials?

(b) Which of the substances in the table above is:
  (i) the strongest oxidising agent,
  (ii) the strongest reducing agent?

(c) Which substance(s) in the table could be used to convert iodide ions to iodine? Write balanced equations for any possible conversions.

(d) A half-cell is constructed by putting a platinum electrode in a solution which is 1.0 mol dm^{-3} with respect to both Fe^{2+} and Fe^{3+} ions. This half-cell is then connected by means of a 'salt bridge' to another half-cell containing an iron electrode in a 1.0 mol dm^{-3} solution of Fe^{2+} ions.
  (i)   What is the e.m.f. of this cell?
  (ii) If the two electrodes are connected externally, what reactions take place in each half-cell?
  (iii) In which direction do electrons flow in the external circuit?

(e) Write an equation for the reaction you would expect to occur when an iron nail is placed in a solution of iron(III) sulphate.

8 When edta (ethylenediaminetetraacetic acid) is titrated against calcium salts, experiments show that one mole of edta reacts with one mole of Ca^{2+}.
This technique is used in clinical analysis to estimate the concentration of calcium in blood serum. Concentrations between 7 and 13 mg of calcium per 100 cm^3 of blood serum are normal.
One cubic centimetre of blood serum was taken from a patient and found to require 2.75 cm^3 of 0.0015 mol dm^{-3} edta solution.

(a) Calculate the concentration of calcium in the patient's blood serum in mg per 100 cm^3 (Ca = 40).

(b) What assumption(s) have you made in your calculation?

(c) Is there an abnormal level of calcium in the patient's blood?

Groups I and II – The Alkali Metals and the Alkaline-Earth Metals

15

15.1 Introduction

Table 15.1
The members of Group I and Group II in the periodic table

Group I		Group II	
Lithium	Li	Beryllium	Be
Sodium	Na	Magnesium	Mg
Potassium	K	Calcium	Ca
Rubidium	Rb	Strontium	Sr
Caesium	Cs	Barium	Ba
The alkali metals		**The alkaline-earth metals**	

The alkali metals and the alkaline-earth metals are members of Groups I and II, respectively, in the periodic table (table 15.1). Whilst you are studying these elements, look for the following important features:

1 The similarities between the elements in Group I.

2 The similarities between the elements in Group II.

3 The differences between the elements in Group I and those in Group II.

4 The gradual trends in properties of the elements and their compounds in Group I and the similar trends in Group II.

In all these metals, the only electrons in their outermost shell occupy an *s*-orbital. Hence these elements are sometimes called the **s-block elements.**

> 1 Write the electronic structures of Na and K in terms of *s, p, d, f* notation.
> (The electronic structure of Li is $1s^2 2s^1$.)
>
> 2 Write the electronic structures of Be, Mg and Ca in terms of *s, p, d, f* notation.

You should now realise that *all Group I metals have one outermost s-electron, whereas Group II metals have two outermost s-electrons.* These s-electrons are situated much further from the nucleus than all other electrons in the metals, so they are only weakly held by the positive nucleus. Consequently, the atoms of Group I metals readily lose their single outermost *s*-electron to form stable ions. These ions have a noble gas electron structure with one positive charge, e.g.

$$Na \rightarrow Na^+ + e^-$$

Similarly, the atoms of Group II metals readily lose two electrons to form stable ions, with a noble gas structure and two positive charges, e.g.

$$Ca \rightarrow Ca^{2+} + 2e^-$$

15.2 Oxidation numbers and s-block elements

Tables 15.2 and 15.3 summarise various physical properties of the elements in Groups I and II, respectively.

Look closely at the first and second ionisation energies for Group I elements. The first ionisation energy is much lower than the second. In the case of Na, it is nine times easier to remove the first electron than the second. The first electron, in an *s*-orbital, can be removed from the element easily. The second electron must be removed from a noble gas core. This is closer to the nucleus and therefore requires much more energy. Thus, sodium readily forms Na^+ ions, but never forms Na^{2+} ions. This means that sodium and the other elements in Group I have only the one oxidation state of +1 in their compounds.

▲ The horizontal bands of rock in the Grand Canyon, Arizona, U.S.A, are limestones and sandstones. These rocks are mainly compounds of calcium and magnesium, two of the alkaline-earth metals in Group II. Limestone is mainly calcium carbonate, and sandstone contains calcium and magnesium silicates.

Table 15.2
Physical properties of the elements in Group I

Element	Li	Na	K	Rb	Cs
Electron structure	$(He)2s^1$	$(Ne)3s^1$	$(Ar)4s^1$	$(Kr)5s^1$	$(Xe)6s^1$
First ionisation energy/kJ mol^{-1}	520	500	420	400	380
Second ionisation energy/kJ mol^{-1}	7300	4600	3100	2700	2440
Atomic radius/nm (metallic radius)	0.15	0.19	0.23	0.25	0.26
Melting point/°C	180	98	64	39	29
Boiling point/°C	1330	892	760	688	690
Density/g cm^{-3}	0.53	0.97	0.86	1.53	1.90
Standard electrode potential, $M^+(aq)/M(s)$/volts	−3.05	−2.71	−2.93	−2.92	−2.92

Table 15.3
Physical properties of the elements in Group II

Element	Be	Mg	Ca	Sr	Ba
Electron structure	$(He)2s^2$	$(Ne)3s^2$	$(Ar)4s^2$	$(Kr)5s^2$	$(Xe)6s^2$
First ionisation energy/kJ mol^{-1}	900	740	590	550	500
Second ionisation energy/kJ mol^{-1}	1800	1450	1150	1060	970
Third ionisation energy/kJ mol^{-1}	14 800	7700	4900	4200	–
Atomic radius/nm (metallic radius)	0.11	0.16	0.20	0.21	0.22
Melting point/°C	1280	650	838	768	714
Boiling point/°C	2770	1110	1440	1380	1640
Density/g cm^{-3}	1.85	1.74	1.55	2.6	3.5
Standard electrode potential, $M^{2+}(aq)/M(s)$/volts	-1.85	-2.37	-2.87	-2.89	-2.91

Look closely at the first, second and third ionisation energies of Group II elements in table 15.3.

1 What is the ratio of the first ionisation energy : second ionisation energy : third ionisation energy for Mg?

2 Why is there a greater increase from the second to the third ionisation energy than from the first to the second?

3 What is the most stable oxidation state for Group II elements?

15.3 Physical properties

The atomic radii of alkali metals are included in table 15.2 and those of the alkaline-earth metals are given in table 15.3. As one might expect, the atomic radii increase with atomic number down the groups. Each succeeding element has electrons in one more shell than the previous element.

The outermost *s*-electrons in these metals are held very weakly by the nucleus. Thus, the outer electrons can drift further from the nucleus than in most other atoms. So, the elements in Groups I and II have larger atomic radii than those elements which follow them in their respective periods. The large atomic size results in weaker forces between neighbouring atoms because there is a reduced attraction of the nuclear charge for the shared mobile outer electrons.

Consequently, the elements in Groups I and II have lower melting points and boiling points than we normally associate with metals. Apart from beryllium, all their melting points are less than 840°C and all the Group I metals melt below 200°C. In contrast, most of the transition metals melt at temperatures above 1000°C.

The reduced interatomic forces in these metals make them relatively soft. All the Group I metals can be cut with a penknife. Although the Group II metals are harder than those in Group I, they are noticeably softer than transition metals.

As we have seen, *s*-block elements have larger atomic radii than transition metals of approximately the same relative atomic mass. This results in lower densities and larger molar volumes. Of the 10 elements in tables 15.2 and 15.3, only barium is more dense than aluminium, a metal normally associated with low density. Their densities vary from approximately 1 g cm^{-3} to 3.5 g cm^{-3} for barium. Most transition metals have a density greater than 7 g cm^{-3}.

15.4 Chemical properties

All the metals in Groups I and II are high in the activity (electrochemical) series. The standard electrode potential (tables 15.2 and 15.3) for the conversion of each metal to its ions is positive and greater than 2.3 volts for all metals except beryllium.

Hence, *these metals are very good reducing agents.* They all react vigorously with water, reducing it to hydrogen.

$$M(s) \rightarrow M^+(aq) + e^-$$
$$H_2O(l) + e^- \rightarrow \tfrac{1}{2}H_2(g) + OH^-(aq)$$

Excluding lithium, which reacts slower than all the other elements of Group I, the reactivities of the elements with water closely follow the values of $E^{\ominus}$.

In Group I, for example, sodium reacts vigorously, fizzing and skating about on the water surface. Potassium reacts even more vigorously. It cracks and pops as the hydrogen explodes and produces a lilac flame as the potassium gets so hot that it starts to burn. Rubidium and caesium explode violently in contact with water.

> 1 Do $E^{\ominus}$ values indicate how fast a reaction will go, or simply the direction in which it should go?
>
> 2 Try to think of a reason or reasons why lithium reacts more slowly with water than other Group I metals, even though it has the largest $E^{\ominus}$?

Although electrode potentials relate only to aqueous solutions and to energy changes rather than reaction rates, they also provide a guide to the general reactivity of a substance. All the s-block elements are good reducing agents. They react with chlorine, bromine, sulphur, hydrogen and oxygen on heating. They all, of course, tarnish rapidly in air, forming a layer of oxide. The reactivity of Group I metals is such that lithium, sodium and potassium are usually stored under oil.

Lithium and beryllium, like other first elements in their group, have some atypical properties.

Lithium compounds are less stable than those of other group I metals. $LiNO_3$, Li_2CO_3 and $LiOH$ all decompose to Li_2O on heating with a bunsen, unlike the corresponding compounds of other alkali metals (section 15.5)

Beryllium differs from other elements in Group II in that its oxide (BeO) and hydroxide ($Be(OH)_2$) are amphoteric (section 13.7) and its chloride ($BeCl_2$) is composed of covalently bonded simple molecules (section 8.6).

15.5 Thermal stability of compounds

The stability of the compound $M^+X^-(s)$ is dependent on the size of its ions and the charge on these ions. The greater the charge, the stronger is the attraction between the ions and the more stable is $M^+X^-(s)$. The smaller the ions become, the closer they can approach each other in the solid crystal and the more stable is $M^+X^-(s)$.

Thus, we might expect M^+X^- to become more stable as ionic charge increases and as ionic radius decreases.

There is, however, one other important factor to bear in mind during any discussion of thermal stability. When large anions such as CO_3^{2-} decompose on heating to form smaller anions such as O^{2-}, the crystal containing the latter will generally be more stable. This is because the charge density on a small O^{2-} ion will be greater than that on a larger CO_3^{2-} ion, and the former will be more strongly attracted to the cations in the crystal. Furthermore, the smaller O^{2-} ions can get closer to the cations than the larger CO_3^{2-} ions.

▼The rocks at Malham Cove in the Yorkshire Dales are mainly limestone. What does this suggest about the temperature of these rocks as the Earth was being formed?

This point is crucial in considering the thermal stability of nitrates, carbonates and hydroxides of the *s*-block elements.

Nitrates

Group I nitrates (except $LiNO_3$) decompose on heating in the bunsen flame to form their corresponding nitrites which are stable:

$$KNO_3(s) \rightarrow KNO_2(s) + \tfrac{1}{2}O_2(g)$$

However, Group II nitrates and $LiNO_3$ decompose on heating to form the oxide:

$$Mg(NO_3)_2(s) \rightarrow MgO(s) + 2NO_2(g) + \tfrac{1}{2}O_2(g)$$
$$2LiNO_3(s) \rightarrow Li_2O(s) + 2NO_2(g) + \tfrac{1}{2}O_2(g)$$

This suggests that the decrease in size from NO_3^- to NO_2^- stabilises the Group I nitrates to heat. The Group II nitrates achieve even greater thermal stability by decomposing from their nitrate to form the much smaller oxide ion.

Carbonates

The pattern of thermal stability amongst the carbonates is very similar to that of the nitrates. The carbonates of Group I (except Li_2CO_3) are stable at the temperature of the bunsen flame, but those of Group II decompose at this temperature to form the corresponding oxide:

$$MgCO_3(s) \rightarrow MgO(s) + CO_2(g)$$

Here again the Group II compounds are achieving thermal stability by forming their oxide.

Hydroxides

Yet again, the pattern is repeated. The hydroxides of Group I (except LiOH) are stable, but those of Group II decompose on heating.

$$Mg(OH)_2(s) \rightarrow MgO(s) + H_2O(g)$$

LiOH starts to decompose at about 650°C, but NaOH does not decompose until temperatures well above this and above the highest temperature obtainable with a bunsen.

Table 15.4
Results obtained when $0.1\,mol\,dm^{-3}$ solutions of Mg^{2+}, Ca^{2+}, Sr^{2+} and Ba^{2+} are treated with various solutions

$0.1\,mol\,dm^{-3}$ solution of Group II cation	Reaction with $1\,mol\,dm^{-3}$ Na_2SO_4		Reaction with saturated $CaSO_4$	Reaction with $1\,mol\,dm^{-3}$ Na_2CO_3		Reaction with $1\,mol\,dm^{-3}$ $Na_2C_2O_4$		Reaction with $1\,mol\,dm^{-3}$ K_2CrO_4		Reaction with $1\,mol\,dm^{-3}$ NaOH	
Mg^{2+}	no ppte		no ppte	white ppte		no ppte		no ppte		thick white ppte	
Ca^{2+}	thin white ppte	Group II sulphates become more insoluble	no ppte	white ppte	Group II carbonates become more insoluble	thin white ppte	Group II ethanedioates (oxalates) become more insoluble	no ppte	Group II chromates(VI) become more insoluble	white ppte	Group II hydroxides become more soluble
Sr^{2+}	white ppte		white ppte	thick white ppte		white ppte		pale yellow ppte		white ppte	
Ba^{2+}	thick white ppte		white ppte	very thick white ppte		thick white ppte		thick pale yellow ppte		thin white ppte	
General equations for precipitations	$M^{2+} + SO_4^{2-}$ $\rightarrow MSO_4(s)$		$M^{2+} + SO_4^{2-}$ $\rightarrow MSO_4(s)$	$M^{2+} + CO_3^{2-}$ $\rightarrow MCO_3(s)$		$M^{2+} + C_2O_4^{2-}$ $\rightarrow MC_2O_4(s)$		$M^{2+} + CrO_4^{2-}$ $\rightarrow MCrO_4(s)$		$M^{2+} + 2OH^-$ $\rightarrow M(OH)_2(s)$	

15 Groups I and II – The Alkali Metals and the Alkaline-Earth Metals

Table 15.5
The solubilities (at 25°C) of some compounds of Group II metals

Compound	Solubility/moles per 100 g water
$MgSO_4$	3600×10^{-4}
$CaSO_4$	11×10^{-4}
$SrSO_4$	0.62×10^{-4}
$BaSO_4$	0.009×10^{-4}
$MgCO_3$	1.3×10^{-4}
$CaCO_3$	0.13×10^{-4}
$SrCO_3$	0.07×10^{-4}
$BaCO_3$	0.09×10^{-4}
$MgCrO_4$	8500×10^{-4}
$CaCrO_4$	870×10^{-4}
$SrCrO_4$	5.9×10^{-4}
$BaCrO_4$	0.011×10^{-4}
$Mg(OH)_2$	0.2×10^{-4}
$Ca(OH)_2$	16×10^{-4}
$Sr(OH)_2$	330×10^{-4}
$Ba(OH)_2$	240×10^{-4}

15.6 The solubility of compounds

Table 15.4 shows the results obtained when 0.1 mol dm^{-3} solutions of Mg^{2+}, Ca^{2+}, Sr^{2+} and Ba^{2+} are treated with various solutions.

Notice that the sulphates, carbonates, ethanedioates (oxalates) and chromates(VI) of the Group II metals become more insoluble as atomic number increases, whereas their hydroxides become more soluble. These results are confirmed by the solubilities of the sulphates, carbonates, chromates(VI), and hydroxides in table 15.5.

When an ionic compound dissolves in water, the following process occurs:

$$M^+X^-(s) + (aq) \rightarrow M^+(aq) + X^-(aq)$$

We can picture this process as the result of two stages. First, the separation of ions in the solid. This is the reverse of the lattice formation process requiring an *input* of the lattice enthalpy.

$$M^+X^-(s) \rightarrow M^+(g) + X^-(g)$$

The second stage involves hydration of these separated ions by water, i.e.

$$M^+(g) + X^-(g) + (aq) \rightarrow M^+(aq) + X^-(aq)$$

As the ionic radii of M^+ and X^- increase, the enthalpy change in both of these stages decreases. The reverse lattice enthalpy process becomes less endothermic and the hydration process becomes less exothermic. Consequently, a decrease in the endothermic process may be cancelled by a decrease in the exothermic process. It is therefore difficult to make predictions about changes in solubility from changes in ionic size (section 12.14).

There are, however, one or two general patterns and trends in the solubility of Group I and Group II salts.

1 All common Group I salts are soluble.

2 Compounds of Group II cations containing anions with a charge of −1 (e.g. chlorides, nitrates) are generally soluble, except for the hydroxides.

3 Salts of Group II cations containing anions with a charge of −2 (e.g. sulphates, carbonates) are generally insoluble (except for some magnesium salts and a few calcium salts).

Furthermore, there is a distinct decrease in solubility as the atomic number of the metal rises. Thus, beryllium and magnesium sulphates are very soluble in water, calcium sulphate is sparingly soluble, but strontium and barium sulphates are virtually insoluble.

Table 15.6
The solubilities (at 25°C) of various magnesium compounds

Compound	Solubility/moles per 100 g water
$MgCl_2$	5.6×10^{-1}
$MgBr_2$	5.5×10^{-1}
$Mg(OH)_2$	2.0×10^{-5}
$Mg(CH_3COO)_2$	5.6×10^{-1}
$Mg(NO_3)_2$	4.9×10^{-1}

Look at the solubilities of the various magnesium compounds (all of which contain an anion with charge −1) in table 15.6.

1 Approximately how many times more soluble are the other compounds in the table than $Mg(OH)_2$?

2 Why are the solubilities given in moles per 100 g of water rather than g/100 g of water?

15.7 Occurrence of the *s*-block elements

The alkali metals occur in nature only as +1 ions. Sodium and potassium are, of course, the commonest Group I metal ions. They are the sixth and seventh most abundant elements in the earth's crust (figure 15.1). Lithium is found in trace amounts in virtually all rocks and clays, but rubidium and caesium are very rare. Francium, the last element in Group I, does not exist naturally. It has only been obtained in tiny amounts during nuclear reactions. Even these amounts disappear rapidly as the radioactive francium disintegrates and forms more stable elements.

Figure 15.1
The relative abundance of elements on the earth. (The figures given indicate the percentage by mass of the particular element in the earth's crust to a depth of 40 kilometres, including the atmosphere.)

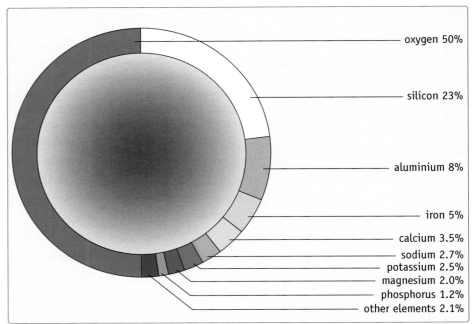

oxygen 50%
silicon 23%
aluminium 8%
iron 5%
calcium 3.5%
sodium 2.7%
potassium 2.5%
magnesium 2.0%
phosphorus 1.2%
other elements 2.1%

Since many of the compounds of alkali metals are soluble in water, it is not surprising that they are found in relatively large quantities in sea water. Compounds of the alkali metals also occur in deposits which have formed by the evaporation of brine. However, many insoluble clays also contain alkali metal ions combined as complex silicates and aluminates.

The alkaline-earth metals occur in nature only as +2 ions. These +2 ions react with −2 ions to form insoluble compounds, so it is not surprising that Group II elements exist in nature principally as their carbonates, silicates and sulphates.

The only beryllium mineral found in any quantity is the silicate beryl ($3BeSiO_3.Al_2(SiO_3)_3$). Emeralds are, in fact, crystals of beryl, coloured green by traces of chromium.

Magnesium is the eighth most abundant element in the earth's crust. It is found extensively as the carbonate in magnesite ($MgCO_3$) and dolomite ($MgCO_3.CaCO_3$). It also occurs as the silicate in minerals such as asbestos ($3MgSiO_3.CaSiO_3$) and as Mg^{2+} ions in sea water.

Calcium is the fifth most abundant element in the earth's crust and the most abundant *s*-block element. Vast quantities of calcium occur as the carbonate in chalk, limestone and marble. These deposits of calcium carbonate have been formed by the effect of high temperature and high pressure on the accumulated skeletons of dead marine animals.

Smaller quantities of calcium occur as the sulphate in anhydrite ($CaSO_4$) and gypsum ($CaSO_4.2H_2O$). These sulphate ores are thought to have resulted from the action of sulphuric acid (produced from the oxidation of sulphide minerals) on limestone.

Strontium and barium are much rarer elements. The commonest ore of strontium is strontianite ($SrCO_3$) and that of barium is barite ($BaSO_4$).

The average abundance of radium (the last element in Group II) in the earth's crust has been estimated at less than 1 part per million million. Radium exists naturally as a radioactive element. It is formed during the nuclear disintegration of heavier elements such as uranium.

▶ A salt worker hacks out salt at the Sambhar Salt Lakes, India.

▶▶ Harvesting salt from lagoons in Brittany, France.

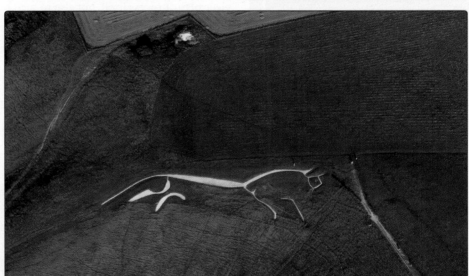

▶The white horse carved in the chalk downland at Uffington, Berkshire.

15.8 Manufacture of the *s*-block elements

In order to extract the *s*-block elements from their naturally occurring compounds, it is necessary to reduce their positive ions, i.e.

$$M^+(aq) + e^- \rightarrow M(s)$$

or

$$M^{2+}(aq) + 2e^- \rightarrow M(s)$$

This reduction can be done either chemically or electrolytically. But the large negative electrode potentials of the *s*-block elements mean that chemical methods are virtually impossible. The processes would require a reducing agent stronger than the alkali or alkaline-earth metals themselves.

Consequently, these elements are usually extracted by electrolysis. The electrolytes are always fused compounds rather than aqueous solutions, since water in the solution would be discharged at the cathode in preference to the *s*-block ion. For example,

$$Na^+(aq) + e^- \rightarrow Na(s) \qquad E^\ominus = -2.71 \text{ volts}$$
$$H_2O(l) + e^- \rightarrow \tfrac{1}{2}H_2(g) + OH^-(aq) \qquad E^\ominus = -0.42 \text{ volts}$$

The more positive (less negative) electrode potential for the lower reaction means that it will occur in preference to the one above.

Sodium and magnesium are in much higher demand than the other metals in Groups I and II. Each of these metals is obtained by electrolysis of its molten chloride.

Look closely at figure 15.2. This shows a simplified diagram of the Downs cell in which sodium is manufactured

1 The electrolyte contains $Na^+(l)$ and $Cl^-(l)$. Write equations for the process at each electrode during the manufacture of sodium.

2 In the Downs cell, the cathode is made of steel, but the anode is made of graphite. Why is the anode made of graphite in spite of the fact that steel would be a better conductor? (Hint: What is produced at the anode?)

3 Why is it necessary to incorporate hoods and a steel gauze cylinder in order to prevent the product at the anode from mixing with the product at the cathode?

The extraction of these *s*-block elements involves fused compounds rather than aqueous solutions. Large quantities of energy are therefore required not only for the electrolytic process itself, but also to melt the electrolyte.

▲ Michael Faraday (1791–1867) carried out many experiments involving electrolysis and discovered the laws of electrolysis and electromagnetic induction.

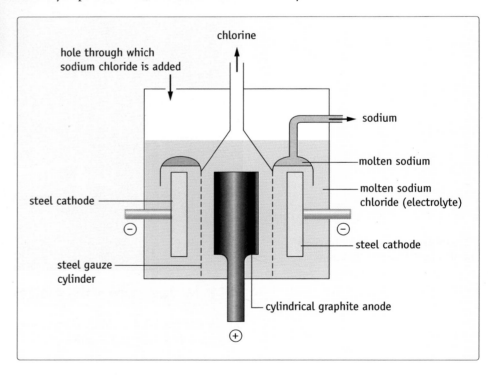

Figure 15.2
The extraction of sodium in the Downs cell.

4 Why do you think that electrolytic plants are often located near supplies of hydroelectricity?

5 In these electrolyses, the naturally occurring raw material is usually mixed with some other compound. For example, in the extraction of sodium, the electrolyte of NaCl(l) contains some calcium chloride. The impurity of calcium chloride causes the mixture to melt at a lower temperature. How does this addition of calcium chloride reduce the energy expenditure?

6 In view of the calcium chloride added to the electrolyte, what substance will contaminate the sodium produced at the cathode?

15 Groups I and II – The Alkali Metals and the Alkaline-Earth Metals

15.9 Uses of the s-block elements and their compounds

In spite of being so reactive, sodium, magnesium and other s-block elements have many important uses. Probably the most important use of sodium is as a liquid coolant in nuclear reactors.

Fast nuclear reactors operate at a temperature of about 600°C. Water, which boils at 100°C, is therefore unsuitable as a cooling liquid in this case. But sodium, which melts at 98°C and boils at 892°C, is ideal for this purpose. Furthermore, sodium is a much better thermal conductor than water and, as a liquid, almost as mobile. It can therefore be circulated around the hot reactor core and will transfer the heat away more effectively than water.

Assuming that the sodium can be prevented from coming into contact with air or moisture, the major problem in its use is its action as a solvent for various metals. These metals might otherwise be used for the reactor core and the cooling system. Fortunately, iron and other transition metals are virtually insoluble in the liquid. So, they can be used in the construction of the reactor core and the cooling pipes.

Another important use of sodium is its alloying with lead in the manufacture of tetraethyllead(IV) from chlorethane.

$$\underbrace{Pb + 4Na}_{\text{lead–sodium alloy}} + 4C_2H_5Cl \rightarrow \underset{\text{tetraethyllead(IV)}}{(C_2H_5)_4Pb} + 4NaCl$$

Tetraethyllead(IV) (TEL) is used as an essential 'anti-knock' additive for petrol. Every gallon of 'leaded' petrol contains about 2 cm^3 of TEL (section 26.7).

Sodium is also used in the reduction of titanium(IV) chloride to titanium by the Goldschmidt process and in the familiar yellow sodium vapour street lamps. In the metals industry, sodium cyanide is used in the extraction of gold and in the hardening of steel.

Sodium hydroxide is the cheapest industrial alkali. Thousands of tonnes are used annually in the manufacture of rayon (cellulose acetate), paper and soap. Sodium polyphosphate ($Na_6P_6O_{18}$) is used as a water softener and sodium nitrite, $NaNO_2$, is used as a preservative in food.

The other metals in Group I are used much less extensively than sodium and its compounds. Potassium is used to make potassium superoxide, KO_2, which reacts with water to give oxygen.

$$4KO_2 + 2H_2O \rightarrow 4KOH + 3O_2$$

KO_2 can therefore be used as an emergency source of oxygen in mines and in submarines. Potassium chloride, KCl, is used as a fertiliser and potassium nitrate is an essential constituent of gunpowder.

Owing to its very low first ionisation energy, caesium is used as the light-sensitive surface in photocells. These convert light signals into electrical signals. Figure 15.3 shows a simplified diagram of a photocell circuit. The evacuated tube contains two electrodes, one of which is coated with caesium or an alloy of caesium, antimony and silver. In the absence of light, the photocell does not conduct electricity. However, when the caesium-coated electrode is struck by light, it emits electrons. These are attracted to the positive electrode and an electric current flows in the circuit. Television pickup devices and photographers' lightmeters incorporate the use of photocells.

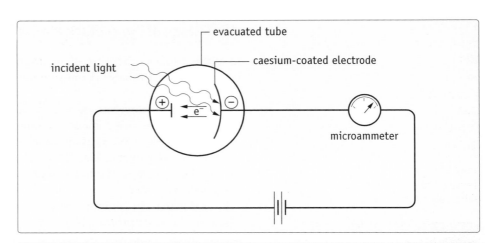

Figure 15.3
A simplified photocell circuit.

▶ Quarrying limestone near Kirkby Steven, Cumbria, UK.

▲ Volatile compounds of calcium, strontium and barium are often used in fireworks.

Magnesium is by far the most commonly used metal in Group II. It is used extensively in the preparation of lightweight alloys with a high tensile strength and particularly in the construction of aircraft parts and household goods. Although pure magnesium has poor structural strength, this can be increased by alloying it with aluminium, zinc and manganese. The aluminium increases the tensile strength of the alloy, the zinc improves its machining properties and the manganese reduces corrosion.

The most commonly used compounds of magnesium are magnesium oxide (MgO), magnesium hydroxide ($Mg(OH)_2$) and hydrated magnesium sulphate ($MgSO_4.7H_2O$). Magnesium oxide is used as a lining for high-temperature furnaces because of its exceptionally high melting point and low reactivity. Magnesium hydroxide is a very weak alkali. Because of this, it is an important constituent of toothpastes. It neutralises acids that would otherwise cause decay. A suspension of $Mg(OH)_2$ in water (commonly called milk of magnesia) is used to treat acid indigestion.

Beryllium is too rare and costly for any large-scale uses, but it is used for hardening alloys and other metals such as copper and nickel. Beryllium–copper alloys are particularly resistant to sea-water corrosion and are therefore used for small marine parts.

Calcium, strontium and barium have very few uses owing to their high reactivity. The affinity of these elements for oxygen results in their use as deoxidisers in steel production. They are also used as 'getters' (i.e. removers of traces of oxygen) inside electron tubes and radio valves. By far the most commercially important compounds of calcium are lime (CaO) and limestone ($CaCO_3$), Limestone, of course, is the source of lime. Many thousands of tonnes of lime are used annually in counteracting soil acidity. Enormous quantities of lime are also used in the manufacture of plaster, mortar and cement.

s-block metal present in salt	Flame colour produced
Lithium	Red
Sodium	Yellow
Potassium	Lilac
Calcium	Red
Strontium	Crimson
Barium	Pale green

Table 15.7
The flame colours of the salts of different s-block elements

The salts of several s-block elements produce vivid flame colours when heated to high temperatures in a hot bunsen flame on the end of a clean nichrome wire (section 6.2). The flame colours of the salts of different s-block elements are listed in table 15.7.

The flame colours of these elements can be used in their identification, and volatile compounds of these elements are used in the production of coloured military signals and fireworks.

▲ Tunnels in an anhydrite mine. Anhydrite is impure calcium sulphate.

▲ Traditional lime kilns are still used in some parts of the world. Limestone (calcium carbonate) is decomposed to lime (calcium oxide) at high temperatures.

Summary

1. In the alkali and alkaline-earth metals, the electrons in the outermost shell occupy an s-orbital.

2. The atoms of these s-block metals readily lose their outermost s-electrons to form stable ions: M^+ if the elements are in Group I, M^{2+} if the elements are in Group II.

3. The metals in Groups I and II show only one oxidation number in their compounds: +1 for Group I elements, +2 for Group II elements.

4. The s-block metals have lower melting points, lower boiling points, lower densities and they are softer than transition metals.

5. The s-block metals are high in the activity series. They will all reduce water to hydrogen and they all tarnish rapidly in air forming a layer of oxide.

6. The compounds of Group I metals are thermally more stable than those of Group II metals. The carbonates and hydroxides of Group I metals (except Li_2CO_3 and LiOH) do not decompose at 1000°C, but those of Group II metals decompose to their oxides on heating.

7. All common Group I salts are soluble.

8. Compounds of Group II cations containing anions with a charge of −1 are generally soluble except for the hydroxides. Salts of Group II cations containing anions with a charge of −2 are generally insoluble.

9. The s-block elements are manufactured by electrolysis of their molten compounds.

10. Sodium and magnesium are the most widely used s-block elements. Sodium is used as a liquid coolant in nuclear reactors and as an alloy with lead in the manufacture of tetraethyllead(IV) ('anti-knock' in petrol). Magnesium is used extensively in the manufacture of low-density, high-strength alloys.

1 (a) How do each of the following properties of the elements in Group II change with increasing atomic number?
  (i)   Atomic radius
  (ii) Ionisation energy
  (iii) Strength as reducing agents
  (iv) Vigour of reaction with chlorine
  (v)   Electropositivity
(b) In each case, explain why the five properties change in the way you have suggested.

2 Group II contains the elements Be, Mg, Ca, Sr, Ba.
(a) Are these elements metals or non-metals? Give one reason for your answer.
(b) What ion or ions are formed by calcium?
(c) Describe briefly five properties that show a general gradation as the group is descended from Be to Ba.
(d) Briefly explain how ionisation energies and redox potentials are related to the reactivity of these elements.
(e) It is easier to form Mg^+ ions than Mg^{2+} ions from magnesium. In spite of this, Mg^+ is not found in magnesium compounds. Why is this?
(f) Group II elements frequently form hydrated salts, while the corresponding compounds of Group I elements are anhydrous. Suggest reasons for this difference.

3 The elements in Group II of the periodic table are barium (Ba), beryllium (Be), calcium (Ca), magnesium (Mg), radium (Ra) and strontium (Sr).
(a) Arrange these elements in order of increasing relative atomic mass.
(b) Write the electronic structures of any two of these elements except beryllium (e.g. Be would be $1s^2 2s^2$).
(c) Draw a sketch graph of ionisation energy against number of electrons removed for the successive ionisation energies of magnesium.
(d) Why do all the metals in Group II have an oxidation number of +2?
(e) Why do the elements in Group II not have an oxidation number of +1 or +3?
(f) How and why does the first ionisation energy of the metals vary as the group is descended?

4 Group I contains the elements Li, Na, K, Rb, Cs.
(a) The first member of a group often shows anomalous properties. Give two respects in which the behaviour of Li is anomalous.
(b) Describe briefly five properties that show a general gradation as the group is descended from Li to Cs.
(c) Explain the meaning of the term 'ionisation energy'.
(d) How will successive ionisation energies of Na vary?
(e) Why is the ion Na^+ formed in normal chemical reactions rather than the ion Na^{2+}?
(f) How are ionisation energies related to the reactivity of these elements?

5 The isotope $^{42}_{19}K$ decays by beta-particle emission to a stable nuclide. The rate of emission of β-particles from $^{42}_{19}K$ was followed by a suitable method.
(a) Draw a sketch graph showing the rate of emission of β-particles (vertically) against time.
(b) 0.02 g of $^{42}_{19}K$ were allowed to decay.
  (i)   Write a nuclear equation (balanced for mass number and atomic number) for the decay process.
  (ii) What is the stable nuclide produced in the decay process?
  (iii) If the half-life of the potassium isotope is 12.5 h, what mass of the stable nuclide will have been formed after 25 h?

6 Magnesium and calcium occur naturally in the mineral dolomite, $MgCO_3.CaCO_3$, a mixture of insoluble magnesium and calcium carbonates. This can be used to produce calcium sulphate and magnesium sulphate. Calcium sulphate is used in the manufacture of building materials such as plaster-board. Magnesium sulphate is used in fireproofing fabrics and as a purgative (Epsom salts).
(a) Describe carefully how you would prepare pure samples of $MgSO_4.7H_2O$ and $CaSO_4$ from dolomite. You may find the following information useful.

Compound	Solubility/g per 100 g water at 20°C
$MgCO_3$	0.01
$CaCO_3$	0.0014
$MgSO_4$	33.0
$CaSO_4$	0.21

(b) How would you obtain pure $MgSO_4$ from crystals of $MgSO_4.7H_2O$?

7 Sodium sulphite (Na_2SO_3) is sometimes added to sausage meat to act as a preservative. The amount of Na_2SO_3 present can be determined by boiling a sample of the meat with acid. The quantity of sulphur dioxide which forms is then determined by titration against iodine.

100 g of sausage meat was boiled with 500 cm^3 of 1 mol dm^{-3} HCl. The sulphur dioxide evolved was dissolved in water. It was found to require 12.00 cm^3 of 0.025 mol dm^{-3} I_2 solution in order to oxidise the SO_2 as in the equation below.

$$SO_2(aq) + 2H_2O(l) + I_2(aq) \rightarrow 4H^+(aq) + SO_4^{2-}(aq) + 2I^-(aq)$$

In order to check the results of the titration, excess barium chloride is added to the final solution after the titration. The resulting precipitate is collected and weighed. (Na = 23, S = 32, O = 16, Ba = 137)

(a) How many moles of SO_2 are evolved from 100 g of the sausage meat?

(b) How many grams of Na_2SO_3 are present in 100 g of the sausage meat?

(c) Government scientists often express the amount of Na_2SO_3 in meat as parts per million (p.p.m.).
(1 p.p.m. = 1 g of Na_2SO_3 in 10^6 g of meat)
Express the amount of Na_2SO_3 in the sausage meat in p.p.m.

(d) Write an equation for the reaction which occurs when $BaCl_2(aq)$ is added to the solution at the end of the titration.

(e) Calculate the mass of precipitate formed when excess $BaCl_2(aq)$ is added to the solution at the end of the titration.

8 Calcium sulphate is found naturally in two forms: anhydrous calcium sulphate ($CaSO_4$ – anhydrite) and hydrated calcium sulphate ($CaSO_4.2H_2O$ – gypsum). When anhydrite is heated with coke, sulphur dioxide is obtained. This can then be used to manufacture sulphuric acid. If gypsum is heated at about 125°C it dehydrates partially forming plaster of Paris, $(CaSO_4)_2.H_2O$. When this is mixed with water it changes back to $CaSO_4.2H_2O$. The paste expands slightly as it hardens and sets quickly to a firm solid.

When gypsum is heated to 200°C, it loses all its water of crystallisation. The anhydrous salt which forms sets only very slowly when mixed with water.

(a) Explain what is meant by the terms:
(i) anhydrous, (ii) hydrated, (iii) water of crystallisation.

(b) Write equations for the reactions which occur when:
(i)   anhydrite is heated with coke,
(ii) plaster of Paris is mixed with water,
(iii) gypsum is heated at 200°C.

(c) Explain why plaster of Paris is so suitable for:
(i)   immobilising broken limbs.
(ii) making models from moulds.

(d) Why is anhydrite not a suitable alternative for the uses mentioned in (c)?

Table 16.1

Group VII (The Halogens)

F
Cl
Br
I
At

16.1 Introduction

The halogens are the elements in Group VII of the periodic table – fluorine (F), chlorine (Cl), bromine (Br), iodine (I) and astatine (At) (table 16.1). They are known as the *halogens* – a name derived from Greek meaning 'salt formers' – because they combine readily with metals to form salts.

Generally speaking, the halogens comprise the most reactive group of non-metals. In this chapter we shall contrast their properties with those of the reactive metals in Groups I and II of the periodic table. We shall find that *the halogens are strong oxidising agents* whilst the alkali and alkaline-earth metals in Groups I and II are strong reducing agents. Furthermore, we shall see that the halogens exhibit various oxidation numbers in their compounds. In contrast, the *s*-block elements have only one oxidation number in their compounds.

Although the halogens show distinct trends in behaviour down Group VII, they also show remarkable similarities. These similarities in their properties and reactions result very largely from their similar electron structures. Each of the halogens has an outer shell containing seven electrons (i.e. ns^2np^5).

1. Write the electron shell structure for fluorine and chlorine. (For example; that for beryllium would be $1s^2\,2s^2$.)

2. Why are fluorine and chlorine strong oxidising agents?

3. What is the most likely oxidation number of these elements in their compounds?

16.2 Sources of the halogens

The halogens are so reactive that they cannot exist free in nature. Indeed, fluorine is reactive enough to combine directly with almost all the known elements including some of the noble gases. Chlorine reacts directly with all elements except carbon, nitrogen, oxygen and the noble gases.

The halogens always occur naturally as compounds with metals. They are present as negative ions: fluoride (F^-), chloride (Cl^-), bromide (Br^-) and iodide (I^-). The last element in the Group, astatine (At), does not occur naturally at all. It is a very unstable, radioactive element which was first synthesised by chemists in the USA in 1940. It has never been obtained in anything other than minute amounts and the most stable isotope, $^{210}_{85}At$, has a half-life of only 8.3 hours. The name astatine is derived from the Greek word *astatos* meaning 'unstable'.

Fluorine and chlorine are by far the most abundant halogens. The most widespread compounds of fluorine are fluorspar (fluorite), CaF_2, such as Derbyshire 'Blue John', and cryolite, Na_3AlF_6. Unfortunately these extensive deposits of fluoride are dispersed very thinly over the earth's surface. Only a few sources can be worked economically.

The commonest chlorine compound is, of course, sodium chloride (NaCl) which occurs in sea water and in rock salt. Each kilogram of sea water contains about 30 g of sodium chloride.

▲ While lying on his side, this miner is drilling for Blue John stone at the Treak Cliff Cavern in Derbyshire. Blue John is still mined in Treak Cliff Cavern and made into jewellery and ornaments. Blue John is calcium fluoride (fluorspar) containing streaks of oil. Over millions of years, the oil has been transformed into beautiful colours in the stone. What do you think caused the changes in the oil?

▲ Iodine can be obtained from laminarian seaweeds. Although seawater contains only traces of iodides, they are concentrated by these seaweeds which can contain up to 800 parts per million.

Bromides and iodides occur in much smaller amounts than either fluorides or chlorides. Sea water contains only small concentrations of bromide, about 70 parts per million by weight, but the extraction of bromine from brine is still economically feasible.

Iodides are even scarcer than bromides. Sea water contains traces of iodide (0.05 parts per million by weight). However, the laminarian seaweeds concentrate iodide from the sea to such an extent that fresh wet weed can contain up to 800 parts per million of iodine. Biologists are still perplexed by this capacity of laminaria to hoard such large concentrations of iodide. Curiously, this is not the only query posed by naturally occurring iodine compounds. The main source of iodine is sodium iodate(v) ($NaIO_3$). This is found only in Chile, mixed with larger proportions of sodium nitrate. Why does iodine occur as iodate(v) in such high concentration in only one part of the world, and why does the iodine exist in such a high oxidation state? At present, no one really knows.

► Chilean workmen drilling holes in layers of caliche (impure sodium iodate(v)) prior to blasting.

16.3 Obtaining the halogens

The halogens are usually obtained by oxidation of halide ions:

$$2Hal^- \rightarrow Hal_2 + 2e^-$$

Fluorine is such a powerful oxidising agent that fluoride ions cannot be oxidised by any of the common oxidising agents such as concentrated H_2SO_4, MnO_2 and $KMnO_4$. This means that fluorine must be obtained commercially by electrolysis. Furthermore, electrolysis cannot be carried out in aqueous solution, since water or OH^- would be discharged in preference to F^-. Even if fluorine were produced it would react with water. The electrolyte is usually potassium fluoride dissolved in liquid anhydrous hydrogen fluoride and the electrodes consist of a graphite anode and a steel cathode.

> 1 Write the symbols, with charges, of all ions present in the electrolyte.
> 2 Write an equation to summarise the formation of fluorine at the anode.
> 3 What problems will the formation of fluorine create?

The other halogens are much less reactive than fluorine. They can be obtained in the laboratory by oxidising the appropriate halide ions using MnO_2 or $KMnO_4$.

Figure 16.1 shows how chlorine, bromine and iodine can be prepared on a small scale. These preparations should be carried out in a fume cupboard. All halogens are poisonous and should be handled with great care. Notice that each preparation is essentially the oxidation of halide ions by MnO_2 in the presence of concentrated H_2SO_4. The apparatus varies from one preparation to the next because chlorine is a gas, bromine is a liquid and iodine is a solid at room temperature.

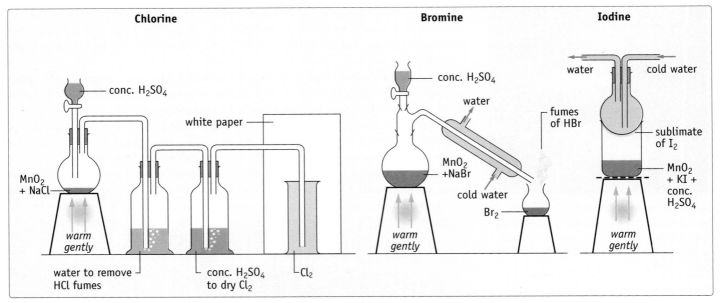

Figure 16.1
Preparing the halogens.

$$2Hal^- \rightarrow Hal_2 + 2e^-$$
$$MnO_2 + 4H^+ + 2e^- \rightarrow Mn^{2+} + 2H_2O$$

1 Why is the acid necessary?

2 Why is concentrated H_2SO_4 used rather than dilute H_2SO_4?

Some hydrogen halide is always evolved from the interaction of the halide and concentrated H_2SO_4. The chlorine, for example, will be contaminated with hydrogen chloride fumes. Pure, dry chlorine can be obtained by passing the gas through water to remove the HCl fumes and then through concentrated H_2SO_4 to dry it. Finally, the gas is collected by downward delivery.

2 Why is the white paper placed behind the gas jar in which Cl_2 collects?

4 The bromine which is collected will be contaminated with fumes of HBr. How could the liquid bromine be purified? (Hint: how are liquids normally purified?)

5 Why is cold water passed through the flask on which solid iodine is deposited?

From an industrial and economic viewpoint, chlorine is by far the most important halogen. At one time, most chlorine was manufactured by the electrolysis of saturated *aqueous* sodium chloride (brine) using cells with a flowing mercury cathode. Serious environmental problems, caused by the escape of relatively small amounts of mercury, have led to the replacement of mercury cells by diaphragm cells (figure 16.2). The name diaphragm arises because the anode and cathode in each cell are separated by a permeable asbestos diaphragm. The process results in the manufacture of hydrogen and sodium hydroxide as well as chlorine.

In the electrolytic process, chlorine is liberated at the titanium anodes and hydrogen is liberated at the steel cathodes.

anode (+) (titanium)	$2Cl^-(aq) \rightarrow Cl_2(g) + 2e^-$
cathode (-) (steel)	$2H_3O^+(aq) + 2e^- \rightarrow H_2(g) + 2H_2O(l)$

As Cl^- ions are removed at the anode, the only ions passing through the permeable asbestos diaphragm into the cathode compartment are Na^+, H^+ and OH^-. H^+ ions are removed at the cathode, leaving Na^+ and OH^- ions (i.e. effectively sodium hydroxide solution) to flow out of the cell.

The permeable diaphragm plays an important part in the process because it keeps the chlorine and sodium hydroxide apart. If the electrolysis were carried out in a cell without a diaphragm, the Cl_2 and NaOH would react immediately (section 16.8).

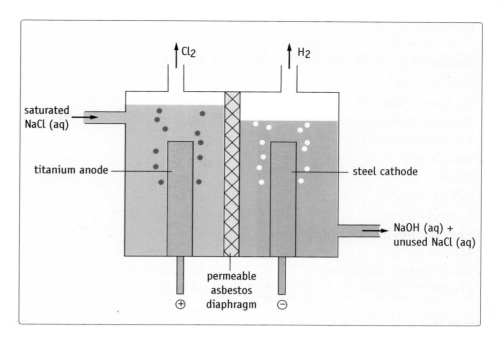

Figure 16.2
A diaphragm cell in which chlorine is manufactured by electrolysis of saturated aqueous sodium chloride.

All three products (chlorine, hydrogen and sodium hydroxide) are valuable materials for the chemical industry. In 1993, the estimated worldwide production of chlorine was 33 million tonnes with 1.5 million tonnes produced in the U.K.

The raw materials used for the process – sodium chloride and water, are readily obtainable and cheap, but the electrolytic process is very energy consuming, making high demands on electrical energy.

Since chlorine is a strong and relatively inexpensive oxidising agent, it is used to manufacture bromine by oxidation of bromide ions in sea water.

$$Cl_2(g) + 2Br^-(aq) \rightarrow 2Cl^-(aq) + Br_2(aq)$$

Unlike chlorine and bromine which normally exist in the −1 oxidation state, iodine occurs naturally in the +5 oxidation state as sodium iodate(V). Consequently, iodine is normally obtained by reduction with sodium hydrogensulphite.

$$(IO_3^-(aq) + 6H^+(aq) + 5e^- \rightarrow \tfrac{1}{2}I_2(aq) + 3H_2O(l)) \times 2$$
$$(HSO_3^-(aq) + H_2O(l) \rightarrow HSO_4^-(aq) + 2H^+(aq) + 2e^-) \times 5$$

A second, somewhat obsolete method of obtaining iodine is to extract the soluble iodides from dried seaweed with water. The iodine is then displaced from the solution by treatment with chlorine.

16.4 Structure and properties of the halogens

All the halogens exist as diatomic molecules. The two atoms are linked by a covalent bond (figure 16.3). These molecules persist in the gaseous, liquid and solid states. Table 16.2 shows some of the physical properties of the halogens.

All the halogens are coloured – the depth of colour increasing with increase in atomic number. Fluorine is pale yellow, chlorine is pale green, bromine is red–brown and iodine is shiny black.

Notice also the change in volatility down the group. Fluorine and chlorine are gases, bromine is a liquid and iodine is a solid. This decreasing volatility is, of course, related to the increasing strength of Van der Waals forces with increasing relative molecular mass. This results in increasing melting points, boiling points and molar enthalpy changes of vaporisation.

All the halogens except fluorine dissolve slightly in water and colour it. Fluorine is such a powerful oxidising agent that it converts water to oxygen. The halogens, being non-polar simple molecules, are more soluble in organic solvents, the colour of the solution depending upon the particular halogen and the solvent.

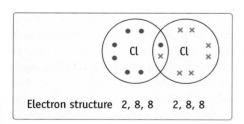

Electron structure   2, 8, 8   2, 8, 8

Figure 16.3
Chlorine is a diatomic molecule. The two chlorine atoms are linked by a covalent bond.

Table 16.2
Physical properties of the halogens

Element	Fluorine	Chlorine	Bromine	Iodine
Atomic number	9	17	35	53
Electron configuration	2, 7	2, 8, 7	2, 8, 18, 7	2, 8, 18, 18, 7
Outer shell electron configuration	$2s^2 2p^5$	$3s^2 3p^5$	$4s^2 4p^5$	$5s^2 5p^5$
Relative atomic mass	19.0	35.5	79.9	126.9
State at 20°C	gas	gas	liquid	solid
Colour	pale yellow	pale green	red–brown	black
Melting point/°C	−220	−101	−7	113
Boiling point/°C	−188	−35	59	183
Molar enthalpy change of vaporisation of liquid /kJ mol^{-1}	+3.3	+10.2	+15	+30
Solubility/g per 100g of water at 20°C	reacts readily with water	0.59 (reacts slightly)	3.6	0.018

You will need to refer to table 16.2 in order to answer the following questions.

Astatine is the halogen element immediately below iodine in the periodic table. Predict the following properties of astatine:

1 Its colour.

2 Its state at room temperature.

3 Its relative atomic mass.

4 Its melting point.

5 Its molar heat of vaporisation.

6 The abundance of its compounds.

In non-polar solvents such as tetrachloromethane and cyclohexane, chlorine is colourless, bromine is red and iodine is violet. In these solvents the elements exist as relatively free molecules as in the gas phase. In polar (electron-donating) solvents such as water, ethanol and propanone (acetone), bromine and particularly iodine give brownish solutions.

The dark blue/black compound which iodine forms with starch is a complex. The iodine molecules are absorbed reversibly within the chains of glucose molecules which make up starch.

16.5 Chemical properties of the halogens

All the halogens have one electron less than the noble gas which follows them in the periodic table. As we might expect, their chemistry is dominated by a tendency to gain a completely filled outermost electron shell.

Consequently, *the halogens react with metals to form ionic compounds containing the Hal⁻ ion. With non-metals and with some metals in high oxidation states, they tend to form simple molecular compounds. In these, the halogen is linked by a single covalent bond (—Hal) to the other element.*

Reactions with metals

The halogens are very reactive with metals. The vigour of reaction depends on

1 the position of the metal in the activity series, and

2 the particular halogen which is reacting.

The reactivity of the halogen decreases with increase in atomic number. Fluorine is, in fact, the most reactive of all non-metals. Because of its great reactivity it was not isolated until 1886. Fluorine combines readily and directly with all metals, whereas iodine reacts very slowly even at high temperatures with metals low in the activity series such as silver and gold. Gold and platinum are readily attacked by fluorine on heating, but some metals, such as copper and nickel alloys, become coated with a superficial layer of fluoride which prevents further reaction. Hence, vessels of these alloys are used for preparing and storing the gas.

Table 16.3
Properties of the halogens and their compounds

Element	Fluorine	Chlorine	Bromine	Iodine
Standard enthalpy change of formation of NaHal/kJ mole^{-1}	−573	−414	−361	−288
Enthalpy change of atomisation of Hal$_2$/kJ per mole of atoms formed	+79	+121	+112	+107
Bond energy E(Hal—Hal) /kJ per mole of bonds	158	242	193	151
Electrode potential Hal$_2$(aq),2Hal$^-$(aq)/Pt/volts	+2.87	+1.36	+1.09	+0.54

The standard enthalpy changes of formation of the sodium halides (table 16.3) provide quantitative evidence for the relative reactivity of the halogens. The heat evolved is greatest for the reaction between sodium and fluorine and least for the reaction between sodium and iodine.

The most reactive halogen is fluorine. Chlorine is the next most reactive, then bromine, then iodine. Thus, reactivity decreases as the relative atomic mass of the halogen increases. A similar decrease in reactivity with increasing atomic mass is found in other groups of non-metals. This is the reverse of the trend in Group I and other groups of metals where reactivity increases as the relative atomic mass increases.

Reactions with non-metals

A similar pattern of reactivity emerges here also. Fluorine reacts directly with all non-metals except nitrogen, helium, neon and argon. It will even react with diamond and xenon on heating.

$$C(diamond) + 2F_2 \rightarrow CF_4$$
$$Xe + 2F_2 \rightarrow XeF_4$$

Fluorine will also attack glass, quartz and silicon dioxide, displacing oxygen.

$$SiO_2 + 2F_2 \rightarrow SiF_4 + O_2$$

If these substances are carefully dried, the reaction is very slow. Thus, fluorine can be studied in dry glass equipment. Any taps must be lubricated with 'fluorocarbon grease' containing fluorinated hydrocarbons rather than grease containing hydrocarbons since fluorine readily attacks organic compounds, displacing hydrogen.

$$—C—H + F_2 \rightarrow —C—F + HF$$

Chlorine and bromine are much less reactive than fluorine. They will not react directly with any of the noble gases, carbon, nitrogen or oxygen. Iodine does not combine with these elements nor with sulphur, but it reacts readily with phosphorus forming the triiodide.

$$\tfrac{1}{4}P_4 + \tfrac{1}{2}I_2 \rightarrow PI_3$$

The relative reactivity of the halogens with non-metals is well illustrated by their reaction with hydrogen. Fluorine explodes with hydrogen even in the dark at −200°C. Chlorine and hydrogen explode in bright sunlight but react slowly in the dark. Bromine reacts with hydrogen only on heating and in the presence of a platinum catalyst. Iodine combines only partially and slowly with hydrogen even on heating.

$$H_2(g) + Cl_2(g) \rightarrow 2HCl(g)$$
$$H_2(g) + I_2(g) \rightleftharpoons 2HI(g)$$

The great reactivity of fluorine in these reactions with non-metals is explained partly in terms of its low atomisation energy (table 16.3). This means that the initial stage (in which the F—F bond breaks) requires little energy. Hence the activation energy for reactions involving fluorine is low. The low atomisation energy of fluorine is reflected in a low F—F bond energy (table 16.3). Fluorine's reactivity is further explained by its ability to form very stable bonds with other elements. Look at table 12.3 and compare the C—F and H—F bond energies with those of other C—Hal and H—Hal bonds respectively.

1 List three factors which influence the reactivity of the halogen elements.

2 How will each of these three factors influence the reactivity of fluorine compared with chlorine?

3 Why is fluorine (F = 19.0) more reactive than chlorine (Cl = 35.5)?

▲ Chlorine acts as an oxidising agent when it bleaches. The piece of printed material on the right was bleached by immersion in chlorine water for a few hours.

16.6 The halogens as oxidising agents

When the halogens combine with metals or non-metals they normally act as oxidising agents. The elements with which they react have positive oxidation numbers in the resultant compounds.

When the halogens combine with metals to form ionic compounds, they gain electrons from the metals to form negative halide ions:

$$2Na(s) + Cl_2(g) \rightarrow 2Na^+Cl^-(s)$$

The halogens accept electrons during these reactions and act as oxidising agents:

$$Hal_2 + 2e^- \rightarrow 2Hal^-$$

Fluorine is the most reactive halogen and the most powerful oxidising agent. *The order of decreasing power as oxidising agents is* $F_2 > Cl_2 > Br_2 > I_2$. The electrode potentials (Hal_2/Hal^-) for the halogens shown in table 16.3 become less positive from fluorine to iodine. This reflects the decreasing oxidising power.

We have already noticed the different oxidising powers of the halogens with metals, non-metals and hydrogen. The following reactions further emphasise their relative oxidising ability.

Fluorine, chlorine and bromine will all oxidise $Fe^{2+}(aq)$ to $Fe^{3+}(aq)$:

$$2Fe^{2+}(aq) \rightarrow 2Fe^{3+}(aq) + 2e^-$$
$$Hal_2(aq) + 2e^- \rightarrow 2Hal^-(aq)$$

Iodine, however, is such a weak oxidising agent that it cannot remove electrons from iron(II) ions to form iron(III) ions. These observations are exactly as we might have predicted from the $E^{\ominus}$ values in table 14.3.

Fluorine and chlorine are such powerful oxidising agents that they can oxidise various coloured dyes to colourless substances. Thus, indicators such as litmus and Universal Indicator are decolorised when exposed to fluorine and chlorine. Chlorine acts as an oxidising agent when it is used for bleaching.

When chlorine water $(Cl_2(aq))$ is added to aqueous KI, the solution becomes brown. This is due to the formation of iodine. When bromine water $(Br_2(aq))$ is used in place of chlorine water, iodine is again liberated from KI(aq).

In these two reactions, iodide ions are oxidised to iodine:

$$2I^-(aq) \rightarrow I_2(aq) + 2e^-$$

The chlorine and bromine act as oxidising agents. They accept electrons from iodide to form chloride and bromide, respectively.

$$Cl_2(aq) + 2e^- \rightarrow 2Cl^-(aq)$$
$$Br_2(aq) + 2e^- \rightarrow 2Br^-(aq)$$

Chlorine and bromine can oxidise iodide to iodine, but iodine cannot oxidise chloride or bromide ions. This is because chlorine and bromine are more reactive than iodine, i.e. they will not release electrons to iodine.

Since chlorine is a stronger oxidising agent than bromine, chlorine can oxidise bromide ions to bromine:

$$Cl_2(aq) + 2Br^-(aq) \rightarrow 2Cl^-(aq) + Br_2(aq)$$

Thus, when chlorine water is added to KBr(aq), yellow–orange bromine is produced.

1 Which halide ions can F_2 oxidise?
2 Which of the following species is the strongest reducing agent?
   I^-, I_2, Cl^-, F_2, F^-.
3 The standard electrode potential for the half-reaction
   $MnO_4^-(aq) + 8H^+(aq) + 5e^- \rightarrow Mn^{2+}(aq) + 4H_2O$ is +1.51 volts.
   Which halide ions could this half-cell oxidise under standard conditions?
   (Refer to the electrode potentials for halogens in table 16.3.)

Explaining the relative oxidising power of the halogens

When halogens act as oxidising agents they undergo the reaction:

$$\tfrac{1}{2}Hal_2 + e^- \rightarrow Hal^-$$

We can picture the reaction in terms of three stages as shown in figure 16.4.

The enthalpy changes for each of these stages for F, Cl, Br and I are also shown in figure 16.4. The order of oxidising power, $F_2 > Cl_2 > Br_2 > I_2$, is due mainly to the relative hydration energies and to a lesser degree to the small atomisation energy of fluorine.

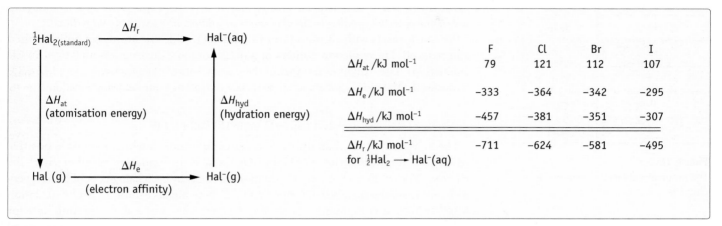

	F	Cl	Br	I
ΔH_{at} /kJ mol^{-1}	79	121	112	107
ΔH_e /kJ mol^{-1}	−333	−364	−342	−295
ΔH_{hyd} /kJ mol^{-1}	−457	−381	−351	−307
ΔH_r /kJ mol^{-1} for $\tfrac{1}{2}Hal_2 \longrightarrow Hal^-(aq)$	−711	−624	−581	−495

Figure 16.4
Stages in the conversion of diatomic halogen elements to aqueous halide ions.

Since the lattice energies of ionic halides also get less exothermic from fluorides to iodides, the order of oxidising power is the same for solid-state reactions as for reactions in aqueous solution.

16.7 The reactions of halogens with water

The electrode potential for the oxidation of water to oxygen,

$$2H_2O(l) \rightarrow O_2(g) + 4H^+(aq) + 4e^-$$

is −1.23 volts. The electrode potentials for the halogens in table 16.3 show that iodine and bromine cannot oxidise water to oxygen, but fluorine and chlorine can do so.

Fluorine reacts with water vapour to form oxygen and ozone:

$$2F_2(g) + 2H_2O(g) \rightarrow 4HF(g) + O_2(g)$$
$$3F_2(g) + 3H_2O(g) \rightarrow 6HF(g) + O_3(g)$$

The reaction of water with chlorine is very slow. In fact, chloric(I) acid (hypochlorous acid) is first formed, which then decomposes to oxygen and hydrochloric acid.

$$Cl_2(g) + H_2O(l) \rightarrow H^+(aq) + Cl^-(aq) + HClO(aq)$$
$$2HClO(aq) \rightarrow 2H^+(aq) + 2Cl^-(aq) + O_2(g)$$

The second reaction is accelerated by sunlight and by catalysts such as platinum and metallic oxides.

16.8 The reactions of halogens with alkalis

Chlorine, bromine and iodine undergo very similar reactions with alkalis. The products depend upon the temperature at which the reaction occurs.

With cold, dilute alkali at 15°C, the halogen reacts to form a mixture of halide (Hal$^-$) and halate(I) (hypohalite, HalO$^-$):

$$Hal_2 + 2OH^-(aq) \rightarrow Hal^-(aq) + HalO^-(aq) + H_2O(l)$$

The HalO$^-$ which is produced in the first reaction then decomposes to form halide and halate(v) (HalO$_3^-$):

$$3HalO^- \rightarrow 2Hal^- + HalO_3^-$$

Oxidation number	Examples
+7	Cl_2O_7, $NaClO_4$
+6	ClO_3
+5	$NaClO_3$
+4	ClO_2
+3	$KClO_2$
+2	
+1	Cl_2O, $NaClO$
0	Cl_2
−1	$NaCl$

Figure 16.5

The range of oxidation numbers for chlorine.

For chlorine, this second reaction is very slow at 15°C, but rapid at 70°C. Therefore, sodium chlorate(I) (sodium hypochlorite) can be obtained by passing chlorine into sodium hydroxide at 15°C. Sodium chlorate(V) ($NaClO_3$) is obtained by carrying out the same reaction at 70°C.

With bromine, both reactions are rapid at 15°C, but decomposition of BrO^- is slow at 0°C.

With iodine, decomposition of IO^- occurs rapidly even at 0°C, so it is difficult to prepare $NaIO$ free from $NaIO_3$.

These two reactions of halogens with alkali involve **disproportionation** – *a change in which one particular molecule, atom or ion is simultaneously oxidised and reduced.*

When Cl_2 reacts with alkali to form Cl^- and ClO^-, chlorine atoms are both oxidised and reduced. The oxidation number of one Cl atom in Cl_2 changes from 0 to −1 in Cl^- (reduction). The oxidation number of the other Cl atom changes from 0 to +1 in ClO^- (oxidation). Oxidation numbers are shown above each atom in the following equation:

$$\overset{0}{Cl_2}(g) + \overset{-2+1}{2OH^-}(aq) \rightarrow \overset{-1}{Cl^-}(aq) + \overset{+1-2}{ClO^-}(aq) + \overset{+1-2}{H_2O}(l)$$

The halogens, other than fluorine, form compounds in which they have positive oxidation numbers up to +7. Figure 16.5 shows an oxidation number chart for chlorine. The exhibition of several oxidation numbers by each halogen is, of course, very different from the metals in Groups I and II. These metals exhibit only one oxidation number in their compounds. As we might expect, the most stable oxidation state for halogens is −1. The relative stability of the −1 state decreases as the group is descended and the electronegativity of the halogen decreases.

Fluorine, the most electronegative element, never exhibits a positive oxidation number. This is because it can never form a compound in which it is the less electronegative element. In this respect, it is significant that fluorine reacts with alkalis differently to the other halogens. Fluorine forms a mixture of fluoride and oxygen difluoride in both of which its oxidation number is −1.

$$2F_2(g) + 2OH^-(aq) \rightarrow OF_2(g) + 2F^-(aq) + H_2O(l)$$

A further consequence of the decreasing electronegativity down Group VII is that the relative stability of the positive oxidation states increases from Cl to I.

Thus, the ease of decomposition of $HalO^-$ into Hal^- and $HalO_3^-$ (halogen in the +5 state) is in the order $IO^- > BrO^- > ClO^-$. This suggests that IO_3^- is relatively more stable than BrO_3^- which in turn is more stable than ClO_3^-. In fact, the positive oxidation state has become sufficiently stable in iodine to permit the existence of iodine cations. Thus, iodine(I) nitrate ($I^+NO_3^-$) and iodine(I) chlorate(VII) ($I^+ClO_4^-$) have been prepared in the presence of pyridine which forms a complex with the I^+ ion. Not surprisingly, these salts react with potassium iodide to give iodine,

$$I^+ + I^- \rightarrow I_2$$

and when electrolysed they give iodine at the cathode. I^+ ions are probably also produced when iodine chloride (ICl) is dissolved in concentrated H_2SO_4.

The existence of I^+ ions is evidence for the slight metallic character of iodine. Other features which show this are the metallic lustre of solid iodine and the small but definite conductivity of liquid iodine.

1 Write an equation for the disproportionation of ClO^- to Cl^- and ClO_3^-.

2 What is the oxidation number of chlorine in each of ClO^-, Cl^- and ClO_3^-?

3 What is the commonest oxidation number for chlorine in its compounds?

4 Write the formula of a compound containing chlorine in a different oxidation state to that in Cl^-, ClO^- or ClO_3^-.

Table 16.4
The electrode potentials of the halogens and some of their oxyanions

Electrode process	$E^{\ominus}$ /volts
$F_2(g) + 2e^- \rightarrow 2F^-(aq)$	+2.87
$2HClO(aq) + 2H^+(aq) + 2e^- \rightarrow Cl_2(aq) + 2H_2O(l)$	+1.59
$2HBrO(aq) + 2H^+(aq) + 2e^- \rightarrow Br_2(aq) + 2H_2O(l)$	+1.57
$2BrO_3^-(aq) + 12H^+(aq) + 10e^- \rightarrow Br_2(aq) + 6H_2O(l)$	+1.52
$Cl_2(aq) + 2e^- \rightarrow 2Cl^-(aq)$	+1.36
$2IO_3^-(aq) + 12H^+(aq) + 10e^- \rightarrow I_2(aq) + 6H_2O(l)$	+1.19
$Br_2(aq) + 2e^- \rightarrow 2Br^-(aq)$	+1.09
$HIO(aq) + H^+(aq) + 2e^- \rightarrow I^-(aq) + H_2O(l)$	+0.99
$ClO^-(aq) + H_2O(l) + 2e^- \rightarrow Cl^-(aq) + 2OH^-(aq)$	+0.89
$I_2(aq) + 2e^- \rightarrow 2I^-(aq)$	+0.54
$IO^-(aq) + H_2O(l) + 2e^- \rightarrow I^-(aq) + 2OH^-(aq)$	+0.49

One sensible deduction that we might make from all this is that the halogens are likely to act as oxidising agents in all their positive oxidation states (table 16.4).

Notice two points from table 16.4.

1 The redox potentials for the various systems involving iodine are smaller than the corresponding systems involving bromine. These in turn are less than those for chlorine (i.e. $F_2 > Cl_2 > Br_2 > I_2$; $HClO > HBrO > HIO$; $BrO_3^- > IO_3^-$; $ClO^- > IO^-$). This is further evidence for the increasing stability of positive oxidation states as Group VII is descended.

2 The $HalO_3^-$ ions are stable in alkali and require acid conditions before they become effective oxidising agents. This is not surprising in view of the fact that each $HalO_3^-$ ion requires six H^+ ions during the reaction.

16.9 Reactions of halide ions

Some reactions of fluorides, chlorides, bromides and iodides in aqueous solution are summarised in table 16.5. All common halides are soluble except all lead halides, AgCl, AgBr and AgI. Notice in table 16.5 that precipitates of these halides are produced when aqueous solutions of halides are treated with either $Pb^{2+}(aq)$ or $Ag^+(aq)$.

$$Ag^+(aq) + Cl^-(aq) \rightarrow AgCl(s)$$
$$Pb^{2+}(aq) + 2Cl^-(aq) \rightarrow PbCl_2(s)$$

Table 16.5
Reactions of aqueous halide ions

Solution added	$F^-(aq)$	$Cl^-(aq)$	$Br^-(aq)$	$I^-(aq)$
$Pb(NO_3)_2(aq)$	white ppte of PbF_2	white ppte of $PbCl_2$	yellow ppte of $PbBr_2$	yellow ppte of PbI_2
$AgNO_3(aq)$	no reaction	white ppte of AgCl	cream ppte of AgBr	yellow ppte of AgI
Solubility of AgHal in				
(a) dil. $NH_3(aq)$	soluble	soluble	insoluble	insoluble
(b) conc. $NH_3(aq)$	soluble	soluble	soluble	insoluble
Effect of sunlight on AgHal	no effect	white AgCl turns purple-grey	cream AgBr turns green-yellow	no effect

Silver nitrate solution followed by sunlight or ammonia solution can be used as a test for halide ions.

F^-(aq) gives no precipitate of AgF.

Cl^-(aq) gives a white precipitate of AgCl which becomes purple-grey in sunlight and dissolves in dilute NH_3.

Br^-(aq) gives a cream precipitate of AgBr which becomes green-yellow in sunlight and dissolves in concentrated NH_3.

I^-(aq) gives a yellow precipitate of AgI which is unaffected by sunlight and which is insoluble in concentrated NH_3.

The colour changes which occur when AgCl and AgBr are exposed to sunlight result from the superficial conversion of these silver halides to silver and halogen.

$$Ag^+Br^-(s) \rightarrow Ag(s) + \tfrac{1}{2}Br_2(g)$$

This photochemical change involving AgBr plays an essential part in black and white photography. Photographic plates and films contain silver bromide. This decomposes to silver on exposure to light. When the photograph is developed, the plate/film is treated with 'hypo' (sodium thiosulphate solution). This removes excess AgBr as a soluble complex ion $[Ag(S_2O_3)_2]^{3-}$, and the silver remains on the plate/film as an opaque shadow.

The reactions of solid halides with concentrated sulphuric acid, with concentrated phosphoric(v) acid and with a mixture of manganese(IV) oxide plus concentrated sulphuric acid are summarised in table 16.6.

Table 16.6
Reactions of solid halides

Reagent added	Fluoride	Chloride	Bromide	Iodide
Conc. H_2SO_4	HF(g) produced	HCl(g) produced	HBr(g) + a little Br_2(g) produced	a little HI(g) + I_2(g) produced
Conc. H_3PO_4	HF(g) produced	HCl(g) produced	HBr(g) produced	HI(g) produced
Conc. H_2SO_4 + MnO_2	HF(g) produced	Cl_2(g) produced	Br_2(g) produced	I_2(g) produced

When concentrated H_2SO_4 is added to solid halides the first product is the hydrogen halide. Being relatively volatile, these hydrogen halides are evolved as gases.

$$Hal^-(s) + H_2SO_4(l) \rightarrow HHal(g) + HSO_4^-(s)$$

However, concentrated sulphuric acid is also an oxidising agent. It is powerful enough to oxidise HBr to Br_2 and HI to I_2, but it cannot oxidise HF and HCl.

$$2HBr(g) + H_2SO_4(l) \rightarrow Br_2(g) + 2H_2O(l) + SO_2(g)$$
$$2HI(g) \; + H_2SO_4(l) \rightarrow I_2(g) \; + 2H_2O(l) + SO_2(g)$$

When concentrated H_2SO_4 is used in the presence of an even stronger oxidising agent, such as MnO_2, the oxidising conditions are sufficient to oxidise HCl to Cl_2, but HF is still not oxidised to F_2.

$$4HCl(l) + MnO_2(s) \rightarrow Cl_2(g) + MnCl_2(aq) + 2H_2O(l)$$

Look at the reactions of solid halides with conc. H_3PO_4 in table 16.6.

1 Write an equation for the reaction of concentrated phosphoric(v) acid (H_3PO_4) with NaBr(s).

2 Why does the conc. H_3PO_4 cause HBr(g) to be evolved during the reaction?

3 Why is no Br_2 produced in this reaction unlike that between conc. H_2SO_4 and NaBr(s)?

PTFE has an extremely low coefficient of friction and anti-stick properties. Because of this, thin layers of it are coated on the running surface of skis.

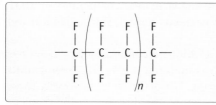

Figure 16.6
The structure of poly(tetrafluoroethene).

Sodium chlorate(I) (sodium hypochlorite) is used as a bleach and as a disinfectant.

16.10 Uses of the halogens

Although the halogen elements have very few uses, their compounds are used extensively in industry, in agriculture, in medicine and in the home. Chlorine is used as a cheap industrial oxidant in the manufacture of bromine (section 16.3).

Iodine dissolved in alcohol, commonly known as 'tincture of iodine', is used as a mild antiseptic for cuts and scratches. Iodine is also mixed with the detergents used in cleaning dairy equipment.

Small quantities of fluorine are used in rocket propulsion. Much larger quantities are used to make uranium(VI) fluoride for the separation of $^{238}_{92}U$ and $^{235}_{92}U$:

$$UF_4(s) + F_2(g) \rightarrow UF_6(s)$$

By allowing gaseous UF_6 to diffuse slowly, it is possible to separate $^{238}_{92}UF_6$ from $^{235}_{92}UF_6$. This allows the separation of isotopes of uranium for use in atomic power stations.

Fluorine is also used to make a wide range of fluorocarbon compounds for use as refrigerants, aerosol propellants, anaesthetics and fire-extinguisher fluids. Many of these compounds are discussed in section 29.3. One of the most important fluorocarbons is poly(tetrafluoroethene), PTFE, frequently sold under the trade name 'Fluon' or 'Teflon' (figure 16.6).

PTFE, like other fully fluorinated hydrocarbons, is very unreactive and resists almost all corrosive chemicals. For this reason, it is used for valves, seals and gaskets in chemical plants and laboratories. It is also an excellent electrical insulator and is used for wire coverings. Furthermore, PTFE has an extremely low coefficient of friction and has anti-stick properties. Because of this, thin layers of it are coated on the cooking surface of non-stick saucepans and on the running surface of skis.

Chlorine is used in the manufacture of many familiar materials. Hydrogen chloride is produced by the reaction between chlorine and hydrogen, itself a by-product in the electrolytic manufacture of chlorine. Hydrogen chloride made in this way is dissolved in water to produce hydrochloric acid:

$$H_2(g) + Cl_2(g) \rightarrow 2HCl(g) \xrightarrow{aq} 2H^+(aq) + 2Cl^-(aq)$$

Hydrochloric acid is the cheapest industrial acid. It is used for removing (de-scaling) rust from steel sheets before galvanising.

The other inevitable by-product of the electrolysis of brine is sodium hydroxide. The sodium hydroxide solution is treated with gaseous chlorine to obtain sodium chlorate(I) (sodium hypochlorite, NaClO).

$$Cl_2(g) + 2NaOH(aq) \rightarrow NaCl(aq) + NaClO(aq) + H_2O(l)$$

The solution produced in this reaction, containing both sodium chlorate(I) and sodium chloride, is sold as liquid bleach. It is used as a bleach in laundries and as a sterilising fluid in dilute solution. Concentrated solutions are used to bleach paper and wood pulp and to treat sewage. The active bleaching agent in the solution is chlorate(I), ClO^-, which acts as a powerful oxidising agent in the bleaching reactions.

On warming, NaClO decomposes to sodium chlorate(V) ($NaClO_3$) which is used as a weed killer.

$$3NaClO(aq) \xrightarrow{heat} 2NaCl(aq) + NaClO_3(aq)$$

In recent years, chlorine compounds have been developed for use as degreasing solvents such as tetrachloromethane and trichloroethene, as plastics such as PVC (section 27.7), as disinfectants and antiseptics such as 'Dettol' and TCP (section 30.5) and as pesticides.

Figure 16.7
The structures of some chlorinated pesticides.

dichlorodiphenyltrichloroethane (DDT)

benzene hexachloride (BHC)

aldrin

dieldrin

During the 1950s and 1960s, peregrine falcons became extinct in some areas of Europe and North America owing to the use of DDT and other toxic insecticides. The birds are now re-establishing themselves in these areas following restrictions on the use of DDT imposed in the 1970s.

A barn owl killed by insecticide poisoning.

During the 1940s and 1950s a range of highly chlorinated aromatic compounds were developed and used as pesticides. Probably the best known of these compounds are DDT (dichlorodiphenyltrichloroethane), BHC (benzene hexachloride), aldrin, dieldrin and heptachlor (figure 16.7).

The spraying of large areas of land with these chlorine-containing pesticides has eliminated many insect-borne diseases, such as malaria. It also led to enormous improvements in the quality and yield of crops. Unfortunately, however, these chlorinated compounds are so stable that they remain unchanged on the crops or accumulate in the soil. Furthermore, these compounds are fat-soluble but not water-soluble. This means that they concentrate, after ingestion, in the fatty tissues of birds and animals, possibly reaching hazardous levels.

In the spring of 1956, large numbers of seed-eating birds were found dead in cereal-growing areas in the UK. Analysis showed that the corpses of these birds contained dieldrin and aldrin which had been sprayed on the spring-sown wheat. In 1964, the International Advisory Committee on Poisonous Substances Used in Agriculture and Food Storage recommended that the use of dieldrin and aldrin should be reserved for the treatment of heavily infected areas. At that time, no restrictions were placed on the use of DDT and BHC, but concern over their use continued to grow. It appears that insects can develop a tolerance to small, non-lethal doses of these chemicals. These insects are eaten by small carnivorous animals and birds which concentrate the DDT in their own fatty tissue. Larger predators who eat these small carnivores concentrate the DDT even further. Ultimately, the DDT may reach toxic levels in animals or birds several stages along the food chain. At one time, for instance, there was some concern that humans might be affected by drinking the milk of cows which had eaten grasses sprayed with DDT. The use of DDT was also thought to be responsible for the decreasing populations of birds of prey and frogs, whose bodies were found to contain the chemical. As a result of concern over its use, the world consumption of DDT fell from 400 000 tonnes in 1963 to only about half that quantity in 1971.

Since 1972, both the British and American Governments have imposed restrictions on the use of DDT, but the need for an effective and cheap substitute is most urgent. Nevertheless, it is sensible to use less toxic insecticides for general purposes and reserve DDT for special control schemes.

Summary

1 The halogens are a group of reactive non-metals. They contrast strongly with the alkali and alkaline-earth elements which are reactive metals.

2 The halogens show remarkable similarities in their properties and reactions. There is also an obvious trend in their behaviour and reactivity with increase in atomic number. As atomic number increases, the halogens become less reactive.

3 The halogens are so reactive that they occur naturally only in compounds.

4 The halogens can be obtained by oxidation of halide ions.

5 The halogens exist as diatomic molecules. Their atoms have seven electrons in the outermost shell. Consequently, the chemistry of the halogens is dictated by a tendency to act as oxidising agents in forming negative halide ions (Hal^-). The most stable oxidation number of the halogens is -1.

6 The order of oxidising power for the halogens is $F_2 > Cl_2 > Br_2 > I_2$.

7 A disproportionation reaction is one in which a particular molecule, atom or ion is simultaneously oxidised and reduced. When halogens react with dilute alkali to form a mixture of Hal^- and $HalO^-$, the halogen is both oxidised and reduced.

8 The action of $AgNO_3(aq)$ followed by sunlight or $NH_3(aq)$ can be used as a test for halide ions.

Review questions

1 (a) Chlorine and sodium hydroxide are manufactured by the electrolysis of brine. Write equations to summarise what happens during the electrolysis when
  (i) precautions are taken to prevent the products mixing with each other,
  (ii) the main products (chlorine and sodium hydroxide) are deliberately mixed with one another.
 (b) Iodine is obtained from iodate(V) by treatment with sodium hydrogensulphite.
  Write an equation (or half-equations) for the reaction of iodate(V) with hydrogensulphite ions in acid solution.
 (c) Fluorine can be obtained by the electrolysis of KF dissolved in liquid HF, but not by the electrolysis of KF dissolved in water. Why not?
 (d) Astatine ($^{211}_{85}At$) has been made by bombarding bismuth with high-energy alpha particles.
  (i) Write a nuclear equation for this reaction.
  (ii) The name 'astatine' is derived from a Greek word meaning unstable. How do you think astatine might decompose? Write a nuclear equation for the decay process.

2 From your knowledge of the halogens, predict what happens in the following situations. Write equations for any reactions which take place. (Ignore the radioactive nature of astatine.)
 (a) Astatine vapour is mixed with hydrogen at 100°C.
 (b) Astatine is added to aqueous sodium hydroxide.
 (c) Concentrated sulphuric acid is added to solid sodium astatide.
 (d) Aqueous silver nitrate is added to aqueous sodium astatide.
 (e) Astatine is added to sodium thiosulphate solution.

3 The halogens (F, Cl, Br and I) form a well-defined group of elements.
 (a) Explain how the following support this statement:
  (i) electron structure,
  (ii) redox behaviour,
  (iii) physical properties of the elements,
  (iv) usual oxidation state
 (b) Describe four specific properties that show a clear gradation as the group is descended from F to I.
 (c) Explain the meaning of the term 'electron affinity'. How does this vary among the halogens, and how is it related to their reactivity?
 (d) Fluorine and fluorides show some properties which are not typical of the rest of the group. Mention three of these properties and suggest a reason or reasons for the difference.

4 Explain the following observations:
 (a) A mass spectrograph of chlorine shows five particles with relative masses of 35, 37, 70, 72 and 74.
 (b) As Group VII of the periodic table is descended, the halogens become weaker oxidising agents.
 (c) In its compounds, fluorine shows only one oxidation state, whereas chlorine shows several.
 (d) Hydrogen fluoride has a higher boiling point than hydrogen chloride and hydrogen iodide.

5 The percentage of copper in a sample of brass was determined as follows. 2.0 g of the brass was converted to 200 cm^3 of a solution of copper(II) nitrate free from nitric or nitrous acid and acidified with ethanoic (acetic) acid. 20.0 cm^3 of this solution liberated sufficient iodine from potassium iodide solution to react with 25.0 cm^3 of 0.1 mol dm^{-3} sodium thiosulphate solution. (Cu = 64; the reaction between Cu^{2+}(aq) and I^-(aq) can be represented as $2Cu^{2+}$(aq) + $4I^-$(aq) → $2CuI$(s) + I_2(aq).)

(a) Write an equation or half-equations for the reaction between iodine and sodium thiosulphate.

(b) How many moles of I_2 were liberated by 20 cm^3 of aqueous Cu^{2+} solution?

(c) How many moles of copper are there in 200 cm^3 of aqueous Cu^{2+} solution?

(d) What is the percentage by mass of copper in the brass?

6 Read the second half of section 16.10 concerning pesticides.

(a) What is the correct systematic name for BHC?

(b) Why are chlorinated pesticides such as BHC and DDT soluble in fat, but not in water?

(c) Seed-eating birds are extremely vulnerable to chlorinated pesticides sprayed on spring crops. Why is there much less danger to these birds when autumn-sown crops are treated with pesticides?

(d) Suggest an explanation for the pesticide action of DDT.

(e) Why can insects develop a tolerance to small doses of chemicals such as DDT?

(f) Suppose you were given the task of synthesising the ideal insecticide for spraying on spring-sown wheat. Make a list of the properties that you would look for in your product.

Group IV – Carbon to Lead, Non-Metal to Metal

17

17.1 Introduction

Table 17.1
The elements in Group IV

C
Si
Ge
Sn
Pb

The similarity between elements in the same family which was so obvious in Groups I, II and VII is much less apparent in Group IV. Here there is a considerable change in the character of the elements as atomic number rises.

Carbon is unquestionably a non-metal. Silicon and germanium are metalloids. Tin and lead show typical metallic properties (table 17.1).

Although carbon occurs naturally as diamonds (in Brazil and South Africa) and as graphite (in Sri Lanka, Germany and USA), it occurs much more abundantly as carbon compounds. These are present in coal, oil and natural gas, in limestone and other carbonates and in living things.

▶ A rough diamond being polished.

▶▶ Large diamond-studded drills being used to bore holes through concrete in the sea wall to improve drainage.

The structure, properties and uses of diamond and graphite were discussed in section 10.6. The high refractive index and dispersive power of diamonds led to their use as jewellery in the very earliest civilisations. The modern industrial uses of diamond almost all result from its hardness – drilling, cutting, grinding and for bearings in precision instruments such as watches.

The uses of graphite are related to its conducting and lubricating qualities. It is used in 'lead' pencils, as a high-temperature lubricant and as inert electrodes in various electrolytic processes. It is also used as a moderator for slowing down neutrons in nuclear reactors.

After oxygen, silicon is the most abundant element on the Earth. It occurs as silicon(IV) oxide (SiO_2) in sand and sandstone and as many forms of silicate in rocks and clays. In comparison with carbon and silicon, germanium, tin and lead are rare elements. Traces of germanium are present in coal and accumulate in flue dust as germanium(IV) oxide, GeO_2. Silicon and germanium are used as semi-conductors in transistorised electronic equipment. These two elements are obtained from their oxides, SiO_2 and GeO_2, by very similar processes.

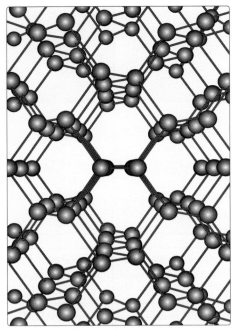

▲ An unusual view through the structure of diamond. Notice the tetrahedral arrangement of carbon atoms.

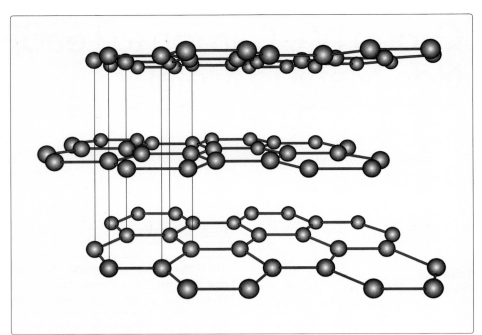

▲ A view of the planes of atoms in the structure of graphite. Within the horizontal planes, the carbon atoms are arranged in interconnected hexagons.

▲ A carbon-filament lamp.

The flow scheme in figure 17.1 shows how pure silicon is obtained industrially. The final zone-refining process allows the production of ultra-pure silicon. This is necessary for use in transistors. The material to be purified is packed in a cylindrical tube and suspended vertically (figure 17.2). At the top of the tube, a short length of the cylinder is surrounded by an electrical heating coil. This melts a narrow band of material within the tube. During the zone-refining operation, the tube is slowly raised through the heating element and the zone of molten material moves down the tube. Once the sample gets above the heating element, it begins to recrystallise and impurities are left in the molten zone. In this way, the impurities collect in the molten phase and end up concentrated at one end of the tube.

1 Write equations for
  (i) the reduction of SiO_2 with carbon,
  (ii) the reaction of Cl_2 with Si.
2 Why is the impure silicon converted to $SiCl_4$?
3 Why can the impure silicon not be zone-refined?

Figure 17.1
A flow scheme showing the industrial manufacture of pure silicon.

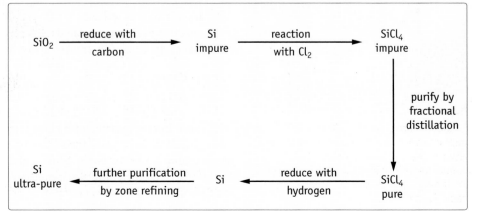

17 Group IV – Carbon to Lead, Non-Metal to Metal

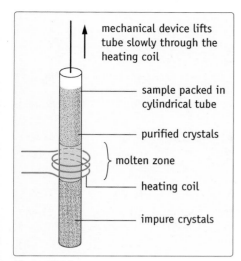

mechanical device lifts
tube slowly through the
heating coil

sample packed in
cylindrical tube

purified crystals

molten zone

heating coil

impure crystals

Figure 17.2
The production of ultra-pure silicon by zone refining.

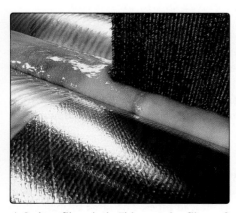

▲ Carbon-fibre cloth. This contains fibres of graphite which can absorb large molecules between the layers of carbon atoms. Carbon-fibre cloth is hung between cooking and dining areas to prevent kitchen smells reaching the diners.

Tin occurs as tin(IV) oxide (SnO_2) in cassiterite (tinstone). Lead occurs as the sulphide ore, galena (PbS). Lead is obtained by roasting the sulphide in air to obtain PbO, which is then reduced with carbon:

$$2PbS(s) + 3O_2(g) \rightarrow 2PbO(s) + 2SO_2(g)$$
$$PbO(s) + C(s) \rightarrow Pb(l) + CO(g)$$

Tin can be obtained by direct reduction of SnO_2 with carbon:

$$SnO_2(s) + 2C(s) \rightarrow Sn(l) + 2CO(g)$$

Tin and lead have important uses as relatively inert metals. Tin is used to tin-plate steel. This is then used to make the so called 'tins' for canning meats, soup, fruit, etc. Lead has been used as an inert material for gas and water pipes, for cable sheathing and for chemical vessels. Lead is also used for the plates of lead–acid accumulators (batteries) and as a screen from radioactivity.

► This speed boat has a hull reinforced with carbon fibre.

Another important use of both tin and lead is in alloying. Indeed, there are several important alloys containing both of these metals. These include

solder (50% Sn, 50% Pb),

pewter (80% Sn, 20% Pb) and

type metal (10% Sn, 75% Pb, 15% Sb).

17.2 Variation in the physical properties of the elements

Table 17.2 shows a list of physical properties for the elements in Group IV. Notice how these properties vary more from one element to the next than with Group I or Group VII.

These changes in property are, of course, related to the *increasing metallic (electropositive) character and the decreasing non-metallic (electronegative) character as atomic number rises.*

The structural changes from giant molecular lattices in carbon and silicon (see section 10.6) to giant metallic structures in tin and lead help to explain the changes in physical properties. Silicon and germanium crystallise in the same structure as diamond. Tin and lead have distorted close-packed metal structures. As the atoms get larger and the atomic radius increases, the interatomic bonding becomes weaker. As a result the attraction of neighbouring nucleii for intervening electrons gets less.

Table 17.2
Various atomic and physical properties of the elements in Group IV

Element	C	Si	Ge	Sn	Pb
Atomic number	6	14	32	50	82
Electron configuration (outer shell only)	$2s^2 2p^2$	$3s^2 3p^2$	$4s^2 4p^2$	$5s^2 5p^2$	$6s^2 6p^2$
Atomic radius/nm	0.077	0.117	0.122	0.141	0.154
Melting point/°C	3730^d	1410	937	232	327
Boiling point/°C	4830^d	2680	2830	2270	1730
Density/g cm^{-3}	2.26^{gr} 3.51^d	2.33	5.32	7.3	11.44
Thermal conductivity /J cm^{-1} s^{-1} K^{-1}	0.24^{gr}	0.84	0.59	0.63	0.35
Conductivity	fairly good^{gr} non-cond.^{d}	semi-cond.	semi-cond.	good	good
Electrical conductivity/ohm^{-1} m^{-1}		1×10^6	2×10^6	8×10^6	5×10^6
Enthalpy change of atomisation/kJ mol^{-1}	+716^{gr}	+456	+376	+302	+195
First ionisation energy/kJ mol^{-1}	1086	787	760	707	715
Type of structure	giant molecular	giant molecular similar to diamond	giant molecular similar to diamond	giant metallic	giant metallic

gr. = graphite   d. = diamond

The weaker interatomic forces result in a change in bonding from covalent to metallic down the Group. Hence there is a decrease in melting point, boiling point, enthalpy change of atomisation and first ionisation energy. At the same time, the increasing metallic character causes a general increase in density and conductivity.

The first ionisation energy decreases considerably from carbon to silicon. After that, it falls relatively little. The reason for this is that, after silicon, there is a larger increase in nuclear charge. This is associated with the filling of '*d*' and '*f*' sub-shells and it counterbalances the increase in atomic radius.

17.3 Variation in the chemical properties of the elements

Information related to the chemical properties of the Group IV elements is given in table 17.3. *The group trends further emphasise the increase in metallic character down the group.* Notice how carbon, the first member of the group, is much more electronegative than the remainder.

In general, chemical reactivity increases from carbon to lead. Electrode potentials show that only tin and lead are strong enough reducing agents to liberate hydrogen from dilute acids.

$$Pb(s) \rightarrow Pb^{2+}(aq) + 2e^- \quad E^{\ominus} = +0.13 \text{ volts}$$

Lead will react very slowly with soft water containing dissolved oxygen to form $Pb(OH)_2$. This reaction has resulted in cases of lead poisoning in areas where householders have drunk hot water directly from lead pipes. The same problem does not arise in hard-water areas. Here the lead piping develops a protective layer of insoluble lead sulphate or lead carbonate.

▲ Leaded windows in Ludlow Parish Church, Shropshire, England. Lead is an ideal material for holding the different-coloured pieces of glass in place on account of its malleability and inertness.

Table 17.3
Various atomic and chemical properties of the elements in Group IV

Element	C	Si	Ge	Sn	Pb
Atomic number	6	14	32	50	82
Electron configuration (outer shell only)	$2s^2 2p^2$	$3s^2 3p^2$	$4s^2 4p^2$	$5s^2 5p^2$	$6s^2 6p^2$
Electronegativity	2.5	1.8	1.8	1.8	1.8
Electrode potential M^{2+}(aq)/M(s)/volts			0.23	−0.14	−0.13
$\Delta H_f^{\ominus}$ (XO_2)/kJ mol^{-1}	−394	−910	−551	−581	−277
$\Delta H_f^{\ominus}$ (XO)/kJ mol^{-1}	−111		−212	−286	−217
$\Delta H_f^{\ominus}$ (XH_4)/kJ mol^{-1}	−75	+ 34	+90	+163	
Bond energy, E(X—H)/kJ mol^{-1}	435	318	285	251	
$\Delta H_f^{\ominus}$ (XCl_4)/kJ mol^{-1}	−136	−640	−544	−511	−320
Bond energy, E(X—Cl)kJ mol^{-1}	327	402	339	314	235

As the enthalpy changes of formation would suggest, all Group IV elements except lead react with oxygen on heating to form the dioxide.

$$C(s) + O_2(g) \rightarrow CO_2(g)$$
$$Sn(s) + O_2(g) \rightarrow SnO_2(s)$$

Lead, however, reacts to form PbO.

The enthalpy changes of formation of hydrides, XH_4, would suggest that only carbon could react directly with hydrogen. In practice, carbon does not react with hydrogen, even at very high temperatures, because of the large activation energy.

All the elements in Group IV react directly on heating with chlorine. They form the tetrachloride, except lead, which forms the dichloride (see $\Delta H_f^{\ominus}$ (XCl_4) values in table 17.3). However, the reaction between carbon and chlorine is so slow that CCl_4 (tetrachloromethane) is manufactured by the reaction between carbon disulphide and chlorine:

$$CS_2 + 3Cl_2 \rightarrow CCl_4 + S_2Cl_2$$

17.4 General features of the compounds

The most striking feature of the compounds of these elements is the existence of two oxidation states, +2 and +4. The existence of compounds in which the elements show different oxidation numbers is typical of the p-block elements.

It is interesting to consider the relative stabilities of the +2 and +4 oxidation states for the different elements. In carbon and silicon compounds, the +4 state is very stable relative to +2. The +2 state is rare and easily oxidised to +4. Thus, CO reacts very exothermically to form CO_2. SiO is too unstable to exist under normal conditions, although it has been obtained at 2000°C.

Germanium forms oxides in both +4 and +2 states. However, GeO_2 is rather more stable than GeO. GeO_2 does not act as an oxidising agent and GeO is readily converted to GeO_2.

In tin compounds, the +4 state is only slightly more stable than the +2 state. Thus, aqueous tin(II) ions are mild reducing agents. They will convert mercury(II) ions to mercury and iodine to iodide:

$$Sn^{2+}(aq) + Hg^{2+}(aq) \rightarrow Sn^{4+}(aq) + Hg(l)$$
$$Sn^{2+}(aq) + I_2(aq) \rightarrow Sn^{4+}(aq) + 2I^-(aq)$$

▲ Through the eye of a needle!
This photograph shows the extremely small size of some silicon chips. The chip shown carries a very complex electrical circuit with 120 components. Silicon chips such as this are used extensively in transistors and other electrical equipment. The material which looks like rope is ordinary sewing cotton.

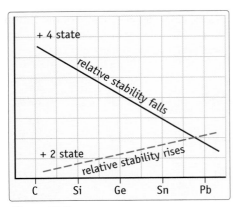

Figure 17.3
Relative stabilities of the +4 and +2 oxidation states for the elements in Group IV.

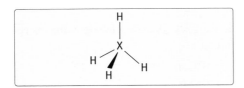

Figure 17.4
The molecular shape of the simple hydrides of Group IV elements.

Table 17.4
Average energies for corresponding carbon and silicon bonds

Bond	Average bond energy $E(X-Y)/kJ\ mol^{-1}$
C—C	346
C—O	360
Si—Si	226
Si—O	464

In lead compounds, however, the +2 state is unquestionably more stable. PbO_2 is a strong oxidising agent, whilst PbO is relatively stable. Thus, PbO_2 can oxidise hydrochloric acid to chlorine and hydrogen sulphide to sulphur.

$$PbO_2(s) + 4HCl(aq) \rightarrow PbCl_2(s) + Cl_2(g) + 2H_2O(l)$$

Notice the steady increase in the stability of the lower oxidation state relative to the higher oxidation state on moving down the group from carbon to lead (figure 17.3).

The greater stability of the +2 oxidation state with respect to the +4 state as the atomic number rises is well illustrated by the standard electrode potentials of the $M^{4+}(aq)/M^{2+}(aq)$ systems for germanium, tin and lead.

$$Ge^{4+} + 2e^- \rightarrow Ge^{2+} \quad E^\ominus = -1.6\ volts$$
$$Sn^{4+} + 2e^- \rightarrow Sn^{2+} \quad E^\ominus = +0.15\ volts$$
$$Pb^{4+} + 2e^- \rightarrow Pb^{2+} \quad E^\ominus = +1.8\ volts$$

As the electrode potentials get more positive from Ge^{4+} to Pb^{4+}, the oxidised form is more readily reduced to the +2 state.

All the Group IV elements have four electrons in their outermost shell (ns^2np^2), so it is not surprising that they show a well-defined oxidation state of +4 (−4 in the hydrides). However, none of the elements forms an M^{4+} cation in its solid compounds. This is due to the high ionisation energies involved in removing four successive electrons from an atom. Consequently, the bonding in the tetravalent compounds is predominantly covalent.

Compounds of tin and lead in which the Group IV element has an oxidation number of +2 (e.g. PbF_2, $PbCl_2$, PbO) are normally regarded as ionic. In these compounds, the Sn^{2+} and Pb^{2+} ions are believed to form by loss of the two 'p' electrons in the outer shell. The two 's' electrons remain relatively stable and unreactive in the filled sub-shell. This is sometimes referred to as the **'inert pair' effect**.

One of the most curious features of Group IV is the unique ability of carbon to form stable compounds containing long chains and rings of carbon atoms. This property is called **catenation**. It results in an enormous range of compounds for carbon. Indeed, there are many thousands of compounds containing only carbon and hydrogen of which methane (CH_4), ethane (C_2H_6), ethene (C_2H_4) and ethyne (C_2H_2) are four of the simplest.

1. What structure and shape do you predict for the hydrides CH_4, SiH_4, GeH_4, SnH_4 and PbH_4?
2. How will the volatility of these hydrides vary from CH_4 to PbH_4?
3. The hydrides become less stable from CH_4 to PbH_4. Why is this?
4. Suggest a reason why there are no compounds of Si, Ge, Sn and Pb analogous to C_2H_4 and C_2H_2.

This ability of carbon to catenate results from the fact that the C—C bond is almost as strong as the C—O bond (table 17.4). This makes the oxidation of carbon compounds to such products as carbon dioxide and water less energetically favourable than in the case of silicon. With silicon, the Si—Si bond is much weaker than the Si—O bond. Consequently, catenated compounds of silicon are energetically unstable with respect to SiO_2 and therefore do not occur naturally. Nevertheless, chemists have succeeded in synthesising a whole series of silicon hydrides, called silanes. Some of these have as many as 11 silicon atoms linked together. In a similar fashion, three hydrides have been synthesised for germanium, but, tin and lead form only one hydride each, SnH_4 and PbH_4. The ability of carbon to catenate is considered further in section 25.1.

Some of the important properties of the hydrides, chlorides and oxides of the Group IV elements are summarised in tables 17.5, 17.6 and 17.7 respectively.

Table 17.5
The preparation and properties of the simple hydrides of Group IV elements

Preparation

1 Reaction of magnesium compound (e.g. magnesium silicide) or magnesium alloy with dilute acid

$$Mg_2Si + 4H^+ \rightarrow 2Mg^{2+} + SiH_4$$

2 Reduction of tetrachloride with lithium aluminium hydride at 0°C in ether

$$SnCl_4 + LiAlH_4 \rightarrow SnH_4 + LiCl + AlCl_3$$

Properties

Structure:	Simple molecular.
Molecular shape:	Tetrahedral (figure 17.4).
Volatility:	Low melting points and boiling points, all are gases at room temperature.
Thermal stability:	The X—H bonds become longer and weaker down the Group (table 17.3). Thus, the hydrides become less stable. CH_4, SiH_4 and GeH_4 are stable up to high temperatures. SnH_4 is unstable and decomposes slowly at room temperature. PbH_4 is so unstable that it cannot be isolated at room temperature.
Reducing powers:	As their ease of decomposition into the element plus hydrogen increases, they become stronger reducing agents. Thermodynamically, the hydrides should all react with oxygen since their enthalpy changes of combustion are all negative. However, CH_4 is kinetically stable due to the high activation energy for its reaction with oxygen.

Table 17.6
The preparation and properties of the tetrachlorides of Group IV elements

Preparation

1 Direct synthesis for $SiCl_4$, $GeCl_4$ and $SnCl_4$

$$Si + 2Cl_2 \rightarrow SiCl_4$$

2 CCl_4 is prepared more easily by the action of Cl_2 on CS_2

$$CS_2 + 3Cl_2 \rightarrow CCl_4 + S_2Cl_2$$

3 $PbCl_4$ is prepared by the reaction of cold conc. HCl with PbO_2.

Properties

Structure:	Simple molecular.
Molecular shape:	Tetrahedral (similar to XH_4 hydrides).
Volatility:	Low melting points and boiling points. All are volatile liquids at room temperature.
Thermal stability:	As the X—Cl bonds become longer and weaker down the group, the tetrachlorides become less stable. Thus, CCl_4, $SiCl_4$ and $GeCl_4$ are stable even at high temperatures. $SnCl_4$ decomposes on heating to form $SnCl_2 + Cl_2$. $PbCl_4$ decomposes readily to form $PbCl_2 + Cl_2$.
Hydrolysis:	All the chlorides (except CCl_4) are readily hydrolysed to form hydroxy compounds + HCl

$$SnCl_4 + 4H_2O \rightarrow Sn(OH)_4 + 4HCl$$

Table 17.7 The preparation and properties of the oxides of Group IV elements

The dioxides	Group IV elements in the +4 oxidation state				
Oxide	CO_2	SiO_2	GeO_2	SnO_2	PbO_2
Preparation	Direct combination of element and oxygen				Electrolytic oxidation of Pb^{2+} in acid solution
Properties					
Boiling point/°C	−78	2590	1200	1900	decomposes on heating
Structure:	Simple molecular	Giant molecular	Intermediate between giant molecular and ionic		
Nature:	Acidic		Amphoteric		

Acidic (left group, CO_2 and SiO_2):

react with aq. alkalis giving XO_3^{2-} salts

$CO_2 + 2OH^- \rightarrow CO_3^{2-} + H_2O$
carbonate

$SiO_2 + 2OH^- \rightarrow SiO_3^{2-} + H_2O$
silicate

Amphoteric (right group, GeO_2, SnO_2, PbO_2):

(i) react with fused alkalis giving XO_3^{2-} salts

$SnO_2 + 2OH^- \rightarrow SnO_3^{2-} + H_2O$
stannate(IV)

$PbO_2 + 2OH^- \rightarrow PbO_3^{2-} + H_2O$
plumbate(IV)

(ii) react with conc. acids forming +4 salts

$SnO_2 + 4H^+ \rightarrow Sn^{4+}(aq) + 2H_2O$

$PbO_2 + 4HCl \rightarrow PbCl_4 + 2H_2O$
conc.

| Thermal stability: | stable even at high temps. | | | | decomposes to PbO on warming $PbO_2 \rightarrow PbO + \frac{1}{2}O_2$ |

The monoxides	Group IV elements in the +2 oxidation state				
Oxide	CO	SiO	GeO	SnO	PbO
Preparation	Reduction of CO_2 with carbon $CO_2 + C \rightarrow 2CO$	Only exists at very high temps.	Reduction of GeO_2 with Ge	Heat appropriate hydroxide or nitrate $Pb(OH)_2 \rightarrow PbO + H_2O$ $Pb(NO_3)_2 \rightarrow PbO + 2NO_2 + \frac{1}{2}O_2$	
Properties					
Boiling point/°C	−191				1470
Structure:	Simple molecular		Predominantly ionic		
Nature:	Neutral oxides		Amphoteric oxides		

Neutral oxides (CO, SiO):

react with neither acids nor alkalis.

Amphoteric oxides (GeO, SnO, PbO):

(i) react with acids to form salts

$PbO + 2H^+ \rightarrow Pb^{2+} + H_2O$

$SnO + 2H^+ \rightarrow Sn^{2+} + H_2O$

(ii) react with alkalis to form salts

$PbO + OH^- + H_2O \rightarrow Pb(OH)_3^-$
trihydroxoplumbate(II)

$SnO + OH^- + H_2O \rightarrow Sn(OH)_3^-$
trihydroxostannate(II)

| Thermal stability: | Readily oxidised to dioxide. (SiO, GeO and SnO revert to dioxide on standing in air.) | | | | Stable. |

Look closely for the following points in these tables:

1 The simple molecular structures of the hydrides and the tetrachlorides.

2 The decreasing thermal stability from C to Pb of the hydrides, tetrachlorides and dioxides.

3 The change in the nature of the dioxides down the group from acidic to amphoteric.

4 The change in the nature of the monoxides down the group from neutral to amphoteric.

▶ Dry ice (solid CO_2) being used to create mist in a production at the Edinburgh International Festival.

Summary

1 The elements in Group IV emphasise the *differences* between one element and the next in a particular group of the periodic table.

2 Catenation is the ability of one element to form chains or rings in which its atoms are bonded to one another.

3 As we go down the group from carbon to lead:

(a) the elements change from non-metallic to metallic. This change in the elements is reflected in their properties and uses.

(b) the +2 oxidation state becomes more stable relative to the +4 oxidation state.

(c) the bonding in compounds changes from covalent to ionic.
Covalent compounds (e.g. hydrides and tetrachlorides) become less stable down the group.
Ionic compounds (e.g. monoxides) become more stable down the group.

(d) the nature of the dioxides changes from acidic to amphoteric.

(e) the nature of monoxides changes from neutral to amphoteric.

▶ Watercolour paintings containing lead(II) compounds as constituents of their pigments darken over many decades. This is caused by the reaction of traces of hydrogen sulphide in the atmosphere with the lead(II) salts. The 'white lead' pigment which contains lead(II) carbonate becomes darker owing to the formation of black lead(II) sulphide. The original colour of the pigment can be restored by oxidation of the black lead(II) sulphide to white lead(II) sulphate using hydrogen peroxide. The photographs on the right show a watercolour painting by Matthew Snelling, in 1674, of Susanna, Lady Dormer before and after restoration. In this case black patches, which had developed on the face and shoulder, were removed.

1 Discuss the following points with respect to the elements in Group IV of the periodic table. Write equations for any reactions which you mention.
(a) The reactivity of the elements.
(b) The state, structure and thermal stability of the hydrides.
(c) The state, structure, thermal stability and hydrolysis of the tetrachlorides.
(d) The acidic/basic character of the dioxides.
(e) The relative stabilities of the +2 and +4 oxidation states.

2 (a) What is the principal ore of tin?
(b) Explain briefly, with an equation, how tin is obtained from its principal ore.
(c) What are the main uses of tin and its alloys?
(d) What are the relative advantages and disadvantages of tin-plating and galvanising (zinc-plating) iron in order to prevent corrosion?

3 (a) How and under what conditions does lead react with:
(i) air (oxygen),
(ii) water (soft and hard),
(iii) dilute nitric acid?
(b) Describe briefly the preparation of tin(II) oxide and tin(IV) oxide from tin.
(c) How (if at all) and under what conditions do tin(II) oxide and tin(IV) oxide react with:
(i) oxygen,
(ii) hydrochloric acid,
(iii) sodium hydroxide?

4 The following passage describes the preparation of tin(IV) iodide.

Add 4.0 g of powdered tin to a solution of 12.7 g of iodine in 100 cm^3 of tetrachloromethane. Reflux the mixture gently until the reaction is complete. Now filter the mixture through a pre-heated funnel. Wash the residue with hot tetrachloromethane, adding the washings to the filtrate. Cool the filtrate in ice until orange crystals of tin(IV) iodide form. Filter and dry the crystals.

(a) Write an equation for the formation of tin(IV) iodide in the above preparation.
(b) Give two reasons for using CCl_4 as the solvent in this preparation.
(c) Calculate the maximum possible yield of tin(IV) iodide. Explain your calculation.
(d) How would you know when the reaction was complete?
(e) Why was the mixture filtered through a pre-heated funnel?
(f) Why was the residue washed with hot CCl_4?
(g) What would you predict for
(i) the structure,
(ii) the thermal stability of tin(IV) iodide?

5 (a) Draw electron 'dot/cross' diagrams for carbon monoxide and carbon dioxide. (You need only show the electrons in the outer shells of the constituent atoms.)
(b) 'Carbon monoxide is iso-electronic with nitrogen.' Explain what is meant by this statement.
(c) Use the standard enthalpy changes of formation of CO_2 and CO in table 17.3 to find the enthalpy changes of the following reactions.

$$CO_2(g) + C(s) \rightarrow 2CO(g)$$
$$CO(g) + \tfrac{1}{2}O_2(g) \rightarrow CO_2(g)$$

Comment on the relative stabilities of CO_2 and CO.

6 (a) State four important similarities in the chemistry of tin and lead or their corresponding compounds.
(b) Explain the following in terms of atomic or electronic properties.
(i) Tin(IV) compounds are more stable than tin(II) compounds.
(ii) Lead(II) compounds are more stable than lead(IV) compounds.
(iii) $PbCl_4$ has a simple molecular structure, whereas $PbCl_2$ is ionic.
(c) Tin and lead have variable oxidation states and form complex ions. Why then are they not classed as transition metals?
(d) Oil and water-colour paintings containing lead(II) compounds in their pigments darken over many decades. This is due to the reaction of traces of hydrogen sulphide in the atmosphere with the lead(II) salts.
(i) Write an equation for the darkening process.
(ii) Explain why dilute solutions of hydrogen peroxide can be used to restore the paintings.

The Transition Metals

18.1 Introduction

The elements from scandium (atomic number 21) to zinc (atomic number 30) in the periodic table form what is generally regarded as the first sequence of transition elements. But, what exactly is a transition element? And why is it that these elements have similarities to each other across their period, unlike other parts of the periodic table?

The answer to these questions concerning the transition metals lies in their electronic structures shown in table 18.1.

Indeed, virtually all the properties of transition elements are related to their electronic structures and the relative energy levels of the orbitals available for their electrons.

Table 18.1

The electron structures of atoms and ions of the elements K to Zn ((Ar) ≡ electron structure of argon)

Element	Symbol	Electron structure of atom	Common ion	Electron structure of ion
potassium	K	$(Ar)4s^1$	K^+	(Ar)
calcium	Ca	$(Ar)4s^2$	Ca^{2+}	(Ar)
scandium	Sc	$(Ar)3d^14s^2$	Sc^{3+}	(Ar)
titanium	Ti	$(Ar)3d^24s^2$	Ti^{4+}	(Ar)
vanadium	V	$(Ar)3d^34s^2$	V^{3+}	$(Ar)3d^2$
chromium	Cr	$(Ar)3d^54s^1$	Cr^{3+}	$(Ar)3d^3$
manganese	Mn	$(Ar)3d^54s^2$	Mn^{2+}	$(Ar)3d^5$
iron	Fe	$(Ar)3d^64s^2$	Fe^{2+}	$(Ar)3d^6$
			Fe^{3+}	$(Ar)3d^5$
cobalt	Co	$(Ar)3d^74s^2$	Co^{2+}	$(Ar)3d^7$
nickel	Ni	$(Ar)3d^84s^2$	Ni^{2+}	$(Ar)3d^8$
copper	Cu	$(Ar)3d^{10}4s^1$	Cu^+	$(Ar)3d^{10}$
			Cu^{2+}	$(Ar)3d^9$
zinc	Zn	$(Ar)3d^{10}4s^2$	Zn^{2+}	$(Ar)3d^{10}$

As the shells of electrons get further and further from the nucleus, successive shells become closer in energy. Thus, the difference in energy between the second and third shells is less than that between the first and second. By the time the fourth shell is reached, there is, in fact, an overlap between the third and fourth shells. In other words, from potassium onwards, the orbitals of highest energy in the third shell (i.e. the $3d$ orbitals) have higher energy than those of lowest energy in the fourth shell (the $4s$ orbitals) (figure 18.1).

Figure 18.1

Relative energy levels of the 3s, 3p, 3d, 4s and 4p orbitals.

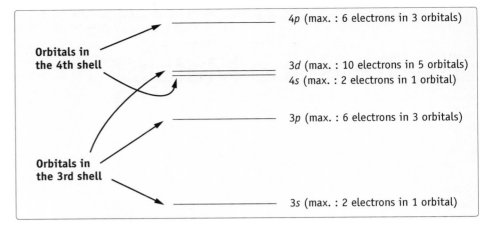

The 3d sub-shell is 'on average' nearer the nucleus than the 4s sub-shell, but at a higher energy level. So, once the 3s and 3p sub-shells are filled at argon, further electrons enter the 4s sub-shell because it is at a lower energy level than the 3d sub-shell. Hence, potassium and calcium have, respectively, one and two electrons in the 4s sub-shell (table 18.1). Once the 4s sub-shell is filled at calcium, electrons enter the 3d level. Hence scandium has the electron structure $(Ar)3d^1 4s^2$, titanium has the electron structure $(Ar)3d^2 4s^2$ and so on (table 18.1 and figure 18.2).

Notice the unexpected electron structures for chromium and copper. The arrangement of electrons in chromium is $(Ar)3d^5 4s^1$ (figure 18.2) not $(Ar)3d^4 4s^2$ which we might have expected. The explanation of this anomaly is that the electron structure $(Ar) 3d^5 4s^1$ with half-filled 3d and 4s sub-shells has a lower energy level than $(Ar)3d^4 4s^2$. The extra stability of a half-filled sub-shell is thought to result from the occupation of each orbital by one electron. This produces an equal distribution of charge around an atom.

Copper atoms have an electron structure $(Ar)3d^{10} 4s^1$ (figure 18.2) rather than $(Ar)3d^9 4s^2$. In this case, it appears that $(Ar)3d^{10} 4s^1$ with a filled 3d sub-shell and a half-filled 4s sub-shell is more stable than $(Ar)3d^9 4s^2$.

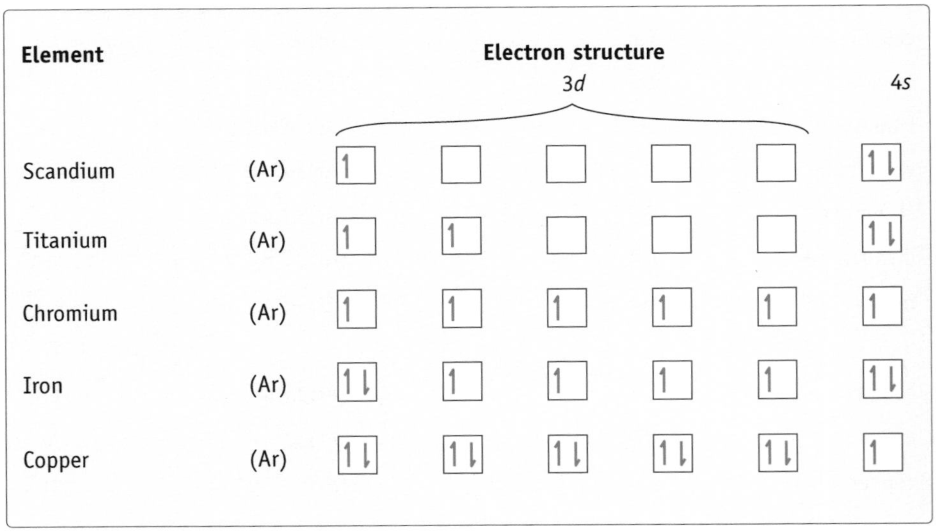

Element		Electron structure					
		3d					4s
Scandium	(Ar)	↑					↑↓
Titanium	(Ar)	↑	↑				↑↓
Chromium	(Ar)	↑	↑	↑	↑	↑	↑
Iron	(Ar)	↑↓	↑	↑	↑	↑	↑↓
Copper	(Ar)	↑↓	↑↓	↑↓	↑↓	↑↓	↑

Figure 18.2

The 'electrons-in-boxes' representation of the electron structures of certain transition metals.

18.2 Ions of the transition metals

Look closely at table 18.1.

1. Which electrons does a calcium atom lose in forming a Ca^{2+} ion?

2. Which electrons would you expect an iron atom to lose in forming an Fe^{2+} ion? Which electrons are lost in practice by the Fe atom?

3. Why do you think Fe^{3+} ions are more stable than Fe^{2+} ions?

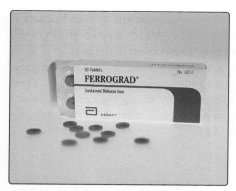

▲ Iron is an important constituent of haemoglobin and people suffering from anaemia may have a diet which is deficient in iron. A supplement of 'iron' tablets containing iron(II) sulphate can improve the problem.

Figure 18.3
The 'electrons-in-boxes' representation of the electronic structures of Fe, Fe^{2+} and Fe^{3+}.

When transition metals form their ions, electrons are lost first from the $4s$ sub-shell rather than the $3d$ sub-shell. Thus, Fe^{2+} ions have the electron structure $(Ar)3d^6$ (figure 18.3) rather than $(Ar)3d^4 4s^2$. V^{3+} ions have the electron structure $(Ar)3d^2$ not $(Ar)4s^2$. At first sight, this seems very odd because prior to occupation by electrons, the $4s$ level is energetically more stable than the $3d$ level. However, once the $3d$ level is occupied by electrons, these repel the $4s$ electrons even further from the nucleus. The $4s$ electrons are therefore pushed to a higher energy level, higher in fact than the $3d$ level now occupied. Consequently, when transition metal atoms form ions, they lose electrons from the $4s$ level before the $3d$ level.

> This means that *all transition metals will have similar chemical properties* which are dictated by the behaviour of the $4s$ electrons in the outermost shell.

This horizontal similarity contrasts sharply with the marked changes across a period of the s- and p-block elements, such as that from sodium to argon.

Atom/Ion		Electron structure					
		3d					4s
Fe	(Ar)	↑↓	↑	↑	↑	↑	↑↓
Fe^{2+}	(Ar)	↑↓	↑	↑	↑	↑	
Fe^{3+}	(Ar)	↑	↑	↑	↑	↑	

18.3 What is a transition element?

The simplest answer to this question is to say that *transition elements are those in the 'd-block' of the periodic table*. The neatness of this definition lies in the fact that it emphasises the four blocks of elements in the periodic table. This division is useful for the s-block elements in Groups I and II which are so alike in their properties. It is much less satisfactory for the elements in the p-block which include metals such as aluminium, reactive non-metals within Group VII and the noble gases. Furthermore, the simple definition of transition metals as 'd-block elements' is rather unsatisfactory because it leads to the inclusion of scandium and zinc as transition elements. As you may have already realised, scandium and zinc show some fairly obvious differences to the elements in the sequence titanium to copper.

- They have only one oxidation state in their compounds (scandium +3, zinc +2), whereas the others have two or more.
- Their compounds are usually white, unlike the compounds of transition metals which are generally coloured.
- They show little catalytic activity.

As scandium and zinc do not show typical transition metal properties, it would seem sensible to find a more satisfactory definition for transition metals. This definition should exclude these two elements, but include all the other elements from titanium to copper.

> In order to achieve this, we can define a transition metal as *an element with at least one ion with a partially filled d sub-shell.*

Look closely at table 18.1 and you will see that neither Sc^{3+} nor Zn^{2+} has a partly filled d sub-shell.

18.4 Trends across the period of transition metals

In building up the elements from sodium to argon, electrons are being added to the outer shell and the nuclear charge is increasing by the addition of protons. The added electrons shield each other only weakly from the extra nuclear charge, so atomic radii decrease sharply from sodium to argon. At the same time the electronegativities and ionisation energies steadily rise. The increasing number of outer-shell electrons also results in major differences in structure and chemical properties from one element to the next.

In moving across the series of metals from scandium to zinc, however, the nuclear charge is also increasing, but electrons are being added to an *inner d* sub-shell. These inner *d* electrons shield the outer 4s electrons from the increasing nuclear charge much more effectively than outer shell electrons can shield each other. Consequently, the atomic radii decrease much less rapidly (table 18.2).

Table 18.2
Electronegativities, ionisation energies and electrode potentials for the elements Sc to Zn

	Sc	Ti	V	Cr	Mn	Fe	Co	Ni	Cu	Zn
Metallic (atomic) radius/nm	0.16	0.15	0.14	0.13	0.14	0.13	0.13	0.13	0.13	0.13
Electronegativity	1.2	1.3	1.45	1.55	1.6	1.65	1.7	1.75	1.75	1.6
First ionisation energy/kJ mol^{-1}	+630	+660	+650	+650	+720	+760	+760	+740	+750	+910
Second ionisation energy/kJ mol^{-1}	+1240	+1310	+1410	+1590	+1510	+1560	+1640	+1750	+1960	+1700
Third ionisation energy/kJ mol^{-1}	+2390	+2650	+2870	+2990	+3260	+2960	+3230	+3390	+3560	+3800
Electrode potential for $M^{2+}(aq) + 2e^- \rightarrow M(s)$/V			−1.20	−0.91	−1.19	−0.44	−0.28	−0.25	+0.34	−0.76
Electrode potential for $M^{3+}(aq) + 3e^- \rightarrow M(s)$/V	−2.1	−1.2	−0.86	−0.74	−0.28	−0.04	+0.40			

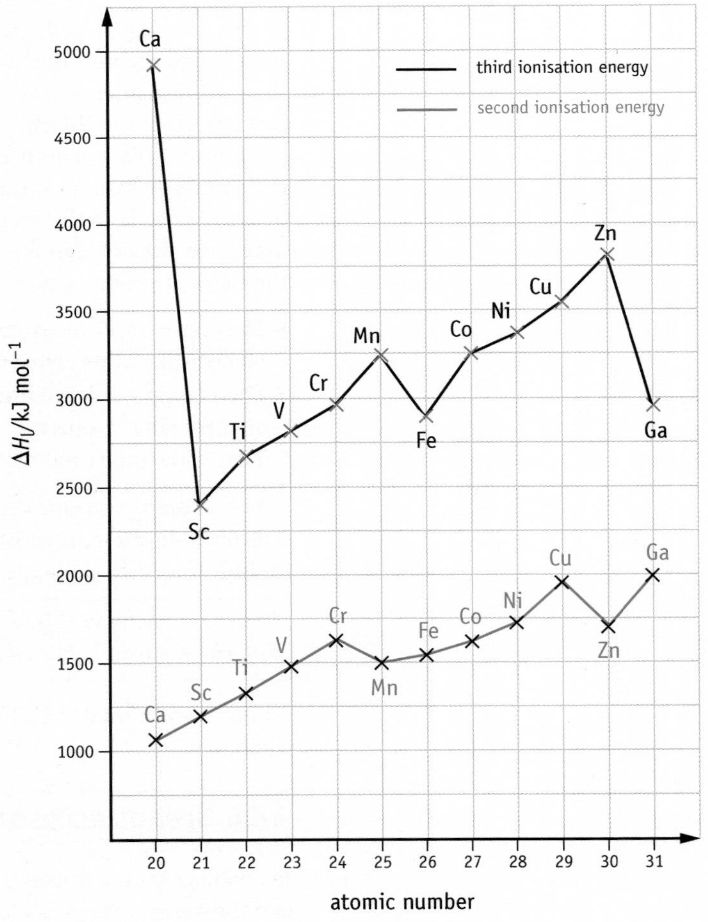

Figure 18.4
Graphs of the second and third ionisation energies of the elements from calcium to gallium.

Similarly, electronegativities and ionisation energies increase from scandium to zinc, but only marginally compared with the increase across period 3 from sodium to argon (table 18.2).

The increasing electronegativity from scandium to copper means that the elements become slightly less metallic. This is reflected in the increasingly positive electrode potentials (sections 14.3 and 14.4) of their M^{2+} and M^{3+} ions.

Look closely at figure 18.4 which shows graphs of the second and the third ionisation energies of the elements from calcium to gallium.

1 Write the electron structures of the following:
(a) Cu (b) Cu^+ (c) Cu^{2+} (d) Cr (e) Cr^+ (f) Mn^{2+} (g) Zn^{2+}
(Use (Ar) to represent the electron structure of argon as we have done before.)

2 By referring to electron structures explain why:
(a) the second ionisation energies of both Cr and Cu are higher than those of the next element.
(b) the third ionisation energies of both Mn and Zn are higher than those of the next element.
Remember that the second ionisation energy refers to the process
$$M^+(g) \rightarrow M^{2+}(g) + e^-$$
and the third ionisation to
$$M^{2+}(g) \rightarrow M^{3+}(g) + e^-$$

18.5 General properties of the first transition series (Sc to Zn)

Most of the transition metals have a close-packed structure in which each atom has twelve nearest neighbours. Furthermore, transition metals have a relatively low atomic radius because the electrons being added to the $3d$ sub-shell are nearer the nucleus than the electrons in the outermost $4s$ orbital. The double effect of close packing and small atomic size results in strong metallic bonds between atoms.

Hence, *the transition metals have higher melting points, higher boiling points, higher densities and higher enthalpy changes of fusion and vaporisation than metals such as potassium and calcium in the s-block of the periodic table* (table 18.3). The strong interatomic bonding in transition metals is also reflected in high tensile strengths and good mechanical properties.

The transition metals are less electropositive than the s-block metals. However, their electrode potentials (table 18.2) indicate that all of them (except copper) should react with dilute solutions of strong acids, such as 1 mol dm^{-3} HCl. In practice, however, many of the metals react only slowly with dilute acids. This is due to protection of the metal from chemical attack by a thin impervious and unreactive layer of oxide. Chromium provides an excellent example of this. Despite its electrode potential, it can be used as a protective, non-oxidising, non-corroding metal owing to the presence of an unreactive layer of Cr_2O_3. This can be compared with the protection of aluminium by a layer of Al_2O_3.

The ions of transition metals are smaller than those of the s-block metals in the same period (table 18.3). Owing to their smaller ionic radii and also to a larger charge in many of their ions, the charge/radius ratios for transition metals are greater than the values for s-block metals. The small, highly charged cations also cause greater polarisation of associated anions. These factors result in the following properties of transition metal compounds compared with those of the s-block metals:

(a) their oxides and hydroxides in oxidation states +2 and +3 are less basic and less soluble;
(b) their salts are less ionic and less thermally stable;
(c) their salts and aqueous ions are more hydrated and more readily hydrolysed forming acidic solutions;
(d) their ions are more easily reduced.

Although the compounds of transition elements in oxidation states +2 and +3 are usually regarded as ionic, polarisation of anions by small, highly charged cations is undoubtedly apparent. Some of the oxides exhibit acidic features and the compounds begin to show covalent bonding. Thus, Cr_2O_3 and Mn_2O_3 are amphoteric and Fe_2Cl_6 is regarded as a molecular solid. Of course, these features begin to predominate in the higher oxidation states. Here the oxides become increasingly acidic and the compounds increasingly molecular. Hence, CrO_3 and Mn_2O_7 are regarded as simple molecular, acidic oxides.

There are relatively small changes in ionic radii from scandium to copper. Because of this, compounds of the simple hydrated +2 and +3 ions have very similar crystalline structures, hydration and solubility. Thus, all the M^{3+} ions form an alum of the type $K_2SO_4.M_2(SO_4)_3.24H_2O$ and all the M^{2+} ions form isomorphous double sulphates of formula $(NH_4)_2SO_4.MSO_4.6H_2O$.

Element	s-block metals		transition metals									
	K	Ca	Sc	Ti	V	Cr	Mn	Fe	Co	Ni	Cu	Zn
Atomic radius/nm	0.24	0.20	0.16	0.15	0.14	0.13	0.14	0.13	0.13	0.13	0.13	0.13
Melting point/°C	64	850	1540	1680	1900	1890	1240	1540	1500	1450	1080	420
Boiling point/°C	770	1490	2730	3260	3400	2480	2100	3000	2900	2730	2600	910
Density/g cm^{-3}	0.86	1.54	3.0	4.5	6.1	7.2	7.4	7.9	8.9	8.9	8.9	7.1
Ionic radius/nm												
M$^+$	0.130											
M^{2+}		0.094		0.090	0.088	0.084	0.080	0.076	0.074	0.072	0.070	0.074
M^{3+}			0.081	0.076	0.074	0.069	0.066	0.064	0.063	0.062		

Table 18.3
Physical properties of the elements K to Zn

18.6 Characteristic properties of transition metals and their compounds

Variable oxidation states

Transition metals have electrons of similar energy in both the $3d$ and $4s$ levels. This means that one particular element can form ions of roughly the same stability by losing different numbers of electrons. Thus, all the transition metals from titanium to copper exhibit two or more oxidation states in their compounds.

The formulas of the common oxides and chlorides of the elements from Sc to Zn are shown in figure 18.5. All the oxidation numbers of the elements are given below these formulas. The more important oxidation states are emphasised by bold print. Notice that both scandium and zinc have only one common oxide, one common chloride, and one oxidation state in their compounds. Notice also, how closely the oxidation states of the elements in their common oxides and chlorides compare with the more important oxidation states of the elements.

The following generalisations emerge from a study of the oxidation states.

(a) *The common oxidation states for each element include +2 or +3 or both.* +3 states are relatively more common at the beginning of the series. +2 states are more common towards the end.

(b) The highest oxidation states from scandium to manganese correspond to the involvement of all the electrons outside the argon core (4 for Ti, 5 for V, 6 for Cr and 7 for Mn). After this, the increasing nuclear charge binds the d electrons more strongly. So, one of the more important oxidation states is that which involves the weakly held electrons in the outer $4s$ shell only (2 for Fe, 2 for Co, 2 for Ni and 1 for Cu).

	Sc	Ti	V	Cr	Mn	Fe	Co	Ni	Cu	Zn
Common oxides	Sc_2O_3	Ti_2O_3 TiO_2	V_2O_3 V_2O_5	Cr_2O_3 CrO_3	MnO MnO_2 Mn_2O_7	FeO Fe_2O_3	CoO Co_2O_3	NiO	Cu_2O CuO	ZnO
Common chlorides	$ScCl_3$	$TiCl_3$ $TiCl_4$	VCl_3	$CrCl_2$ $CrCl_3$	$MnCl_2$ $MnCl_3$	$FeCl_2$ Fe_2Cl_6	$CoCl_2$	$NiCl_2$	$CuCl$ $CuCl_2$	$ZnCl_2$
Oxidation numbers that occur in compounds	3	4 3 2 1	5 4 3 2 1	6 5 4 3 2 1	7 6 5 4 3 2 1	6 5 4 3 2 1	5 4 3 2 1	4 3 2 1	3 2 1	2

Figure 18.5
Oxidation states of the elements Sc to Zn. (Common oxidation numbers are in bold print.)

(c) The transition metals usually exhibit their highest oxidation states in compounds with oxygen or fluorine. These are the two most electronegative elements.

(d) Ti, V, Cr and Mn never form simple ions in their highest oxidation state since this would result in ions of extremely high charge density. Hence the compounds of these elements in their highest oxidation state are either covalently bonded (e.g. TiO_2, V_2O_5, CrO_3, Mn_2O_7) or contain complex ions (e.g. VO_3^-, CrO_4^{2-}, MnO_4^-).

One of the most beautiful and effective demonstrations of the range of oxidation states of a transition metal can be shown by shaking a solution containing a compound of vanadium(V) with zinc and dilute acid. The solution of vanadium(V) can be made by dissolving about 3 g of ammonium vanadate(V) (NH_4VO_3) in 40 cm^3 of 2 mol dm^{-3} NaOH and then adding 80 cm^3 of 1 mol dm^{-3} H_2SO_4. This solution is a yellow colour. It contains dioxovanadium(V) ions, VO_2^+, in acid solution.

$$VO_3^-(aq) + 2H^+(aq) \rightarrow VO_2^+(aq) + H_2O(l)$$

When the yellow solution is shaken with granulated zinc or zinc amalgam, it changes gradually through green to blue oxovanadium(IV) ions, VO^{2+}(aq). It then changes to green vanadium(III) ions, V^{3+}(aq), and finally to violet vanadium(II) ions, V^{2+}(aq) (figure 18.6).

▲ This photograph shows the common oxidation states in chromium compounds very colourfully. From left to right, the solutions contain violet chromium(III) chloride, green chromium(III) nitrate, yellow potassium chromate(VI) and orange potassium dichromate(VI).

Figure 18.6
The oxidation states of vanadium.

Oxidation state	+5	+4	+3	+2
Colour in aqueous solution	yellow	blue	green	violet
Ion	VO_2^+	VO^{2+}	V^{3+}	V^{2+}
Name	dioxovanadium(V) ion	oxovanadium(IV) ion	vanadium(III) ion	vanadium(II) ion

All the transition metals from titanium to copper exhibit oxidation numbers of +3 and +2 in their compounds. But what are the relative stabilities of the +3 and +2 states for the different elements? Why, for example, is manganese more stable in the +2 state than the +3 state, whilst the reverse is true for iron? Look closely at figure 18.7 which shows the electrode potentials of the $M^{3+}(aq)/M^{2+}(aq)$ systems for the elements from titanium to cobalt. There is a fairly steady rise in $E^{\ominus}$ values across the series. This is interrupted by what appears to be an abnormally high value for manganese and a low value for iron.

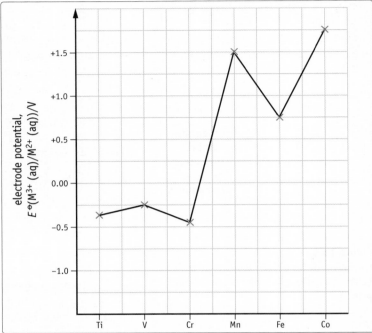

Figure 18.7
Redox potentials of the M^{3+}/M^{2+} systems for transition metals.

The more positive the value for $E^{\ominus}$, the more will the aqueous M^{3+} ion become reduced to M^{2+}.

$$M^{3+}(aq) + e^- \rightarrow M^{2+}(aq)$$

Thus, low values of $E^{\ominus}$ for Ti, V, Cr and Fe indicate that the 3+ oxidation state is relatively more stable in these elements than in Mn and Co which have higher values of $E^{\ominus}$

The relative stabilities of the +3 and +2 states in manganese and iron can be interpreted using an 'electrons-in-boxes' representation of their electronic structures (figure 18.8)

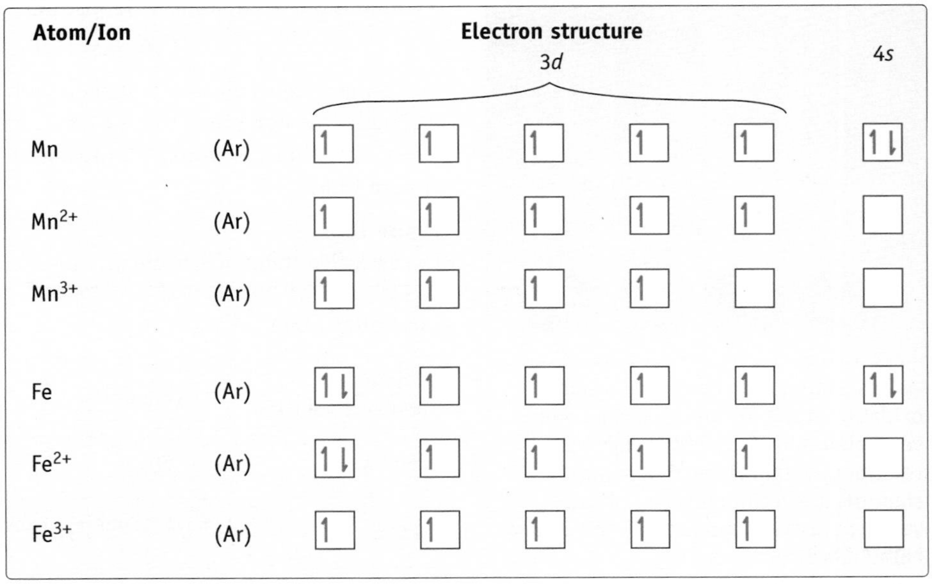

Atom/Ion		Electron structure					
		3d					4s
Mn	(Ar)	↑	↑	↑	↑	↑	↑↓
Mn^{2+}	(Ar)	↑	↑	↑	↑	↑	
Mn^{3+}	(Ar)	↑	↑	↑	↑		
Fe	(Ar)	↑↓	↑	↑	↑	↑	↑↓
Fe^{2+}	(Ar)	↑↓	↑	↑	↑	↑	
Fe^{3+}	(Ar)	↑	↑	↑	↑	↑	

Figure 18.8
Electron structures of manganese, iron and some of their respective ions.

1 Write the electron structures of Cu, Cu$^+$ and Cu^{2+} using the 'electrons-in-boxes' notation.

2 In which oxidation state would you expect copper to be more stable, +1 or +2?

3 From experience you will know that copper compounds usually exist in the +2 state. What factors might increase the relative stability of the +2 state over the +1 state for copper?

Mn^{2+} and Fe^{3+} each have half-filled $3d$ orbitals. This makes them more stable than Mn^{3+} and Fe^{2+}, respectively. Hence, in manganese the +2 state is more stable than +3, whereas in iron the +3 state is more stable than +2.

Formation of complex ions

Complex ions are composed of a central metal ion surrounded by anions or molecules, called **ligands** (section 14.2). In transition metal complexes, non-bonded pairs of electrons on the ligand form co-ordinate bonds to the central ion. The unshared electron pairs are donated into vacant orbitals of the transition metal ion. The number of co-ordinate bonds from ligands to the central ion is known as the **co-ordination number** of the central ion. Generally, the co-ordination number of a particular ion is the same whatever the ligand. Thus, Cu^{2+} ions have a co-ordination number of four in $[Cu(H_2O)_4]^{2+}$, in $[Cu(NH_3)_4]^{2+}$, in $[CuCl_4]^{2-}$ and in $[Cu(H_2NCH_2CH_2NH_2)_2]^{2+}$. In contrast, Fe^{3+} ions usually have a co-ordination number of six as in $[Fe(H_2O)_6]^{3+}$, $[FeF_6]^{3-}$, $[Fe(CN)_6]^{3-}$ and $[Fe(edta)]^-$. Ag$^+$ ions normally have a co-ordination number of two ($[Ag(NH_3)_2]^+$, $[Ag(CN)_2]^-$).

In the case of the transition metals, the most common co-ordination number is six, but examples of four and two are not uncommon. In aqueous solution, transition metal ions exist as hydrated complexes with water molecules $[M(H_2O)_6]^{2+}$(aq) and $[M(H_2O)_6]^{3+}$(aq). However, the high charge density on the central metal ion causes the hydrated complexes to dissociate.

$$[Fe(H_2O)_6]^{3+}(aq) \rightleftharpoons [Fe(H_2O)_5OH]^{2+}(aq) + H^+(aq)$$

This dissociation process involves hydrolysis of the water molecules in the complex ion. The high charge density of the metal ion draws electrons in the O—H bonds of the water molecules towards itself, enabling the water molecules to become proton donors. This explains why solutions of FeCl$_3$ and CrCl$_3$ are acidic.

Consequently, the aqueous solutions of most transition metal compounds such as CuSO$_4$(aq), FeCl$_3$(aq) and Co(NO$_3$)$_2$(aq) are acidic. In oxidation states higher than +3, the polarising power of the central ion is even greater. This causes a release of even more protons and loss of water molecules resulting in the formation of oxyanions. For example, neither $[Cr(H_2O)_6]^{6+}$ nor $[Mn(H_2O)_6]^{7+}$ exist in aqueous solution. The loss of two water molecules and eight protons from each of these ions results in the formation of CrO$_4^{2-}$ and MnO$_4^-$, respectively:

$$[Cr(H_2O)_6]^{6+} \rightarrow 2H_2O + 8H^+ + CrO_4^{2-}$$

$$[Mn(H_2O)_6]^{7+} \rightarrow 2H_2O + 8H^+ + MnO_4^-$$

In complexes with a **co-ordination number of six**, the ligands usually occupy **octahedral positions**. The six electron pairs around the central metal ion are repelled as far as possible from each other (figure 18.9).

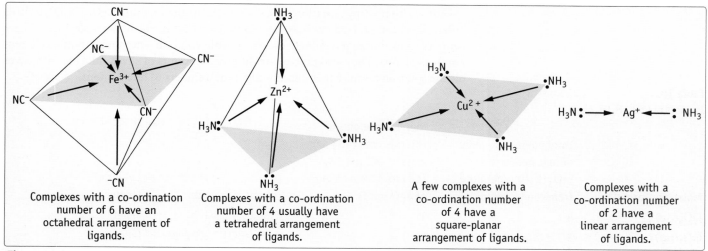

Complexes with a co-ordination number of 6 have an octahedral arrangement of ligands.

Complexes with a co-ordination number of 4 usually have a tetrahedral arrangement of ligands.

A few complexes with a co-ordination number of 4 have a square-planar arrangement of ligands.

Complexes with a co-ordination number of 2 have a linear arrangement of ligands.

Figure 18.9
The stereochemistry of complex ions.

In complexes with a **co-ordination number of four**, the ligands usually occupy **tetrahedral positions** although a few four-co-ordinated complexes (such as $[Cu(H_2O)_4]^{2+}$ and $[Cu(NH_3)_4]^{2+}$) have a square-planar structure (figure 18.9).

Complexes with a **co-ordination number of two** usually have a linear arrangement of ligands (figure 18.9).

Owing to the stereochemical positions of the ligands, isomers can occur in four co-ordinated and six-co-ordinated complexes. Isomers are compounds with the same molecular formula, but different arrangements of their constituent atoms. (See section 25.8.)

The neutral complex $PtCl_2(NH_3)_2$ in which the Cl^- ions and NH_3 molecules act as ligands to the Pt^{2+} ion, has two isomers.

1. Does $PtCl_2(NH_3)_2$ have a tetrahedral or a square-planar structure? Explain your answer.
2. The isomers of $PtCl_2(NH_3)_2$ are described as *cis* and *trans*. Draw these isomers. Indicate which is the *cis* form and which is the *trans* form.

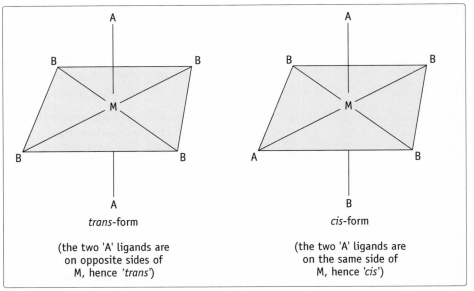

trans-form

(the two 'A' ligands are on opposite sides of M, hence '*trans*')

cis-form

(the two 'A' ligands are on the same side of M, hence '*cis*')

Figure 18.10
Isomers of complexes of the type MA_2B_4.

Octahedral complexes of the type MA_2B_4 also exhibit *cis/trans* (geometric) isomerism. Figure 18.10 shows the two isomers for MA_2B_4.

One of the most striking examples of this type of isomerism occurs with tetraamminedichlorocobalt(III) chloride ($[Co(NH_3)_4Cl_2]^+Cl^-$). In this compound, the two isomers have different-coloured crystals. The *cis* form is blue–violet in colour, whereas the *trans* form is green.

Another interesting situation arises with salts of the formula $CrCl_3(H_2O)_6$. Three differently coloured isomers have been isolated with this formula (table 18.4). These isomers show different conductivities in aqueous solution since they produce different numbers of ions. They also precipitate different amounts of silver chloride with silver nitrate solution because they contain different numbers of free Cl^- ions.

Table 18.4
Isomers of $CrCl_3(H_2O)_6$

Molecular formula	Total number of moles of ions per mole of $CrCl_3(H_2O)_6$ (deduced from conductivity)	Number of moles of chloride ions (Cl^-) per mole of $CrCl_3(H_2O)_6$ (deduced from amount of AgCl pptd)	Structural formula	Colour of isomer
$CrCl_3(H_2O)_6$	4	3	$[Cr(H_2O)_6]^{3+}.3Cl^-$	violet
$CrCl_3(H_2O)_6$	3	2	$[Cr(H_2O)_5Cl]^{2+}.2Cl^-.H_2O$	light green
$CrCl_3(H_2O)_6$	2	1	$[Cr(H_2O)_4Cl_2]^+.Cl^-.2H_2O$	dark green

Coloured compounds

Most of the compounds of transition elements are coloured. The colour of these compounds can often be related to incompletely filled d-orbitals in the transition metal ion.

In general, when light hits a substance, part is absorbed, part is transmitted (if the substance is transparent) and part may be reflected.

If all the incident radiation is absorbed then the substance looks *'black'*. If all the incident radiation is reflected, then the substance looks *'white'*. On the other hand, if only a very small proportion of the incident white light is absorbed and if all the radiations in the visible region of the spectrum are transmitted equally, then the substance will appear *'colourless'*.

However, many substances only absorb the light photons in certain areas of the visible spectrum. In this case, the transmitted or reflected light is relatively richer in the radiations of the remaining regions and the substance has a characteristic colour. Suppose a material absorbs all radiations in the yellow–orange–red region of the spectrum. It will appear blue in white light because only the radiations in the blue region are reflected towards our eyes.

When light energy is absorbed by a substance, an electron in the substance is promoted from an orbital of lower to one of higher energy. The atom or ion absorbing the radiation changes from its **ground state** (i.e. stable state) to an **excited state** (section 6.2). The different electron transitions involve the absorption of radiation of different frequencies (i.e. different quanta of energy). If the absorbed frequencies are in the visible region of the spectrum, then the material appears coloured. Why are solutions of transition metal compounds usually coloured, whereas solutions of compounds of non-transition metals are usually colourless?

Consider first a solution of a non-transition metal compound such as aqueous sodium chloride. The sodium ion has the electronic structure $1s^2 2s^2 2p^6$. The energy difference between the $2p$ orbitals and the next available orbital, $3s$, is very large. To undergo the electron transition, $1s^2 2s^2 2p^6$ (ground state) $\rightarrow 1s^2 2s^2 2p^5 3s^1$ (excited state), a sodium ion must absorb a photon of high-energy radiation well beyond the range of visible light. Consequently, none of the photons of visible light has sufficient energy to promote electrons in Na^+ ions to even the lowest possible excited state. Hence, $Na^+(aq)$ does not absorb any photons of visible light and so solutions of sodium salts are usually colourless. The $Cl^-(aq)$ ion is also colourless for the same reason.

Now let us consider an aqueous solution of titanium(III) chloride. This contains the octahedral complex ions, $[Ti(H_2O)_6]^{3+}$, and $Cl^-(aq)$ ions. A solution of titanium(III) chloride is violet. An absorption spectrum shows that the solution absorbs most effectively in the green–yellow region of the spectrum (figure 18.11).

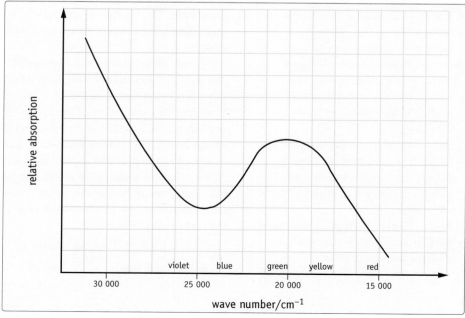

Figure 18.11
The absorption spectrum of $[Ti(H_2O)_6]^{3+}$.

Red, blue and violet radiations are absorbed less efficiently, so the solution looks violet. Both water and $Cl^-(aq)$ are colourless, so the $Ti^{3+}(aq)$ ion must be responsible for the colour of the solution.

The Ti^{3+} ion has the electron structure $(Ar)3d^1$. When a solution of titanium(III) chloride absorbs photons of visible light, we might expect electrons in the Ti^{3+} ion to be promoted from $(Ar)3d^1$ (ground state) to $(Ar)4s^1$ (excited state). However, calculations show that this excitation of Ti^{3+}, like that for Na^+, requires photons of radiation well beyond the visible region of the spectrum.

How then do we explain the absorption of green light by $Ti^{3+}(aq)$? Which electron transition is responsible for the absorption of green light? The answer lies in electron transitions *within* the set of five $3d$ orbitals.

All five $3d$ orbitals have exactly the same energy level in the isolated gaseous Ti^{3+} ion (figure 18.12(a)). However, when the Ti^{3+} ion is surrounded by ligands, the $3d$ orbitals are no longer symmetrically arranged. Orbitals closer to the ligands are pushed to a slightly higher energy level than those orbitals further away. This splitting of the $3d$ orbitals in the octahedral $[Ti(H_2O)_6]^{3+}$ ion is shown in figure 18.12(b).

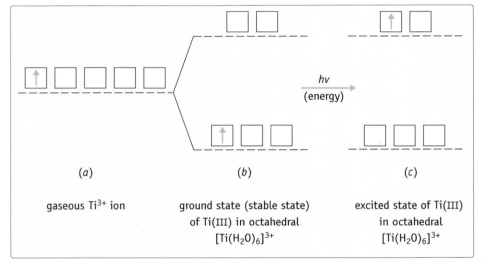

hv
(energy)

(a)

(b)

(c)

gaseous Ti^{3+} ion

ground state (stable state) of Ti(III) in octahedral $[Ti(H_2O)_6]^{3+}$

excited state of Ti(III) in octahedral $[Ti(H_2O)_6]^{3+}$

Figure 18.12
Relative energy levels for the five $3d$ orbitals of the gaseous and hydrated Ti^{3+} ion.

Assume that the $3d$ orbitals of Cu(II) in the aqueous Cu^{2+} ion are split into two levels as in figure 18.12.

1 The electron structure for a copper atom is $(Ar)3d^{10}4s^1$. Write the electron structure for a Cu^{2+} ion.

2 Draw a diagram similar to figure 18.12(b) to show how electrons occupy the higher and lower $3d$ orbitals in the ground state for $Cu^{2+}(aq)$.

3 Draw a diagram similar to figure 18.12(c) to show how electrons occupy the higher and lower $3d$ orbitals in the excited state for $Cu^{2+}(aq)$.

4 What colour(s) of light do you think the $Cu^{2+}(aq)$ ion absorbs when the d–d transition occurs?

In the ground (stable) state, the single d electron occupies one of the d orbitals of slightly lower energy as shown in figure 18.12(b). It is now possible for the Ti^{3+} ion in $[Ti(H_2O)_6]^{3+}$ to absorb sufficient energy for the $3d$ electron to be promoted from a lower to a higher $3d$ orbital (figure 18.12(c)). The difference in energy between the two sets of $3d$ orbitals in $[Ti(H_2O)_6]^{3+}$ is relatively small. It coincides with the wavelength for green light. Thus, when aqueous $TiCl_3$ is exposed to white light, it absorbs photons of green light but transmits those of red and blue light. Hence $TiCl_3(aq)$ appears purple or violet.

In considering the colour of aqueous $TiCl_3$, the situation is relatively simple because the $[Ti(H_2O)_6]^{3+}$ ion has only one d electron. The situation is much more complex for most other transition metal ions since many more d–d transitions are possible. In principle, however, the explanation of the colour of other transition metal compounds is similar to that for $TiCl_3(aq)$. The five d orbitals are split into two or more slightly different energy levels. Thus, the promotion of an electron from the lower to the higher of these d orbitals just happens to require energies within the range of visible light.

The colour of a transition metal complex depends mainly on the central cation. It is influenced much less by the co-ordinating ligands. For example the colours of most Cu^{2+} complexes are blue or violet as in the reaction scheme below. The complexes of Cu^+ are usually colourless.

$[Cu(H_2O)_4]^{2+}$ $\xrightarrow{NH_3(aq)}$ $[Cu(NH_3)_4]^{2+}$ $\xrightarrow{H_2NCH_2CH_2NH_2(aq)}$ $[Cu(H_2NCH_2CH_2NH_2)_2]^{2+}$
pale blue   dark blue   violet

Since different oxidation states have different numbers of d electrons, the colours of complexes often change when the oxidation state of the transition metal changes. For example, when potassium manganate(VII) is heated with 50% KOH it is reduced through the following colour changes.

Substance present in alkaline soln	MnO_4^-	MnO_4^{2-}	MnO_3^-	MnO_2	$Mn(OH)_3$	$Mn(OH)_2$
Oxidation state	+7	+6	+5	+4	+3	+2
Colour	purple	green	blue	dark brown	brown	pink

▲ Virgin desert scrub with reclaimed land in the middle ground. This barren land in Israel was made productive simply by the application of minute quantities of transition metals to the soil. This made good the mineral deficiencies that had prevented previous settlement of the area.

Catalytic properties

Transition metals and their compounds are important **catalysts** in industry and in biological systems. Many transition metal ions are required by humans and other living things in minute but definite quantities. Elements required in such small amounts are called **trace elements**. These trace elements include copper, manganese, iron, cobalt, nickel and chromium. They are essential for the effective catalytic activity of various enzymes (section 9.8). One of the most important enzymes containing copper is cytochrome oxidase. This enzyme is involved when energy is obtained from the oxidation of food. In the absence of copper, cytochrome oxidase is completely inhibited and the animal or plant is unable to metabolise food effectively.

A large number of industrial catalysts are either transition metals or their compounds. Table 18.5 shows a list of some of the more important examples.

Chemists believe that the catalytic activity of transition metals depends on their ability to exist in different states of oxidation or co-ordination. For example, in the **Contact Process**, vanadium compounds in the +5 state (either V_2O_5 or VO_3^-) are used to oxidise sulphur dioxide to sulphur(VI) oxide:

$$SO_2 + \tfrac{1}{2}O_2 \xrightarrow{V_2O_5} SO_3$$

It is thought that the actual oxidation process involves two stages. In the first of these, V^{5+} in the presence of oxide ions converts SO_2 to SO_3. At the same time, V^{5+} is reduced to V^{4+}:

$$2V^{5+} + O^{2-} + SO_2 \rightarrow 2V^{4+} + SO_3$$

Table 18.5
Transition metals and their compounds used as catalysts in industry

Transition element	Substance used as catalyst	Reaction catalysed
Ti	$TiCl_3/Al_2(C_2H_5)_6$	$nC_2H_4 \longrightarrow \left(-\overset{\mid}{\underset{\mid}{C}}-\overset{\mid}{\underset{\mid}{C}}-\right)n$ Polymerisation of ethene → poly(ethene)
V	V_2O_5 or vanadate (VO_3^-)	$2SO_2 + O_2 \rightarrow 2SO_3$ Contact process
Fe	Fe or Fe_2O_3	$N_2 + 3H_2 \rightarrow 2NH_3$ Haber process
Ni	Ni	$RCH {=\!=} CH_2 + H_2 \rightarrow RCH_2CH_3$ Hardening of vegetable oils (e.g. manufacture of margarine)
Cu	Cu or CuO	$CH_3CH_2OH + \tfrac{1}{2}O_2 \rightarrow CH_3CHO + H_2O$ Oxidation of ethanol to ethanal
Pt	Pt	$2SO_2 + O_2 \rightarrow 2SO_3$ Contact process
Pt	Pt	$4NH_3 + 5O_2 \rightarrow 4NO + 6H_2O$ then $NO \rightarrow NO_2 \rightarrow HNO_3$ Manufacture of nitric acid from ammonia

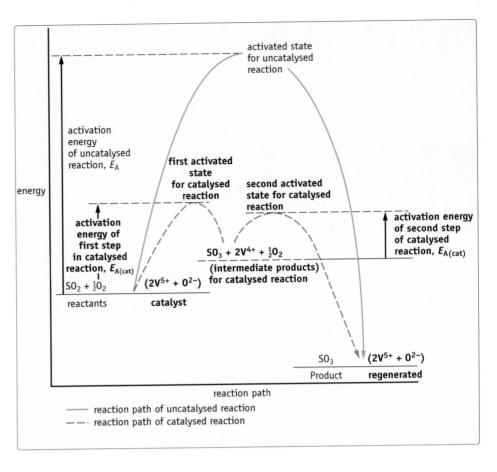

Figure 18.13
The activation energies of catalysed and uncatalysed reactions.

In the second stage, V^{5+} is regenerated from V^{4+} by oxygen:

$$2V^{4+} + \tfrac{1}{2}O_2 \rightarrow 2V^{5+} + O^{2-}$$

The overall process is, of course, the sum of these two stages:

$$2\cancel{V}^{5+} + \cancel{O}^{2-} + SO_2 \rightarrow 2\cancel{V}^{4+} + SO_3$$

$$2\cancel{V}^{4+} + \tfrac{1}{2}O_2 \rightarrow 2\cancel{V}^{5+} + \cancel{O}^{2-}$$

$$\underline{\text{Sum:} \quad SO_2 + \tfrac{1}{2}O_2 \rightarrow SO_3}$$

Transition metals and their compounds catalyse reactions because they introduce an entirely new reaction mechanism. This new reaction mechanism has a lower activation energy (section 24.10) than the uncatalysed reaction. Since the activation energy of the catalysed reaction is lower, the reaction rate is faster. This is shown diagrammatically in figure 18.13.

During the course of the reaction, the catalyst undergoes changes in oxidation state. It is, however, regenerated in its original form when the reactants form the products.

An excellent demonstration of the change in oxidation state of a transition metal during catalysis can be shown using cobalt(II) ions with a mixture of 2,3-dihydroxybutanedioate (tartrate) ions and hydrogen peroxide.

When aqueous solutions of 2,3-dihydroxybutanedioate (tartrate) ions and hydrogen peroxide are mixed, there is little or no evolution of gas from the decomposition of hydrogen peroxide. If cobalt(II) chloride solution is added, the solution becomes pink and there is still little evolution of gas. After an induction period, the solution begins to darken, going brown and then green due to the presence of cobalt(III). At the same time, effervescence from the solution gradually increases and the maximum evolution of gas occurs when the solution is green. As the reaction subsides, the green colour of the cobalt(III) complex fades to be replaced by the original pink of cobalt(II) ions.

Summary

1 A transition metal is an element which forms at least one ion with a partially filled d sub-shell.

2 Unlike elements in other parts of the periodic table, the elements in a transition series have very similar properties to one another.

3 Many of the properties of transition elements can be attributed to their electronic structures and the relative energy levels of the orbitals available for their electrons.

4 Prior to occupation by electrons, the $4s$ level is energetically more stable than the $3d$ level. Once the $3d$ level is occupied by electrons, the latter repel the $4s$ electrons further from the nucleus to a higher energy level than the $3d$ level. Thus, transition metals lose electrons from the $4s$ level before the $3d$ level in forming ions.

5 Transition metals have higher melting points, higher boiling points and higher densities than other metals.

6 Transition metals show the following characteristic properties:
   (a) variable oxidation states,
   (b) formation of complex ions,
   (c) coloured compounds,
   (d) catalytic properties.

7 The more important oxidation states for each transition element include +2 or +3 or both.

8 Complex ions are composed of a central cation surrounded by a cluster of anions or molecules known as ligands. Non-bonded pairs of electrons on the ligand form co-ordinate bonds to the central metal cation.

9 The number of co-ordinate bonds from ligands to the central cation is known as the co-ordination number of the central ion. Generally, the co-ordination number of a particular ion is the same whatever the ligand.

10 In complexes with a co-ordination number of six, the ligands usually occupy octahedral positions. In complexes with a co-ordination number of four the ligands occupy tetrahedral or square-planar positions. Complexes with a co-ordination number of two usually have a linear arrangement of ligands.

11 Transition metals and their compounds are important catalysts in industry and in biological systems.

12 Many enzymes rely on transition metals for their catalytic activity.

Review questions

1 (a) By reference to copper or its compounds, give three properties that are characteristic of transition metals.
   (b) Explain the following:
   (i) Aqueous copper(II) sulphate turns blue litmus paper red.
   (ii) The addition of ammonia solution to aqueous copper(II) sulphate gives a pale blue precipitate at first and then a deep blue solution when more ammonia solution is added.

2 (a) Give the name and formula of:
   (i) one complex cation containing a transition metal,
   (ii) one complex anion containing a transition metal.
   (b) Describe the overall shape (octahedral, tetrahedral, etc.) of the two ions in part (a).
   (c) Outline the electron structures of the ions that you chose in (a), restricting yourself to a consideration of the outer shell of each atom.

3 (a) How do the electron structures of transition elements differ from those of the elements in the main groups of the periodic table?
   (b) Describe the electron structure of either vanadium or chromium. How are the important oxidation states of the element determined by this electron structure?
   (c) Give three characteristic features of transition metals or their compounds (other than variable oxidation states).

Illustrate your answer with reference to vanadium or chromium.

4 Complex ions can sometimes exhibit isomerism.
   (a) The compound $[Co(NH_3)_5Br]^{2+}SO_4^{2-}$ is isomeric with the compound $[Co(NH_3)_5SO_4]^+Br^-$.
   (i) What ions will these two isomers yield in solution?
   (ii) How would you confirm which isomer was which? (You are required to describe a simple positive test for each isomer.)
   (iii) What is (i) the oxidation state,
                  (ii) the co-ordination number,
       of cobalt in each complex ion?
   (iv) Draw the structure of the $[Co(NH_3)_5Br]^{2+}$ ion indicating its shape and the co-ordinate bonds involved.
   (b) The compound $NiCl_2(NH_3)_2$ has cis–$trans$ isomers. These have a complex non-ionic structure.
   (i) Does $NiCl_2(NH_3)_2$ have a tetrahedral or a square-planar structure? Explain your answer.
   (ii) Draw the cis and $trans$ isomers for $NiCl_2(NH_3)_2$.

5 Ascorbate oxidase is a metallo-protein enzyme in plants. The enzyme consists of a protein associated with a copper(II) ion.
   (i) The enzyme protein alone has no catalytic activity.
   (ii) Copper(II) ions alone can act as a catalyst for the reaction, but much less efficiently than the metal–protein combination.

(iii) When egg albumen is added to aqueous copper(II) ions, the mixture shows greater catalytic activity than copper(II) ions alone. But, this activity is not so good as with the specific metal–protein in ascorbate oxidase.

    (a) Write the electron structure of copper(II) ions.

    (b) Explain how copper(II) ions might act as catalysts.

    (c) What explanation can you offer for observations (i), (ii) and (iii) above?

6 Early in the 20th century, the German scientist Werner succeeded in clarifying the situation concerning the five compounds of $PtCl_4$ and ammonia. The properties of these compounds are listed in the table below.

Compound	Empirical formula	Total no. of ions in the empirical formula	No. of Cl⁻ ions in the empirical formula
A	$PtCl_4.6NH_3$	5	4
B	$PtCl_4.5NH_3$	4	3
C	$PtCl_4.4NH_3$	3	2
D	$PtCl_4.3NH_3$	2	1
E	$PtCl_4.2NH_3$	0	0

(a) What is the oxidation state of Pt in each of the compounds, A–E?

(b) The co-ordination number of Pt in each compound is six. Write a formula for each of the five compounds. Show the complex ion and the other ions and/or molecules present.

(c) Each of the compounds forms an octahedral complex ion. Draw structures for the complex ions in A, B, C and D.

(d) Which of the complex ions in (c) have isomers?

(e) Draw structures to show the various structural isomers of A, B, C and D.

Metals and the Activity Series

19

▲ A sample of 'kidney' iron ore. This is a form of haematite (impure iron(III) oxide).

How do transition metals compare and contrast with the metals in Groups I and II in their:

1 melting points and boiling points,
2 density,
3 reactivity with water,
4 reactivity with dilute HCl,
5 number of oxidation states,
6 colour of compounds,
7 formation of complex ions?

▲ Roman iron objects. Successful methods for smelting and casting iron existed even in Roman times. The collection includes a knife, a hammer-head, a large nail and a spear-head.

19.1 Iron – a typical transition metal

Iron typifies transition metals in many respects.

(a) It has *two* relatively stable **oxidation states** in its compounds (+2 and +3).
(b) It forms a variety of **complex ions** (such as $[Fe(H_2O)_6]^{2+}$, $[Fe(H_2O)_6]^{3+}$, $[Fe(CN)_6]^{4-}$ and $[Fe(CN)_6]^{3-}$).
(c) It has characteristically **coloured compounds**: pale green for iron(II) salts and yellow or brown for iron(III) salts.
(d) It has important **catalytic properties**. For example, it is used to catalyse the synthesis of ammonia in the Haber process.

The typical properties of transition metals were considered in detail in the last chapter.

19.2 The occurrence of iron

Apart from its presence in certain meteorites, iron does not occur as the element in the earth's crust. Nevertheless, iron is the fourth most abundant element in the earth's crust (figure 15.1). Its compounds are abundant. The commonest ores are iron pyrites (FeS_2), haematite (Fe_2O_3), magnetite (Fe_3O_4) and limonite or brown iron ore ($Fe_2O_3.H_2O$).

Iron is also present in silicates, sands and clays, and in all living matter. It is essential for the production of haemoglobin in blood and chlorophyll in plants. Many pregnant women must take iron tablets (usually in the form of iron(II) sulphate). These tablets enable them to produce sufficient haemoglobin for themselves and their growing babies without becoming anaemic.

Iron has considerable economic and industrial importance because:
(a) its ores are abundant in the earth's crust,
(b) it is easily manufactured from its ores,
(c) it has very useful mechanical properties.

19.3 The manufacture of iron

At the end of the nineteenth century, the annual production of iron in the UK was only about 20 000 tonnes. Nowadays, its annual production is about 20 000 000 tonnes.

The most important raw materials from which iron is obtained are haematite (Fe_2O_3) and brown iron ore ($Fe_2O_3.H_2O$). In most parts of the world, there are still large quantities of high quality iron ores from which iron can be obtained.

The first stage in the manufacture of iron involves preparing the ore for reduction in a blast furnace. This involves crushing the ore to produce lumps about the size of one's fist. These lumps are then pre-heated using hot gases from the furnace. This drives off water and other volatile impurities.

The prepared iron ore contains between 30% and 95% iron oxides. The main impurities are silicon(IV) oxide (silica, SiO_2) and aluminium oxide (alumina, Al_2O_3).

The second stage in the manufacture of iron involves reducing the iron oxides. This is done using carbon (coke) in the presence of limestone (calcium carbonate) in a blast furnace (figure 19.1).

▲ This photo shows iron ore being crushed and transported to the blast furnace prior to reduction.

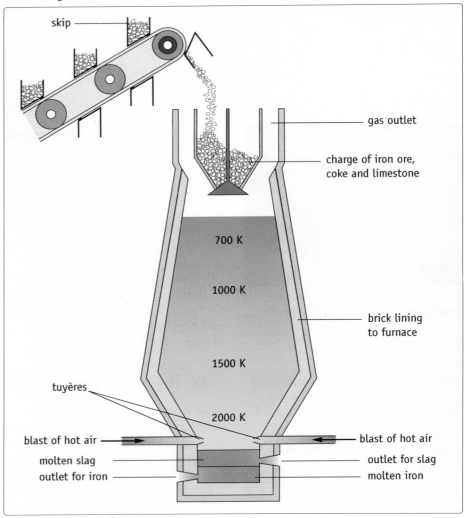

Figure 19.1
A blast furnace.

▲ A workman taps off the molten slag which has collected on top of molten iron in a blast furnace.

The furnace itself is a tapered cylindrical tower, about 30 metres in height. It is made of steel and lined with refractory bricks. The furnace is fed mechanically so that no gas escapes during the charging. A mixture of iron ore, coke and limestone is added at the top of the furnace. At the same time, blasts of hot air at about 1000 K are driven into the furnace through small holes or 'tuyères' near its base. These blasts of air give the furnace its name.

As the air enters the furnace, it reacts with coke. A highly exothermic reaction occurs forming carbon dioxide:

$$C(s) + O_2(g) \rightarrow CO_2(g) \qquad \Delta H = -394 \, kJ$$

This raises the temperature at the bottom of the furnace to about 2000 K. As the CO_2 rises up the furnace, it reacts with more coke forming carbon monoxide:

$$CO_2(g) + C(s) \rightarrow 2CO(g) \qquad \Delta H = +173 \, kJ$$

The carbon monoxide then reduces the iron oxides to iron in the upper parts of the furnace. Here the temperature is between 750 K and 1000 K.

$$Fe_2O_3(s) + 3CO(g) \rightarrow 2Fe(s) + 3CO_2(g)$$

19 Metals and the Activity Series

As the iron falls towards the base of the furnace, it eventually melts (m.pt. = 1812 K) and flows to the bottom of the furnace where it collects and is tapped off every few hours. A typical blast furnace produces about 1000 tonnes of iron every 24 hours.

So far, we have said nothing about the use of limestone in the blast furnace. The limestone plays an essential part in the extraction of impurities from the iron.

At the high temperatures inside the furnace, the limestone decomposes into calcium oxide and carbon dioxide:

$$CaCO_3(s) \rightarrow CaO(s) + CO_2(g)$$

The calcium oxide then reacts with impurities in the iron such as sand (silicon(IV) oxide) and alumina (aluminium oxide). It forms calcium silicate and calcium aluminate, respectively:

$$CaO(s) \quad + \quad SiO_2(s) \quad \rightarrow \quad CaSiO_3(l)$$
$$CaO(s) \quad + \quad Al_2O_3(s) \quad \rightarrow \quad CaAl_2O_4(l)$$

The mixture of calcium silicate and calcium aluminate which remains molten at the furnace temperature is known as **slag**. This flows to the bottom of the furnace where it floats on the molten iron. It is tapped off at a different level, separately from the iron. The slag is not wasted. It is used for road-making materials, for cement manufacture and for lightweight building materials.

The iron obtained from the blast furnace is far from pure. It is known generally as **pig iron**. It is hard, but brittle, and melts at about 1500 K.

It is re-melted, mixed with scrap steel and cooled in moulds to form **cast iron**. This has much the same properties as pig iron and can be used for articles such as gates, pipes and lamp posts where cheapness is more important than strength.

19.4 The conversion of iron into steel

Most of the pig iron manufactured nowadays is used for the production of **steel**. This is an alloy of iron with carbon and other elements such as manganese, nickel, chromium, tungsten and vanadium.

The main impurities in pig iron and in mild steel are shown in table 19.1.

▲ An attractive cast iron gate.

▶ When steel was first produced it was an expensive alloy. Thus, the first metal bridge, which was erected at Coalbrookdale in 1799, was made of 400 tonnes of cast iron not steel.

Table 19.1
The main impurities in pig iron and in mild steel

Impurity	% impurity in pig iron	% impurity in mild steel
Carbon	3–5	0.15
Silicon	1–2	0.03
Sulphur	0.05–0.10	0.05
Phosphorus	0.05–1.5	0.05
Manganese	0.5–1.0	0.50

The carbon impurity comes from coke used during the reduction of iron ore. The silicon, sulphur and phosphorus are present as a result of the reduction of impurities of silicates, sulphates and phosphates by carbon and carbon monoxide during the blast furnace processes. The manganese arises from the reduction of manganese compounds in the iron ore.

In order to obtain steel from pig iron, a large proportion of these impurities must be removed.

During the last 50 years, the techniques of **steel production** have undergone vast changes in scale. New processes have also been developed to keep pace with the demands of quantity and quality. At present, there are only three major steel-making processes; **Open Hearth**, **Electric Arc**, and **Basic Oxygen**.

All of these processes share the same general principle. This entails removing the impurities of carbon, silicon, sulphur and phosphorus from molten pig iron by oxidation. Known quantities of other elements are then added to obtain steel of the desired composition and properties.

During steel production, carbon and sulphur impurities in the pig ion are removed by oxidation to CO_2 and SO_2 which then escape as gases. On the other hand, silicon and phosphorus are oxidised to less volatile oxides, phosphorus(V) oxide and silicon(IV) oxide. These combine with the lime added to the furnace to form slag.

$$6CaO(s) + P_4O_{10}(s) \rightarrow 2Ca_3(PO_4)_2(l)$$
calcium phosphate(V)

$$CaO(s) + SiO_2(s) \rightarrow CaSiO_3(l)$$
calcium silicate

} slag

The Basic Oxygen Process

In the Basic Oxygen Process, no external heating is required because the reactions inside the furnace are very exothermic. Initially the furnace is charged with hot, molten pig iron and lime.

Oxygen is blown onto the surface of the metal at great speed through water-cooled pipes. The oxygen penetrates into the melt and oxidises the impurities rapidly.

Nowadays, over 80% of the steel produced in the UK is obtained by the Basic Oxygen Process. This process is increasingly used because it is so fast. A large furnace can convert 300 tonnes of pig iron into steel in only 40 minutes.

► Charging one of the basic oxygen furnaces with hot molten pig iron at a Japanese steel works.

The Open Hearth Process

In the Open Hearth Process, heat is provided by burning gas or oil in a mixture of air and oxygen.

Oxygen is injected into the furnace through water-cooled pipes. This oxidises the impurities in the iron.

About 10% of the steel produced in the UK is obtained by the Open Hearth Process. The process is rather slow. A large furnace takes 10 hours to convert 300 tonnes of pig iron to steel. This process is being replaced by the very much faster Basic Oxygen Process.

By adding different metals to the steel it is possible to impart specific properties, such as hardness, resistance to corrosion or high tensile strength. 'Manganese steel' containing as much as 13% manganese can be made incredibly tough by heating to 1000°C and 'quenching' (rapid cooling) in water. It is used for parts of rock-breaking machinery and for railway cross-overs.

Stainless steel, containing about 20% chromium and 10% nickel, is used for cutlery, surgical instruments and bathroom taps.

'High-speed' steels used as the cutting edges on lathes for metalwork, contain about 18% tungsten and 5% chromium. The tungsten hardens the steel and makes it less brittle. Using high-speed steels, metal parts can be machined up to twenty times faster than ordinary steel.

19.5 The reactions of iron and its aqueous ions

Several reactions of iron and the aqueous Fe^{2+} and Fe^{3+} ions are summarised in figure 19.2.

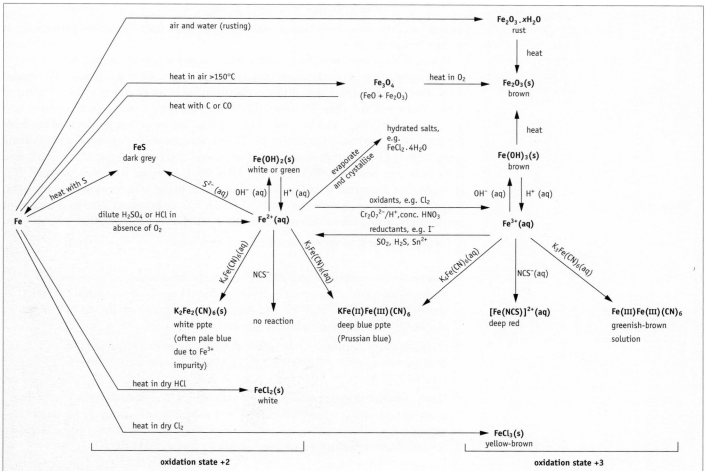

Figure 19.2
The reactions of Fe, Fe^{2+}(aq) and Fe^{3+}(aq).

Iron combines with most non-metals on heating, forming compounds such as Fe_3O_4, FeS, $FeCl_3$ and FeI_2. Those non-metals which are strong oxidising agents, such as chlorine and oxygen, form products containing iron in the more oxidised +3 state.

The electrode potentials involving iron in its +2 and +3 states are given below.

In 1.0 mol dm^{-3} acid solution

$$Fe(s) \xrightarrow{+0.44} Fe^{2+}(aq) \xrightarrow{-0.77} Fe^{3+}(aq)$$

In 1.0 mol dm^{-3} alkaline solution

$$Fe(s) \xrightarrow{+0.89} Fe(OH)_2(s) \xrightarrow{+0.56} Fe(OH)_3(s)$$

The figures show that in acid solution the more stable state of iron is +2 and in alkaline solution the more stable state of iron is +3.

In alkaline solution, oxygen in the air or oxygen dissolved in the solution will oxidise $Fe(OH)_2$ to $Fe(OH)_3$:

$$2Fe(OH)_2(s) + 2OH^-(aq) \rightarrow 2Fe(OH)_3(s) + 2e^- \qquad +0.56 \text{ V}$$
$$\tfrac{1}{2}O_2(g) + H_2O(l) + 2e^- \rightarrow 2OH^-(aq) \qquad +0.40 \text{ V}$$

As expected, the reaction has a relatively large overall positive potential (+0.96 V), indicating that it is energetically feasible.

So, freshly prepared pale green $Fe(OH)_2$ rapidly turns dirty green and finally red–brown on contact with the air as it is oxidised to $Fe(OH)_3$.

In acid solution and under standard conditions, $Fe^{2+}(aq)$ is not oxidised to $Fe^{3+}(aq)$ by oxygen in the air or by oxygen dissolved in the solution.

$$2Fe^{2+}(aq) \rightarrow 2Fe^{3+}(aq) + 2e^- \qquad -0.77 \text{ V}$$
$$2H^+(aq) + O_2(g) + 2e^- \rightarrow H_2O_2(aq) \qquad +0.68 \text{ V}$$

In this case, the overall reaction potential is negative (−0.09 V) and the process is not energetically feasible.

Thus, $Fe^{2+}(aq)$ ions are stable in acid solution under standard conditions.

Look again at the electrode potentials given at the beginning of this section. The figures show that iron will react with dilute acids (in the absence of air) to yield solutions of iron(II) salts and hydrogen. However, dilute nitric acid and concentrated sulphuric acid yield a mixture of Fe^{2+} and Fe^{3+} ions together with the reduction products of the acids (mainly NH_4^+ and SO_2, respectively). This is because these two acids are oxidising agents. On the other hand, concentrated nitric acid and concentrated chromic(VI) acid have only a momentary effect on iron. After this, the metal becomes 'passive'. It is thought that the initial attack of the acid coats the metal with an exceptionally tight, impervious and unreactive layer of oxide. This layer prevents further attack. Not surprisingly, iron is sometimes treated in this way to prevent corrosion.

Try to answer these questions about the chemistry of iron.

1 Aqueous iron(II) salts are only slightly acidic whereas iron(III) salts are strongly acidic. Why is this? (Hint: your explanation should involve a consideration of the charge density on Fe^{2+} and Fe^{3+} and the possible hydrolysis of $[Fe(H_2O)_6]^{2+}$ and $[Fe(H_2O)_6]^{3+}$ in aqueous solution. See section 18.6)

2 Draw structural formulas to show the octahedral stereo-structures of the hexacyanoferrate(III) ion ($[Fe(CN)_6]^{3-}$) and the hexacyanoferrate(II) ion ($[Fe(CN)_6]^{4-}$).

3 Iron(III) compounds are more covalent than iron(II) compounds. Why is this? (For example, anhydrous iron(III) chloride is very similar to anhydrous aluminium chloride. It sublimes readily. It is soluble in alcohol and ether. It has a co-ordinately bonded dimer structure, Fe_2Cl_6. See section 9.2 if you need help.)

4 Use figure 19.2 to tabulate the reactions of Fe^{2+}(aq) and Fe^{3+}(aq) with:
  (a) OH^-(aq)
  (b) NCS^-(aq)
  (c) $[Fe(CN)_6]^{4-}$(aq)
  (d) $[Fe(CN)_6]^{3-}$(aq)
  Which of the above ions would be most effective in:
  (i) testing for Fe^{2+}(aq)
  (ii) testing for Fe^{3+}(aq)
  (iii) distinguishing Fe^{2+}(aq) from Fe^{3+}(aq)?

19.6 The corrosion of iron and its prevention

Rusting costs us millions of pounds each year. The costs arise on two counts: the need to protect iron and steel objects and the replacement of rusted articles.

Rusting *requires the presence of both oxygen and water* but other factors also influence the rate of rusting. These factors include impurities in the iron, availability of dissolved oxygen and electrolytes in the solution in contact with the iron.

When rusting occurs, a hydrated form of iron(III) oxide with variable composition ($Fe_2O_3.xH_2O$) is produced. This oxide is very permeable to both air and water. It cannot protect the metal from further corrosion, which continues unhindered below the rusted surface. The process is represented schematically in figure 19.3.

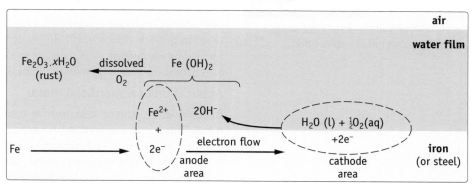

Figure 19.3
The chemical reactions involved in rusting.

In our climate, iron or steel objects open to the atmosphere are often wet. The layer of water on the surface of the metal dissolves oxygen and carbon dioxide from the air. In addition to these, the water will also contain dissolved sulphur dioxide in any industrialised area and sodium chloride from salt spray near the sea.

Look closely at figure 19.3.

1 Why are oxygen and water both required for rusting?

2 Why do electrolytes, particularly acids, accelerate rusting?

3 Which substances, present in the atmosphere, can provide the H^+ ions to accelerate rusting? (Hint: why is rusting accelerated in polluted areas?)

In the initial stages of rusting, iron(II) ions pass into solution at the anode area. At the same time, reduction of atmospheric oxygen to hydroxide ions takes place at the cathode area.

		$E^{\ominus}$/volts
anode (+)	$Fe(s) \rightarrow Fe^{2+}(aq) + 2e^-$	+0.44
cathode (−)	$H_2O(l) + \frac{1}{2}O_2(aq) + 2e^- \rightarrow 2OH^-(aq)$	+0.40

The electrode potentials for the two half reactions indicate that the overall reaction is energetically feasible.

▲ A wrecked and rusting tanker.

The Fe^{2+} and OH^- ions produced diffuse away from the 'electrodes' and precipitate as iron(II) hydroxide. This is then oxidised by dissolved oxygen to form rust (see section 19.5)

Notice that the rust is formed by a secondary process *in the solution* as Fe^{2+} and OH^- ions diffuse away from the metal surface. Thus, the rust does not usually form as a protective layer in contact with the iron surface. However, if the solution contains a relatively high concentration of dissolved oxygen, Fe^{2+} ions are converted into rust more rapidly. In this case a protective layer may be formed on the iron surface which retards further rusting.

Acid conditions (low pH) accelerate rusting because the iron reacts with H^+ ions to form Fe^{2+}. Electrolytes in the water and impurities in the iron also assist the process. They increase the conductivity of the solution and initiate cell action, respectively.

Two general methods are used to protect iron and steel from corrosion: the **application of a protective layer** and the **application of a sacrificial metal**.

Application of a protective layer

Many different substances are used to protect iron and steel objects from rusting.

In the motor car industry, the chassis of vehicles are painted with phosphoric(V) acid. This reacts with the iron to form an insoluble, tenacious film of iron(III) phosphate(V). Engine parts, like other machinery and tools, are often protected by a film of oil or grease. However, the most common method of rust prevention is painting. This often incorporates the use of dilead(II), lead(IV) oxide (red lead) or zinc chromate(VI) as protective priming coats.

Alternatively, iron and steel articles can be coated with non-rusting metals. Buckets, dustbins and gates are coated with a layer of zinc ('galvanised'). This is done by dipping them in molten zinc at 450°C or by depositing the zinc by electrolysis. 'Tin plate' is formed by immersing steel sheeting in molten tin. Other articles, such as car bumpers, are chromium-plated by making them the cathode in an electrolytic cell.

Application of a sacrificial metal

In this case, the iron or steel must be connected to a more reactive, more electropositive metal. This is then oxidised in preference to iron, thus preventing the formation of Fe^{2+} ions and hence rust. 'Galvanising' is really an example of a sacrificial method of rust prevention. When the 'galvanised' surface is undamaged, zinc is protected from any reaction by a thin, firmly adhering layer of zinc oxide. When the galvanised surface is scratched and the iron is exposed, zinc ions pass into solution rather than iron(II) ions. This occurs because zinc is more electropositive. Electrode potentials confirm this fact.

$$
\begin{array}{ll}
 & E^{\ominus}/\text{volts} \\
Fe(s) \rightarrow Fe^{2+}(aq) + 2e^- & +0.44 \\
Zn(s) \rightarrow Zn^{2+}(aq) + 2e^- & +0.76
\end{array}
$$

Thus, zinc is sacrificed in the protection of iron. Tin plating, however, is not so effective as galvanising. Provided the coating is undamaged, the tin surface protects the iron below. But when the tin plate is scratched rusting occurs. Iron is more electropositive than tin, so Fe^{2+} ions pass into solution and then form rust.

$$
\begin{array}{ll}
 & E^{\ominus}/\text{volts} \\
Fe(s) \rightarrow Fe^{2+}(aq) + 2e^- & +0.44 \\
Sn(s) \rightarrow Sn^{2+}(aq) + 2e^- & +0.14
\end{array}
$$

Sacrificial protection is used to prevent corrosion in underground pipelines made of steel. Pieces of zinc or manganese alloy are connected to the buried pipeline at intervals. In oil refineries, magnesium alloys are bolted inside the distillation and cracking plant. These metals will then corrode in preference to steel between regular maintenance inspections.

One other general method is used in the fight against rusting. This is **alloying**. Iron can be alloyed with chromium, nickel or manganese to produce corrosion-resistant steels.

▲ Hot galvanised sheets of steel being carefully inspected at a steel works.

▲ Blocks of magnesium on the hull of a large sailing yacht act as a sacrificial metal.

19 Metals and the Activity Series

19.7 Copper – extraction and manufacture

Although the metal occurs naturally in a few places, most copper is obtained from its naturally occurring compounds. The most important ores are copper pyrites ($CuFeS_2$) and copper glance (Cu_2S). These two sulphides provide 80% of the element.

Most copper ores contain only a few per cent of the metal, so the first stage in obtaining copper is to crush the ore and concentrate it by froth flotation.

The second stage involves reduction of the sulphide ore to the metal. Initially, the concentrate is roasted in air to form copper(I) sulphide, iron(II) oxide and gaseous sulphur dioxide:

$$2CuFeS_2(s) + 4O_2(g) \rightarrow Cu_2S(s) + 2FeO(s) + 3SO_2(g)$$

The product is then heated with silica in a closed furnace. Most of the iron(II) oxide reacts with the silica to form a molten slag which floats on the molten copper(I) sulphide and can be tapped off separately:

$$FeO(s) + SiO_2(s) \rightarrow FeSiO_3(l)$$
$$\text{iron(II) silicate (slag)}$$

► The open-cast copper mine at Chuquicamata, in the Atacama Desert, Chile. This is the largest copper-producing mine in the world.

The impure copper(I) sulphide is then heated in air so that part of it reacts, forming copper(I) oxide:

$$2Cu_2S(s) + 3O_2(g) \rightarrow 2Cu_2O(s) + 2SO_2(g)$$

The copper(I) oxide, mixed with unchanged copper(I) sulphide, is now heated strongly in the absence of air. This forms copper and sulphur dioxide:

$$2Cu_2O(s) + Cu_2S(s) \rightarrow 6Cu(s) + SO_2(g)$$

The product is known as 'blister copper'. It releases bubbles of SO_2 as it solidifies and therefore gets a blistered appearance. It still contains 2 or 3% of impurities, mainly iron and sulphur.

The final stage in the manufacture of copper involves purification of the blister copper. This is usually achieved by electrolysis of copper(II) sulphate solution. A thin sheet of pure copper is used as the cathode and the impure blister copper is the anode. During the electrolytic purification, copper dissolves away from the anode, whilst a thickening deposit of pure copper appears on the cathode.

1 Write equations for the processes at the anode and the cathode during this electrolytic purification of blister copper.
2 If the impure blister copper contains metals above copper in the electrochemical series, these metals will dissolve from the anode in preference to copper. Why do these metals, which get into solution, not deposit on the pure copper cathode?
3 Suppose the impure blister copper contains metals below copper in the electrochemical series such as silver. Why do these metals not contaminate the copper which deposits on the cathode?

▲ Purifying crushed copper ore by froth flotation.

▶ Sparks fly during the smelting of copper. The copper ore is heated to a high temperature in air. The ore reacts with oxygen to produce copper(I) sulphide, iron(II) oxide and sulphur dioxide in a very exothermic reaction.

▲ Electrolytic refining of copper. Pure copper has deposited on these sheets of the metal.

In practice, silver and gold can be obtained as valuable by-products from the impure 'blister copper'. They help to pay for the copper-refining process.

19.8 The properties and uses of copper

The uses of copper result from its malleability and ductility, its high electrical and thermal conductivity, its resistance to corrosion and its ability to form alloys. The pure metal is used extensively as electrical conducting wires and cables. It is also used as pipes and radiators in central heating systems.

Copper provides a wide range of different alloys with other metals. Alloying copper with zinc produces various types of brass which are harder and stronger than pure copper. Alloying copper with tin produces bronze, an alloy which has greater tensile strength and is more readily cast into moulds than pure copper. Copper also alloys readily with nickel, aluminium and gold. For example, 75% copper and 25% nickel are used in our present 'silver coins', whilst 9 carat gold is about two-thirds copper and one-third gold.

19.9 Aluminium – extraction and manufacture

Although aluminium is not found free in nature, its compounds are widespread. Aluminium is the third most abundant element (after oxygen and silicon) in the Earth's crust (figure 15.1).

Economically, the most important ore is bauxite (impure hydrated aluminium oxide, $Al_2O_3.2H_2O$) from which the metal is obtained. There are also huge, widespread deposits of aluminosilicates in clays, mica and other minerals, and deposits of cryolite (sodium hexafluoroaluminate(III), Na_3AlF_6) are found in Greenland.

Aluminium is manufactured by the electrolysis of molten aluminium oxide which is obtained from bauxite.

The main impurities in bauxite are iron(III) oxide and silicon(IV) oxide. The first stage in the manufacture of aluminium is therefore the production of pure aluminium oxide from bauxite (figure 19.4). The purification process is based on the amphoteric nature of aluminium oxide (section 13.7).

When the impure bauxite is treated with concentrated sodium hydroxide, Al_2O_3 and SiO_2 dissolve, but Fe_2O_3 and other basic materials remain insoluble and are removed by filtration.

$$Al_2O_3(s) + 2OH^-(aq) + 3H_2O(l) \rightarrow 2[Al(OH)_4]^-(aq)$$
$$SiO_2(s) + 2OH^-(aq) \rightarrow SiO_3^{2-}(aq) + H_2O(l)$$

The solution is then treated with weak acid (e.g. CO_2 to produce carbonic acid) in order to precipitate aluminium hydroxide.

$$[Al(OH)_4]^-(aq) + H^+(aq) \rightarrow Al(OH)_3(s) + H_2O(l)$$

Finally, the aluminium hydroxide is removed by filtration and is heated to obtain pure Al_2O_3:

$$2Al(OH)_3(s) \rightarrow Al_2O_3(s) + 3H_2O(g)$$

▲ Rows of electrolytic cells at Alcan's aluminium smelter at Kitimat, British Columbia. What is the white powder on the floor?

Figure 19.4
The production of pure aluminium oxide from bauxite.

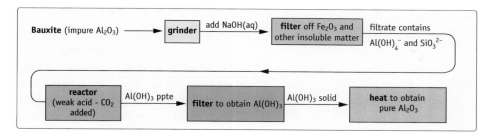

Many attempts have been made to obtain aluminium oxide from the vast, ubiquitous deposits of clays throughout the world. Unfortunately, its separation and purification from large quantities of impurity are very difficult and, at present, uneconomical.

During the nineteenth century, attempts were made to reduce aluminium oxide chemically. However, the oxide is so stable that it required such unconventional and undesirable reducing agents as sodium or potassium.

The answer to the problem of aluminium production was obviously electrolysis, but this also presented problems. Aluminium is so reactive that the electrolytic process must use molten solids rather than aqueous solutions. This in turn requires very high temperatures in order to maintain the molten state.

Fortunately, the electrolytic process (figure 19.5) can be performed by dissolving the oxide in molten cryolite (Na_3AlF_6). This has a much lower melting temperature than aluminium oxide. Even so, the temperature of the molten electrolyte must be raised to 850 °C and maintained at this temperature by the current through the electrolyte. Obviously, the amount of electrical energy required for this process is enormous, and it is usually carried out where electrical energy is cheap and plentiful.

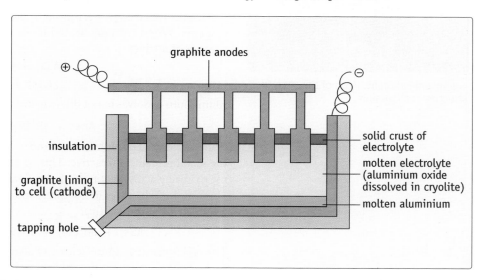

Figure 19.5
The electrolytic manufacture of aluminium.

Aluminium is discharged at the graphite lining of the cell which acts as the cathode:

$$Al^{3+}(l) + 3e^- \rightarrow Al(l)$$

The molten aluminium collects at the bottom of the cell and is tapped off periodically. Oxygen is evolved at the anodes:

$$2O^{2-}(l) \rightarrow O_2(g) + 4e^-$$

The oxygen reacts with the carbon of the anode to form oxides of carbon at the high temperatures involved. As a result, the anodes are gradually burnt away and must be replaced from time to time.

19.10 The properties and uses of aluminium

The position of aluminium in the electrochemical series and its electrode potential suggest that it should react readily with oxygen. It should also dissolve rapidly in dilute acids, liberating hydrogen:

$$Al(s) \rightarrow Al^{3+}(aq) + 3e^- \qquad E^{\ominus} = +1.66 \text{ V}$$
$$3H^+(aq) + 3e^- \rightarrow \tfrac{3}{2}H_2(g) \qquad E^{\ominus} = 0.00 \text{ V}$$

However, the rapid formation of a thin layer of oxide on the surface of the metal prevents further attack by oxygen. This layer also retards reaction of the aluminium with dilute acids. Furthermore, the oxide layer is virtually non-porous to water, so the aluminium is completely protected from further oxidation. This is very different from iron, which forms a porous oxide layer (rust). This is readily penetrated by water, allowing the process of corrosion to continue beneath the superficial layer of rust.

Most of the uses of aluminium are only possible because of the protective oxide coating – saucepans, aircraft and vehicle bodywork, tent frames, etc.

In order to protect the aluminium even more, it is possible to increase the thickness of the oxide layer by a process known as **anodising**. The aluminium is anodised by making it the anode during the electrolysis of sulphuric acid. Oxygen, released at the anode, combines with the aluminium and thickens the oxide layer. By carrying out the electrolytic anodising process in the presence of dyes which are absorbed by the oxide layer, the anodised material can be coloured attractively for use in domestic and personal articles.

> 1 What properties of aluminium (besides its resistance to corrosion) make it suitable for use as: (i) milk bottle tops, (ii) baking foil, (iii) aircraft bodywork, (iv) electricity cables, (v) kettles and pans?
> 2 Why does aluminium not corrode away like iron?
> 3 Aluminium and iron are the two most widely used metals. What advantages are there in using iron rather than aluminium?

Once the oxide layer is removed or penetrated, aluminium reacts readily with many reagents. When strongly heated in air, it will burn to form the oxide and a smaller amount of nitride:

$$2Al(s) + \tfrac{3}{2}O_2(g) \rightarrow Al_2O_3(s)$$
$$2Al(s) + N_2(g) \rightarrow 2AlN(s)$$

Aluminium dissolves in hot dilute hydrochloric and sulphuric acids, evolving hydrogen:

$$Al(s) + 3H^+(aq) \rightarrow Al^{3+}(aq) + \tfrac{3}{2}H_2(g)$$

With concentrated sulphuric acid and both concentrated and dilute nitric acid, however, the metal remains unreactive. This so-called 'passivity' of the aluminium is caused by the oxidising acids reacting with the metal to form an inert, impervious oxide layer.

Aluminium dissolves in hot dilute sodium hydroxide, forming hydrogen and sodium tetrahydroxoaluminate(III):

$$Al(s) + OH^-(aq) + 3H_2O(l) \rightarrow [Al(OH)_4]^-(aq) + \tfrac{3}{2}H_2(g)$$

The reactions with dilute acids and alkalis are usually very slow at first. They become increasingly vigorous as the oxide layer is removed and the solutions become hotter.

▲ Concorde in flight. Much of the aircraft's structure is aluminium.

19.11 The reactions of metals and the activity series

Many of the reactions of metals can be summarised using either the periodic table or the activity series. The reactions of some common metals with oxygen, chlorine, water, steam and dilute acids are summarised in table 19.2. The metals are written in order of their position in the activity (electrochemical) series.

The heats evolved when each metal reacts with one mole of oxygen and with one mole of chlorine are also included in the table as are the electrode potentials of the metals. Notice that there is a similar order of reactivity of these metals for each series of reactions.

Metals act as reducing agents in all their reactions. They donate electrons to non-metals, water or H^+ ions.

$$M \xrightarrow[\text{reducing agent}]{\text{oxidation}} M^{2+} + 2e^-$$

Of the metals listed in table 19.2, potassium is the strongest reducing agent whereas gold is the weakest. Potassium donates its electrons most readily, gold least readily. The ease with which a metal loses electrons is, of course, directly linked to the value of its electrode potential and hence its position in the electrochemical series.

19.12 The extraction of metals from their ores

Table 19.3 gives some information concerning the methods of extraction, the major sources, the annual productions and the uses of nine common metals.

Some of these extraction processes have been considered in this and previous chapters. References to these sections are given in the final column of table 19.3.

The extraction of all metals from their naturally occurring ores involves three stages.

(A) Purification and concentration of the ore

Very few metal ores are sufficiently concentrated to be used without purification. The main impurities are earthy materials such as rocks, clays and sand. In many cases, the ore is separated from useless impurities before transportation. This reduces costs.

Usually the material is first crushed and ground into small pieces. Magnetic material can then be separated by using strong electromagnets. Alternatively, and more commonly, differences in density between the ore and its impurities will allow separation of crushed material on an agitated sloping table. Jets of water play across the line of inclination of the table. This causes the denser material to accumulate at the lower end.

Another common method of purifying metal ores is **froth flotation**. In this process, the finely ground ore is added to water which contains special oils known as 'frothing agents'. These include pine oil, creosote and xanthates. Separation is possible because of the different densities of the materials and their different wetting characteristics with the frothing agent and water.

When air is blown through the mixture, bubbles adhere to certain minerals, but not to others. The 'frothing agent' is carefully chosen so that particles of the ore become attached to air bubbles. The particles then float to the surface while earthy particles sink to the bottom. The ore particles collect as a froth and are removed periodically.
The process of froth flotation is particularly suitable for concentrating dense ores such as galena (lead sulphide) and zinc blende (zinc sulphide).

(B) Reduction of the purified ore to the metal

The various methods of reducing metal ores are discussed in section 19.14. The method used for a particular metal depends mainly on the position of the metal in the activity series.

1 Why is iron used in much larger quantities than any other metal?
2 Which compounds do metals occur as most frequently in their ores?
3 As a general rule, the most electropositive metals occur in nature in combination with the most electronegative anions such as oxide and chloride. The less electropositive metals are found in combination with the less electronegative anions such as sulphide. Do you think this is a good generalisation?
4 How do very poor metals such as gold and platinum occur in nature?
5 Why is there a clear division between those metals discovered in ancient times and those discovered in the eighteenth and nineteenth centuries?

Table 19.2

Summary of the reactions of metals in the activity series (M is used as the general symbol for a metal and assumed to be divalent)

Metal	Reaction with $O_2(g)$ on heating	Heat evolved when metal reacts with 1 mole of O_2 to form oxide shown /kJ		Reaction with cold water	Reaction with steam	Reaction with dilute strong acids
K	form oxides (e.g. Na_2O in limited supplies of O_2, but peroxides (e.g. Na_2O_2) with excess O_2	K_2O	723	displace $H_2(g)$ from cold water with decreasing reactivity (K, violently; Mg, very slowly)	displace $H_2(g)$ from steam with decreasing vigour (K, very violently; Fe, very slowly)	displace $H_2(g)$ from dilute strong acids with decreasing vigour (K, explosively; Mg, very vigorously; Fe, steadily; Pb, very slowly)
Na		Na_2O	832			
Ca		CaO	1272			
Mg	burn with decreasing vigour to form oxides	MgO	1204			
Al		Al_2O_3	1114	do *not* displace $H_2(g)$ from cold water		
Zn		ZnO	697			
Fe		Fe_2O_3	548			
Pb		PbO	436			
Cu	do *not* burn, but only form a surface layer of oxide	CuO	311		do *not* displace $H_2(g)$ from steam	do *not* displace $H_2(g)$ from dilute strong acids
Hg		HgO	182			
Ag	do *not* burn or oxidise on surface	Ag_2O	61			
Pt						
Au		Au_2O_3	54			
General eqn	$2M + O_2 \rightarrow 2M^{2+}O^{2-}$			$M + H_2O \rightarrow$ $M^{2+}O^{2-} + H_2$	$M + H_2O \rightarrow$ $M^{2+}O^{2-} + H_2$	$M + 2H^+ \rightarrow$ $M^{2+} + H_2$
Half-eqn	$2M \rightarrow 2M^{2+} + 4e^-$ $O_2 + 4e^- \rightarrow 2O^{2-}$			$M \rightarrow M^{2+} + 2e^-$ $H_2O + 2e^- \rightarrow$ $H_2 + O^{2-}$	$M \rightarrow M^{2+} + 2e^-$ $H_2O + 2e^- \rightarrow$ $H_2 + O^{2-}$	$M \rightarrow M^{2+} + 2e^-$ $2H^+ + 2e^- \rightarrow H_2$
Comment				The oxides which form react with water to form hydroxides $O^{2-} + H_2O \rightarrow 2OH^-$	Oxides of the top four metals react with water to form their hydroxides $O^{2-} + H_2O \rightarrow 2OH^-$	

Reaction with Cl_2 on heating	Heat evolved when metal reacts with 1 mole of $Cl_2(g)$ to form the chloride shown/kJ		Strength as reducing agents	Reactions with other aqueous cations	Standard electrode potential, $M(s)$ /$M^{n+}(aq)$/volts	Metal
	KCl	873	↓ decreasing strength as reducing agents (i.e. electrons are donated less readily by the metal)		+2.92	K
	NaCl	826			+2.71	Na
	$CaCl_2$	795			+2.87	Ca
	$MgCl_2$	642			+2.37	Mg
all the metals react with $Cl_2(g)$ on heating to form their chloride with decreasing vigour	Al_2Cl_6	464		any metal will displace from solution ions of a metal below it in the activity series	+1.66	Al
	$ZnCl_2$	416			+0.76	Zn
	$FeCl_3$	270			+0.44 ($Fe(s)/Fe^{2+}(aq)$)	Fe
	$PbCl_2$	359			+0.13	Pb
	$CuCl_2$	206			−0.34	Cu
	$HgCl_2$	230			−0.79 ($Hg(l)/Hg_2^{2+}(aq)$)	Hg
	AgCl	255			−0.80	Ag
	$PtCl_3$	142			−1.20	Pt
	$AuCl_3$	79				Au
$M + Cl_2 \rightarrow M^{2+}(Cl^-)_2$				$M_h + M_l^{2+} \rightarrow M_h^{2+} + M_l$		General eqn
$M \rightarrow M^{2+} + 2e^-$ $Cl_2 + 2e^- \rightarrow 2Cl^-$				$M_h \rightarrow M_h^{2+} + 2e^-$ $M_l^{2+} + 2e^- \rightarrow M_l$	$M(s) \rightarrow M^{2+}(aq) + 2e^-$	Half-eqn
Metals with 2 oxidation states form the chloride of the metal in the highest state (e.g. $FeCl_3$ not $FeCl_2$)						Comment

Table 19.3
The extraction and uses of metals

Metal	Date of discovery	Main ore from which metal is obtained	Main method of extraction	Annual World Production /1000 tonnes (1998)	Price per tonne (1999)	Main uses	Reference to other sections in the book
Sodium	1807	Rock salt, NaCl	Electrolysis of molten NaCl	400	—	Manufacture of tetraethyllead(IV), sodium vapour lamps, titanium	15.7–15.9, 19.14
Magnesium	1808	Magnesite, $MgCO_3$ and Mg^{2+} ions in sea water	Electrolysis of molten $MgCl_2$	450	£1540	Low-density alloys, e.g. in aircraft and nuclear reactors	15.7–15.9, 19.14
Aluminium	1827	Bauxite Al_2O_3	Electrolysis of Al_2O_3 in molten cryolite (Na_3AlF_6)	16 400	£920	Kitchen utensils, packaging foil, structural alloys in aircraft, automobiles	19.9–19.10, 19.14
Zinc	1746	Zinc blende, ZnS	Heat sulphide in air → oxide. Dissolve oxide in H_2SO_4, electrolyse	8 000	£720	Galvanising iron, die castings, alloys (brass)	19.14
Iron	Ancient	Haematite, Fe_2O_3	Reduce Fe_2O_3 with carbon monoxide	780 000 (steel)	£130	Most important structural metal (as steel), vehicles, engines, tools	19.1–19.6, 19.14
Tin	Ancient	Tinstone, SnO_2	Reduce SnO_2 with carbon	138	£3400	Tin plate (coating iron), alloys (e.g. solder, pewter, bronze, etc.)	17.1–17.2
Lead	Ancient	Galena, PbS	Heat sulphide in air → oxide. Reduce oxide with carbon	6000	£310	Roof and cable covering, battery plates, alloys (e.g. solder, printing metals)	17.1–17.2
Copper	Ancient	Copper pyrites, $CuFeS_2$ (CuS + FeS)	Controlled heating with correct amount of air → Cu + SO_2	10 100	£1080	Electrical wires, cables, etc., water pipes, alloys (e.g. brass, bronze)	19.7–19.8, 19.14
Mercury	Ancient	Cinnabar, HgS	Heat in air → Hg + SO_2	3.5	£3770	Scientific equipment (e.g. barometers, thermometers, etc.), mercury vapour lamps, chlor-alkali cells	19.14

(C) Purification of the metal

The metal obtained by reduction of the ore is usually contaminated with impurities. These include the unchanged ore, other metals present in the ore and non-metals from the anions in the ore. For example, the 'blister copper' obtained by reduction of copper ores contains small amounts of copper sulphide, iron and sulphur. The purification of blister copper is described in section 19.7.

▲ An Egyptian worker smelting metal ore. Why is he using a blow pipe?

▲ Egyptian workmen casting molten metal.

Impure pig iron obtained from the blast furnace contains about 8% impurities. These include carbon, silicon, sulphur, phosphorus and manganese. The removal of these impurities from pig iron is described in section 19.4.

19.13 Factors influencing the choice of method used to reduce a metal ore

Chemical factors

Here, the most important question to ask is 'How easily can the ore be reduced to the metal?' This will dictate the cost of the method used and is closely related to the position of the metal in the activity series.

Other factors which must be considered are:

▲ A small working tin mine in South America.

- the accessibility of the ore,
- the ease with which the ore can be purified,
- the scale on which the metal is required, and
- the most suitable type of furnace or electrolytic technique to be used.

The enormous quantities of iron required for the production of steels could not be extracted by any method other than carbon reduction. This is the cheapest method available. In contrast, a much wider choice of techniques is possible for metals which command higher prices and those required in smaller quantities. Thus, titanium can be obtained economically by reduction of titanium(IV) chloride using sodium in an inert atmosphere.

$$TiCl_4 + 4Na \rightarrow Ti + 4NaCl$$

Economic factors

The final choice of extraction method will usually depend on cost. In this respect the quality of the ore is obviously important. Low-grade ores may not be economically worth exploitation. In this respect, it is interesting that some Cornish tin mines, which closed down some decades ago, re-opened in the 1980s. The escalating prices of tin made their operation economically profitable for a few years.

The demand for a particular metal affects its scale of production. Large-scale operations require plentiful and cheap materials. These may include a cheap reducing agent, such as coke, and possibly a cheap supply of electricity.

Consequently, most iron works are located near coalfields and aluminium smelters are usually situated near a source of cheap hydro-electric power.

▲ A derelict tin mine in Cornwall, England.

Another crucial economic factor in the choice of extraction method relates to the value of any by-products. During the 1950s, a world surplus of zinc arose. Prices sank to an almost uneconomic level. Nevertheless, extraction of zinc continued because the process yielded sulphur dioxide. This was needed for the manufacture of sulphuric acid which was in short supply at the time.

19.14 Metal extractions and the electrochemical series

Electrolysis of fused compounds for metals at the top of the electrochemical series

This method is used to extract potassium, sodium, calcium, magnesium and aluminium. Chemical reduction of the oxides of these metals by carbon or carbon monoxide is not feasible. The temperature required for reduction is too high to make the process economical or practical.

Futhermore, these metals cannot be obtained by electrolysis of their aqueous solutions. Water in the solution would be discharged at the cathode in preference to the metal ions.

As neither chemical reduction nor aqueous electrolysis is possible, the only viable method of extraction is electrolysis of their fused compounds. This usually involves the chlorides.

In practice, the cathode at which the metal ions are discharged is usually made of steel. The anode is made of graphite.

cathode(−) (steel)	$M^{2+}(l) + 2e^- \rightarrow M(l)$
anode (+) (graphite)	$2Cl^-(l) \rightarrow Cl_2(g) + 2e^-$

Large quantities of electricity are required not only to electrolyse the chloride but also to maintain the molten state. Suitable impurities are normally added to the chloride. This reduces its melting point and less electrical energy is then needed to keep the electrolyte molten.

Chemical reduction of compounds for metals in the middle of the electrochemical series

Metals in the middle of the activity series (such as zinc, iron, tin, lead and copper) can be extracted from their oxides and sulphides by chemical reduction. Very often, the sulphides are converted to oxides by heating in air before reduction to the metal.

$$2PbS(s) + 3O_2(g) \rightarrow 2PbO(s) + 2SO_2(g)$$
galena

When metals are extracted from sulphide ores, sulphur dioxide is produced. Precautions must therefore be taken to prevent pollution from this sulphur dioxide.

(a) Reduction of oxides by carbon

The oxides are reduced by coke in a closed furnace. This forms the metal and carbon monoxide. This method is used to obtain lead by heating a mixture of powdered lead oxide and coke at 1400°C.

$$PbO(s) + C(s) \rightarrow Pb(l) + CO(g)$$

One drawback to the use of coke (carbon) as a reductant is the pollution that can arise from the toxic carbon monoxide which is produced.

(b) Reduction of oxides by carbon monoxide

Reduction of iron ore (Fe_2O_3) by carbon monoxide takes place in a blast furnace.

$$Fe_2O_3(s) + 3CO(g) \rightarrow 2Fe(l) + 3CO_2(g)$$

(c) Self-reduction of sulphide ores

Self-reduction plays an essential part in the extraction of copper from its sulphide ores.

Part of the sulphide is first converted to oxide by roasting in air.

$$2Cu_2S(s) + 3O_2(g) \rightarrow 2Cu_2O(s) + 2SO_2(g)$$

1 Why is the anode made of graphite rather than steel, even though steel is a better conductor than graphite? (Hint: What is the product at the anode?)

2 Why must the metal liberated at the cathode be protected from the chlorine produced at the anode?

3 In the electrolytic manufacture of sodium, calcium chloride is added to sodium chloride electrolyte. This reduces the melting point of the sodium chloride. Will this result in any impurity in the sodium produced? Explain.

The supply of air is then cut off and the temperature is raised. The rest of the sulphide now reacts with oxide forming the metal and sulphur dioxide.

$$2Cu_2O(s) + Cu_2S(s) \rightarrow 6Cu(s) + SO_2(g)$$

(d) Reduction of compounds by more reactive metals

This method is particularly useful for those metals such as chromium and titanium which are expensive. Usually they are required in only small quantities.

For example titanium is obtained by heating titanium(IV) chloride to 850°C with magnesium or sodium in an atmosphere of argon.

$$TiCl_4 + 2Mg \rightarrow Ti + 2MgCl_2$$
$$TiCl_4 + 4Na \rightarrow Ti + 4NaCl$$

1 Why is it necessary to use an atmosphere of argon?

2 Chromium is obtained from chromium(III) oxide by heating with aluminium powder. Write an equation for this process.

Heating the ore alone or displacement from aqueous solution for metals at the bottom of the electrochemical series

(a) Heating the ore alone

The compounds of mercury and silver are so unstable that they decompose to the metal on heating. Consequently, mercury can be extracted from the ore, cinnabar (HgS), by heating in air.

$$HgS(s) + O_2(g) \rightarrow Hg(l) + SO_2(g)$$

The mercury distils over at the temperature of the furnace. It is condensed in water-cooled receivers.

(b) Displacement from aqueous solution

In this case, the metal is precipitated from a solution of its ions. This requires a metal higher in the electrochemical series.

Silver can be extracted from very low-grade ore by this method. The insoluble silver sulphide ore is first dissolved in a solution of cyanide ions. It reacts with this to form dicyanoargentate(I) ions.

$$Ag_2S(s) + 4CN^-(aq) \rightarrow 2[Ag(CN)_2]^-(aq) + S^{2-}(aq)$$

Silver is then precipitated from the solution of dicyanoargentate(I) ions by the addition of powdered zinc dust.

$$2[Ag(CN)_2]^-(aq) + Zn(s) \rightarrow [Zn(CN)_4]^{2-}(aq) + 2Ag(s)$$

19.15 Identifying cations

Many common cations can be identified from the colour they produce during flame tests (section 6.2) and/or the colour and solubility of their hydroxides.

Flame tests

Flame tests are carried out by heating the substance under test mixed with concentrated hydrochloric acid on the end of a clean nichrome wire in a roaring bunsen flame. Certain cations emit distinct colours when they are heated in this way (table 19.4).

The colour and solubility of hydroxides

The hydroxides of all metals (except those in group I and barium) are insoluble in water. These insoluble hydroxides form as precipitates when aqueous sodium hydroxide, containing OH^- ions, is added to a solution of metal ions. For example,

$$Zn^{2+}(aq) + 2OH^-(aq) \rightarrow Zn(OH)_2(s)$$

white ppte
zinc hydroxide

Table 19.4
The flame colours frome some cations

Cation present in the substance tested	Flame colour
Li^+	red
Na^+	yellow
K^+	lilac
Mg^{2+}	no flame colour
Ca^{2+}	red
Ba^{2+}	pale green
Cu^{2+}	blue–green

Some of these insoluble hydroxides, including zinc hydroxide, dissolve when excess sodium hydroxide is added.

$$Zn(OH)_2(s) + 2OH^-(aq) \rightarrow Zn(OH)_4^{2-}(aq)$$

clear solution
tetrahydroxozincate(II)

Table 19.5 shows what happens when
(i) a little sodium hydroxide solution, NaOH(aq), and then
(ii) excess sodium hydroxide solution is added to solutions of some common cations.

1 Which metal cations give no precipitate with NaOH(aq)?

2 How can these three cations be distinguished by flame tests?

3 Which cations give a white precipitate with NaOH(aq) which remains with excess NaOH(aq)?

4 How can these two cations be distinguished using flame tests?

5 Which cations give a white precipitate with NaOH(aq) which dissolves in excess? Can you think of a way in which these three cations can be distinguished?

Table 19.5
Identifying cations with sodium hydroxide solution

Cation in solution	A little (3 drops) of NaOH(aq) added to 3 cm^3 of solution of cation	Excess (10 cm^3) of NaOH(aq) added to 3 cm^3 of solution of cation
K^+	No precipitate	No precipitate
Na^+	No precipitate	No precipitate
NH_4^+	No precipitate	No precipitate
Ba^{2+}	No precipitate	No precipitate
Ca^{2+}	White ppte. of $Ca(OH)_2$ with high $[Ca^{2+}(aq)]$ $Ca^{2+}(aq) + 2OH^-(aq) \rightarrow Ca(OH)_2(s)$	White precipitate remains
Mg^{2+}	White precipitate of $Mg(OH)_2$ forms $Mg^{2+}(aq) + 2OH^-(aq) \rightarrow Mg(OH)_2(s)$	White precipitate remains
Al^{3+}	White precipitate of $Al(OH)_3$ forms $Al^{3+}(aq) + 3OH^-(aq) \rightarrow Al(OH)_3(s)$	White precipitate dissolves to give clear solution $Al(OH)_3(s) + OH^-(aq) \rightarrow Al(OH)_4^-(aq)$
Zn^{2+}	White precipitate of $Zn(OH)_2$ forms $Zn^{2+}(aq) + 2OH^-(aq) \rightarrow Zn(OH)_2(s)$	White precipitate dissolves to give clear solution $Zn(OH)_2(s) + 2OH^-(aq) \rightarrow Zn(OH)_4^{2-}(aq)$
Pb^{2+}	White precipitate of $Pb(OH)_2$ forms $Pb^{2+}(aq) + 2OH^-(aq) \rightarrow Pb(OH)_2(s)$	White precipitate dissolves to give clear solution $Pb(OH)_2(s) + 2OH^-(aq) \rightarrow Pb(OH)_4^{2-}(aq)$
Cr^{3+}	Grey-green precipitate of $Cr(OH)_3$ forms $Cr^{3+}(aq) + 3OH^-(aq) \rightarrow Cr(OH)_3(s)$	Grey-green ppte. dissolves partly to give dark green solution $Cr(OH)_3(s) + OH^-(aq) \rightarrow Cr(OH)_4^-(aq)$
Fe^{3+}	Brown precipitate of $Fe(OH)_3$ forms $Fe^{3+}(aq) + 3OH^-(aq) \rightarrow Fe(OH)_3(s)$	Brown precipitate remains
Fe^{2+}	Dark green precipitate of $Fe(OH)_2$ forms $Fe^{2+}(aq) + 2OH^-(aq) \rightarrow Fe(OH)_2(s)$	Dark green precipitate remains
Ni^{2+}	Pale green precipiate of $Ni(OH)_2$ forms $Ni^{2+}(aq) + 2OH^-(aq) \rightarrow Ni(OH)_2(s)$	Pale green precipitate remains
Cu^{2+}	Pale blue precipitate of $Cu(OH)_2$ forms $Cu^{2+}(aq) + 2OH^-(aq) \rightarrow Cu(OH)_2(s)$	Pale blue precipitate remains

19 Metals and the Activity Series

One way to distinguish between Al^{3+}, Zn^{2+} and Pb^{2+} is to filter the precipitates of $Al(OH)_3$, $Zn(OH)_2$ and $Pb(OH)_2$ and heat the residue (precipitate) to dryness.

$$2Al(OH)_3(s) \xrightarrow{\text{heat}} Al_2O_3(s) + 3H_2O(g)$$
$$\text{white}$$

$$Zn(OH)_2(s) \xrightarrow{\text{heat}} ZnO(s) + H_2O(g)$$
$$\text{yellow when hot,}$$
$$\text{white when cold}$$

$$Pb(OH)_2(s) \xrightarrow{\text{heat}} PbO(s) + H_2O(g)$$
$$\text{yellow/orange}$$

1 Which cations give distinctive coloured (non-white) precipitates with NaOH(aq)?

2 Can all these five metals be identified from the colour of their hydroxides?

The simple test with sodium hydroxide solution, NaOH(aq), does not easily distinguish between Cr^{3+}(aq) and Ni^{2+}(aq) which both give pale greenish precipitates. The two cations can, however, be distinguished using ammonia solution which contains aqueous NH_3 molecules, undissociated NH_4OH and dissociated NH_4^+ and OH^- ions.

$$NH_3(aq) + H_2O(l) \rightleftharpoons NH_4OH(aq) \rightleftharpoons NH_4^+(aq) + OH^-(aq)$$

When a little ammonia solution is added to solutions of Cr^{3+} and Ni^{2+}, they both give pale greenish precipitates of their hydroxides.

$$Cr^{3+}(aq) + 3OH^-(aq) \rightarrow Cr(OH)_3(s)$$
$$\text{grey–green}$$

$$Ni^{2+}(aq) + 2OH^-(aq) \rightarrow Ni(OH)_2(s)$$
$$\text{pale green}$$

With excess NH_3 solution, the precipitate of $Cr(OH)_3$ remains but that of $Ni(OH)_2$ dissolves to form a green solution of tetraamminenickel(II) ions.

$$Ni(OH)_2(s) + 4NH_3(aq) \rightarrow [Ni(NH_3)_4]^{2+}(aq) + 2OH^-(aq)$$

Two other cations which react in a similar fashion to Ni^{2+}(aq) are Cu^{2+}(aq) and Zn^{2+}(aq).

$$Cu^{2+}(aq) \xrightarrow[\text{NH}_3\text{(aq)}]{\text{little}} Cu(OH)_2(s) \xrightarrow[\text{NH}_3\text{(aq)}]{\text{excess}} [Cu(NH_3)_4]^{2+}(aq)$$

| pale blue solution | pale blue ppte copper hydroxide | dark blue solution tetraamminecopper(II) ion |

$$Zn^{2+}(aq) \xrightarrow[\text{NH}_3\text{(aq)}]{\text{little}} Zn(OH)_2(s) \xrightarrow[\text{NH}_3\text{(aq)}]{\text{excess}} [Zn(NH_3)_4]^{2+}(aq)$$

| clear solution | white ppte zinc hydroxide | clear solution tetraamminezinc(II) ion |

Test for ammonium ions

A more positive test for ammonium ions, NH_4^+, is to add an equal volume of sodium hydroxide solution to the solution of NH_4^+ ions and then heat the mixture. If NH_4^+ ions are present, they react with OH^- ions to produce ammonia which is released as a gas.

$$NH_4^+(aq) + OH^-(aq) \rightarrow NH_3(g) + H_2O(l)$$
$$\text{ammonia}$$

Ammonia has a distinct smell and turns damp red litmus paper blue.

19.16 Recycling

When metal articles are finished with or worn out, the metal can be reclaimed and used again. This is called **recycling**.

Our supplies of metal ores will not last for ever. These ores, like coal, oil and natural gas, are a **finite resource** and geologists can estimate how long the known **reserves** of these materials will last (figure 19.6).

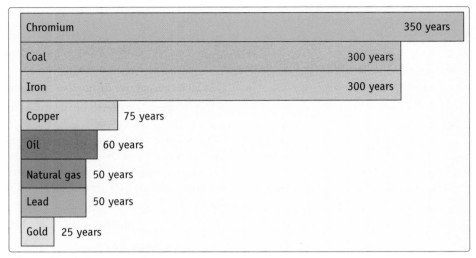

Figure 19.6
How long will the reserves of different materials last?

Recycling is important because:

- **It saves money**. The high costs of extracting aluminum by electrolysis and concentrating low grade tin ores makes the recycling of these two metals very economical and worthwhile.
- **It saves energy and fuels**. Think of the enormous amounts of electricity needed to extract aluminium and the amount of coke used to extract iron.
- **It saves the environment** because we dig up less metal ore, mine less coal and drill for less oil.
- **It saves the reserves of materials** for the future.
- **It solves the problem of waste disposal**. Recycling metal articles prevents them causing litter.

Sometimes recycling is not economical because of the cost of sorting, collecting and processing the waste material. The higher the value of the material, the more economical it is to recycle. Thus, virtually all gold is recycled, but only 40% of aluminium.

Recycling plastics presents a particular problem because of the difficulty of identifying the type of polymer used. It is easy to separate iron from copper in recycling, but pvc and polythene are much more difficult to distinguish and separate. One way round this is to indicate the type of polymer by means of a special symbol (figure 19.7).

Figure 19.7
These recycling symbols are already used on some plastic bottles, particularly in the USA.
PET stands for polyethene terephthalate (polyester).
HDPE stands for high density polyethene and V stands for polyvinyl chloride.
What do you think LDPE, PP and PS stand for?

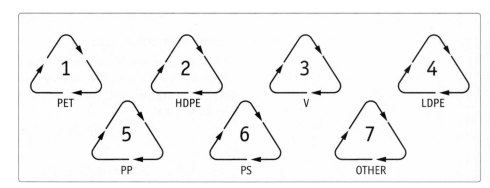

19 Metals and the Activity Series

Summary

1 The world production of iron far exceeds that of any other metal. The reasons for this are that iron ores are abundant, rich, readily accessible and easily reduced to iron.

2 Iron can be alloyed easily with other elements to produce different steels with a great variety of properties and uses.

3 In many respects, iron is a typical transition metal. It produces coloured ions. It forms stable complexes. It shows variable oxidation state. The element and its compounds show catalytic activity.

4 Rusting requires the presence of both oxygen and water. During rusting, iron is first oxidised to Fe^{2+} ions by dissolved oxygen in the water. Iron(II) hydroxide then precipitates. This is further oxidised by dissolved oxygen to rust (hydrated iron(III) oxide, $Fe_2O_3.xH_2O$).

5 Two general methods are used to protect iron and steel from corrosion: the application of a protective layer or the application of a sacrificial metal.

6 The manufacture of copper from its ores involves three stages.
   (a) Conversion of the sulphide ore to Cu_2O by roasting in air.
   (b) Reduction of the Cu_2O to copper by heating with more sulphide ore in the absence of air.
   (c) Purification of the impure copper by electrolysis.

7 Aluminium is obtained industrially by the electrolysis of pure aluminium oxide in molten cryolite. The pure oxide is obtained from bauxite.

8 Aluminium is a reactive metal, but its reactivity is suppressed by a thin coating of inert oxide.

9 Metals at the top of the electrochemical series (e.g. K, Na, Ca, Mg, Al) are usually obtained by electrolysis of their fused compounds.

   Metals in the middle of the electrochemical series (e.g. Zn, Fe, Sn, Pb, Cu) are usually obtained by chemical reduction of their oxides with carbon or carbon monoxide or by self-reduction of sulphide ores.

   Metals at the bottom of the electrochemical series (e.g. Cu, Hg, Ag, Au) are usually obtained by heating the ore alone or by displacement from aqueous solution.

10 Many common cations can be identified from the colour they produce during flame tests and/or the colour and solubility of their hydroxides.

11 Recycling is important in our conservation of finite resources of fossil fuels and metal ores.

Review questions

1 (a) What do you understand by the term 'transition metal'?
  (b) Which of the following do you regard as transition metals? Explain your answer.
     (i) Scandium
     (ii) Iron
     (iii) Zinc
  (c) Although the salts of transition elements are usually coloured, there are several copper(I) compounds which are white. Suggest an explanation for this.
  (d) The densities of transition elements in the same period gradually increase with relative atomic mass (table 18.3). Why is this?

2 Comment upon and explain the following observations.
  (a) An iron pipe buried in the earth is protected from rusting by joining it to a bar of magnesium.
  (b) A transition metal, M, forms only two isomers of formula $[M(NH_3)_4Cl_2]^+Cl^-$.
  (c) Copper does not react appreciably with hydrochloric acid, but in the presence of strong complexing agents, it will do so, evolving hydrogen steadily.
  (d) Iron immersed in copper(II) sulphate solution is rapidly coated with copper. This does not occur when the iron has previously been dipped in concentrated nitric acid.

3 Metals are usually isolated from their ores by reduction. Describe briefly the methods of reduction for the following metals. Suggest reasons for the method chosen for each metal.
  (a) Zinc from zinc blende (ZnS)
  (b) Magnesium from magnesium chloride extracted from sea water
  (c) Iron from haematite (Fe_2O_3)
  (d) Silver from silver glance (Ag_2S)

4 10 g of an impure iron(II) salt were dissolved in water and made up to 200 cm^3 of solution. 20 cm^3 of this solution acidified with dilute H_2SO_4 required 25 cm^3 of 0.03 mol dm^{-3} $KMnO_4$(aq) before a faint pink colour appeared.
  (a) Write a balanced equation (or half-equations) for the reaction of acidified manganate(VII) ions with iron(II) ions.
  (b) How many moles of iron(II) ions react with 1 mole of MnO_4^- ions?
  (c) How many moles of Fe^{2+} react with 25 cm^3 of 0.03 mol dm^{-3} $KMnO_4$(aq)?

(d) How many grams of Fe^{2+} are there in the 200 cm^3 of original solution? (Fe = 56)

(e) What is the percentage by mass of iron in the impure iron(II) salt?

5 (a) Most metals exist naturally in a combined state. They are too reactive to occur native (uncombined). However, a few metals do occur native. Give examples of:
(i) two metals which never occur native,
(ii) two metals which occur both native and combined,
(iii) two metals which almost always occur naturally in an uncombined form.

(b) Before extraction of the metal can begin, ores must be purified and concentrated. Mention two different processes by which this is done.

(c) Why do metal extractions often produce slag?

(d) In many cases, the primary reduction of an ore does not produce metal of sufficient purity. The metal requires further purification. Outline two different processes by which further purification is carried out.

6 Economic considerations are important in deciding the method by which a metal is extracted from its ore. How may economic considerations affect:
(a) the nature of the ore used,
(b) the nature of the reducing agent used,
(c) the location of the industrial plant at which extraction is carried out?
Give examples wherever possible to illustrate the point you make.

7 The method by which any metal is extracted is dependent upon the position of that metal in the electrochemical series.

(a) Highly reactive metals must be extracted by electrolysis.
(i) Give a brief outline of the extraction process for one such metal. Mention the electrolyte used, the materials of which the electrodes are made and the reactions taking place at each electrode.
(ii) Why are such reactive metals not obtained by chemical methods of reduction?

(b) A few metals are obtained by reduction of their purified ores with a second more reactive metal.
(i) Outline the chemical principles of one such extraction process.
(ii) Why is this type of process fairly uncommon?

(c) Less reactive metals can be extracted by reduction of their ores with carbon.
(i) Outline the chemical principles of the extraction of two different metals using carbon.
(ii) What are the disadvantages of reduction with carbon?

Equilibria

20.1 Introduction

Have you ever been in the difficult position of having your loyalty and responsibility pulled in two directions at the same time? If you have, you will know that very often you must settle for a compromise (or balance) between the alternatives.

In the same way, many physical and chemical processes also exist in a position of balance. In chapter 14, we studied the competition processes taking place in acid–base, redox and complexing systems. In these and other reactions, the starting materials are not always completely converted to the products. Sometimes, they reach an intermediate position or **equilibrium** in which both reactants and products are present.

In real life, compromise in difficult circumstances is usually dictated by several factors – responsibility, loyalty, honesty, age, parents, finance, etc. Fortunately, the factors which dictate the balance position in chemical reactions are fewer and less complicated than those in real-life situations.

► The tendency for materials to exist in a state of disorder is a key factor in dictating the position of equilibrium. Disorder is, in fact, a more natural state than order. Even so, it may be difficult to convince your parents of this when you are next told to tidy your room.

20.2 Equilibria in physical processes

Liquid–vapour equilibrium: vapour pressure

When liquid bromine is shaken in a stoppered flask, some of the liquid evaporates, forming an orange gas. Gradually, the gas becomes thicker. Eventually, the intensity of the gas does not change any more. However much we shake the flask, its colour remains constant. Provided we took more than a tiny volume of liquid bromine, some liquid remains in the flask. The constant colour of the gas in the flask suggests that a position of balance has been reached. Some of the bromine has formed a vapour and some of it remains as a liquid. A position of equilibrium has been reached between bromine liquid and bromine gas. This equilibrium can be summarised as:

$$Br_2(l) \rightleftharpoons Br_2(g)$$

The equilibrium sign ($\rightleftharpoons$) is used to show that both bromine liquid and bromine gas are present in the flask. But, what is happening to the molecules in the flask at equilibrium? Do all the gas molecules remain as gas while all the liquid molecules remain as liquid (a **static equilibrium**)? Or, are some gas molecules becoming liquid while an equal number of liquid molecules become gas (a **dynamic equilibrium**)?

Experiments show that liquid and gas molecules move around rapidly and randomly. So, it would seem likely that molecules in the flask are in a dynamic rather than a static equilibrium. Thus, the rate at which molecules leave the liquid surface and enter the vapour is equal to the rate at which other molecules in the vapour return to the liquid. Random molecular activity occurs even after all the external signs of change have disappeared. But, we cannot see the molecules moving. Nor can we measure the rate at which they enter or leave the vapour.

The differences between a static and a dynamic equilibrium are illustrated very neatly in figure 20.1.

▲ One of the key factors which dictates the position of equilibrium in any physical or chemical process is the tendency for materials to exist in the lowest energy state. Will equilibrium ever be achieved in the waterfall?

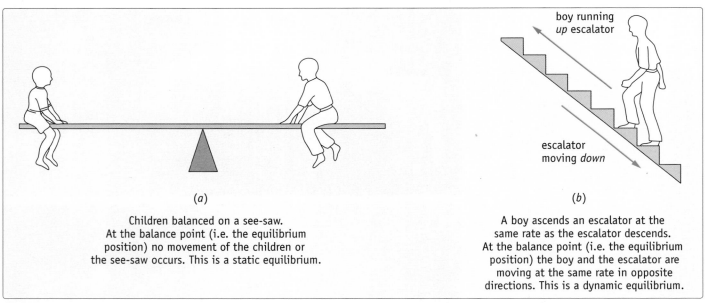

(a)

Children balanced on a see-saw. At the balance point (i.e. the equilibrium position) no movement of the children or the see-saw occurs. This is a static equilibrium.

(b)

boy running *up* escalator

escalator moving *down*

A boy ascends an escalator at the same rate as the escalator descends. At the balance point (i.e. the equilibrium position) the boy and the escalator are moving at the same rate in opposite directions. This is a dynamic equilibrium.

Figure 20.1
The difference between a static and a dynamic equilibrium.

Consider the equilibrium between bromine liquid and bromine vapour once again. Figure 20.2(*a*) shows a perfect balance between the rate of evaporation and the rate of condensation at equilibrium.

Now, suppose that some of the vapour is suddenly removed without affecting the system in any other way (figure 20.2(*b*)).

1 Will the rate of condensation of gas molecules still equal the rate of evaporation of liquid molecules? Explain.

Figure 20.2
Liquid–vapour equilibria.

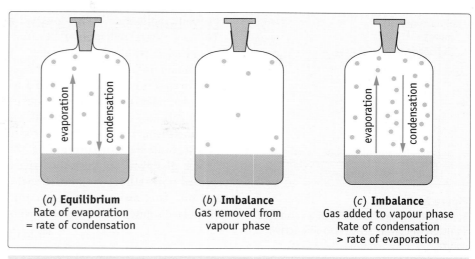

(a) **Equilibrium**
Rate of evaporation
= rate of condensation

(b) **Imbalance**
Gas removed from
vapour phase

(c) **Imbalance**
Gas added to vapour phase
Rate of condensation
> rate of evaporation

2 What happens to the concentration of molecules in the gas phase as time goes on? Explain.

3 How and when is equilibrium attained once more?

Suppose, now, the equilibrium is disturbed by injecting an excess of bromine vapour into the flask (figure 20.2(c)). The concentration of gas molecules suddenly rises and the rate of condensation increases. Condensation occurs faster than evaporation. Thus the concentration of gas molecules in the vapour phase decreases. Gradually, the rate of condensation falls until finally the rate of condensation equals the rate of evaporation. The system is in equilibrium once more and the vapour pressure of the bromine is the same as it was before the equilibrium was disturbed.

Notice that equilibrium can be reached from either direction. We can start with either bromine liquid or with excess bromine vapour. *Provided we have a closed container,* equilibrium will eventually be obtained.

Solute–solution equilibrium: solubility

When one teaspoon of sugar is added to a cup of tea, all the sugar dissolves forming a solution. As more sugar is added, this also dissolves at first, but a stage is eventually reached when no more will dissolve. The solution (tea) is now saturated with solute (sugar) at the temperature involved. Solute particles in the undissolved sugar are in equilibrium with solute particles in the solution.

$$sugar(s) \rightleftharpoons sugar(aq)$$

Provided the system is closed, no solvent can escape and the amounts of dissolved and undissolved solute remain constant. As in the liquid–vapour equilibrium, macroscopic properties have become constant. In this case, the constant macroscopic properties are the concentration of the solution and the amount of undissolved solute.

Investigating the nature of equilibrium

What are the molecules doing when equilibrium is established between a solute and its saturated solution? No changes are apparent at a macroscopic level, but what is happening at a molecular (microscopic) level?

It would seem reasonable to predict that the equilibrium in this case is also dynamic. This would involve particles leaving the undissolved solute and entering the solution at the same rate as dissolved particles rejoin the solid.

In order to check these ideas about a dynamic equilibrium, we can 'label' some of the particles in either the saturated solution or in the undissolved solute. This is done by using a radioactive isotope.

If solid radioactive $^{212}PbCl_2$ is added to a saturated solution of non-radioactive $PbCl_2(aq)$, no increase in the amount of dissolved lead chloride can occur. Nevertheless, if our ideas about a dynamic equilibrium are correct, there will be an interchange of Pb^{2+} and Cl^- ions between the undissolved solid and the saturated solution.

▲When the atmosphere is saturated with water vapour, water is in equilibrium with its vapour.

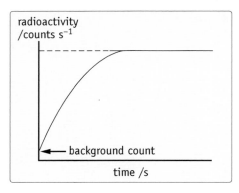

Figure 20.3
Following the radioactivity of a saturated solution in contact with its radioactive solute.

☐ Figure 20.3 shows that the radioactivity of the saturated solution eventually reaches a constant value. Why is this?

As a result, radioactive $^{212}Pb^{2+}$ ions will appear in the solution.

The background radiation of the saturated aqueous lead chloride is first measured. It is then shaken with radioactive solid $^{212}PbCl_2$ for 10 minutes. After this, the saturated solution and undissolved solute are separated by centrifuging and the radioactivity of the saturated solution is measured a second time. The mixing and centrifuging are repeated at intervals and the radioactivity of the saturated solution is measured at each stage. Figure 20.3 shows how the radioactivity of the solution changes with time.

The increasing radioactivity of the saturated solution suggests that solid radioactive $^{212}PbCl_2$ must have dissolved. But, the solution was already saturated before any $^{212}PbCl_2$ dissolved. So, the radioactive particles dissolving from the solid must have replaced non-radioactive particles which have crystallised onto the solid from the solution. This means that a dynamic equilibrium is taking place between the undissolved solute and the saturated solution:

$$PbCl_2(s) \rightleftharpoons PbCl_2(aq)$$

20.3 Characteristic features of a dynamic equilibrium

The last section has highlighted four important features of dynamic equilibria.

1 *At equilibrium, macroscopic properties are constant* under the given conditions of temperature, pressure and initial amounts of substances.

2 *At equilibrium, microscopic (molecular-scale) processes continue but these are in balance. This means that no overall macroscopic (large-scale) changes occur.* The particles participate in both forward and reverse processes. The rate of the forward process is equal to the rate of the reverse process so that no net change results.

3 *The equilibrium can be attained from either direction* beginning with only the materials on one side of the change. Changes of this kind are described as **reversible**.

4 *Equilibrium can only be achieved in a closed system.* A closed system is one in which there is no loss or gain of materials to or from the surroundings. An open system may allow matter to escape or to enter. This cannot reach equilibrium.

20.4 Equilibria in chemical reactions

The decomposition of calcium carbonate

When calcium carbonate is heated strongly, it decomposes forming calcium oxide and carbon dioxide:

$$CaCO_3(s) \xrightarrow{\text{heat}} CaO(s) + CO_2(g)$$

Normally, CO_2 escapes well away from the solid CaO, so their recombination to form $CaCO_3$ never occurs in an open container. The system can never reach equilibrium.

If, however, a few grams of $CaCO_3(s)$ are heated at 800°C in a *closed* evacuated container, only part of the solid is decomposed no matter how long it is heated. The pressure of CO_2 inside the container rises, and then remains steady at a constant value. As long as the temperature stays at 800°C, the pressure remains constant at 25 kPa (0.25 atm). The reaction appears to have reached an equilibrium. Constant macroscopic properties have been achieved in a closed system.

We can check that the system is at equilibrium by showing that it reaches the same macroscopic composition when approached from the opposite direction.

Solid CaO is first placed in the reaction vessel. This is then evacuated and refilled with CO_2 at a pressure well above 25 kPa. Finally, the container is closed and maintained at 800°C. The pressure begins to fall and becomes steady at 25 kPa (0.25 atm). The same equilibrium pressure of CO_2 results whether we start with $CaCO_3$ or with CaO and CO_2. This is further evidence that the materials are in equilibrium. This equilibrium can be summarised as

$$CaCO_3(s) \rightleftharpoons CaO(s) + CO_2(g)$$

A modern rotatory lime kiln. The kiln is viewed from the lower heated end. Raw limestone slurry is pumped in at the higher end. As the slurry flows down the kiln, it is first dried and then decomposed to lime at the lower heated end. The kiln is inclined at 1 in 30. It makes one revolution per minute.

In this case, $CaCO_3$ is decomposing to form $CaO + CO_2$ at the same rate as $CaO + CO_2$ are reforming $CaCO_3$.

Although this reaction will come to equilibrium in a closed vessel, the industrial process is never allowed to do so. Millions of tonnes of calcium oxide are produced annually by heating limestone in large open kilns. The calcium oxide is used in liming the soil and as a constituent of plaster and mortar.

The iodine chloride–iodine trichloride equilibrium

When dry chlorine is passed over brown liquid iodine chloride (ICl), yellow crystals of iodine trichloride (ICl_3) form and the brown liquid disappears:

$$\begin{array}{ccccc} Cl_2(g) & + & ICl(l) & \rightarrow & ICl_3(s) \\ \text{pale green} & & \text{brown} & & \text{yellow} \end{array}$$

If the supply of chlorine is stopped, the ICl_3 decomposes if it is left in an open container. The ICl_3 decomposes to a brown liquid (ICl) and a pale green gas (Cl_2):

$$\begin{array}{ccccc} ICl_3(s) & \rightarrow & ICl(l) & + & Cl_2(g) \\ \text{yellow} & & \text{brown} & & \text{pale green} \end{array}$$

This is another example of a reversible reaction. But what happens when a mixture of $Cl_2(g)$ and ICl(l) is left in a closed vessel from which the air has been removed? Yellow crystals of ICl_3 form, but not all the reactants are used up. As the reaction proceeds, the amounts of ICl and Cl_2 fall and the amount of ICl_3 rises (figure 20.4).

After some time, the composition of the mixture becomes constant and the amount of each substance remains steady. The reaction seems to have reached an equilibrium because the macroscopic properties of the closed system have become constant.

1 What is happening to the molecules of ICl_3, ICl and Cl_2 inside the container at equilibrium?

2 How is the rate of formation of ICl_3 related to the rate of decomposition of ICl_3 at equilibrium?

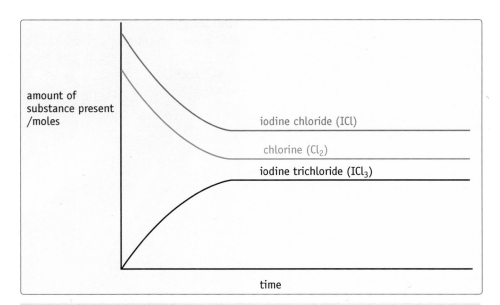

Figure 20.4
A sketch graph showing the amounts of substances present as equilibrium is approached in the reaction between ICl and Cl_2.

Now look closely at figure 20.4

3 What is the relationship between the decrease in amount of ICl, the decrease in the amount of Cl_2 and the increase in amount of ICl_3?

4 Why is there always a constant difference between the amounts of ICl and Cl_2?

5 Why do the three curves become horizontal at the same time?

20.5 The equilibrium of a solute between two immiscible solvents – the partition coefficient

The following experiment investigates the equilibrium of a solute between two immiscible solvents. Dissolve a small crystal of iodine in 5 cm³ of aqueous potassium iodide. A brown solution forms. Then carefully add 5 cm³ of tetrachloromethane to the mixture. The tetrachloromethane is clear at first, but when the mixture is shaken gently, iodine dissolves from the KI solution into the tetrachloromethane. A purple solution of iodine forms in the latter. Now shake vigorously. The density of the purple colour in the tetrachloromethane darkens whilst the brown colour of the KI(aq) becomes paler. Eventually, the colours of the two solutions remain constant. The iodine has distributed itself between the two solvents and an equilibrium has been attained. No matter how much the mixture is shaken, no further changes in colour intensity occur. The concentrations of iodine in the two solutions remain constant.

Other solutes which are soluble in two immiscible solvents will also distribute themselves between both solvents when the three substances are shaken together. These are further examples of equilibrium systems. But, how do the concentrations of the solute in the different solvents depend on the initial amount of solute taken? Is there a relationship between the equilibrium concentrations of the solute in the two solvents?

Investigating the partition of butanedioic acid (succinic acid) between water and ethoxyethane (diethyl ether)

One gram of butanedioic acid was added to a separating funnel containing 25 cm³ of water and 25 cm³ of ethoxyethane (ether). This was shaken until all the solid had dissolved and equilibrium had been established. The mixture was then left to stand for some time. This allowed the organic and aqueous layers to separate as fully as possible.

The two layers were then separated. The concentration of butanedioic acid in each layer was determined by titration against sodium hydroxide solution of known molarity using phenolphthalein indicator.

Further experiments were carried out using different initial amounts of butanedioic acid between 0.5 and 1.5 grams. The results obtained are shown in table 20.1.

Table 20.1
Equilibrium concentrations of butanedioic acid in ether and in water

Experiment number	Equilibrium concentration of butanedioic acid in ether layer/mol dm^{-3}	Equilibrium concentration of butanedioic acid in water layer/mol dm^{-3}
1	0.023	0.152
2	0.028	0.182
3	0.036	0.242
4	0.044	0.300
5	0.052	0.358
6	0.055	0.381

When these equilibrium concentrations in table 20.1 are plotted graphically, a straight line is obtained (figure 20.5).

1. Would you expect the graph to go through the origin? Explain.
2. What does the straight line graph suggest about the ratio of the equilibrium concentrations of butanedioic acid in the two solvents?
3. How does the ratio of concentrations depend upon the amount of butanedioic acid taken?

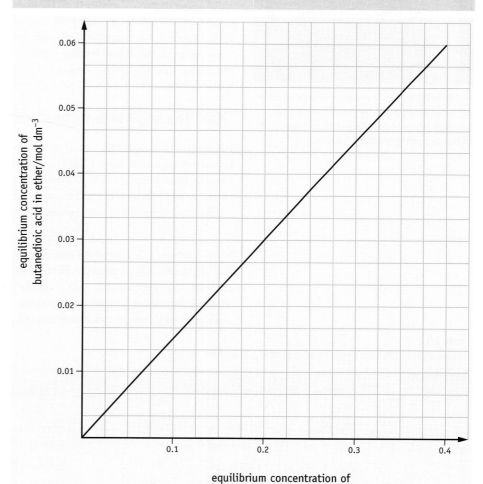

Figure 20.5
A graph of the equilibrium concentration of butanedioic acid in ether against that in water.

The results in figure 20.5 show that the ratio:

$$\frac{\text{concentration of butanedioic acid in ether}}{\text{concentration of butanedioic acid in water}} \text{ is constant.}$$

Using square brackets, [ ], to denote the concentrations, we can deduce from the graph in figure 20.5 that

$$\frac{[\text{butanedioic acid (ether)}]_{eq}}{[\text{butanedioic acid (water)}]_{eq}} = 0.15$$

This ratio is known as the **partition coefficient** (or the **distribution ratio**) for butanedioic acid between ether and water. *The partition coefficient is independent of the amount of solute taken and also independent of the volumes of the solvents used.* Notice that it is *the ratio of concentrations and not the ratio of masses* of solute which matter. Other investigations of the distribution of solutes between immiscible liquids at equilibrium give similar results. In each case, the partition coefficient is constant provided:

(i)   the temperature is constant,
(ii) the solvents are immiscible and do not react with each other,
(iii) the solute does not react, associate or dissociate in the solvents.

20.6 Solvent extraction

The most important application of partition is in solvent extraction. Organic compounds are generally more soluble in non-polar solvents such as methylbenzene (toluene), tetrachloromethane and ethoxyethane (ether) than in water. These solvents are themselves immiscible with water. Thus, organic compounds can be extracted from aqueous solutions or suspensions by shaking with a non-polar organic solvent and then separating the two layers. The pure organic compound can then be obtained by distilling off the solvent. For example, penicillin is extracted from a dilute aqueous solution using trichloromethane. Phenylamine (aniline) can be reclaimed from a mixture with water by using ether. Iodine can be extracted from an aqueous mixture using a hydrocarbon solvent such as cyclohexane.

▲ *Penicillium* mould growing in a petri dish on a nutrient jelly. The penicillin can be washed into aqueous solution and then extracted using trichloromethane.

20.7 The equilibrium constant

Earlier in this chapter we obtained a constant value for the ratio of concentrations of a solute between two immiscible solvents at equilibrium. Thus, it would seem sensible to investigate whether there is a similar relationship between the concentrations of reactants and products at equilibrium in chemical reactions.

Table 20.2 shows information concerning the following reaction at 731 K:

$$H_2(g) + I_2(g) \rightleftharpoons 2HI(g)$$

In experiments 1, 2, 3 and 4 the sealed reaction vessel contains gaseous hydrogen and gaseous iodine initially. After a time the mixture in the flask remains unchanged and equilibrium is attained. In experiments 5 and 6, equilibrium is approached from the opposite direction. The reaction vessel contains only gaseous hydrogen iodide initially. Table 20.2 shows the initial and equilibrium concentrations of hydrogen, iodine and hydrogen iodide.

Table 20.2
Initial and equilibrium concentrations of H_2, I_2 and HI

Experiment number	Initial concentrations/mol dm^{-3}			Equilibrium concentrations/mol dm^{-3}		
	$[H_2(g)]$	$[I_2(g)]$	$[HI(g)]$	$[H_2(g)]_{eq}$	$[I_2(g)]_{eq}$	$[HI(g)]_{eq}$
1	2.40×10^{-2}	1.38×10^{-2}	0	1.14×10^{-2}	0.12×10^{-2}	2.52×10^{-2}
2	2.40×10^{-2}	1.68×10^{-2}	0	0.92×10^{-2}	0.20×10^{-2}	2.96×10^{-2}
3	2.44×10^{-2}	1.98×10^{-2}	0	0.77×10^{-2}	0.31×10^{-2}	3.34×10^{-2}
4	2.46×10^{-2}	1.76×10^{-2}	0	0.92×10^{-2}	0.22×10^{-2}	3.08×10^{-2}
5	0	0	3.04×10^{-2}	0.345×10^{-2}	0.345×10^{-2}	2.35×10^{-2}
6	0	0	7.58×10^{-2}	0.86×10^{-2}	0.86×10^{-2}	5.86×10^{-2}

1 According to the equation

1 According to the equation

$$H_2(g) + I_2(g) \rightleftharpoons 2HI(g)$$

one mole of H_2 reacts with one mole of I_2 forming two moles of HI. Check the data from experiments 1 and 2 to see whether:

no. of moles H_2 reacted = no. of moles I_2 reacted = $\frac{1}{2} \times$ no. of moles HI formed.

2 Why is $[H_2(g)]_{eq} = [I_2(g)]_{eq}$ in experiments 5 and 6?

3 How could you show that the system involving H_2, I_2 and HI is in dynamic equilibrium?

Comparing the reaction $H_2(g) + I_2(g) \rightleftharpoons 2HI(g)$ with a partition equilibrium, it is reasonable to calculate the ratio of product and reactant concentrations as

$$\frac{[HI(g)]_{eq}}{[H_2(g)]_{eq}[I_2(g)]_{eq}}$$

For each set of equilibrium concentrations, we want to see if the expression is constant. The computed values are shown in the second column of table 20.3. The ratios are far from constant.

If, however, we calculate values for the ratio

$$\frac{[HI(g)]_{eq}^2}{[H_2(g)]_{eq}[I_2(g)]_{eq}}$$

we obtain the results in column three of table 20.3. The values obtained this time are constant, within experimental limits. Therefore, we can write

$$\frac{[HI(g)]_{eq}^2}{[H_2(g)]_{eq}[I_2(g)]_{eq}} = \text{a constant} = 46.8 \text{ at } 731 \text{ K}$$

Notice that the ratio of concentrations which gives a constant value is related to the number of moles of each substance in the balanced equation.

$$H_2(g) + I_2(g) \rightleftharpoons 2HI(g)$$

The ratio uses the *second* power of the [HI] in the equilibrium expression and the first power for both $[H_2]$ and $[I_2]$. Thus, *the power to which we raise the concentration of a substance in the equilibrium expression is the same as its coefficient in the balanced equation.*

Table 20.3
Possible equilibrium constants for the reaction $H_2 + I_2 \rightleftharpoons 2HI$

Experiment number	$\dfrac{[HI(g)]_{eq}}{[H_2(g)]_{eq}[I_2(g)]_{eq}}$	$\dfrac{[HI(g)]_{eq}^2}{[H_2(g)]_{eq}[I_2(g)]_{eq}}$
1	1840	46.4
2	1610	47.6
3	1400	46.7
4	1520	46.9
5	1970	46.4
6	790	46.4

The value of this ratio of concentrations at equilibrium is known as the **equilibrium constant**. The equilibrium constant is represented by the symbol K_c. Thus, for the reaction $H_2(g) + I_2(g) \rightleftharpoons 2HI(g)$

$$K_c = \frac{[HI(g)]_{eq}^2}{[H_2(g)]_{eq}[I_2(g)]_{eq}}$$

Very often, the 'eq' subscripts are omitted from the concentration terms because it is assumed that the concentrations in the expression for K_c are equilibrium ones. Hence, the expression for K_c is shortened to

$$K_c = \frac{[HI(g)]^2}{[H_2(g)][I_2(g)]}$$

The subscript 'c' indicates that K_c is expressed in concentrations.

20.8 The equilibrium law

The equilibria in many other chemical reactions have also been studied. In each case the equilibrium constant relates to the balanced equation in a similar manner to that for the hydrogen, iodine and hydrogen iodide system.

For example, the reaction

$$Fe^{3+}(aq) + NCS^-(aq) \rightleftharpoons Fe(NCS)^{2+}(aq)$$

has been studied by colorimetry. The reaction has an equilibrium constant in which

$$K_c = \frac{[Fe(NCS)^{2+}(aq)]}{[Fe^{3+}(aq)][NCS^-(aq)]}$$

> These observations lead to the general statement known as the **Equilibrium Law** or the **Law of Chemical Equilibrium**.
>
> If an equilibrium mixture contains substances A, B, C and D related by the equation
>
> $$aA + bB \rightleftharpoons cC + dD$$
>
> it is found experimentally that
>
> $$\frac{[C]^c[D]^d}{[A]^a[B]^b} = K_c$$
>
> K_c, the equilibrium constant, is constant at a given temperature.

In writing expressions for the equilibrium constants of reactions, there are important conventions. Concentrations of substances on the right-hand side of the equation are written in the numerator. Concentrations of substances on the left-hand side are written in the denominator.

Thus, it is essential to relate any numerical value for an equilibrium constant to the particular equation concerned. Suppose the equilibrium constant for the reaction $H_2(g) + I_2(g) \rightleftharpoons 2HI(g)$

$$K_c = \frac{[HI]^2}{[H_2][I_2]} = x$$

The equilibrium constant for the reverse reaction $2HI \rightleftharpoons H_2 + I_2$ at the same temperature is

$$K'_c = \frac{[H_2][I_2]}{[HI]^2} = \frac{1}{x}$$

As you would expect, $K'_c = \dfrac{1}{K_c}$

On the other hand, the equilibrium constant for the reaction

$$\tfrac{1}{2}H_2(g) + \tfrac{1}{2}I_2(g) \rightleftharpoons HI(g)$$

$$\text{is } K''_c = \frac{[HI]}{[H_2]^{\frac{1}{2}}[I_2]^{\frac{1}{2}}} = \sqrt{x}$$

1 What is the relation between K_c and K''_c?

K_c has no units in reactions with equal numbers of particles on both sides of the equation. This is because the concentration units cancel out in the expression for K_c, i.e. for the reaction

$$H_2(g) + I_2(g) \rightleftharpoons 2HI(g)$$

$$K_c = \frac{[HI(g)]^2}{[H_2(g)][I(g)]} \qquad \text{units} = \frac{(mol\,dm^{-3})^2}{(mol\,dm^{-3})(mol\,dm^{-3})}$$

For reactions in which the numbers of reactant and product particles are not equal, K_c will, of course, have units.

What are the units of K_c for the following reactions?

1 $N_2(g) + 3H_2(g) \rightleftharpoons 2NH_3(g)$,
2 $2NH_3(g) \rightleftharpoons N_2(g) + 3H_2(g)$,
3 $2NO(g) + O_2(g) \rightleftharpoons 2NO_2(g)$,
4 $NO(g) + \tfrac{1}{2}O_2(g) \rightleftharpoons NO_2(g)$.

Determination of equilibrium constants

The essential stages in determining an equilibrium constant are listed below.

(i)   Write the balanced equation.
(ii) Mix known molar amounts of either the reactants or the products.
(iii) Allow the mixture to reach equilibrium.
(iv) Determine the equilibrium concentration of at least one substance in the equilibrium mixture. This analysis might be carried out using colorimetry, pressure measurements, titration or some other method.
(v)   Deduce the equilibrium concentrations of the other materials in the mixture.
(vi) Substitute the equilibrium concentrations in the expression for K_c.
(vii) Repeat the determination of K_c using different initial concentrations.

In determining and using equilibrium constants it is important to realise the following points:

- *The Equilibrium Law only applies to systems at equilibrium.*
- K_c *is constant only so long as the temperature remains constant.* If the temperature changes, the value of K_c will change.
- *The numerical value of* K_c *is unaffected by changes in concentration of either reactants or products.* Obviously, when more reactant is suddenly added to a system in equilibrium, more of the products will tend to form. Eventually the system will adjust itself at a new equilibrium position in which the concentrations of reactants and products give the same numerical value for K_c.
- *The magnitude of* K_c *provides a useful indication of the extent of a chemical reaction.* A large value for K_c indicates a high proportion of products to reactants (i.e. an almost complete reaction). A low value for K_c indicates that only a small fraction of reactants have been converted to products.
- *The equilibrium constant for a reaction indicates the extent of a reaction. It gives no information about the rate of reaction.* K_c tells us *how far*, but *not how fast* the reaction goes. In fact, the extent and the rate of a reaction are quite independent. For example, the conversion of sulphur dioxide and oxygen to sulphur trioxide at 450°C occurs very slowly but almost completely. In contrast, the conversion of nitrogen oxide and oxygen to nitrogen dioxide at the same temperature occurs rapidly but only partially.

The following worked example will help you to understand the ideas we have covered so far. It will show you how equilibrium constants and equilibrium concentrations can be calculated.

Calculation of an equilibrium constant

When 1 mole of hydrogen iodide is allowed to dissociate in a 1.0 dm^3 vessel at 440°C, only 0.78 moles of HI are present at equilibrium. What is the equilibrium constant at this temperature for the reaction, $2HI(g) \rightleftharpoons H_2(g) + I_2(g)$?

	2HI $\rightleftharpoons$	H$_2$ +	I$_2$	
No. of moles initially	1	0	0	
No. of moles at equilibrium	0.78	0.11	0.11	Since 0.22 moles of HI have decomposed, 0.11 moles of H$_2$ and 0.11 moles of I$_2$ have formed at equilibrium
Concentration at equilibrium/mol dm^{-3}	$\dfrac{0.78}{1}$	$\dfrac{0.11}{1}$	$\dfrac{0.11}{1}$	

$$K_c = \frac{[H_2(g)][I_2(g)]}{[HI(g)]^2} = \frac{0.11 \times 0.11}{(0.78)^2} = \frac{1}{50} = 0.02$$

$$\Rightarrow K_c \text{ for the reaction} = 0.02$$

Calculation of concentrations at equilibrium

Suppose 2 moles of hydrogen and 1 mole of iodine are mixed together in a 1.0 dm^3 vessel at 440°C. How many moles of HI, H_2 and I_2 will be present at equilibrium?

$$2HI \rightleftharpoons H_2 + I_2$$

	2HI	H_2	I_2	
No. of moles initially	0	2	1	
No. of moles at equilibrum	$2x$	$2-x$	$1-x$	We have assumed that $2x$ moles of HI have formed at equilibrium. Hence, x moles of H_2 and x moles of I_2 must have disappeared.
Concentration at equilibrium/mol dm^{-3}	$\dfrac{2x}{1}$	$\dfrac{2-x}{1}$	$\dfrac{1-x}{1}$	

$$\Rightarrow K_c = \frac{(2-x)\ (1-x)}{(2x)^2} = 0.02$$

$$\therefore \frac{2 - 3x + x^2}{4x^2} = 0.02$$

$$\Rightarrow 0.92x^2 - 3x + 2 = 0$$

Solving this quadratic equation using the formula,

$$x = \frac{-b \pm \sqrt{b^2 - 4ac}}{2a}$$

we get

$$x = 0.935$$

Hence, at equilibrium:
number of moles of HI $= 2x = 1.870$
number of moles of $H_2 = (2 - x) = 1.065$
number of moles of $I_2 = (1 - x) = 0.065$

See if you can now answer the following questions.

1 The equilibrium constant for the reaction

$$2NO_2(g) \rightleftharpoons N_2O_4(g)$$

at 298 K is 200 mol^{-1} dm^3.

(a) Write an expression for the equilibrium constant for the reaction.
(b) If the concentration of $N_2O_4(g)$ in the equilibrium mixture at 298 K is 2×10^{-2} mol dm^{-3}, what is the concentration of $NO_2(g)$?
(c) Calculate the equilibrium constant at 298 K for the reaction.

$$\tfrac{1}{2}N_2O_4(g) \rightleftharpoons NO_2(g)$$

The answer to part (b) in the question above is 10^{-2} mol dm^{-3}

2 The equilibrium constants for the synthesis of hydrogen chloride, hydrogen bromide and hydrogen iodide at a particular temperature are given below.

	K_c
$H_2(g) + Cl_2(g) \rightleftharpoons 2HCl(g)$	10^{17}
$H_2(g) + Br_2(g) \rightleftharpoons 2HBr(g)$	10^9
$H_2(g) + I_2(g) \rightleftharpoons 2HI(g)$	10

(a) What do the values of K_c tell you about the extent of each reaction?
(b) Which of these reactions would you regard as virtually complete conversions?

20.9 Equilibrium constants in gaseous systems

Equilibrium constants are normally expressed in terms of concentrations using the symbol, K_c. For reactions involving gases, however, it is usually more convenient to express the amount of gas present in terms of its partial pressure rather than its molar concentration.

Using the ideal gas equation

$$pV = nRT$$

$$\Rightarrow p = \frac{n}{V} RT$$

In this equation, p is the pressure in atmospheres, n is the number of moles of gas, V is the volume in cubic decimetres and T is the temperature in kelvins. In this case, R, the gas constant, has units of $atm\ dm^3\ K^{-1}\ mol^{-1}$.

$$\therefore p = [gas]RT$$

where [gas] is the concentration of the gas in $mol\ dm^{-3}$. Thus, at a constant temperature, the pressure of a gas is proportional to its concentration, i.e.

$$p \propto [gas]$$

This means that for the equilibrium

$$H_2(g) + I_2(g) \rightleftharpoons 2HI(g)$$

we can write either

$$K_c = \frac{[HI(g)]^2}{[H_2(g)][I_2(g)]}$$

or

$$K_p = \frac{(P_{HI})^2}{(P_{H_2})(P_{I_2})}$$

Now, since

$$P_{HI} = [HI(g)]RT$$

$$P_{H_2} = [H_2(g)]RT$$

and

$$P_{I_2} = [I_2(g)]RT$$

it follows that

$$K_p = \frac{(P_{HI})^2}{(P_{H_2})(P_{I_2})} = \frac{[HI(g)]^2 \cancel{(RT)^2}}{[H_2(g)]\cancel{RT}[I_2(g)]\cancel{RT}} = \frac{[HI(g)]^2}{[H_2(g)][I_2(g)]} = K_c$$

In this particular example, $K_p = K_c$. Neither K_p nor K_c has any units, but this is not always the case. Just consider the reaction

$$N_2(g) + 3H_2(g) \rightleftharpoons 2NH_3(g)$$

$$K_p = \frac{(P_{NH_3})^2}{(P_{N_2})(P_{H_2})^3} = \frac{[NH_3(g)]^2(RT)^2}{[N_2(g)]RT[H_2(g)]^3(RT)^3}$$

$$= \frac{[NH_3(g)]^2}{[N_2(g)][H_2(g)]^3} (RT)^{-2} = K_c(RT)^{-2}$$

In this case, $K_p = K_c(RT)^{-2}$.

Can you see that the numerical value of K_p is the same as that of K_c only when there are the same number of moles on each side of the balanced equation?

In general,

$$K_p = K_c(RT)^{\Delta n}$$

where Δn = number of moles on the right of the equation – number of moles on the left. In calculating values for K_p, pressure should be expressed in pascals and *not* in atmospheres.

$$1\ pascal\ (Pa) = 1\ N\ m^{-2}$$

$$1\ atm \approx 10^5\ Pa = 100\ kPa$$

The equilibrium constant, K_c, for the reaction

$$N_2(g) + 3H_2(g) \rightleftharpoons 2NH_3(g)$$

at 620 K is $2\ mol^{-2}\ dm^6$.

1 What is the value of K_p for this reaction at 620 K? ($R = 8.31\ J\ K^{-1}\ mol^{-1}$)
2 The units of K_p for this reaction are $(Pa)^{-2}$. Check that you agree with this.

20.10 Heterogeneous equilibria

Most of the equilibria that we have discussed so far may be described as **homogeneous**. In these systems, all the reactants and products co-exist in the *same phase* at equilibrium. Either they are all gases or they are all mixed together in aqueous solution.

In this section, we shall be looking at **heterogeneous equilibria** in more detail. In these systems, *two or more phases are present* at equilibrium. We have already discussed three types of heterogeneous equilibria:

(i) liquid in equilibrium with its saturated vapour;
(ii) solid solute in equilibrium with its saturated solution;
(iii) solute partitioned between two immiscible solvents.

Data concerning the equilibrium of water with its vapour at various temperatures are given in table 20.4.

The important feature in these results is that the vapour pressure exerted by the water at a particular temperature is independent of the mass of water present. How can this be explained?

For equilibrium between water and its vapour

$$H_2O(l) \rightleftharpoons H_2O(g)$$

we can write an equilibrium constant as

$$K_c = \frac{[H_2O(g)]}{[H_2O(l)]} \qquad \text{Equation (1)}$$

Table 20.4
Saturation vapour pressure of water at various temperatures

Mass of water taken/g	Pressure/Pa		
	20°C	40°C	60°C
1	23.4×10^2	73.8×10^2	199×10^2
2	23.4×10^2	73.8×10^2	199×10^2
10	23.4×10^2	73.8×10^2	199×10^2
50	23.4×10^2	73.8×10^2	199×10^2

The value of $[H_2O(l)]$ is, however, effectively constant, whatever the amount of water taken. We can, in fact, calculate its value from the density of water:

$$\text{density of water} = 1 \text{ g cm}^{-3}$$
$$= 1000 \text{ g dm}^{-3}$$
$$\Rightarrow [H_2O(l)] = \frac{1000}{18} = 55.56 \text{ mol dm}^{-3}$$

Using equation (1) above we can write

$$K_c[H_2O(l)] = K'_c = [H_2O(g)]$$

where K'_c may be described as a modified equilibrium constant. Instead of writing $K'_c = [H_2O(g)]$, we can write

$$K'_p = P_{H_2O}$$

This last equation confirms the results in table 20.4 which show that the saturation vapour pressure of water is constant at a particular temperature.

Similar results are obtained when other liquids or solids are used in place of water. In each case, the results confirm that *the concentration of a pure liquid or a pure solid is constant. This concentration in mol dm^{-3} is also independent of the amount of liquid or solid present.*

In other words, $[X(s)]$ and $[X(l)]$ are constant whatever the amount of X taken, but $[X(g)]$ and $[X(aq)]$ will vary as the amount of X in a given volume varies.

When solid NH_4HS is allowed to dissociate in an evacuated vessel it forms ammonia and hydrogen sulphide:

$$NH_4HS(s) \rightleftharpoons NH_3(g) + H_2S(g)$$

If excess NH_4HS is used at 25°C, the total pressure at equilibrium is 6×10^4 Pa.

1. Write an expression for the modified equilibrium constant for the system, K'_p.
2. What are the partial pressures of ammonia and hydrogen sulphide at equilibrium?
3. What is the value of K'_p at 25°C?

An interesting example of heterogeneous chemical equilibrium involves the thermal dissociation of calcium carbonate:

$$CaCO_3(s) \rightleftharpoons CaO(s) + CO_2(g)$$

From this equation, we can write

$$K_c = \frac{[CaO(s)][CO_2(g)]}{[CaCO_3(s)]}$$

but $[CaCO_3(s)]$ and $[CaO(s)]$ are both constant, so the modified equilibrium constant for the system becomes

$$K'_c = [CO_2(g)]$$

or

$$K'_p = P_{CO2}$$

This shows that at a particular temperature, there is a constant pressure (or concentration) of CO_2 in equilibrium with $CaO(s)$ and $CaCO_3(s)$. It does not matter what masses of these two solids are present. This prediction is confirmed by experiment.

Thus, at 800°C, the pressure of CO_2 in equilibrium with $CaCO_3(s)$ and $CaO(s)$ is 2.5×10^4 Pa. Hence, the equilibrium constant at 800°C is:

$$K'_p = P_{CO2} = 2.5 \times 10^4 \text{ Pa}$$

Summary

1. A dynamic equilibrium is characterised by the following features:
   (a) Constant macroscopic properties.
   (b) Continuing microscopic processes. Particles participate in both forward and reverse processes, but the rate of the forward process is equal to the rate of the reverse process. Thus, no overall change occurs.
   (c) Equilibrium attained from either direction.
   (d) Equilibrium achieved only in a closed system.

2. When a dissolved solute, X, distributes itself between two immiscible solvents A and B, at equilibrium

   $$\frac{\text{concentration of X in solvent A}}{\text{concentration of X in solvent B}} = \text{a constant}$$

   This constant is called the distribution ratio or the partition coefficient. It remains constant provided
   (a) the temperature is constant,
   (b) the solvents are immiscible and do not react with each other.
   (c) the solute neither reacts, nor associates, nor dissociates in the solvents.

3. The most important application of partition is in solvent extraction both industrially and in the laboratory.

4. If an equilibrium mixture contains substances A, B, C and D related by the equation

   $$aA + bB \rightleftharpoons cC + dD$$

   it is found experimentally that

   $$\frac{[C]^c[D]^d}{[A]^a[B]^b} = K_c$$

   K_c is called the Equilibrium Constant. It is constant at a given temperature. This experimental result is known as the Equilibrium Law.

5. The only factor which affects the value of an equilibrium constant is temperature. The values of K_c, K_p, K'_c and K'_p are unaffected by changes in concentration or changes in pressure provided equilibrium is established at the same temperature.

6. The partial pressure of a gas is proportional to its concentration. So, we can express the equilibrium constant of a gaseous reaction either in terms of partial pressures or in terms of concentrations.

   This means that for the equilibrium,
   $H_2(g) + I_2(g) \rightleftharpoons 2HI(g)$ we can write either

   $$K_c = \frac{[HI]^2}{[H_2][I_2]} \quad \text{or} \quad K_p = \frac{(P_{HI})^2}{(P_{H_2})(P_{I_2})}$$

7. The concentration of a solid or a pure liquid is constant. So, we can write modified equilibrium constants for heterogeneous equilibria excluding these constant concentrations. Thus, for the equilibrium

   $$AgCl(s) + 2NH_3(aq) \rightleftharpoons Ag(NH_3)_2^+(aq) + Cl^-(aq)$$

   $[AgCl(s)]$ is constant, so we can write a modified equilibrium constant:

   $$K'_c = \frac{[Ag(NH_3)_2^+(aq)][Cl^-(aq)]}{[NH_3(aq)]^2}$$

1 The Mogul Oil Company is worried about the impurity, M, in its four-star petrol. One dm^3 of petrol contains 5 g of M. In an effort to reduce the concentration of M in the petrol, Mogul have discovered the secret solvent, S. The partition coefficient of M between petrol and S is 0.01.

(a) What is meant by the term 'partition coefficient'?

(b) Explain the principles of solvent extraction.

(c) Calculate the total mass of M removed from 1 dm^3 of petrol by shaking it with $100 cm^3$ of solvent, S.

2 5 moles of ethanol, 6 moles of ethanoic acid, 6 moles of ethyl ethanoate and 4 moles of water were mixed together in a stoppered bottle at 15°C.

After equilibrium had been attained the bottle was found to contain only 4 moles of ethanoic acid.

(a) Write an equation for the reaction between ethanol and ethanoic acid to form ethyl ethanoate and water.

(b) Write an expression for the equilibrium constant, K_c, for this reaction.

(c) How many moles of ethanol, ethyl ethanoate and water are present in the equilibrium mixture?

(d) What is the value of K_c for this reaction?

(e) Suppose 1 mole of ethanol, 1 mole of ethanoic acid, 3 moles of ethyl ethanoate and 3 moles of water are mixed together in a stoppered flask at 15°C. How many moles of:

(i) ethanol,   (ii) ethanoic acid,

(iii) ethyl ethanoate,   (iv) water

are present at equilibrium?

3 At a certain temperature and a total pressure of 10^5 Pa, iodine vapour contains 40% by volume of I atoms:

$$I_2(g) \rightleftharpoons 2I(g)$$

(a) Calculate K_p for the equilibrium.

(b) At what total pressure (without temperature change) would the percentage of I atoms be reduced to 20%?

4 (a) Deduce the relationship between K_p and K_c for the gaseous equilibria:

(i) $2NO(g) + O_2(g) \rightleftharpoons 2NO_2(g)$,

(ii) $NO(g) + \frac{1}{2}O_2(g) \rightleftharpoons NO_2(g)$.

(b) What are the units of K_p and K_c for the two equilibria referred to in (a)?

5 Consider the following reaction:

$$H_2(g) + I_2(g) \rightleftharpoons 2HI(g)$$

(a) Write an expression for the equilibrium constant in terms of partial pressures. At a certain temperature, analysis of an equilibrium mixture of the gases yielded the following results:

$P_{H_2} = 2.5 \times 10^4$ Pa

$P_{I_2} = 1.6 \times 10^4$ Pa

$P_{HI} = 4.0 \times 10^4$ Pa

(b) Calculate the equilibrium constant for the reaction. What are its units?

(c) In a second experiment at the same temperature, iodine and hydrogen iodide were mixed together. Each gas had a partial pressure of 3×10^4 Pa. What are the partial pressures of hydrogen, iodine and hydrogen iodide at equilibrium?

(d) In a third experiment at the same temperature, pure hydrogen iodide was injected into the flask at a pressure of 6×10^4 Pa. What are the partial pressures of hydrogen, iodine and hydrogen iodide at equilibrium?

Factors Affecting Equilibria

21

21.1 The effect of concentration changes on equilibria

When a system in equilibrium is suddenly disturbed, it will respond until the equilibrium is eventually restored.

Consider the equilibrium

$$Fe^{3+}(aq) \quad + \quad NCS^-(aq) \quad \rightleftharpoons \quad Fe(NCS)^{2+}(aq)$$
$$\text{pale yellow} \qquad \text{colourless} \qquad \text{deep red}$$

When 10^{-3} mol dm^{-3} iron(III) nitrate solution is added to an equal volume of 10^{-3} mol dm^{-3} potassium thiocyanate, a red solution is produced. The red colour is due to the formation of thiocyanatoiron(III) ions. The system forms an equilibrium mixture containing unreacted Fe^{3+}, unreacted NCS^- and the product $Fe(NCS)^{2+}$. But what happens to the equilibrium when one of the concentrations is suddenly changed?

If a soluble iron(III) salt is added to an equilibrium solution containing $Fe^{3+}(aq)$, $NCS^-(aq)$ and $Fe(NCS)^{2+}(aq)$, the colour of the solution becomes darker (figure 22.1).

A new state of equilibrium is quickly attained. In this equilibrium the concentration of $Fe(NCS)^{2+}(aq)$ is obviously greater than before the addition of Fe^{3+}. Increasing the concentration of Fe^{3+} has increased the concentration of $Fe(NCS)^{2+}(aq)$. In the same way, the concentration of $Fe(NCS)^{2+}(aq)$ also rises when a soluble thiocyanate is added to the system. On the other hand, removal of Fe^{3+} or NCS^- from the equilibrium mixture causes the solution to become paler. A decrease in the concentration of Fe^{3+} or NCS^- results in the conversion of some $Fe(NCS)^{2+}$ into Fe^{3+} and NCS^- in an attempt to replace the substance removed.

The results of these experiments can be summarised by the following statement:

> *If the concentration of one of the substances in a reversible equilibrium is altered, the equilibrium will shift to oppose the change in concentration.*

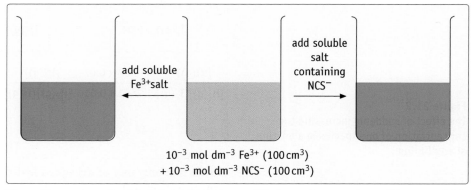

Figure 21.1
The effect of Fe^{3+} and NCS^- on the equilibrium, $Fe^{3+}(aq) + NCS^-(aq) \rightleftharpoons Fe(NCS)^{2+}(aq)$.

add soluble Fe^{3+}salt

add soluble salt containing NCS$^-$

10^{-3} mol dm^{-3} Fe^{3+} (100 cm^3)
+ 10^{-3} mol dm^{-3} NCS$^-$ (100 cm^3)

Thus, if a reactant is added to a system in equilibrium, that reaction will occur which uses up the added reactant. Conversely, if a reactant is removed, that reaction will occur which replenishes the removed reactant. The statement in italics above is a specific application of an important generalisation known as **Le Chatelier's Principle**.

The Frenchman, Henri Louis Le Chatelier, was one of the first chemists to investigate the effects of different factors such as temperature, pressure and concentration on equilibria. After studying data concerning equilibria, Le Chatelier proposed the following generalisation:

> *If a system in equilibrium is subjected to a change, processes will occur which tend to counteract the change imposed.*

Although the concentration of individual substances in an equilibrium may vary, the equilibrium constant is always the same at one particular temperature. This, of course, is the crucial point of the equilibrium law. We can emphasise this further by considering the effect of suddenly increasing the concentration of hydrogen in an equilibrium mixture of $H_2(g)$, $I_2(g)$ and $HI(g)$. In the initial equilibrium mixture (figure 21.2),

$$[HI(g)] = 0.07 \text{ mol dm}^{-3}, [H_2(g)] = 0.01 \text{ mol dm}^{-3} \text{ and } [I_2(g)] = 0.01 \text{ mol dm}^{-3}$$

$$\therefore K_c = \frac{[HI(g)]^2}{[H_2(g)][I_2(g)]} = \frac{(0.07)^2}{(0.01) \times (0.01)} = 49$$

When the concentration of $H_2(g)$ is suddenly doubled,

$$\Rightarrow \frac{[HI(g)]^2}{[H_2(g)][I_2(g)]} = \frac{(0.07)^2}{(0.02) \times (0.01)} = 24.5 < K_c$$

The system is no longer in equilibrium. In order to restore the equilibrium, the concentration of $HI(g)$ must rise, whilst those of $H_2(g)$ and $I_2(g)$ must fall. This is achieved by an *overall* conversion of *some* of the hydrogen and iodine in the mixture to hydrogen iodide:

$$H_2(g) + I_2(g) \rightarrow 2HI(g)$$

When equilibrium is restored once more (figure 21.2), we find that

$$[HI(g)] = 0.076 \text{ mol dm}^{-3}, [H_2(g)] = 0.017 \text{ mol dm}^{-3} \text{ and } [I_2(g)] = 0.007 \text{ mol dm}^{-3}$$

$$\Rightarrow \frac{[HI(g)]^2}{[H_2(g)][I_2(g)]} = \frac{(0.076)^2}{(0.017) \times (0.007)} = 49 = K_c$$

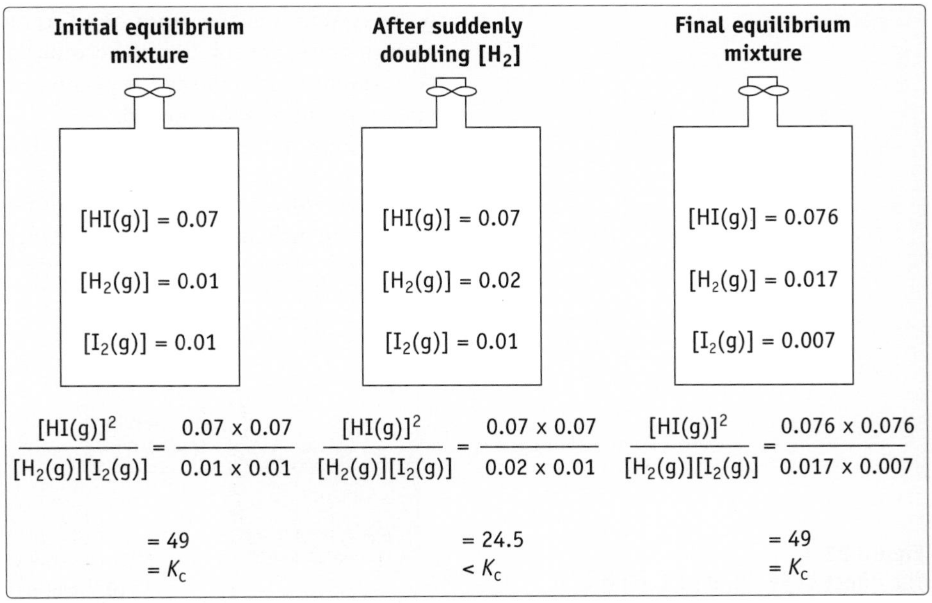

Figure 21.2
The effect of suddenly increasing the concentration of one species in a mixture at equilibrium.

Notice that only part of the added hydrogen is used up in restoring equilibrium. The concentration of $H_2(g)$ was suddenly doubled from 0.01 mol dm^{-3} in the initial equilibrium to 0.02 mol dm^{-3}. When equilibrium is achieved once more the final concentration of hydrogen is not 0.01 mol dm^{-3} but 0.017 mol dm^{-3}. Obviously, $[HI(g)]$ in the final equilibrium is greater than that in the initial equilibrium whilst $[I_2(g)]$ in the final equilibrium is less than that initially.

21.2 The effect of pressure changes on equilibria

If the partial pressure of *only one* of the gases in an equilibrium mixture is changed, the overall effect can be predicted like those involving changes in concentration in the last section. But what happens when the *total* pressure of a gaseous system at equilibrium is suddenly increased or decreased? In this case, the partial pressures of *all* the gases increase or decrease. Consider, first, the reaction

$$N_2(g) + 3H_2(g) \rightleftharpoons 2NH_3(g)$$

for which

$$K_p = \frac{(P_{NH3})^2}{P_{N2}(P_{H2})^3}$$

Now, suppose that the equilibrium partial pressures of nitrogen, hydrogen and ammonia are a, b and c atm, respectively. Thus:

$$K_p = \frac{(P_{NH3})^2}{P_{N_2}(P_{H_2})^3} = \frac{c^2}{ab^3} \; atm^{-2}$$

What happens when the total pressure is suddenly doubled? Let us follow the arguments in the last section. What do the equilibrium law and Le Chatelier's principle predict will happen after a change in pressure?

Using the equilibrium law to predict the results of a change in pressure

When the total pressure is suddenly doubled, all of the partial pressures are doubled. Hence

$$P_{N_2} = 2a \; atm,$$
$$P_{H_2} = 2b \; atm,$$
$$P_{NH_3} = 2c \; atm.$$

$$\Rightarrow \quad \frac{(P_{NH3})^2}{P_{N_2} \cdot (P_{H_2})^3} = \frac{(2c)^2}{2a \cdot (2b)^3} = \frac{4c^2}{2a.8b^3} = \frac{1}{4} \frac{c^2}{ab^3} \; atm^{-2}$$

Momentarily, 'the equilibrium constant ratio' is reduced to one-quarter of its value at equilibrium. So, nitrogen and hydrogen react to form ammonia until equilibrium is restored once more. Table 21.1 shows how the percentage of ammonia in the equilibrium mixture rises as the total pressure on the system increases.

Table 21.1
The effect of pressure on the equilibrium percentage of ammonia in the system
$N_2 + 3H_2 \rightleftharpoons 2NH_3$

Total pressure/atm	1	50	100	200
Equilibrium percentage of NH₃ at 723 K	0.24	9.5	16.2	25.3

Using Le Chatelier's principle to predict the results of a change in pressure

When the total pressure is suddenly increased, the molecules are crowded closer together. The additional pressure can be relieved if the molecules are able to react and reduce the number of molecules present.

In the reaction we are considering, one molecule of nitrogen reacts with three molecules of hydrogen to form two molecules of ammonia. In other words, four molecules of gas are reacting to form only two molecules of gas. This reduction in the total number of gas molecules results in a reduction in the total pressure. Hence, any increase in pressure in the $N_2/H_2/NH_3$ system at equilibrium can be relieved by a conversion of nitrogen and hydrogen to ammonia. Conversely, a decrease in pressure will favour the formation of nitrogen and hydrogen. This results in an increase in the number of molecules present, thereby counteracting the pressure reduction.

In general, for gaseous reactions an increase in pressure favours the reaction which produces fewer molecules. A decrease in pressure favours the gaseous reaction which produces more molecules.

On the other hand, pressure has no effect on a gaseous reaction if there is no change in the number of molecules. Consider the reaction

$$H_2(g) + I_2(g) \rightleftharpoons 2HI(g)$$

Suppose the equilibrium partial pressures of hydrogen, iodine and hydrogen iodide are x, y and z atm, respectively, at a particular temperature.

1. Write an expression for the equilibrium constant, K_p, in terms of x, y and z.
2. Suppose the overall pressure is halved. What are the partial pressures of hydrogen, iodine and hydrogen iodide now?
3. What is the value of the equilibrium constant ratio

$$\frac{(P_{HI})^2}{(P_{H_2})(P_{I_2})}$$

when the overall pressure is suddenly halved?
4. Use the equilibrium law to explain why reducing the pressure has no effect on this system.
5. How does Le Chatelier's principle explain why pressure changes have no effect on this system?

21.3 The effect of catalysts on equilibria

The equilibrium constant expression includes *only* those substances shown in the overall balanced equation. Catalysts do not appear in the overall equation for a reaction. Therefore, it is not surprising that they have no effect on the equilibrium position.

Experiments show that catalysts can increase the *rates* of both forward and backward reactions in an equilibrium. So they enable equilibrium to be achieved much more rapidly, but they do not alter the concentrations of reacting substances at equilibrium.

21.4 The effect of temperature changes on equilibria

Although the equilibrium concentrations of reactants and products can vary over a wide range, the numerical value of the equilibrium constant remains constant at one particular temperature. This means that K_c and K_p are unaffected by catalysts or by changes in pressure and concentration.

The equilibrium constant does, however, vary with temperature. Table 21.2 shows the values of K_p at different temperatures for three important reactions. The corresponding enthalpy changes for the complete conversion of reactants to products are also shown.

Table 21.2
Values of K_p for three different reactions at various temperatures

$N_2(g) + 3H_2(g) \rightleftharpoons 2NH_3(g)$ $\Delta H^{\ominus} = -92$ kJ		$N_2O_4(g) \rightleftharpoons 2NO_2(g)$ $\Delta H^{\ominus} = +57$ kJ		$2SO_2(g) + O_2(g) \rightleftharpoons 2SO_3(g)$ $\Delta H^{\ominus} = -197$ kJ	
T/K	$K_p = \dfrac{(P_{NH_3})^2}{P_{N_2}(P_{H_2})^3}$ atm^{-2}	T/K	$K_p = \dfrac{(P_{NO_2})^2}{P_{N_2O_4}}$ atm	T/K	$K_p = \dfrac{(P_{SO_3})^2}{(P_{SO_2})^2 P_{O_2}}$ atm^{-1}
400	1.0×10^2	200	1.9×10^{-6}	600	3.2×10^3
500	1.6×10^{-1}	300	1.7×10^{-1}	700	2.0×10^2
600	3.1×10^{-3}	400	5.1×10	800	3.2×10
700	6.3×10^{-5}	500	1.5×10^3	900	6.3
800	7.9×10^{-6}	600	1.4×10^4	1000	2.0

Notice that the two exothermic reactions have K_p values which decrease with an increase in temperature. In contrast, the endothermic reaction has K_p values which increase as the temperature rises. Evidence from other investigations confirms this pattern of results. In general, it is found that:

(i) equilibria in which the **forward reaction is exothermic (i.e. $\Delta H^{\ominus}$ negative)** have equilibrium constants that decrease as temperature rises;

(ii) equilibria in which the **forward reaction is endothermic (i.e. $\Delta H^{\ominus}$ positive)** have equilibrium constants that increase as temperature rises.

Look at the information in table 21.2.

1 How does the proportion of ammonia in the $N_2/H_2/NH_3$ system change as temperature increases?

2 What is the value of $\Delta H^{\ominus}$ for the reaction

$$2NH_3(g) \rightleftharpoons N_2(g) + 3H_2(g)?$$

What is the value of K_p for this reaction at 400 K?

3 Predict the effect of increasing temperature on K_p for the reaction

$$2NH_3(g) \rightleftharpoons N_2(g) + 3H_2(g)$$

Le Chatelier's principle can be used once again to predict the effect of temperature on chemical systems in equilibrium. Changing the temperature of a system in equilibrium provides a constraint. The system will try to remove the constraint. Hence, increase in temperature favours the endothermic process which will absorb the additional heat. Alternatively, decrease in temperature favours the exothermic process. This may be summarised as:

$$A+B \quad \underset{\substack{\textit{endothermic process favoured}\\ \textit{by increase in temperature}}}{\overset{\substack{\textit{exothermic process favoured}\\ \textit{by decrease in temperature}}}{\rightleftharpoons}} \quad C+D \qquad \Delta H = -x\,\text{kJ}$$

The following demonstration illustrates the effect of temperature on the position of an equilibrium. Three identical sealed tubes are prepared containing dark brown nitrogen dioxide (NO_2) in equilibrium with pale yellow dinitrogen tetraoxide (N_2O_4).

$$N_2O_4(g) \quad \rightleftharpoons \quad 2NO_2(g)$$
$$\text{pale yellow} \qquad\qquad \text{dark brown}$$

Initially, all three tubes contain the same amounts of NO_2 and N_2O_4 and they have the same brown appearance.

One tube is now placed in iced water. A second tube is left at room temperature and the third tube is placed in hot water (figure 21.3).

The tube in cold water becomes much paler whilst that in hot water turns dark brown. This shows that the equilibrium in the reaction is displaced towards the formation of darker NO_2 at higher temperatures. This is the endothermic direction. The equilibrium moves towards the formation of paler N_2O_4 at lower temperatures.

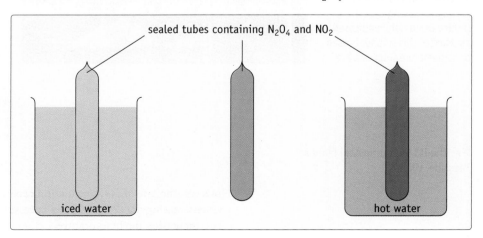

Figure 21.3
The effect of temperature on the equilibrium, $N_2O_4 \rightleftharpoons 2NO_2$.

21.5 Applying the principles of reaction rates and equilibria to industrial processes

The principles of reaction rates and chemical equilibria play an important part in the design and conditions of industrial processes. The economic and commercial competitiveness of any process rests on the speed, efficiency and economy with which products can be obtained from starting materials. Many people (including managers, economists and engineers) will be involved in decisions about the methods and materials in any industrial process, but the major problems confronting the chemist are to convert reactants into products:

(i) as quickly as possible and
(ii) as completely as possible.

The first of these problems is clearly a kinetic one. It involves obtaining the maximum rate of reaction and product formation.

The second problem is one of equilibrium. It involves the choice of those conditions which maximise the proportion of product in the equilibrium mixture.

The solution to each of these problems requires a careful choice of reaction conditions such as temperature, pressure and catalyst. In this respect, Le Chatelier's principle can be used to predict the conditions for maximum yield of product at equilibrium.

The manufacture of sulphuric acid: the contact process

The essential stages in the manufacture of sulphuric acid are shown diagrammatically in figure 21.4. Sulphur dioxide is first obtained by burning sulphur or by roasting sulphide ores in air. SO_2 is then mixed with excess air and thoroughly purified. This prevents 'poisoning' of the catalyst by dust and other impurities. The SO_2 is now ready for further oxidation to sulphur trioxide using the Contact Process. Finally, the SO_3 is combined with water to form sulphuric acid.

The bottleneck in the production of sulphuric acid is the slow third stage. This is the conversion of SO_2 to SO_3 in the Contact Process:

$$2SO_2(g) + O_2(g) \rightleftharpoons 2SO_3(g) \qquad \Delta H = -197 \text{ kJ}$$

▶ Piles of sulphur at a chemical plant near Vancouver, Canada. This sulphur has been obtained from underground deposits by melting it with superheated steam and forcing it to the surface using hot compressed air. Most of this sulphur is used to manufacture sulphuric acid.

▶▶ The ICI Sulphuric Acid Plant at Teesside, UK.

However, the rate of SO_3 production can be improved by
(i) increasing the concentration (pressure) of O_2 and/or SO_2,
(ii) increasing the temperature,
(iii) employing a catalyst.

21 Factors Affecting Equilibria

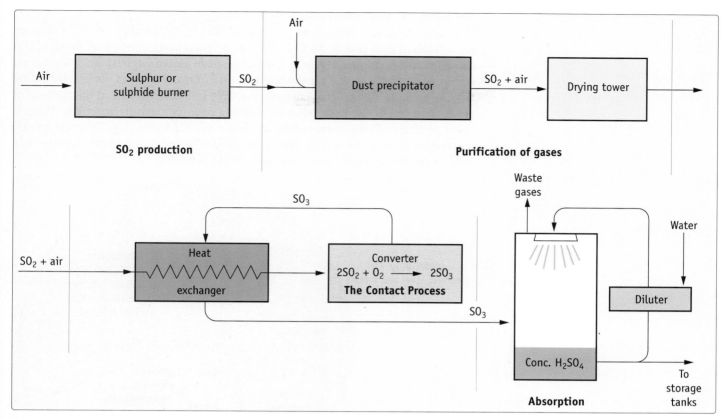

Figure 21.4
Essential stages in the manufacture of sulphuric acid.

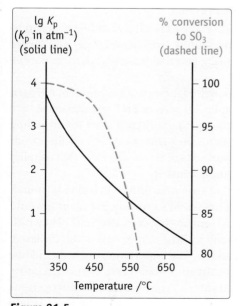

lg K_p
(K_p in atm^{-1})
(solid line)

% conversion to SO_3
(dashed line)

Temperature /°C

Figure 21.5
The effect of temperature on the equilibrium constant and the percentage conversion to SO_3 at equilibrium for a typical mixture of SO_2 and O_2 at one atmosphere pressure.

Both vanadium compounds and platinum have been used as catalysts for the Contact Process (section 18.6). The vanadium catalysts (incorporating either VO_3^- or V_2O_5) are less efficient than platinum. They are, however, cheaper and less susceptible to poisoning. Very few platinum catalyst plants have been built since 1945.

As the reaction is exothermic and three moles of reactants form two moles of products, Le Chatelier's principle predicts that the maximum yield of SO_3 at equilibrium will be obtained:

(i) at high pressure,
(ii) at low temperature.

Notice how kinetic and equilibrium considerations conflict in the choice of reaction temperature. The greatest yield of sulphur trioxide would be obtained at low temperatures, but under these conditions the reaction rate would be very slow. In practice, a compromise temperature of 450°C is chosen. This is the lowest that can be used without reducing the reaction rate to an unacceptable level. There are two other important reasons for keeping the temperature as low as possible. Fuel costs and corrosion of reaction chambers increase rapidly with rising temperature. Figure 21.5 shows the effect of temperature on the equilibrium constant and the percentage conversion to SO_3 at equilibrium for a typical gas mixture. Notice how the percentage conversion to SO_3 falls rapidly above about 450°C. At 450°C, conversion to SO_3 is 97%, but at 550°C conversion to SO_3 falls to 86%. The high conversion to SO_3 at 450°C, even at atmospheric pressure, makes it unnecessary to carry out the process at increased pressure.

There is one other aspect of the Contact Process worth consideration. As the reaction proceeds, the heat evolved in the exothermic reaction moves the system to a higher temperature. At this higher temperature the percentage conversion to SO_3 is much reduced (figure 21.5). Thus, it is necessary to cool gases between successive beds of catalyst. By clever use of heat exchangers to heat the incoming gases, the operating temperature can be maintained at 450°C without external heating.

After passing through the heat exchange system, the product gases pass into an absorption tower. Here SO_3 dissolves in concentrated H_2SO_4. Direct absorption in water is unsatisfactory because the heat evolved vaporises the H_2SO_4. The H_2SO_4 vapour would condense as a fog of tiny droplets which is slow to settle out.

The uses and importance of sulphuric acid

Sulphuric acid is one of the most widely used chemicals. About $2\frac{1}{2}$ million tonnes are manufactured each year in the UK. The annual world production of sulphuric acid is about 110 million tonnes. Figure 21.6 shows the main uses of sulphuric acid. Notice that sulphuric acid is required for the manufacture of many basic materials, including fertilisers, paints, fibres, detergents, plastics, dyes and steel. Because of this, its level of production can be used as a reliable guide to a country's industrial activity. What is more, problems of bulk storage of the acid mean that production must respond quickly to any changes in consumption. Hence, the economic and technical development of a country can be estimated from its sulphuric acid production.

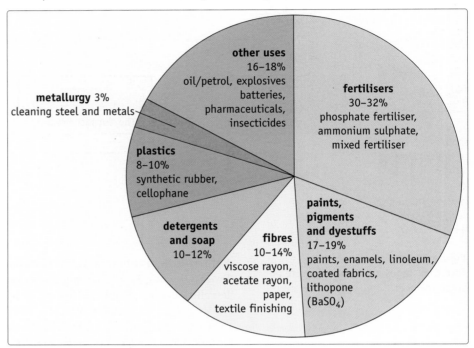

Figure 21.6
The uses of sulphuric acid in the UK.

The manufacture of ammonia: the Haber process

During the latter part of the nineteenth century, the agricultural industry began to require rapidly increasing quantities of nitrogenous fertilisers. Larger and larger supplies of food were being needed to feed growing populations, particularly in Europe and North America.

At the same time, the chemical industry also required increasing quantities of nitrogen compounds to make nitric acid for dyes and explosives such as TNT and dynamite.

Consequently, agriculture and industry were competing with each other for dwindling supplies of nitrogenous raw material. By 1900, Peruvian guano, a valuable fertiliser from the droppings of sea-birds, had already been worked out. It was also clear that supplies of sodium nitrate from Chile would soon become exhausted.

An alternative supply of nitrogen in the form of ammonia or nitrate had to be found, or the chemical industry would stagnate and the world's growing population would starve. Ironically, it was the war preparations in Germany between 1909 and 1914 which solved the problem. Military leaders in Germany realised that once war was declared, their country would be subjected to a strict blockade. The importation of goods and raw materials from the rest of unoccupied Europe and America would cease. German industry, therefore, had to be capable of meeting its country's requirements for nitrogenous fertilisers and the tremendous demand for explosives from nitric acid that a war would create.

In 1909, the leading German chemical company, Badische Anilin- und Soda-Fabrik (BASF), turned its research expertise and financial resources towards the problem. They started to investigate the possible manufacture of ammonia from atmospheric nitrogen.

In the previous year, a young German research chemist, Fritz Haber, had discovered that nitrogen and hydrogen would form an equilibrium mixture containing ammonia. The reaction needed a suitable catalyst at 600°C and a pressure of 200 atmospheres.

$$N_2(g) + 3H_2(g) \rightleftharpoons 2NH_3(g); \qquad \Delta H^{\ominus} = -92 \text{ kJ}$$

1 What conditions did Haber employ to increase the reaction rate?

2 Haber's experiments yielded an equilibrium mixture containing only 8% by volume of ammonia. What conditions of temperature and pressure does Le Chatelier's principle predict for maximum yield of ammonia at equilibrium?

3 Why do you think Haber employed a heat exchanger in his equipment?

▲ Fritz Haber (1868–1934). Haber was the son of a merchant of Breslau. After studying chemistry, Haber obtained a post as lecturer at the technical college in Karlsruhe. It was here that he discovered a method of synthesising ammonia, a discovery for which he was awarded the Nobel Prize for Chemistry in 1918.

BASF bought from Haber the rights to his ammonia process. Then, they spent more than £1m in transforming Haber's simple pilot process into a giant industrial plant capable of producing 10 000 tonnes of ammonia per year. By 1913, German production of nitrogen compounds had reached 120 000 tonnes per year. Without this effort, Germany would have run out of food and explosives and the war would certainly have ended before 1918.

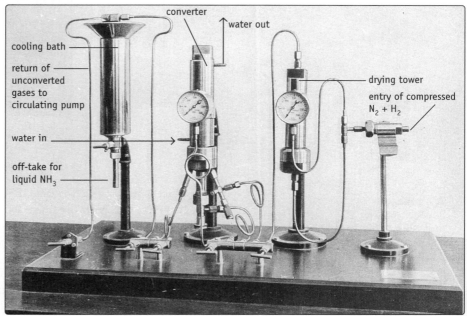

▶ Haber's apparatus for the synthesis of ammonia.

Figure 21.7 shows a flow diagram for the modern Haber process. Nitrogen can be obtained by the fractional distillation of liquid air, whilst hydrogen is obtained from naphtha or natural gas. Naphtha is a mixture of hydrocarbons containing 5–9 carbon atoms. The production of hydrogen involves either catalysis with steam:

$$C_6H_{14}(g) + 6H_2O(g) \rightarrow 6CO(g) + 13H_2(g)$$
in naphtha

$$CH_4(g) + H_2O(g) \rightarrow CO(g) + 3H_2(g)$$
in natural gas

or partial oxidation with oxygen:

$$C_6H_{14}(g) + 3O_2(g) \rightarrow 6CO(g) + 7H_2(g)$$
in naphtha

$$CH_4(g) + \tfrac{1}{2}O_2(g) \rightarrow CO(g) + 2H_2(g)$$
in natural gas

Thorough purification of both the nitrogen and hydrogen is necessary. This removes carbon monoxide from the hydrogen and also sulphur compounds, water vapour and carbon dioxide. These impurities would poison the catalyst in the converter.

Figure 21.7
A flow diagram for the Haber process.

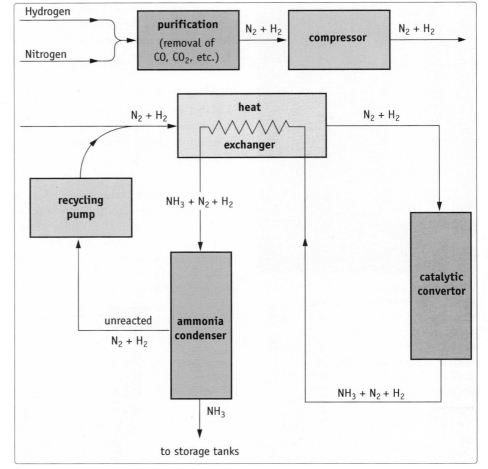

Hot product gases are used to warm up the purified nitrogen and hydrogen in a heat exchanger before they enter the converter.

Le Chatelier's principle suggests that increase in pressure and decrease in temperature will increase the proportion of ammonia at equilibrium. These predictions are borne out by the results in figure 21.8.

High pressure obviously gives a higher yield of ammonia, but the higher the pressure the greater the cost and maintenance of equipment. Although pressures up to 600 atm have been used, the favoured pressure nowadays is 250 atm.

In contrast to pressure, the temperature must be low to give a high yield of ammonia. But at low temperature the rate of reaction is so slow that it makes the process uneconomical. In practice, the operating temperature is usually about 450°C.

In addition to temperature and pressure, the catalyst is a vitally important variable in any industrial process. A more efficient catalyst permits lower operating temperatures.

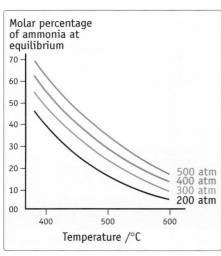

Figure 21.8
The percentage of ammonia in the equilibrium mixture obtained from a 1:1 mixture of N_2 and H_2 at different temperatures and pressures.

▶ The world's first ammonia manufacturing plant. This was built by BASF at Oppau in Germany and opened in 1912.

Experience has shown that the best catalyst is iron mixed with small amounts of promoters such as potassium oxide and aluminium oxide. These improve its catalytic activity (table 21.3).

The hot gases leaving the converter pass through the heat exchange system and are then cooled to −50°C. The ammonia liquifies (b.pt. −33°C) and collects in the storage vessels. The unreacted nitrogen and hydrogen are recycled.

Table 21.3
The effect of promoters on the efficiency of iron as a catalyst for the Haber process
(at 200 atm and 400°C)

Catalyst	Promoter	% ammonia in exit gases
Fe	nil	3–5
Fe	K_2O	8–9
Fe	$K_2O + Al_2O_3$	13–14

The uses and importance of ammonia

Ammonia forms the basis of the nitrogen industry. It will react with acids to give ammonium salts. It can also be oxidised to nitric acid which in turn can give nitrates. Both ammonium salts and nitrates are used as fertilisers, which provide the outlet for about 85% of ammonia. Of the remaining 15%, one-third is used in the production of nylon.

21.6 From ammonia to nitric acid

The manufacture of nitric acid from ammonia involves three stages.

1 Catalytic oxidation of ammonia to nitrogen oxide (NO).

$$4NH_3(g) + 5O_2(g) \rightleftharpoons 4NO(g) + 6H_2O(g) \qquad \Delta H^\ominus = -950 \text{ kJ}$$

2 Oxidation of nitrogen oxide (NO) to nitrogen dioxide (NO_2).

$$2NO(g) + O_2(g) \rightarrow 2NO_2(g) \qquad \Delta H^\ominus = -114 \text{ kJ}$$

3 Reaction of nitrogen dioxide with water to form nitric acid.

$$3NO_2(g) + H_2O(l) \rightarrow 2HNO_3(aq) + NO(g) \qquad \Delta H^\ominus = -117 \text{ kJ}$$

▲ Carl Bosch (1874–1940). Bosch was the son of a plumber. After studying chemistry at university, Bosch joined BASF in 1899 and quickly gained a reputation as a brilliant chemical engineer. Bosch was responsible for developing an industrial plant to manufacture ammonia from Haber's laboratory process. Bosch was awarded the Nobel Prize for Chemistry in 1931 for his work on high-pressure reactions.

Look closely at stage 1 above.

1 What conditions of temperature and pressure would give the maximum yield of nitrogen oxide (NO) at equilibrium?

2 In practice, the first stage is carried out by passing dry ammonia and air at 7 atm over a platinum gauze catalyst at 900°C. Explain why the conditions for the process differ from those predicted in your answer to the last question.

After stage 1, the product gases are cooled to 25°C. They are then mixed with more air so that nitrogen oxide is immediately oxidised to red–brown nitrogen dioxide:

$$2NO(g) + O_2(g) \rightarrow 2NO_2(g)$$

In the third stage, nitrogen dioxide reacts with water in large absorption towers. These are designed to ensure thorough mixing of the ascending gases and the descending solution. The final product contains about 60% nitric acid. More concentrated acid can be obtained by distilling the 60% solution with concentrated sulphuric acid.

► This technician is holding a sheet of the platinum-alloy catalyst used in stage 1 of the manufacture of nitric acid from ammonia. The sheet of catalyst is a gauze woven from wire of the platinum-alloy.

21.7 Fertilisers and explosives from nitric acid

Nitric acid plays an important part in the production of fertilisers and explosives which consume 75% and 15% of its production, respectively.

Fertilisers

Nitrogen is an essential element for plant growth. It is required for the formation of proteins, chlorophyll and nucleic acids. Plants suffering from nitrogen deficiency become stunted with yellow leaves.

> 1 Why does a deficiency of nitrogen cause plants to become stunted?
>
> 2 Why does a deficiency of nitrogen cause yellowing of the leaves?

The intensive cropping of agricultural land means that nitrogen removed from the soil must be replaced by the application of fertilisers, otherwise the soil will become barren and infertile. Nitrogenous fertilisers are usually nitrates or ammonium salts. Indeed, ammonium nitrate (e.g. ICI's 'NITRAM') is the most widely used fertiliser in many countries, because of its relatively high percentage of nitrogen.

> 1 What is the percentage by mass of nitrogen in ammonium nitrate, NH_4NO_3?
>   ($N = 14$, $H = 1$, $O = 16$)
>
> 2 How do you think ammonium nitrate is obtained industrially?

Calcium ammonium nitrate contains about 26% nitrogen. It consists of ammonium nitrate crystals coated with chalk (calcium carbonate). It is used less frequently than 'NITRAM', but it is non-deliquescent (unlike ammonium nitrate crystals). It also provides a convenient way of liming the soil at the same time.

Explosives

Although 90% of ammonium nitrate is used in fertilisers, a large proportion of the remaining 10% goes towards the production of explosives. Nowadays, virtually all explosives are manufactured by processes which use concentrated nitric acid.

The formulas of some of these explosives are shown in figure 21.9. All of these substances undergo rapid chemical reactions on explosion. The explosions produce large amounts of gas and heat and a sudden increase in pressure. Carbon and hydrogen in the explosives undergo combustion with oxygen from the nitro-groups forming carbon dioxide and water. For example, the explosion of 'nitroglycerine' (made by the action of concentrated nitric acid on glycerol (propane-1,2,3-triol)) can be summarised by the equation:

$$4C_3H_5(NO_3)_3(l) \rightarrow 12CO_2(g) + 10H_2O(g) + 6N_2(g) + O_2(g)$$

cellulose nitrate
(nitrocellulose, gun cotton)

propane-1,2,3-triyl trinitrate
(nitroglycerine)

methyl-2,4,6-trinitrobenzene
(trinitrotoluene, TNT)

2,4,6-trinitrophenol
(picric acid)

Figure 21.9
Explosives manufactured from nitric acid.

Notice that in this case, there is more than enough oxygen within the 'nitroglycerine' molecule to ensure complete combustion to CO_2 and water. This is not the case with aromatic nitro compounds such as TNT and picric acid. These explosives are frequently blended with compounds containing a high percentage of oxygen, such as chlorates and nitrates, to ensure complete oxidation during explosion.

Dynamite is a general term for explosives which contain both nitroglycerine and nitrocellulose. Cordite and gelignite are each different forms of dynamite.

The manufacture of explosives requires elaborate and very special safety precautions. You *should never* attempt to make any explosive yourself.

▶Controlled demolition, using explosives, of a block of flats at Hackney, East London, UK.

Summary

1. If the concentration of one of the substances in a reversible equilibrium is altered, the equilibrium will shift in such a way as to oppose the change in concentration.

2. The influence of different factors, such as temperature, pressure and concentration, on a system in equilibrium can be predicted using Le Chatelier's principle. This says:
   If a system in equilibrium is subjected to a change, processes occur which tend to counteract the change imposed.

3. The values of K_c and K_p are unaffected by changes in pressure or concentration.

4. Temperature is the only factor which influences the values of K_c and K_p.

5. Changes in concentration or pressure may, however, result in changes in the concentration or partial pressure of substances in the equilibrium mixture.

6. Catalysts affect neither the position of equilibrium nor the values of K_c and K_p.

7. Equilibria for which the forward reaction is exothermic (i.e. $\Delta H^{\ominus}$ negative) have equilibrium constants that decrease as temperature rises.
   Equilibria for which the forward reaction is endothermic (i.e. $\Delta H^{\ominus}$ positive) have equilibrium constants that increase as temperature rises.

Review questions

1. Nowadays, hydrogen can be obtained from natural gas by partial oxidation with steam. This involves the following endothermic reaction.
$$CH_4(g) + 3H_2O(g) \rightleftharpoons CO(g) + 3H_2(g)$$
   (a) Write an expression for K_p for this reaction.
   (b) How will the value of K_p be affected by
      (i) increasing the pressure,
      (ii) increasing the temperature,
      (iii) using a catalyst?
   (c) How will the composition of the equilibrium mixture be affected by:
      (i) increasing the pressure,
      (ii) increasing the temperature,
      (iii) using a catalyst?

2. The first step in the manufacture of nitric acid from ammonia involves the exothermic oxidation of ammonia to nitrogen oxide (NO) and steam.
   (a) Write an equation for the reaction of ammonia with oxygen to form nitrogen oxide and steam.
   (b) Predict, qualitatively, the conditions of temperature and pressure for maximum yield of nitrogen oxide in the equilibrium mixture.
   (c) The industrial manufacture of nitrogen oxide from ammonia employs high temperature and a pressure of 7 atm. How and why are these industrial conditions different from those you predicted in (b) for maximum yield of nitrogen oxide at equilibrium?
   (d) Describe, with equations, how nitrogen oxide produced by this process is converted to nitric acid.

3. At 25°C, the value of K_c for the following system is 10^{10}:
$$Sn^{2+}(aq) + 2Fe^{3+}(aq) \rightarrow Sn^{4+}(aq) + 2Fe^{2+}(aq)$$
   (a) Write an expression for K_c for this reaction.
   (b) Explain why K_c has no units.
   (c) What is the value of K_c for:
      (i) $Sn^{4+}(aq) + 2Fe^{2+}(aq) \rightarrow Sn^{2+}(aq) + 2Fe^{3+}(aq)$,
      (ii) $2Sn^{2+}(aq) + 4Fe^{3+}(aq) \rightarrow 2Sn^{4+}(aq) + 4Fe^{2+}(aq)$?

4. At 200°C, K_c for the reaction
$$PCl_5(g) \rightleftharpoons PCl_3(g) + Cl_2(g) \quad \Delta H^{\ominus} = +124 \text{ kJ}$$
   has a numerical value of 8×10^{-3}.
   (a) Write an expression for K_c for this reaction.
   (b) What are the units of K_c?
   (c) What is the value of K_c for the reverse reaction at 200°C and what are its units?
   (d) How will the amounts of PCl_5, PCl_3 and Cl_2 in the equilibrium mixture change if (i) more PCl_5 is added, (ii) the pressure is increased, (iii) the temperature is increased?
   (e) What would be the effect on K_c if (i) more PCl_5 is added, (ii) the pressure is increased, (iii) the temperature is increased?
   (f) A sample of pure PCl_5 was introduced into an evacuated vessel at 200°C. When equilibrium was obtained, the concentration of PCl_5 was 0.5×10^{-1} mol dm^{-3}.
   What are the concentrations of PCl_3 and Cl_2 at equilibrium?

5 At 488 K the equilibrium constant (K_p) for the reaction
$$COCl_2(g) \rightleftharpoons CO(g) + Cl_2(g)$$
is 0.2 Pa (2×10^{-6} atm).
(a) Assuming that the total pressure at equilibrium is P and the degree of dissociation of $COCl_2$ is α, deduce a relationship between K_p, α and P.
(Degree of dissociation (α) is the fraction dissociated.)
(b) Calculate the degree of dissociation
  (i) at 10^5 Pa (1 atm) pressure assuming the temperature remains constant,
  (ii) at 2×10^5 Pa (2 atm) pressure assuming the temperature remains constant.
  (Hint: when α is very small, $(1 - \alpha) \simeq 1$; $(1 + \alpha) \simeq 1$)

6 State and explain what happens to the concentrations of hydrogen, carbon monoxide and methanol which are in equilibrium according to the reaction
$$CO(g) + 2H_2(g) \rightleftharpoons CH_3OH(g) \qquad \Delta H = -92 \text{ kJ}$$
when
(a) the volume in which they are contained is suddenly reduced to half,
(b) the temperature is increased,
(c) the partial pressure of hydrogen is suddenly doubled,
(d) a catalyst is added,
(e) an inert gas is added to the system.

7 The densities of diamond and graphite are 3.5 and 2.3 g cm^{-3}, respectively. The change from graphite to diamond is represented by the equation
$$C(graphite) \rightleftharpoons C(diamond) \qquad \Delta H = +2 \text{ kJ}$$
Is the formation of diamond from graphite favoured by
(a) high or low temperature,
(b) high or low pressure?
Explain your answers.

8 (a) Describe in outline the manufacture of *one* of the following industrial chemicals:
  (i) sulphuric acid, (ii) nitric acid, (iii) ammonia.
(b) Discuss the chemical principles which determine the optimum operating conditions for the process which you describe.
(c) For the substances you have chosen, mention:
  (i) two large-scale uses,
  (ii) two important features of its chemistry. Give reactions and equations to illustrate the points you make.

22 Ionic Equilibria in Aqueous Solution

22.1 Introduction

Equilibria involving ions in aqueous solution are important in industrial, analytical and biological processes. The principles and characteristics of these ionic equilibria are very similar to those in other systems in chemical equilibrium.

In this chapter, we shall concentrate on two fundamental types of ionic equilibria:

(i) the equilibrium between an undissolved solid solute and its dissolved products in solution, i.e. **solubility equilibria**;
In most cases, these solubility equilibria involve sparingly soluble ionic solids in water.
(ii) the equilibrium between a dissolved undissociated molecule and its dissociated ions, i.e. **dissociation equilibria**.
In most cases, these dissociation equilibria involve either acids or bases.

22.2 The solubility of sparingly soluble ionic solids in water

When increasing quantities of a sparingly soluble ionic solid are added to water, a saturated solution is eventually formed. Ions in the saturated solution are in equilibrium with the excess undissolved solute:

$$MX(s) \rightleftharpoons M^+(aq) + X^-(aq)$$

In this section, we need to explore the relationship between the concentrations of the aqueous ions and the undissolved solute at equilibrium.

Table 22.1 shows the equilibrium concentrations of $Ag^+(aq)$ and $BrO_3^-(aq)$ in contact with undissolved $AgBrO_3$ when different initial volumes of $0.1\ mol\ dm^{-3}$ $AgNO_3$ and $0.1\ mol\ dm^{-3}$ $KBrO_3$ are added to $200\ cm^3$ of distilled water at 16°C.

The concentrations of BrO_3^- were obtained as follows. A measured volume of the aqueous solution was pipetted. Acid was then added followed by excess potassium iodide. The liberated iodine was then titrated against sodium thiosulphate solution of known concentration.

▲ As a coral reef grows, the concentration of Ca^{2+} and CO_3^{2-} ions from reef-building organisms in the sea-water around it must be sufficient to precipitate calcium-carbonate which forms the reef.

Table 22.1
Concentrations of $Ag^+(aq)$ and $BrO_3^-(aq)$ ions in contact with undissolved $AgBrO_3$ when different initial volumes of $0.1\ mol\ dm^{-3}$ $AgNO_3$ and $0.1\ mol\ dm^{-3}$ $KBrO_3$ are added to $200\ cm^3$ of distilled water

Initial volume of $0.1\ mol\ dm^{-3}$ $AgNO_3/cm^3$	Initial volume of $0.1\ mol\ dm^{-3}$ $KBrO_3/cm^3$	Concentration of $Ag^+(aq)$ at equilibrium $/mol\ dm^{-3}$	Concentration of $BrO_3^-(aq)$ at equilibrium $/mol\ dm^{-3}$	$[Ag^+]_{eq} \times [BrO_3^-]_{eq}$ $/mol^2\ dm^{-6}$
40	10	0.0144	0.0024	3.45×10^{-5}
30	20	0.0081	0.0041	3.32×10^{-5}
25	25	0.0058	0.0058	3.36×10^{-5}
20	30	0.0042	0.0082	3.44×10^{-5}
10	40	0.0033	0.0102	3.37×10^{-5}

Once we have determined the concentration of $BrO_3^-(aq)$, we can calculate the concentration of $Ag^+(aq)$.

Notice that the product of the concentrations of $Ag^+(aq)$ and $BrO_3^-(aq)$ is constant (column 5 in table 22.1). So, for this equilibrium we can write

$$[Ag^+(aq)][BrO_3^-(aq)] = \text{a constant at a given temperature}$$
$$= 3.39 \times 10^{-5} \text{ mol}^2 \text{ dm}^{-6} \text{ at } 16°C.$$

In other words, the product of the concentrations of Ag^+ and BrO_3^- is independent of the amount of $AgBrO_3$ present, provided there is some undisssolved $AgBrO_3$ in contact with the solution. This situation is comparable to other heterogeneous systems involving solid–liquid or solid–gas equilibria (section 20.10).

How can we explain the constant value for the product $[Ag^+(aq)][BrO_3^-(aq)]$?

When equilibrium between pure $AgBrO_3$ and its solution is reached, we have

$$AgBrO_3(s) \rightleftharpoons Ag^+(aq) + BrO_3^-(aq)$$

Hence we can write an equilibrium constant expression as

$$K_c = \frac{[Ag^+(aq)][BrO_3^-(aq)]}{[AgBrO_3(s)]}$$

But, $[AgBrO_3(s)]$, which represents the concentration of a pure solid, is constant (see section 20.10). So,

$$[Ag^+(aq)][BrO_3^-(aq)] = K_c[AgBrO_3(s)] = \text{a new constant}$$

This new constant is known as the **solubility product** and is given the symbol, $K_{s.p.}$.

Using the general formula A_xB_y for a sparingly soluble salt, we can deduce a general expression for the solubility product as follows.

At equilibrium,

$$A_xB_y(s) \rightleftharpoons xA^{y+}(aq) + yB^{x-}(aq)$$

$$\text{Hence } K_c = \frac{[A^{y+}(aq)]^x[B^{x-}(aq)]^y}{[A_xB_y(s)]}$$

But $[A_xB_y(s)]$ is constant. Therefore

$$[A^{y+}(aq)]^x[B^{x-}(aq)]^y = K_{s.p.}, \text{ the solubility product of } A_xB_y$$

The solubility products of some common compounds are given in table 22.2.

The solubility product of a salt is usually obtained from its solubility. The following example shows how this is done.

A saturated solution of silver chloride contains 1.46×10^{-3} g dm^{-3} at 18°C. What is the solubility product of silver chloride at this temperature?

$$\text{The solubility of silver chloride at } 18°C = 1.46 \times 10^{-3} \text{ g dm}^{-3}$$

$$= \frac{1.46 \times 10^{-3}}{143.5} \text{ mol dm}^{-3}$$

$$= 1 \times 10^{-5} \text{ mol dm}^{-3}$$

Write an expression for the solubility product of:

1 Bi_2S_3,
2 $AgCl$,
3 PbI_2.

Table 22.2
The solubility products of some common compounds at 25°C

Compound	Solubility product	Compound	Solubility product
Barium sulphate	1.0×10^{-10} mol^2 dm^{-6}	Lead(II) sulphate	1.6×10^{-8} mol^2 dm^{-6}
Calcium carbonate	5.0×10^{-9} mol^2 dm^{-6}	Lead(II) sulphide	1.3×10^{-28} mol^2 dm^{-6}
Calcium fluoride	4.0×10^{-11} mol^3 dm^{-9}	Nickel sulphide	4.0×10^{-21} mol^2 dm^{-6}
Calcium sulphate	2.0×10^{-5} mol^2 dm^{-6}	Silver bromide	5.0×10^{-13} mol^2 dm^{-6}
Copper(II) sulphide	6.3×10^{-36} mol^2 dm^{-6}	Silver chloride	2.0×10^{-10} mol^2 dm^{-6}
Lead(II) bromide	3.9×10^{-5} mol^3 dm^{-9}	Silver iodide	8.0×10^{-17} mol^2 dm^{-6}
Lead(II) chloride	2.0×10^{-5} mol^3 dm^{-9}	Zinc sulphide	1.6×10^{-24} mol^2 dm^{-6}
Lead(II) iodide	7.1×10^{-9} mol^3 dm^{-9}		

Now at equilibrium,

$$AgCl(s) \rightleftharpoons Ag^+(aq) + Cl^-(aq)$$
$$\therefore [Ag^+(aq)] = 1 \times 10^{-5} \text{ mol dm}^{-3}$$
$$\text{and } [Cl^-(aq)] = 1 \times 10^{-5} \text{ mol dm}^{-3}$$
$$\Rightarrow K_{s.p.} (AgCl) = [Ag^+(aq)][Cl^-(aq)] = 1 \times 10^{-5} \times 1 \times 10^{-5}$$
$$= 10^{-10} \text{ mol}^2 \text{ dm}^{-6}$$

so, the solubility product of silver chloride at 18°C is $10^{-10} \text{ mol}^2 \text{ dm}^{-6}$.

Conversely, the solubility of a salt can be obtained from its solubility product. The following calculation shows how this is done.

The solubility product of silver carbonate at 20°C is $8 \times 10^{-12} \text{ mol}^3 \text{ dm}^{-9}$. What is its solubility at this temperature?

Let us suppose the solubility is s mol dm^{-3}. At equilibrium,

$$Ag_2CO_3(s) \rightleftharpoons 2Ag^+(aq) + CO_3^{2-}(aq)$$
$$\therefore \text{ if the solubility of } Ag_2CO_3 \text{ is } s \text{ mol dm}^{-3}$$
$$[Ag^+(aq)] = 2s \text{ and } [CO_3^{2-}(aq)] = s$$
$$\therefore K_{s.p.} (Ag_2CO_3) = [Ag^+(aq)]^2[CO_3^{2-}(aq)] = 8 \times 10^{-12} \text{ mol}^3 \text{ dm}^{-9}$$
$$= (2s)^2 s = 8 \times 10^{-12}$$
$$\therefore 4s^3 = 8 \times 10^{-12}$$
$$\Rightarrow s^3 = 2 \times 10^{-12}$$
$$\therefore s = 1.25 \times 10^{-4} \text{ mol dm}^{-3}$$

i.e. the solubility of silver carbonate at 20°C is $1.25 \times 10^{-4} \text{ mol dm}^{-3}$.

Notice that the concentration of Ag^+ is both doubled and squared in the solubility product expression relative to the solubility of Ag_2CO_3. This is because the balanced solubility equation shows that the concentration of Ag^+ will be double the solubility (i.e. $2s$ not s) and the expression for $K_{s.p.}$ requires this concentration of Ag^+ to be squared (i.e. $(2s)^2$ not $2s$).

22.3 Limitations to the solubility product concept

The solubility product concept is valid only for saturated solutions in which the total concentration of ions is no more than about 0.01 mol dm^{-3}. For concentrations greater than this, the value of $K_{s.p.}$ is no longer constant. This means that it is quite inappropriate to use the solubility product concept for soluble compounds such as NaCl, $CuSO_4$ and $AgNO_3$. As a consequence of this, the numerical values of solubility products are always very small, rarely exceeding 10^{-4}. For substances of extremely low solubility, $K_{s.p.}$ may be less than 10^{-40}.

The solubility product of a sparingly soluble salt is essentially a modified equilibrium constant. Like other equilibrium constants, its value will change with temperature. Consequently, the temperature at which a solubility product is measured should always be specified unless it relates to the selected standard temperature of 298 K.

22.4 Using the solubility product concept

The common ion effect

Although the solubility product of a salt is constant at constant temperature, the concentrations of the individual ions may vary over a very wide range. When a saturated solution is obtained by dissolving the pure salt in water the concentrations of the ions are in a ratio determined by the formula of the compound. For example, the concentrations of Ca^{2+} and F^- ions in pure saturated calcium fluoride solution must be in the ratio 1:2.

However, when a saturated solution is obtained by mixing two solutions containing a common ion, there may be a big difference in the concentration of the ions of any sparingly soluble electrolyte. In these cases, solubility products can be used to determine the concentration of ions in the solution. We can illustrate this by considering the solubility of $BaSO_4$, first in water and then in 0.1 mol dm^{-3} sodium sulphate solution ($K_{s.p.}(BaSO_4) = 1 \times 10^{-10}$ mol^2 dm^{-6}).

Solubility of $BaSO_4$ in water

Suppose the solubility of $BaSO_4$ in water $= s$ mol dm^{-3}

$$BaSO_4(s) \rightleftharpoons Ba^{2+}(aq) + SO_4^{2-}(aq)$$
$$\therefore K_{s.p.}(BaSO_4) = [Ba^{2+}][SO_4^{2-}]$$
$$\Rightarrow 1 \times 10^{-10} = s \times s = s^2$$
$$\Rightarrow s = 10^{-5} \text{ mol dm}^{-3}$$

$\therefore$ solubility of $BaSO_4$ in water $= 10^{-5}$ mol dm^{-3}.

Solubility of $BaSO_4$ in 0.1 mol dm^{-3} Na$_2$SO$_4$

Suppose the solubility of $BaSO_4$ in 0.1 mol dm^{-3} Na$_2$SO$_4 = s'$ mol dm^{-3}

$$BaSO_4(s) \rightleftharpoons Ba^{2+}(aq) + SO_4^{2-}(aq)$$
$$Na_2SO_4(s) \rightarrow 2Na^+(aq) + SO_4^{2-}(aq)$$

In this case, $[Ba^{2+}(aq)] = s'$ mol dm^{-3}
but, $[SO_4^{2-}(aq)] = (s' + 0.1)$ mol dm^{-3}

$$\Rightarrow K_{s.p.} = [Ba^{2+}][SO_4^{2-}] = s'(s' + 0.1)$$
$$\Rightarrow s'(s' + 0.1) = 1 \times 10^{-10}$$

Now since $s' \ll 0.1$; $(s' + 0.1) \simeq 0.1$

$$\Rightarrow s' \times 0.1 = 1 \times 10^{-10}$$
$$\Rightarrow s' = 10^{-9} \text{ mol dm}^{-3}$$

$\therefore$ solubility of $BaSO_4$ in 0.1 mol dm^{-3} Na$_2$SO$_4 = 10^{-9}$ mol dm^{-3}

Calculate the solubility of silver chloride

1 in water

2 in 0.1 mol dm^{-3} NaCl.
($K_{s.p.}(AgCl) = 2.0 \times 10^{-10}$ mol^2 dm^{-6})

This calculation illustrates the important generalisation known as the **common ion effect**. This says that *in the presence of either A$^+$ or B$^-$ from a second source, the solubility of the salt AB is reduced.*

Predicting precipitation

Another important application of solubility products is that they enable chemists to predict the maximum concentrations of ions in a solution. Hence, it is possible to tell whether or not precipitation will occur.

Suppose we mix a 10^{-3} mol dm^{-3} solution of Ca^{2+} ions with an equal volume of a 10^{-3} mol dm^{-3} solution of SO$_4^{2-}$ ions at 25°C. Will a precipitate of CaSO$_4$ form?

The solubility product for calcium sulphate is 2×10^{-5} mol^2 dm^{-6} at 25°C.

$$\text{So, } K_{s.p.}(CaSO_4) = [Ca^{2+}][SO_4^{2-}] = 2 \times 10^{-5} \text{ mol}^2 \text{ dm}^{-6}$$

Immediately after mixing equal volumes of the two solutions and before any precipitation has occurred,

$$[Ca^{2+}] = [SO_4^{2-}] = \frac{10^{-3}}{2} \text{ M} = 5 \times 10^{-4} \text{ mol dm}^{-3}$$

(The concentration of each ion is halved since each solution is diluted by mixing with the other.) Hence, the ionic product for CaSO$_4$ immediately after mixing,

$$[Ca^{2+}][SO_4^{2-}] = 5 \times 10^{-4} \times 5 \times 10^{-4}$$
$$= 25 \times 10^{-8} = 2.5 \times 10^{-7} \text{ mol}^2 \text{ dm}^{-6}$$

This ionic product is less than the value of $K_{s.p.}$ for CaSO$_4$, so no precipitate will form.

▲ Stalactites growing from the roof of caverns at Marianna, Florida, USA. The stalactites form as calcium carbonate precipitates from a saturated solution dripping from the roof.

▲ Stag's horn coral growing off the shore of the Seychelle Islands in the Indian Ocean.

Let us now suppose that we mix equal volumes of 10^{-2} mol dm^{-3} solutions.

Immediately after mixing,

$$[Ca^{2+}] = [SO_4^{2-}] = 5 \times 10^{-3} \text{ mol dm}^{-3}$$

and the ionic product

$$= [Ca^{2+}][SO_4^{2-}]$$
$$= 5 \times 10^{-3} \times 5 \times 10^{-3}$$
$$= 25 \times 10^{-6}$$
$$= 2.5 \times 10^{-5} \text{ mol}^2 \text{ dm}^{-6}$$

In this case, the ionic product is greater than the solubility product. Therefore precipitation of CaSO$_4$ occurs. The concentrations of aqueous Ca^{2+} and SO$_4^{2-}$ ions are lowered by the reaction

$$Ca^{2+}(aq) + SO_4^{2-}(aq) \rightarrow CaSO_4(s)$$

The product $[Ca^{2+}][SO_4^{2-}]$ falls from 2.5×10^{-5} to 2.0×10^{-5}.

The precipitation of solids from aqueous solution is of great importance in nature and industry. Stalagmites and stalactites precipitate slowly from water in which the concentrations of Ca^{2+}(aq) and CO$_3^{2-}$(aq) have an ionic product greater than the solubility product of calcium carbonate. Coral reefs grow in a similar fashion. In this case, the concentration of Ca^{2+} and CO$_3^{2-}$ ions around the coral must be large enough to precipitate calcium carbonate from the surrounding sea water.

Selective precipitation

The different solubilities of salts can be used as a means of separating different substances from each other by carefully selected precipitation reactions. Just suppose that we have a solution containing magnesium chloride, calcium chloride and barium chloride. How can a separation of the metal ions be achieved? Both magnesium and calcium chromate(VI) are soluble but barium chromate(VI) is insoluble. So, addition of a solution of K$_2$CrO$_4$(aq) will precipitate BaCrO$_4$(s), which can then be removed by filtration.

The remaining solution now contains Mg^{2+} and Ca^{2+} ions, However, MgSO$_4$ is soluble, while CaSO$_4$ is insoluble. Thus, addition of Na$_2$SO$_4$(aq) to the mixture will precipitate CaSO$_4$ and leave Mg^{2+}(aq) in solution. Finally, the Mg^{2+}(aq) can be removed as solid MgCO$_3$ by adding Na$_2$CO$_3$ solution.

Notice that the order in which reagents are added is important. If we added Na$_2$SO$_4$ solution before adding K$_2$CrO$_4$ solution, a mixture of BaSO$_4$ and CaSO$_4$ would be precipitated. On the other hand, if Na$_2$CO$_3$ solution was added to the mixture of the three cations, MgCO$_3$, CaCO$_3$ and BaCO$_3$ would all be precipitated. Thus, both the precipitating reagents and their order of addition must be chosen carefully.

The results in table 22.3 show how three similar cations (Cu^{2+}, Zn^{2+} and Ni^{2+}) can be separated by careful, selective precipitation.

When hydrogen sulphide is bubbled through a solution containing these cations, the following equilibrium is established:

$$H_2S(aq) \rightleftharpoons 2H^+(aq) + S^{2-}(aq)$$

In acid solution, H$^+$ ions will displace the equilibrium to the left. This reduces the concentration of S^{2-}(aq) to very low values. Hence, only those metal sulphides with very

▲ As an oyster grows, it must adjust conditions so that the concentration of carbonate ions and calcium ions is large enough to precipitate calcium carbonate from sea water to form its shell.

Table 22.3
Selective precipitation of CuS, ZnS and NiS

Cation under test	H$_2$S bubbled into acidic solution of cation	H$_2$S bubbled into neutral solution of cation	H$_2$S bubbled into alkaline solution of cation
Cu^{2+}(aq)	Black ppte of CuS $Cu^{2+}(aq) + S^{2-}(aq) \rightarrow CuS(s)$	Black ppte of CuS	Black ppte of CuS
Zn^{2+}(aq)	No ppte	White ppte of ZnS $Zn^{2+}(aq) + S^{2-}(aq) \rightarrow ZnS(s)$	White ppte of ZnS
Ni^{2+}(aq)	No ppte	No ppte	Black ppte of NiS $Ni^{2+}(aq) + S^{2-}(aq) \rightarrow NiS(s)$

22 Ionic Equilibria in Aqueous Solution

low solubility products (such as CuS, $K_{s.p.}$ (CuS) $= 6.3 \times 10^{-36}$) will be precipitated. So, in acid solution;

$$[Cu^{2+}][S^{2-}] > K_{s.p.}(CuS) = 6.3 \times 10^{-36}$$

but
$$[Zn^{2+}][S^{2-}] < K_{s.p.}(ZnS) = 1.6 \times 10^{-24}$$

and
$$[Ni^{2+}][S^{2-}] < K_{s.p.}(NiS) = 4 \times 10^{-21}$$

Thus CuS is precipitated, but ZnS and NiS do not precipitate.

In neutral solution, the concentration of S^{2-}(aq) from H_2S will be higher and this enables ZnS with a higher solubility product ($K_{s.p.}$ (ZnS) $= 1.6 \times 10^{-24}$) to precipitate as well as CuS.

When the solution is made slightly alkaline with ammonia, the H_2S equilibrium is displaced further to the right and the concentration of S^{2-}(aq) is even higher. Under these conditions, NiS is precipitated as well as CuS and ZnS.

Thus, by careful adjustment of the pH before passing in H_2S, it is possible to separate a mixture of Cu^{2+}, Zn^{2+} and Ni^{2+}.

22.5 The strengths of acids and bases

The strengths of different acids and bases can be compared using conductivity measurements.

Strong electrolytes, such as hydrochloric acid and sodium hydroxide, are virtually completely dissociated into ions in aqueous solutions. Therefore, they are better conductors than weak electrolytes, such as ethanoic (acetic) acid, which are only partially dissociated:

$$HCl(aq) \rightleftharpoons H^+(aq) + Cl^-(aq) \qquad \text{almost complete dissociation}$$
$$CH_3COOH(aq) \rightleftharpoons H^+(aq) + CH_3COO^-(aq) \qquad \text{only partial dissociation}$$

The simple descriptive terms 'strong' and 'weak' are much too limited and inaccurate as a method of comparing the strengths of electrolytes. So, chemists looked for a more accurate and quantitative comparison. In the case of acids, relative strengths can be compared by measuring the concentration of H^+ ions or by measuring their **pH**. The 'p' in pH comes from the German word 'potenz' meaning power and the 'H' from $[H^+]$.

> The pH of a solution is the negative logarithm to base ten of the molar hydrogen ion concentration, i.e.,
>
> $$pH = -lg[H^+(aq)]$$

(Notice that the accepted abbreviation for logarithm to base 10 is lg not $\log_{10}$.)

Hydrogen ion concentrations in aqueous solution range from about 10^{-15} to 10 mol dm^{-3}, so it is convenient to have a scale that is both negative and logarithmic to show the relative strengths of acids. The negative sign produces positive pH values for almost all solutions encountered in practice. The logarithmic scale reduces the extremely wide variation in $[H^+(aq)]$ to a narrow range of pH from about 15 to −1.

The following examples show how pH can be obtained from $[H^+(aq)]$ and vice versa.

(a) What is the pH of 10^{-1} mol dm^{-3} HCl?
Since HCl is fully dissociated and monobasic,
$$[H^+(aq)] \text{ in } 10^{-1} \text{ mol dm}^{-3} \text{ HCl} = 10^{-1} \text{ mol dm}^{-3}$$
$$\therefore pH = -lg[H^+(aq)] = -lg[10^{-1}] = -(-1) = +1$$

(b) What is the pH of 10^{-3} mol dm^{-3} H_2SO_4?
Since H_2SO_4 is fully dissociated and dibasic,
$$[H^+(aq)] \text{ in } 10^{-3} \text{ mol dm}^{-3} \text{ H}_2SO_4 = 2 \times 10^{-3} \text{ mol dm}^{-3}$$
$$\therefore pH = -lg[H^+(aq)] = -lg(2 \times 10^{-3})$$
$$= -(+0.30 - 3.00)$$
$$= -(-2.70)$$
$$= +2.70$$

What are the pH values of the following solutions:

1 10^{-3} mol dm^{-3} HCl,
2 1.0 mol dm^{-3} HCl,
3 3 mol dm^{-3} HX which is only 50% dissociated?

(c) The pH of pure water at 25°C is 7. What is its hydrogen ion concentration?

$$pH = -lg[H^+(aq)]$$
$$\Rightarrow 7 = -lg[H^+(aq)]$$
$$\therefore lg[H^+(aq)] = -7$$
$$\therefore [H^+(aq)] = 10^{-7} \text{ mol dm}^{-3}$$

22.6 The dissociation of water

When water is purified by repeated distillation, its conductance falls to a constant, low value. Even the purest water has a tiny electrical conductivity. This is further evidence that water dissociates to form ions, i.e.

$$H_2O(l) \rightleftharpoons H^+(aq) + OH^-(aq)$$

Obviously, the concentration of ions is very small, as the pH of pure water shows. The equilibrium in this reaction lies far to the left. Nevertheless, we can write an equilibrium constant for the dissociation of water as

$$K_c = \frac{[H^+][OH^-]}{[H_2O]}$$

Now, as only a minute trace of the water is ionised, each cubic decimetre of water will contain virtually 1000 g of H_2O. Furthermore, as 1 mole of water weighs 18 g, we can say,

$$[H_2O] \text{ in water} = \frac{1000}{18} = 55.56 \text{ mol dm}^{-3}, \text{ which is constant.}$$

Thus, we can incorporate this constant for $[H_2O]$ in the value of K_c, just as we did with the concentration of undissolved solute in considering the solubility product of a sparingly soluble salt, i.e.

$$K_c[H_2O] = [H^+][OH^-] = \text{a new constant, } K_w$$

constant   constant

This overall constant, K_w, is called the **ionic product for water**.

At 25°C, $[H^+] = [OH^-] = 10^{-7}$ M
hence $K_w = [H^+][OH^-] = 10^{-7} \times 10^{-7} = 10^{-14}$ mol^2 dm^{-6}

1 Why is $[H^+] = [OH^-]$ in pure water?
2 At 25°C, $K_w = 10^{-14}$. What is the value of K_c for the reaction

$$H_2O(l) \rightleftharpoons H^+(aq) + OH^-(aq) \text{ at 25°C?}$$

3 Look at the information in table 22.4. How does the value of K_w change with an increase in temperature?

4 Explain the effect of temperature on K_w using Le Chatelier's principle
$(H_2O(l) \rightleftharpoons H^+(aq) + OH^-(aq); \quad \Delta H = +58 \text{ kJ})$.

Table 22.4
Values of K_w at various temperatures

Temperature/°C	K_w/mol^2 dm^{-6}
0	0.11×10^{-14}
10	0.30×10^{-14}
20	0.68×10^{-14}
25	1.00×10^{-14}
50	5.47×10^{-14}
100	51.3×10^{-14}

22.7 The pH scale

In pure water and in neutral solutions such as sodium chloride, H$^+$ and OH$^-$ ions arise only from the ionisation of water.

Hence in water and in neutral solutions,

$$[H^+] = [OH^-]$$

At 25°C, $[H^+][OH^-] = 10^{-14}$

$$\therefore [H^+] = [OH^-] = 10^{-7} \therefore pH = 7$$

Notice, however, in table 22.4 that K_w, which equals $[H^+][OH^-]$, rises with temperature.

At 50°C, for example,

$$K_w = [H^+][OH^-] = 5.47 \times 10^{-14} \text{ mol}^2 \text{ dm}^{-6}$$

So, at 50°C,

$$[H^+] \text{ in neutral solution} = (5.47 \times 10^{-14})^{\frac{1}{2}}$$
$$= 2.34 \times 10^{-7} \text{ mol dm}^{-3}$$

$$\therefore \text{ pH of neutral solutions at 50°C}$$
$$= -\lg[2.34 \times 10^{-7}]$$
$$= 6.6$$

This shows that the pH of neutral solutions varies with temperature. It is 7.0 only at 25°C. At 0°C, the pH of a neutral solution is 7.5, at 50°C it is 6.6 and at 100°C it is 6.1.

So far, we have been considering neutral solutions in which H^+ and OH^- ions can only come from water. In acidic and alkaline solutions H^+ and OH^- ions may arise from sources other than water. Nevertheless, the system

$$H_2O(l) \rightleftharpoons H^+(aq) + OH^-(aq)$$

is still in equilibrium. So, the product $[H^+][OH^-]$ remains constant for all solutions at the same temperature. Thus, it is possible to determine both the $[H^+]$ and the $[OH^-]$ in any solution. For example, in 10^{-2} mol dm^{-3} HCl,

$$[H^+] = 10^{-2} \text{ and } \therefore \text{ pH} = 2$$

but since $[H^+][OH^-] = 10^{-14}$ for this solution,

$$[OH^-] \text{ in } 10^{-2} \text{ mol dm}^{-3} \text{ HCl} = 10^{-12}$$

Likewise, in 10^{-1} mol dm^{-3} NaOH,

$$[OH^-] = 10^{-1}$$

but since $[H^+][OH^-] = 10^{-14}$ for this solution,

$$[H^+] \text{ in } 10^{-1} \text{ mol dm}^{-3} \text{ NaOH} = 10^{-13} \text{ and } \therefore \text{ pH} = 13$$

These results will help you to appreciate the following important generalisations.

For neutral solutions,
$[H^+] = [OH^-] = 10^{-7}$ mol dm^{-3} and pH = 7 at 25°C.

For acidic solutions,
$[H^+] > [OH^-]$ and pH < 7 at 25°C.

For basic (alkaline) solutions,
$[H^+] < [OH^-]$ and pH > 7 at 25°C.

Figure 22.1 relates the pH scale to the hydrogen ion concentration and to changing acidity and alkalinity.

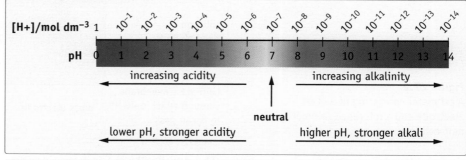

Figure 22.1
The pH scale.

22.8 The measurement of hydrogen ion concentration and pH

The most obvious method of measuring the hydrogen ion concentration of a solution is by using the hydrogen half-cell (hydrogen electrode) described in section 14.3. Under standard conditions, the hydrogen half-cell is assigned an electrode potential of 0.00 volts. Thus, when the standard hydrogen electrode is combined with another electrode to form a complete cell, the numerical value of the overall e.m.f. will depend on the second electrode. However, if the concentration of H^+ ions in the hydrogen half-cell changes, then the electrode potential of this cell is no longer 0.00 volts and the overall e.m.f. will also change. A calibration curve can be constructed showing the overall e.m.f. against the $[H^+]$ in the hydrogen half-cell using various solutions of known $[H^+]$. It is then possible to place the hydrogen electrode in any solution and obtain its pH from the resulting e.m.f.

Unfortunately, the hydrogen half-cell is awkward and inconvenient to use. It requires the use of a bulky hydrogen cylinder, and it is difficult to adjust the gas pressure so as to provide a steady and satisfactory flow rate. It also takes time to reach equilibrium and the platinum electrode is easily 'poisoned' by impurities in the gas or in the solution.

Consequently, more convenient electrodes were sought, of which the so-called '**glass electrode**' is the most widely used.

The glass electrode consists of a silver/silver chloride electrode (a silver wire coated with AgCl) in a solution of fixed acidic pH. The electrode is placed inside a thin glass membrane permeable only to H^+ ions. The acid solution inside the glass electrode is 0.1 mol dm^{-3} HCl. This half-cell is combined with a reference half-cell which is usually a calomel electrode (mercury in contact with mercury(I) chloride in saturated potassium chloride solution). This arrangement is attached to a sensitive voltmeter to form a **pH meter** (figure 22.2).

The electrode potential of the glass electrode is dependent on the $[H^+]$ of the solution in which it is placed. In practice, the voltmeter is calibrated to give a direct reading of the pH of the solution rather than the e.m.f. of the cell. The calibration is done using solutions of known pH. In most commercial pH meters, the glass electrode and the reference half-cell are combined in a single unit. This unit can be dipped into the solution under test.

The whole cell (glass electrode + calomel electrode) can be written as

$$Ag(s)/AgCl(s)/HCl\,(0.1\,mol\,dm^{-3})/glass/test\,solution \vdots KCl(sat)/Hg_2Cl_2(s)/Hg(l)$$

glass electrode   calomel electrode

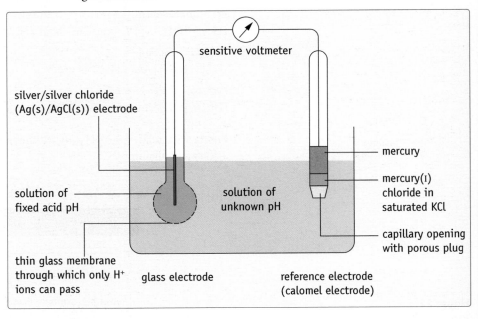

Figure 22.2
A pH meter consisting of a glass electrode and a reference calomel half-cell.

The overall cell e.m.f. is determined by

- the electrode potential of the Ag(s)/AgCl(s) electrode, E_1, which is constant;
- the potential set up at the thin glass membrane which separates two solutions of different $[H^+]$, E_2,
- the electrode potential of the calomel electrode which is constant, E_3.

▲ pH meters with protected probes, such as this one, allow the accurate determination of the pH of any solution.

Now since E_1 and E_3 are constant and since the [H$^+$] *inside* the glass bulb is constant, the overall e.m.f. depends only on the [H$^+$] of the test solution in which the glass electrode is placed.

22.9 Dissociation constants of acids and bases

The pH of a solution is sometimes used to indicate the strength of a constituent acid or base. The use of pH is, however, very limited in this context since its value will change as the concentration changes. Consequently, chemists looked for a more useful means of representing the strengths of acids and bases. They found this by considering the dissociation of these substances in aqueous solution.

When the weak acid HA is dissolved in water, we can write

$$HA(aq) \rightleftharpoons H^+(aq) + A^-(aq)$$

Hence, $K_c = \dfrac{[H^+(aq)][A^-(aq)]}{[HA(aq)]}$

In dealing with acids, K_c is usually replaced by the symbol, K_a, which is known as the **dissociation constant** of the acid. By eliminating the (aq) state symbols, we can simplify the last equation to

$$K_a = \dfrac{[H^+][A^-]}{[HA]}$$

Since dissociation constants are effectively equilibrium constants, they are unaffected by concentration changes. They are influenced only by changes in temperature. Hence, the numerical value of K_a provides an accurate measure of the extent to which an acid is dissociated (i.e. the strength of the acid).

The stronger the acid, the greater the extent of dissociation, the greater are [H$^+$] and [A$^-$] and the larger is K_a. The values of K_a for some acids are given in table 22.5.

Table 22.5
The values of K_a for some acids

Acid	Equilibrium in aqueous solution	K_a at 25°C /mol dm^{-3}
Sulphuric acid	$H_2SO_4 \rightleftharpoons H^+ + HSO_4^-$	very large
Nitric acid	$HNO_3 \rightleftharpoons H^+ + NO_3^-$	40
Trichloroethanoic acid	$CCl_3COOH \rightleftharpoons H^+ + CCl_3COO^-$	2.3×10^{-1}
Dichloroethanoic acid	$CHCl_2COOH \rightleftharpoons H^+ + CHCl_2COO^-$	5.0×10^{-2}
Sulphurous acid	$H_2SO_3 \rightleftharpoons H^+ + HSO_3^-$	1.6×10^{-2}
Chloroethanoic acid	$CH_2ClCOOH \rightleftharpoons H^+ + CH_2ClCOO^-$	1.3×10^{-3}
Nitrous acid	$HNO_2 \rightleftharpoons H^+ + NO_2^-$	4.7×10^{-4}
Methanoic acid	$HCOOH \rightleftharpoons H^+ + HCOO^-$	1.6×10^{-4}
Benzoic acid	$C_6H_5COOH \rightleftharpoons H^+ + C_6H_5COO^-$	6.4×10^{-5}
Ethanoic acid	$CH_3COOH \rightleftharpoons H^+ + CH_3COO^-$	1.7×10^{-5}
Hydrated aluminium ion	$[Al(H_2O)_6]^{3+} \rightleftharpoons H^+ + [Al(H_2O)_5OH]^{2+}$	1.0×10^{-5}
Carbonic acid	$H_2CO_3 \rightleftharpoons H^+ + HCO_3^-$	4.5×10^{-7}
Hydrogen sulphide	$H_2S \rightleftharpoons H^+ + HS^-$	8.9×10^{-8}
Boric acid	$H_3BO_3 \rightleftharpoons H^+ + H_2BO_3^-$	5.8×10^{-10}
Hydrogen peroxide	$H_2O_2 \rightleftharpoons H^+ + HO_2^-$	2.4×10^{-12}
Water	$H_2O \rightleftharpoons H^+ + OH^-$	1.8×10^{-16}

decreasing acid strength

Determination of dissociation constants

The value of K_a for a weak acid can be determined using the equation above if we know the concentration of the acid and its pH. The following example shows how this is done.

The pH of 0.01 mol dm^{-3} ethanoic(acetic) acid (CH$_3$COOH) is 3.40 at 25°C. What is the dissociation constant of ethanoic acid at this temperature?

$$CH_3COOH(aq) \rightleftharpoons H^+(aq) + CH_3COO^-(aq)$$

$$\Rightarrow K_a = \frac{[H^+][CH_3COO^-]}{[CH_3COOH]}$$

Since the concentration of H$^+$ ions arising from the water is much smaller than the concentration of those from the acid, we can say that

$$[H^+] \simeq [CH_3COO^-]$$

and
$$[CH_3COOH] = 0.01 - [H^+] \simeq 0.01$$

as $[H^+] \ll 0.01$, because ethanoic acid is a weak acid.

$$pH = -lg[H^+] = 3.40$$
Hence
$$lg[H^+] = -3.40 = (-4.00 + 0.60)$$
$$\therefore [H^+] = 4.0 \times 10^{-4} \text{ mol dm}^{-3}$$
and
$$[CH_3COO^-] = 4.0 \times 10^{-4} \text{ mol dm}^{-3}$$

Now, assuming $[CH_3COOH] = 0.01$ mol dm^{-3},

$$K_a = \frac{[H^+][CH_3COO^-]}{[CH_3COOH]} = \frac{4 \times 10^{-4} \times 4 \times 10^{-4}}{0.01}$$

$$K_a = 1.60 \times 10^{-5} \text{ mol dm}^{-3}$$

By reversing this calculation, it is possible to predict the pH of a solution of a weak acid. We must, however, know the molarity of the acid and its dissociation constant.

> The following questions will help you to calculate the pH of 1.0 mol dm^{-3} benzoic acid (C$_6$H$_5$COOH). K_a(C$_6$H$_5$COOH) = 6.4 × 10^{-5} mol dm^{-3}
>
> 1 Write an equation for the dissociation of C$_6$H$_5$COOH.
>
> 2 Using this equation, write an expression for the dissociation constant of benzoic acid.
>
> 3 Assuming $[H^+] = [C_6H_5COO^-]$ and $[C_6H_5COOH] \gg [H^+]$, calculate $[H^+]$ for 1.0 mol dm^{-3} benzoic acid.
>
> 4 What is the pH of 1.0 mol dm^{-3} benzoic acid?

The K_a values of weak acids vary from 10^{-2} to 10^{-10}. In order to have and compare more accessible and more manageable values, the term pK_a is often used. p$K_a = -lg\, K_a$ (i.e. $-$ log to base 10 of K_a). This gives pK_a values from 2 to 10.

Just as acid dissociation constants can be used to compare the strengths of different acids, we can also use base dissociation constants for bases. If the base BOH is in equilibrium with its dissociated ions as

$$BOH \rightleftharpoons B^+ + OH^-$$

then
$$K_b = \frac{[B^+][OH^-]}{[BOH]}$$

where K_b is known as the dissociation constant of the base.

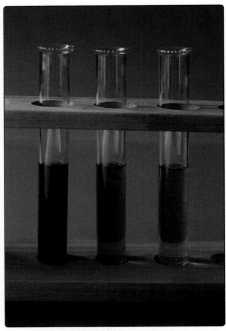

▲ The colouring material in red cabbage can be used as an indicator. The red coloured extract shown in the middle tube is yellow with acid (right-hand tube) and blue/purple with alkali (left tube).

Acid–base indicators such as methyl orange, phenolphthalein and bromothymol blue *are substances which change colour according to the hydrogen ion concentration of the solution to which they are added* (figure 22.3). Consequently, they are used to test for acidity and alkalinity and to detect the end point in acid–base titrations.

Most indicators can be regarded as weak acids of which either the undissociated molecule or the dissociated anion, or both, are coloured. If we take methyl orange as our example and write the undissociated molecule as HMe,

$$HMe \rightleftharpoons H^+ + Me^-$$
$$\text{red} \qquad \text{colourless} \quad \text{yellow}$$

Addition of acid (i.e. H^+ ions) displaces this equilibrium to the left. When this happens $[HMe] \gg [Me^-]$ and the solution becomes red.

On the other hand, when alkali (containing OH^- ions) is added to methyl orange it reacts with H^+ ions forming water. The equilibrium in the above system moves to the right in order to replace some of the H^+ ions. In this case, $[Me^-] \gg [HMe]$, and the methyl orange becomes yellow.

> The dissociation of phenolphthalein in aqueous solution can be represented as
>
> $$HPh \rightleftharpoons H^+ + Ph^-$$
>
> 1 What colour is phenolphthalein in
>   (a) strongly acid solution,
>   (b) strongly alkaline solution?
>
> 2 What are the colours of
>   (a) HPh
>   (b) Ph^- for phenolphthalein?

Figure 22.3
The formulas of two common indicators.

Indicators can be regarded as weak acids, so it is possible to determine their dissociation constants. Using HIn for the undissociated form of the indicator, we can write

$$HIn(aq) \rightleftharpoons H^+(aq) + In^-(aq)$$

$$\Rightarrow K_a(HIn) = \frac{[H^+][In^-]}{[HIn]}$$

The numerical values of these dissociation constants can, of course, be obtained by measuring the pH of a solution of known molarity for each indicator (see section 22.9). The dissociation constants for some indicators are shown in table 22.6.

Table 22.6
The values of K_a for some indicators

Indicator	K_a at 25°C /mol dm^{-3}
Phenolphthalein	7×10^{-10}
Bromothymol blue	1×10^{-7}
Litmus	3×10^{-7}
Methyl orange	2×10^{-4}

The end point of an indicator

The aim of any titration is to determine the volumes of two solutions which just react with each other. The **end point** is the point at which the titration is stopped. This must coincide with the **equivalence point** for the two reacting solutions. In order to achieve this, the indicator should change colour sharply at the equivalence point on adding a single drop of either acid or alkali. At the exact equivalence point of the titration, the colour of the indicator will be mid-way between the acid colour of HIn and the alkaline colour of In^-, i.e. $[HIn] = [In^-]$.

Now, $$K_a(HIn) = \frac{[H^+][In^-]}{[HIn]}$$

Table 22.7
The pH at the end point for some indicators

Indicator	pH at end point $= -\lg K_a$ (HIn)
Phenolphthalein	9.1
Bromothymol blue	7.0
Litmus	6.5
Methyl orange	3.7

But, at the end point, $[\text{HIn}] = [\text{In}^-]$

$\therefore$ At the end point, K_a (HIn) $= [\text{H}^+]$

$\Rightarrow$ pH at end point $= -\lg [\text{H}^+] = -\lg K_a$ (HIn).

The pHs at the end point for the four indicators previously mentioned are shown in table 22.7.

The range of an indicator

The colour change of an indicator is due to the change from one coloured form to another. Near the end point, both coloured forms will be present in appreciable quantities. It is not possible to say precisely when the two forms are at equal concentrations. Experience shows that our eyes cannot judge the exact end point, and indicators effectively change colour over a range of about 2 pH units.

The range of an indicator is the pH range over which it changes colour. The values in tables 22.6 and 22.7 show that the dissociation constants of indicators differ widely. Consequently, they change colour over widely differing ranges of pH. This last point is illustrated more clearly in figure 22.4.

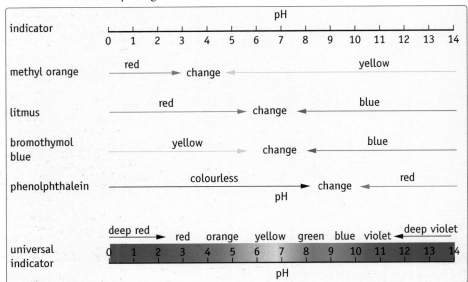

Figure 22.4
The pH ranges of some indicators.

> Look closely at figure 22.4.
>
> 1 What is the pH of pure water?
>
> 2 What colours will the following indicators give in pure water:
> (a) methyl orange,
> (b) phenolphthalein,
> (c) bromothymol blue?
>
> 3 When too much phenolphthalein is added to water, it forms a cloudy suspension. Why is this?
> (*Hint*: Phenolphthalein is an organic compound of high relative molecular mass.)

Some indicators, like phenolphthalein, are only slightly soluble in water. They are therefore often used as solutions in alcohol or in a mixture of alcohol and water. Because of this, the colour of an indicator may be confusing before it is added to an aqueous solution.

Notice that the end point of each indicator is in the centre of its pH range. Figure 22.4 also shows that many indicators change colour over a range of pH well away from 7. This is an important point. Many students misunderstand the use of indicators and expect them all to change colour at pH 7. In fact it is because different indicators change colour at different pH values that acid–base titrations have such a wide application in industry and in the laboratory.

22 Ionic Equilibria in Aqueous Solution

During any acid–base titration there is a change in pH as alkali is added to acid or vice versa. The pH must change sharply by several units at the equivalence point if it is to be identifiable using an indicator.

The actual change in pH during the course of a titration depends largely upon the strength of the acid and alkali used.

Titrating strong acid against strong alkali

The graph in figure 22.5 shows how the pH changes during the titration of 50 cm^3 of 0.1 mol dm^{-3} HCl with 0.1 mol dm^{-3} NaOH. As the alkali is added, the pH changes slowly at first. It does, however, change very rapidly from about 3.5 to 9.5 at the equivalence point. Thus, any indicator which changes colour between pH 3.5 and 9.5 will identify the equivalence point. This means that any one of methyl orange, bromothymol blue or phenolphthalein could be used.

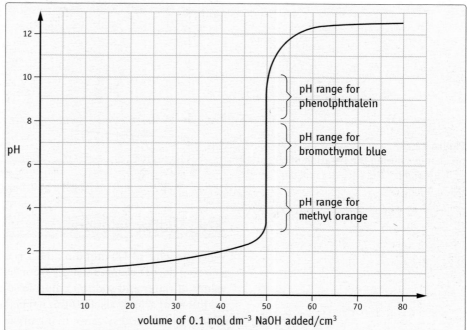

Figure 22.5
pH changes during the titration of 50 cm^3 of 0.1 mol dm^{-3} HCl with 0.1 mol dm^{-3} NaOH.

▶ The cells of animals and plants are protected against small amounts of acid or alkali by buffers. However serious damage results when too much acid or alkali is involved. The conifers in this photo have been damaged by acid rain.

Titrating strong acid against weak alkali

The graph in figure 22.6 shows how the pH changes when 0.1 mol dm^{-3} NH$_3$ solution (weak alkali) is added to 50 cm^3 of 0.1 mol dm^{-3} HCl. As before, there is little variation in pH when the alkali is first added. But, at the equivalence point the pH changes rapidly from about 3.5 to 7.0. Thus, any indicator which changes colour between 3.5 and 7.0 will identify the equivalence point accurately. In this case, methyl orange is a very suitable indicator. Bromothymol blue may also be used, but phenolphthalein is useless because it does not begin to change colour until about pH 8.

The pH changes during the titration of a weak acid with a strong alkali and a weak acid with a weak alkali are discussed in review question 10 at the end of this chapter.

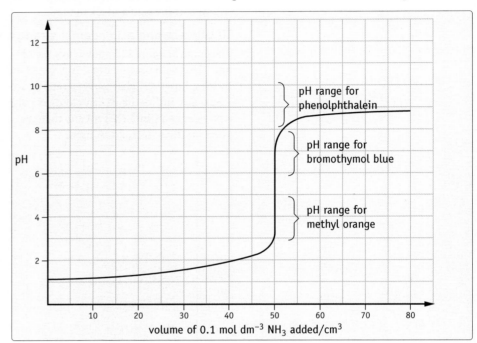

Figure 22.6
pH changes during the titration of 50 cm^3 of 0.1 mol dm^{-3} HCl with 0.1 mol dm^{-3} NH$_3$.

22.12 Buffer solutions

When 0.1 cm^3 of 1.0 mol dm^{-3} HCl is added to 1 dm^3 of water or sodium chloride solution, the pH changes sharply from 7.0 to 4.0 (i.e. by 3 units of pH). Clearly, the pH values of water and sodium chloride solution are extremely sensitive to even small additions of acid or alkali. If this happened when small amounts of acid or alkali were added to biological systems, living organisms would be killed instantly. Fortunately, animals and plants are protected against sharp changes in pH by the presence of **buffers**.

> *Buffers are solutions which resist changes in pH on addition of acid or alkali.*

Buffers are also important in many industrial processes where the pH must not deviate very much from an optimum value. Furthermore, many synthetic and processed foods must be prepared in a buffered form so that they may be eaten and digested in our bodies without undue change in pH. The pH of blood is normally 7.4. Under most circumstances, a change of only 0.5 units in the pH of blood would be fatal.

How does a buffer act? How does it resist changes in pH when acid or alkali is added?

Buffer solutions usually consist of:
either a solution of a weak acid in the presence of one of its salts (e.g. ethanoic acid and sodium ethanoate, carbonic acid and sodium hydrogencarbonate);
or a solution of a weak base in the presence of one of its salts (e.g. ammonia solution and ammonium chloride).

In order to understand how a buffer works, we can consider the hypothetical weak acid, HA, in a solution with its salt MA. In this solution, HA will be slightly dissociated whilst MA is fully dissociated:

$$HA \rightleftharpoons H^+ + A^-$$
$$MA \rightarrow M^+ + A^-$$

Hence the mixture contains a relatively high concentration of un-ionised HA (an acid) and a relatively high concentration of A^- (a base).

If an acid is suddenly added to this system, the H^+ ions in the acid combine with A^- ions to form un-ionised HA. Provided there is a large reservoir of A^- ions in the buffer, nearly all the added H^+ ions are removed. Thus, $[H^+]$ changes very little and the pH is only slightly altered.

If an alkali, such as sodium hydroxide, is suddenly added to the system, the OH^- ions combine with H^+ ions to form water. This reduces the concentration of H^+ ions in the buffer, so more HA dissociates to restore the equilibrium and the $[H^+]$ rises almost to its original value. Provided there is a large reservoir of HA in the buffer, the $[H^+]$ changes very little and again the pH is only slightly altered. By having these reserves of both HA and A^- in the buffer mixture, changes in the pH resulting from the addition of acid or alkali can be minimised.

Essentially, the stable pH of a buffer is due to
(a) a high $[A^-]$ which traps added H^+ ions, and
(b) a high [HA] which can supply H^+ ions to trap added OH^- ions.

Solutions of this kind can minimise the effects of adding acid or alkali.

Calculating the pH of buffer solutions

In a buffer composed of the weak acid HA and its salt MA,

$$HA \rightleftharpoons H^+ + A^-$$

and

$$MA \rightarrow M^+ + A^-$$

We can write an expression for the dissociation constant of HA as

$$K_a = \frac{[H^+][A^-]}{[HA]}$$

$$\therefore [H^+] = K_a \frac{[HA]}{[A^-]}$$

In the buffer mixture, [HA] is effectively the concentration of acid taken ([acid]) because the acid will be only very slightly dissociated. $[A^-]$ is effectively the concentration of salt taken ([salt]) because the salt is fully dissociated into ions.

Thus we can write

$$[H^+] \simeq K_a \times \frac{[acid]}{[salt]}$$

This last equation also explains why the $[H^+]$, and therefore the pH, of a buffer, is affected very little by dilution. This is because the ratio [acid]/[salt] will remain constant on dilution.

Using the last equation, notice the special case which applies when [acid] = [salt], i.e.

$$[H^+] \simeq K_a$$

The following example will help you to understand the points which have just been discussed.

A buffer solution was made by adding 3.28 g of sodium ethanoate to 1 dm^3 of 0.01 mol dm^{-3} ethanoic acid. What is the pH of this buffer?

$$(K_a(CH_3COOH) = 1.7 \times 10^{-5} \text{ mol dm}^{-3})$$

$$K_a = \frac{[H^+][CH_3COO^-]}{[CH_3COOH]}$$

$$\therefore [H^+] = K_a \frac{[CH_3COOH]}{[CH_3COO^-]}$$

But, $[CH_3COOH] = [acid] = 0.01 \text{ mol dm}^{-3}$

and $[CH_3COO^-] = [salt] = \dfrac{3.28}{82} = 0.04 \text{ mol dm}^{-3}$

$$\therefore [H^+] = 1.7 \times 10^{-5} \times \dfrac{0.01}{0.04} = 4.25 \times 10^{-6} \text{ mol dm}^{-3}$$

$$\therefore \text{pH of the buffer} = -lg\,[H^+] = -lg\,(4.25 \times 10^{-6})$$
$$= -(+0.63 - 6.00) = -(-5.37)$$
$$\Rightarrow \underline{\text{pH of the buffer} = 5.37}$$

In order to appreciate the action of a buffer let us first consider the change in pH when 1 cm^3 of 1.0 mol dm^{-3} NaOH is added to 1 dm^3 of the buffer in the last example. Then we should consider the change in pH when 1 cm^3 of 1.0 mol dm^{-3} NaOH is added to 1 dm^3 of 0.01 mol dm^{-3} ethanoic acid.

1 *The change in pH when 1 cm^3 of 1.0 mol dm^{-3} NaOH is added to 1 dm^3 of buffer in the last example*

$$\text{pH of buffer initially} = 5.37$$

When NaOH is added to the buffer it reacts with CH$_3$COOH and forms CH$_3$COONa (i.e.[CH$_3$COOH] falls and [CH$_3$COO$^-$] rises). 1 cm^3 of 1.0 mol dm^{-3} NaOH contains 0.001 moles of OH$^-$. This removes 0.001 moles of CH$_3$COOH and forms 0.001 moles of CH$_3$COONa.

∴ After adding 0.001 moles of NaOH,

$$[CH_3COOH] = 0.010 - 0.001 = 0.009$$

$$[CH_3COO^-] = 0.040 + 0.001 = 0.041$$

$$\Rightarrow [H^+] = K_a \dfrac{[acid]}{[salt]} = 1.7 \times 10^{-5} \times \dfrac{0.009}{0.041}$$

$$[H^+] = 3.73 \times 10^{-6}$$
$$\therefore \underline{\text{pH after adding NaOH} = 5.43}$$

Therefore, the pH of the buffer changes by only 0.06 units.

2 *The change in pH when 1cm^3 of 1.0 mol dm^{-3} NaOH is added to 1 dm^3 of 0.01 mol dm^{-3} ethanoic acid. Before the addition of alkali,*

$$[H^+] = [CH_3COO^-] \ll [CH_3COOH]$$

$$\therefore \text{using } K_a = \dfrac{[H^+][CH_3COO^-]}{[CH_3COOH]}$$

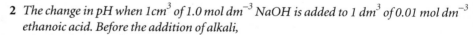

$$[H^+]^2 = K_a \times [CH_3COOH] = 1.7 \times 10^{-5} \times 10^{-2} = 17 \times 10^{-8}$$
$$\therefore [H^+] = 4.12 \times 10^{-4}$$
$$\therefore \underline{\text{pH of 0.01 mol dm}^{-3} \text{ ethanoic acid} = 3.39}$$

On adding 0.001 moles of NaOH,

$$[CH_3COOH] = 0.010 - 0.001 = 0.009$$

$$\text{and } [CH_3COO^-] = 0.001$$

$$\therefore [H^+] = K_a \times \dfrac{[acid]}{[salt]} = 1.7 \times 10^{-5} \times \dfrac{0.009}{0.001} = 1.53 \times 10^{-4}$$

$$\Rightarrow \underline{\text{pH after adding NaOH} = 3.82}$$

Therefore, the pH of the 0.01 mol dm^{-3} ethanoic acid changes by 0.43 units on adding alkali. This is more than seven times the pH change of the buffer. It illustrates the pH-stabilising nature of the buffer very nicely.

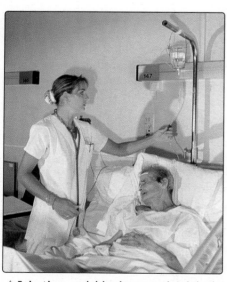

▲ Injections and drips into a patient's body must be carefully buffered so that the pH values of body fluids do not change too much.

22 Ionic Equilibria in Aqueous Solution

The main use of buffers in the laboratory is in preparing solutions of known and constant pH. Buffer solutions cannot be made by simply preparing acid or alkaline solutions of a given concentration. The pH of such solutions will vary slightly as gases from the atmosphere, such as CO_2, dissolve in them, or as traces of alkali dissolve from the glass vessel.

Buffer solutions are also important in medicine and in agriculture because the pH values of living systems must be maintained at certain critical values. Because of this, intravenous injections must be carefully buffered so as not to change the pH of the blood from its normal value of 7.4. In the same way, most fermentation processes must also be buffered, otherwise relatively small changes in the pH will cause the death of the fermenting organisms. In living systems, the buffering action is usually provided by H_2CO_3 and HCO_3^-, by $H_2PO_4^-$ and HPO_4^{2-} and by various proteins which can both accept and donate H^+ ions.

Summary

1   For the sparingly soluble strong electrolyte, A_xB_y, in contact with excess undissolved solute,

$$[A^{y+}]^x[B^{x-}]^y = K_{s.p.}$$

$K_{s.p.}$, known as the solubility product of A_xB_y, is constant at constant temperature.

2   The solubility product concept is valid only for saturated solutions in which the total concentration of ions is no more than about 0.01 mol dm^{-3}. It is therefore inappropriate for soluble salts such as NaCl, KNO_3 and $CuSO_4$.

3   In the presence of either A^{y+} or B^{x-} from a second source, the solubility of the salt A_xB_y is reduced. This generalisation is known as the common ion effect.

4   Precipitation of an insoluble salt occurs when

ionic product > solubility product`

In this case, precipitation occurs until

ionic product = solubility product

5   The pH of a solution provides a quantitative measure of its acidity or alkalinity.

$$pH = -lg[H^+]$$

6   At 25°C, the product, $[H^+][OH^-] = 10^{-14}$ for all aqueous solutions.

For neutral solutions, $[H^+] = [OH^-] = 10^{-7}$ mol dm^{-3} and pH = 7.

For acidic solutions, $[H^+] > [OH^-]$ and pH < 7.

For alkaline solutions, $[H^+] < [OH^-]$ and pH > 7.

7   For the weak acid, HA, dissolved in water

$$K_a = \frac{[H^+][A^-]}{[HA]}$$

K_a, known as the dissociation constant of the acid, is constant at constant temperature.

The numerical value of K_a provides a quantitative indication of the extent to which an acid is dissociated (i.e. the strength of the acid). The greater the extent of dissociation, the greater are $[H^+]$ and $[A^-]$, the larger is K_a and the stronger the acid.

8   K_a, like $K_{s.p.}$, is a modified equilibrium constant.

9   Acid–base indicators change colour according to the $[H^+]$ of the solution to which they are added. They are used to test for acidity and alkalinity and to detect the end point in acid–base titrations.

10   The range of an indicator is the pH range over which it changes colour.

11   Buffers are solutions which resist changes in pH on dilution or on addition of acid or alkali. Animals and plants are protected against changes in pH by the presence of buffers.

Review questions

1   The solubility of silver bromide in water is 7×10^{-7} mol dm^{-3} at 25°C. Calculate its solubility product.

2   (a) Calculate the solubility of silver ethanedioate (oxalate) ($Ag_2C_2O_4$) in water.
($K_{s.p.}$ ($Ag_2C_2O_4$) = 5×10^{-12} mol^3 dm^{-9})

(b) How would you expect the value of a solubility product to vary with temperature? Explain your answer.

3 The solubility product of lead(II) sulphate, $PbSO_4$, in water, is 1.6×10^{-8} mol^2 dm^{-6}.
  (a) Calculate the solubility of lead(II) sulphate in
      (i)   pure water,
      (ii) 0.1 mol dm^{-3} $Pb(NO_3)_2$ solution,
      (iii) 0.01 mol dm^{-3} Na_2SO_4 solution.
  (b) Why is lead(II) sulphate more soluble in water than in any solution containing either Pb^{2+} or SO_4^{2-} ions?
  (c) Use your understanding of solubility product to explain the so-called 'common ion' effect.
      Illustrate your answer with an example.

4 Suppose the acid, HX, is a weak electrolyte.
  (a) What happens to the degree of dissociation of HX in aqueous solution if:
      (i)   water is added,
      (ii) gaseous HCl is added,
      (iii) solid NaX is added?
  (b) Assume that $K_a(HA) = 10^{-6}$, $K_a(HB) = 10^{-8}$ and $K_a(HC) = 10^{-10}$ mol dm^{-3}, respectively.
      (i)   Which solution has the highest $[H^+]$, 1.0 mol dm^{-3} HA, 1.0 mol dm^{-3} HB or 1.0 mol dm^{-3} HC?
      (ii) What is the $[H^+]$ in the solution with the highest value of $[H^+]$?

5 Chloric(I) (hypochlorous) acid, HClO, is a weak acid.
  $K_a(HClO) = 3.2 \times 10^{-8}$ mol dm^{-3}
  (a) Calculate the $[H^+]$ and $[OH^-]$ in 1.25×10^{-2} mol dm^{-3} HClO.
  (b) What is the pH of 1.25×10^{-2} mol dm^{-3} HClO?

6 Explain the following observations:
  (a) When 1.0 mol dm^{-3} hydrochloric acid is diluted, the pH rises. Eventually, the pH reaches a static value and does not change on further dilution.
  (b) A solution of iron(III) chloride is acid to litmus.
  (c) Phenolphthalein can be used to determine the equivalence point in the titration of ethanoic acid with sodium hydroxide, but methyl orange is no use in this case.
  (d) The pH of 10^{-8} mol dm^{-3} HCl is not 8.

7 Benzoic acid, C_6H_5COOH, is a weak monobasic acid.
  ($K_a = 6.4 \times 10^{-5}$ mol dm^{-3})
  (a) Explain how a mixture of benzoic acid and sodium benzoate can act as a buffer on the addition of small amounts of either HCl(aq) or NaOH(aq).
  (b) What is the $[H^+]$ in 0.02 mol dm^{-3} benzoic acid?
  (c) What is the pH of 0.02 mol dm^{-3} benzoic acid?
  (d) What is the pH of a solution containing 7.2 g of sodium benzoate in 1 dm^3 of 0.02 mol dm^{-3} benzoic acid?
  (e) By how much will the pH change if 1 cm^3 of 1.0 mol dm^{-3} NaOH is added to the buffer in part (d)?

8 Assuming that the pH of blood is maintained at 7.4 by the acid $H_2PO_4^-$ and its salt, HPO_4^{2-}, calculate the ratio of the concentration of $H_2PO_4^-$ to that of HPO_4^{2-} in blood. ($K_a(H_2PO_4^-) = 6.4 \times 10^{-8}$ mol dm^{-3})

9 Calculate the pH of
  (a) 10^{-4} mol dm^{-3} HCl,
  (b) 10^{-4} mol dm^{-3} $Ba(OH)_2$,
  (c) 1.0 mol dm^{-3} H_2X which is only 50% dissociated,
  (d) 0.01 mol dm^{-3} propanoic acid, ($K_a = 1.45 \times 10^{-5}$ mol dm^{-3})
  (e) 1.0 mol dm^{-3} NH_4OH. ($K_b = 1.7 \times 10^{-5}$ mol dm^{-3})

10 Figure 22.7 shows how the pH changes when 0.1 mol dm^{-3} CH_3COOH is titrated against 0.1 mol dm^{-3} NaOH (curve a). Curve b shows what happens when 0.1 mol dm^{-3} CH_3COOH is titrated against 0.1 mol dm^{-3} NH_3.
  (a) The pH of 0.1 mol dm^{-3} HCl is 1.0. Why is the pH of 0.1 mol dm^{-3} CH_3COOH about 2.8 rather than 1.0?
  (b) Is methyl orange a suitable indicator to use when titrating 0.1 mol dm^{-3} CH_3COOH against 0.1 mol dm^{-3} NaOH? Explain.
  (c) Is phenolphthalein a suitable indicator to use when titrating 0.1 mol dm^{-3} CH_3COOH against 0.1 mol dm^{-3} NaOH? Explain.
  (d) No indicator will detect the equivalence point accurately when titrating 0.1 mol dm^{-3} CH_3COOH against 0.1 mol dm^{-3} NH_3. Why is this?
  (e) In view of the statement in (d), how could you determine the equivalence point in titrating a weak acid against a weak alkali?

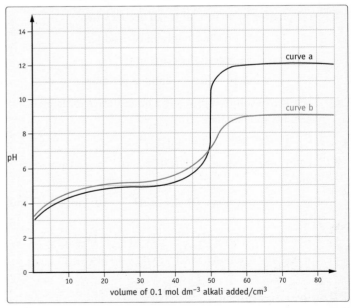

Figure 22.7
pH changes during the titration of 50 cm^3 of 0.1 mol dm^{-3} CH_3COOH with 0.1 mol dm^{-3} NaOH (curve a) and during the titration of 50 cm^3 of 0.1 mol dm^{-3} CH_3COOH with 0.1 mol dm^{-3} NH_3 (curve b).

23 Entropy and Free Energy

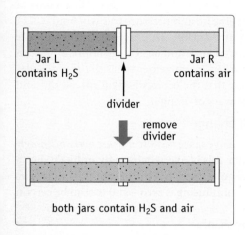

Figure 23.1
Mixing hydrogen sulphide and air.

Jar L contains H₂S — Jar R contains air — divider — remove divider — both jars contain H₂S and air

23.1 Introduction

Why do things happen the way they do? If you put a block of ice over the flame of a bunsen burner, it melts. Why? Why does the water not refreeze over the bunsen? Take the bunsen burner itself for that matter, and the methane gas burning in it. Why do methane and oxygen readily turn to carbon dioxide and water, heating the surroundings? Why do carbon dioxide and water never form methane and oxygen of their own accord?

Chemists are always interested in the direction of change. The answers to the questions above may seem obvious, but there are many changes for which it is not at all easy to predict the direction in which a reaction will go. Consider, for example, the Haber process (page 312). Will nitrogen and hydrogen form ammonia of their own accord or vice versa? Here it is very important to choose the right conditions – temperature and pressure in particular. Much of industrial chemistry involves choosing the right conditions in order to get as much of the product as possible. By understanding the factors that make a reaction go in a particular direction, it is possible to predict the conditions which might allow a reluctant reaction to go in the way you want.

23.2 A simple example – the diffusion of gases

Hydrogen sulphide, H₂S, is a very smelly gas. If you opened a gas jar of H₂S in one corner of a room, it would not be long before everyone in the room could smell it. But suppose you wanted to remove the smell from the room. It would be pointless to open a new, empty gas jar and hope the H₂S would go back in. It just would not. Why? The answer, of course, is that all gases diffuse naturally to fill any available space. Why do they do this? To answer this question, let us simplify the problem a little and reduce the room to the size of a second gas jar (figure 23.1).

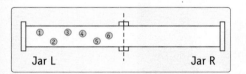

Figure 23.2
Two gas jars – there are six molecules of hydrogen sulphide in jar L, and jar R is empty.

Look at figure 23.1.

1 Why will the gases never unmix?

Simplify things still further and imagine there are only six molecules of H₂S in jar L, and that jar R is completely empty (figure 23.2). We can represent this arrangement of the six molecules in jar L as LLLLLL. The molecules move around in jar L completely at random, colliding haphazardly with each other and with the walls of the container and not caring where they go. They continue like this after the divider has been removed, but now some of them can pass into jar R. To start with, perhaps only one molecule might go into jar R (figure 23.3). This arrangement could be represented as LLLLLR. Of course, there are many other possible arrangements of the six molecules. In fact, each of the six molecules can be arranged in one of two ways, L or R. There are two arrangements for each molecule, and we can calculate the total number of possible arrangements by multiplying the number of individual arrangements together.

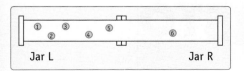

Figure 23.3
Two gas jars – the divider has been removed and some of the molecules can now pass from jar L into jar R.

For the six molecules, this is

$$2 \times 2 \times 2 \times 2 \times 2 \times 2 = 2^6 = 64$$

Only one of these 64 arrangements is

LLLLLL

– the arrangement at the beginning. So there is a 1 in 64 chance that the molecules will all end up in jar L. However, it is much more likely (a 63 in 64 chance) that the six molecules will be distributed in some other way.

So far we have considered only six molecules. For 100 molecules, the number of arrangements will be 2^{100} – about 10^{30}. Only one of these arrangements will have all 100 molecules in jar L – a 1 in 10^{30} chance. It is extremely likely that the gas molecules will arrange themselves between the two jars.

In real life, the numbers of molecules involved are much, much bigger. The volume of a gas jar might be 0.5 dm^3, and at room temperature and pressure this will hold 0.5/24 moles – about 1/50 moles of gas, containing approximately 10^{22} molecules. The number of ways these molecules can be arranged between jar L and jar R is therefore $2^{10^{22}}$ – an unimaginably huge number. Only one of these $2^{10^{22}}$ arrangements has all the molecules in jar L. Therefore it is overwhelmingly likely that the molecules will spread out and occupy both jars – there are so many different ways of doing it.

Chance, probability and the direction of change

We have just seen that gases diffuse into any available space *because they are overwhelmingly likely to do so*. They never 'undiffuse' because this is overwhelmingly unlikely. For a single molecule, the chance of 'undiffusion' is quite good, but the numbers of molecules are so huge that the probability of *all* of them undiffusing is zero – and we can be certain that they will stay diffused.

Diffusion is not the only process dictated by chance.

All changes, physical and chemical, are governed by the laws of chance and probability. Quite simply, the most likely thing will always happen. If you appreciate this, it will be very helpful in predicting changes.

▲ The cream spreads through the coffee because there are more ways it can be spread out than stay all in one place.

▶ The smell of cooking soon spreads. The laws of chance and probability say it must.

23.3 Entropy

The examples in the last section show that probability is vital in deciding whether a physical or chemical change will occur. The more ways a change can happen, the higher the chance of its happening. Clearly, the 'number of ways' in which a change can take

▲ Below 0°C, ice does not melt spontaneously, even though this would increase its entropy. Why not?

place is of vital importance in deciding whether the change occurs.

When you are dealing with just a few molecules, the number of ways in which the change can occur, W, is quite small and the figures are quite easy to cope with. But in real situations the numbers of molecules are huge. We have already seen that when 1/50 mole of gas (about 10^{22} molecules) diffuses from one gas jar to fill two jars, the increase in the number of ways of arranging these molecules is

$$W = 2^{10^{22}}$$

This is an enormous number and very difficult to handle. Fortunately, W itself is not usually used. Instead, a quantity called **entropy**, S, is used. Entropy is closely related to the number of ways, W.

Entropy, $S = k \ln W$

(ln means the natural logarithm, $\log_e$.) k is a constant, called the **Boltzmann Constant**. Its value is 1.38×10^{-23} J K^{-1}.

Using this expression, we can calculate the value of S corresponding to $W = 2^{10^{22}}$

$$S = k \ln W = 1.38 \times 10^{-23} \ln 2^{10^{22}}$$
$$= 0.096 \text{ or about } 0.1 \text{ J K}^{-1}$$

This is a much easier number to handle.

$k \ln W$ is an extremely important quantity. It shows, in a 'scaled-down' form, the number of ways of achieving a particular physical or chemical situation, and therefore indicates the probability of that situation arising.

We have already seen that changes that happen of their own accord are those with the greatest probability, i.e. the greatest number of ways, W, of happening. So, for all changes that happen spontaneously,

W must increase

But entropy $S = k \ln W$. So for all spontaneous changes

entropy, S, must increase

or

entropy change, ΔS, must be positive

This is one way of stating the *Second Law of Thermodynamics*.

At the beginning of this section we asked the question 'why does ice melt spontaneously?' Using the ideas that have been developed, we can begin to see an answer.

Look at figure 23.4.

1 Are there more ways of arranging water molecules in an ordered, regular ice crystal or in the disordered liquid state?

2 Which has the higher entropy – solid water (ice) or liquid water?

water molecules in an ice crystal → water molecules in the liquid state

Figure 23.4
Changes in the arrangement of water molecules when ice melts.

This simple argument suggests that ice will tend to melt spontaneously because there are more arrangements for liquid water than for solid water. Liquid water has a higher entropy than solid water, and ΔS for the melting process is positive.

But if you were reading this at the South Pole, you would disagree – ice does not melt spontaneously below 0°C – indeed, at that temperature liquid water *freezes* spontaneously. Clearly we have not yet considered the whole picture.

Entropy and energy

For spontaneous changes, the number of arrangements, W, must always increase. Furthermore, because entropy $S = k \ln W$, entropy must always increase too. So far, we have considered only the number of ways, W, that molecules can be arranged in space. But there are other kinds of arrangement to consider – the ways *energy* can be distributed among the molecules.

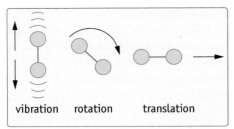

Figure 23.5
Distribution of energy in a diatomic molecule.

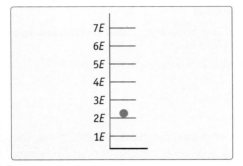

Figure 23.6
Vibrational energy levels of hydrogen. The molecule represented here possesses two energy quanta. It vibrational energy is 2E.

Figure 23.7
Transfer of one quantum of vibrational energy from one H$_2$ molecule to another.

You should already be familiar with the idea of energy levels. Electrons in atoms can exist in certain energy levels, but not between these levels. Their energy comes in packets, or **quanta**, and the number of quanta they possess defines the energy level they occupy.

When the electrons in an atom or molecule move between energy levels, energy changes occur. But atoms and molecules can change their energy states in other ways. Molecules can vibrate, rotate, and translate (move about) as well. All these movements involve energy (figure 23.5). *Energy comes in quanta for each of these changes too.*

A diatomic molecule such as hydrogen has definite vibrational energy levels spaced rather like the rungs of a ladder. At any instant, the molecule will have a definite quantity of vibrational energy, depending on how many quanta it possesses. The molecule represented by the energy level diagram in figure 23.6 has two quanta, each of energy *E*. Now suppose there are two hydrogen molecules, each with energy 2*E*, i.e. two quanta (figure 23.7(*a*)). These molecules might exchange one quantum, giving the arrangement in figure 23.7(*b*).

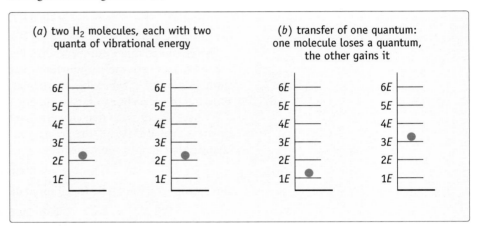

There are, of course, other ways in which these two molecules could share their four energy quanta. The full list of arrangements is

Number of quanta	
Molecule 1	**Molecule 2**
2	2
1	3
3	1
0	4
4	0

– five different ways. If the molecules were given one extra quantum to share, there would be six ways. Try to list these six arrangements in a table similar to the one above. The more quanta there are to share, the greater the number of ways of sharing them. And the greater the number of molecules involved, the greater still the number of ways of sharing the quanta (table 23.1).

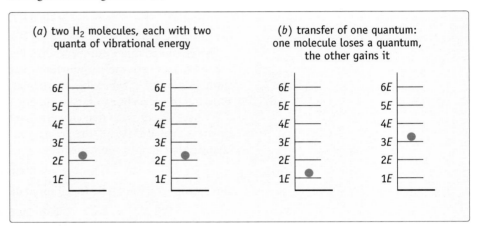

Figure 23.8 (hot metal / cold metal)

Figure 23.8
Hot and cold metal in contact. Each piece of metal contains 100 atoms.

Table 23.1
Number of ways of sharing quanta between molecules

Number of molecules	Number of quanta	Number of different ways of sharing the quanta
10	100	~10^{12}
100	10	~10^{13}
100	100	~10^{60}
200	110	~10^{86}

Consider two small pieces of metal in contact, each containing 100 atoms. One piece of metal is hot and has 100 energy quanta, the other is cold and has only 10 quanta (figure 23.8 on previous page).

Use table 23.1 on the previous page to answer these questions.

1. How many ways can the quanta be distributed in (a) the hot piece of metal, and (b) the cold piece?

2. What is the total number of arrangements of quanta for the hot piece and the cold piece?

3. Suppose all the quanta are now distributed between the two pieces of metal. What is the number of ways this could be done?

4. Which is the more probable arrangement – to have 100 quanta in one piece and 10 in the other, or to distribute the quanta evenly between both pieces?

5. What will happen to the temperatures of the pieces of metal when they are placed in contact like this?

You should now see *that sharing energy quanta increases the number of energy states and therefore the number of ways in which atoms and molecules can arrange themselves and their energy.* Thus, when entropy changes are worked out, using $S = k \ln W$, we must include in W the number of ways of distributing the available energy among the particles, as well as the number of ways of arranging the particles in space.

Increasing the temperature of a substance will increase the number of energy quanta available, and so increase the entropy.

Some entropy values

Using a variety of experimental methods, it is possible to work out the total entropy of elements and compounds under different conditions. Table 23.2 lists the standard entropies of some substances.

Table 23.2

Standard entropies, $S^{\ominus}$, of various substances at 298 K

Substance	State at 298 K	$S^{\ominus}$/J K^{-1} mol^{-1}
diamond	s	2.4
argon	g	154.7
carbon dioxide	g	213.6
aluminium	s	28.3
iron	s	27.2
water (solid)	s	48.0
water (liquid)	l	70.0
water (gas)	g	188.7
sodium chloride	s	72.4
quartz (SiO$_2$)	s	41.8

Examine the table carefully.

1. Try to formulate a general rule concerning which types of substance have low standard entropies, and which have high ones.

2. What will happen to all these entropy values when the temperature is increased?

In general, gases have higher standard entropies than solids. This is reasonable because there are more ways of arranging molecules and energy quanta in a disordered gas like carbon dioxide than in an ordered solid like diamond. Indeed, entropy is sometimes loosely described as 'disorder'. This is valid to an extent, but remember that in calculating entropy it is not enough to think simply of the arrangement of the particles in space – the distribution of energy among them must be considered as well.

Table 23.3 gives some more standard entropy values.

1. What conclusions can you draw concerning the effect of molecular complexity on the entropy of a substance?

2. Try to explain this effect.

3. Why is the trend of steadily rising $S^{\ominus}$ values broken by pentane?

Table 23.3

Standard entropies of some alkanes at 298 K

Substance	State at 298 K	$S^{\ominus}$/J K^{-1} mol^{-1}
methane	g	186.2
ethane	g	229.5
propane	g	269.9
butane	g	310.1
pentane	l	261.1

▲ Which has the higher entropy: the copper sulphate crystals, or the same crystals dissolved in water?

We can work out the entropy change, ΔS, for a reaction using tabulated standard entropy values. However, we can get a qualitative idea of the entropy change by inspecting the equation for the reaction and then using the following simple rules.

1 Gases generally have higher entropies than solids or liquids, so if the number of molecules of gas increases during a reaction, it is likely that the overall entropy will also increase.

2 Ions and molecules in solution generally have higher entropies than solids.

3 Substances with larger, complex molecules have higher entropies than those with smaller, simpler ones.

4 If a large molecule breaks down into smaller ones, entropy increases since there are more ways of arranging several small molecules than one large one.

Can entropy ever decrease?

For spontaneous changes *entropy*, S, *must always increase*.

We do not have to look far, however, to find spontaneous changes which *seem* to involve an entropy *decrease*. For example, at room temerature, steam spontaneously condenses to water:

$$H_2O(g) \longrightarrow H_2O(l)$$

steam   water

$$S^\ominus = 188.7 \text{ J K}^{-1} \text{ mol}^{-1} \qquad S^\ominus = 70.0 \text{ J K}^{-1} \text{ mol}^{-1}$$

For this change, ΔS is negative – the entropy has decreased by $(188.7 - 70) = 118.7 \text{ J K}^{-1} \text{ mol}^{-1}$. How can this be reconciled with the rule that entropy must always *increase*?

To find the answer, we must look a little further than the water alone, and remember the importance of *energy* in deciding entropy changes.

When steam condenses, energy is given out.

$$H_2O(g) \rightarrow H_2O(l); \qquad \qquad \Delta H^\ominus = -44.1 \text{ kJ mol}^{-1}$$

Where does this energy go? If the water condenses on a window, it goes to the glass. If it condenses as droplets in the air, it goes to the air molecules. The atoms in the glass, or the air molecules, or whatever – call them the **surroundings** – gain energy quanta and this increases the number of ways in which they can share energy (figure 23.9). Thus the entropy of the surroundings increases, whereas the entropy of the water *decreases*. There are two entropy changes to consider – the entropy change of water, ΔS_{water}, and the entropy change of the surroundings, $\Delta S_{surroundings}$.

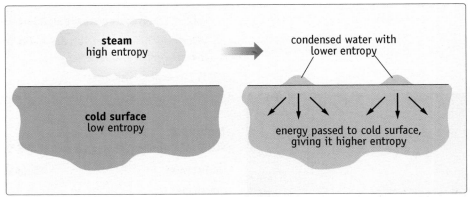

$$\Delta S_{total} = \Delta S_{water} + \Delta S_{surroundings}$$

As we shall see shortly, the positive $\Delta S_{surroundings}$ outweighs the negative ΔS_{water}, so ΔS_{total} is positive and the *total* entropy increases. The Second Law of Thermodynamics has not been broken after all.

Figure 23.9
Entropy changes when water condenses.

---

1 For each of the following reactions, say whether you would expect the entropy change of the chemicals to be

A positive
B negative
C approximately zero

(a) $2Mg(s) + O_2(g) \rightarrow 2MgO(s)$
(b) $CaCO_3(s) \rightarrow CaO(s) + CO_2(g)$
(c) $N_2(g) + 3H_2(g) \rightarrow 2NH_3(g)$
(d) $Zn(s) + Cu^{2+}(aq) \rightarrow Zn^{2+}(aq) + Cu(s)$

Calculating the entropy change in the surroundings

In any change in which energy is passed to or taken from the surroundings, it is vital to consider the entropy change in the surroundings, $\Delta S_{\text{surroundings}}$, as well as the entropy change in the chemicals or system, ΔS_{system}. ΔS_{system} is easily worked out from tabulated values of S. But $\Delta S_{\text{surroundings}}$ cannot be calculated in the same way, because the surroundings cannot be accurately defined.

Fortunately, there is a simple way of calculating $\Delta S_{\text{surroundings}}$. When a quantity of energy Q is passed to a body whose temperature is T, the change in entropy, ΔS, of the body is just

$$\Delta S = Q/T$$

In a chemical reaction for which the enthalpy change is ΔH, the quantity of energy Q passed to the surroundings is $-\Delta H$. (A minus sign is needed because if the surroundings gain energy, the chemicals must lose it.) Therefore

$$\Delta S_{\text{surroundings}} = -\Delta H/T$$

This enables us to calculate $\Delta S_{\text{surroundings}}$ very easily. Going back to the condensation of steam,

$$H_2O(g) \rightarrow H_2O(l); \qquad\qquad \Delta H^{\ominus} = -44.1 \text{ kJ mol}^{-1}$$

$$\Delta S^{\ominus}_{\text{system}} = -118.7 \text{ J K}^{-1} \text{ mol}^{-1}$$

At 298 K,

$$\Delta S^{\ominus}_{\text{surroundings}} = \frac{-\Delta H^{\ominus}}{T} = \frac{44\,100}{298} = +148.0 \text{ J K}^{-1} \text{ mol}^{-1}$$

(ΔH must be in joules, T in kelvins). Hence

$$\Delta S^{\ominus}_{\text{total}} = \Delta S^{\ominus}_{\text{system}} + \Delta S^{\ominus}_{\text{surroundings}}$$

$$\Delta S^{\ominus}_{\text{total}} = -118.7 + 148.0 = +29.3 \text{ J K}^{-1} \text{mol}^{-1}$$

When the surroundings as well as the system are taken into account, ΔS is positive

The condensation of steam is an example of a situation where ΔS_{system} is negative and unfavourable, but more than compensated for by a positive $\Delta S_{\text{surroundings}}$.

For many changes, however, ΔS_{system} is positive. For example,

$$C(s) + O_2(g) \rightarrow CO_2(g), \qquad\qquad \Delta S^{\ominus}_{\text{system}} = +6.31 \text{ J K}^{-1} \text{ mol}^{-1}$$

In such cases, $\Delta S_{\text{surroundings}}$ need not be positive. Indeed it could be negative, and the change will still be thermodynamically favourable so long as ΔS_{system} is bigger so that ΔS_{total} remains positive. Table 23.4 sums up the various possibilities.

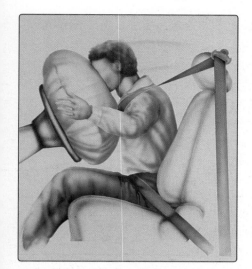

▲ A car airbag is inflated by the sudden decomposition of sodium azide, NaN_3, producing nitrogen gas. In this decomposition reaction, both ΔS_{system} and $\Delta S_{\text{surroundings}}$ have large positive values, so the reaction is spontaneous.

Table 23.4
Possible combinations of ΔS_{system} and $\Delta S_{\text{surroundings}}$

ΔS_{system}	$\Delta S_{\text{surroundings}}$	ΔS_{total}	Is the reaction thermodynamically favourable? Will it go?	Example
+	+	must be +	yes	$C + O_2 \rightarrow CO_2$
+	−	may be +	maybe	liquids evaporating; solids melting; solids dissolving
−	+	may be +	maybe	gases condensing; solutions crystallising; solids freezing
−	−	must be −	no	$CO_2 \rightarrow C + O_2$

23.5 Entropy changes and free energy changes

Entropy changes are very useful to the chemist, because they help us to predict whether a particular reaction will 'go', and if not, what can be done to make it go. However, the *total* entropy change, including both system and surroundings, must always be considered. This is rather a nuisance, because chemists are always more interested in what is going on *inside* a test tube or *inside* a reaction vessel than what is happening to the vessel itself and its surroundings. Fortunately there is a convenient way round this problem, by using *free energy changes*.

We know that

$$\Delta S_{total} = \Delta S_{surroundings} + \Delta S_{system}$$

and

$$\Delta S_{surroundings} = \frac{-\Delta H}{T}$$

therefore

$$\Delta S_{total} = \frac{-\Delta H}{T} + \Delta S_{system}$$

Multiplying through by T

$$T\Delta S_{total} = -\Delta H + T\Delta S_{system}$$

or

$$-T\Delta S_{total} = \Delta H - T\Delta S_{system}$$

The quantity $(-T\Delta S_{total})$ is called the **free energy change, ΔG**

$$\Delta G = \Delta H - T\Delta S_{system}$$

(The symbol G for free energy is in honour of the American chemist Willard Gibbs.)

▲ Using ΔG enables chemists to concentrate on the reaction system, and forget the surroundings.

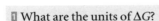 **1** What are the units of ΔG?

For changes that happen spontaneously, ΔS_{total} must be positive – the total entropy must increase.

Now, since $\Delta G = -T\Delta S_{total}$, for ΔS_{total} to be positive, ΔG must be negative. *For all spontaneous changes, ΔG must be negative.* This is another way of stating the Second Law of Thermodynamics.

Free energy changes, ΔG, are extremely helpful to chemists, because using them means we can ignore the surroundings. ΔG values are tabulated in data books in much the same way as ΔH values. ΔG has the same units as ΔH: J mol^{-1} or kJ mol^{-1}. In particular, the free energy of formation, ΔG_f, is known and tabulated for a wide variety of compounds. Look up a few examples in your own data book.

ΔG and ΔH

In many chemical reactions, $T\Delta S_{system}$ is a lot smaller than ΔH, at least at room temperature, and in these cases it is fairly safe to say that if ΔH is negative, ΔG will also be negative. This is why ΔH values can often be used to predict whether or not reactions could go spontaneously. But while it is true that for all spontaneous changes ΔG must be negative, a negative ΔH does not *always* mean a spontaneous reaction.

If ΔS_{system} is large and negative, $T\Delta S_{system}$ may exceed ΔH, making ΔG positive. This is particularly likely when T is large, i.e. at high temperatures. Table 23.5 on the next page gives some values of $\Delta G^{\ominus}$ and $\Delta H^{\ominus}$ for comparison.

Table 23.6 on the next page summarises the different possible combinations of ΔH and ΔS. Compare the table with table 23.4.

Look at the figures in table 23.5, then answer these questions.

1 For which two changes do ΔG and ΔH have opposite signs?

2 Why does ammonium chloride dissolve spontaneously under standard conditions even though ΔH for the process is positive?

3 Why does liquid water not freeze under standard conditions even though ΔH for the process is negative?

4 For the reaction of zinc with copper ions, why are the values of ΔG and ΔH so close?

23 Entropy and Free Energy

Table 23.5

Standard enthalpy changes and free energy changes (note the units of $\Delta G^{\ominus}$ and $\Delta H^{\ominus}$ are the same, kJ mol^{-1}. Note too that these values refer to standard conditions: 298 K and 1 atm)

Reaction	Equation	$\Delta G^{\ominus}$/kJ mol^{-1}	$\Delta H^{\ominus}$/kJ mol^{-1}
burning carbon	$C(s) + O_2(g) \rightarrow CO_2(g)$	−394.4	−393.5
condensing steam	$H_2O(g) \rightarrow H_2O(l)$	−8.6	−44.1
freezing water	$H_2O(l) \rightarrow H_2O(s)$	+0.6	−6.01
burning magnesium	$Mg(s) + \frac{1}{2}O_2(g) \rightarrow MgO(s)$	−1138.8	−1204
reaction of zinc with copper ions	$Zn(s) + Cu^{2+}(aq) \rightarrow Zn^{2+}(aq) + Cu(s)$	−212.1	−216.7
making ammonia by the Haber process	$\frac{1}{2}N_2(g) + \frac{3}{2}H_2(g) \rightarrow NH_3(g)$	−33.4	−92.0
combining hydrogen and oxygen to make water	$H_2(g) + \frac{1}{2}O_2(g) \rightarrow H_2O(g)$	−457.2	−483.6
dissolving ammonium chloride in water	$NH_4Cl(s) \rightarrow NH_4^+(aq) + Cl^-(aq)$	−6.7	+16

Table 23.6

Possible combinations of ΔH and ΔS_{system}

ΔH	ΔS_{system}	ΔG	Result	Example
0	+	−	change occurs spontaneously (if rate is reasonable)	mixing of gases
0	−	+	change does not occur spontaneously	unmixing of gases
−	+	−	change occurs spontaneously (if rate is reasonable)	exothermic reactions producing gases, e.g. burning most fuels
−	−	+ or −	change may occur spontaneously depending on conditions. Most likely at low temperatures	exothermic reactions in which number of moles of gas decreases, e.g. $N_2(g) + 3H_2(g) \rightarrow 2NH_3(g)$, or condensing of a gas
+	+	+ or −	change may occur spontaneously depending on conditions. Most likely at high temperatures	endothermic reactions in which number of moles of gas increases, e.g. liquids evaporating
+	−	+	change does not occur spontaneously	endothermic reactions in which number of moles of gas decreases, e.g. $6C(s) + 3H_2(g) \rightarrow C_6H_6(g)$, or photosynthesis

▲ ΔG for the photosynthesis reaction is positive. Photosynthesis cannot occur spontaneously: it needs the input of sunlight.

How to make reluctant reactions go

A reaction for which ΔG is positive will not happen spontaneously, because its total entropy change will be negative. This does not mean it can *never* happen – if this were the case, photosynthesis would never occur, and it would be impossible to make ice cubes in warm climates.

To make a reluctant reaction take place we need to alter the conditions, and this usually means altering the temperature. Consider the problem of making ice cubes in a warm climate (figure 23.10). To make the water freeze we put it in the freezer compartment of a refrigerator. At room temperature (298 K), ΔG for the change

$$H_2O(l) \rightarrow H_2O(s)$$

is +546 J mol^{-1} and freezing does not occur. At the temperature of the freezer (263 K), ΔG is −224 J mol^{-1} and the water freezes.

Why does the water freeze at 263 K but not at 298 K? When freezing occurs, energy equivalent to 6010 J ($-\Delta H$) passes to the surroundings, and this increases the entropy of these surroundings by 6010/T J K^{-1} mol^{-1}, where T is the temperature of the surroundings. The quantity of energy passed to the surroundings is the same at both temperatures, but at 263 K this energy has a bigger effect on the colder surroundings than at 298 K when the surroundings are already quite warm.

At 263 K,

$$\Delta S_{\text{surroundings}} = (6010/263) = 22.85 \text{ JK}^{-1} \text{ mol}^{-1}, \text{ but at } 298 \text{ K,}$$
$$\Delta S_{\text{surroundings}} = (6010/298) = 20.16 \text{ JK}^{-1} \text{ mol}^{-1}$$

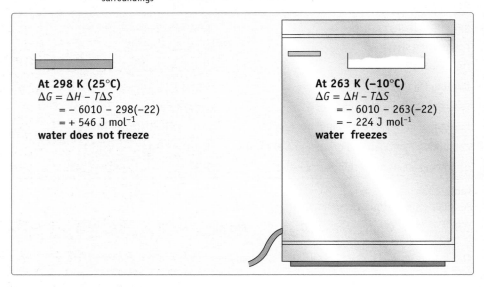

At 298 K (25°C)
$$\Delta G = \Delta H - T\Delta S$$
$$= -6010 - 298(-22)$$
$$= +546 \text{ J mol}^{-1}$$
water does not freeze

At 263 K (−10°C)
$$\Delta G = \Delta H - T\Delta S$$
$$= -6010 - 263(-22)$$
$$= -224 \text{ J mol}^{-1}$$
water freezes

Figure 23.10
Making ice cubes.

Another reluctant reaction which can be made to occur by changing the conditions is the production of quicklime, CaO, from limestone, $CaCO_3$:

$$CaCO_3(s) \rightarrow CaO(s) + CO_2(g); \qquad \Delta H^{\ominus} = +178 \text{ kJ mol}^{-1}$$
$$\Delta S^{\ominus} = +165 \text{ J K}^{-1} \text{ mol}^{-1}$$

Answer these questions.

1. Use $\Delta G^{\ominus} = \Delta H^{\ominus} - T\Delta S^{\ominus}$ to calculate $\Delta G^{\ominus}$ (in J mol^{-1}) for this reaction at 298 K.
2. Will the reaction happen spontaneously at 298 K?
3. Will (a) raising the temperature, (b) lowering the temperature, help to make ΔG more negative, and hence help the reaction to go?
4. At what temperature will ΔG for this reaction be zero?
5. What is the minimum temperature to which $CaCO_3$ must be heated to make it decompose?

You should have worked out that under standard conditions at 298 K,

$$\Delta G^{\ominus} = 128\,830 \text{ J mol}^{-1}$$

This is a positive value, so $CaCO_3$ does not decompose at this temperature. Raising the temperature makes ΔG less positive, until at 1078 K, $\Delta G = 0$. At any temperature above this, ΔG will be negative and the $CaCO_3$ will decompose.

6. What other change in the conditions would encourage decomposition?

You should now be able to see how useful ΔG can be, not only in telling whether a reaction will go, but also in deciding what can be done to make it go if it is reluctant to do so.

Limitations of ΔG

There are many reactions for which ΔG is negative, and which should therefore go spontaneously at room temperature. Surprisingly, some of these reactions do not happen spontaneously.

Fuels provide an example. ΔG values suggest that fuels should burn spontaneously at room temperature, yet they do not. This is because they have a high *activation energy*, and until they are heated (i.e. ignited) the reaction is infinitely slow. So beware: ΔG *provides information only about the thermodynamic or energetic feasibility* of a reaction. In practice, a reaction may not occur because of *kinetic stability*. To find out about kinetic factors, we need to know the activation energy for the reaction, and this is quite independent of ΔG. The kinetics or *rates* of reaction are considered in chapter 24.

▲ ΔG values suggest that matches should burn spontaneously at room temperature. Fortunately, a high activation energy stops them burning before they are struck.

What's 'free' about free energy?

Reactions which give out heat can be made to do useful work. For example, the energy from burning coal can be used in a power station to drive a generator. At first sight, we might expect that the energy available to do useful work would be given by ΔH. It turns out, in fact, that ΔG, not ΔH, gives the maximum quantity of energy available. ΔG, the free energy change, tells us the quantity of chemical energy *free* to do useful work, and that is how it gets its name. In reality, the actual quantity of work obtained is always less than ΔG, so the free energy change represents the *maximum* amount of useful work available.

23.6 Free energy, electrode potentials and equilibrium constants

The free energy change, ΔG, indicates whether a reaction is likely to proceed spontaneously. There are, however, other criteria that can be used to answer this question.

Electrode potentials

A useful way of deciding whether a redox reaction will take place is to look at standard electrode potentials. In general, if $E^{\ominus}$ for a reaction is positive, it will go, provided the reaction is kinetically feasible. We might expect $\Delta G^{\ominus}$ and $E^{\ominus}$ to be related, and this relationship is quite simple.
Under standard conditions,

$$\Delta G^{\ominus} = -zFE^{\ominus}$$

z is the number of moles of electrons transferred during the cell reaction; F is the charge on one mole of electrons = 96 500 coulombs.
This equation applies quite generally and shows the close relationship between $E^{\ominus}$

Consider the Daniell cell

$$Zn \,|\, Zn^{2+} \;\vdots\; Cu^{2+} \,|\, Cu; \; E^{\ominus} = 1.1\,V$$

1 Calculate the value of $\Delta G^{\ominus}$ for the cell reaction.

$$Zn(s) + Cu^{2+}(aq) \rightarrow Zn^{2+}(aq) + Cu(s)$$

and $\Delta G^{\ominus}$.

Equilibrium constants

The extent of a reaction can be judged directly by looking at its equilibrium constant. There is a simple relationship between K and $\Delta G^{\ominus}$.

$$\Delta G^{\ominus} = -RT \ln K$$

R is the gas constant $= 8.31\,J\,K^{-1}\,mol^{-1}$.

If K is large and positive the reaction is expected to be spontaneous, providing it is kinetically feasible. This equation is quite general and applies to all reactions, whether or not they involve redox cells. It shows the close relationship between $\Delta G^{\ominus}$ and K.

2 Use the value of $\Delta G^{\ominus}$ for the Zn/Cu^{2+} reaction which you derived above to calculate the equilibrium constant for the reaction at 298 K.

You should have found, for the Daniell Cell reaction, that $\Delta G^{\ominus} = -212\,300\,J\,mol^{-1}$ and $K = 1.7 \times 10^{37}$. This indicates that the reaction should certainly go under these conditions. However, even with this big number, a tiny amount of Zn and Cu^{2+} will be left unreacted at the end. If it was not, K would be infinite.

Comparison of $\Delta G^{\ominus}$, $E^{\ominus}$ and K

Table 23.7 gives the corresponding values of $\Delta G^{\ominus}$, $E^{\ominus}$ and K for certain types of reaction. Note that a 'complete' reaction has been defined arbitrarily as one for which $K > 10^{10}$, and a reaction which 'doesn't go' as one with $K < 10^{-10}$. These figures correspond to $\Delta G^{\ominus}$ values of -60 kJ mol^{-1} and $+60$ kJ mol^{-1}, respectively. Up to now we have simply said that any reaction with a negative ΔG will be spontaneous, but table 23.7 shows how ΔG can be extended to see just how far a reaction goes.

But remember, ΔG only provides information about *thermodynamic* or *energetic* feasibility. Many reactions appear to be feasible on thermodynamic grounds, but in fact do not go because of kinetic stability: the reaction *rate* is too slow. Reaction rates are covered in the next chapter.

Table 23.7
Values of K, $\Delta G^{\ominus}$ and $E^{\ominus}$ for certain types of reaction

K	$\Delta G^{\ominus}$/kJ mol^{-1}	$E^{\ominus}$/volt	Extent of reaction
$> 10^{10}$	< -60	$> +0.6$	reaction 'complete'
1	0	0	reaction balanced between reactants and products
$< 10^{10}$	$> +60$	< -0.6	reaction 'doesn't go'

23.7 A final point

In this chapter we have asked the question 'Why do things happen the way they do?' The answer has been 'Because that is the most probable way'. There are, however, other, quite valid answers. For example, we have asked the question 'Why does ice melt when heated?' The answer given is 'Because there are more ways of melting than unmelting.' Nevertheless, it would have been equally correct to say 'Because heating provides the energy needed to break bonds between water molecules in the solid state.'

Entropy and thermodynamics provide one way of explaining natural phenomena, but there are other approaches. Nevertheless, the Second Law of Thermodynamics has proved to be one of the most valuable and universally applicable scientific laws.

Summary

1 Spontaneous changes occur if they result in a more probable arrangement of molecules and energy, i.e. an increase in the number of ways, W, in which molecules and energy can be distributed.

2 Entropy, S, is equal to $k \ln W$. In spontaneous changes entropy always increases.

3 In general, the entropy values for gases and aqueous solutions are greater than those for solids.

4 When considering entropy changes, it is important to consider the entropy change of the surroundings as well as that of the chemical system:

$$\Delta S_{surroundings} = -\Delta H / T$$

5 Free energy changes, ΔG, provide a way of accounting for the surroundings as well as the system:

$$\Delta G = \Delta H - T\Delta S_{system}$$

For all spontaneous changes, ΔG must be negative.

6 Reactions for which ΔG is positive can sometimes be made to go by altering conditions, particularly temperature.

7 ΔG indicates the energetic feasibility of a reaction. It says nothing about its kinetic feasibility.

8 $\Delta G^{\ominus}$ is closely related to K and $E^{\ominus}$.

$$\Delta G^{\ominus} = -RT \ln K$$
$$\Delta G^{\ominus} = -zFE^{\ominus}$$

1 Consider the equilibrium between solid iodine and iodine dissolved in CCl_4:

$$I_2(s) \rightleftharpoons I_2(CCl_4)$$

This equilibrium lies well to the right: iodine is soluble in CCl_4.

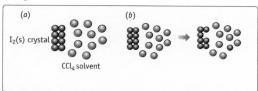

(a) (b)

$I_2(s)$ crystal

CCl_4 solvent

Figure 23.11
(a) A lattice containing eight iodine molecules, surrounded by ten CCl_4 molecules. *(b)* One iodine molecule has dissolved.

(a) Consider the simple situation of a lattice containing eight iodine molecules, surrounded by ten CCl_4 molecules (figure 23.11(a)). Suppose one iodine molecule dissolves, thereby occupying the space previously occupied by a CCl_4 molecule (figure 23.11(b)). How many different CCl_4 molecules could this I_2 molecule displace? In how many different ways could one I_2 molecule arrange itself among the CCl_4 molecules?

(b) Suppose now that the I_2 molecule returned to its vacant place in the $I_2(s)$ crystal. How many ways are there of doing this?

(c) Which is the more probable arrangement for the I_2 molecule – in the crystal or among the CCl_4 molecules?

(d) Suppose this example is now scaled up and we have one mole of $I_2(s)$ surrounded by one mole of CCl_4. If just one I_2 molecule dissolves in the CCl_4, how many different arrangements are possible?

(e) Suppose the molecule now returns to its original place in the lattice. In how many ways can it do this? Is it likely to?

(f) Use these answers to explain why iodine is soluble in CCl_4.

(g) (Hard) If iodine is steadily added to CCl_4, the solution eventually reaches saturation, and no more iodine dissolves. Suggest a reason why.

2 Consider the entropy changes involved in the burning of magnesium.

$$2Mg(s) + O_2(g) \rightarrow 2MgO(s); \quad \Delta H^\ominus = -1204 \text{ kJ mol}^{-1}$$

$$S^\ominus(Mg) = 32.7 \text{ J K}^{-1} \text{ mol}^{-1}$$

$$S^\ominus(O_2) = 204.9 \text{ J K}^{-1} \text{ mol}^{-1}$$

$$S^\ominus(MgO) = 26.8 \text{ J K}^{-1} \text{ mol}^{-1}$$

(a) What is the standard entropy of the reactants, two moles of magnesium and one mole of oxygen?

(b) What is the standard entropy of the products, two moles of magnesium oxide?

(c) What is the standard entropy change for the reaction as written in the equation?

(d) Will this entropy change alone favour the reaction?

(e) Use the value of $\Delta H^\ominus$ to calculate $\Delta S^\ominus_{surroundings}$ at 298 K. (Remember to work in joules.)

(f) What is the overall entropy change, $\Delta S^\ominus_{total}$?

(g) Is this a favourable entropy change?

3 For each of the following reactions, say whether you think the entropy change of the *chemicals* (the system) will be

A positive
B negative
C approximately zero

(a) $NH_4Cl(s) \rightarrow NH_3(g) + HCl(g)$
(b) $CH_4(g) + 2O_2(g) \rightarrow CO_2(g) + 2H_2O(g)$
(c) $C_3H_8(g) + 5O_2(g) \rightarrow 3CO_2(g) + 4H_2O(g)$
(d) $H_2(g) + I_2(g) \rightarrow 2HI(g)$
(e) $NaCl(s) \xrightarrow{aq} Na^+(aq) + Cl^-(aq)$
(f) $H^+(aq) + OH^-(aq) \rightarrow H_2O(l)$
(g) amino acids $\rightarrow$ proteins
(h) starch $\rightarrow$ glucose

4 Some standard entropy values are given in the table below.

Substance	Formula	$S^\ominus$ /J K^{-1} mol^{-1}
helium	He(g)	126
neon	Ne(g)	146
argon	Ar(g)	155
krypton	Kr(g)	164
xenon	Xe(g)	167
hydrogen	H$_2$(g)	131
water (gaseous)	H$_2$O(g)	189
water (liquid)	H$_2$O(l)	70
water (solid)	H$_2$O(s)	48

(a) What pattern do you notice in the standard entropies of the noble gases? Suggest a reason for the pattern.

(b) Compare the standard entropy values of hydrogen and helium. Suggest a reason for any difference.

(c) Compare the standard entropy values of steam, water and ice. What do the values suggest about the three states of matter?

5 (a) If you have not already done so, answer the questions in section 23.5 relating to the decomposition of calcium carbonate.

(b) Repeat the calculations for magnesium carbonate instead of calcium carbonate, using these data:
$$MgCO_3(s) \rightarrow MgO(s) + CO_2(g); \Delta H^\ominus = +100.3 \text{ kJ mol}^{-1}$$
$$\Delta S^\ominus = +174.8 \text{ J K}^{-1} \text{mol}^{-1}$$

(c) Use your results to compare the thermal stability of magnesium carbonate with that of calcium carbonate.

(d) Suggest a reason for the difference in thermal stability between the two compounds.

24 Reaction Rates

▲ Why does a pressure cooker reduce the time needed to cook food?

► This huge limestone lion outside Leeds Town Hall, England, has been slowly weathered by acidic gases in the atmosphere.

▲ Finding the optimum temperature, pressure and catalyst for an industrial process is crucial. It could mean the difference between success and failure on the commercial market. This workman is checking the temperature in a cheese-making process.

24.1 Introduction

The rates of chemical reactions are just as important to you as they are to industrialists and chemical engineers.

At home you might be interested in the rate at which you can boil an egg or bake a cake. Out-of-doors you might be interested in the rate at which your car is rusting, the rate at which your lettuces are growing and possibly the rate at which the stonework of buildings is being weathered by acidic gases in the atmosphere.

In industry, engineers and other workers are concerned with the rates of chemical reactions in industrial, engineering and farming processes. These might include the rate at which ammonia can be obtained from nitrogen and hydrogen, the rate at which concrete sets or the rate of growth of fruit and vegetable crops.

Industrialists and chemical engineers are not satisfied with merely turning one substance into another. In most cases, they want to obtain products rapidly, easily and as cheaply as possible. Time and money are important in industry. It is often necessary to accelerate reactions so that they are economically worthwhile.

At normal temperatures and pressures and in the absence of a catalyst, ammonia cannot be obtained from nitrogen and hydrogen. Fortunately, chemical engineers have found that a reasonable reaction rate results when the process is carried out at 250 atmospheres pressure and 450°C in the presence of an iron catalyst (section 21.5).

$$N_2(g) + 3H_2(g) \rightarrow 2NH_3(g)$$

Biological reactions also rely on the presence of catalysts. Almost every chemical reaction in your body is controlled by one or more catalysts. Catalysts are involved in simple reactions like the hydrolysis of starch to sugars. They are also involved in highly complex reactions like the replication of DNA which forms the genes in the nuclei of your cells. These biological catalysts, called **enzymes**, are proteins.

Reaction rates are also important in archaeology. Archaeologists can estimate the age of rocks, fossils or prehistoric remains by a process known as radioactive dating. This entails measuring the concentration of a decaying radioactive isotope such as $^{14}_{6}C$ in the object under scrutiny.

> 1 Why does a pressure cooker enable vegetables to be cooked more rapidly?
> 2 What conditions and processes are used to slow down the rate at which perishable foods deteriorate?
> 3 How do gardeners accelerate the growth of their crops?

24.2 The concept of reaction rate

▲ Why are these carcasses stored in a refrigerator? What other foods must be stored in refrigerated containers?

During a chemical reaction, reactants are being converted to products. The reaction rate tells us how fast the reaction is taking place by indicating how much of a reactant is consumed or how much of a product forms in a given time. Hence,

$$\text{reaction rate} = \frac{\text{change in amount (or concentration) of a substance}}{\text{time taken}}$$

Thus, *we can define reaction rate as the rate of change of amount or concentration of a particular reactant or product.*

When acidified hydrogen peroxide is added to a solution of potassium iodide, iodine is formed.

$$H_2O_2 + 2I^- + 2H^+ \rightarrow 2H_2O + I_2$$

The concentration of iodine rises from 0 to 10^{-5} mol dm^{-3} in 10 seconds.

$$\therefore \text{ reaction rate} = \frac{\text{change in concentration of iodine}}{\text{time taken}}$$

Using the symbol Δ to represent the change in a particular quantity, we can write

$$\frac{\Delta[I_2]}{\Delta t} = \frac{10^{-5} \text{ mol dm}^{-3}}{10 \text{ s}}$$
$$= 10^{-6} \text{ mol dm}^{-3} \text{ s}^{-1}$$

▲ ICI spent millions of pounds in the production of a suitable biodegradable plastic. Their product 'Biopol' is used in this plastic bag. The rate at which 'Biopol' degrades is important.

Strictly speaking, this result gives the average reaction rate during the 10 seconds that it took for the concentration of iodine to become 10^{-5} mol dm^{-3}. By measuring the change in concentration (or amount) over shorter and shorter time intervals we obtain an increasingly accurate estimate of the reaction rate at any moment. The disadvantages of 'clock' techniques such as this one are illustrated more effectively in figure 24.1. Using 'clock' techniques, the rate is obtained as the inverse of the time for a certain proportion of the reaction to occur. Provided the reaction has gone only a little way towards completion, very little error is introduced, but serious errors result if the 'end point' is, say, half-way to completion.

Ideally we should make the time interval almost zero, and then we obtain what is effectively the reaction rate at a particular instant, i.e.

$$\frac{\Delta[I_2]}{\Delta t} = \frac{d[I_2]}{dt}$$
$$\Delta t \rightarrow 0$$

In practice, it is usual to plot a graph of the concentration or amount of a particular substance against time. The reaction rate can then be obtained at particular times by drawing tangents to the resulting curve. This technique is illustrated in section 24.5.

Normally it is convenient to express reaction rates in mol dm^{-3} s^{-1} or in mol s^{-1}, but occasionally it is more convenient to use minutes or even hours as the unit of time.

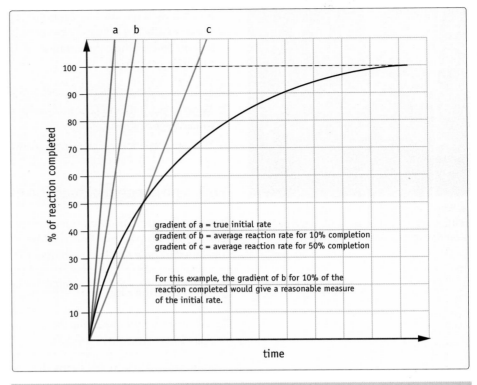

Figure 24.1
Errors in the 'clock technique' for measuring reaction rates.

In the reaction of acidified hydrogen peroxide with potassium iodide

$$H_2O_2(aq) + 2I^-(aq) + 2H^+(aq) \rightarrow 2H_2O(l) + I_2(aq)$$

the rate of formation of iodine was found to be $10^{-6}\,mol\,dm^{-3}\,s^{-1}$

1 What is the rate of consumption of H_2O_2 in this reaction?

2 What is the rate of consumption of I^- in this reaction?

▲ A researcher lifts a container of frozen animal tissue from its storage in dry ice.

24.3 Factors affecting the rate of a reaction

Our studies have already indicated several factors which can influence the rate of a reaction.

The availability of reactants and their surface area

Anyone who has camped knows that it is easier to start a fire using sticks rather than logs. Similarly, magnesium powder will react much more rapidly than magnesium ribbon with dilute sulphuric acid. In general, the smaller the size of reacting particles, the greater is the total surface area exposed for reaction and, consequently, the faster the reaction. In the case of heterogeneous systems, in which the reactants are in different states, the area of contact between the reacting substances will influence the reaction rate considerably. In homogeneous systems, reacting substances normally occur in their *maximum* state of subdivision. They are usually present as individual particles in the gaseous or aqueous phase. In this case, the idea of surface area becomes meaningless.

The concentration of reactants

Increasing the concentration of a reactant normally causes an increase in the rate of a reaction, but this is not always the case. Furthermore, the different reactants can affect the rate of a particular reaction in different ways. For example, when nitrogen oxide reacts with oxygen

$$2NO(g) + O_2(g) \rightarrow 2NO_2(g)$$

the reaction rate doubles when the oxygen concentration doubles. But doubling the concentration of NO quadruples the rate of reaction. The effect of concentration on reaction rates is considered further in sections 24.5 and 24.6.

The temperature of the reactants

Perishable foods like milk 'go bad' much more rapidly in summer than in winter. In summer, the chemical reactions in the deterioration processes occur more rapidly at the higher temperatures. In general, increasing the temperature increases the rate of chemical reactions.

Figure 24.2
Graphs of reaction rate against temperature.

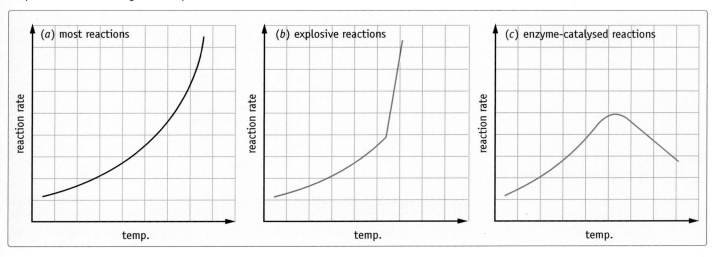

The effect of temperature on reaction rates is considered further in sections 24.8 and 24.9.

Catalysts

Catalysts are substances which alter the rate of chemical reactions without undergoing any overall chemical change themselves. Although catalysts are not used up in reactions, they participate by forming intermediate compounds in the conversion of the reactants to products (section 18.6).

Normally, catalysts are used to accelerate reactions. Certain catalysts can, however, slow reactions down. For example, propane-1,2,3-triol (glycerine) is sometimes added to hydrogen peroxide as a negative catalyst in order to slow down its rate of decomposition.

Catalysts play an important part in biological and industrial processes by enabling reactions to take place which would never occur in their absence. Many industrial processes, including the manufacture of ammonia, sulphuric acid, nitric acid, ethene, polythene and polystyrene, rely heavily on the use of catalysts. Living things rely even more heavily on catalysts because every chemical reaction in animals, plants and micro-organisms requires its own enzyme (catalyst). Some enzymes are so specific that they can only catalyse the reaction of one particular substrate (reactant). Other enzymes are less specific.

Light

Photosynthesis and photography both involve light-sensitive reactions. The leaves of plants contain a green pigment called chlorophyll. This can absorb radiation in the visible region of the electromagnetic spectrum and use this energy to synthesise chemicals and provide food for the plant.

During photosynthesis, plants transform carbon dioxide and water into oxygen and sugars such as glucose:

$$6CO_2(g) + 6H_2O(l) \xrightarrow{h\nu} C_6H_{12}O_6(aq) + 6O_2(g) \quad \Delta H = +2820 \text{ kJ}$$

In the absence of sunlight, energy is no longer provided and photosynthesis ceases.

White silver chloride turns purple and finally dark grey when it is exposed to sunlight. Sunlight provides the energy required to decompose the silver chloride.

$$AgCl(s) \xrightarrow{h\nu} Ag(s) + \tfrac{1}{2}Cl_2(g)$$

The use of silver salts in photography depends on photosensitivity of this kind.

The reactions of halogens with hydrogen and with alkanes are further examples of photochemical reactions.

Thus, chlorine reacts slowly with hydrogen or methane in diffused daylight. When the reaction is exposed to intense ultraviolet radiation it becomes explosive. The effect of sunlight results from its ability to split chlorine molecules into highly reactive single atoms. These are known as **free radicals** (section 26.5) which contain an unpaired electron (figure 24.3).

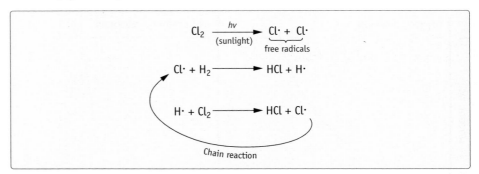

Figure 24.3
The reaction of chlorine with hydrogen.

24.4 Measuring reaction rates

The rate of a chemical reaction can be obtained by following a property which alters as the reaction occurs. By following this property and analysing the reaction mixture at suitable intervals, it is possible to determine the concentration of both reactants and products. Hence, we can obtain a measure of the reaction rate (i.e. the rate at which the concentration of a particular substance changes with time).

The method used to analyse the reaction mixture depends on the reaction under consideration. The following techniques illustrate four possible approaches.

Titrimetric analysis

This method is particularly suitable for reactions in solution such as that between iodine and propanone catalysed by acid:

$$CH_3COCH_3(aq) + I_2(aq) \xrightarrow{H+} CH_2ICOCH_3(aq) + H^+(aq) + I^-(aq)$$

Titration techniques can also be used to analyse the alkaline hydrolysis of an ester (such as methyl methanoate):

$$HCOOCH_3(l) + OH^-(aq) \rightarrow HCOO^-(aq) + CH_3OH(aq)$$

In these cases, the reaction can be followed by removing and analysing small portions of the reaction mixture at intervals. Very often, the removed portion must be added to a reagent which will 'quench' the reaction (i.e. stop it). This prevents further changes in concentration before the analysis is carried out. For example, in studying the reaction between iodine and propanone, portions of the reaction mixture can be pipetted into sodium hydrogencarbonate solution. This 'quenches' the reaction by neutralising the acid catalyst. The quenched mixture can then be analysed carefully by titrating the unreacted iodine against a standard solution of sodium thiosulphate.

Colorimetric analysis

This method is especially convenient for systems in which one of the substances is coloured (e.g. the reaction of iodine with propanone or the reaction of bromine with methanoic (formic) acid).

The intensity of colour can be followed during the reaction using a photoelectric colorimeter (figure 24.4). From these measurements the concentration of the coloured substance can be obtained at different times.

In a colorimeter, a narrow beam of light passes through the solution under test towards a sensitive photocell. In many colorimeters, it is possible to select the most appropriate colour of light by choosing a particular filter or by adjusting a diffraction grating.

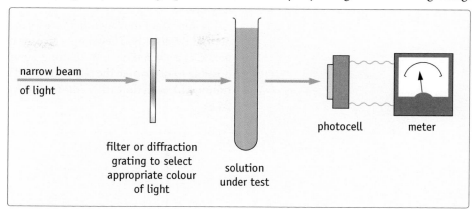

Figure 24.4
A simplified diagram of a colorimeter.

The current generated in the photocell is, of course, proportional to the amount of light transmitted by the solution. This, in turn, depends upon the depth of colour of the substance under test. Thus, the current from the photocell will be greatest when the light transmitted by the solution is the greatest. This occurs when the coloured substance is most dilute. However, the meter is usually calibrated to show not the fraction of light transmitted but the fraction of light absorbed. This will be proportional to the concentration of the coloured substance in the test solution.

Pressure measurements

This technique is particularly suitable for gaseous reactions which involve changes in pressure. For example, the gaseous decomposition of 2-methyl-2-iodopropane can be followed conveniently by measuring the pressure at suitable time intervals.

$$H_3C - \underset{\underset{I}{\overset{\overset{CH_3}{|}}{|}}{C}} - CH_3(g) \longrightarrow \underset{H_3C}{\overset{H_3C}{>}} C = C \underset{H(g)}{\overset{H}{<}} + HI(g)$$

Conductimetric analysis

Many reactions in aqueous solution involve changes in the ions present as the reaction proceeds. Consequently the electrical conductivity of the solution will change during the reaction. This can be used to determine the changing concentrations of reactants and products with time. Essentially, this method involves immersing two inert electrodes in the reaction mixture and then following the change in electrical conductivity of the solution with time (figure 24.5).

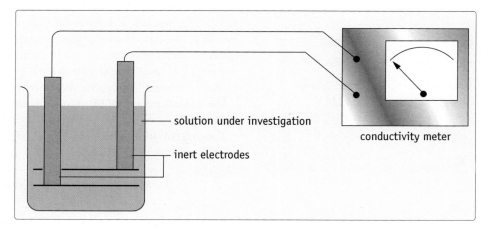

Figure 24.5
Following the change in electrical conductivity of a solution with time.

The last three methods have one advantage over titrimetric analysis. They do not require the removal of samples from the reacting mixture. In these three cases, the extent of the reaction is determined at regular intervals without disturbing the reaction mixture.

It is important to realise that measurements on a reacting system do not give the rate of reaction directly. They simply give the concentration of a particular reactant or product, X, at a given time, t. By plotting a graph of the concentration of X against time, it is possible to determine the reaction rate. This involves measuring the change in concentration of X with time ($d[X]/dt$) from the gradient of the tangent at a given point (figure 24.6).

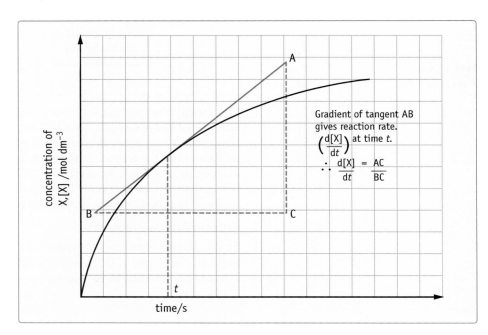

Figure 24.6
Obtaining the rate of reaction at a given time from a graph of concentration against time.

24.5 Investigating the effect of concentration on the rate of a reaction

We must now consider the influence of concentration on reaction rates in more detail. A convenient reaction to study is that between bromine and methanoic acid (formic acid) in aqueous solution.

The reaction is catalysed by acid:

$$Br_2(aq) + HCOOH(aq) \xrightarrow{H+} 2Br^-(aq) + 2H^+(aq) + CO_2(g)$$

The reaction can be followed colorimetrically by measuring the intensity of the red–brown bromine at suitable time intervals. By plotting a calibration curve of known bromine concentrations against colorimeter readings, it is possible to deduce the concentrations of bromine from the colorimeter readings obtained in the experiment. Some typical results are shown in table 24.1. The concentration of methanoic acid was virtually constant throughout the experiment because it was present in large excess.

The concentrations of bromine in table 24.1 are plotted graphically against time in figure 24.7.

The concentration of bromine ($[Br_2]$) falls during the course of the reaction. So, the rate of the reaction can be expressed in terms of the rate at which the bromine concentration changes.

Reaction rate = – rate of change of bromine concentration

$$= - \frac{d[Br_2]}{dt}$$

Table 24.1
Results for the kinetic study of the reaction between bromine and methanoic acid

Time/s	$[Br_2]$/mol dm^{-3}
0	0.0100
30	0.0090
60	0.0081
90	0.0073
120	0.0066
180	0.0053
240	0.0044
360	0.0028
480	0.0020
600	0.0013

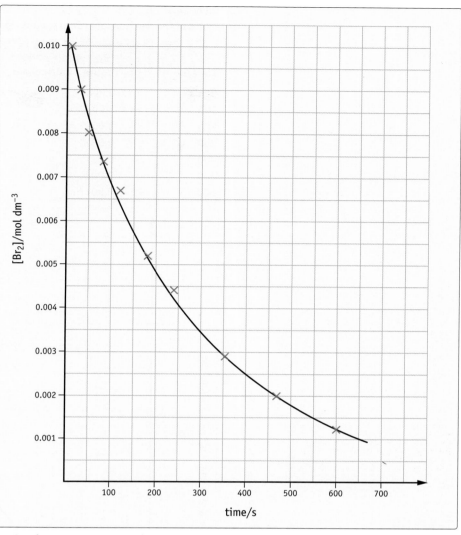

Figure 24.7
The variation of bromine concentration with time in the reaction between methanoic acid and bromine.

Notice the negative sign in the last expression. $d[Br_2]$ is negative because the bromine is being used up. So, the negative sign is necessary to give the rate of reaction, $-d[Br_2]/dt$, a positive value.

In order to obtain the reaction rate at any given time, we must draw a tangent to the curve at this particular time and measure its gradient. Values of the reaction rate corresponding to different bromine concentrations at different times are shown in table 24.2. These values of reaction rate are plotted vertically against bromine concentration in figure 24.8.

1 How does the bromine concentration change with time?
2 How does the reaction rate change with time?
3 Is the rate of reaction affected by the bromine concentration?
4 How does the rate of reaction depend on the bromine concentration?
5 Write a mathematical expression relating reaction rate to bromine concentration.

Table 24.2
Values of the reaction rate corresponding to different bromine concentrations

Time/s	[Br$_2$] obtained from figure 24.7/mol dm^{-3}	Reaction rate ($-d$[Br$_2$]/dt) obtained from gradients in figure 24.7/mol dm^{-3} s^{-1}
50	0.0085	2.9×10^{-5}
100	0.0072	2.4×10^{-5}
200	0.0050	1.7×10^{-5}
250	0.0042	1.5×10^{-5}
300	0.0035	1.2×10^{-5}
400	0.0024	0.8×10^{-5}
500	0.00175	0.6×10^{-5}

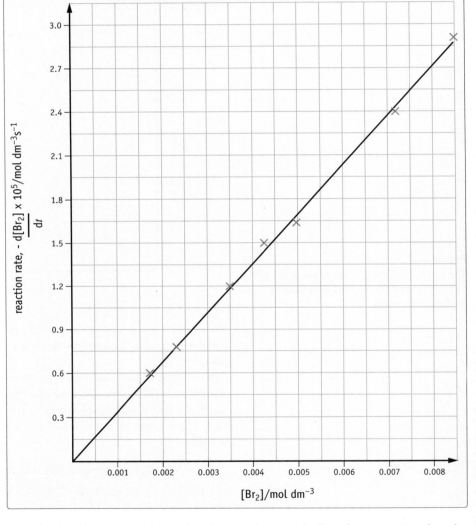

Figure 24.8
Variation of reaction rate with bromine concentration.

The graph in figure 24.8 shows that the reaction rate is directly proportional to the bromine concentration, i.e.

$$\text{reaction rate} \propto [\text{Br}_2]$$
$$\Rightarrow \text{reaction rate} = k\,[\text{Br}_2]$$

where k is a constant, known as the **rate constant** for the reaction.

24.6 Order of reaction and rate equations

Experiments show that the rates of most reactions can be related to the concentrations of individual reactants by an equation of the form

$$\text{Rate} = k\,[\text{X}]^n$$

This expression, in which X is the reactant under consideration and n is usually 0, 1 or 2, is known as a **rate equation**. The value of n gives the **order of the reaction with respect to X.**

When $n = 0$, the reaction rate is said to be **zero order** with respect to X, i.e.

$$Rate = k\,[X]^0$$
but, since $[X]^0 = 1$,
$$Reaction\ rate = k$$

In other words, the reaction rate is independent of the concentration of X. This means that changing the concentration of X will not affect the rate of the reaction (figure 24.9).

When $n = 1$, the reaction rate is proportional to $[X]^1$ and the reaction is said to be **first order** with respect to X (figure 24.9).

When $n = 2$, the reaction rate is proportional to $[X]^2$ and the reaction is said to be **second order** with respect to X (figure 24.9).

> 1 What is the order with respect to Br_2 for the reaction we studied in the last section?

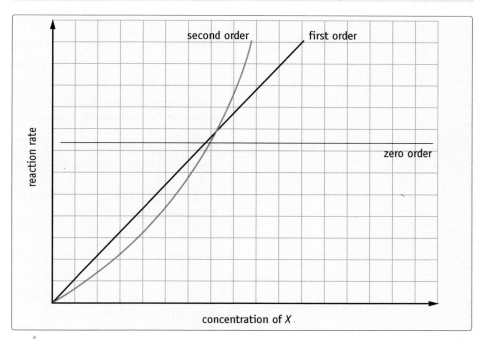

second order   first order

zero order

reaction rate

concentration of X

Figure 24.9
The variation of reaction rate with concentration for reactions which are zero, first and second order.

Look at the information in table 24.3. This relates to the reaction between hydrogen and nitrogen oxide at 800°C:

$$2H_2(g) + 2NO(g) \rightarrow 2H_2O(g) + N_2(g)$$

> 1 What happens to the initial rate, when the initial concentration of hydrogen is doubled in experiment 2 compared with experiment 1?
>
> 2 What happens to the initial rate, when the initial concentration of hydrogen is trebled in experiment 3 compared to experiment 1?
>
> 3 How does the reaction rate depend on $[H_2]$?
>
> 4 What is the order of the reaction with respect to hydrogen?

In experiments 1, 2 and 3 the initial concentration of nitrogen oxide is the same. In experiments 4, 5 and 6 the initial concentration of hydrogen is the same. When [NO] is doubled in experiment 5 compared with experiment 4, the reaction rate is *not* doubled, but quadrupled (i.e. 2^2).

Table 24.3

Information concerning the rate of reaction between hydrogen and nitrogen oxide at 800°C

Experiment number	Initial concentration of nitrogen oxide /mol dm^{-3}	Initial concentration of hydrogen /mol dm^{-3}	Initial rate of production of nitrogen /mol dm^{-3} s^{-1}
1	6×10^{-3}	1×10^{-3}	3×10^{-3}
2	6×10^{-3}	2×10^{-3}	6×10^{-3}
3	6×10^{-3}	3×10^{-3}	9×10^{-3}
4	1×10^{-3}	6×10^{-3}	0.5×10^{-3}
5	2×10^{-3}	6×10^{-3}	2.0×10^{-3}
6	3×10^{-3}	6×10^{-3}	4.5×10^{-3}

Similarly, when [NO] is trebled (experiment 6 compared with experiment 4), the reaction rate is *not* trebled, but increases nine-fold (i.e. 3^2). These results show

$$\text{Reaction rate} \propto [NO]^2 \text{ and}$$
$$\text{Reaction rate} \propto [H_2]$$

∴. The order of reaction with respect to NO is 2 and the order of reaction with respect to H_2 is 1.

We can combine these results in a single rate equation as

$$\text{Reaction rate} = k[NO]^2[H_2]$$

k, *the rate constant, is constant for a given reaction at a particular temperature.* It is important to realise that *the rate equation can only be obtained by experiment; it cannot be deduced either theoretically or from the balanced equation.*

> ⚊ Using the concentrations of NO and H_2 and the initial rate of reaction in experiment 1, determine the value of k from the rate equation.
>
> $$\text{Rate} = k[NO]^2[H_2]$$
>
> ⚋ What are the units of k in this case?

In experiment 2

$[NO] = 6 \times 10^{-3}$ mol dm^{-3},
$[H_2] = 2 \times 10^{-3}$ mol dm^{-3} and
rate = 6×10^{-3} mol dm^{-3} s^{-1}.

Substituting these values in

$$\text{Rate} = k[NO]^2[H_2], \text{ we get}$$
$$6 \times 10^{-3} = k(6 \times 10^{-3})^2 \times 2 \times 10^{-3}$$
$$\Rightarrow k = \frac{6 \times 10^{-3}}{(6 \times 10^{-3})^2} \times \frac{1}{2 \times 10^{-3}} = \frac{10^6}{12} = 8.33 \times 10^4$$

By substituting the units for reaction rate and concentrations in the rate equation, we can determine the units of k. For example, in the NO/H_2 reaction

$$k = \left(\frac{\text{rate}}{[NO]^2[H_2]}\right) \quad \frac{\cancel{\text{mol dm}^{-3}} \text{ s}^{-1}}{(\text{mol dm}^{-3})^2 \cdot \cancel{(\text{mol dm}^{-3})}}$$

$$= \left(\frac{\text{rate}}{[NO]^2[H_2]}\right) \quad \text{mol}^{-2} \text{ dm}^6 \text{ s}^{-1}$$

As the rate constant is constant at a fixed temperature, our results show that $k = 8.33 \times 10^4$ mol^{-2} dm^6 s^{-1} at 800°C.

Suppose we take the general case of a reaction,

$$x\text{A} + y\text{B} \rightarrow \text{products}$$

with a rate equation which can be expressed as

$$Rate = k[A]^m[B]^n$$

The indices m and n are known as the **orders of the reaction with respect to A and B**, respectively. The **overall order** of the reaction is described as $(m + n)$.

It is therefore important to be clear whether we are discussing the overall order of the reaction or the order with respect to an individual reactant.

> *The order of a reaction with respect to a given reactant is the power of that reactant's concentration in the experimentally determined rate equation.*
>
> *The overall order of the reaction is the sum of the powers of the concentration terms in the rate equation.*

The reaction between propanone and iodine in aqueous solution,

$$CH_3COCH_3(aq) + I_2(aq) \rightarrow CH_2ICOCH_3(aq) + H^+(aq) + I^-(aq)$$

is catalysed by H^+ ions. Experiments show that the rate can be expressed as:

$$Rate = k[CH_3COCH_3][H^+]$$

> **1** What is the order of the reaction with respect to
>
> (a) propanone,
> (b) iodine,
> (c) H^+ ions?
>
> **2** What is the overall order of the reaction?

This particular reaction shows very clearly that the balanced equation tells us nothing about the rate equation. I_2 appears in the balanced equation, but it does not feature in the rate equation. Furthermore, H^+ ions which feature in the rate equation do not appear as reactants in the balanced equation.

24.7 The half-life for a first-order reaction and for radioactive decay

Look closely at figure 24.7 once more.

> **1** How many seconds does it take for the concentration of bromine to fall from 0.010 mol dm^{-3} to 0.005 mol dm^{-3}?
>
> **2** How many seconds does it take for the concentration of bromine to fall from 0.005 mol dm^{-3} to 0.0025 mol dm^{-3}?
>
> **3** How many seconds does it take for the concentration of bromine to fall from 0.0025 mol dm^{-3} to 0.00125 mol dm^{-3}?

At any time during the reaction it takes 200 seconds for the concentration of bromine to halve (i.e. to fall from a concentration of say x mol dm^{-3} to $\frac{x}{2}$ mol dm^{-3}).

> The time taken for the concentration of a reactant to fall to half its original value is called the **half-life of the reaction**.
>
> *All first-order reactions have constant half-lives.*

Thus, the kinetics of a first-order reaction with its constant half-life are similar to the decay of a radioactive isotope, which also has a constant half-life. (See figure 24.10 and section 7.4.)

Indeed, the constant half-lives of radioactive isotopes form the basis of archaeological dating. Archaeological dating can be carried out with organic remains containing radioactive $^{14}_{6}C$ or with mineral remains containing isotopes such as radioactive $^{238}_{92}U$.

Half-lives range from minute fractions of a second (such as 10^{-21} seconds for $^{5}_{2}He$) to thousands of millions of years (such as 4.5×10^9 years for $^{238}_{92}U$). The half-life of a radioactive isotope is a fundamental property unaffected by chemical and physical changes.

▶ A mutant blue spruce damaged by radioactive fallout from the Chernobyl disaster in April 1986 when the core of a nuclear reactor, near Kiev in the Ukraine, exploded. Some of the ejected radioactive isotopes have long half-lives and they will contaminate the area for many decades. Some of the tree's needles are growing longer and are lighter in colour than normal as a result of mutations.

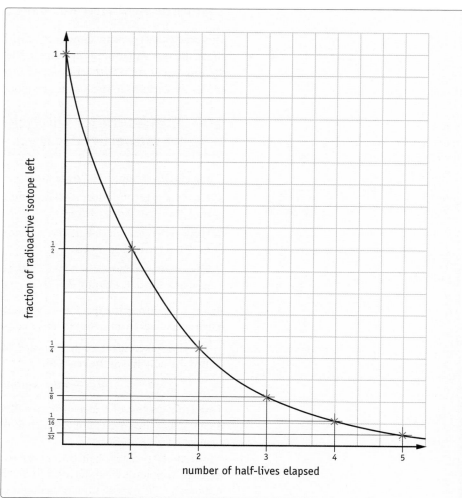

Figure 24.10
The fraction of a radioactive isotope remaining after 1, 2, 3, 4 and 5 half-lives.

▲ Research workers examining fragments of the Dead Sea Scrolls. The age of the Dead Sea Scrolls was checked by carbon dating. The process showed that the scrolls were probably authentic.

Dating geological remains

From the half-life of a radioactive isotope, we can deduce the time it will take for a certain proportion of the isotope to decay. Thus, the decay of $^{238}_{92}U$ to $^{206}_{82}Pb$ provides a method of dating rocks in the earth's crust. We begin by assuming that the uranium-bearing rocks in the earth's crust originally contained $^{238}_{92}U$ but no $^{206}_{82}Pb$. The present ratio of $^{238}_{92}U : ^{206}_{82}Pb$ in the rocks can then be used to calculate the time which has elapsed since the rocks formed. Using this technique, scientists have found that the ages of different rocks vary from forty million to four thousand million years. Geologists often take the larger of these two values to be the age of the earth.

Dating the remains of living things

The remains of living things are often dated from the amount of carbon-14 in them. The basis of carbon dating (section 7.5) is the simultaneous production and disintegration of radioactive $^{14}_{6}C$. As a result of this simultaneous formation and decay of $^{14}_{6}C$, all living things have a constant proportion of their carbon in the form of $^{14}_{6}C$. However, when the animal or plant dies, replacement of $^{14}_{6}C$ ceases while decay of $^{14}_{6}C$ continues.

Suppose that carbon-14 makes up $x\%$ of the carbon in living things and that the half-life of $^{14}_{6}C$ is 5800 years.

1 What percentage of $^{14}_{6}C$ will the remains of a plant contain 5800 years after it has died?

2 How long will it take for the percentage of $^{14}_{6}C$ in a dead plant to fall from $\frac{x}{2}\%$ to $\frac{x}{4}\%$?

3 Approximately how old is an object which has $\frac{x}{8}\%$ of $^{14}_{6}C$?

By comparing the content of $^{14}_{6}C$ in archaeological specimens with that in similar materials living at the present time, it is possible to estimate the age of the specimens. Radiocarbon dating has been widely used to establish Egyptian chronology and to check the authenticity of ancient remains such as the Dead Sea scrolls.

24.8 Investigating the effect of temperature on the rate of a reaction

When dilute hydrochloric acid is added to sodium thiosulphate solution ($Na_2S_2O_3(aq)$), the mixture becomes cloudy. The cloudiness is caused by a precipitate of sulphur. As the precipitate gradually thickens, its yellow colour becomes apparent.

$$S_2O_3{}^{2-}(aq) + 2H^+(aq) \rightarrow S(s) + SO_2(g) + H_2O(l)$$

How can we measure the reaction rate in this case?

Add 10 cm³ of 1.0 mol dm⁻³ HCl to 50 cm³ of 0.05 mol dm⁻³ $Na_2S_2O_3(aq)$ and mix the contents thoroughly. Place the flask above an ink cross on white paper. Measure the interval between the addition of HCl and the obscuring of the ink cross as the precipitate thickens.

The results in table 24.4 were obtained when the experiment was carried out at different temperatures.

Notice that a rise in temperature of about 10 K causes the reaction rate to double. For example, the cross disappears in 119 seconds at 298 K, but in about half the time (62 s) at 307 K. The cross disappears in 90 seconds at 302 K and in half the time (45 s) at 311 K.

The question we must now ask is 'Why does a small rise in temperature cause such a large percentage increase in the reaction rate?'

Table 24.4
Investigating the effect of temperature on the reaction between HCl(aq) and $Na_2S_2O_3(aq)$

Temperature /K	Time for ink cross to disappear/s
296	135
298	119
302	90
307	62
311	45
317	37
320	34
326	24
332	20

24.9 Explaining the increase in reaction rate with temperature – the collision theory

As temperature rises, the average speeds of reacting particles increase. At high temperatures, there are more collisions per second and this results in an increase in the rate of reaction.

The increase in collision frequency accounts for an increase in reaction rate as temperature rises; but does it also explain how rapidly the rate increases – the rates of many reactions double for a temperature rise of only 10 K?

From the kinetic theory, we can predict the relative increase in collisions when the temperature rises by 10 K.

The kinetic energy of a particle is proportional to its absolute temperature:

$$\tfrac{1}{2}mV^2 \propto T$$

but the mass of a given particle remains constant

$$\Rightarrow V^2 \propto T$$

$$\therefore \frac{V_1^2}{V_2^2} = \frac{T_1}{T_2} \quad \text{(equation 1)}$$

where V_1 is the velocity at temperature T_1, and V_2 is the velocity at temperature T_2.

Now, suppose that the average speed of a particle is V at 300 K. What will its average speed be at 310 K?

Substituting in equation 1,

$$\frac{V_1^2}{V^2} = \frac{310}{300}$$

$$\therefore V_1 = \sqrt{\frac{310}{300}}\, V = \sqrt{1.033}V = 1.016V$$

$\therefore$ The average speed at 310 K is only 1.016 times greater than that at 300 K, i.e. it has increased by only 1.6%.

Assuming that the frequency of collisions depends on the average speed of the particles, we might expect the rate of collisions and hence the reaction rate to be 1.6% greater at 310 K than at 300 K. In practice, the reaction rate roughly doubles between 300 and 310 K, i.e. it increases by approximately 100%.

Clearly, the *simple* collision theory cannot account fully for the increase in reaction rate as temperature rises. How then do we explain the relatively large increase in reaction rate with temperature?

During a chemical reaction, bonds are first broken and others are then formed. Consequently, energy is required to break bonds and start this process, whether the overall reaction is exothermic or endothermic. Therefore, it is reasonable to assume that *particles do not always react when they collide*. They may not have sufficient energy for the necessary bonds to be broken.

A reaction will only occur if the colliding particles possess more than a certain minimum amount of energy. This minimum energy for a reaction to occur is known as the **activation energy**, E_A. The activation energy enables chemical bonds to stretch and break, and rearrangements of atoms, ions and electrons to occur as the reaction proceeds.

The reaction can be imagined to proceed as shown in figure 24.11.

The diagrams show the relationship between the activation energy, E_A, and the enthalpy change of reaction, ΔH, for an exothermic and an endothermic reaction. E_A is related to the rate constant of a reaction (see later in this section). It gives an indication of *how fast* the reaction will occur. If the activation energy is very large, only a small

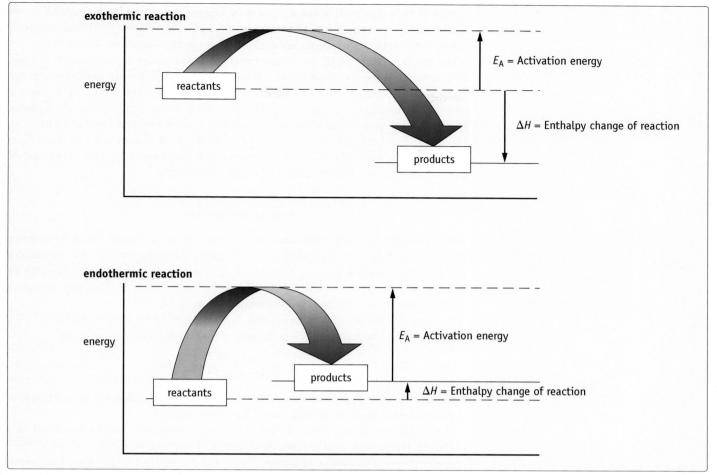

Figure 24.11
The progress of an exothermic and an endothermic reaction according to the collision theory.

proportion of molecules have enough energy to react, so the reaction proceeds very slowly. If, however, the activation energy is very small, most of the molecules have sufficient energy to react and the reaction proceeds very fast.

Unlike E_A, ΔH is related to the equilibrium constant of a reaction (section 21.4) rather than to the rate constant. ΔH gives information concerning *how far* a reaction goes towards completion.

The fact that a certain minimum energy is needed to initiate most reactions is well illustrated by fuels and explosives. These usually require a small input of energy to get started even though they are highly exothermic.

The idea of activation energies leads to the next important question.

What fraction of the particles have more than the activation energy – the minimum energy required for reaction? In the case of a gas, the energy of particles is largely kinetic

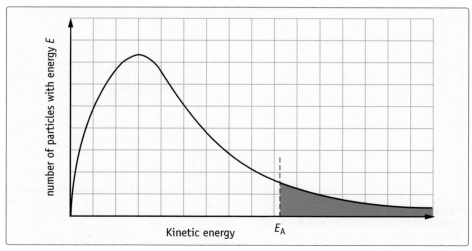

Figure 24.12
Distribution of the kinetic energies of particles in a gas.

energy ($\frac{1}{2}mv^2$). For particles of a given mass this is determined by their speed. As we discovered in section 11.7, the distribution of molecular speeds (and hence energies) can be determined by an apparatus similar to that used by Zartmann (figure 11.3).

The graph in figure 24.12 shows how the energies of particles are distributed in a gas. The spread of energies in the graph is sometimes called the Maxwell–Boltzmann distribution in honour of the two scientists who predicted these results. The graph is essentially a histogram. It shows the number of particles in each small range of kinetic energy. So, the area beneath the curve is proportional to the total number of particles involved. Furthermore, the number of particles with energy greater than E_A is proportional to the red area beneath the curve at energies above E_A. Hence the *fraction* of particles with energy greater than E_A is given by the ratio

$$\left(\frac{\text{red area beneath curve}}{\text{total area beneath curve}} \right)$$

Using probability theory and the kinetic theory of gases, Maxwell and Boltzmann derived equations for the distribution of kinetic energies amongst the molecules of a gas. From their equations, they calculated that the fraction of molecules with an energy greater than $E_A \text{ J mol}^{-1}$ is given by $e^{-E_A/RT}$. R is the gas constant ($8.3 \text{ J K}^{-1} \text{ mol}^{-1}$) and T is the absolute temperature.

This suggests that at a given temperature, T, the reaction rate is proportional to $e^{-E_A/RT}$. Now, since the rate constant, k, is a measure of the reaction rate, we can write

$$k \propto e^{-E_A/RT}$$
$$\Rightarrow k = Ae^{-E_A/RT}$$

This last expression is sometimes called the **Arrhenius equation** because it was first predicted by the Swedish chemist Svante Arrhenius in 1889.

In the Arrhenius equation, A (the Arrhenius constant), can be regarded as a **collision frequency and orientation factor** in the reaction rate, whilst $e^{-E_A/RT}$ represents an **activation state factor**. Thus, A is determined by the total number of collisions per unit time and the orientation of molecules when they collide (collision geometry). In contrast $e^{-E_A/RT}$ is determined by the fraction of molecules with sufficient energy to react.

If we take logarithms to base e in $k = Ae^{-E_A/RT}$

$$\ln k = \ln A + \ln e^{-E_A/RT}$$
$$\Rightarrow \ln k = \ln A - \frac{E_A}{R} \times \frac{1}{T}$$

Comparing the last equation with

$$y = c + mx$$

a graph of $\ln k$ against $1/T$ should be a straight line with gradient $-E_A/R$ (figure 24.13).

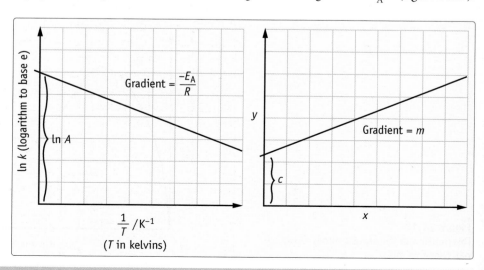

Figure 24.13
Comparing the equation $\ln k = \ln A - E_A/RT$ with $y = c + mx$.

E_A is the activation energy in J mol^{-1} and R is the gas constant (8.3 J K^{-1} mol^{-1}). Thus, by plotting a graph of the logarithm to base e of velocity constants against the reciprocal of the absolute temperature, we can measure the gradient ($-E_A/R$). From this gradient, we can calculate the activation energy (E_A).

Using the concept of activation energy, we can now return to our original question concerning the relatively large increase in reaction rate with temperature.

Look closely at figure 24.14. This shows how the kinetic energies of the molecules in a gas might be distributed at T K and ($T + 10$) K. Let us suppose that colliding molecules must have a kinetic energy of E_A before a reaction takes place. Notice in figure 24.14 that only a small fraction of molecules (indicated by the blue area) have sufficient energy to react at T K. However, when the temperature rises by 10 K, the fraction of molecules with sufficient energy to react (indicated by the red vertically lined area) roughly doubles. This causes the reaction rate to double as well.

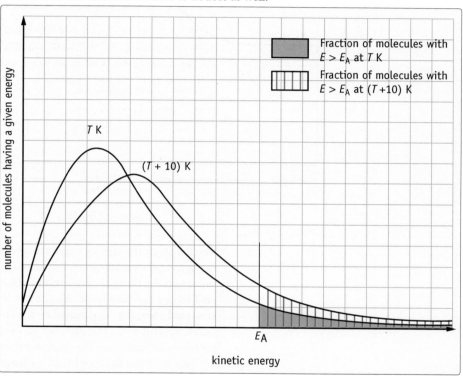

Figure 24.14
Distribution of the kinetic energies of gas molecules at T K and ($T + 10$) K.

24.10 Catalysis

In a chemical reaction, existing bonds must break and new bonds must form as reactants are converted to products. In order to do this, energy is needed, usually in the form of heat. The 'energy barrier' (as we might call the activation energy) for the reaction between N_2 and H_2 to produce NH_3 is shown in figure 24.15. The horizontal axis of the diagram shows the progress of the reaction from initial reactants to final products. The curve showing the energy of the materials throughout the reaction is usually referred to as an **energy profile**.

Let us follow the progress of the reaction from left to right along the energy profile in figure 24.15. As the molecules approach, there is little change in their total energy until they get close to each other. Hence the flat part on the left of the energy profile. When the molecules are within a few nanometres of each other, repulsions between their nuclei and between their negative electron clouds begin to operate. Thus, the molecules must have sufficient speed and kinetic energy (668 kJ per mole of N_2) if they are to overcome these repulsive forces as the molecules get closer.

The peak of the energy profile represents an 'energy barrier'. This has to be surmounted before bonds are stretched and sufficiently broken for products to form.

Figure 24.15
An energy profile for the reaction, $N_2 + 3H_2 \rightarrow 2NH_3$

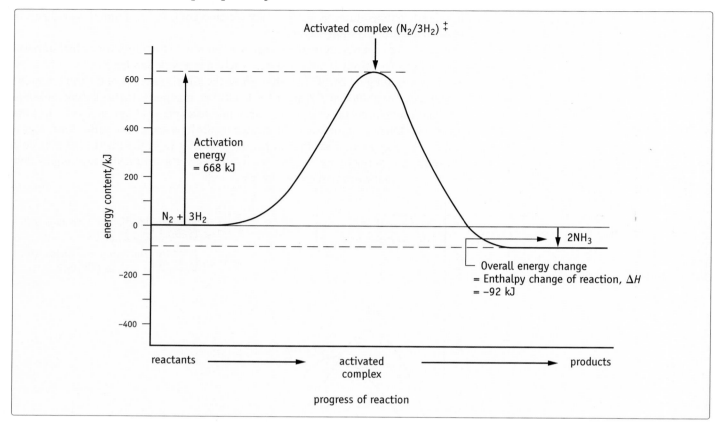

Activated complex $(N_2/3H_2)^{\ddagger}$

Activation energy = 668 kJ

$N_2 + 3H_2$

$2NH_3$

Overall energy change = Enthalpy change of reaction, ΔH = −92 kJ

energy content/kJ

reactants — activated complex — products

progress of reaction

The height of the energy barrier above the original height of the reactant molecules is the activation energy for the reaction. Only those molecules with enough energy to surmount the energy barrier will be able to form products. At the summit of the energy profile, the reactant molecules have a high energy content. They are described as the **activated complex** (designated by the superscript ‡). This activated complex can either break up and form the product molecules down the right-hand side of the energy profile or separate into the original molecules.

When nitrogen is mixed with hydrogen, no detectable reaction occurs even at high temperatures and high pressures. As the molecules approach each other, they have insufficient kinetic energy to overcome their mutual repulsions. So they never reach the activated state. They rise part of the way up the left-hand side of the energy profile, repel one another and separate again.

We can, however, speed up the reaction by using a **catalyst** (sections 18.6 and 21.3).

> *A catalyst can be defined as a substance which alters the rate of a reaction without itself undergoing any permanent chemical change.*

Thus, a small amount of catalyst is capable of catalysing an infinite amount of reaction. Usually, the catalysed reaction is much faster than the uncatalysed reaction.

When the reaction

$$N_2(g) + 3H_2(g) \rightarrow 2NH_3(g)$$

is catalysed by tungsten, the activation energy is much lower than in the uncatalysed reaction (figure 24.16).

In the presence of tungsten, a greater proportion of molecules have sufficient energy to overcome the lower activation energy barrier. Therefore the reaction goes much faster (figure 24.17). Notice in figure 24.16 that the energy levels of reactants and products are the same in the catalysed reaction as in the uncatalysed reaction.

▲ Catalytic converters like this contain platinum and rhodium or transition metal oxides such as chromium(III) oxide as the catalysts. In car exhaust systems, catalytic converters can oxidise carbon monoxide to carbon dioxide and reduce oxides of nitrogen (NO and NO_2) to nitrogen.

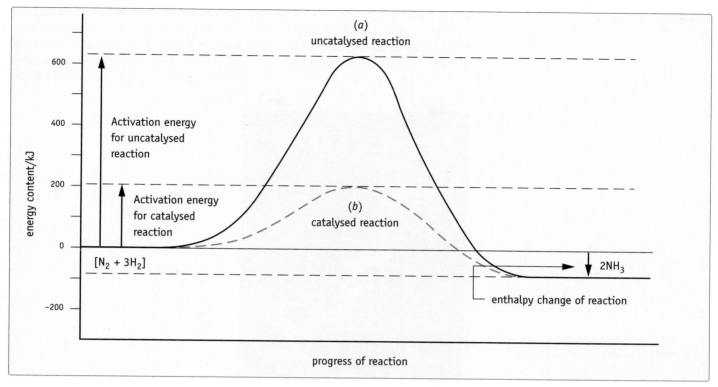

Figure 24.16
Energy profiles for the reaction,
$N_2 + 3H_2 \rightarrow 2NH_3$ (a) uncatalysed and (b)
catalysed by tungsten.

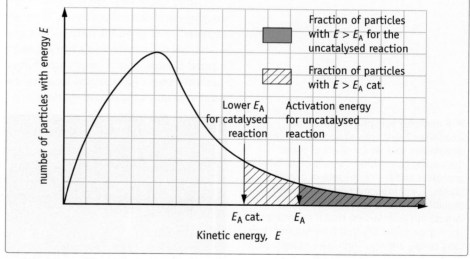

Figure 24.17
Distribution of the kinetic energies of
reacting particles and the activation
energies for catalysed and uncatalysed
reactions. Notice the greater proportion of
molecules which have energies greater than
the activation energy for the catalysed
reaction.

1 Name the catalyst used in each of the
following industrial processes,
(a) the Haber Process,
(b) the Contact Process,
(c) the oxidation of ammonia to
nitrogen oxide in the
manufacture of nitric acid.

2 Do these processes involve
heterogeneous or homogeneous
catalysis?

The catalyst has not supplied any extra energy for the reactants, yet the reaction has been speeded up.

The catalyst has, in fact, provided a new reaction path for the breaking and rearrangement of bonds. The new reaction path has a larger rate constant and lower activation energy, so that many more molecules can pass over the energy barrier. The situation in a catalysed reaction can be compared to a pole-vaulting event in which the bar has been lowered so that many more athletes can get over.

The reaction between nitrogen and hydrogen involves gases passing over a solid catalyst. Reactions of this kind in which the reactants are in a different physical state to the catalyst are said to involve **heterogeneous catalysis**.

In contrast to this, catalysed reactions in which reactants and catalyst are mixed together in the same state are said to involve **homogeneous catalysis**.

Many important industrial processes involve heterogeneous catalysis. In contrast, the action of enzymes in biological systems usually involves homogeneous catalysis.

Heterogeneous catalysis can reduce the atmospheric pollution from car exhaust fumes. The toxic chemicals in exhaust fumes include oxides of nitrogen, carbon monoxide and unburnt petrol. These are normally blown away without causing any problems. Under certain conditions, however, and particularly in traffic-dense areas these fumes cause a great deal of atmospheric pollution. Consequently, industrial chemists have developed solid catalysts which can be used in the exhaust systems of all vehicles. These catalyse the conversion of these polluting chemicals to harmless exhaust products. This is discussed further in section 26.8.

► A strong pole-vaulter gets sufficient height (potential energy) to clear the cross bar (activation energy barrier).

Another interesting development in the field of catalysis has been the increasing use of enzymes in industrial processes. These include the use of papain (a protein-hydrolysing enzyme) to remove the haze in beer and to tenderise beef. The manufacture of fruit juices, beer, vitamins and pharmaceutical products also uses enzymes extracted from animal tissues, plants, yeasts and fungi. The so-called 'biological' washing powders contain enzymes which attack animal and plant tissues. These are especially useful for removing the stains caused by biological materials such as food and blood.

► A gardener spraying the selective weedkiller 2,4-D onto a lawn. Selective weedkillers such as 2,4-D act as plant hormones, controlling the rate of metabolic processes and hence the rate of growth in plants. 2,4-D is such a powerful growth promoter in broad-leaved plants, such as dandelions and daisies, that it can make them grow so rapidly that they exhaust their food supply and die. Thin-leaved plants such as grass are unaffected. The full name of 2,4-D is 2,4-dichlorophenoxyethanoic acid. Its formula is

Many catalysts are highly specific and the details of their operation are still not fully understood. The final choice of catalyst for many industrial processes may often be a combination of scientific deduction, trial and error and simply inspired guesswork. Indeed, the precise composition of an industrial catalyst is often a closely guarded secret.

24.11 The importance of reaction rate studies

Reaction rate studies give us useful information about the rates of chemical reactions. This is important for industrial processes in which time and the efficient use of resources are crucial. It is also useful for biological processes, archaeological dating and many other reactions, such as rusting and burning, which affect our everyday lives. Nevertheless, the most exhaustive studies of reaction rates have usually been done on industrial processes. In these cases chemists, engineers and economists endeavour to obtain maximum product from the minimum amount of raw material, using minimum fuel in the minimum possible time. When viewed in this light, it is not surprising that catalysts play such an important part in industrial processes.

Rate studies also allow us to interpret reactions on a molecular level. By considering the order of a reaction with respect to the different reactants, we can speculate about the sequence in which bonds break and atoms rearrange. From these ideas, it is possible to suggest a reaction sequence.

As an example, let us consider the reaction between oxygen and hydrogen bromide at 700 K:

$$4HBr(g) + O_2(g) \rightarrow 2H_2O(g) + 2Br_2(g)$$

This equation indicates that four HBr molecules react with one O_2 molecule. If the reaction were to take place in a single step, these five molecules would need to collide with each other simultaneously. This is extremely improbable. Since the reaction occurs quite rapidly at 700 K, it is likely that it proceeds by a sequence of steps rather than by a single step involving the simultaneous collision of five molecules. In fact, *most chemical reactions which proceed at a measurable rate are believed to take place in a series of simple steps. This series of simple reactions is known as the* **reaction mechanism**.

Quantitative studies of the reaction between HBr and O_2 show that the reaction rate is proportional to both the concentration of HBr and the concentration of O_2. This means that the reaction is first order with respect to both HBr and O_2. We can summarise this information in the rate equation

$$Rate = k[HBr][O_2]$$

Notice that the balanced equation involves HBr and O_2 in molar proportions of 4 : 1, yet the rate equation suggests proportions of 1:1. How can we explain this?

The overall reaction must take place in a series of simple steps. These must satisfy both the rate equation and the balanced equation.

The following mechanism has been proposed for the reaction:

	$HBr + O_2$	$\rightarrow$	$HBrOO$	Step 1	Slow
$HBrOO +$	HBr	$\rightarrow$	$2HBrO$	Step 2	Fast
$HBrO$	$+ HBr$	$\rightarrow$	$H_2O + Br_2$	Step 3	Fast
$HBrO$	$+ HBr$	$\rightarrow$	$H_2O + Br_2$	Step 4	Fast
Overall	$4HBr + O_2$	$\rightarrow$	$2H_2O + 2Br_2$		

Notice that each step in the reaction mechanism involves the collision of only two molecules. This is a much more likely event than the simultaneous collision of five molecules. Notice also the suggestion that the first step in the mechanism is slow whilst the others are fast. This explains why the reaction rate is proportional to both [HBr] and $[O_2]$.

The first step producing HBrOO is very much a 'bottleneck' in the oxidation of hydrogen bromide. HBrOO forms, but it is immediately consumed in the fast second step by reaction with HBr. Although the second, third, and fourth steps are very rapid, they produce water and Br_2 only as fast as the slowest stage in the sequence. Hence, those factors that determine the rate of formation of HBrOO determine the overall rate of reaction. The formation of HBrOO is the one step which dictates the rate because it is the slowest stage in the reaction mechanism. The slowest stage in a mechanism is called the **rate-determining step** or the **rate-limiting step**.

Iodine reacts with propanone in acid solution as follows:

$$I_2(aq) + CH_3COCH_3(aq) \rightarrow CH_2ICOCH_3(aq) + H^+(aq) + I^-(aq)$$

The reaction is first order with respect to CH_3COCH_3, zero order with respect to I_2 and first order with respect to H^+.

1 Write a rate equation for the reaction.

2 What is the overall order of reaction?

3 Which substances are probably involved in the slow, rate-determining step of the reaction?

The rate-determining step in the reaction involves propanone and H^+, but not iodine. Hence the suggested mechanism is:

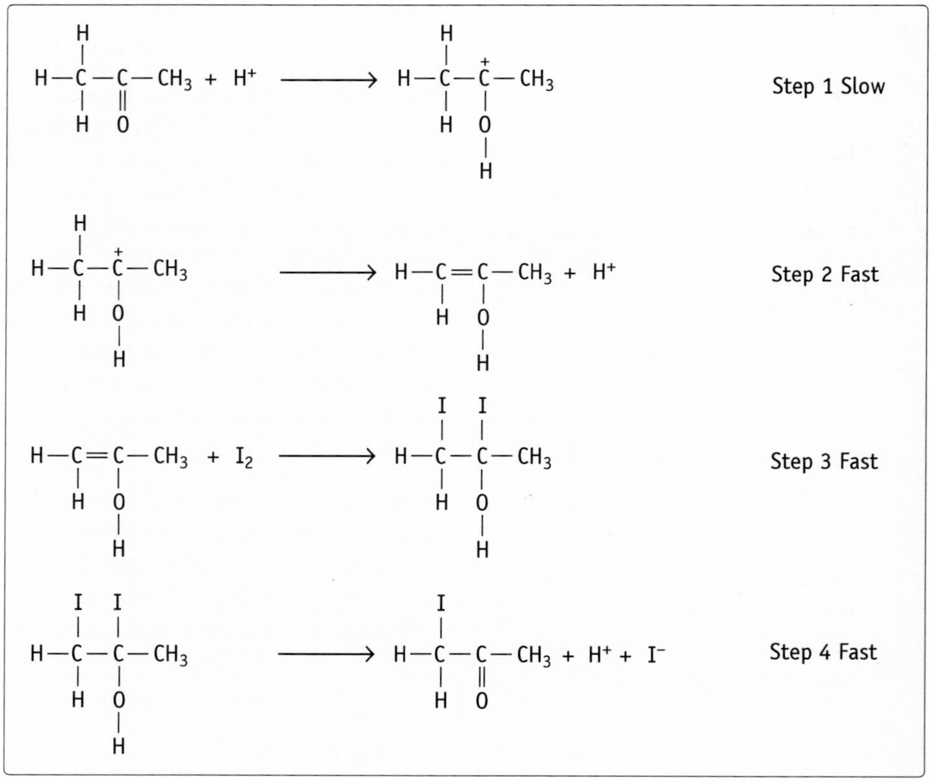

The reaction rate is dictated by the first, slow stage which requires only the participation of CH_3COCH_3 and H^+. Once this first step is completed, the remaining steps take place rapidly. Consequently, the reaction rate is independent of the concentration of iodine and $[I_2]$ does not feature in the rate equation. If this mechanism is correct, then the reaction of Br_2 with propanone should take place at a similar rate to iodination. This is found to be so.

Both reaction mechanisms that we have considered so far involve an initial slow step. This is not always the case. In order to illustrate this point, consider the reaction between bromide and bromate(v) ions in acid solution:

$$5Br^-(aq) + BrO_3^-(aq) + 6H^+(aq) \rightarrow 3Br_2(aq) + 3H_2O(l)$$

Kinetic studies show that the reaction is fourth order overall; first order with respect to bromide, first order with respect to bromate(v) and second order with respect to H^+ ions.

$$\text{Rate} = k\,[Br^-][BrO_3^-][H^+]^2$$

The immediate deduction from this is that the rate-determining step involves one Br^-, one BrO_3^- and two H^+ ions. But, the simultaneous collision of four ions is most unlikely. A more likely explanation is that the slow rate-determining step is preceded by faster reactions. Hence, the suggested mechanism involves the initial formation of HBr and $HBrO_3$ in two fast reactions, followed by a reaction between these two substances in the slow rate-determining step.

$$H^+ + Br^- \quad \rightarrow HBr \qquad \text{Step 1 \quad Fast}$$
$$H^+ + BrO_3^- \quad \rightarrow HBrO_3 \qquad \text{Step 2 \quad Fast}$$
$$HBr + HBrO_3 \xrightarrow{\text{rate-determining step}} HBrO + HBrO_2 \quad \text{Step 3 \quad Slow}$$

The HBrO and $HBrO_2$, produced in the slow step, then react rapidly with more HBr, forming bromine and water:

$$HBrO_2 + HBr \quad \rightarrow 2HBrO \qquad \text{Step 4 \quad Fast}$$
$$HBrO + HBr \quad \rightarrow H_2O + Br_2 \qquad \text{Step 5 \quad Fast}$$

Summary

1 Reaction rate =
$$\frac{\text{change in amount (or concentration) of a substance}}{\text{time taken}}$$

2 The rate of a reaction can be affected by:

(a) the availability of reactants and their surface area,
(b) concentration (partial pressure for gases),
(c) temperature,
(d) catalysts,
(e) light.

3 For the hypothetical reaction

$$a\text{A} + b\text{B} + c\text{C} \rightarrow \text{products}$$

it is found experimentally that the reaction rate can be expressed as

$$\text{Reaction rate} = k[\text{A}]^{\alpha}[\text{B}]^{\beta}[\text{C}]^{\gamma}.$$

This is known as the rate equation for the reaction. k is known as the rate constant and its value is constant for a given reaction at a particular temperature. When all reactant concentrations are 1.0 mol dm^{-3}, k, the rate constant, is numerically equal to the reaction rate.

4 The rate equation can only be obtained experimentally. It cannot be deduced from the balanced equation.

5 The order of a reaction with respect to a given reactant is the power of that reactant's concentration in the rate equation. For the hypothetical reaction just mentioned,

order of reaction with respect to A = α,
order of reaction with respect to B = β,
order of reaction with respect to C = γ.

The overall order of a reaction is the sum of the powers of the concentration terms in the rate equation. Hence, the overall order of the hypothetical reaction = $\alpha + \beta + \gamma$.

6 The half-life of a first-order reaction is constant. The kinetics of a first-order reaction with its constant half-life are similar to the decay of a radioactive isotope. This also has a constant half-life.

7 A reaction will only occur if the colliding particles possess more than a certain minimum amount of energy known as the activation energy.

8 A catalyst is a substance which alters the rate of a reaction without itself undergoing any permanent chemical change. Catalysts usually speed up a reaction. They do this by introducing an entirely different reaction mechanism with a lower activation energy than the uncatalysed reaction.

9 Most chemical reactions proceed by a sequence of simple steps, each of which involves only one or two particles. This series of simple steps is known as the reaction mechanism.
The slowest step in the reaction mechanism dictates the overall reaction rate. This slowest step is usually known as the rate-determining step.

10 The mechanism for a reaction is related to the rate equation. It cannot be deduced from the balanced equation.

1 (a) Give one example in each case of a reaction which takes place:
  (i)   instantaneously,
  (ii) at a moderate rate,
  (iii) rapidly at a high temperature, but not at all at room temperature.
  (b) Explain why the reactions you chose in (a) (i) and (a) (iii) behave as they do.

2 For the gaseous reaction.

$A(g) + B(g) \rightarrow C(g) + D(g)$ it is found that,

Reaction rate = $k[A]^2[B]$

How many times does the rate increase or decrease if:
(a) the partial pressures of both A and B are doubled,
(b) the partial pressure of A doubles, but that of B remains constant,
(c) the volume of the reacting vessel is doubled,
(d) an inert gas is added, which doubles the overall pressure whilst the partial pressures of A and B remain constant.
(e) the temperature rises by 30°C?

3 (a) Draw a sketch graph of the percentage reactant remaining against time, for a zero-order and a first-order reaction. (Assume, in each case, that it takes 10 minutes for the amount of reactant to fall from 100% to 50%.)
  (b) The half-life of radioactive $^{238}_{92}U$ is 4.5×10^9 years. It takes about 4.5×10^9 years for half of a given amount of uranium to disintegrate by radioactive emission and turn into lead. When uranium and lead are found together in rocks, the age of the rock may be deduced. In a particular sample, uranium and lead are found in molar proportions of 1:3.
  Estimate the age of the rock, stating any assumptions which you make.

4 (a) Explain the following terms:
  (i)   order of reaction,
  (ii) rate constant,
  (iii) half-life,
  (iv) activation energy,
  (v)   activated state.
  (b) Rate constants (k) for the decomposition of hydrogen iodide at different temperatures are given in the table below.
  (i) Plot a graph of ln k ($lg_e k$) against $1/T$.
  (ii) Use this graph to obtain a value for the activation energy for the decomposition of hydrogen iodide. (The gas constant, $R = 8.3$ J K^{-1} mol^{-1}.)

Rate constant, k /mol^{-1} dm^{-3} s^{-1}	Temperature/K
3.75×10^{-9}	500
6.65×10^{-6}	600
1.15×10^{-3}	700
7.75×10^{-2}	800

5 Hydrogen peroxide reacts with iodide ions in acid solution according to the following equation:

$H_2O_2(aq) + 2I^-(aq) + 2H^+(aq) \rightarrow I_2(aq) + 2H_2O(l)$

The rate of the reaction can be calculated by measuring the time for the first appearance of I_2 in the solution. When iodine first appears the concentration of I_2 is 10^{-5} mol dm^{-3}.
(a) For a particular experiment, the initial concentrations are $[H_2O_2] = 0.010$ mol dm^{-3}, $[I^-] = 0.010$ mol dm^{-3} and $[H^+] = 0.10$ mol dm^{-3}.
  Calculate the reaction rate if I_2 first appears after 6 seconds.
(b) In a second experiment, the initial concentrations are $[H_2O_2] = 0.005$ mol dm^{-3}, $[I^-] = 0.010$ mol dm^{-3} and $[H^+] = 0.10$ mol dm^{-3}.
  Calculate the reaction rate if I_2 first appears after 12 seconds.
(c) From these calculations show that the reaction is first order with respect to H_2O_2.
(d) Given the further information that the rate law is, Reaction rate = $k[H_2O_2][H^+][I^-]$, calculate the rate constant, k.
(e) What are the units of k?
(f) Predict the rate of reaction when $[H_2O_2] = 0.05$ mol dm^{-3}, $[H^+] = 0.10$ mol dm^{-3} and $[I^-] = 0.02$ mol dm^{-3}.

6 In an experiment to study the acid-catalysed reaction of propanone (CH_3COCH_3) with iodine, 50 cm^3 of 0.02 mol dm^{-3} I_2 were mixed with 50 cm^3 of acidified 0.25 mol dm^{-3} propanone solution. 10 cm^3 portions of the reaction mixture were removed at 5 minute intervals and added rapidly to excess $NaHCO_3(aq)$. The remaining iodine was then titrated against $Na_2S_2O_3(aq)$. The graph in figure 24.18

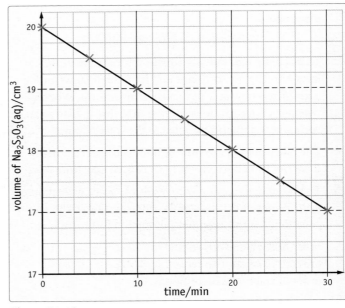

Figure 24.18
Volume of $Na_2S_2O_3(aq)$ required to react with the remaining iodine at different times during the reaction of iodine with propanone.

shows the volume of $Na_2S_2O_3(aq)$ required to react with the remaining iodine at different times from the start of the reaction.

(a) Why are the 10 cm^3 portions of the reaction mixture added rapidly to excess $NaHCO_3(aq)$ before titration with $Na_2S_2O_3(aq)$?

(b) What is the rate of reaction in terms of cm^3 of $Na_2S_2O_3(aq)$ min^{-1}?

(c) How does the *rate* of change of iodine concentration vary during the experiment?

(d) Is the reaction rate dependent on the concentration of iodine?

(e) What is the order of reaction with respect to iodine?

(f) Write an equation for the reaction between $S_2O_3{}^{2-}$ and I_2.

(g) What is the concentration of I_2 in the 100 cm^3 of reaction mixture at time = 0 min?

(h) Use the graph to predict the volume of $Na_2S_2O_3(aq)$ which reacts with 10 cm^3 of the reaction mixture at time = 0 min.

(i) What is the molarity of the $Na_2S_2O_3(aq)$ used in the titrations?

(j) Suppose the reaction is first order with respect to propanone. What would be the rate of reaction (in cm^3 $Na_2S_2O_3(aq)$ min^{-1}) if 0.50 mol dm^{-3} propanone were used in place of 0.25 mol dm^{-3}?

7 Two gases, X and Y, react according to the equation
$$X(g) + 2Y(g) \rightarrow XY_2(g)$$
Experiments were performed at 400 K in order to determine the order of this reaction and the following results were obtained.

Experiment number	Initial concentration of X/mol dm^{-3}	Initial concentration of Y/mol dm^{-3}	Initial rate of formation of XY$_2$/mol dm^{-3}s^{-1}
1	0.10	0.10	0.0001
2	0.10	0.20	0.0004
3	0.10	0.30	0.0009
4	0.20	0.10	0.0001
5	0.30	0.10	0.0001

(a) What is the order of this reaction with respect to (i) X, (ii) Y?

(b) Write a rate equation for the reaction of X with Y.

(c) Using the rate equation, predict a possible mechanism for this reaction.

(d) Using the results from experiment 1, calculate the numerical value of the rate constant, k.

(e) What are the units of k?

(f) What further experiments would you carry out to find the activation energy of the reaction between X and Y?

(g) Why are chemists interested in obtaining orders of reaction and rate equations?

8 Suggest experimental means by which the rates of the following reactions could be followed.

(a) $CaCO_3(s) \rightarrow CaO(s) + CO_2(g)$

(b) $2NO(g) + 2H_2(g) \rightarrow N_2(g) + 2H_2O(g)$

(c) $Cl_2(aq) + 2Br^-(aq) \rightarrow Br_2(aq) + 2Cl^-(aq)$

9 (a) Draw an energy profile curve for the reaction
$$H_2(g) + I_2(g) \rightarrow 2HI(g) \qquad \Delta H^{\ominus} = -10 \text{ kJ}$$
Put labelled arrows on your diagram to indicate the activation energy of the reaction and the enthalpy change of the reaction.

(b) The rate of this reaction is given by
Rate = $k[H_2][I_2]$

(i) What is the order of the reaction with respect to iodine?

(ii) What is the overall order of the reaction?

(c) When 0.1 moles of hydrogen and 0.2 moles of iodine were mixed at 400°C in a 1 dm^3 vessel, the initial rate of formation of hydrogen iodide was 2.3×10^{-5} mol dm^{-3} s^{-1}.

(i) What is the value of k at 400°C?

(ii) What are the units of k?

10 The following results were obtained from a study of the isomerisation of cyclopropane to propene in the gas phase at 433°C.

Time/hours	0	2	5	10	20	30
% of cyclopropane remaining	100	91	79	63	40	25

(a) Write an equation for the reaction involved.

(b) Show that the reaction is first order with respect to cyclopropane.

25 Introduction to Carbon Chemistry

25.1 Carbon – a unique element

The number of compounds containing carbon and hydrogen whose formulas are known to chemists is over ten million. This is far more than the number of compounds of all the other elements put together. Why does carbon have this unique ability to form an enormous number of compounds? How can we hope to begin to study more than a tiny fraction of them?

There are three important properties of carbon that enable it to form so many stable compounds.

(a) Carbon has a fully shared octet of electrons in its compounds

In methane the outer-shell electrons are shared as shown in figure 25.1. This means that the carbon atoms have no lone pairs or empty orbitals in their outer shells, so they are unable to form dative bonds. The inability of carbon to bond in this way once it has an octet of electrons is responsible for the kinetic stability of its compounds (see below).

(b) Carbon can form strong single, double and triple bonds to itself

The stability of the single C—C bond can be seen by comparing the bond energies in table 25.1.

It is worth comparing carbon and silicon here since we might expect silicon to show similarities to carbon, being the next member of group IV. Note the strength of the C—C bond compared with that of the Si—Si bond. Note too the high strength of the C—H bond: all but a handful of the vast range of carbon compounds also contain hydrogen. Of course, it is not enough to consider simply the strength of bonds between carbon atoms and hydrogen atoms. If carbon compounds are to be stable they must be stable under normal conditions, and that means in the presence of air. In fact, compounds containing carbon and hydrogen are *not* stable relative to their oxidation products, carbon dioxide and water. We would therefore expect them to react with oxygen exothermically, which, of course, they do. For example, methane:

$$CH_4(g) + 2O_2(g) \rightarrow CO_2(g) + 2H_2O(g); \Delta H^{\ominus} = -890 \text{ kJ mol}^{-1}$$

Most people are familiar with this reaction: it occurs in most gas-fired appliances in Britain, since North Sea gas is largely methane. It is a familiar fact that although methane is energetically unstable relative to its combustion products, it does not react with air until heated to quite high temperatures. In other words, it needs lighting before it burns. This is because the reaction between methane and oxygen has a high activation energy which must be supplied before the reaction will proceed. Thus methane, like most compounds containing carbon and hydrogen, is energetically unstable in the presence of air, but kinetically stable.

Compounds containing silicon and hydrogen are also energetically unstable relative to their combustion products:

$$SiH_4(g) + 2O_2(g) \rightarrow SiO_2(s) + 2H_2O(g); \Delta H^{\ominus} = -1428 \text{ kJ mol}^{-1}$$

This reaction is much more exothermic than the corresponding one for methane, largely because of the very high energy of the Si—O bond. Unlike methane, silane is not

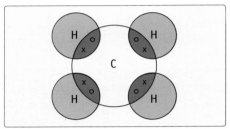

Figure 25.1
Electron sharing in methane.

Table 25.1
Some average bond energies

Bond	Bond energy /kJ mol^{-1}
C—C	346
C=C	610
C≡C	835
Si—Si	226
Si=Si	318 (estimated)
S—S	272
C—O	360
Si—O	464
C—H	413
Si—H	318

▲ In the presence of air, hydrocarbons are energetically unstable, but kinetically stable. They can, therefore, be safely stored at room temperature ...

▲ ... but at high temperatures they burn rapidly.

▶ Like carbon atoms, these skydivers can form four bonds each. This means they can form chains and rings.

kinetically stable in the presence of oxygen. The activation energy of the above reaction is quite low, and silane bursts into flame spontaneously in air. Other compounds containing silicon and hydrogen behave similarly, so there is no huge range of silicon compounds to compare with that of carbon. However, the high strength of the Si—O bond relative to the Si—Si and Si—H bonds means that silicon exists naturally as highly stable silicon(IV) oxide (sand) and silicates.

The ability of carbon to form strong bonds to itself means that it can form chains and rings of varying size. This is called **catenation**. These chains and rings are the basis of carbon's many stable compounds.

The kinetic stability of hydrocarbons in air is very important to society. It means they can be stored and, barring accidents, the energy of their oxidation can be released when it is required. This makes hydrocarbons, of which oil is our major source, the most important modern fuels.

(c) Carbon can form four covalent bonds

The bond energies given in table 25.1 suggest that, like carbon, sulphur should be able to form reasonably stable bonds to itself. However, sulphur forms only two bonds, so a chain of sulphur atoms cannot have side-groups attached to it. By contrast, carbon can form *four* bonds. This means a chain of carbon atoms can have many different groups attached, leading to a wide diversity of compounds.

25.2 Organic chemistry

The diversity of carbon chemistry is responsible for the diversity of life itself. The ability to form a virtually unlimited range of compounds has led to an almost unlimited range of living organisms constructed out of molecules containing carbon. You yourself are a unique individual because you contain unique DNA. Only carbon could form a range of compounds diverse enough to provide a different one for every individual.

The major source of compounds containing carbon and hydrogen is living or once-living material: animals, plants, coal, oil and gas. For this reason, it was originally thought that only living organisms could produce these compounds. This has since been shown to be untrue, but the name **organic** is still applied to that branch of chemistry concerned with the study of compounds containing C—H bonds. This includes the vast majority of carbon compounds, but traditionally compounds such as CO, CO_2 and carbonates have been considered to belong to the field of inorganic chemistry.

The position of carbon as the basis of the molecules of life means that the study of organic chemistry is of central importance in understanding the chemistry, and therefore the biology, of living systems. (The chemical study of living systems is called **biochemistry**.) A knowledge of organic chemistry enables chemists to develop and

▲ These identical twins are made from identical sets of carbon compounds.

manufacture medicines, agricultural chemicals, anaesthetics and other chemicals whose effects on life processes are important to humans.

Many other organic chemicals are of prime importance to modern society. These include, for example, the many polymers (polythene, nylon) whose properties of flexibility and elasticity are a direct result of carbon's unique ability to form chains.

It is clearly important to understand organic chemistry, but with so many compounds to consider, we need some means of simplifying and systematising our study.

25.3 Functional groups

The ability of carbon to form strong bonds to itself and to hydrogen leads to the formation of stable compounds. Hydrocarbons contain *only* hydrogen and carbon. The simplest hydrocarbons, containing only single bonds, are the **alkanes.** Butane is an example:

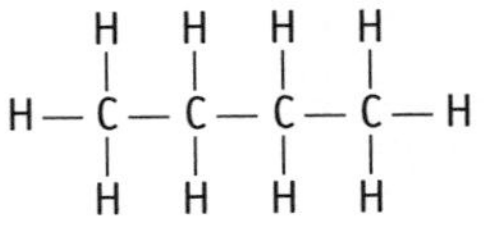

But consider another compound, butan-1-ol:

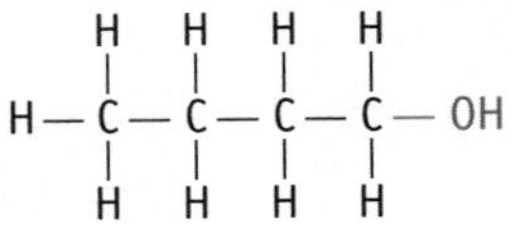

Butane and butan-1-ol, despite their structural similarities, have very different properties. Butane is a gas, butan-1-ol is a liquid. Butane has no effect on sodium, but butan-1-ol reacts with the evolution of hydrogen. Clearly, the —OH group in butan-1-ol has a big effect on the properties of the unreactive butane skeleton to which it is attached.

The —OH group in butan-1-ol is an example of a **functional group.** A given functional group, such as —OH, has much the same effect whatever the size and shape of the hydrocarbon skeleton it is attached to. This greatly simplifies the study of organic compounds because all molecules containing the same functional group can be considered as members of the same family, with similar properties. As the hydrocarbon chain gets bigger it increasingly dominates the properties of the compound. Because of this, members of the family show a steady gradation of physical and chemical properties as the size of the hydrocarbon portion increases.

A family of compounds containing the same functional group is called a **homologous series.** Butan-1-ol is a member of the homologous series of **alcohols,** all of which contain the —OH group. The first two alcohols are methanol and ethanol:

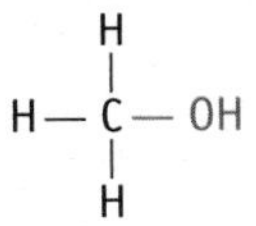

methanol 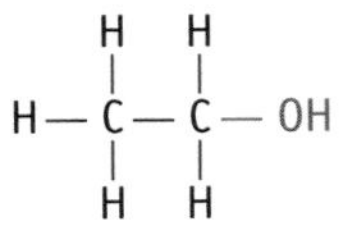   ethanol

For any homologous series we can write a general formula in terms of the number of carbon atoms present. For example, the general formula of the alcohols is $C_nH_{2n+1}OH$.

The main functional groups and homologous series considered in this book are shown in table 25.2. The methods of naming the different compounds will be explained as we go along.

The idea of functional groups can also be applied to compounds containing more than one group. Thus, the properties of the molecule as a whole can be predicted by considering the effect of each functional group.

Look at table 25.2 and answer these questions:

1 To what homologous series does propanal, CH_3CH_2CHO, belong?

2 To what homologous series does ethoxyethane, $CH_3CH_2OCH_2CH_3$ belong?

Table 25.2
Common functional groups

Functional group	Name of homologous series	Example
—OH	alcohols	CH_3OH methanol
—NH_2	amines	CH_3NH_2 methylamine
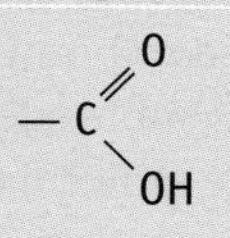 $-C\overset{\displaystyle O}{\underset{\displaystyle OH}{}}$	carboxylic acids	CH_3COOH ethanoic acid
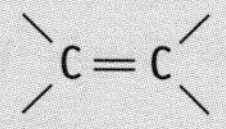 $C=C$	alkenes	$H_2C\equiv CH_2$ ethene
—Halogen	halogeno compound	CH_3Cl chloromethane
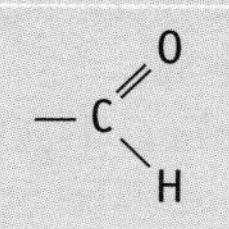 $-C\overset{\displaystyle O}{\underset{\displaystyle H}{}}$	aldehydes	CH_3CHO ethanal
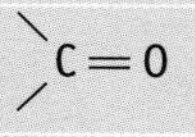 $C=O$	ketones	CH_3COCH_3 propanone
$-O-$	ethers	CH_3OCH_3 methoxymethane

25.4 Finding the formulas of organic compounds

To predict the properties of a compound, we need to know its **structural formula,** showing the position and nature of its functional groups.

Finding empirical formulas

Chapter 1 explains how the empirical formula of a compound may be found from its percentage composition. The composition of organic compounds can be found by **combustion analysis.** A known mass of a compound is burned and the carbon dioxide and water formed are collected and measured. Other elements that may be present, such as nitrogen and halogens, can also be estimated. From the masses of the combustion products the empirical formula can be calculated. In modern laboratories this composition analysis is performed automatically by machines.

Example

A compound X containing only carbon, hydrogen and oxygen was subjected to combustion analysis. 0.1 g of the compound on complete combustion gave 0.228 g of carbon dioxide and 0.0931 g of water. Calculate the empirical formula of the compound.

First, calculate the mass of C and H in 0.1 g of the compound.

$$44 \text{ g of } CO_2 \text{ contains 12 g of C} \Rightarrow \text{mass of C in 0.1 g of X} = \frac{12}{44} \times 0.228 = 0.0621 \text{ g}$$

$$18 \text{ g of } H_2O \text{ contains 2 g of H} \Rightarrow \text{mass of H in 0.1 g of X} = \frac{2}{18} \times 0.0931 = 0.0103 \text{ g}$$

$$\text{Mass of C} + \text{H in 0.1 g of X} = 0.0621 + 0.0103 = 0.0724 \text{ g}$$

The remainder of the 0.1 g of X must have been oxygen.

$$\Rightarrow \text{Mass of O in 0.1 g} = 0.1 - 0.0714 = 0.0276 \text{ g}$$

$$\therefore \text{Ratio by mass C : H : O is} \quad 0.0621 \quad : \quad 0.0103 \quad : \quad 0.0276$$

$$\Rightarrow \text{Ratio by moles C : H : O is} \quad \frac{0.0621}{12} \quad : \quad \frac{0.0103}{1} \quad : \quad \frac{0.0276}{16}$$

$$= \quad 0.00518 : \quad 0.0103 \quad : \quad 0.00173$$

$$= \quad 3 \quad : \quad 6 \quad : \quad 1$$

$$\therefore \text{Empirical formula of X is } C_3H_6O.$$

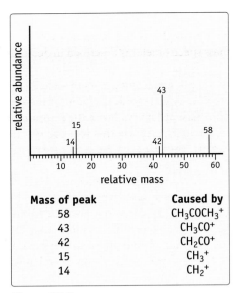

Mass of peak	Caused by
58	$CH_3COCH_3^+$
43	CH_3CO^+
42	CH_2CO^+
15	CH_3^+
14	CH_2^+

Figure 25.2
Mass spectrum of compound **X**, molecular formula C_3H_6O. The table identifies the fragments causing the major peaks. (There are several smaller peaks, which have been omitted.)

The structural formula of ethanoic acid is given in table 25.2 on page 381.

1 What is its empirical formula?

2 What is its molecular formula?

3 Write one other structure with the same molecular formula.

Finding molecular formulas

Once we have found the empirical formula, we can find the molecular formula of the compound, provided we know its relative molecular mass. Several methods are available for doing this, but nowadays it is usually done by mass spectrometry (see section 1.2).

When the molecules of an organic compound pass through a mass spectrometer, they get broken up into fragments. This means that the mass spectrum of the compound contains several peaks, corresponding to the different fragments. But a few molecules pass through intact. These intact molecules give a peak showing the relative molecular mass of the compound.

The mass spectrum obtained for the compound **X** in the example above is shown in figure 25.2.

The relative mass of the heaviest particle recorded in the spectrum is 58. We can assume that this corresponds to the intact molecule with a single positive charge, i.e. **X**$^+$, the **molecular ion.** This means that the relative molecular mass of **X** must be 58.

With an empirical formula of C_3H_6O, **X** could have molecular formula C_3H_6O, $C_6H_{12}O_2$, $C_9H_{18}O_3$ and so on. But since its relative molecular mass is known to be 58, the only possible molecular formula is C_3H_6O.

Section 25.5 explains how the fragments can tell you the molecule's structure.

Finding structural formulas

The molecular formula of a compound gives the number of atoms of the different elements in one molecule of the compound, but it gives no information about the way the atoms are arranged. A compound with molecular formula C_2H_6O, for example, could have one of two structural formulas:

$$CH_3OCH_3 \qquad\qquad CH_3CH_3OH$$

methoxymethane   ethanol

To find the exact structural formula of a compound, we need more information to decide which functional groups it contains. This information can be found by various methods.

(a) Instrumental methods

These are described in the next section.

(b) Physical properties of the compound, such as boiling point

Physical properties are dependent on structure and provide information that can help us work out the structural formula. For example methoxymethane is a gas at room temperature, its boiling point being 248 K. In contrast, ethanol, with the same molecular formula, is a liquid with boiling point 341 K. All members of the homologous series of alcohols tend to have high boiling points relative to other compounds of comparable relative molecular mass.

(c) Chemical properties of the compound

Each functional group has certain chemical characteristics. For example, alcohols such as ethanol react with sodium, giving hydrogen. Ethers such as methoxymethane do not. Thus a knowledge of the chemical properties of a compound can provide clues about the functional groups it contains.

25.5 Instrumental methods of analysis

Modern chemistry laboratories use sophisticated instruments to find the structure of organic chemicals. These instruments are very sensitive and they usually need only very small amounts of the chemical to work on. Laboratories in industry and universities are equipped with a range of instruments, but they tend to be very expensive, so you don't find many of them in schools.

Spectroscopy involves using instruments to examine the radiation emitted or absorbed by chemicals, giving information about their molecular structure.

Spectroscopy is covered in section 25.6.

Mass spectrometry

The mass spectrometer is an instrument which turns atoms and molecules into ions and measures their mass. The working of the mass spectrometer is described in detail in section 1.2.

When an organic compound passes through a mass spectrometer, its molecules get broken into positively-charged fragments. These fragments provide useful information.

Each fragment gives a corresponding line in the mass spectrum. From the position of the line, we can find the relative mass of the fragment, and use this to work out its formula. By piecing together the fragments we can then deduce the structure of the parent molecule.

Look again at figure 25.2. The mass spectrum for compound **X**, molecular formula C_3H_6O, has prominent peaks at 15 and 43, and these give strong clues to the compound's structure. They correspond to CH_3^+ and CH_3O^+, respectively, and this suggests the structural formula CH_3COCH_3 (propanone). The other possible structure for **X**, CH_3CH_2CHO (propanal) is ruled out because this would give a strong peak at mass 29, corresponding to the two fragments $CH_3CH_2^+$ and CHO^+, both of which we would expect CH_3CH_2CHO to form.

Certain fragments are very common in mass spectra. They include

mass	fragment
15	CH_3^+
28	CO^+ or $C_2H_4^+$
29	$CH_3CH_2^+$
77	$C_6H_5^+$ (benzene ring)

We can find out a lot by looking at the differences between the masses of peaks in the spectrum. For example, in the mass spectrum for **X** (figure 25.2), the peaks at mass 58 and 43 differ in mass by $(58 - 43) = 15$. This suggests that the peak at 43 has been formed by the loss of a CH_3 group from the molecular ion.

Look at figure 25.3, which shows the mass spectrum of butanone, $CH_3CH_2COCH_3$.

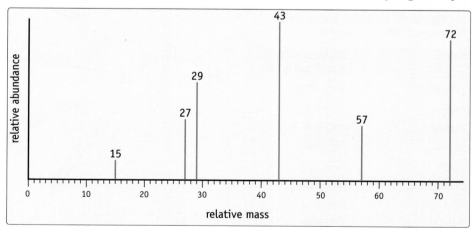

Figure 25.3
The mass spectrum of butanone.

1 Identify the molecular ion peak. What is the relative molecular mass of butanone?

2 Identify the fragments responsible for the peaks at 15 and 29.

3 Work out the difference in mass between the peaks at 72 and 57.
  What fragment must have been lost from the molecular ion to create the peak at 57?

4 Work out the difference in mass between the peaks at 72 and 43.
  What fragment must have been lost from the molecular ion to create the peak at 43?

5 Identify the fragment responsible for each peak, and work out how each has been formed by the fragmentation of butanone.

The yellow colour of sodium vapour street lamps is due to the visible emission spectrum of sodium.

25.6 Spectroscopy

When electromagnetic radiation, such as light or infrared, shines on a chemical, the chemical may interact with the radiation in some way. The commonest example is colour: colour is produced when chemicals emit or absorb visible light of a particular frequency.

The way a particular chemical interacts with radiation can tell chemists about its molecules. A **spectroscope** is an instrument which allows radiation to interact with a sample of a chemical, then analyses the changes.

Different kinds of radiation interact with chemicals in different ways (figure 25.4). The effects are summarised in table 25.3. In this section we will look at three types of spectroscopy that are particularly useful to chemists: ultraviolet/visible, infrared and nuclear magnetic resonance.

Table 25.3
How different types of radiation interact with chemicals

Type of radiation	Frequency range /Hz	Effect on molecule	Type of spectroscopy
ultraviolet	10^{15}–10^{17}	excites the electrons	ultraviolet/visible spectroscopy (see below)
visible light	10^{14}–10^{15}	excites the electrons	ultraviolet/visible spectroscopy (see below)
infrared	10^{11}–10^{12}	makes bonds vibrate	infrared spectroscopy (page 385)
microwaves	10^{9}–10^{11}	makes molecules rotate	microwave spectroscopy
radio waves	10^{6}–10^{8}	changes the magnetic alignment of the nuclei of some atoms	nuclear magnetic resonance (page 387)

▶ **Figure 25.4**
The effect of different types of radiation on the water molecule.

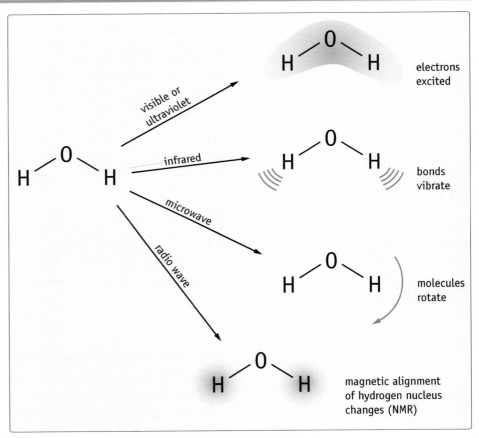

electrons excited

bonds vibrate

molecules rotate

magnetic alignment of hydrogen nucleus changes (NMR)

Table 25.4
The approximate wavelengths of visible radiation of different colours

Colour	Approximate wavelength/nm
[infrared]	above 700
red	620–700
orange	600–620
yellow	580–600
green	520–580
blue-green	490–520
blue	440–490
indigo	420–440
violet	400–420
[ultraviolet]	below 400

Table 25.4 gives the approximate wavelengths of visible radiation of different colours. Look at the visible/UV spectrum of methylene blue in figure 25.5.

1 What colours of visible light does methylene blue absorb?

2 Explain why methylene blue has a blue colour.

Visible and ultraviolet spectroscopy

When you heat sodium chloride in a flame, it gives a yellow flame colour. This is an example of a **visible emission spectrum**, where a substance emits certain visible frequencies when its electrons have been excited by heating or by an electrical discharge. The radiation may be in the ultraviolet region of the spectrum as well as the visible.

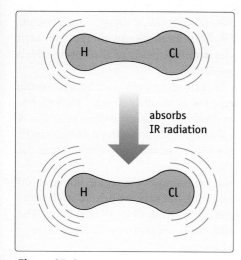

Figure 25.5 The visible/UV spectrum of methylene blue.

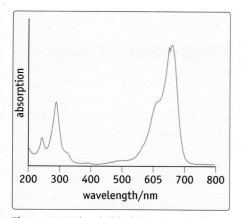

Figure 25.6
When an HCl molecule absorbs infrared radiation, it vibrates more energetically. The frequency of radiation absorbed is 7.21×10^{13} Hz, and this frequency is characteristic of the H—Cl bond.

Visible and ultraviolet emission spectra are covered in detail in section 6.3. By examining the frequencies emitted, chemists can get information about the substance. The emission spectrum of hydrogen gave chemists the first clues about the energy levels of electrons in an atom.

When white light shines on methylene blue (an organic dye), some of the electrons in the dye's molecules become excited. The electrons absorb certain frequencies of light radiation and change their energy level. The radiation they absorb happens to be in the red end of the spectrum. Removing red from white light makes it look blue – but you can only tell the colour of methylene blue if light is shining on it. This is an example of a **visible absorption spectrum**, where a chemical absorbs certain frequencies of visible radiation. Ultraviolet radiation can be absorbed in a similar way: it is not visible, but can be detected by instruments. By finding which frequencies have been absorbed, chemists can get information about the chemical. You can read about visible spectra and the way they give rise to colour in transition metal compounds on page 259.

In an ultraviolet/visible spectroscope, radiation consisting of a mixture of ultraviolet and visible frequencies is shone through a sample of a compound. A detector measures the frequencies of radiation that have been absorbed, and the spectrometer prints out an absorption spectrum. Figure 25.5 shows the ultraviolet/visible absorption spectrum of methylene blue. Notice the axes: the horizontal axis shows the wavelength of radiation absorbed, in nanometres, nm, and the vertical axis shows the relative absorbance: how strongly it is absorbed.

The ultraviolet/visible spectrum of an organic compound is characteristic of that compound. The compound can be identified by comparing its spectrum with the spectra of known compounds. However, other types of spectroscopy, particularly infrared and nuclear magnetic resonance, are more useful for piecing together the structure of an organic compound.

Infrared spectroscopy

Infrared spectroscopy makes use of the fact that molecules absorb infrared (IR) radiation which has a wavelength longer than visible light, between about 2500 nm and 25 000 nm. The energy of the radiation is absorbed in making the bonds vibrate (figure 25.6). When the molecule absorbs the radiation, the bonds vibrate more energetically.

Different bonds absorb radiation of different frequencies, and the frequency is characteristic of the particular bond concerned. We can use IR absorption to identify the bonds, and therefore the functional groups, in an organic molecule.

The spectrometer produces an **infrared spectrum** on a chart recorder, and an example is shown in figure 25.7 – this is the IR spectrum of propanone, CH_3COCH_3.

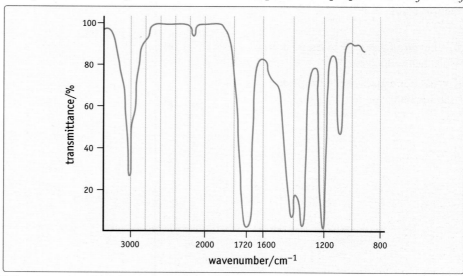

Figure 25.7
The infrared spectrum of propanone.

Notice that the horizontal axis shows 'wavenumber': this is the conventional way of representing the frequency of radiation absorbed. It is the reciprocal of the wavelength measured in cm.

Table 25.5
Characteristic IR absorptions of some common bonds

Bond	Compound it is in	Absorption/cm^{-1}	Intensity (M = medium, S = strong)
C—H	alkanes, alkenes, arenes	2840 to 3095	M/S
C=C	alkenes	1610 to 1680	M
C=O	aldehydes, ketones, acids, esters	1680 to 1750	S
C—O	alcohols, ethers, esters	1000 to 1300	S
C≡N	nitriles	2200 to 2280	M
C—Cl	chloro compound	700 to 800	S
O—H	'free'	3580 to 3670	S
	hydrogen-bonded in alcohols, phenols	3230 to 3550	S (broad)
	hydrogen-bonded in acids	2500 to 3300	M (broad)
N—H	primary amines	3100 to 3500	S

You will see that the spectrum is quite complicated, even though propanone is a simple molecule with only three types of bond. The complexity arises because the bonds can each vibrate in a number of different ways, and the vibrations can interact with each other. Nevertheless, it is possible to see some characteristic peaks of absorption which we can use to identify functional groups in the molecule.

Table 25.5 gives the characteristic absorptions of some common bonds. In the spectrum of propanone in figure 25.7, the strong peak at about 1720 cm^{-1} corresponds to the C=O bond. The weaker absorption at 3000 cm^{-1} corresponds to the C—H bond. This peak is weaker even though there are more H atoms in the molecule: in IR spectroscopy the strength of the peak is a characteristic of the bond itself, not of the number of bonds present.

Most of the interesting parts of an IR spectrum are found in the region above about 1500 cm^{-1}. The peaks outside this region are less useful, but they are helpful in showing the **fingerprint** of the compound: the characteristic pattern of its IR spectrum. The fingerprint can be used to compare the compound's spectrum with IR spectra of known compounds given in standard reference books.

Figure 25.8 shows the IR spectrum of ethanol, CH_3CH_2OH. Look closely at the spectrum.

1 What bond gives rise to the peak at just below 3000 cm^{-1}?

2 What bond gives rise to the peak at about 3400 cm^{-1}?

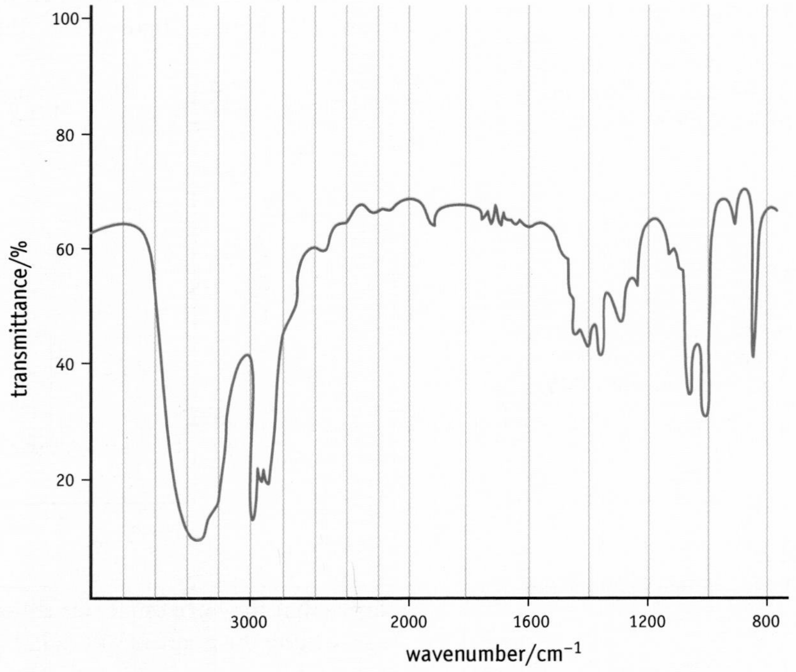

Figure 25.8
The infrared spectrum of ethanol, CH_3CH_2OH.

Infrared spectra are very useful: they are particularly helpful for identifying the functional groups in an unknown compound. In the IR spectrum for ethanol in figure 25.8, the peak just below 3000 cm^{-1} is from the C—H bonds, and the peak at about 3400 cm^{-1} is from the O—H bond.

Nuclear magnetic resonance (NMR)

When certain atoms are placed in a strong magnetic field, their nuclei behave like tiny bar magnets and align themselves with the field. Electrons behave like this too, and for this reason both electrons and nuclei are said to possess 'spin', since any spinning electric charge has an associated magnetic field.

Just as electrons with opposite spin pair up together (section 6.3), a similar thing happens with the protons and neutrons in the nucleus. If a nucleus has an even number of protons and neutrons (e.g. $^{12}_{6}C$), their magnetic fields cancel out and it has no overall magnetic field. But if the number of protons and neutrons is odd (e.g. $^{13}_{6}C$, $^{1}_{1}H$), the nucleus has a magnetic field.

▲ A chemist using a nuclear magnetic resonance spectrometer. NMR is probably the most valuable technique available to organic chemists.

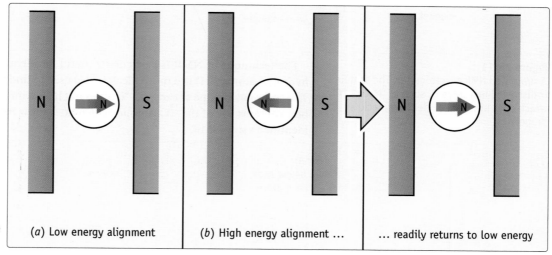

Figure 25.9
Alignment of a compass needle in a magnetic field.

(a) Low energy alignment   (b) High energy alignment ...   ... readily returns to low energy

If the substance is now placed in an external magnetic field, the nuclear 'magnets' line up with the field, in the same way as a compass needle lines up with a magnetic field (figure 25.9).

The nuclear 'magnet' can have two alignments, of low and high energy (figure 25.10). To make the nucleus change to the high energy alignment, energy must be supplied.

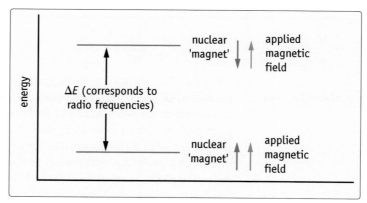

Figure 25.10
Two alignments of the nuclear 'magnet' in an external magnetic field. The energy difference between the two orientations is the basis of the technique of NMR.

It happens that the energy absorbed corresponds to radio frequencies. The precise frequency of energy depends on the environment of the nucleus, in other words, on the other nuclei and electrons in its neighbourhood. So by placing the sample being examined in a strong magnetic field and measuring the frequencies of radiation it absorbs, information can be obtained about the environments of nuclei in the molecule. The technique is called **nuclear magnetic resonance (NMR)**. Figure 25.11 shows a simplified diagram of an NMR spectrometer.

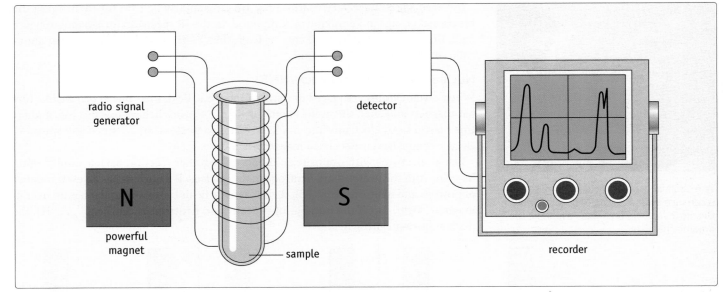

Figure 25.11
A simplified NMR spectrometer. The sample is dissolved in a solvent, such as ^2H_2O or CCl_4 which does not have nuclear magnetic properties.

The technique of NMR is particularly useful for identifying the number and type of hydrogen atoms (^1H) in a molecule. It is also used to find the positions of carbon atoms. The common isotope of carbon, ^{12}C, does not have a nuclear 'magnet', but natural carbon contains 1% of the ^{13}C isotope, which does show magnetic behaviour and can be identified using NMR.

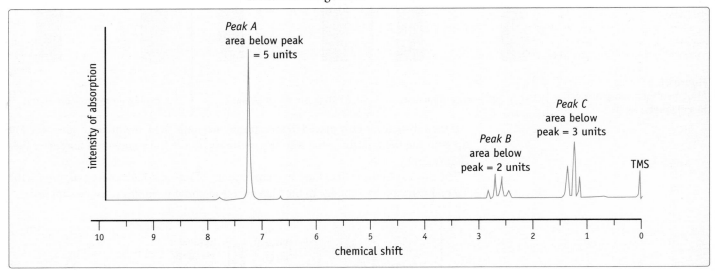

Figure 25.12
The proton NMR spectrum of ethylbenzene, $C_6H_5CH_2CH_3$.

Figure 25.12 shows the NMR spectrum of ethylbenzene, $C_6H_5CH_2CH_3$. This is a **proton NMR spectrum:** the frequencies correspond to the absorption of energy by ^1H nuclei, which are protons. Notice that there are three major peaks, of differing heights. Each peak corresponds to an H atom in a different molecular environment. The area under each peak is proportional to the number of that type of H atom in the molecule. The largest peak (A) corresponds to the 5 H atoms in C_6H_5, the benzene ring. The second largest (C) corresponds to the 3 H atoms in the —CH_3 group, and the third peak (B) corresponds to the 2 H atoms in the CH_2 group.

The H atoms in a particular type of environment have similar positions in the NMR spectrum. Normally, this position is measured as a **chemical shift**, δ, from a fixed reference point. The reference point normally used is the absorption of a substance known as TMS. The chemical shift of TMS is set at zero.

TMS stands for tetramethylsilane, $Si(CH_3)_4$. This non-toxic and unreactive substance is chosen as the NMR reference because its protons give a single peak that is well separated from the peaks found in the NMR spectra of most organic compounds.

Table 25.6 gives the chemical shifts for some common proton environments.

25 Introduction to Carbon Chemistry

▲ Nuclear magnetic resonance is the principle behind Magnetic Resonance Imaging (MRI) body scanners, which detect protons in water molecules in the body. Unlike X-ray investigation, the technique is believed to be completely harmless to the patient. The patient lies inside the tube, surrounded by a powerful magnetic field.

1 Use table 25.6 to identify each of the peaks in figure 25.13.

2 Explain the relative areas under the peaks

Table 25.6 Chemical shifts for some types of protons

Type of proton	Chemical shift, δ, in region of
R—CH_3	0.9
R—CH_2—R	1.3
R—CH—R (with R above)	2.0
—C—CH_2— (with O double bonded)	2.3
—O—CH_3	3.8
—O—CH_2—R	4.0
—O—H	5.0
(benzene ring)—H	7.5
—C=O with H	9.5
—C=O with O—H	11.0

Figure 25.13 shows a simplified proton NMR spectrum for ethanol, CH_3CH_2OH. It has been simplified by removing some of the detail, so the peaks appear single. Notice that it includes the peak for TMS. Notice too that an **integrated trace** is shown: this gives the relative areas under each of the peaks.

Figure 25.13
A simplified proton NMR spectrum for ethanol, CH_3CH_2OH. The integrated trace shows the areas under the peaks are in the ratio 1:2:3.

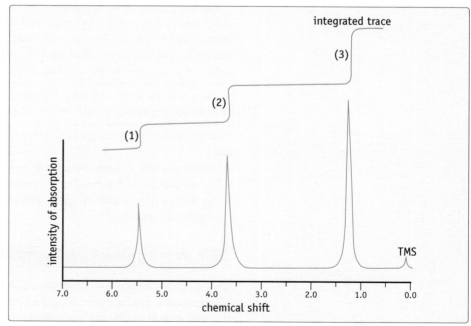

Spin-spin coupling

The simplified NMR spectrum of ethanol shown in figure 25.13 shows three single peaks. The smallest peak corresponds to the single OH proton; the middle peak corresponds to the two CH_2 protons and the largest peak corresponds to the three CH_3 protons. A detailed, high-resolution spectrum of ethanol shows that the CH_2 and CH_3 peaks are in fact split into a number of subsidiary peaks (figure 25.14). This splitting is caused by **spin-spin coupling** between protons on neighbouring carbon atoms.

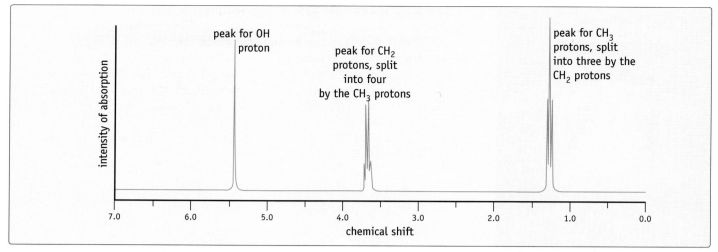

Figure 25.14
A detailed high-resolution NMR spectrum for ethanol.

Here is what happens. One of the carbon atoms in ethanol has two protons on it (CH_2), the other has three (CH_3). Consider the CH_3 protons, which act like tiny magnets. When the external magnetic field is applied, these three tiny magnets can arrange themselves in four different ways.

1 All three aligned *with* the magnetic field

2 All three aligned *against* the magnetic field

3 Two aligned with the field and one aligned against it

4 One aligned with the field and two aligned against it

Each of these four arrangements gives a slightly different overall magnetic field. Each different field interacts with the neighbouring CH_2 protons slightly differently, so these CH_2 protons give four different peaks, very close to one another. These peaks are called a **quartet**.

Similarly, the two protons on the CH_2 group can arrange themselves differently in the external magnetic field. This time there are three different arrangements – see if you can work out what they are. Each of these different fields interacts with the neighbouring CH_3 protons slightly differently, so we see three different peaks close to one another. These three peaks are called a **triplet**.

> As a general rule: **a group carrying *n* protons will cause the protons on a neighbouring group to split into *n* + 1 peaks.**

When you are interpreting a high-resolution spectrum like the one in figure 25.14, you can

• Use the position of each overall peak to identify the type of proton causing the peak
• Use the integrated trace to find the numbers of each type of proton
• Use the '*n* + 1' rule above to get information about the numbers of protons on neighbouring groups.

25.7 Writing structural formulas

The structural formula of a compound shows how the atoms are joined together. This information is crucial in deciding the properties of the compound. The actual shape of the molecule also affects its properties. It is helpful if the structural formula can give some indication of the shape or **stereochemistry** of the molecule.

The methane molecule is tetrahedral in shape (section 8.7). This tetrahedral arrangement of bonds is common to all saturated carbon atoms (that is, carbon atoms bonded to four other groups). Unfortunately the tetrahedral shape creates a problem in representing structural formulas, because it is difficult to show a three-dimensional shape on two-dimensional paper.

25 Introduction to Carbon Chemistry

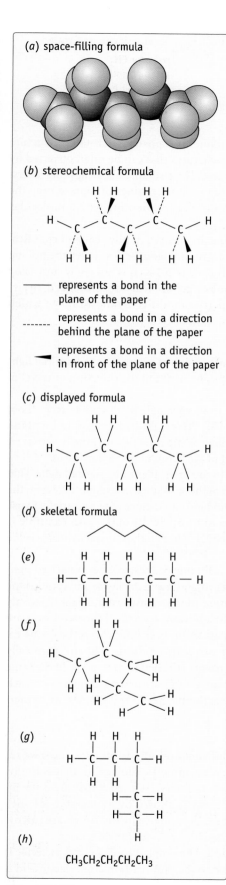

(a) space-filling formula

(b) stereochemical formula

—— represents a bond in the plane of the paper

------ represents a bond in a direction behind the plane of the paper

▬◄ represents a bond in a direction in front of the plane of the paper

(c) displayed formula

(d) skeletal formula

(e)

(f)

(g)

(h)

CH₃CH₂CH₂CH₂CH₃

Figure 25.15
Representations of the structural formula of pentane.

Figure 25.15 shows eight ways of representing the structural formula of pentane. Possibly the most accurate representation is the **space-filling model** (figure 25.15(a)) and it would be useful if you could try building this model yourself. Space-filling models show the extent of the electron cloud of each atom accurately.

Figure 25.15(b) attempts to represent the tetrahedral carbon atom by showing the directions of the bonds. This is called a **stereochemical** formula. This system is of limited use, and it can be simplified to figure 25.15(c), which is called a **displayed** formula.

Sometimes a **skeletal formula** (figure 25.15(d) is used, in which only carbon–carbon bonds and functional groups are shown.

A common and easy way of writing formulas is shown in figure 25.15(e) but this has drawbacks because it represents the bond angles at each carbon atom as 90° instead of 109°. This type of diagram shows its limitations when an attempt is made to represent rotation about a single bond: compare figure 25.15(g) with figure 25.15(f). It is easy to see that (f) can be obtained from (c) by rotating about the bond between the second and third carbon atoms. On the other hand, (g) appears to be a different compound to (e), though both in fact represent pentane.

Finally, figure 25.15(h) shows a shortened way of writing a structural formula.

These different ways of representing structural formulas will be used as appropriate in different parts of this book.

1. Write a displayed formula for $CH_3CH_2CHOHCH_2CH_3$
2. Write a skeletal formula for

$$
\begin{array}{c}
\overset{\displaystyle H}{|} \\
H-C-H \\
\end{array}
$$

$$
H-\overset{\overset{\displaystyle H}{|}}{\underset{\underset{\displaystyle H}{|}}{C}}-\overset{\overset{\displaystyle H}{|}}{\underset{\underset{\displaystyle H}{|}}{C}}-\overset{\overset{\displaystyle H}{|}}{\underset{\underset{\displaystyle H}{|}}{C}}-\overset{\overset{\displaystyle H}{|}}{\underset{\underset{\displaystyle H}{|}}{C}}-H
$$

25.8 Isomerism

We saw in section 25.4 that a compound of molecular formula C_2H_6O could have two possible structural formulas, CH_3OCH_3 (methoxymethane) or CH_3CH_2OH (ethanol). Compounds such as these, possessing the same molecular formula but with their atoms arranged in different ways, are called **isomers**. Isomerism is very common in organic chemistry. All carbon compounds with four or more carbon atoms, and many with less, show isomerism. There are 4.11×10^9 isomers of molecular formula $C_{30}H_{62}$. Isomerism also occurs, though less commonly, in inorganic chemistry. In this book we are concerned with three main types of isomerism: structural, optical and *cis–trans*.

Structural isomerism

Structural isomers differ in which atom is attached to which. Methoxymethane and ethanol are structural isomers. In methoxymethane the oxygen atom is attached to two carbon atoms while in ethanol it is attached to a carbon and a hydrogen. Structural isomers can be members of different homologous series, like the pair just mentioned, or they may be members of the same series like the following two structural isomers of molecular formula C_4H_{10}, which are both alkanes:

butane

methylpropane

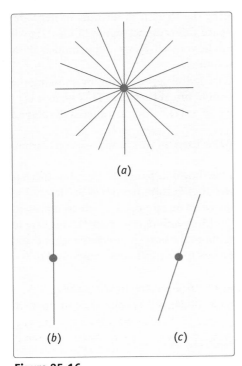

Figure 25.16
(a) Normal light ray travelling towards the observer: each line represents a wave seen 'end-on' (b) Plane-polarised light: only waves in a single plane are present. (c) Plane-polarised light from (b) after passing through an optically active solution.

☐ 1 How many isomers are there of molecular formula C_5H_{12}?

☐ 2 Write the structural formula of all the alcohols (i.e. the compounds with the —OH functional group) of molecular formula $C_4H_{10}O$.

▲ The different flavours of oranges and lemons are caused by the two different optical isomers of the same molecule, limonene.

Note that

$$H-\overset{\overset{\displaystyle H}{|}}{\underset{\underset{\displaystyle H}{|}}{C}}-\overset{\overset{\displaystyle H}{|}}{\underset{\underset{\displaystyle H}{|}}{C}}-\overset{\overset{\displaystyle H}{|}}{\underset{\underset{\displaystyle H-\overset{\overset{\displaystyle H}{|}}{\underset{\underset{\displaystyle H}{|}}{C}}-H}{|}}{C}}-H$$

is *not* a third isomer of C_4H_{10}, because it can be formed from butane simply by rotating about a single bond. It is possible to rotate a structure freely about any single carbon-carbon bond. Structures that can be interconverted by rotating about a bond in this way are not isomeric. The structure above appears at first sight to be a separate isomer, but this is due to the shortcomings of representing the carbon bond angles as 90° instead of 109°. A little experimenting with a molecular model will make this point clear.

Structural isomers usually show considerable differences in physical and chemical properties, even if they are members of the same homologous series. Thus, the boiling point of butane is 273 K while that of methylpropane is 261 K. It is not really surprising that structural isomers differ. Any functional group will be influenced by its environment and will therefore have its properties modified by the atoms to which it is attached.

Optical isomerism

Light is a form of electromagnetic radiation and consists of waves. A ray of normal light has waves which vibrate in many directions at right angles to the direction of travel of the ray (figure 25.16(a)).

Certain materials have the ability to remove from normal light all waves except those vibrating in a single plane (see figure 25.16(b)). The light is then said to be **plane polarised.** It is rather like the light being combed as it passes through the polariser. A well known polariser is polaroid, which is used in the lenses of some sunglasses.

Certain chemical substances have the ability to rotate the plane of polarised light. That is, if polarised light is shone through a solution of the substance in a suitable solvent, the plane in which the polarised light vibrates will be rotated clockwise or anticlockwise (figure 25.16(c)). These substances are said to be **optically active.** An example is 2-hydroxypropanoic acid (lactic acid, $CH_3CHOHCOOH$), the substance responsible for the sour taste in sour milk.

Chemists have isolated two forms of lactic acid: one of these rotates polarised light clockwise and is called the (+) isomer, the other rotates it anticlockwise and is called the (−) isomer. For a given concentration of solution the two forms rotate light to exactly the same extent, but in opposite directions. In addition, crystals of the two forms are found to be mirror images of one another. Apart from these differences, the forms are physically and chemically identical. Since the two forms of lactic acid are chemically identical, they must contain exactly the same groups attached to each other in the same way. Their difference can only lie in the way the groups are arranged relative to each other in space. Furthermore they must be exact opposites in this respect since their effect on polarised light is equal and opposite.

Chiral molecules

When we look at the molecular structure of different optically active substances we see they have one thing in common. They all have asymmetric molecules, that is to say, their molecules have no centre, axis or plane of symmetry. As a result they are different from their own mirror images. Optical activity is shown by all substances with asymmetric molecules. Such substances have two isomeric forms which are mirror images of one another and which rotate polarised light in opposite directions.

Objects that are asymmetric are said to be **chiral.** The simplest type of chiral molecule is one in which four different groups are attached to the same carbon atom; lactic acid is such a molecule. The asymmetric carbon atom is called a **chiral centre.** Figure 25.17 shows the two forms, called (+)-lactic acid and (−)-lactic acid, but you should try making models of each of the forms to satisfy yourself that they are indeed different molecules and that they are mirror images. You can show they are different by trying to

Figure 25.17
The enantiomeric forms of lactic acid. The two isomers are images of one another reflected in an imaginary mirror placed between them.

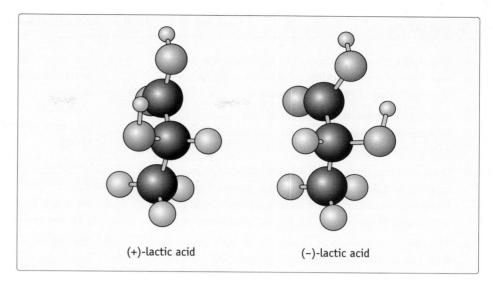

(+)-lactic acid   (−)-lactic acid

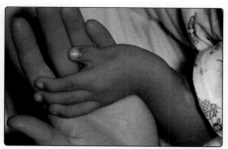

▲ The thalidamide tragedy in the 1960s was caused by a drug that was prescribed to pregnant mothers to treat morning sickness. One of the enantiomers of thalidamide is a safe and effective medicine; unfortunately the other enantiomer is a powerful foetus-deformer. If chemists had known this at the time, they might have been able to prevent the tragedy.

1. Would you expect (a) CH_2ClF (b) CHClBrF to show optical isomerism?

2. Which of the isomeric alcohols whose structures you wrote earlier in this section would show optical isomerism? Write the structures of the enantiomers.

3. Write the structure of the first alkane to show optical isomerism.

superimpose the two molecules on one another: it is impossible to arrange them so that all the groups correspond in position. Note, though, that in both (+)- and (−)-lactic acid all the groups are attached together in the same way, and that the spacings of the various groups are the same in each isomer. This is why the two forms of lactic acid have identical chemical properties.

The relationship between optical isomers such as (+)- and (−)-lactic acid is like the relationship between your right and your left hand. The lengths of the different fingers and the distances between them are the same for both hands, but they are mirror images of one another and cannot be superimposed on each other – try putting your right glove on the left hand. Structures that are mirror images of one another are called **enantiomers.**

Many organic compounds and some inorganic ones have chiral molecules and therefore show optical isomerism. Optically active compounds are very common in nature. Almost all amino acids are optically active, as are all the sugars. The interesting thing is that most naturally occurring optically active compounds occur as one isomer only. Thus all naturally occurring glucose is (+)-glucose. The enzymes that produce and break down these substances are able to recognise them because of their asymmetric molecules. The active site of the enzyme is also asymmetric, and exactly fits the asymmetric substrate. This is why enzymes are so specific in their action.

Many optically active compounds have complex structures and it is difficult to tell quickly whether or not their molecules are chiral. For simpler molecules it is safe to say that if the molecule contains a carbon atom to which four different groups are attached, it will show optical isomerism. Such a carbon atom is called a **chiral centre.**

Lactic acid can be extracted from natural sources such as sour milk or muscle tissue and the acid from these sources always shows optical activity. It is also possible to prepare lactic acid in the laboratory from fairly simple starting materials, but the acid prepared in this way shows no optical activity at all. This is because most laboratory methods for preparing lactic acid produce the (+)- and (−)-isomers in equal amounts. The two isomers cancel each other out in their effect on polarised light. This kind of mixture of optical isomers containing equal molar quantities of (+) and (−) isomers is called a **racemic mixture** or a **racemate.** Because of the identical properties of the isomers it is very difficult to separate them from a racemic mixture, though methods do exist for doing so. The process of separation is called **resolution.** It is important in the preparation of pure pharmaceutical products.

Cis–trans isomerism

This type of isomerism is described in section 27.4.

Summary

1 The ability of carbon atoms to form strong bonds to four other atoms, including other carbon atoms, results in a huge range of organic compounds.

2 Organic compounds, though kinetically stable in air, are energetically unstable relative to their combustion products.

3 Organic compounds can be regarded as having a basic hydrocarbon skeleton with functional groups attached. Compounds containing the same functional group form a family with similar properties called a homologous series.

4 The empirical formula of an organic compound can be found by analysis of its combustion products.

5 The molecular formula of a compound can be found from the empirical formula once the relative molecular mass is known.

6 The structural formula shows the precise arrangement of atoms. It can be found from a knowledge of some of the properties of the compound or by using instrumental methods, particularly mass spectrometry, infrared (IR) spectroscopy and nuclear magnetic resonance (NMR).

7 Compounds with the same molecular formula but with their atoms arranged in different ways are called isomers.

8 Structural isomers differ in the way the atoms are attached to one another.

9 Compounds with chiral (asymmetric) molecules exist in two forms whose molecules are mirror images or enantiomers. The two forms rotate polarised light in opposite directions and are called optical isomers.

Review questions

1 An organic compound was subjected to combustion analysis. 1.0 g of the compound formed 1.37 g carbon dioxide, 1.12 g water and no other products.
  (a) Calculate the percentage by mass of carbon and of hydrogen in the compound.
  (b) What other element must be present?
  (c) Calculate the empirical formula of the compound.
  (d) The mass spectrum of the compound is shown in figure 25.18.
    (i) Use this to find the relative molecular mass of the compound, and thus its molecular formula.
    (ii) Using the fragments shown in the mass spectrum, deduce the structural formula of the compound.
    (iii) Give the formula of the fragment to which each peak can be attributed.

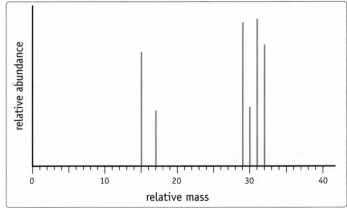

Figure 25.18

2 **A**, **B** and **C** are isomeric compounds of molecular formula C_3H_8O. Two of the compounds are members of the same homologous series. The table gives some data about **A**, **B** and **C**.

	A	B	C
Boiling point/K	370	284	356
Density/g cm^{-3}	0.80	0.72	0.79

  (a) Which two compounds are members of the same homologous series?
  (b) Write the structural formulas of all three possible isomers of C_3H_8O.
  (c) Use table 25.2 (page 381) to decide to which homologous series each of the isomers you have drawn belongs.
  (d) Is it possible from the data given to say which of the isomers you have drawn is **A**, which is **B** and which is **C**? Explain your answer.

3 Use table 25.2 (page 381) to decide to which homologous series each of the following compounds belongs.
  (a) $CH_3CH_2CH_2OH$
  (b) $CH_3CH_2COCH_3$
  (c) CH_3CH_2Cl
  (d) CH_3CH_2COOH
  (e) $CH_3CH{=}CHCH_3$
  (f) H_2NCH_2COOH
  (g) $CH_3CH_2OCH_2CH_2CH_3$

4 Write the full structural formulas of all the isomers of the following, stating which type of isomerism is involved:
  (a) C_3H_7Cl
  (b) C_6H_{14}
  (c) $C_2H_3Cl_2Br$

5 Silicones are unreactive polymers with the structure

$$\begin{array}{ccccccccc}
& CH_3 & & CH_3 & & CH_3 & & CH_3 & & CH_3 \\
& | & & | & & | & & | & & | \\
-\,Si & -\,O\, & -\,Si & -\,O\, & -\,Si & -\,O\, & -\,Si & -\,O\, & -\,Si\,- \\
& | & & | & & | & & | & & | \\
& CH_3 & & CH_3 & & CH_3 & & CH_3 & & CH_3
\end{array}$$

25 Introduction to Carbon Chemistry

Look at the bond energies given in table 25.1 (page 378). Suggest a reason why silicones are inert and unreactive while silanes (compounds analogous to alkanes but containing silicon atoms instead of carbon atoms) are unstable and ignite spontaneously in air.

6 Fluorocarbons are compounds analogous to alkanes but containing fluorine instead of hydrogen, e.g. CF_4, CF_3CF_3, etc. They are extremely unreactive, being quite stable in air even at high temperatures. Long-chain fluorocarbons such as PTFE ($—CF_2—CF_2—CF_2—CF_2—CF_2—$) are used as unreactive corrosion-resistant materials for gaskets and protective coatings.

Use the bond energies given below to explain why fluorocarbons are stable in air while hydrocarbons are energetically unstable.

Bond	Bond energy/kJ mol^{-1}
C—H	413
C—F	485
H—O	463
F—O	234

7 A compound containing carbon, hydrogen and nitrogen only was analysed. 0.1 g of the compound on combustion gave 0.228 g of carbon dioxide and 0.124 g of water. On reduction of the compound, all the nitrogen in it was converted to ammonia and it was found that 0.1 g of the compound gave ammonia equivalent to 17.2 cm^3 of 0.1 mol dm^{-3} hydrochloric acid when titrated.
(a) How many moles of carbon, hydrogen and nitrogen are there in 0.1 g of the compound?
(b) What is the empirical formula of the compound?
(c) The relative molecular mass of the compound was found to be 116. What is its molecular formula?
(d) The infrared spectrum of the compound indicated the presence of the —NH$_2$ group. Suggest one possible structure for the compound.

8 The following structural formulas represent only three different substances. Some of the formulas are equivalent to others. Which formulas are equivalent to which?

A

H H H H
| | | |
H—C—C—C—C—H
| | |
H H H
      |
   H—C—H
      |
   H—C—H
      |
      H

B

C $CH_3CH_2CH_2CH_2CH_2CH_3$

D

H
|
H—C—H

H H H   H H
| | | | |
H—C—C—C—C—C—C—H
| | | | | |
H H H H H H

E

F $CH_3(CH_2)_5CH_3$

G $CH_3CH_2CH_2CHCH_2CH_3$
                |
              CH_3

H

H   H   H
| | |
H—C———C———C—H
| | |
H—C—H H H—C—H
| |
H—C—H   H—C—H
| |
H   H

9 The earth's crust contains only 0.036% by mass of carbon, compared with 49% oxygen and 26% silicon. Despite its relatively low abundance on earth, carbon is the element that forms the basis of all living things. What are the properties of carbon that make it so suitable for this role?

10 Figure 25.19 shows the simplified NMR spectrum of 1-phenylbutan-2-one.

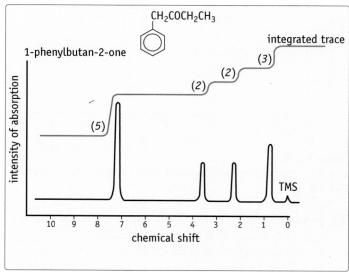

Figure 25.19
The simplified NMR spectrum of 1-phenylbutan-2-one.

(a) Table 25.6 on page 389 shows the chemical shifts for some types of protons (H atoms). Use the table to identify the protons responsible for the peaks at
(i) 7.2   (ii) 2.3   (iii) 0.9
(b) Using your answer to (a), and a process of elimination, identify the protons responsible for the peak at 3.6.

11 Ethoxyethane, $CH_3CH_2OCH_2CH_3$, and butan-1-ol, $CH_3CH_2CH_2CH_2OH$ are isomers.
Figure 25.20 shows the infrared spectra of these two compounds. The spectra are labelled A and B.
(a) Use table 25.5, page 386 (*Characteristic IR absorptions of some common bonds*) to decide which spectrum belongs to which compound.
(b) Identify the bonds responsible for the peaks marked **1**, **2**, **3** and **4** on figure 25.20.

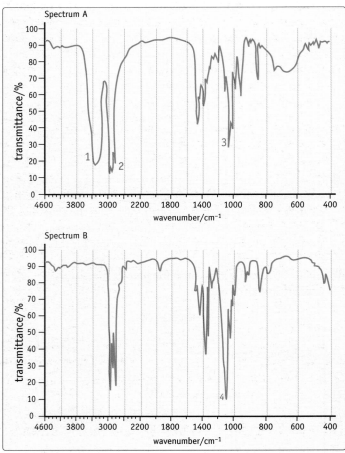

Figure 25.20
The IR spectra for ethoxyethane and butan-1-ol (not necessarily in that order).

12 Phenolphthalein is an organic chemical, molecular formula $C_{20}H_{14}O_3$. Figure 25.21(*a*) shows the visible/ultraviolet absorption spectrum of phenolphthalein at pH 6. Figure 25.21(*b*) shows the spectrum of the same chemical at pH 10. Table 25.4 on page 384 gives the approximate wavelengths of visible radiation of different colours.
(a) Which colours does phenolphthalein absorb at pH 6?

(b) What colour do you predict that a solution of phenolphthalein will be at pH 6?
(c) Which colours does phenolphthalein absorb at pH10?
(d) What colour do you predict that a solution of phenolphthalein will be at pH 10?
(e) Explain how you could use phenolphthalein in an acid–base titration?

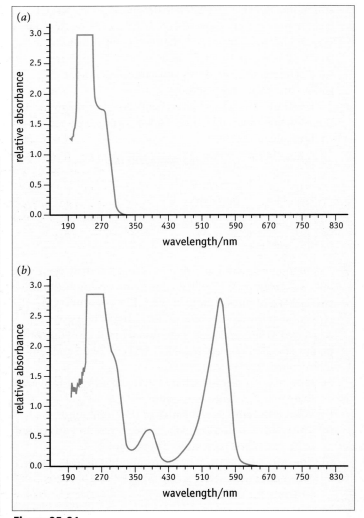

Figure 25.21
The visible/ultraviolet spectrum of phenolphthalein (*a*) at pH 6, (*b*) at pH 10.

25 Introduction to Carbon Chemistry

Petroleum and Alkanes

26.1 Crude oil

As well as supplying a large part of our energy needs, crude oil is the source of 70% of Britain's organic chemicals. It is the most important of all modern raw materials.

Crude oil was formed from the remains of small marine animals and plants that were buried in the beds of the seas millions of years ago. The decay of these remains under layers of overlaying rock formed the liquid known as crude oil or petroleum (from Greek words meaning 'rock oil'). Similar conditions led to the formation of natural gas. Natural gas is often found associated with crude oil as well as in deposits on its own, such as those in the North Sea.

Crude oil is a complex mixture of hydrocarbons. It has no uses in its raw form: to provide useful products its components must be partly separated and if necessary modified. Once crude oil has been found and extracted, it must therefore be transported to a refinery where it is processed. The fundamental process is primary distillation, the details of which are summarised in diagrammatic form in figure 26.1.

Fractions from primary distillation

Refinery gas (1–2% of crude oil) is similar in composition to natural gas. It contains those hydrocarbons that are gases at normal temperatures. These include the alkanes with one to four carbon atoms in their molecules, with methane as the major component. The main use of refinery gas is as a gaseous fuel, but like natural gas it can also be used as a starting point for making petrochemicals, since most organic chemicals are built up from small molecules containing one, two or three carbon atoms.

Gasoline (15–30%) is a complex liquid mixture of hydrocarbons containing mainly C_5–C_{10} compounds whose boiling points range from 40°C to 180°C. The major use of gasoline is of course as a fuel in internal combustion engines (see section 26.6). In the USA, where the motor car is the most popular method of transport, fuel taxes are low and cars have traditionally been built to consume large quantities of fuel. Gasoline is therefore in

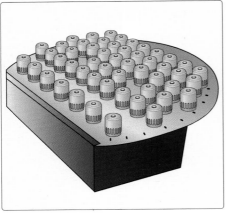

▲ Bubble tray in a primary distillation column

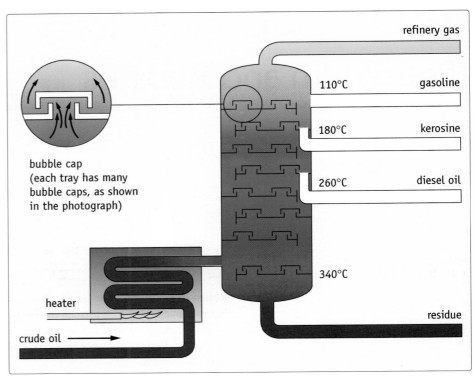

Figure 26.1
Primary distillation of crude oil.

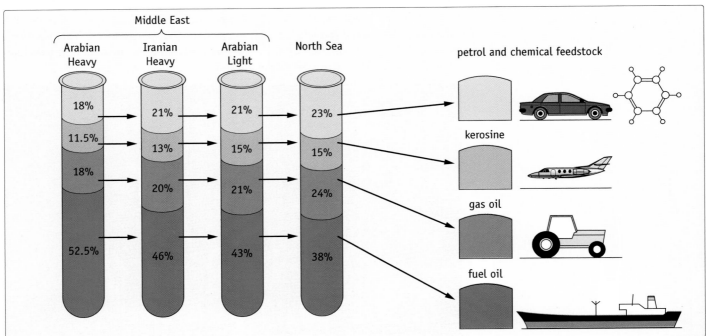

Figure 26.2
Variation of composition of crude oil from different sources.

Different crudes contain varying proportions of the different fractions. The output of the refinery can be arranged to suit market demands by blending different crudes and by converting heavy fractions into lighter ones by cracking.

great demand and American refineries use most of the gasoline fraction for blending in motor fuels. In Western Europe and most of the rest of the world, the demand for gasoline is less. A considerable proportion of this fraction is used in the manufacture of chemicals, after the cracking processes described in section 26.7. The part of the gasoline fraction used to produce chemicals is called **naphtha.**

Kerosine (10–15%) consists mainly of C_{11} and C_{12} hydrocarbons, with boiling points from 160°C to 250°C. It is used as a fuel in jet engines and for domestic heating. It can also be cracked to produce extra gasoline (see section 26.7).

Diesel oil or **gas oil**, (15–20%) containing C_{13}– C_{25} compounds, boils in the range 220°C–350°C. It is used in diesel engines, where the fuel is ignited by compression instead of by a spark and also in furnaces for industrial heating purposes. Like kerosine, it can be cracked to produce extra gasoline.

26 Petroleum and Alkanes

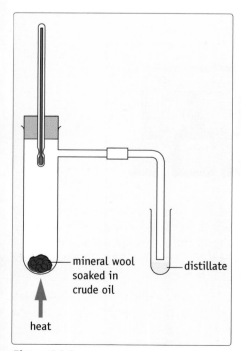

▶ Refuelling a jet in mid-flight. Jet fuel is made from kerosine.

Residue (40–50%) The residual oil from the primary distillation boils above 350°C and is a highly complex mixture of involatile hydrocarbons. Most of it is used as fuel oil in large furnaces such as those in power stations or big ships. A proportion of it, however, is used to make **lubricating oils** and **waxes**. Both these materials contain C_{26}–C_{28} hydrocarbons. When pure these hydrocarbons are solid. However, lubricating oil is a complex mixture, in which each member depresses the melting point of the others so that the mixture is a liquid. In paraffin wax the components are similar enough in structure to form a solid.

To obtain lubricating oil and paraffin wax from the residue, the appropriate hydrocarbons must be distilled off. The distillation is done in a vacuum to avoid the high temperatures that would be needed for distillation at atmospheric pressure, since such temperatures would tend to crack the hydrocarbons. Paraffin wax is separated from lubricating oil by solvent extraction. The solid left after vacuum distillation is an involatile tarry material called **bitumen** or **asphalt**. It is used to surface roads and to waterproof materials.

26.2 The composition of crude oil

The distillation of crude oil can be carried out on a small scale in the laboratory: you will probably be familiar with the experiment represented in figure 26.3. Table 26.1 gives the properties of the fractions collected over different temperature ranges in this laboratory experiment.

Notice that all these properties show a steady gradation: there are no sudden changes in the properties of the different fractions. This suggests that crude oil is a mixture of many components and that the components have similar properties, many of them perhaps belonging to the same homologous series.

Hydrocarbons are the main and the most important components of crude oil.

Figure 26.3
The laboratory distillation of crude oil. Fractions are collected over several different boiling ranges.

Table 26.1
Properties of fractions obtained in the laboratory distillation of crude oil

Property	Room temp. to 70°C	70°C–120°C	120°C–170°C	170°C–220°C
Colour	pale yellow	yellow	dark yellow	brown
Viscosity	runny	fairly runny	fairly viscous	viscous
Behaviour when ignited	burns readily: clean yellow flame	quite easily ignited: yellow flame, some smoke	harder to ignite: quite smoky flame	hard to ignite: smoky flame

Three different homologous series of hydrocarbons are present: **alkanes**, which we will look at in more detail later in this chapter, **cycloalkanes**, considered in section 26.7, and **aromatics** (compounds containing benzene-type rings), considered in chapter 28. The proportions of the different types of compounds present depend on the source of the oil. Alkanes and cycloalkanes form the large majority, with aromatics making only about 10% of the hydrocarbon total.

All the volatile fractions of crude oil are made up of hydrocarbons. The increasing boiling ranges of the fractions correspond to the increasing size of the alkane molecules they contain. The involatile tarry residue left after vacuum distillation consists of compounds with large molecules called resins and asphalts. These contain mainly carbon and hydrogen, with some oxygen and sulphur.

Look again at table 26.1.

1 Why are the first fractions easiest to ignite?
2 Why do the higher fractions burn with the smokiest flames?
3 Suggest a reason why the fractions become increasingly viscous.

We will now consider in detail the largest group of compounds present in crude oil, the alkanes.

26.3 Naming alkanes

The alkanes are **saturated hydrocarbons**. 'Saturated' means that they contain the maximum content of hydrogen possible, with no double or triple bonds between carbon atoms. The general formula of the alkanes is C_nH_{2n+2}. It is possible to have alkanes with straight or with branched chains, for example

$$CH_3—CH_2—CH_2—CH_2—CH_3 \text{ straight chain}$$

$$CH_3—CH_2—CH—CH_3 \text{ branched chain}$$
$$\underset{|}{}$$
$$CH_3$$

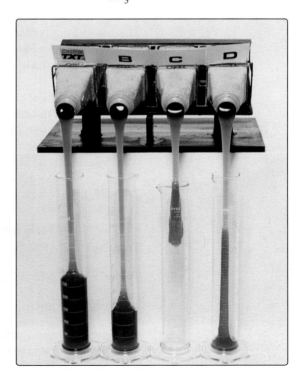

▶ The viscosity of oil depends on the length of the alkane chains in it.

Table 26.2
Straight-chain alkanes

Formula	Name
CH_4	methane
CH_3CH_3	ethane
$CH_3CH_2CH_3$	propane
$CH_3CH_2CH_2CH_3$	butane
$CH_3(CH_2)_3CH_3$	pentane
$CH_3(CH_2)_4CH_3$	hexane
$CH_3(CH_2)_8CH_3$	decane
$CH_3(CH_2)_{18}CH_3$	eicosane

Table 26.2 gives the names of some simple straight-chain alkanes.

You can see that after the first four alkanes the name is formed by adding the suffix **-ane** to the Greek root indicating the number of carbon atoms in the molecule (e.g. *pent-* five, *hex-* six). The first part of the name indicates the number of carbon atoms; the ending -ane indicates that it is an alkane. All compounds based on alkane skeletons are named by this method, with a stem to indicate the number of carbon atoms and a suffix to indicate the functional group.

Branched-chain alkanes are named by considering them as straight-chain alkanes with side groups attached. For example:

$$CH_3 — CH_2 — \underset{\underset{CH_3}{|}}{CH} — CH_3$$

is regarded as butane with CH_3 attached to the *second* atom. It is therefore called 2-methylbutane. The name **methyl** indicates the CH_3— group, which is just methane with a hydrogen atom removed so it can be attached to another atom. Similarly CH_3CH_2— is called **ethyl**, $CH_3CH_2CH_2$— **propyl**, and so on. Side groups of this kind are called **alkyl groups**; their general formula is C_nH_{2n+1}. The symbol **R** is often used to represent a general alkyl group.

Another example of a branched-chain alkane is:

$$CH_3 — CH_2 — CH — \underset{\underset{CH_2 — CH_3}{|}}{\overset{\overset{CH_3}{|}}{CH}} — CH_3$$

This molecule can be regarded as a five-carbon chain with a methyl and an ethyl group attached to the second and third carbon atoms, respectively. It is therefore called 3-ethyl-2-methylpentane, the side groups being arranged in alphabetical order. Note that if the carbon atoms were numbered from the left, the name would be 3-ethyl-4-methylpentane. This name is not used because the convention is to use the name that includes the lowest numbers. Note also that for each side group the name includes a number showing the carbon atom to which the group is attached. Thus

$$CH_3 — \underset{\underset{CH_3}{|}}{\overset{\overset{CH_3}{|}}{C}} — CH_2 — CH_3$$

is called 2,2-dimethylbutane, and

$$CH_3 — \underset{\underset{CH_3}{|}}{\overset{\overset{CH_3}{|}}{C}} — \underset{\underset{CH_2 — CH_3}{|}}{CH} — CH_2 — CH_2 — CH_2 — CH_3$$

is called 2,2-dimethyl-3-ethylheptane.

To name an alkane from its structural formula:

1 Look for the longest unbranched chain in the molecule.

2 Look for the side groups attached to the main chain and the numbers of the carbon atoms to which they are attached.

3 The name then consists of the name of the longest unbranched chain, prefixed by the names of the side groups and the numbers of the carbon atoms to which they are attached.

The system of nomenclature described above is called the IUPAC system (IUPAC stands for International Union of Pure and Applied Chemistry). The system can be extended to apply to all organic compounds. Using this system, it is possible to write the structural formula of any compound from its name.

The IUPAC system is gradually replacing the older, less systematic way of naming organic compounds. IUPAC names will be used wherever possible in this book. The only disadvantage of the system is that for complicated molecules the name becomes

1 Name the following alkanes:

$$CH_3 — \underset{\underset{CH_3}{|}}{CH} — \underset{\underset{CH_3}{|}}{CH} — \underset{\underset{CH_3}{|}}{\overset{\overset{CH_3}{|}}{C}} — CH_3$$

$$CH_3 — \underset{\underset{CH_3}{|}}{\overset{\overset{CH_3}{|}}{C}} — CH_3 \qquad CH_3(CH_2)_6CH_3$$

2 Write the formulas of the following alkanes:

2,2,4-trimethylhexane, methyl-propane, 5-ethyldecane.

very cumbersome. In such cases systematic nomenclature is often abandoned and a less informative but more easily spoken name is used. Thus the name 2,3,4,5,6-pentahydroxyhexanal is dropped in favour of 'glucose'.

26.4 Physical properties of alkanes

The alkanes form a homologous series. As with all homologous series the members show a gradual change in physical properties as the number of carbon atoms in their molecules increases. We can see this in the distillation of crude oil, which is largely composed of alkanes. Successive fractions from crude oil have successively higher boiling ranges and steadily increasing viscosity. The fractions range from gas (refinery gas), through liquids (gasoline, kerosine) to solids (paraffin wax, bitumen).

Some properties of individual straight-chain alkanes are shown in table 26.3. Notice that these properties show a gradual steady change as the number of carbon atoms in the molecules increases. Figure 26.4 on the next page shows graphically the smooth increase of boiling point with increasing number of carbon atoms.

Such steady variation in physical properties is characteristic of all homologous series. This means we can predict the properties of a compound from the properties of other members of its homologous series.

Table 26.3
Physical properties of straight-chain alkanes

Number of carbon atoms	Formula	Name	State (at 298 K)	Boiling point/K	Melting point/K	Density /g cm^{-3}
1	CH_4	methane	g	112	90	0.424
2	C_2H_6	ethane	g	184	101	0.546
3	C_3H_8	propane	g	231	85	0.501
4	C_4H_{10}	butane	g	273	138	0.579
5	C_5H_{12}	pentane	l	309	143	0.626
6	C_6H_{14}	hexane	l	342	178	0.657
7	C_7H_{16}	heptane	l	371	182	0.684
8	C_8H_{18}	octane	l	399	216	0.703
9	C_9H_{20}	nonane	l	424	219	0.718
10	$C_{10}H_{22}$	decane	l	447	243	0.730
11	$C_{11}H_{24}$	undecane	l	469	247	0.740
12	$C_{12}H_{26}$	dodecane	l	489	263	0.749
15	$C_{15}H_{32}$	pentadecane	l	544	283	0.769
20	$C_{20}H_{42}$	eicosane	s	617	310	0.785

Use the data in table 26.3 to predict:

1 The density of tridecane ($C_{13}H_{28}$).

2 The boiling point of tetradecane ($C_{14}H_{30}$).

3 The formula of the first alkane to be solid at 20°C.

As table 26.3 shows, the first four straight-chain alkanes (C_1–C_4) are gases. The next twelve (C_5–C_{16}) are liquids, and the remainder are solids. Solid alkanes of high molecular mass are similar in nature to polythene, which is effectively composed of long-chain alkane molecules.

Alkanes are colourless when pure. The viscosity of liquid alkanes increases with increasing molecular mass. Alkanes are all less dense than water and therefore float on it. This can cause major environmental hazards when oil-carrying ships run into trouble (see photographs).

The steady change in physical properties of alkanes is caused by the steadily increasing molecular size. The increase in boiling point is due to the increasing forces of attraction between molecules of increasing size (see section 9.4). The higher viscosity of the higher alkanes is due to the tendency of the long molecules to become 'tangled up' with one another.

The trends in physical properties considered so far have been those among the straight-chain alkanes. Branched-chain alkanes do not show the same steady gradation of properties as straight-chain alkanes. The variation in molecular structures of branched alkanes is too great for trends to be clear. On the whole, the effect of branching is to increase volatility and reduce density. Thus the boiling point of pentane is 309 K while that of its highly branched isomer 2,2-dimethylpropane is 283 K.

▲ When the freighter the *New Carlissa* foundered off the coast off Oregon, US Navy demolition experts blew up its fuel tanks and burned off the engine oil, to stop it polluting nearby beaches.

▲ These sea birds were coated in oil from the oil tanker *Braer* which ran aground in Shetland in 1993.

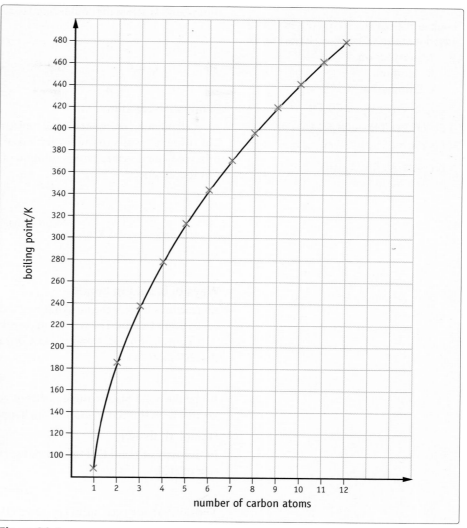

Figure 26.4
Variation of boiling point of straight-chain alkanes with number of carbon atoms.

26.5 Reaction mechanisms in organic chemistry

One of the things that makes organic chemistry enjoyable is the fact that although there are many, many different organic compounds, the reactions of these compounds are quite simply classified.

In the next section we will look at the characteristic reactions of alkanes, but in this section we will look at some of the general principles behind organic reactions. Although some organic reactions go in a single step, most reactions involve a series of steps. The sequence of steps in a reaction is called the **reaction mechanism.**

Types of organic reaction

In this book we will be concerned with three important types of organic reaction.

Substitution reactions

In this type of reaction one atom or group of atoms is substituted by another. For example, when bromoethane reacts with ammonia, bromine atoms are replaced by —NH_2 groups:

$$CH_3CH_2Br + NH_3 \rightarrow CH_3CH_2NH_2 + HBr$$
bromoethane   ethylamine

Addition reactions

In these reactions, two substances react together to form a single substance. A double or triple bond is usually involved. For example, when bromine reacts with ethene, the bromine adds across the double bond to form a single product, 1,2-dibromoethane:

$$\underset{\substack{H \\ | \\ H}}{\overset{\substack{H \\ |}}{C}} = \underset{\substack{| \\ H}}{\overset{\substack{H \\ |}}{C}} \quad + \quad Br_2 \quad \longrightarrow \quad \underset{\substack{| \\ Br}}{\overset{\substack{H \\ |}}{H - C}} - \underset{\substack{| \\ Br}}{\overset{\substack{H \\ |}}{C - H}}$$

Elimination reactions

In an elimination reaction, a small molecule is removed, or eliminated, from a larger one. This usually results in the formation of a double or triple bond. For example, when bromoethane is heated with a solution of hydroxide ions in ethanol, HBr is eliminated, forming ethene:

$$\underset{\substack{| \quad | \\ H \quad Br}}{\overset{\substack{H \quad H \\ | \quad |}}{H - C - C - H}} \xrightarrow{\text{\, }^-OH \text{ (ethanol)}} \underset{\substack{H}}{\overset{\substack{H}}{C}} = \underset{\substack{H}}{\overset{\substack{H}}{C}} \quad + \quad HBr$$

Ways of breaking bonds

All reactions involve the breaking, and remaking, of bonds. Breaking bonds is sometimes called **bond fission.** The way bonds break has an important influence on reactions.

In a covalent bond, a pair of electrons is shared between two atoms. For example, in the HCl molecule:

$$H \div Cl$$

When the bond breaks, these electrons get redistributed between the two atoms. There are two ways this redistribution can happen.

Homolytic fission

In this type of bond fission, one of the two shared electrons goes to each atom. In the case of HCl:

$$H \div Cl \rightarrow H\cdot + Cl\cdot$$

The dot · beside each atom represents the unpaired electron that the atom has gained from the shared pair in the bond. The atoms have no overall electric charge, because each has equal numbers of protons and electrons. But the atoms are highly reactive, because the unpaired electron has a strong tendency to pair up with another electron from another substance (figure 26.5). These highly reactive atoms or groups of atoms with unpaired electrons are called **free radicals.**

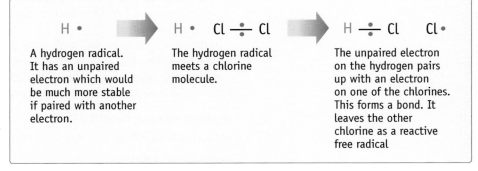

H •	H • Cl $\div$ Cl	H $\div$ Cl Cl•
A hydrogen radical. It has an unpaired electron which would be much more stable if paired with another electron.	The hydrogen radical meets a chlorine molecule.	The unpaired electron on the hydrogen pairs up with an electron on one of the chlorines. This forms a bond. It leaves the other chlorine as a reactive free radical

Figure 26.5
Why free radicals are reactive.

Another example of free radical formation is when a C—H bond in methane is broken:

$$\underset{\substack{| \\ H}}{\overset{\substack{H \\ |}}{H - C - H}} \longrightarrow \underset{\substack{| \\ H}}{\overset{\substack{H \\ |}}{H - C\cdot}} \quad + \quad H\cdot$$

methane   methyl radical   hydrogen radical

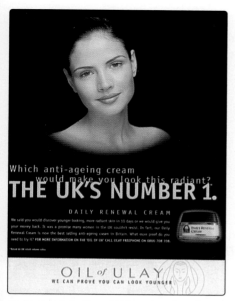

▲ Ageing is believed to be related to the action of free radicals on the body.

Free radicals are most commonly formed when the bond being broken has electrons that are fairly equally shared. When the electrons are unequally shared, the bond is polar and heterolytic fission is more likely.

Heterolytic fission

In heterolytic bond fission, when the bond breaks, *both* of the shared electrons go to just *one* of the atoms. This atom becomes negatively charged, because it has one more electron than it has protons. The other atom becomes positively charged. In the case of HCl:

$$H \div Cl \rightarrow H^+ + :Cl^-$$

Heterolytic fission is more common where a bond is already polar. For example, the bromoalkane shown below contains a polar C—Br bond, and under certain conditions this can break heterolytically:

$$CH_3 - \overset{\overset{\displaystyle CH_3}{|}}{\underset{\underset{\displaystyle CH_3}{|}}{C}} \overset{\delta+}{-} Br^{\delta-} \longrightarrow CH_3 - \overset{\overset{\displaystyle CH_3}{|}}{\underset{\underset{\displaystyle CH_3}{|}}{C^+}} + Br^-$$

Notice that the first ion contains a *positively* charged carbon atom. It is an example of a **carbocation**. An ion containing a *negatively* charged carbon atom is called a **carbanion**. Both these types of ions tend to be unstable and highly reactive, so they only exist as short-lived reaction intermediates.

Electrophiles and nucleophiles

Like all ions, carbocations and carbanions are attracted to groups which carry an opposite electric charge. Carbocations are **electrophiles** ('electron-lovers'): they are short of electrons, so they are attracted to groups which can donate electron pairs. Carbanions are **nucleophiles** ('nucleus-lovers'): they are electron-rich, so they are attracted to groups which can accept electron pairs. This idea is illustrated in figure 26.6.

1. Look at the groups listed below. Classify each as
   A carbocation
   B carbanion
   C free radical
   D none of these

   1 $CH_3CH_2 \cdot$ 2 $CH_3CH_2CH_2^+$
   3 CH_3^+ 4 $CH_3CH_2^-$
   5 ^-OH 6 $\cdot OH$

2. Which type of bond fission, homolytic or heterolytic, would you expect for (a) a bond between identical atoms, (b) a bond between atoms whose electronegativities differ widely, (c) a bond between atoms whose electronegativities are similar?

Carbocations are **electrophiles**. They are attracted to groups which can donate electron pairs.

Carbanions are **nucleophiles**. They are attracted to groups which can accept electron pairs.

Figure 26.6

The terms electrophile and nucleophile don't apply just to ions. Many organic compounds are polar: they carry partial charges, though not full positive or negative charges. These partial charges can also make a group electrophilic or nucleophilic, like this:

$$\overset{\delta+}{CH_3} - \overset{\delta-}{Br}$$

electrophilic   nucleophilic

1. Look at the ions and groups below. Classify each as
   A electrophile
   B nucleophile
   C could act as electrophile or nucleophile

   1 Cl^- 2 NO_2^+ 3 CH_3^+
   4 HI 5 ^-CN 6 H^+
   7 CH_3NH_2 8 CH_3OH

Describing reaction mechanisms

We can use the ideas and terms outlined above to describe the mechanisms of the important classes of organic reactions that we will meet in this book. You will find them in the following sections:

Nucleophilic substitution: section 29.4

Electrophilic addition: section 27.5

Electrophilic substitution: section 28.6

Nucleophilic addition: section 31.1

Free radical substitution: section 26.6.

Curly arrows

When reaction mechanisms are being described, a 'curly arrow' is sometimes used to show the **movement of a pair of electrons.** The beginning of the arrow shows where the electron pair starts from and the arrow head shows where the pair ends up. Figure 26.7 shows an example.

The arrow shows a pair of electrons moving from the Br^- ion to the region between the bromine and the carbon, where it forms a covalent bond between the two atoms.

The same reaction is shown again below, with all the bonding electrons indicated.

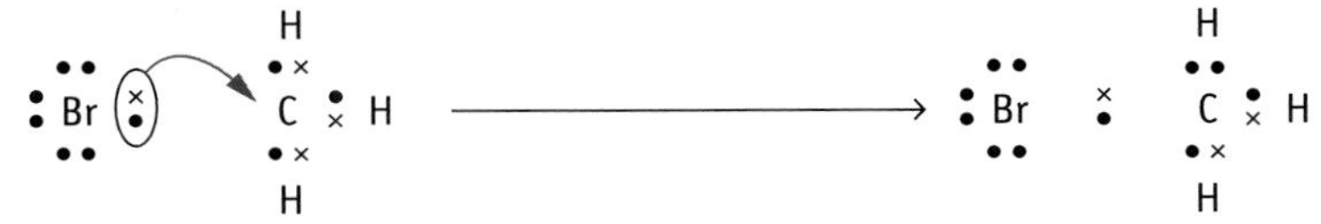

Figure 26.7
Using curly arrows.

A *half-arrow* is used to show the *movement of a single electron* in reactions involving free radicals. The beginning of the arrow shows where the single electron starts from, and the half-arrow head shows where it ends up.

For example, the reaction

$$H^\bullet \ + \ Cl {-\!\!-} Cl \longrightarrow H {-\!\!-} Cl \ + \ Cl^\bullet$$

would be shown as

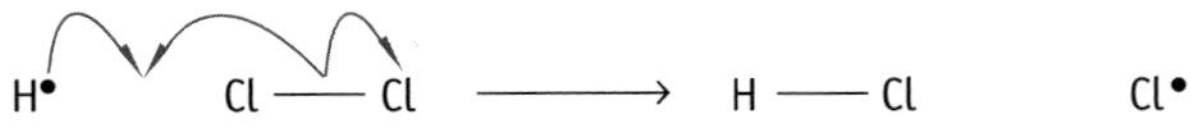

This is shown again below, with all the bonding electrons indicated.

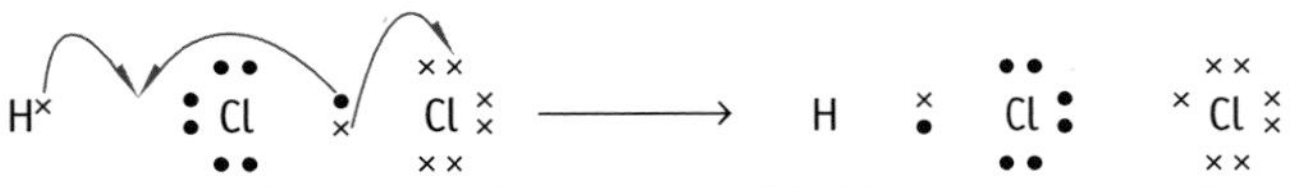

Figure 26.8 summarises the way curly arrows and half-arrows are used.

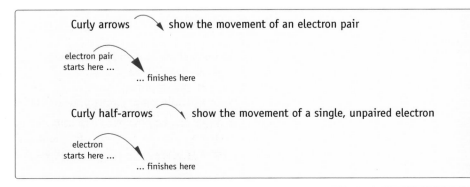

Figure 26.8

The effects of various reagents on hexane, a typical alkane, are shown in table 26.4.

Clearly hexane is unreactive. It is unaffected by acids, alkalis, dehydrating agents and aqueous oxidising agents. A close look at the substances in the table with which hexane *does* react, namely bromine and oxygen, shows two things. First, neither of these reagents has any centre of electrical charge in their molecule: they are non-polar. This is in contrast with the other substances in the table, which are all polar reagents. Second, before reaction can occur, energy must be supplied: heat in the case of oxygen, light in the case of bromine.

Table 26.4
The effect of common reagents on hexane

Reagent	Effect
air	no effect cold. Burns when heated
sodium hydroxide solution	no effect hot or cold
concentrated sulphuric acid	no effect hot or cold
potassium manganate(VII) (potassium permanganate) solution	no effect hot or cold
bromine	no effect in dark. Bromine is slowly decolorised in sunlight

Ions and polar molecules do not have any effect on alkane molecules, because of the non-polarity of the C—H bond. Carbon and hydrogen are very close in electronegativity, so the electron pair in the covalent bond between carbon and hydrogen is fairly evenly shared: consequently there is little polarity in the C—H bond. The electron pair in the C—C bond is also evenly shared, so this bond is non-polar too. Thus there are no polar bonds in alkane molecules, and so no centres of electrical charge to act as electrophiles or nucleophiles and attract normally reactive species such as ^-OH, H^+ and MnO_4^-.

Alkanes, then, are unreactive towards polar or ionic reagents, but will react with non-polar substances such as oxygen or bromine. The mechanism of these reactions involves free radicals.

Free-radical reactions of alkanes

Hexane and bromine react in sunlight. A similar reaction occurs between methane and chlorine and we will use this as a simple example to look at in more detail.

Methane and chlorine do not react at all in the dark, but in sunlight an explosive reaction occurs, forming chloromethane and hydrogen chloride:

$$CH_4(g) + Cl_2(g) \xrightarrow{light} CH_3Cl(g) + HCl(g) \qquad \Delta H = -98 \text{ kJ mol}^{-1}$$

This is a substitution reaction. Like all reactions between covalent molecules, it involves breaking some bonds, for which energy must be supplied, and making new bonds, when energy is released.

The reaction between methane and chlorine does not proceed in the dark at room temperature because there is not enough energy available to break bonds and start the reaction.

$$Cl_2 \rightarrow Cl^{\bullet} + Cl^{\bullet} \qquad \Delta H = +242 \text{ kJ mol}^{-1}$$

$$CH_4 \rightarrow CH_3^{\bullet} + H^{\bullet} \qquad \Delta H = +435 \text{ kJ mol}^{-1}$$

These figures show that the Cl—Cl bond is easier to break than the C—H bond. When light is shone on the reaction mixture, chlorine molecules are supplied with the energy necessary to split them into atoms. This stage is called **initiation.**

The chlorine atoms, being free radicals, are highly reactive. When they collide with a methane molecule they combine with one of its hydrogen atoms, forming a new free radical:

$$CH_4 + Cl^{\bullet} \rightarrow CH_3^{\bullet} + HCl \qquad \Delta H = +4 \text{ kJ mol}^{-1}$$

Using curly arrows, we can show the movement of the single electrons in this reaction:

$$H-\overset{\overset{\textstyle H}{|}}{\underset{\underset{\textstyle H}{|}}{C}}-H \quad Cl^{\bullet} \longrightarrow H-\overset{\overset{\textstyle H}{|}}{\underset{\underset{\textstyle H}{|}}{C}}^{\bullet} \quad H-Cl$$

The $CH_3^{\bullet}$ free radical then reacts with another chlorine molecule:

$$CH_3^{\bullet} + Cl_2 \rightarrow CH_3Cl + Cl^{\bullet} \qquad \Delta H = -97 \text{ kJ mol}^{-1}$$

and so the process continues. These two reactions enable a chain reaction to occur: they are **propagation** steps. Note that each propagation step involves both the breakage and the formation of a bond: the net energy change is therefore relatively small.

The reaction chain ends when two free radicals collide and combine; this is called **termination** and is highly exothermic:

$$Cl^{\bullet} + Cl^{\bullet} \rightarrow Cl_2 \qquad \Delta H = -242 \text{ kJ mol}^{-1}$$

$$CH_3^{\bullet} + Cl^{\bullet} \rightarrow CH_3Cl \qquad \Delta H = -339 \text{ kJ mol}^{-1}$$

$$CH_3^{\bullet} + CH_3^{\bullet} \rightarrow C_2H_6 \qquad \Delta H = -346 \text{ kJ mol}^{-1}$$

Each chain may go through 100 to 10 000 cycles before termination occurs. The processes are extremely rapid, hence the explosive nature of the reaction.

The net result of such reactions is the formation of large amounts of CH_3Cl and HCl and small amounts of C_2H_6. (In the presence of excess chlorine, further substitution may occur, forming CH_2Cl_2, $CHCl_3$ and CCl_4, as explained in the next section.) The overall energy change of -98 kJ mol^{-1} represents the difference between the energy released in forming C—Cl and H—Cl bonds and the energy absorbed in breaking C—H and Cl—Cl bonds. The energy supplied by the light which initiates the reaction is the activation energy.

Practically all the reactions of alkanes proceed by free-radical mechanisms, characterised by high activation energies and a tendency to proceed rapidly in the gas phase. Some of these reactions will now be considered.

26.7 Important reactions of alkanes

Alkanes show little reactivity towards all the common polar and ionic reagents. Indeed the alkanes were once known as the **paraffins**, from the Latin words *parum* (little) and *affinitas* (affinity). Hence there are only a few reactions of alkanes, but these are of great importance.

Burning

Alkanes are kinetically stable in the presence of oxygen, but they are energetically unstable with respect to their oxidation products. When the necessary activation energy is supplied, in other words when they are ignited, combustion occurs. The reaction has a free-radical mechanism, and it occurs rapidly in the gas phase. Because it is a gas-phase reaction, liquid and solid alkanes must first be vaporised. This is why less volatile alkanes burn less readily. If the oxygen supply is plentiful, the combustion products are carbon dioxide and water. For example:

$$C_7H_{16}(g) + 11O_2(g) \longrightarrow 7CO_2(g) + 8H_2O(l); \qquad \Delta H = -485 \text{ kJ mol}^{-1}$$

If the oxygen supply is limited, the products may include carbon monoxide and carbon (soot).

The combustion of hydrocarbons in general and alkanes in particular is of immense importance. It occurs in power stations, furnaces, domestic heaters, candles, gas heaters, internal combustion engines and many other devices essential to a technological society. An understanding of the nature of the combustion process is vital to the design of such devices.

▲ The combustion of alkanes is a gas-phase reaction. Before it can burn, candle wax must vaporise. The wick provides a surface from which the molten wax vaporises.

1. Write an equation for the complete combustion of octane, C_8H_{18}.

2. Estimate the enthalpy change of combustion of octane given that the enthalpy changes of combustion of hexane and heptane are $-4195 \text{ kJ mol}^{-1}$ and $-4854 \text{ kJ mol}^{-1}$, respectively.

3. Write an equation for the combustion of octane to form carbon monoxide and water.

An important example is the **petrol engine.** In the cylinder of a motor-car engine a mixture of gasoline vapour (which contains mostly C_5 to C_{10} alkanes) and air is ignited by an electric spark. This produces an explosive reaction which drives the piston down. The rate at which this reaction occurs and the ease with which it is initiated are very important for the efficiency of the engine. If the explosion occurs too rapidly, heat will be dissipated instead of being converted to useful kinetic energy. If the explosion starts too early the pistons are subjected to harmful jarring. 'Knocking' or 'pinking' of the engine then occurs. This is more likely to occur in the engines of high-performance cars, where high compression ratios tend to cause premature ignition.

Reducing the knocking problem

Gasoline mixtures that are rich in straight-chain alkanes such as heptane ignite very readily and explode rapidly, causing 'knocking' and inefficient combustion. The combustion of branched-chain alkanes such as 2,2,4-trimethylpentane (iso-octane) is much smoother and more controlled, so gasoline mixtures rich in branched-chain alkanes are more efficient fuels and less likely to cause knocking. The **octane rating** of 2,2,4-trimethylpentane is set at 100, and that of heptane at 0. Gasoline mixtures rich in branched-chain alkanes have high octane numbers and burn smoothly and efficiently in high-performance engines.

There are two ways of meeting the demand of modern high-performance engines for fuels with high octane numbers. One is to produce artificial gasoline mixtures that are rich in branched-chain alkanes (see below). The other is to add an anti-knock compound to gasoline. The anti-knock compound normally used is tetraethyllead(IV), $Pb(CH_3CH_2)_4$. When burned, this compound produces small particles of lead oxide which tend to combine with free radicals produced in the chain reaction of combustion. This slows the reaction down and makes it smoother.

To prevent lead accumulating in the engine, 1,2-dibromoethane (CH_2BrCH_2Br) is also added to the gasoline. This results in the formation of lead bromide, which is volatile and is swept away in the car exhaust. The unfortunate effect of this is to add lead to the atmosphere, whence it can be inhaled. Lead is a neurotoxin, or nerve poison. It can cause damage to the brain and nervous system, particularly in young children.

Most countries are now phasing out the use of leaded petrol and replacing it with unleaded. The production of unleaded petrol relies on cracking, isomerisation and reforming reactions (see below) to produce hydrocarbons with a high octane rating. Unleaded petrol also contains special high-octane additives such as 'methyl tertiary butyl ether' (MTBE) (see section 30.9). These high-octane ingredients are expensive, so unleaded petrol costs more to produce than leaded. However, in the UK it is taxed at a lower rate to encourage people to use it, so its overall price is lower.

Section 26.8 has more about the environmental effects of burning motor fuels.

Reaction with halogens

The reaction of methane with chlorine has already been considered in some detail (page 407). This reaction produces several products:

$$CH_4 + Cl_2 \rightarrow CH_3Cl + HCl$$
chloromethane

$$CH_3Cl + Cl_2 \rightarrow CH_2Cl_2 + HCl$$
dichloromethane

$$CH_2Cl_2 + Cl_2 \rightarrow CHCl_3 + HCl$$
trichloromethane (chloroform)

$$CHCl_3 + Cl_2 \rightarrow CCl_4 + HCl$$
tetrachloromethane (carbon tetrachloride)

The proportions of different products formed depend on the proportions of chlorine and methane used. All four products are useful in industry, but they have to be separated from one another by distillation. Chlorination of alkanes is therefore of limited use for preparing chloroalkanes, particularly in the case of the higher alkanes where the number of possible products is very large.

Similar reactions occur between alkanes and fluorine, chlorine and bromine. The reaction with fluorine occurs in the absence of sunlight and is violent.

Cracking

When alkanes are heated to high temperatures their molecules vibrate strongly enough to break and form smaller molecules. One of these molecules is usually an alkane. Such reactions are known as **cracking**. For example:

$$C_{11}H_{24} \longrightarrow C_9H_{20} + H_2C = CH_2$$
$$\text{undecane} \qquad \text{nonane} \qquad \text{ethene}$$

1. In which petroleum fraction would $C_{11}H_{24}$ be found?
2. In which fraction would C_9H_{20} be found?
3. Why must one of the products of cracking always be unsaturated?

By using a catalyst, cracking can be made to occur at fairly low temperatures. This is known as **catalytic cracking** (figure 26.9).

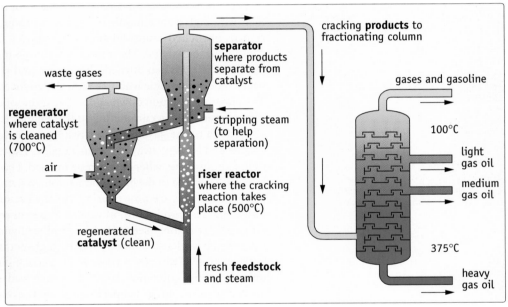

Figure 26.9
A catalytic cracking unit. The diagram alongside explains the working of the unit. The reaction actually occurs in the central unit: the catalyst is regenerated in the left-hand unit and the products are separated in the distillation tower on the right.

Cracking involves *free radical* reactions. Heating alkanes makes them break up into free radicals. These radicals then react with one another to give a mixture of products, including branched-chain alkanes and alkenes.

Cracking is very important in the petroleum industry. It is used to convert lower value, heavy fractions into higher value products. In particular, it is used for the following.

(a) To provide extra gasoline

The example above shows how undecane, a member of the kerosine fraction, can be cracked to produce nonane, a component of gasoline. Thus, heavier fractions can be cracked to produce extra gasoline. Furthermore, cracking tends to produce branched-chain rather than straight-chain alkanes, so the gasoline produced this way has a high octane rating. Processes similar to cracking can be used to convert low-grade gasoline to high-grade fuel. **Reforming** involves converting straight-chain alkanes into ring molecules: arenes and cycloalkanes. **Isomerisation**, as its name implies, involves breaking up straight-chain alkanes and reassembling them as branched-chain isomers. Both of these processes are important in the production of unleaded gasoline.

(b) As a source of alkenes

Because they are so unreactive, alkanes are not a good starting point from which to make the many organic chemicals derived from crude oil. Alkenes, with their reactive double bonds, are a more suitable starting point. The petrochemical industry uses vast quantities of ethene and propene as units for building larger organic molecules.

Cracking reactions produce alkenes, and under the right conditions large yields of ethene and propene can be obtained. Alkenes are thus produced as by-products from cracking heavy fractions (such as $C_{11}H_{24}$) and by cracking some of the gasoline fraction (which is called naphtha when used as a source of alkenes). Alkenes are also made by cracking gaseous alkanes (ethane, propane and butane).

The reactions of alkenes and the products obtained from them are discussed in chapter 27.

26.8 The environmental impact of motor vehicle fuels

Table 26.5 gives the composition of typical petrol engine exhaust for cars with and without catalytic converters.

1 What single gas omitted from table 26.5 makes up most of the remainder of car exhaust?
2 Why does exhaust contain carbon monoxide?
3 Why does exhaust contain oxides of nitrogen?
4 Why does exhaust contain hydrocarbons?

Table 26.5
Gases in average petrol engine exhaust, percentage by volume

Gas	Non-catalyst cars	Catalyst cars
carbon dioxide	14	15
oxygen	0.7	0
hydrogen	0.25	0
carbon monoxide	1.0	0.2
hydrocarbons	0.06	0.01
aldehydes	0.004	0.002
nitrogen oxides	0.2	0.02
sulphur dioxide	0.005	0.005
lead (as solids if leaded petrol is used)	4 mg m^{-3}	not applicable

▲ Car exhausts are analysed as part of the annual MOT test. Emissions must fall below certain limits if the car is to pass.

Effects on humans

The most dangerous pollutants in car exhaust, apart from lead, are probably carbon monoxide, hydrocarbons and oxides of nitrogen. Carbon monoxide and hydrocarbons are present in the exhaust because of incomplete combustion of the fuel. Carbon monoxide is very toxic because it forms a stable compound with haemoglobin in the blood, making the haemoglobin unable to transport oxygen. In a confined space the carbon monoxide in exhaust gases can be fatal. In the open the danger is less, though the carbon monoxide concentration in a busy street is undesirably high. It is worth noting, though, that the carbon monoxide level in a railway compartment full of smokers is twice as high as in a busy London street.

Acid deposition

Acid deposition describes all the different ways in which acid materials get deposited on the earth, including rain, snow, mist and solid particles. Acid deposition has serious environmental effects, including the destruction of life in acidified lakes, the destruction of trees and the erosion of stonework on buildings. Acid deposits mainly contain sulphuric acid, nitric acid, sulphates and nitrates, and automobile exhausts are thought to play a major part in their formation. Acidic oxides of sulphur and nitrogen are heavily involved.

Sulphur oxides, SO_2 and SO_3, collectively known as SO_x, are formed when sulphur compounds in petrol are oxidised as the petrol burns. Sulphur oxides are also produced by power stations, especially those that burn coal, and when sulphur ores are smelted to produce metals such as copper and zinc.

Nitrogen oxides, NO and NO_2, collectively known as NO_x, are formed inside the combustion chambers of motor vehicles. At the high temperature of combustion, nitrogen and oxygen from the air combine to form NO and NO_2.

$$N_2(g) + O_2(g) \rightarrow 2NO(g)$$

$$2NO(g) + O_2(g) \rightarrow 2NO_2(g)$$

These oxides are formed wherever combustion occurs on a large scale: power stations are another major source.

The chemical reactions leading to the formation of acid in the atmosphere are very complex and involve ozone and hydrocarbons as well as SO_x and NO_x. An important part of the process is the reaction in which SO_2 is oxidised to SO_3 by NO_x. SO_2 is not oxidised spontaneously by O_2, but it is oxidised readily by NO_2:

$$SO_2(g) + NO_2(g) \rightarrow NO(g) + SO_3(g)$$

Then

$$2NO(g) + O_2(g) \rightarrow 2NO_2(g)$$

The NO_2 is constantly regenerated by the second reaction, and thus acts as a catalyst in the oxidation of SO_2.

SO_3 reacts readily with water in the atmosphere to form sulphuric acid, which eventually reaches the ground in rain, snow or mist.

$$SO_3(g) + H_2O(l) \rightarrow H_2SO_4(aq)$$

Exhaust gases are also involved in the formation of **photochemical smog**, in cities such as Los Angeles, where the strong sunlight interacts with pollutant gases from millions of automobiles, causing a complex series of photochemical reactions which produce a choking mixture of ozone, NO_x and other gases.

Removing pollutant gases from car exhaust

The most significant pollutant gases in car exhaust are CO, NO_x and unburnt hydrocarbons, C_xH_y. As figure 26.10 on the next page shows, governments (particularly the US) have passed laws requiring that these gases are progressively reduced in vehicle exhausts.

There are several ways that the exhaust pollution problem can be tackled.

Using oxygenated fuels

'Oxygenates' are fuels containing oxygen in their molecules. The most important are the alcohols methanol and ethanol (chapter 30). Adding these oxygenates to petrol makes combustion more complete and reduces the quantities of CO and C_xH_y that are produced.

Controlling the quantity of air mixed with the fuel

Before entering the combustion chamber, petrol is mixed with air in the carburettor or fuel-injection system. The 'stoichiometric' quantity of air is the quantity that is exactly enough to combine with the fuel, with none left over. A 'rich' mixture has an excess of fuel: this is what you get when the engine is 'choked' for a cold start. Rich mixtures produce large quantities of CO and C_xH_y so cold winter mornings are a bad time for exhaust pollution. A 'lean' mixture has an excess of air, and it produces less CO and C_xH_y. 'Lean burn' engines are one way that car manufacturers can meet the demands for reduced emissions of these gases.

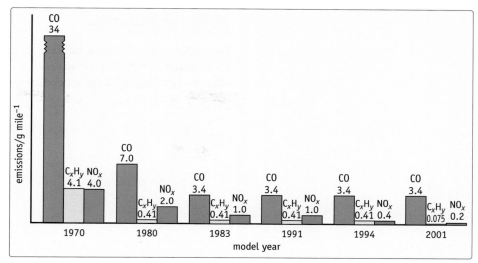

Figure 26.10
Exhaust emission limits set by the US Federal government for new vehicles. (From 2001, the hydrocarbon limit C_xH_y is for hydrocarbons other than methane.)

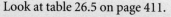

Look at table 26.5 on page 411.

1. Which exhaust gases are *significantly* lower in catalyst cars than in non-catalyst cars?

2. Which exhaust gas is *higher* in catalyst cars than in non-catalyst cars?

3. Suggest a reason for your answer in 2.

Using catalytic converters

A catalytic converter removes pollutant gases from the exhaust by oxidising or reducing them. The exhaust gases pass through a converter containing a precious metal catalyst, usually an alloy of platinum and rhodium. Several reactions may take place. NO_x and CO may take part in a redox reaction which neatly removes both of them at the same time: NO_x oxidises CO to CO_2, and is reduced to harmless nitrogen gas.

$$2NO(g) + 2CO(g) \rightarrow N_2(g) + 2CO_2(g)$$

CO and C_xH_y are oxidised by air:

$$2CO(g) + O_2(g) \rightarrow 2CO_2(g)$$

$$C_7H_{16}(g) + 11O_2(g) \rightarrow 7CO_2(g) + 8H_2O(g)$$

(using C_7H_{16} to represent a typical hydrocarbon).

For all three of these reactions to happen, it is necessary to use a 'three-way converter', and to have an oxygen monitor fitted to the engine: this checks the quantity of oxygen going into the engine to make sure there is enough to carry out the oxidation reactions.

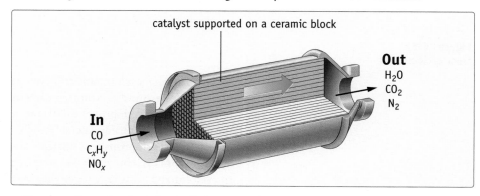

Figure 26.11
A three-way catalytic converter.

The overall result of passing exhaust gases through this kind of catalyst system is to convert CO, NO_x and C_xH_y to relatively harmless N_2, CO_2 and H_2O (figure 26.11). The catalytic reactions do not start working until the catalyst has reached a temperature of about 200°C, so they are not effective until the engine has warmed up.

Catalyst systems of this type cost several hundred pounds, mainly because of the high cost of the precious metal they contain. The catalyst is 'poisoned' by lead, so unleaded fuel must always be used.

The alkanes considered so far have all had open-chain molecules, that is molecules which come to an end at some point. It is also possible for alkane molecules to form rings: such compounds are named by using the prefix **cyclo-**. Some cycloalkanes are shown in figure 26.12.

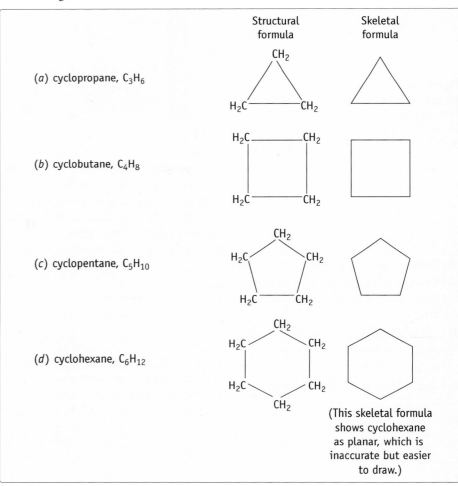

	Structural formula	Skeletal formula

(a) cyclopropane, C_3H_6

(b) cyclobutane, C_4H_8

(c) cyclopentane, C_5H_{10}

(d) cyclohexane, C_6H_{12}

(This skeletal formula shows cyclohexane as planar, which is inaccurate but easier to draw.)

Figure 26.12
Some cycloalkanes.

Cyclopropane, cyclobutane and cyclopentane all have planar molecules. The natural bond angles around a saturated carbon atom are 109° (the angle at the centre of a tetrahedron). In cyclopropane the bond angle is only 60°, since the carbon atoms within the molecule form an equilateral triangle. Hence there is considerable **ring strain** in this molecule, so cyclopropane is unstable and reactive, tending to break open its ring. The bond angle in cyclobutane is 90°, so the ring strain is less, but nevertheless considerable, so cyclobutane is more reactive than open-chain alkanes. Cyclopentane, with a bond angle of 108°, has little ring strain and is therefore stable.

Cyclohexane and the higher cycloalkanes can relieve ring strain by puckering, so that their rings are no longer planar. These cycloalkanes are therefore stable and very similar in properties to open-chain alkanes. Figure 26.13 shows the puckered structure of cyclohexane, in which all the carbon atoms have an unstrained bond angle of 109°. You should try building this molecule with models to see exactly how ring strain is eliminated.

The general formula of cycloalkanes is C_nH_{2n}, compared with C_nH_{2n+2} for open-chain alkanes. Although cyclopentane, cyclohexane and higher cycloalkanes behave similarly to open-chain alkanes their melting and boiling points tend to be somewhat higher.

Cycloalkanes are present in considerable quantities in crude oil. Industrially the most important is cyclohexane, used in the manufacture of nylon.

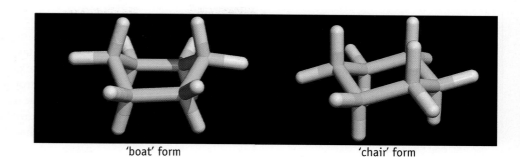

Figure 26.13
Computer-generated models of the two molecular forms of cyclohexane.

'boat' form   'chair' form

Summary

1 Crude oil is a complex mixture of hydrocarbons consisting of alkanes, cycloalkanes and aromatic compounds. It is separated into useful fractions by fractional distillation.

2 Alkanes are saturated hydrocarbons with general formula C_nH_{2n+2}.

3 Straight-chain alkanes are named by adding the suffix -*ane* to a prefix indicating the number of carbon atoms in the molecule.

4 Branched-chain alkanes are named by considering their molecules as straight-chain alkanes with side groups attached.

5 Alkanes show a steady gradation in physical properties with increasing molecular size.

6 When a covalent bond breaks, it may form either uncharged free radicals (homolytic fission) or positive and negative ions (heterolytic fission).

7 Electrophiles are electron-deficient and are attracted to negatively charged groups which can donate lone pairs. Nucleophiles are electron-rich and are attracted to positively charged groups.

8 Alkanes are chemically unreactive towards polar or ionic reagents, but they react with reagents such as oxygen or halogens by free-radical mechanisms.

9 Alkanes burn in a plentiful supply of oxygen to form carbon dioxide and water. They undergo substitution reactions with halogens in sunlight.

10 The combustion of fossil fuels produces oxides of nitrogen and sulphur, which contribute to acid deposition.

11 Alkane molecules break down to smaller molecules at high temperatures: this process is called cracking and is important in the petroleum industry.

12 Cycloalkanes are saturated hydrocarbon rings with general formula C_nH_{2n}.

Review questions

1 Name the following alkanes

(a) $CH_3-CH_2-CH_2-CH_2-CH_2-CH_3$

(b) $CH_3-CH-CH_2-CH-CH_2-CH_3$
           | |
           CH_3 CH_2CH_3

(c)

(d) $CH_3-CH_2-CH_2-CH_2$
                          CH_2
                          CH_3

(e) CH_2
      | $CH-CH_3$
   CH_2

(f)

2 Consider the following alkanes

A $CH_3(CH_2)_{18}CH_3$ B

C $CH_3(CH_2)_6CH_3$ D $CH_3CH(CH_2)_4CH_3$
                                                     |
                                                     CH_3

E $CH_3CH_2CH_3$

(a) In which crude oil fraction would each be found?

(b) Which two alkanes are isomers?

(c) Which alkane might have been formed by cracking another alkane shown above? (Identify *both* alkanes.)

(d) Arrange the alkanes in order of increasing boiling point.

(e) Which alkane is a solid at 298 K?

(f) Which alkane's formula does *not* fit the general formula C_nH_{2n+2}?

3 Write (a) structural formulas; (b) molecular formulas for the following alkanes:

    (i)   ethylcyclohexane

    (ii) 1,2-dimethylcyclopentane

    (iii) 2,2,3-trimethylbutane

    (iv) 3,4-diethyl-2,2-dimethylheptane.

4 Study the table, which gives the enthalpy changes of combustion and the relative molecular masses of some alkanes.

Alkane	No. of carbon atoms, n	Relative molecular mass	Enthalpy change of combustion ΔH_c/kJ mol^{-1}
methane	1	16	−890
ethane	2	30	−1560
propane	3	44	−2220
butane	4	58	−2877
pentane	5	72	−3509
methylbutane	5	72	−3503
2,2-dimethylpropane	5	72	−3517
hexane	6	86	−4195

(a) Plot a graph of ΔH_c against n for the first six straight-chain alkanes.

(b) Does ΔH_c increase by approximately the same amount for each extra carbon atom in an alkane chain?

(c) What is the average increase in ΔH_c per carbon atom?

(d) Your answer to (c) represents the enthalpy change of combustion of which structural group?

(e) Compare the ΔH_c values for the three isomeric alkanes with five carbon atoms. Why are they so similar?

(f) Work out the enthalpy change of combustion per gram of butane, pentane and hexane, respectively. Comment on your result.

5 For the purposes of this question, assume that petrol has density of 0.75 g cm^{-3} and is a mixture of hydrocarbons all of formula C_8H_{18}. Assume 0.5 cm^3 of tetraethyllead(IV) are added to every dm^3 (litre) of leaded petrol. Tetraethyl-lead(IV), $Pb(C_2H_5)_4$, is a liquid of density 1.6 g cm^{-3}. (H=1, C=12, Pb = 207)

(a) Why is tetraethyllead(IV) added to petrol?

(b) Write an equation for the complete combustion of petrol in excess air.

(c) How many moles of C_8H_{18} are there in 1 dm^3 of petrol?

(d) Calculate the volume of air needed to burn completely 1 dm^3 of petrol (assume air is 20% oxygen by volume and that 1 mole of a gas occupies 24 dm^3 under ordinary conditions).

(e) Calculate the volume of nitrogen emitted as exhaust when 1 dm^3 of petrol burns. (Assume all the nitrogen in the air intake passes into the exhaust.)

(f) Calculate the mass of nitrogen emitted as exhaust when 1 dm^3 of petrol burns.

(g) Calculate the mass of carbon dioxide produced when 1 dm^3 of petrol burns.

(h) Assuming that all the lead in the tetraethyllead(IV) passes out in the exhaust, calculate the mass of lead passed out in the exhaust when 1 dm^3 of leaded petrol burns.

6 This question is about octane (C_8H_{18}) and cyclooctane (C_8H_{16}).

(a) Draw the structural formula of octane.

(b) Draw the skeletal formula of cyclooctane.

(c) Suppose octane and cyclooctane were each reacted with chlorine so that only one hydrogen atom in each molecule was substituted by chlorine. Draw structural formulas for all the products you would expect in each case.

7 The table shows the enthalpy changes of combustion of some cycloalkanes.

Name	Formula	Enthalpy change of combustion, ΔH_c/kJ mol^{-1}
cyclopropane	$(CH_2)_3$	−2090
cyclobutane	$(CH_2)_4$	−2740
cyclopentane	$(CH_2)_5$	−3320
cyclohexane	$(CH_2)_6$	−3948
cycloheptane	$(CH_2)_7$	−4635
cyclooctane	$(CH_2)_8$	−5310
cyclononane	$(CH_2)_9$	−5980
cyclodecane	$(CH_2)_{10}$	−6630

(a) For each compound, calculate the enthalpy change of combustion per CH_2 group.

(b) Would you expect the enthalpy change of combustion per CH_2 group to be constant among: (i) open-chain alkanes; (ii) unstrained cycloalkanes; (iii) cycloalkanes with ring strain?

(c) Explain your answers and comment on the figures you obtained in (a).

8 If a few drops of bromine are added to hexane a deep red solution is formed. No reaction occurs if the mixture is kept in the dark, but in sunlight the red colour slowly disappears and a misty gas is given off.

(a) Identify the misty gas formed in this reaction.

(b) Write an overall equation for the reaction of one mole of bromine with one mole of hexane under these conditions.

(c) Why does the reaction only occur in sunlight?

(d) This is a free-radical chain reaction. Write equations for

    (i)   the initiation stage,

    (ii) a propagation stage,

    (iii) a termination stage.

9 Write the structural formulas of all the products you would expect to be formed when ethane reacts with excess chlorine in sunlight.

10 For each of the following, state whether or not you would expect reaction to occur to any significant extent. Write equations for any reactions which you think will occur.
(a) Chlorine is bubbled through hexane in sunlight.
(b) Sodium metal is added to warm hexane.
(c) Hexane is boiled with acidified potassium dichromate(VI) solution.
(d) Chlorine and hexane vapour are heated in the dark.
(e) Hexane and hydrogen are heated in sunlight.
(f) Hexane vapour is heated strongly on its own, in the absence of air.

11 Methane can be used as an alternative to gasoline as a fuel for motor cars. What modifications would be necessary to an ordinary car to enable it to run on methane? What would be the advantages of using methane instead of gasoline?

12 Natural gas from the Texas Panhandle has the percentage composition by volume shown in the table.

Component	Percentage by volume
Methane	80.9
Ethane	6.8
Propane	2.7
Butane and higher alkanes	1.6
Nitrogen	7.9
Carbon dioxide	0.1

The natural gas is processed in the following way. Carbon dioxide is first removed, then ethane, propane, butane and higher alkanes are separated off and mainly used to make alkenes, especially ethene.
(a) How might the carbon dioxide be removed from the gas?
(b) How might ethane and the higher alkanes be separated off?
(c) What would the remaining natural gas be used for?
(d) Why is it not necessary to remove the nitrogen from the gas?
(e) Why is ethene an important industrial chemical?
(f) Ethene is produced form ethane by a cracking reaction that eliminates hydrogen. Write an equation for this reaction. What conditions would be needed to make the reaction occur?
(g) In Europe ethene is usually manufactured from naphtha (part of the gasoline fraction from petroleum distillation) rather than ethane. Suggest a reason for this difference.

27 Unsaturated Hydrocarbons

27.1 Why alkenes are important

What do PVC raincoats and antifreeze have in common? Or polythene bottles and adhesives? Like many other things in everyday use, they are made from ethene, the simplest alkene and the most versatile organic compound in use today. In the UK, more than a million tonnes of ethene are produced and consumed every year. Ethene, $CH_2\text{=}CH_2$, with its reactive double bond, can be used as a building block to prepare complex organic molecules. Propene, $CH_3CH\text{=}CH_2$, is used in a similar way, though on a smaller scale. Large quantities of these alkenes are manufactured by cracking processes described in chapter 26.

27.2 Naming alkenes

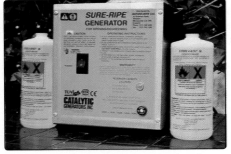

▲ Ethene is used to speed up the ripening of fruit such as bananas. This 'Ethylene Gas Generator' produces ethene by the dehydration of ethanol over a heated catalyst. What might the catalyst be?

Ethene and propene are the first two members of the homologous series of **alkenes.** All members of this series contain a double carbon–carbon bond, $C\text{=}C$. They have two atoms of hydrogen less than the corresponding alkane and their general formula is C_nH_{2n}. Because they have less than the maximum content of hydrogen they are said to be **unsaturated**.

Physically, alkenes are similar to alkanes, with boiling points generally a little lower (e.g. ethane 185 K, ethene 169 K; propane 231 K, propene 225 K).

Alkenes are named using the same general rules described for alkanes in chapter 26. The suffix **-ene** is used instead of -ane, together with a number indicating the position of the double bond in the chain. Thus the molecule $CH_3CH\text{=}CHCH_3$ is named but-2-ene. Note that although the double bond joins carbon atoms 2 and 3, the number 2 is used because this is the lower. Table 27.1 gives the formulas and names of some more alkenes.

Table 27.1 Nomenclature of alkenes

Formula	Name
$CH_3CH_2CH_2CH\text{=}CH_2$	pent-1-ene
$CH_3CH_2CH\text{=}CHCH_3$	pent-2-ene
$CH_3C\text{=}CHCH_3$ $\quad\vert$ $\quad CH_3$	2-methylbut-2-ene
$CH_2\text{=}CHCH_2CH\text{=}CH_2$	penta-1,4-diene
⬡ (cyclohexene ring with double bond)	cyclohexene

Name these alkenes:

(a) CH_2=$CHCH_2CH_3$

(b) CH_2=C—CH_3
       |
      CH_3

(c)

2 Write structural formulas for these alkenes:
(a) hex-2-ene
(b) buta-1,3-diene
(c) 2,3-dimethylbut-2-ene.

27.3 The nature of the double bond

When ethene is bubbled through bromine, the bromine is decolorised. No sunlight is needed. A colourless liquid, immiscible with water, is formed, but no hydrogen bromide is produced. This reaction is clearly different in nature to the reaction of bromine with alkanes such as ethane, in which sunlight is needed to make the reaction occur and hydrogen bromide is produced. The reaction between ethene and bromine is an **addition** reaction. One of the two carbon–carbon bonds breaks, enabling bonds to be formed to bromine:

1,2-dibromoethane
(a colourless liquid)

Addition reactions are characteristic of all alkenes, which are much more reactive than alkanes. This reactivity is perhaps a little surprising at first: we might expect a double bond to be stronger than a single one, and therefore more stable.

Table 27.2
Bond energies and bond lengths of C—C, C=C, and C≡C

	C—C (in ethane)	C=C (in ethene)	C≡C (in ethyne)
bond energy/kJ mol^{-1}	346	598	813
bond length/nm	0.154	0.134	0.121

Look at the bond energies in table 27.2. You can see that the bond energy of a double bond is greater than that of a single bond, though not twice as great. This suggests that the two bonds in C=C may not be identical. In fact, two kinds of covalent bond are involved.

(a) A bond situated symmetrically between the two carbon atoms, formed by the overlap of two orbitals. This is called a **sigma (σ) bond** (see figure 27.1(a)).
(b) A bond formed by the 'sideways' overlap of two 2p orbitals. Because each p orbital has two lobes, this bond has two regions, one above and one below the plane of the molecule. It is called a **pi (π) bond** (see figure 27.1(b)).

The two electrons of the π-bond are not situated axially between the carbon atoms. This means they are not 'on average' as close to the nuclei of the atoms as the electron pair in the σ-bond. Therefore they do not attract the nuclei so strongly, so the π-bond is not as strong as the σ-bond. Nevertheless, the π-bond and σ-bonds together are stronger than the single σ-bond that links the carbon atoms in ethane. Consequently, the C=C bond in ethene is stronger and shorter than the C—C bond in ethane.

If double bonds are stronger than single bonds, why are alkenes more reactive than alkanes? When bromine adds across the double bond in ethene the π-bond is broken: this requires energy. The energy used in breaking the one π-bond, however, is more than repaid by the energy released when two new bonds are made to bromine atoms. Hence alkenes are *energetically* unstable relative to their products in an addition reaction. They are also *kinetically* unstable, because the high electron density in the double bond tends to attract electron-deficient groups (electrophiles), thus initiating addition reactions. There is more about the mechanism of addition reactions in section 27.5.

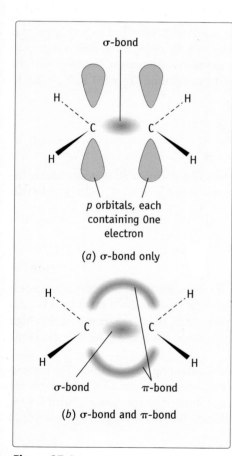

Figure 27.1
σ– and π–bonds in ethene. Note that the plane of the molecule is perpendicular to the page.

27.4 *Cis–trans* isomerism

X-ray diffraction evidence shows that the ethene molecule is planar. This fits in with our ideas on the shapes of molecules (section 8.6), since ethene has no lone pairs. The three groups round each carbon atom are arranged trigonally, at approximately 120° to one another, so the ethene molecule is drawn thus:

$$
\begin{array}{ccc}
\text{H} & & \text{H} \\
& \diagdown \quad \diagup & \\
& \text{C} = \text{C} & \\
& \diagup \quad \diagdown & \\
\text{H} & & \text{H}
\end{array}
$$

Unlike the CH_3 groups in ethane the CH_2 groups in the ethene molecule cannot be rotated about the bond between the carbon atoms. It is possible to rotate about a σ-bond, because this does not affect the orbital overlap. With a double bond, though, rotation would involve breaking the π-bond and this requires more energy than is available at normal temperatures.

Now consider the compound 1,2-dibromoethene, $BrCH{=}CHBr$. This can be made by addition of bromine to ethyne ($HC{\equiv}CH$). The product is a colourless non-polar liquid, boiling point 381 K, melting point 266 K and immiscible with water.

However, another compound is known with the same formula $BrCH{=}CHBr$. This second substance is also a colourless liquid, immiscible with water, but its boiling point is 383 K, its melting point is 220 K and its molecule is quite polar. These two compounds exhibit a kind of isomerism called ***cis–trans*** or **geometric** isomerism. It arises from the lack of free rotation about a double bond. The first compound ($T_b = 381$ K, $T_m = 266$ K, non-polar) has the structure

$$
\begin{array}{ccc}
\text{H} & & \text{Br} \\
& \diagdown \quad \diagup & \\
& \text{C} = \text{C} & \\
& \diagup \quad \diagdown & \\
\text{Br} & & \text{H}
\end{array}
$$

It is called *trans*-1,2-dibromoethene because the bromine atoms are on opposite sides of the double bond (*trans*, Latin: opposite). The second compound ($T_b = 383$ K, $T_m = 220$ K, polar) has the structure

$$
\begin{array}{ccc}
\text{H} & & \text{H} \\
& \diagdown \quad \diagup & \\
& \text{C} = \text{C} & \\
& \diagup \quad \diagdown & \\
\text{Br} & & \text{Br}
\end{array}
$$

It is called *cis*-1,2-dibromoethene (*cis*, Latin: on the same side), both bromine atoms being on the same side of the double bond. Since free rotation about the double bond is prevented, these isomers cannot readily be interconverted, unless the temperature is high enough to make available the energy needed to break the π-bond. Figure 27.2 shows the three-dimensional shapes of the two isomers.

Cis–trans isomerism is common in compounds containing double bonds. The isomers normally have similar chemical properties but often their physical properties are markedly different. *Cis–trans* isomerism is not limited to compounds with C=C double bonds. It can arise wherever rotation about a bond is restricted, for example in ring compounds.

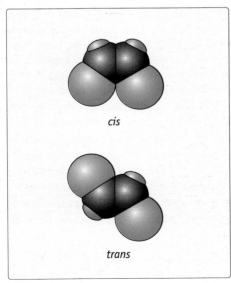

cis

trans

Figure 27.2
Computer-generated models showing three-dimensional shapes of *cis*- and *trans*-dibromoethene.

1 Why is *cis*-1,2-dibromoethene polar while its *trans* isomer is not?

2 Suggest a reason for the large difference in melting point between the two isomers.

3 Would you expect but-2-ene, $CH_3CH{=}CHCH_3$, to have *cis* and *trans* isomers? If so, write their structures.

4 Why are
$$
\begin{array}{cc}
\text{H} & \text{Br} \\
| & | \\
\text{H—C—C—H} \\
| & | \\
\text{Br} & \text{H}
\end{array}
$$
and
$$
\begin{array}{cc}
\text{H} & \text{H} \\
| & | \\
\text{H—C—C—H} \\
| & | \\
\text{Br} & \text{Br}
\end{array}
$$
not considered to be isomers?

27.5 Mechanism of addition to a double bond

Ethene and bromine undergo an addition reaction to form 1,2-dibromoethane. The reaction occurs in the dark at room temperature, which suggests that a free-radical mechanism like that described in section 26.6 is *not* involved. The mechanism involves heterolytic rather than homolytic fission.

The ethene molecule has a region of high electron density caused by the π-electrons in its double bond. As the bromine molecule approaches the ethene, it becomes polarised by this negative charge:

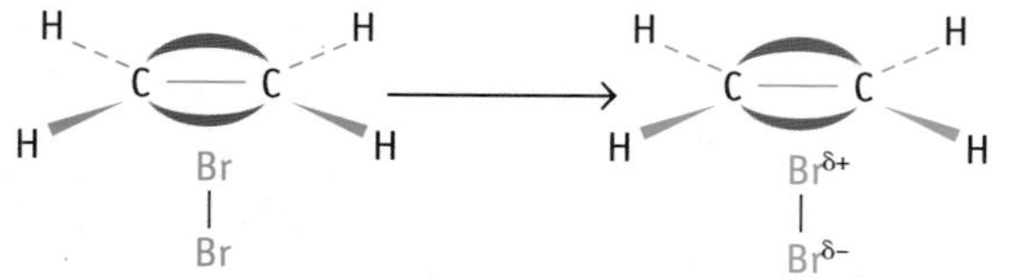

A loose association forms between the ethene and bromine molecules. Negative charge moves from the double bond towards the positively charged bromine atom. At the same time, electrons in the Br—Br bond are repelled towards the negatively charged Br atom. The result is the formation of a C—Br bond and the production of two ions, a carbocation and a Br⁻ ion:

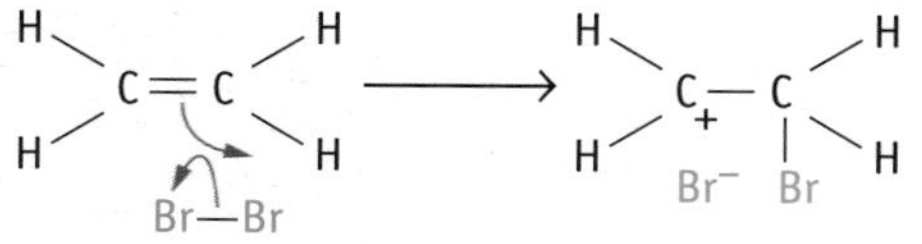

Notice the curly arrows in this equation. They are used to show the movement of electron pairs (section 26.5).

The carbocation is very unstable and quickly combines with the Br⁻ to form 1,2-dibromoethane:

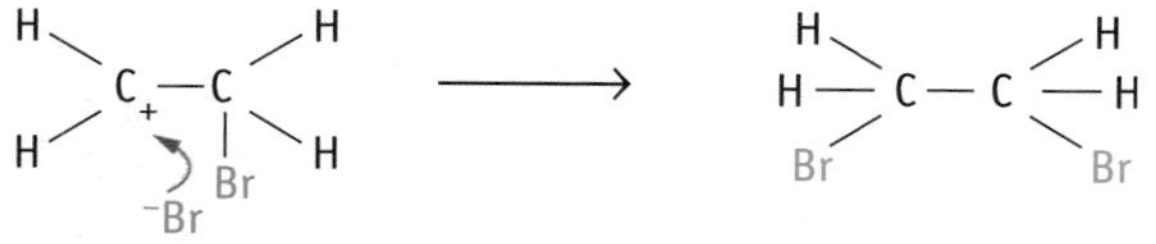

This mechanism is accepted by most chemists, though the exact structure of the intermediate carbocation is not altogether clear. The mechanism is described as **electrophilic addition** because the Br_2 molecule acts as an electrophile when it is attracted to the electron-rich double bond.

Evidence for this mechanism comes from observing what happens when the reaction is carried out in the presence of Cl⁻ ions. When ethene reacts with bromine in the presence of Cl⁻ ions, it is found that 1-chloro-2-bromoethane, $ClCH_2$—CH_2Br, is formed as well as 1,2-dibromoethane. This suggests that the intermediate carbocation has indeed been formed and has reacted with Cl⁻ as well as Br⁻ ions.

1 What product would you expect from an addition reaction between ethene and hydrogen chloride molecules? Write its formula.

2 In the reaction between ethene and HCl, which end of the HCl molecule would you expect to attack the double bond initially?

3 What would be the structure of the intermediate carbocation in this reaction?

27.6 Important reactions of alkenes

Nearly all the important reactions of alkenes are addition reactions. Many have mechanisms similar to that described for the reaction with bromine. The product of an addition reaction can often be predicted. You need to look carefully at the molecule which is being added to ethene. Try to decide how it will split into a positive and negative part. These two parts will then join on either side of the double bond.

> ▣ Predict the structure of the products of additional reactions between ethene and:
>
> (a) H_2O (b) $HOCl$ (c) H_2 (d) H_2SO_4
>
> *Hint*: think of H_2SO_4 as
>
>
> You can check your answers below.

The reactions given below all involve ethene, which is industrially the most important alkene. The reactions of ethene are in any case typical of alkenes in general, for the alkenes are a well-graded homologous series.

Many of these reactions can be carried out in the laboratory, though some require special conditions normally only available on an industrial scale. Industrially, ethene is produced by cracking light hydrocarbon fractions. In the laboratory it can be conveniently produced by the dehydration of ethanol (see section 30.7).

Reaction with halogens

We have already seen that bromine reacts with ethene under normal conditions to form 1,2-dibromoethane. This compound is added to gasoline as a scavenger for lead (see section 26.7). As you might expect, chlorine also reacts with ethene:

$$CH_2{=}CH_2 + Cl_2 \rightarrow CH_2Cl{-}CH_2Cl$$
$$\text{1,2-dichloroethane}$$

The reaction occurs under ordinary conditions. The product, 1,2-dichloroethane, is used to manufacture chloroethene ('vinyl chloride'), from which PVC is made.

Fluorine reacts explosively with ethene, but the reaction with iodine is rather slow.

Decolorisation of bromine water is a useful test-tube reaction to detect a double bond.

Reaction with hydrogen

Hydrogen and ethene do not react under normal conditions, but in the presence of a finely divided metal catalyst, usually nickel at about 140°C, ethane is produced:

$$H_2C{=}CH_2 + H_2 \xrightarrow{\text{Ni}} CH_3CH_3$$
$$\text{ethane}$$

This reaction is useful when analysing organic compounds. By measuring the number of moles of hydrogen absorbed by one mole of a hydrocarbon, the number of double (or triple) bonds in its molecule can be found.

Hydrogenation of double bonds is used to convert edible oils into margarine. Vegetable oils such as palm oil consist of esters of long-chain carboxylic acids (see section 32.6) which contain double bonds. Treatment of these oils with hydrogen in the presence of a nickel catalyst saturates the carbon chains. This raises the melting point of the oil so that it becomes solid, i.e. a fat, at room temperature. By controlling the degree of hydrogenation, the margarine can be made as soft or hard as required.

There is now considerable medical evidence that saturated fats, i.e. those containing few double bonds, are dangerous to health. In particular, they are thought to contribute to heart and circulatory disease. Animal fats such as cream, butter and pork fat tend to have a high proportion of saturated fats as well as containing cholesterol, which is also thought to be dangerous in large amounts. Vegetable fats, including soft margarine made from vegetable oil, contain more unsaturated fats and less cholesterol. They are therefore thought to be healthier.

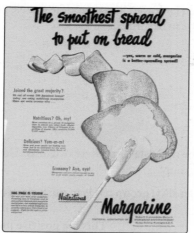

► In 1948, margarine was something of a novelty.

1. Predict the formula of the product of the reaction between propene and chlorine.
2. Predict the formula of the product of the reaction between but-2-ene and hydrogen.

Reaction with hydrogen halides

Ethene reacts with concentrated aqueous solutions of hydrogen halides in the cold:

$$CH_2 \!\!=\!\! CH_2 + HX \rightarrow CH_3CH_2X$$

Hydrogen ions act as electrophiles, attacking the double bond and forming an intermediate carbocation:

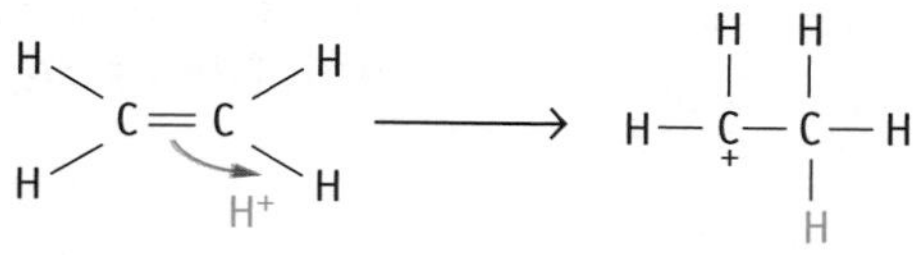

This ion then reacts with halide ions to form the product:

The reaction of hydrogen chloride with ethene produces chloroethane. When reacted with a sodium–lead alloy chloroethane produces tetraethyllead(IV), the anti-knock additive for leaded gasoline.

When propene reacts with a hydrogen halide such as HCl, there are two possible products:

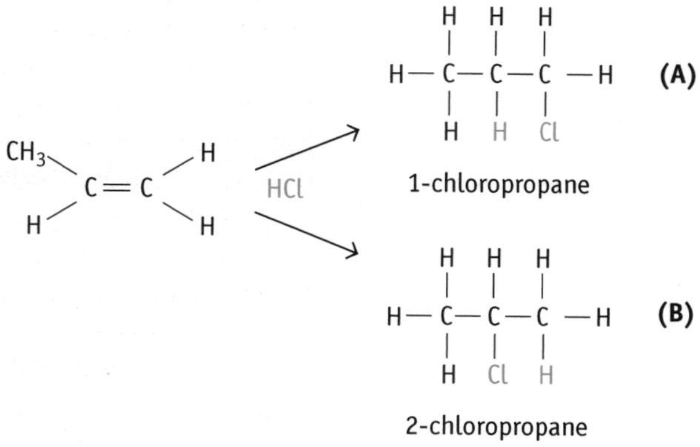

(A) 1-chloropropane

(B) 2-chloropropane

1. When 2-methylpropene

   $(CH_3\!\!-\!\!\overset{\displaystyle |}{\underset{\displaystyle CH_3}{C}}\!\!=\!\!CH_2)$ reacts with HCl,

   what are the structures of the two possible intermediate carbocations?

2. Which of these two ions is the more stable?

3. What will be the major product of the reaction between 2–methylpropene and HCl?

When this reaction is actually carried out, it is found that much more **B** is formed than **A**. We can find a reason for this by looking at the mechanism of the reaction.

When hydrogen ions attack the double bond in propene, two different carbocations, **C** and **D**, can be formed:

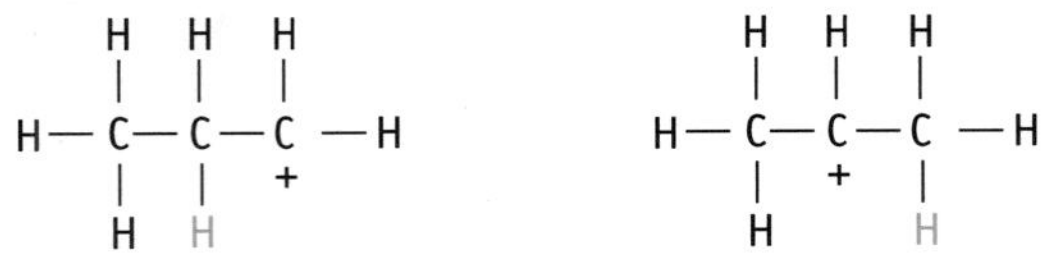

(C)   (D)

D is the more stable of these two ions. This is because of a very important property of an alkyl group:

> **an alkyl group tends to push electrons slightly towards any carbon atom to which it is attached.**
> **This is a positive inductive effect (see box).**

In ion **D** there are two methyl groups pushing electrons onto the positively charged carbon atom; in **C** there is only one ethyl group doing so. As a result the positive charge is stabilised slightly more in **D** than in **C**, because the donated electrons tend to cancel out the charge. **D** is therefore the more stable of the two possible intermediate carbocations, though still very unstable. It therefore tends to persist longer, making it more likely to combine with Cl⁻ to form the product **B**. This means that **B** is the major product, though a certain amount of **A** is also formed.

The major products of asymmetric addition reactions can be predicted by a useful general rule, known as **Markovnikoffs Rule**. It says that:

> **when a molecule HA adds to an asymmetric alkene, the major product is the one in which the hydrogen atom attaches itself to the carbon atom already carrying the larger number of hydrogen atoms.**

Thus in the case of 2-methylpropene,

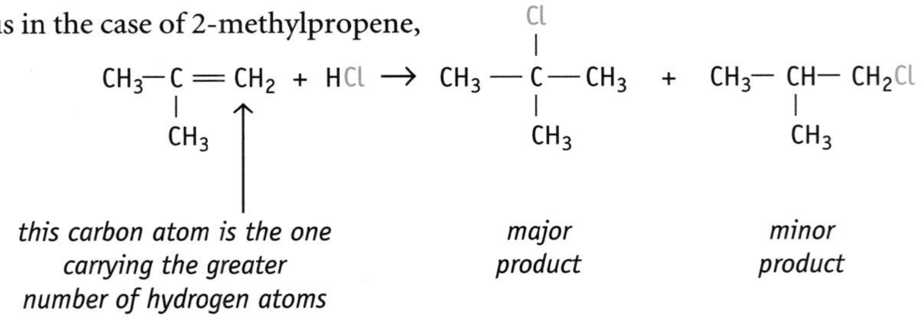

$$CH_3\!-\!\underset{\underset{\displaystyle CH_3}{|}}{C}\!=\!CH_2 + HCl \longrightarrow CH_3\!-\!\underset{\underset{\displaystyle CH_3}{|}}{\overset{\overset{\displaystyle Cl}{|}}{C}}\!-\!CH_3 + CH_3\!-\!\underset{\underset{\displaystyle CH_3}{|}}{CH}\!-\!CH_2Cl$$

this carbon atom is the one major minor
carrying the greater product product
number of hydrogen atoms

1. Predict the major products of reactions between

   (a) $CH_3CH_2CH\!=\!CH_2$ and HBr
   (b) $CH_3\underset{\underset{\displaystyle CH_3}{|}}{C}\!=\!CHCH_3$ and HI

Reaction with sulphuric acid: hydration

Sulphuric acid is another strong acid with which ethene undergoes addition. Concentrated sulphuric acid reacts with ethene in the cold:

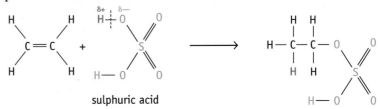

sulphuric acid

ethyl hydrogensulphate

When added to water and warmed, ethyl hydrogensulphate is converted to ethanol. This is **hydrolysis**: breaking up a compound by reacting it with water.

ethyl hydrogensulphate   ethanol   sulphuric acid

The overall effect of these two reactions is to combine ethene with water to form ethanol. The sulphuric acid is regenerated. At one time this was the most important method of manufacturing ethanol from ethene. Nowadays most ethanol is manufactured by the direct catalytic hydration of ethene in the vapour phase:

$$CH_2{=}CH_2(g) + H_2O(g) \xrightarrow[330°C,\ 60\ atm]{H_3PO_4} CH_3CH_2OH(g)$$
ethanol

The reaction with sulphuric acid is still used industrially to produce propan-2-ol from propene.

Reaction with potassium manganate(VII)

Ethene will decolorise acidified dilute potassium manganate(VII) (potassium permanganate). This reaction is a useful test for a double bond.

The reaction with manganate(VII) is complicated and involves both addition and oxidation. The product is ethane-1,2-diol, $HOCH_2CH_2OH$, which is itself further oxidised if excess manganate(VII) is present.

The equation for this reaction, like all redox reactions in organic chemistry, can be written using the half-equation method described in section 3.4:

$$CH_2{=}CH_2 + 2H_2O \rightarrow HOCH_2CH_2OH + 2H^+ + 2e^- \qquad (1)$$
ethane-1,2-diol

$$MnO_4^- + 8H^+ + 5e^- \rightarrow Mn^{2+} + 4H_2O \qquad (2)$$

Multiplying (1) by 5 and (2) by 2 and adding, we get

$$5CH_2{=}CH_2 + 2H_2O + 2MnO_4^- + 6H^+ \rightarrow 5HOCH_2CH_2OH + 2Mn^{2+}$$

Potassium manganate(VII) is much too expensive for this process to be useful industrially for making ethane-1,2-diol. In industry this important compound is made from epoxyethane (see section 30.9).

Reaction with oxygen

Like all hydrocarbons, ethene burns in air. This is an unimportant reaction – it is a waste of such a useful compound.

In the presence of a finely divided silver catalyst at 180°C, ethene can be made to combine with oxygen much more fruitfully, forming epoxyethane:

epoxyethane

Epoxyethane, with its strained three-membered ring, is unstable and can be readily converted to a number of useful products. There is more about its reactions in section 30.9.

Reaction with itself

Perhaps the most important industrial reaction of ethene is with itself, to form poly(ethene). This is described in section 27.8.

All the reactions above are applicable to alkenes in general, not just to ethene.

Predict the outcome of the following reactions:

1. $CH_3CH=CH_2 + H_2$ over Ni catalyst
2. $CH_3CH=CHCH_3 + H_2SO_4$ (concentrated)
3. excess $CH_3CH=CH_2 + KMnO_4$ in acid
4. $CH_3CH=CHCH_3 + Br_2$

27.7 Polymers

About 80% of the world's output of organic chemicals is used to make polymers. The most important synthetic polymers are addition polymers, described in the next section. In this section we will look at the general properties of polymers.

The use of polymers by humans is not new: wood, cotton, wool and rubber are all naturally occurring polymeric materials. However, the manufacture and use of *synthetic* polymers has only really developed since the Second World War. Since then their use has increased very rapidly and they are steadily replacing traditional natural materials. Synthetic polymers are usually cheaper than natural materials. They are often better suited to their particular function, since chemists are now able to produce a polymer to suit most specifications.

One drawback of the increased use of synthetic polymers is the problem of their disposal. Synthetic polymer molecules, not being of natural origin, cannot be broken down by the enzymes in bacteria. They are said to be **non-biodegradable**. What is more, polymers are difficult to recycle economically because of the difficulty of separating the different types of polymer in plastic waste (see section 19.16). For these reasons their disposal presents a considerable problem: plastic containers are among the worst forms of modern litter.

Polymers are long-chain molecules made by joining together many small molecules. The small molecules from which the polymer is built are called **monomers**. Polymers may be natural or synthetic. Table 27.3 gives some examples of both types. Natural polymers include many important materials in common use. Many biochemicals are polymeric, including the proteins in your hair, the DNA in your genes and the starch in your food. The structures of natural polymers are considered in other sections of this book (proteins, sections 9.8(D) and 33.8; nucleic acids, section 9.8(D); carbohydrates, sections 9.8(D), 31.7 and 31.8).

Polymers vary widely in physical properties such as strength, flexibility and softening temperature. Some, such as polyesters, are very strong and not readily stretched. They are therefore suitable for use as **fibres**. Others, such as polythene, are more easily deformed and are classed as **plastics**.

Table 27.3
Examples of natural and synthetic polymers

	Polymer	Monomer	Where you find it
Natural	protein	amino acids	wool, silk, muscle, etc.
	starch	glucose	potato, wheat, etc.
	cellulose	glucose	paper, wood, dietary fibre
	DNA	nucleotides	chromosomes, genes
Synthetic	poly(ethene)	ethene	bags, washing-up bowls, etc.
	poly(chloroethene) (PVC)	chloroethene	fabric coatings, electrical insulation, etc.
	poly(phenylethene) (polystyrene)	phenylethene	toys, expanded polystyrene
	polyester	ethane-1,2-diol and benzene-1,2-dicarboxylic acid	skirts, shirts, trousers

▲ Recycling PVC. This woman separates fragments of PVC from other plastics by tipping them into water, The denser PVC sinks; the other plastics float.

Properties of polymers

The properties of a polymer depend on the properties of the molecular chains it contains. Four factors are particularly important.

(i) Chain length

The characteristic polymer properties of flexibility and tensile strength arise from the chain structure of polymer molecules. Strength and melting point increase with chain length. Typical polymer properties become apparent for average chain lengths of 50 monomer units and upwards. Mechanical strength increases up to about 500 units, after which it changes only slightly. It is necessary to talk in terms of *average* chain lengths where polymers are concerned, since individual chains vary in length even in a pure sample of polymer.

(ii) Intermolecular forces

If the intermolecular forces between chains are high, the polymer will tend to be strong and difficult to melt. The size of the intermolecular forces depends on the nature of the side groups on the chain.

(iii) Branching

Highly branched polymer chains cannot pack together so regularly as straight chains. Highly branched polymers therefore tend to have lower tensile strength and to melt more easily (see figure 27.3). In most polymers there are regions in which the chains are regularly packed. These are called **crystalline** regions. The regions in which packing is irregular are called non-crystalline regions.

(iv) Cross-linking

Some polymers have extensive cross-linking between chains. This forms a rigid network, making a hard material (figure 27.3). The more cross-linking, the more rigid and brittle the polymer. Such polymers are called *thermosetting* (see section 30.8).

Forces between polymer molecules are considered further in section 9.4.

(a) Polymer with few branched chains, e.g. high-density polythene.
Chains regularly packed, with extensive crystalline regions. High tensile strength, high melting point.

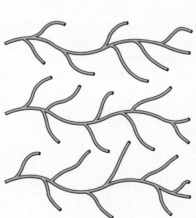

(b) Polymer with many branched chains, e.g. low-density polythene.
Chains irregularly packed: largely amorphous with few crystalline regions. Lower tensile strength, lower melting point.

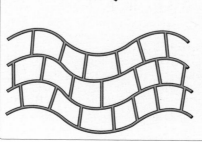

(c) Polymer with much cross-linking, e.g. bakelite. Rigid and hard, fairly brittle.

Figure 27.3
Physical properties of polymers.

Addition and condensation polymerisation

When a polymer forms, the monomer molecules join together. This joining together can happen in two ways: by addition and by condensation.

(a) *Addition polymerisation* involves **addition** reactions between monomer molecules. Usually only a single type of monomer molecule is involved. Most addition polymers are synthetic: for example, ethene forms an important synthetic polymer, poly(ethene). This type of polymerisation is considered further in the next section.

(b) *Condensation polymerisation* involves a **condensation** reaction between monomers. In this type of reaction, a small molecule, usually a water molecule, is eliminated from between two larger molecules. In the process a bond is formed between the two larger molecules:

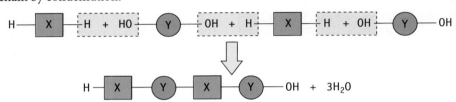

Condensation polymerisation usually involves two different monomers joined in a chain by condensation:

Condensation polymerisation can sometimes be reversed by hydrolysis. The condensation polymer is heated for a long period with aqueous acid or alkali. This reverses the condensation reaction, putting water molecules back into the polymer and re-forming the monomers. Addition polymerisation cannot be reversed in this way. Most natural polymers are condensation polymers, and so are some synthetics. Table 27.4 gives examples of both types. There is more about condensation polymerisation on page 507.

27.8 Addition polymerisation

Molecules containing double bonds are particularly useful monomers as they can usually be made to undergo addition reactions among themselves. The double bond in each molecule can be thought of as breaking open, enabling the free bonds to link with one another forming a chain.

Thus with ethene:

Pure PVC is rigid, because of strong intermolecular forces. It can be made more flexible by adding a plasticiser.

▲ (*a*) Guttering made from unplasticised PVC.

▲ (*b*) Clothing made from plasticised PVC.

Table 27.4
Addition and condensation polymers

Addition	Condensation
polythene	polyester
polystyrene	nylon
PVC	proteins
poly(propene)	starch
PTFE	cellulose
acrylic	nucleic acids

▲ These two bottles, one made from high-density polythene and the other from low-density polythene, were heated in an oven. Which is which?

The product, **poly(ethene)** or **polythene**, is the most commonly used synthetic polymer. Note that the empirical formula of the polymer is the same as that of the monomer, CH_2. This is always the case with addition polymerisation, but is not so with condensation polymerisation. Details of some important addition polymers are now given.

Polythene

Polythene was discovered by ICI in 1933. The method they used to make it involved heating ethene at about 200°C and a pressure of 1200 atm in the presence of traces of oxygen. This is a free-radical reaction, in which oxygen acts as an initiator. Organic peroxides, R—O—O—R, are also used to initiate the reaction (figure 27.4).

The polythene produced in this way has branched chains which cannot pack closely together. It is therefore fairly readily melted and easily deformed. The polymer melts at about 105°C and softens in boiling water. The majority of polythene manufactured today is of this form: it is called **low-density polythene** (LDPE). It is used for making film and sheeting for bags and wrappers and for making moulded articles such as washing-up bowls and 'squeezy' bottles.

Another method of making polythene was developed by Ziegler in the 1950s. This process uses catalysts at low temperatures and pressure (about 60°C and 1 atm). The molecules produced have few branches and can pack closely together. This form of polythene, called **high-density polythene** (HDPE), is therefore more rigid and melts at a higher temperature (about 135°C) than the low-density form. High-density polythene is used for moulding rigid articles such as bleach bottles and milk-bottle crates.

Initiation

A free-radical initiator $\boxed{RO}{}^{\bullet}$ starts the reaction.
($RO^{\bullet}$ is formed by the breakdown of an organic peroxide R—O—O—R.)

Propagation

$\boxed{RO}{}^{\bullet}$ reacts with an ethene molecule . . .

$\boxed{RO}{}^{\bullet} \ CH_2{=}CH_2 \longrightarrow \boxed{RO}—CH_2—CH_2{}^{\bullet}$

. . . the chain grows by reacting with other ethene molecules . . .

$\boxed{RO}—CH_2—CH_2{}^{\bullet} + CH_2{=}CH_2 \longrightarrow \boxed{RO}—CH_2—CH_2—CH_2—CH_2{}^{\bullet}$ etc.

. . . the chains grow fast, at up to 10,000 monomers per second. Sometimes the growing chain 'back-bites' and attacks itself. . .

$$\boxed{RO}—CH_2—CH_2—CH_2—\overset{\bullet}{CH_2}\!\!\overset{\displaystyle {}^{\bullet}CH_2{>}CH_2}{}\text{ etc.} \rightarrow \boxed{RO}—CH_2—CH_2—\overset{\bullet}{CH}—CH_2—CH_2—CH_3$$

. . . the chain now grows from the middle, giving a branch-chain.

Termination

The reaction ends when two radicals react together, e.g.

$\boxed{RO}—CH_2—CH_2—CH_2—CH_2—CH_2—CH_2{}^{\bullet} + \boxed{RO}{}^{\bullet}$

$\longrightarrow \boxed{RO}—CH_2—CH_2—CH_2—CH_2—CH_2—CH_2—\boxed{RO}$

Figure 27.4
Mechanism for the polymerisation of ethene to produce low-density polythene.

Poly(propene) (polypropylene)

Propene, which like ethene is readily available from petroleum, can be polymerised by the Ziegler process:

▲ Polypropene is used to make rope. Unlike rope made from natural fibres, it does not rot and it can be any colour.

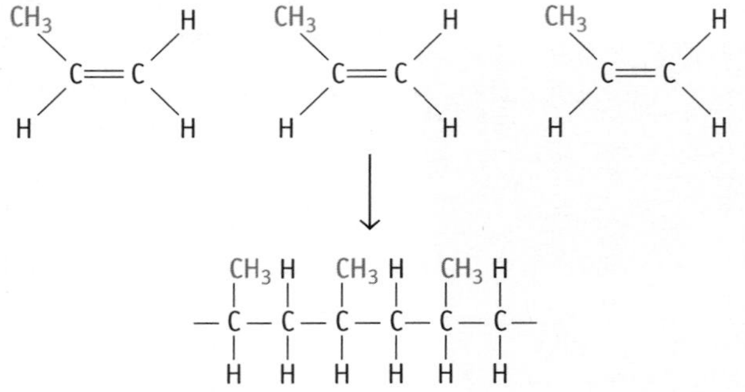

Poly(propene) produced in this way has the —CH$_3$ side groups arranged in a highly regular fashion. Its chains pack together closely, producing a material similar to high-density polythene. It is used in mouldings and film and can be made into a fibre.

Poly(chloroethene) (PVC)

Chloroethene (vinyl chloride) polymerises to form poly(chloroethene) or polyvinyl chloride, PVC:

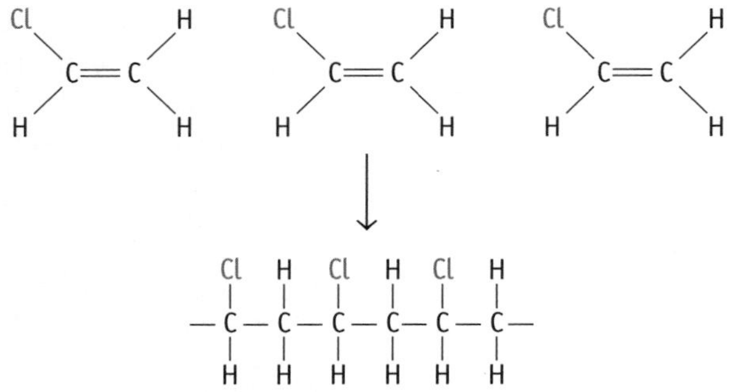

The polar C—Cl bond results in considerable intermolecular attraction between the polymer chains, making PVC a fairly strong material. Its best feature is its versatility. On it own, PVC is a fairly rigid plastic, but additives called plasticisers can be used to make it highly flexible. It is used among other things for coating fabrics and for covering wires and cables.

PVC is a versatile polymer, but its disposal can present problems. Burning waste PVC produces hydrogen chloride gas, which is toxic. It must be removed from the combustion gases before they are released into the atomsphere, by reacting the acidic HCl(g) with a base.

▶ Furniture made from Perspex.

27 Unsaturated Hydrocarbons

Table 27.5
Addition polymers

(d) Other addition polymers

Table 27.5 shows some other addition polymers, their monomers and their uses.

Polymer: systematic name	poly(phenylethene)	poly(tetrafluoroethene)	poly(methyl-2-methyl-propenoate)	poly(propenenitrile)
Polymer: common name	polystyrene	PTFE, 'Teflon', etc.	'Perspex', acrylic	'Acrilan', etc.
Monomer: systematic name	phenylethene	tetrafluoroethene	methyl 2-methyl-propenoate	propenenitrile
Monomer: formula	$C_6H_5CH{=}CH_2$	$CF_2{=}CF_2$	CH_3 \\ $C{-}COOCH_3$ // CH_2	$CH_2{=}CHCN$
Monomer: common name	styrene	tetrafluoroethylene	methyl methacrylate	acrylonitrile
Properties	brittle (but cheap)	very stable. Low friction, anti-stick properties	transparent	strong, fibre properties
Uses	expanded polystyrene for insulation. Plastic toys, etc.	non-stick coatings on pans. Insulators	as a substitute for glass	making textiles (wool substitute)

Use table 27.5 to answer these questions.

1. Draw a section of the chain structures of poly(phenylethene) and poly(propenenitrile).

2. Suggest a reason why poly(phenylethene) is brittle.

3. Suggest a reason why poly(propenenitrile) is a strong material, suitable for making fibres.

27.9 Rubber – a natural addition polymer

Rubbers are **elastomers** – polymers with elastic properties (figure 27.5).

Many forms of synthetic rubber are now available, many of them cheaper and more useful than natural rubber, but the natural polymer is still much used. Most naturally occurring polymers are condensation polymers, but natural rubber can be regarded as an *addition* polymer. The monomer is 2-methylbuta-1,3-diene (isoprene), whose formula is shown in figure 27.5.

When dienes of this sort undergo addition polymerisation, one of the double bonds is retained in the resulting polymer. The structure of poly(2-methylbuta-1,3-diene) is also shown in figure 27.5.

Because of the double bond, *cis–trans* isomerism is possible. Two isomeric polymers are known – natural rubber (*cis* form) and **gutta-percha** (*trans* form).

The structural difference between rubber and gutta-percha gives them different

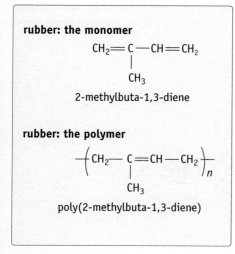

rubber: the monomer

$$CH_2{=}C{-}CH{=}CH_2$$
$$|$$
$$CH_3$$

2-methylbuta-1,3-diene

rubber: the polymer

$$-\left(CH_2{-}C{=}CH{-}CH_2\right)_n$$
$$|$$
$$CH_3$$

poly(2-methylbuta-1,3-diene)

Figure 27.5
Why rubber is elastic.

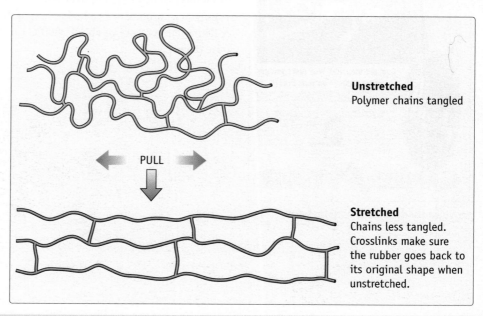

Unstretched
Polymer chains tangled

PULL

Stretched
Chains less tangled. Crosslinks make sure the rubber goes back to its original shape when unstretched.

▲ Tapping latex from a rubber tree.

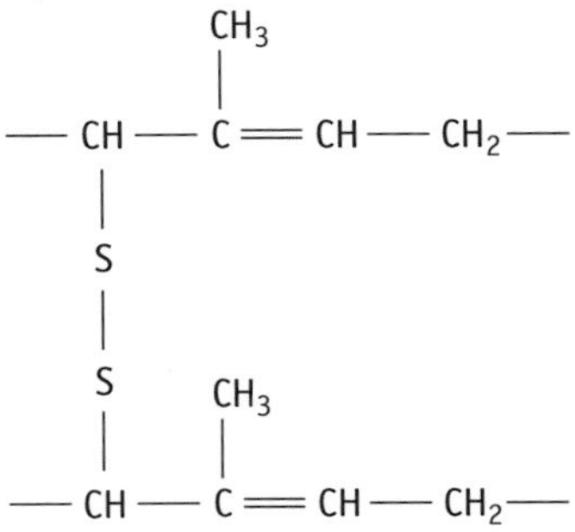

Cross-link between rubber chains

properties. Gutta-percha is much less elastic than rubber. The tough outer cover of golf balls is often made of gutta-percha.

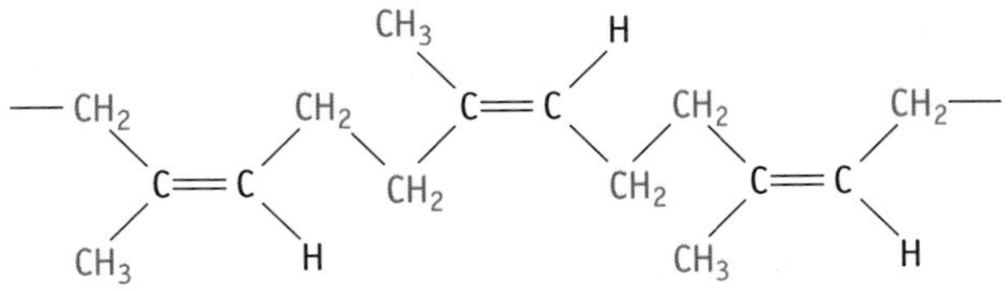

Natural rubber – *cis*

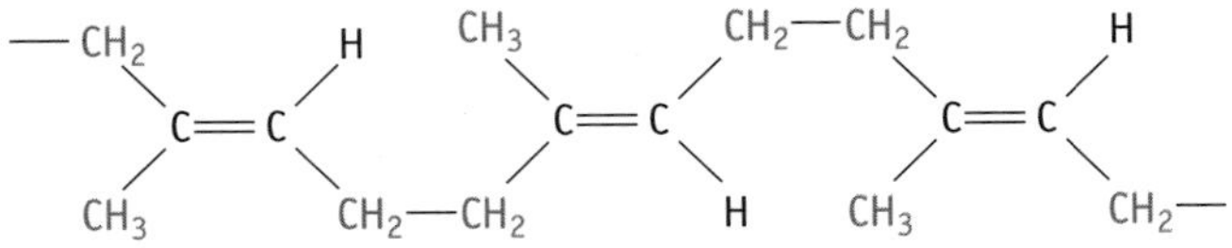

Gutta-percha – *trans*

Natural rubber comes from the Para rubber tree. When the outer bark is cut, a milky liquid called **latex** oozes out. This contains small globules of rubber suspended in water. Some adhesives (e.g. Copydex) are made from latex with ammonia added as a preservative. Raw rubber is readily precipitated from latex, but is sticky and soft. In 1832 Charles Mackintosh used rubber to waterproof cotton coats, but these became sticky in hot weather.

It was not until Charles Goodyear discovered vulcanisation in 1839 that rubber became really useful for footwear, tyres, hoses, foams, etc. When raw rubber is heated with sulphur it becomes harder, tougher and less temperature-sensitive. It is said to be **vulcanised**. Sulphur atoms form cross-links between rubber chains, as shown in the margin.

In ordinary soft rubber about 4% sulphur is used. This gives cross-links between 5–10% of the polymer units. By using more sulphur, harder rubbers can be made.

27.10 Alkynes

Alkynes are hydrocarbons containing triple bonds. The simplest is **ethyne,** commonly called acetylene, H—C≡C—H. Alkynes are named in a similar manner to alkenes, using the suffix **–yne** instead of **–ene.**

Ethyne used to be a very important material for the manufacture of organic chemicals, but it has lost this position to ethene, which is now a great deal cheaper to produce.

Ethyne is important as a fuel in 'oxy-acetylene' torches for cutting and welding metal. A flame temperature of about 2800°C can be obtained by the combustion of ethyne in pure oxygen.

▲ Vulcanised rubber is used for tyres.

Summary

1 Alkenes are unsaturated hydrocarbons containing a carbon–carbon double bond in their molecules. Their general formula is C_nH_{2n} and they are named using the suffix -*ene* together with a number to indicate the position of the double bond.

2 Double bonds consist of a σ-bond situated axially between the carbon atoms and a weaker non-axial π-bond.

3 *Cis* and *trans* isomers differ in the way their substituents are arranged about a double bond.

4 The characteristic reaction of alkenes is electrophilic addition. In general:

$$
\begin{array}{c}
\diagdown \\
\diagup
\end{array} C = C
\begin{array}{c}
\diagup \\
\diagdown
\end{array}
+ \;\; X-Y \longrightarrow
\begin{array}{cc}
X & Y \\
| & | \\
-C & -C- \\
| & |
\end{array}
$$

5 The mechanism of electrophilic addition involves initial attack on the π-electrons, often leading to the formation of an intermediate carbonium ion.

6 Some important reactions of a double bond are shown in figure 28.5.

7 There are two types of polymerisation: addition polymerisation and condensation polymerisation.

8 The properties of polymers are determined by the properties of the chains they contain, in particular chain length, branching and cross-linking.

9 Many compounds containing double bonds readily undergo addition polymerisation. The most important case is ethene, which polymerises to polythene.

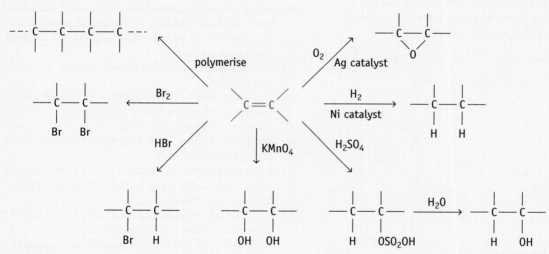

Figure 27.6
Reactions of a double bond.

The importance of alkenes in the synthesis of organic compounds is discussed in chapter 34.

Review questions

1 (a) Name the following compounds:

  (i) $CH_3CH_2CH_2CH{=}CH_2$

  (ii)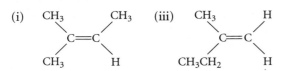

  (iii)

  (iv) $CH_3C{\equiv}CCH_3$

  (v)

(b) Write structural formulas for these compounds:
  (i) propene
  (ii) *cis*-pent-2-ene
  (iii) cycloocta–1,3,5,7-tetraene
  (v) 2-methylpent-2-ene.

2 1,1-Dichloroethene, $Cl_2C{=}CCH_2$ readily undergoes addition polymerisation.
(a) What is the systematic name of the polymer?
(b) Write the structure of a section of the polymer chain.
(c) Would you expect the polymer to have a higher or a lower melting point than poly(chloroethene) (PVC)? Explain your answer.

3 (a) For each of the following compounds, say whether you would expect them to show *cis–trans* isomerism. For the compounds that show *cis–trans* isomerism, draw the structures of all the possible isomers.

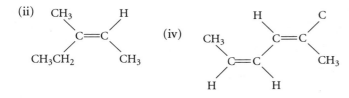

(b) 1,2-Dichlorocyclopropane, CH_2

  ClCH—CHCl

has *cis* and *trans* isomers.
Explain why this is so, and draw the structure of the two isomers.

4 A hydrocarbon **A** contains 87.8% carbon and 12.2% hydrogen by mass. Its relative molecular mass is 82. **A** decolorises bromine water and in the presence of a nickel catalyst it reacts with hydrogen to form **B**. 0.1 g of **A** was found to absorb 27.3 cm^3 of hydrogen (measured at s.t.p.). **B** does not decolorise bromine water.
The molar volume of a gas at s.t.p. is 22.4 dm^3
(a) What is the empirical formula of **A**?
(b) What is the molecular formula of **A**?

(c) How many moles of hydrogen react with 1 mole of **A**?
(d) How many double bonds does **A** have in its molecule?
(e) What is the molecular formula of **B**?
(f) Suggest structural formulas for **A** and **B**.

5 Propan-2-ol is an important alcohol. It is used as a solvent, as a de-icing fluid and for manufacturing propanone (acetone). Propan-2-ol is manufactured as follows: Impure propene, containing traces of ethene, is passed up an absorption tower down which 85% sulphuric acid trickles. The propene reacts to form 2-propyl hydrogensulphate:

$$CH_3CH{=}CH_2 + H_2SO_4 \rightarrow CH_3CHCH_3$$
$$OSO_2OH$$

Virtually no 1-propyl hydrogensulphate is produced, and hardly any of the ethene present as impurity reacts. The 2-propyl hydrogensulphate produced in this way is added to water, when a highly exothermic reaction occurs. 2-Propyl hydrogensulphate is hydrolysed to propan-2-ol:

$$CH_3{-}CH{-}CH_3 + H_2O \rightarrow CH_3{-}CH{-}CH_3 + H_2SO_4$$
$$OSO_2OH \qquad\qquad OH$$

(a) How is propene obtained industrially? Why is it contaminated with ethene?
(b) When propene reacts with sulphuric acid, initial attack is by H^+ on the π-electrons of the double bond. Write the structures of the two possible carbocations formed as a result of this attack.
(c) Which of the two carbocations formed in (b) is the more stable?
(d) Use the answer to (c) to explain why virtually no 1-propyl hydrogensulphate is formed in the reaction of propene with sulphuric acid.
(e) Write the structure of the carbocation formed when ethene is attacked by H^+.
(f) Is the carbocation in (e) more or less stable than the more stable carbocation formed by propene (refer back to your answer to (c))?
(g) Use your answer to (f) to explain why hardly any of the ethene contaminating the propene reacts with sulphuric acid, whereas the propene reacts readily.
(h) Why is it important that 2-propyl hydrogensulphate is virtually the only product of the reaction of the gases with sulphuric acid?

6 (a) How would you distinguish between hex-1-ene and cyclohexane, using simple test-tube reactions?
(b) Predict the structures of the products, if any, of the following reactions.

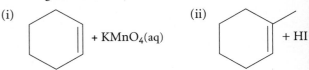

  (iii) $CH_3CH{=}CHCH_3$ reacts with itself, i.e. polymerises.

7 Some bond energies are given below.
C≡C (in ethene) 598 kJ mol^{-1}
C—C (in ethane) 346 kJ mol^{-1}
C—H (general) 413 kJ mol^{-1}
H—H 436 kJ mol^{-1}
Consider the reaction of ethene with hydrogen:

(a) What conditions are needed for this reaction to occur?
(b) How much energy must be supplied to break the π-bonds in a mole of ethene?
(c) How much energy must be supplied to split a mole of hydrogen molecules into hydrogen atoms?
(d) How much energy is released when two moles of new C—H bonds are formed by a mole of ethene?
(e) Calculate the energy change for the reaction of a mole of ethene with a mole of hydrogen.

8 Alkenes such as ethene and propene have been described as the building blocks of the organic chemical industry. Discuss this statement, giving examples. What particular features of the chemistry of alkenes make them suitable for this rôle and why are alkanes less suitable?

9 Consider the following compounds.
A hex-2-ene $CH_3CH_2CH_2CH$=$CHCH_3$
B hex-1-ene, $CH_3CH_2CH_2CH_2CH$=CH_2
C hexane $CH_3CH_2CH_2CH_2CH_2CH_3$
D cyclohexane

The answers to the following questions may be one or more than one of the compounds **A–D**.
(a) Which would decolorise bromine in the absence of sunlight?
(b) Which would react with chlorine, but only when heated or exposed to light?
(c) Which would absorb 1 mole of hydrogen per mole of the compound in the presence of a nickel catalyst?
(d) Which has *cis* and *trans* isomers?
(e) Which are unsaturated?

10 Study figure 27.4 on page 429.
(a) Explain why ethene needs an initiator to make it polymerise.
(b) Compare a polythene molecule with a long-chain alkane molecule.
    (i) In what ways are they similar?
    (ii) In what ways are they different?
(c) Explain how 'back-biting' leads to chain branching.
(d) Give one way that termination of the chain reaction could happen.

28 Aromatic Hydrocarbons

28.1 Aromatic hydrocarbons

In the last chapter we considered unsaturated compounds with double and triple bonds. There is another important class of unsaturated compounds that we need to look at separately because their properties are so different from the alkenes and alkynes.

This class of compounds is called the **aromatic hydrocarbons** and its simplest and most important member is **benzene, C_6H_6**. The name 'aromatic' was originally used because some derivatives of these hydrocarbons have pleasant smells. It is now known that just as many of them smell unpleasant, and in any case many of the aromatic vapours are toxic, so it is unwise to smell them. Benzene itself is particularly toxic, and long-term exposure to benzene may cause certain kinds of cancer. The name aromatic has been retained to indicate certain chemical characteristics rather than odorous properties.

28.2 The structure of benzene

Since 1834 the molecular formula of benzene has been known to be C_6H_6. The exact structural formula, however, posed a problem for many years. With such a high C : H ratio, benzene must clearly be highly unsaturated. A possible structure might be hexatetraene, e.g. $CH_2\!=\!C\!=\!CH\!-\!CH\!=\!C\!=\!CH_2$. Such a structure would be expected to have two isomeric forms of C_6H_5Cl:

$$ClCH\!=\!C\!=\!CH\!-\!CH\!=\!C\!=\!C\!=\!CH_2 \quad \text{and} \quad CH_2\!=\!C\!=\!CCl\!-\!CH\!=\!C\!=\!CH_2.$$

Only one form of chlorobenzene has ever been isolated. The structure of benzene must therefore be one in which all six hydrogen atoms occupy equivalent positions.

The problem was solved by Friedrich August Kekulé in 1865. He proposed a ring structure in which alternate carbon atoms were joined by double bonds:

Here is Kekulé's description of how this structure occurred to him.

'I turned my chair to the fire and dozed. Again the atoms were gambolling before my eyes. This time the smaller groups kept modestly in the background. My mental eye, rendered more acute by repeated visions of this kind, could now distinguish larger structures, of manifold conformation; long rows, sometimes more closely fitted together; all twining and twisting in snakelike motion. But look. What was that? One of the snakes had seized hold of its own tail, and the form whirled mockingly before my eyes. As if by a flash of lightning I awoke.'

This structure, called the Kekulé or cyclohexatriene structure, explains many of the properties of benzene and was accepted for a long time. But even this structure leaves some problems, particularly concerning bond length and thermochemistry.

Bond lengths

X-ray diffraction studies show that benzene is planar: this is to be expected from the Kekulé structure. X-ray diffraction also shows that all the C—C bonds in benzene are the same length:

carbon–carbon bond length in all bonds in benzene, 0.139 nm
carbon–carbon single bond length in cyclohexane, 0.154 nm
carbon–carbon double bond length in cyclohexene, 0.133 nm

The Kekulé model would suggest unequal carbon–carbon lengths, alternating between double and single bond values. In fact we find a constant bond length, somewhere between the value for a single and a double bond.

Thermochemistry of benzene

It is revealing to work out a theoretical value for the enthalpy change of formation of benzene on the basis of the Kekulé model, and to compare it with the experimental value obtained from the enthalpy change of combustion.

The enthalpy change of formation of gaseous benzene is the enthalpy change when a mole of gaseous benzene is formed from its elements:

$$6C(s) + 3H_2(g) \rightarrow C_6H_6(g)$$

Relevant data are:

enthalpy change of atomisation of C(s)	$+715$ kJ (mol of C atoms)$^{-1}$
enthalpy change of atomisation of $H_2(g)$	$+218$ kJ (mol of H atoms)$^{-1}$
bond energy of C=C (average)	610 kJ mol^{-1}
bond energy of C—C (average)	346 kJ mol^{-1}
bond energy of C—H (average)	413 kJ mol^{-1}

1 Work out the enthalpy change of formation of benzene by the following stages.

(a) Calculate the energy needed to produce
  (i) six moles of gaseous carbon atoms from C(s)
  (ii) six moles of gaseous hydrogen atoms from $H_2(g)$
(b) Calculate the energy released when
  (i) three moles of C—C bonds are formed from gaseous atoms
  (ii) three moles of C=C bonds are formed from gaseous atoms
  (iii) six moles of C—H bonds are formed from gaseous atoms.
(c) Use your answers to 1 and 2 to calculate the total energy change when a mole of gaseous benzene is formed from its elements.
(d) Compare your answer with the experimental value of $+82$ kJ mol^{-1}.
(e) Do your results suggest that 'real' benzene is more or less stable than the Kekulé structure?

Using the calculation above, we find that the theoretical enthalpy change of formation of gaseous benzene based on the Kekulé structure is $+252$ kJ mol^{-1}. This is about 170 kJ mol^{-1} greater (i.e. more endothermic) than the experimental value of $+82$ kJ mol^{-1}. This implies that the actual structure of benzene is considerably more stable than the Kekulé structure. The result agrees reasonably closely with the stabilisation energy of benzene obtained using the enthalpy changes of hydrogenation in section 12.11.

Electron delocalisation in benzene

The extra stability of benzene and the fact that its C—C bonds are all of equal length can be explained using the following model, which is currently accepted by most chemists.

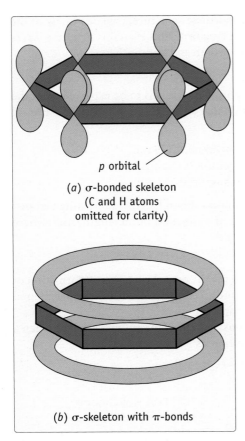

p orbital

(*a*) σ-bonded skeleton
(C and H atoms
omitted for clarity)

(*b*) σ-skeleton with π-bonds

Figure 28.1
Bonding in benzene. Note that the plane of the molecule is perpendicular to the paper.

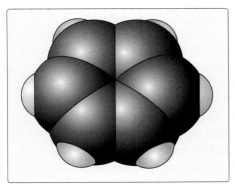

Figure 28.2
Computer-generated model of the benzene molecule. Note the planar, symmetrical shape.

The carbon atoms in the ring are bonded to one another and to their hydrogen atoms by σ-bonds. This leaves one unused *p* orbital on each carbon, each containing a single electron. These *p* orbitals are perpendicular to the plane of the ring, with one lobe above and one below this plane (figure 28.1(*a*)). Each *p* orbital overlaps sideways with the two neighbouring orbitals to form a single π-bond that extends as a ring of charge above and below the plane of the molecule (figure 28.1(*b*)).

The electrons in the π-bond cannot be said to 'belong to' any particular carbon atom. Each electron is free to move throughout the entire π system, so the electrons are said to be **delocalised.** It is this delocalisation that gives benzene its extra stability: any system in which electron delocalisation can occur is stabilised. The reason for this is not hard to see. Electrons tend to repel one another, so a system in which they are as far apart from one another as possible will involve minimum repulsion and will therefore be stabilised.

To conform with this model the structural formula of benzene is nowadays often written as ⬡ rather than ⬡ . As we shall see in section 28.5, the delocalisation of π-electrons has a profound effect on the chemical properties of benzene.

The word **aromatic** is used to describe any system that is stabilised by a ring of delocalised π-electrons. Non-aromatic compounds (such as alkanes and alkenes) are called **aliphatic.**

The structure described above means that the benzene molecule is planar, symmetrical and non-polar (figure 28.2). Its lack of polarity results in benzene being a liquid at room temperature and makes it immiscible with water. Its boiling point is 80°C and its melting point 6°C. The surprisingly high melting point is due to the ease with which highly symmetrical benzene rings can pack into a crystal lattice. Compare it with that of the structurally similar but less symmetrical compound methylbenzene $C_6H_5CH_3$, which melts at −95°C.

28.3 Naming aromatic compounds

Benzene is not the only aromatic hydrocarbon. Many others exist, either substituted forms of benzene or compounds containing different ring systems. Some other aromatic hydrocarbons are discussed in section 28.9. Aromatic hydrocarbons are sometimes known as **arenes.**

The name 'benzene' comes from *gum benzoin*, a natural product containing benzene derivatives. The name *phene*, derived from the Greek *pheno* 'I bear light', was at one time suggested as an alternative to 'benzene' (benzene was originally isolated from illuminating gas by Michael Faraday). It was not adopted, but it survives in the word **phenyl** used for the C_6H_5— group. The phenyl group is an example of an **aryl** group.

The hydrogen atoms in a benzene ring can be substituted by other atoms and groups, as we shall see in section 28.5. Some examples are given in table 28.1.

Table 28.1
Some derivatives of benzene

Substituent group	Systematic name	Other name
methyl —CH_3	methylbenzene	toluene
chloro —Cl	chlorobenzene	—
nitro —NO_2	nitrobenzene	—
hydroxy —OH	phenol	—
amino —NH_2	phenylamine	aniline
carboxylic acid —COOH	benzenecarboxylic acid	benzoic acid

Where more than one hydrogen atom is substituted, numbers are used to indicate which of the six possible ring positions are concerned. Thus:

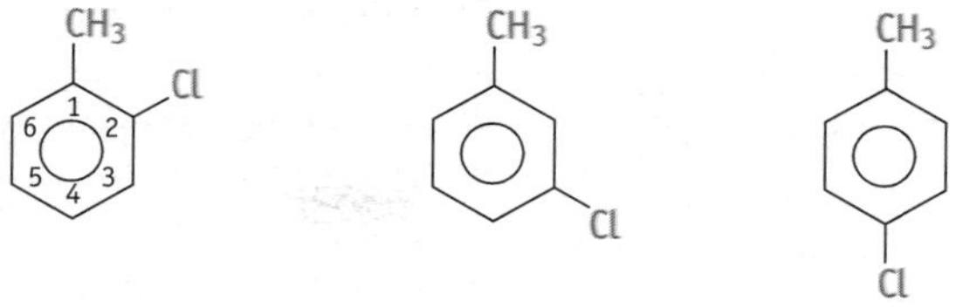

methyl-2-chlorobenzene   methyl-3-chlorobenzene   methyl-4-chlorobenzene

The methyl group is regarded as occupying the 1 position and the ring is numbered clockwise as shown in the first formula. Two more examples are

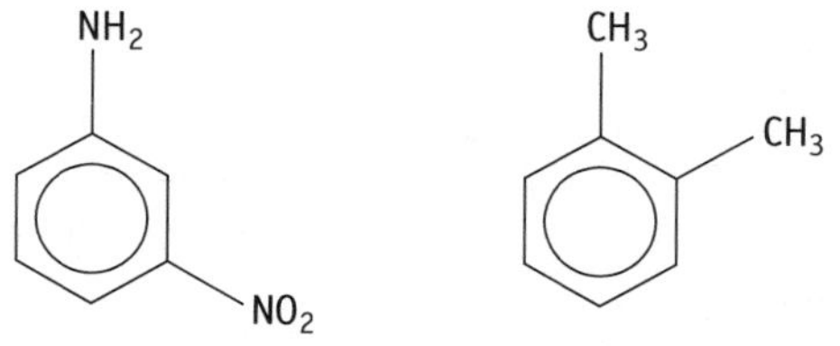

3-nitrophenylamine   1,2-dimethylbenzene

Note that other numbers could be used (e.g. 3,4-dimethylbenzene instead of 1,2-dimethylbenzene), but the numbers actually employed are the lowest ones possible.

1 Use these rules and table 28.1 to write the structural formulas of the following compounds:

(a) ethylbenzene
(b) 2-methylphenol
(c) 1,3-dinitrobenzene.

2 Name the following:

(a) Br

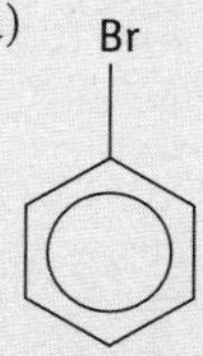

(b) COOH, CH₃

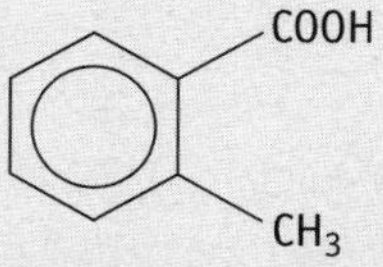

(c) CH₃, O₂N, NO₂, NO₂

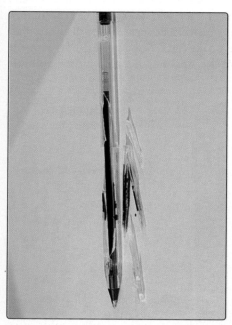

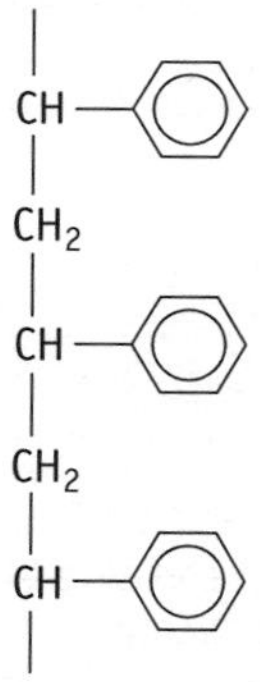

▲ Polystyrene contains alkane chains with benzene rings as side groups. The bulky benzene rings make it difficult for the polymer chains to slide over each other so polystyrene is brittle, not flexible like polythene.

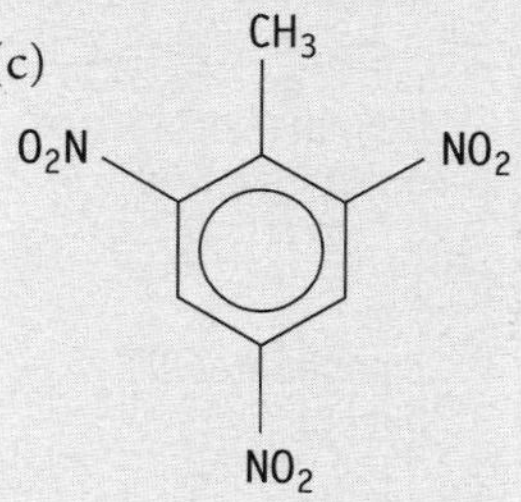

You may come across the terms *ortho*, *meta* and *para* in the names of disubstituted benzene derivatives, particularly in older books. Under this system, 1,2 derivatives are given the prefix *ortho-*, 1,3 derivatives are *meta-* and 1,4 derivatives are *para-*. Thus the three methylchlorobenzenes shown on the previous page become *ortho*-methylchlorobenzene, *meta*-methylchlorobenzene and *para*-methylchlorobenzene, respectively.

28.4 The importance of benzene

Detergents, polystyrene, nylon and insecticides can all be made from benzene, which is industrially the most important arene. Most benzene is normally manufactured from oil by **catalytic reforming**. In the presence of a catalyst, C_6–C_8 hydrocarbons from the gasoline fraction rearrange their molecules, producing a variety of aromatic hydrocarbons, including benzene. In the past, benzene was produced as a by-product of the destructive distillation of coal. Figure 28.3 shows a small-scale version of this process which can be carried out in the laboratory.

Coal is heated in the absence of air and the gases produced are bubbled through water. Coal tar condenses and floats on the water, leaving a gaseous fuel called coal gas. The solid residue is coke. On an industrial scale this process was once the basis of the gas industry in Britain, and also produced coke for boilers and for steel-making. The tar obtained contains a wide range of aromatic compounds which can be separated by distillation. At one time all aromatic hydrocarbons were produced from coal tar, but with the development of catalytic reforming and the decline of coal, most aromatic hydrocarbons now come from oil.

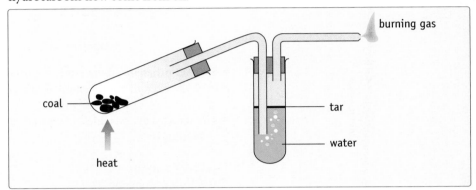

Figure 28.3
The destructive distillation of coal. Aromatic compounds such as benzene are present in coal tar and coal gas.

28.5 Chemical characteristics of benzene

Table 28.2 compares some reactions of cyclohexane, cyclohexene and benzene.

Table 28.2
Some reactions of cyclohexane, cyclohexene and benzene

Reagent	cyclohexane	cyclohexene	benzene
bromine (in dark)	no reaction	bromine decolorised, no HBr evolved	no reaction with bromine alone. In presence of iron filings, bromine decolorised and HBr fumes evolved
acidified potassium manganate(VII) (potassium permanganate)	no reaction	manganate(VII) decolorised	no reaction
hydrogen over very finely divided nickel catalyst	no reaction	one mole absorbs one mole of hydrogen at room temperature	one mole absorbs three moles of hydrogen at 150°C
mixture of concentrated nitric and concentrated sulphuric acids	no reaction	oxidised	substitution reaction: yellow oil formed

1 Does the evidence suggest that (a) cyclohexene, (b) benzene undergo addition with bromine in the dark?
2 Does benzene undergo (i) addition, (ii) substitution with bromine in the presence of iron filings?
3 Does benzene undergo catalytic hydrogenation as readily as cyclohexene?

As table 28.2 shows, benzene undergoes addition reactions far less readily than we might expect for such an unsaturated compound. In fact, *substitution* reactions are more characteristic of benzene than addition reactions. You can see this by looking at the reaction of benzene with bromine in the presence of iron filings. HBr is evolved, which implies a substitution reaction.

If we remember the delocalisation of π-electrons in benzene, it is quite easy to see why addition reactions are difficult. For example, if a molecule of benzene underwent addition with a molecule of bromine, the ring of delocalised π-electrons would be broken:

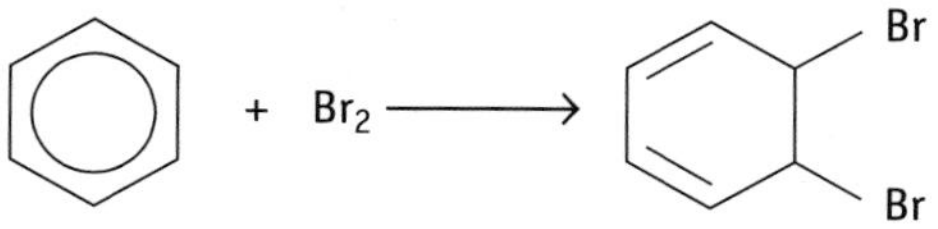

This would require considerably more energy than is needed to break the one double bond in cyclohexene. In its reactions, therefore, benzene has a tendency to maintain its π-electron ring intact and to undergo substitution rather than addition (see review question 10 at the end of this chapter).

28.6 Mechanism of substitution reactions of benzene

The substitution reactions of benzene involve **electrophilic substitution.** The high density of negative charge in the delocalised electron system of the benzene ring tends to attract electrophiles.

Consider as an example the nitration of benzene. Benzene reacts with a mixture of concentrated nitric acid and concentrated sulphuric acid (called a **nitrating mixture**) at 50°C. The product is nitrobenzene:

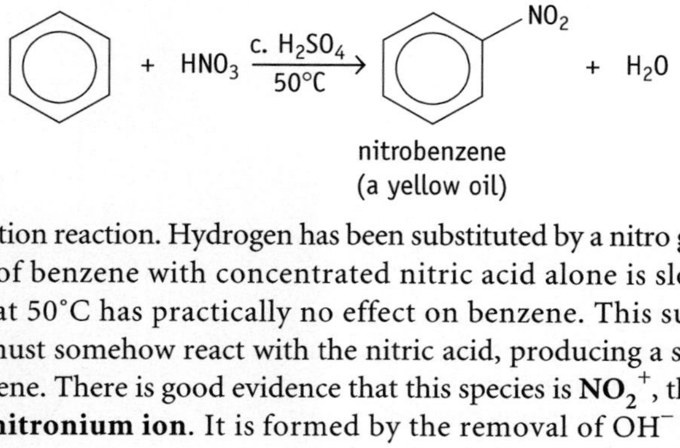

nitrobenzene
(a yellow oil)

This is a substitution reaction. Hydrogen has been substituted by a nitro group, —NO_2.

The reaction of benzene with concentrated nitric acid alone is slow, whilst pure sulphuric acid at 50°C has practically no effect on benzene. This suggests that the sulphuric acid must somehow react with the nitric acid, producing a species that then reacts with benzene. There is good evidence that this species is NO_2^+, the **nitryl cation,** also called the **nitronium ion.** It is formed by the removal of OH^- from HNO_3 by sulphuric acid:

$$HNO_3 + 2H_2SO_4 \rightarrow NO_2^+ + 2HSO_4^- + H_3O^+$$
nitryl cation

Here nitric acid is acting as a base in the presence of the stronger sulphuric acid.

The NO_2^+ ion is a strong electrophile and tends to attack the negative π-electron system in benzene. First a loose association is formed:

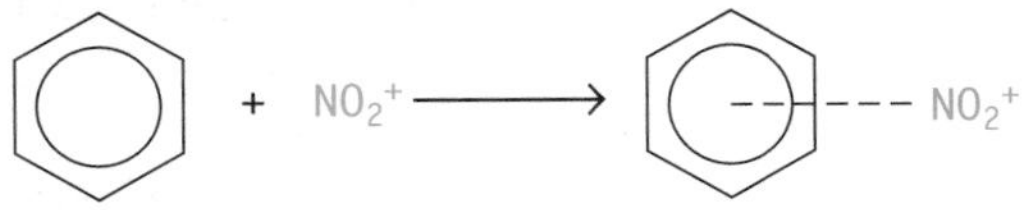

The NO_2^+ then attacks one of the carbon atoms of the ring, forming a bond to it and disrupting the delocalised π system:

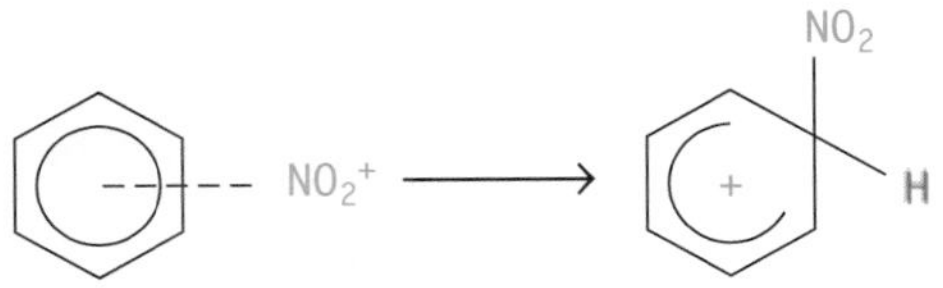

The formation of this intermediate requires the input of considerable energy to break the delocalised π system, so the reaction has a fairly high activation energy.

The intermediate cation then breaks down, either reforming benzene or producing nitrobenzene. The delocalised π system re-forms, releasing energy:

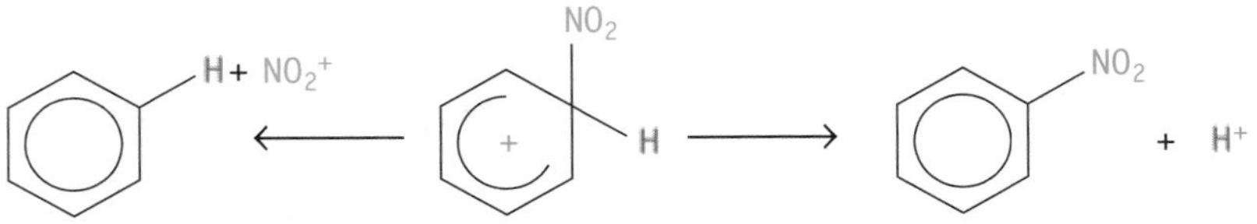

Notice that the first two stages of this mechanism, up to the formation of the intermediate cation, are similar to the stages in addition to an alkene double bond. The difference is that the aromatic cation loses its charge by loss of H^+, thus regaining aromatic character. With alkenes there is no delocalisation energy to be lost or gained, so the intermediate cation combines with an anion to form an addition product.

► Many food colourings contain 'coal tar dyes', so called because they were originally made from aromatic compounds found in coal tar. These dyes are rigorously checked for safety before they can be used in foods.

28.7 Important electrophilic substitution reactions of benzene

Nitration

The detailed mechanism of this reaction has already been considered in the last section. It is used industrially to manufacture nitrobenzene, from which phenylamine (aniline), $C_6H_5NH_2$, is produced by reduction. Phenylamine is used to manufacture dyes (see chapter 33).

Sulphonation

If benzene and concentrated sulphuric acid are refluxed together for several hours benzenesulphonic acid is formed:

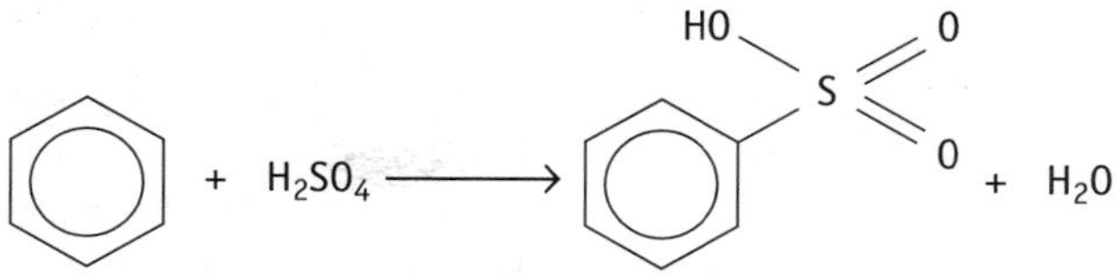

benzenesulphonic acid

This reaction is known as **sulphonation**. The electrophile that initially attacks the benzene ring is thought to be SO_3, which carries a large partial positive charge on the sulphur atom.

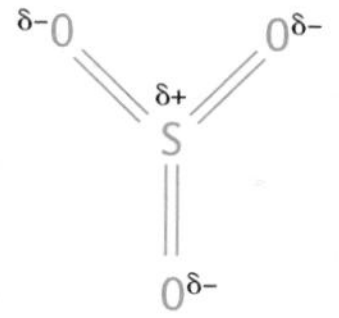

This theory is borne out by the fact that benzene is sulphonated in the cold by 'fuming sulphuric acid', which is a solution of SO_3 in concentrated sulphuric acid, but not by concentrated sulphuric acid alone unless heated.

Benzenesulphonic acid is important industrially as an intermediate in the manufacture of phenol (section 30.8). Phenol, C_6H_5OH, is a simple derivative of benzene, but it cannot be made by the direct reaction of ^-OH with benzene. This is because the high electron density in the π system makes the ring unreactive towards negatively charged groups. When the sulphonate group is attached to a benzene ring, however, the electron-withdrawing effect of the group gives rise to a positive charge on the carbon atom to which it is attached.

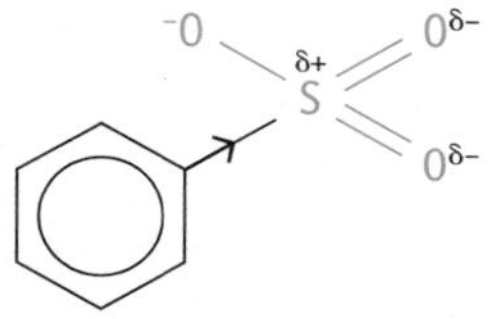

This makes the carbon atom susceptible to attack by negatively charged groups such as ^-OH, which can displace the $-SO_3^{2-}$ group. Consequently when benzenesulphonic acid is heated with molten sodium hydroxide, phenol is formed:

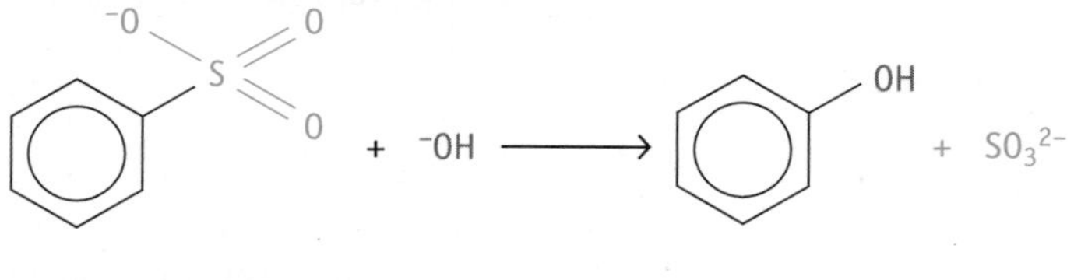

benzenesulphonate ion   phenol

This is one of several methods used today to manufacture phenol. What other group might have a similar effect to the sulphonate group in making the carbon atom it is attached to susceptible to attack by negatively charged groups?

Halogenation

Benzene does not react with chlorine, bromine or iodine on their own in the dark. This is because the non-polar halogen molecule has no centre of positive charge to initiate electrophilic attack on the benzene ring. However, in the presence of a catalyst such as iron filings, iron(III) bromide or aluminium chloride, benzene is substituted by chlorine or bromine:

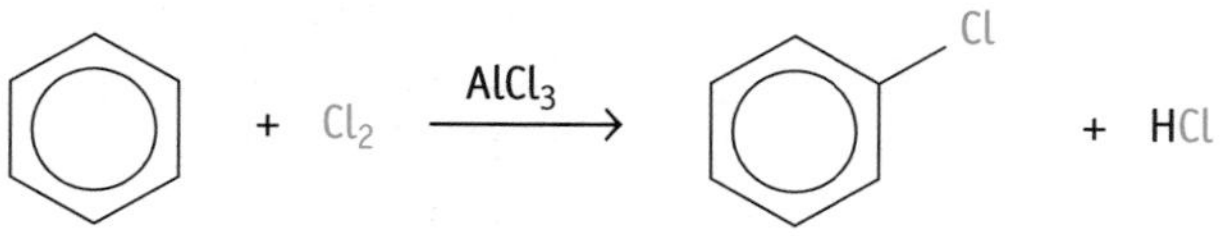

chlorobenzene

The catalyst (called a **halogen carrier**) is thought to work by accepting a lone pair from one of the halogen atoms. This induces polarisation in the halogen molecule:

$$Cl—Cl: \rightarrow AlCl_3$$
$$\delta+ \quad \delta-$$

The positively charged end of the halogen molecule is now electrophilic and attacks the benzene ring.

The electron-withdrawing tendency of the —Cl group on chlorobenzene has a similar effect to that of the —SO_3H group in benzenesulphonic acid. It makes the carbon atom to which it is attached susceptible to nucleophilic attack. For example, when chlorobenzene is boiled with hot, aqueous sodium hydroxide under pressure, it undergoes nucleophilic substitution to form phenol:

chlorobenzene   phenol

This reaction provides another method for the manufacture of phenol from benzene. Other reactions of chlorobenzene are described in chapter 29.

Interhalogen compounds such as BrCl will substitute benzene without a halogen carrier, since they are already polarised. BrCl reacts with benzene to form bromobenzene.

In the presence of sunlight, chlorine and bromine undergo *addition* reactions with benzene (see section 28.8).

Alkylation and acylation: Friedel–Crafts reaction

Aluminium chloride can be used as a catalyst to polarise halogen molecules and cause them to substitute a benzene ring. The same catalyst can be used to bring about the substitution of a benzene ring by halogenoalkanes. For example, if benzene is warmed with chloromethane and aluminium chloride under anhydrous conditions, a substitution reaction occurs and methylbenzene is formed:

methylbenzene

As in the reaction with halogens, the aluminium chloride accepts an electron pair from the chlorine atom, polarising the chloromethane molecule:

$$\delta+ \quad \delta-$$
$$CH_3—Cl: \rightarrow AlCl_3$$

The positively charged methyl group attacks the benzene ring and electrophilic substitution occurs. This is an example of a **Friedel–Crafts** reaction. Such reactions occur between aromatic hydrocarbons and any combination of reagents that can give rise to a positively charged carbon atom. These include alkenes and alcohols as well as halogenoalkanes. For example, ethene and benzene undergo a Friedel–Crafts reaction in the presence of hydrogen chloride and aluminium chloride, to form ethylbenzene:

ethylbenzene

This reaction is used industrially to manufacture ethylbenzene, from which phenylethene (styrene) is made by catalytic dehydrogenation:

ethylbenzene   phenylethene

---

1 What would you expect to be the product of a substitution reaction between benzene and bromine(I)chloride, BrCl? (*Hint:* think about the way this molecule is polarised.)

2 Iodine is too unreactive to substitute benzene even in the presence of a halogen carrier. Quite good yields of iodobenzene can however be obtained by reacting benzene with iodine(I) chloride, ICl. Explain why.

▲ Most solid detergents contain benzenesulphonate derivatives, made by reacting arenes with sulphuric acid (see section 32.3).

$CH_3CH_2CH_2CH_2CH_2CH_2CH_2CH_2CH_2CH_2CH_2$ —⬡— $SO_3^-Na^+$

A synthetic detergent

1 What important plastic is manufactured from phenylethene?
2 Ethylbenzene could be manufactured by a Friedel–Crafts reaction between chloroethane and benzene. Why is ethene used in preference to chloroethane?

Dodecylbenzene, important in the manufacture of detergents (see section 32.3) is made by a Friedel–Crafts reaction between benzene and dodecene:

⬡ + $CH_3(CH_2)_9$ $CH == CH_2$ $\xrightarrow[HCl]{AlCl_3}$ ⬡ $CH_2CH_2(CH_2)_9CH_3$

dodecene   dodecylbenzene

A similar reaction between benzene and propene is used to make (1-methylethyl)-benzene, which is used in the manufacture of phenol (see review question 7). The importance of electrophilic substitution reactions in the synthesis of simple aromatic compounds is shown in chapter 34.

Friedel–Crafts acylation

Friedel–Crafts reactions can also be used to join an acyl group, RCO—, to a benzene ring. For example:

⬡ + CH_3COCl $\xrightarrow[heat]{AlCl_3}$ ⬡ $COCH_3$ + HCl

ethanoyl chloride   phenylethanone

28.8 Other important reactions of benzene

The characteristic reactions of benzene and other aromatic hydrocarbons involve electrophilic substitution, because this type of reaction retains the delocalised π-electron system. Reactions involving the disruption of this system do occur, however.

Addition reactions

(i) With hydrogen

Alkenes undergo addition with hydrogen in the presence of a nickel catalyst. Benzene also gives this reaction, but considerably higher temperatures are required than for aliphatic compounds. The higher temperature is needed because extra energy must be supplied to break up the delocalised π-electron system.

$3H_2$ + ⬡ $\xrightarrow[150°C]{Raney\ nickel}$ ⬡

cyclohexane

The catalyst, Raney nickel, is a form of nickel with an extremely high surface area, and is very active.

If one mole of benzene and one mole of hydrogen are reacted in this way, one-third of the benzene is converted into cyclohexane and the remainder is left unreacted.
1 We might expect some cyclohexadiene,   and cyclohexene,

to be formed, but this does not in fact happen. Can you suggest a reason why?

The catalytic hydrogenation of benzene is important industrially in the manufacture of cyclohexane, from which nylon is made (section 33.9).

(ii)With chlorine

Benzene can undergo addition as well as substitution reactions with chlorine. In ultraviolet light, chlorine adds to benzene to form 1,2,3,4,5,6-hexachlorocyclohexane. The need for light suggests a free-radical mechanism (sections 26.5 and 26.6).

There are eight geometric isomers for 1,2,3,4,5,6-hexachlorocyclohexane due to the restricted rotation about the C—C bonds. Try drawing them or making models of them. One of these eight forms is a very effective insecticide, known commercially as Gammexane or BHC.

Burning

Benzene burns in air with a sooty, smoky flame. This sort of flame is characteristic of all hydrocarbons containing a high percentage of carbon.

Benzene has a high octane rating, and is used as a component of motor fuel. However, its use is strictly limited because of its toxicity.

28.9 Other arenes

Some examples of arenes other than benzene are shown in table 28.3. Naphthalene and anthracene are examples of **fused ring** arenes, in which two or more rings are joined.

Table 28.3
Some arenes

Systematic name	Other name	Molecular formula	Structural formula
methylbenzene	toluene	C_7H_8	
1,4-dimethylbenzene	*para*-xylene	C_8H_{10}	
naphthalene	—	$C_{10}H_8$	
anthracene	—	$C_{14}H_{10}$	

Methylbenzene

The reforming reactions used to manufacture benzene from petroleum (see section 28.4) also produce considerable quantities of methylbenzene (commonly called toluene). This substance is used in the manufacture of plastics and explosives (see below), but much more is produced than can be used in this way. The majority of the methylbenzene produced is added to motor fuel to increase its octane rating: it is much less toxic than benzene.

Reactions of methylbenzene

The properties of aromatic compounds are very different from those of aliphatic ones. Methylbenzene has an aromatic portion (the benzene ring) and an aliphatic portion (the —CH₃ group). These two portions make different contributions to the properties of methylbenzene and have a modifying effect on one another.

(i) The —CH_3 group

The —CH_3 group shows some reactions we would expect of an alkyl group: for example, it can be substituted by chlorine. This reaction occurs when chlorine is bubbled into boiling methylbenzene in sunlight:

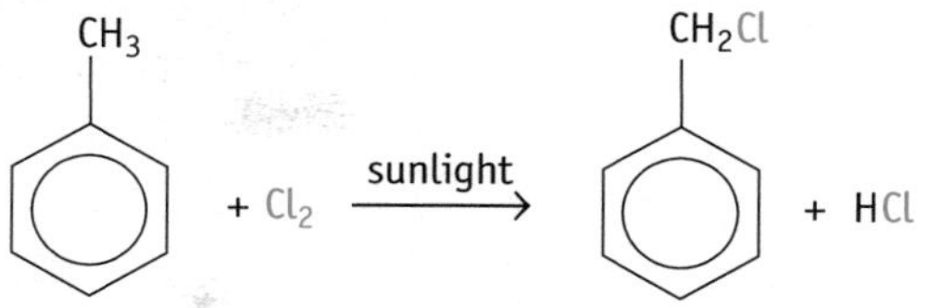

$$CH_3 \text{-benzene} + Cl_2 \xrightarrow{\text{sunlight}} CH_2Cl\text{-benzene} + HCl$$

The reaction has a free-radical mechanism similar to the reaction of methane with chlorine described in section 26.6.

> **1** Write the structure of two other compounds that might form when the —CH_3 group is substituted by chlorine.
> **2** Under certain conditions chlorine will substitute the *ring* in methylbenzene instead of the —CH_3 group. What will these conditions be?

The —CH_3 group in methylbenzene does not behave as a typical alkyl group in all its reactions. The benzene ring with its regions of high electron density has a modifying effect on any group that is attached to it. A given functional group behaves differently depending on whether it is attached to an aliphatic or to an aromatic molecule.

One example of the way the ring modifies the properties of the —CH_3 group in methylbenzene is the reaction with potassium manganate(VII). Alkanes are inert to oxidation, but the alkyl group in methylbenzene can be oxidised by alkaline manganate(VII) to give benzoic acid. The manganate(VII) is reduced to green manganate(VI).

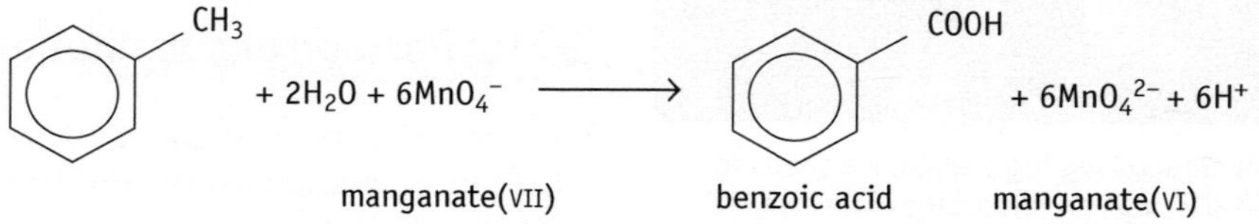

$$CH_3\text{-benzene} + 2H_2O + 6MnO_4^- \longrightarrow COOH\text{-benzene} + 6MnO_4^{2-} + 6H^+$$

manganate(VII)　　　　　benzoic acid　　manganate(VI)

Note that the ring is not affected. This is another indication of its stability.

(ii) The aromatic ring

If chlorine is bubbled through methylbenzene in the absence of sunlight and in the presence of a halogen carrier such as $AlCl_3$, the *ring* is substituted instead of the CH_3 side chain. This reaction proceeds by the electrophilic substitution mechanism described in section 28.6. A mixture of two isomers is obtained:

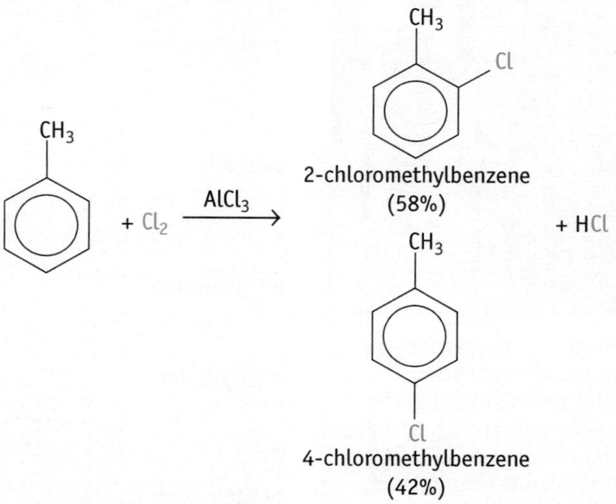

$$CH_3\text{-benzene} + Cl_2 \xrightarrow{AlCl_3}$$

2-chloromethylbenzene
(58%)

+ HCl

4-chloromethylbenzene
(42%)

Virtually none of the other possible isomer, 3-chloromethylbenzene, is produced. Furthermore, the reaction proceeds at a considerably higher rate than the corresponding reaction of chlorine with benzene. Clearly the —CH$_3$ group influences the aromatic ring, making it more susceptible to electrophilic substitution and dictating the positions in which it is substituted. This effect is discussed further in the next section.

An important substitution reaction of methylbenzene is nitration. When methylbenzene is heated with a nitrating mixture of concentrated nitric acid and concentrated sulphuric acid, it is substituted by one, two or three —NO$_2$ groups, depending on the conditions. The main products are

2-nitromethylbenzene   2,4-dinitromethylbenzene

4-nitromethylbenzene   2,4,6-trinitromethylbenzene

Note the effect of the —CH$_3$ group in determining that the main positions substituted are 2, 4 and 6 rather than 3 or 5. 2,4,6-Trinitromethylbenzene (trinitrotoluene or TNT) is an important high explosive. It is fairly resistant to shock and so can be used in shells without risk of explosion under the shock of firing a gun. When detonated it decomposes forming large volumes of CO, H$_2$O and N$_2$ at high temperature. The sudden expansion of these gases gives TNT its explosive force. TNT has the advantage that it is a solid which melts below 100°C, so it can be melted with steam and poured into its container.

28.10 Position of substitution in benzene derivatives

The methyl group in methylbenzene activates the ring towards electrophilic substitution. It also favours substitution in positions 2, 4 and 6 rather than 3 or 5. In fact, *any substituent group* attached to a benzene ring affects the rate and the position at which further substitution occurs. Table 28.4 shows the main products of mononitration (i.e. substitution by one nitro group) of different benzene derivatives, and whether they are nitrated faster or slower than benzene.

Table 28.4
Mononitration products of benzene derivatives

Compound	Main products of mononitration			Rate of nitration relative to benzene
methylbenzene	CH$_3$	CH$_3$ NO$_2$	CH$_3$ NO$_2$	Faster
phenol	OH	OH NO$_2$	OH NO$_2$	Faster
nitrobenzene	NO$_2$	NO$_2$ NO$_2$		Slower
phenylamine	NH$_2$	NH$_2$ NO$_2$	NH$_2$ NO$_2$	Faster
benzoic acid	COOH	COOH NO$_2$		Slower

▲ Manufacturing high explosive shells in 1940. TNT was an important explosive in both world wars.

▲ One of the earliest high explosives was nitroglycerine. It is made by heating glycerol with a nitrating mixture, but the reaction must be watched carefully to make sure it does not get out of control. In this print of early nitroglycerine manufacture, the operator minding the reaction is sitting on a one-legged stool to prevent him nodding off to sleep.

Look at the table and answer these questions.

1. Which groups tend to direct substitution to the 2 or 4 position?
2. Which groups tend to direct substitution to the 3 position?
3. Is there any correlation between the *position* to which a group directs substitution and the *rate* at which it causes the ring to substitute?

It can be seen from table 28.4 that benzene derivatives fall into two classes as far as further substitution is concerned:

- Those which substitute faster than benzene and in which the new substituent is directed to the 2 or 4 position, a mixture of the two isomers being obtained. Functional groups causing this behaviour include:
  —CH_3 and all alkyl groups, —OH, —NH_2, —OCH_3.
- Those which substitute more slowly than benzene and in which the new substituent is directed to the 3 position. Functional groups causing this behaviour include:
  —COOH, —SO_3H, —NO_2.

In practice, a mixture of all possible isomers is obtained, but these rules give the *main* products. Note that the rules apply whatever the nature of the new substituent, not just to nitration. Note too, that as far as monosubstitution is concerned, the 6 position is equivalent to the 2 position and the 5 to the 3.

4. Write structural formulas for the main products you would expect from the following substitution reactions. Assume monosubstitution occurs in each case.

(a) CH_3 ... $\xrightarrow[AlCl_3]{CH_3Cl}$

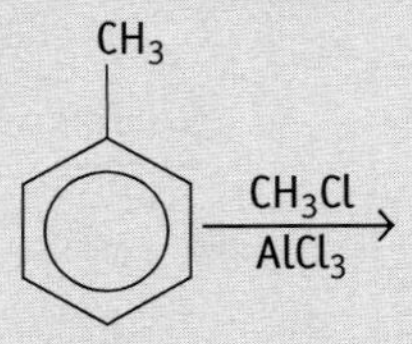

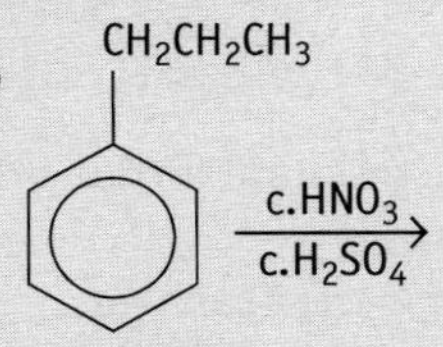

(b) NO_2 ... $\xrightarrow[AlCl_3]{Cl_2}$

(c) $CH_2CH_2CH_3$ ... $\xrightarrow[c.H_2SO_4]{c.HNO_3}$

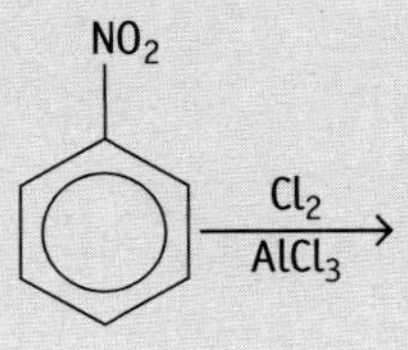

(d) COOH ... $\xrightarrow{c.H_2SO_4}$

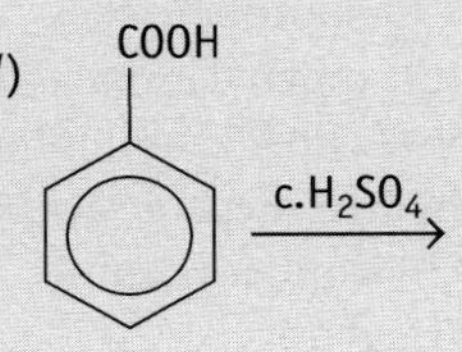

Summary

1. Aromatic hydrocarbons (arenes) are ring compounds stabilised by electron delocalisation.

2. Benzene, C_6H_6, has a symmetrical planar ring of six carbon atoms with a ring of delocalised π-electrons above and below the plane of the molecule.

3. Benzene is manufactured by catalytic reforming of petroleum fractions.

4. Disubstituted benzene derivatives are named by referring to numbered positions on the ring.

5. The C_6H_5— group is called phenyl.

6. Benzene undergoes addition considerably less readily than alkenes.

7. The characteristic reaction type of benzene is electrophilic substitution. Some important substitution reactions are shown on the following diagram:

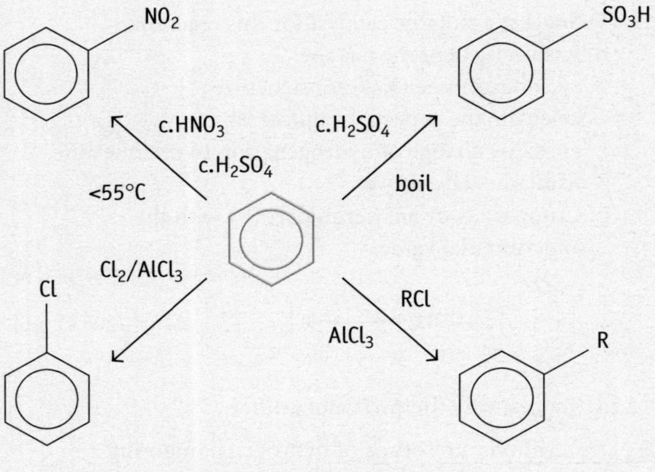

8. A benzene ring strongly influences the properties of functional groups attached to it.

9. A functional group influences the position and rate of substitution of the benzene ring to which it is attached.

10. Methylbenzene shows some properties of an alkane and some of the arene, but the properties of the CH_3— group are substantially modified by the benzene ring.

1 Name the following benzene derivatives:

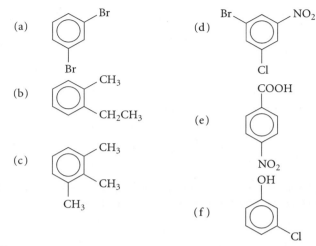

(a)

(b)

(c)

(d)

(e)

(f)

2 Write structural formulas for the following benzene derivatives:
(a) 2,4,6-trinitrophenol
(b) 1,4-dichlorobenzene
(c) 4-nitrophenylamine
(d) 2-methylbenzenesulphonic acid
(e) 2-hydroxybenzoic acid
(f) 2-chlorophenylamine.

3 Consider the catalytic hydrogenation of cyclohexene:

$$+ H_2 \longrightarrow \qquad \Delta H = -120 \text{ kJ mol}^{-1}$$

(a) Suggest a suitable catalyst for this reaction.
(b) Assuming benzene has the cyclohexatriene (Kekulé) structure calculate the expected value of its enthalpy change of hydrogenation to cyclohexane, using the data above.
(c) Compare your answer in part (b) with the experimental value:

$$+ 3H_2 \longrightarrow \qquad \Delta H = -208 \text{ kJ mol}^{-1}$$

(d) Suggest why the two values differ.

4 Deuterium is an isotope of hydrogen containing a neutron in its nucleus in addition to the single proton present in normal hydrogen. The symbol for deuterium is $^2_1$H or D, and its chemical properties are almost identical to those of normal hydrogen, $^1_1$H. If deuterium chloride, DCl, is dissolved in methylbenzene, no reaction occurs. However, if anhydrous aluminium chloride is added to the solution, hydrogen atoms on the aromatic ring are rapidly substituted by deuterium atoms. If excess DCl is

present, all five ring hydrogens are substituted, forming:

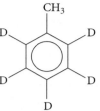

(a) Write the structure of the complex formed between AlCl₃ and DCl.
(b) What is the effect of the AlCl₃ on the polarisation of the DCl molecule?
(c) Why is AlCl₃ effective in causing DCl to substitute the aromatic ring?
(d) Why are none of the hydrogen atoms on the CH₃ side chain substituted, even in the presence of excess DCl?

5 How would you distinguish between the members of the following pairs of compounds, using simple chemical tests?

(a)

(b)

(c)

(d)

6 Predict the major products of the following reactions.

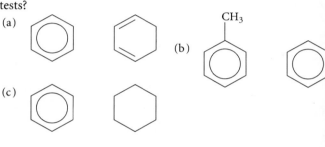

(a) CH₃ ... $\xrightarrow{\text{H}_2/\text{Ni}}{300°C}$

(b) SO₃H ... $\xrightarrow[\text{150°C}]{\text{c.HNO}_3 \ \text{c.H}_2\text{SO}_4}$

(c) CH₃ ... CH₃ ... $\xrightarrow[\text{warm}]{\text{alkaline KMnO}_4(aq)}$

(d) NO₂ ... NO₂ ... $\xrightarrow[\text{120°C}]{\text{c.HNO}_3 \ \text{c.H}_2\text{SO}_4}$

(e) ... $\xrightarrow[\text{AlCl}_3,\text{warm}]{\text{CH}_3\text{CHClCH}_3}$ (mono-substituted product only)

(f) CH₃ ... $\xrightarrow[\text{cold, in dark}]{\text{Br}_2/\text{FeBr}_3}$

7 (1-Methylethyl)benzene, commonly called cumene, has the structure

CH₃—CH—CH₃

(benzene ring structure)

It is an important intermediate in a process used to manufacture phenol and propanone (acetone).

(1-Methylethyl)benzene is manufactured by a Friedel–Crafts-type reaction between propene and benzene in the presence of an acid catalyst:

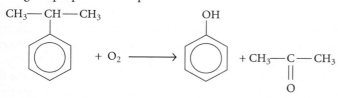

The reaction is believed to proceed via a carbocation intermediate.

(a) Write the structures of the two possible carbocations formed by the attack of H⁺ on the double bond of propene. (Sections 27.5 and 27.6 may help you.)

(b) Which is the more stable of these two carbocations?

(c) Show how the carbocation you identified in (b) can attack and substitute the benzene ring. Explain why (1-methylethyl)benzene is virtually the only product of this reaction, hardly any propylbenzene being produced.

Treatment of the product with air followed by dilute acid gives propanone and phenol:

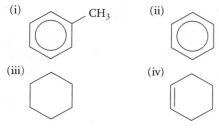

(d) Suggest a reason why this is the most economic of the several methods available for manufacturing phenol.

8 (a) Write structural formulas for all the compounds of molecular formula C_8H_{10} containing one benzene ring.

(b) For each of these compounds, write the formulas of all the possible mononitration products (not just the ones you would expect for the major product).

(c) For one of the compounds, state which of the mononitration products you would expect to be produced in the majority.

9 Consider the following compounds,

(i) (benzene with CH₃)   (ii) (benzene ring)

(iii) (cyclohexane ring)   (iv) (cyclohexene ring)

(v) $CH_3CH_2CH\!=\!CHCH_2CH_3$

(a) Which are aromatic hydrocarbons?
(b) Which are cyclic compounds?
(c) Which are unsaturated hydrocarbons?
(d) Which have a planar ring in their molecule?
(e) Which would decolorise bromine water in the dark?
(f) Which would evolve fumes of HBr when treated with bromine and iron filings?
(g) Which would react with alkaline potassium manganate(VII) solution?

10 Consider two possible reactions of chlorine with benzene.

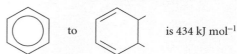

A (benzene (g)) + 2Cl₂(g) ⟶ (dichlorobenzene) + 2HCl(g)

B (benzene (g)) + Cl₂(g) ⟶ (addition product with Cl)

Some relevant bond energies (in kJ mol⁻¹) are: Cl—Cl 242; C—Cl (general value) 339; C—H (in benzene) 430; H—Cl 431.

(a) Which of the two reactions is addition and which is substitution?

(b) Calculate the energy change in reaction **A** by the following stages.
  (i) How much energy is needed to split two moles of chlorine molecules into atoms?
  (ii) How much energy is needed to break two moles of C—H bonds in benzene?
  (iii) How much energy is released when two moles of new C—Cl bonds form?
  (iv) How much energy is released when two moles of HCl molecules are formed?
  (v) What is the total energy change for reaction **A**?

(c) Calculate the energy change in reaction **B** by the following stages, assuming that the energy needed to break one C—C bond in benzene, i.e. convert

(benzene) to (cyclohexadiene) is 434 kJ mol⁻¹

  (i) How much energy is needed to split one mole of chlorine molecules into atoms?
  (ii) How much energy is released when two moles of C—Cl bonds form?
  (iii) What is the total energy change in reaction **B**?

(d) Which reaction is more likely to occur between chlorine and benzene, **A** or **B**? Give reasons for your answer.

29 Organic Halogen Compounds

29.1 Anaesthetics

Before the advent of anaesthetics, surgery was a savage and primitive affair. It was agony for the patient, and surgeons were therefore only prepared to operate if it was absolutely essential, for example the amputation of a damaged limb that would otherwise become gangrenous. Anaesthetics enabled surgery to develop from crude carpentry to its present sophisticated form.

Three of the most important early anaesthetics were nitrous oxide (dinitrogen oxide, N_2O), ether (ethoxyethane, $CH_3CH_2OCH_2CH_3$) and chloroform (trichloromethane, $CHCl_3$). Nitrous oxide is non-toxic and non-flammable, but it only produces light anaesthesia. Ether is an effective anaesthetic but it is highly flammable and therefore dangerous. Chloroform produces deep anaesthesia and is non-flammable, but it is toxic and carries the risk of liver damage.

The ideal inhalant anaesthetic must be a gas or volatile liquid, so that it can be inhaled and absorbed via the lungs. It must be non-flammable. It must produce deep anaesthesia but must be non-toxic. In 1951, ICI began the search for a new anaesthetic. They decided to look at **halogenoalkanes**.

It was known that the substitution of chlorine atoms into an alkane increased its anaesthetic properties, but also made it toxic. Thus dichloromethane, CH_2Cl_2, is a fairly weak anaesthetic with little toxicity, trichloromethane (chloroform), $CHCl_3$, is stronger and more toxic, and tetrachloromethane is a very strong anaesthetic and also very toxic.

The introduction of halogen atoms into an alkane skeleton also tends to make it non-flammable. Fluorine is useful in this respect as the C—F bond is very stable. Fluoroalkanes are inert, non-flammable and non-toxic. ICI therefore looked for a short-chain halogenoalkane containing fluorine for inertness and chlorine for anaesthetic properties, with a suitable boiling point in the range 40°C to 60°C. They concentrated on two-carbon molecules, and after many trials produced the compound

The bromine atom was introduced to produce a substance with a high enough boiling point to be conveniently stored. This substance, 2-bromo-2-chloro-1,1,1-trifluoroethane, was given the name Halothane. Following its discovery in 1956 it became widely used in hospitals.

Halothane illustrates several of the important properties of halogenoalkanes. It shows the increasing reactivity of the C—Hal bond as we go from F to I; the decreasing volatility of R—Hal as we move in the same direction; and the effect of halogen atoms in reducing the flammability of a hydrocarbon.

Although organic halogen compounds are uncommon in nature a study of their properties is important to chemists as they have many uses in industry and in the laboratory.

▲ A tiger having a false tooth fitted under Halothane anaesthetic.

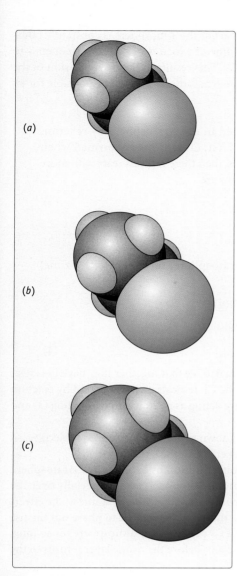

(a)

(b)

(c)

Figure 29.1
Molecules of (a) chloroethane,
(b) bromoethane and (c) iodoethane.

29.2 Naming halogen compounds

Organic halogen compounds contain the **halogeno** functional group: —F, —Cl, —Br, —I. They are named using the prefixes **fluoro-**, **chloro-**, **bromo-** and **iodo-**. Numbers are used if necessary to indicate the position of the halogen atom in the molecule. Thus,

CH_3CH_2Cl chloroethane,
$CH_3CHBrCH_3$ 2-bromopropane.

If the molecule contains more than one halogen atom of the same kind, the prefixes **di-**, **tri-**, etc., are used. Thus,

CH_2ClCH_2Cl 1,2-dichloroethane,
$CHCl_2CHClCH_3$ 1,1,2-trichloropropane.

1 Name these compounds.

    (a) $CH_3CH_2CHICH_3$
    (b) CH_3CHCl_2

2 Write formulas for these compounds.

    (c)   1,3,5-tribromobenzene
    (d)   1,2-dibromo-3-chloropropane

29.3 The nature of the carbon–halogen bond

Unreactive halogenoalkanes: the C—F bond

The C—F bond is very strong: compare its bond energy of 485 kJ mol^{-1} with 435 kJ mol^{-1} for C— H and 327 kJ mol^{-1} for C—Cl. The C—F bond is thus very unreactive. Consequently, fluorocarbons, compounds containing fluorine and carbon only, are extremely inert (see review question 6, chapter 25). The C—Cl bond is more reactive than C—F, but highly chlorinated compounds such as CCl_4 and $CHCl_3$ are, nevertheless, fairly inert. In particular they are non-flammable.

An important group of compounds is the **chlorofluorocarbons**, or **CFCs**. Some examples are given in table 29.1

Table 29.1
Some important CFCs and their uses

formula	name	typical use
CCl_3F	trichlorofluoromethane	'blowing agent' for making foam plastics
CCl_2F_2	dichlorodifluoromethane	refrigerant fluid, aerosol propellant
CCl_2F—$CClF_2$	1,1,2-trichloro-1,2,2-trifluoroethane	solvent for degreasing

At one time, CFCs were used widely as refrigerants, solvents and aerosol propellants. They are useful in these applications because they are non-flammable, non-toxic and, being very unreactive, they do not affect other materials, such as the contents of the aerosol pack.

It is now known that, unfortunately, CFCs are responsible for serious damage to the stratospheric ozone layer which filters out most of the Sun's harmful ultraviolet radiation before it can reach us on Earth.

CFCs and the ozone layer

The trouble is that CFCs are *too* unreactive. Once they have entered the atmosphere, from an aerosol spray, an old refrigerator or whatever, they stay there and do not break down for many years. Slowly the molecules of CFCs diffuse upwards until they reach

▲ This large plastic ballon being filled with helium will carry instruments for measuring the ozone concentration over the Arctic.

▲ In Australia and New Zealand, ultraviolet radiation from the sun is particularly intense. Most people keep themselves covered up even in the sea.

the upper atmosphere, or stratosphere. This part of the atmosphere is constantly bombarded with powerful ultraviolet radiation, some of which is of the right frequency to break the relatively weak C—Cl bond in CFCs. This results in homolytic fission of the C—Cl bond, producing Cl· radicals. It is these free radicals that are responsible for all the damage to the ozone layer.

This is what happens.

Ozone, O_3, is continuously being formed in the stratosphere by reaction between O· radicals and O_2 molecules. First, O• radicals are formed when ultraviolet radiation splits up O_2 molecules, then these radicals react with more O_2 molecules to make O_3. The two reactions are

$$O_2(g) \quad \rightarrow \quad 2O·(g) \tag{1}$$

Then

$$O·(g) \quad + \quad O_2(g) \quad \rightarrow \quad O_3(g) \tag{2}$$

When Cl· radicals are present in the stratosphere, they react with ozone, forming O_2 and ClO radicals.

$$O_3(g) \quad + \quad Cl·(g) \quad \rightarrow \quad O_2(g) \quad + \quad ClO·(g) \tag{3}$$

The ClO· radicals then react with O· radicals

$$ClO·(g) \quad + \quad O·(g) \quad \rightarrow \quad Cl·(g) \quad + \quad O_2(g) \tag{4}$$

If you look closely at reactions (3) and (4), you will see that together they have the effect of removing both O_3 and O·. What is more, the Cl· radicals get regenerated by reaction (4): in effect they never get used up, so they are acting as a catalyst in converting O_3 and O· to O_2.

In fact, one Cl· radical can catalyse the breakdown of about 1 million O_3 molecules, so you can see why CFCs have such a devastating effect on the ozone layer.

During the 1980s, measurements of the concentration of ozone in the stratosphere showed that the ozone layer was thinning and developing a 'hole', especially over the Antarctic. Chemists became increasingly convinced that this 'hole' was due to the effects of CFCs, and in 1990 more than 60 countries signed an agreement to phase out the use of CFCs by the year 2000. Even so, because CFCs persist in the atmosphere for so long, scientists estimate that it will be well into the 2000s before the ozone layer returns to the condition it was in before the trouble started.

Replacing CFCs

Few compounds have CFCs' combination of non-flammability, non-toxicity and inertness, and for uses such as refrigeration and aerosols it is necessary to find compounds with exactly the right boiling point.

Some of the important replacements for these uses are the **hydrofluorocarbons, HFCs** (also called hydrofluoroalkanes, or HFAs). An example is 1,1,1,2-tetrafluoroethane, CF_3CH_2F, which is used as a refrigerant. HFCs have the advantage that they contain no Cl atoms, so they do not release damaging Cl· radicals in the stratosphere. Moreover, their molecules include C—H bonds, which are relatively reactive, which means that these compounds break down in the atmosphere more quickly than CFCs, so they do not persist for so long.

Reactive halogenoalkanes: the C—Br bond

Compounds containing C—F bonds are not typical of the halogenoalkanes as a whole, which tend to be much more reactive than fluoroalkanes. To get some idea of the nature of the C—Hal bond, we will consider some bromine-containing compounds.

Table 29.2 on the next page compares the effect of adding silver nitrate solution to sodium bromide, 1-bromobutane and bromobenzene. In the case of the two organic liquids, vigorous shaking is necessary since they do not mix appreciably with the aqueous phase.

Table 29.2
Effect of aqueous silver nitrate on bromine-containing compounds at room temperature

Sodium bromide	1-Bromobutane	Bromobenzene
pale yellow precipitate appears immediately	no reaction at first: faint precipitate appears after several minutes	no reaction even after several hours

Look at table 29.2 and answer these questions.

1 What is the pale yellow precipitate produced in the reaction between silver nitrate and sodium bromide?
2 Write an ionic equation for this reaction.
3 Why does silver nitrate produce no immediate precipitate with 1-bromobutane, even though it contains bromine?
4 Suggest a reason why a precipitate appears after several minutes.
5 Compare the reactivity of bromobenzene with that of 1-bromobutane.

The C—Br bond in 1-bromobutane is covalent. 1-Bromobutane therefore contains no Br^- ions, so it does not produce a precipitate of silver bromide with silver nitrate. The slow appearance of a precipitate of silver bromide suggests that Br^- ions are slowly being produced. Why? To explain, we need to look at the nature of the C—Br bond.

Bromine is more electronegative than carbon and the C—Br bond is thus polar:

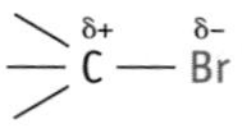

$$\overset{\delta+}{C} - \overset{\delta-}{Br}$$

The partial positive charge on the carbon atom tends to attract nucleophiles, such as NH_3, ^-OH and H_2O, with their lone pairs of electrons. Water molecules from the aqueous silver nitrate can act as nucleophiles since their oxygen atoms carry two lone pairs and a partial negative charge:

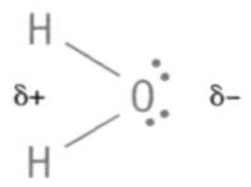

$$\overset{\delta+}{H}\underset{\overset{|}{H}}{\overset{}{}} \overset{}{O:} \ \delta-$$

The water attacks the partially positive carbon atom in 1-bromobutane and a substitution reaction takes place, releasing bromide ions:

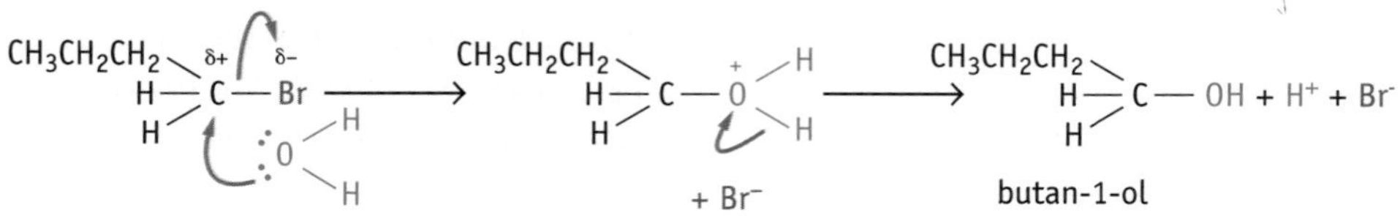

This is an example of a **nucleophilic substitution** reaction. Such reactions are typical of halogenoalkanes and their mechanism is discussed further in the next section. The failure of bromobenzene to react under these conditions suggests that the C—Br bond is stronger in aromatic than in aliphatic compounds.

Notice that nucleophilic substitution involves *heterolytic fission* of the C—Br bond: the bond breaks to form oppositely charged ions, rather than breaking homolytically to form free radicals. Heterolytic fission is characteristic of most reactions of halogenoalkanes, though in the presence of ultraviolet radiation homolytic fission is possible (see 'CFCs and the ozone layer' above).

Physical properties of halogenoalkanes

The C—Hal bond is polar, but not polar enough to have an appreciable effect on the physical properties of organic halogen compounds. They are all immiscible with water. As table 29.3 shows, their volatility is determined more by the size and number of halogen atoms they contain than by the polarity of the bonds. Thus iodomethane is a liquid at room temperature while chloromethane is a gas, even though chloromethane has a more polar molecule.

Table 29.3
Boiling points of some halogen compounds

Compound	State at 298 K	Boiling point/K
CH_3F	g	195
CH_3Cl	g	249
CH_3Br	g	277
CH_3I	l	316
CH_2Cl_2	l	313
$CHCl_3$	l	335
CCl_4	l	350
C_6H_5Cl	l	405

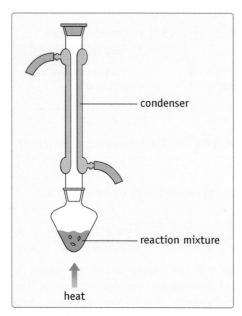

Figure 29.2
Reflux apparatus used for boiling an organic reaction mixture. The condenser prevents escape of volatile reagents.

Table 29.4
Average bond energies. The weaker the bond between the carbon and halogen atoms, the easier it is to hydrolyse.

Bond	Average bond energy/kJ mol^{-1}
C—F	484
C—Cl	338
C—Br	276
C—I	238

29.4 Nucleophilic substitution

Water can act as a nucleophile towards halogenoalkanes. In general:

$$RHal + H_2O \rightarrow ROH + HHal$$

This reaction is slow, but if ^-OH is used as a nucleophile instead of H_2O it is quicker. Thus bromoethane forms ethanol quite rapidly when heated under reflux with aqueous sodium hydroxide (figure 29.2).

$$\underset{\text{bromoethane}}{CH_3CH_2Br} + {^-OH} \rightarrow \underset{\text{ethanol}}{CH_3CH_2OH} + Br^-$$

The mechanism of the reactions between halogenoalkanes and hydroxide ions is discussed later in this section.

The substitution reaction outlined above between CH_3CH_2Br and ^-OH is general for all halogenoalkanes. It can be used as a method for preparing alcohols:

$$RHal + {^-OH} \rightarrow ROH + Hal^-$$

For a given alkyl group —R, the iodo compound reacts most readily, the bromo compound less so and the chloro compound reacts least readily. This is because the C—Hal bond becomes progressively stronger passing from I to Cl (table 29.4).

If the halogen is attached to a benzene ring, substitution is more difficult. It is thought that lone pairs of electrons on the halogen atom interact with the delocalised π-electron system of the benzene ring, strengthening the C—Hal bond. In addition, the high electron density on the aromatic ring tends to repel the approaching negatively charged ^-OH ion. Consequently, chlorobenzene reacts with aqueous sodium hydroxide at a reasonable rate only at 300°C under a pressure of 200 atmospheres. (This reaction provides one way of manufacturing phenol – see section 28.7.) Compare this with the behaviour of chlorobutane, which is readily converted to butanol by refluxing with aqueous sodium hydroxide under ordinary conditions.

> **1** Predict the organic product of the reactions when aqueous sodium hydroxide is boiled with
> (a) 2-chloropropane
> (b) 1,2-dibromoethane
>
> (c)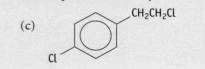

The mechanism of nucleophilic substitution

The mechanism of reactions between halogenoalkanes and hydroxide ions has been extensively studied by chemists. It involves nucleophilic attack by the ^-OH ion, e.g.

$$H-\overset{\overset{\displaystyle H}{|}}{\underset{\underset{\displaystyle H}{|}}{C}}-\overset{\overset{\displaystyle H}{|}}{\underset{\underset{\displaystyle H}{|}}{\overset{\delta+}{C}}}\overset{\delta-}{Br} \longrightarrow H-\overset{\overset{\displaystyle H}{|}}{\underset{\underset{\displaystyle H}{|}}{C}}-\overset{\overset{\displaystyle H}{|}}{\underset{\underset{\displaystyle H}{|}}{C}}-OH + Br^-$$

The actual mechanism is more complex than the simplified version shown here. Chemists believe that there are two different mechanisms possible:
- a step-by-step, two-stage mechanism known as S_N1
- a single step mechanism, known as S_N2.

The evidence for the proposed mechanisms, and the reasons for the terms 'S_N1' and 'S_N2', are explained below (see 'How do we know?').

We will look at the two mechanisms in turn, using a generalised substitution reaction:

$$R^2-\overset{\overset{\displaystyle R^1}{|}}{\underset{\underset{\displaystyle R^3}{|}}{C}}-Br + {^-OH} \longrightarrow R^2-\overset{\overset{\displaystyle R^1}{|}}{\underset{\underset{\displaystyle R^3}{|}}{C}}-OH + Br^-$$

(R^1, R^2 and R^3 are alkyl groups or hydrogen atoms.)

Step-by-step mechanism (S_N1)

In this case, it is proposed that a two-stage mechanism is involved:

$$R^2 - \overset{\overset{\displaystyle R^1}{|}}{\underset{\underset{\displaystyle R^3}{|}}{C}} - Br \xrightarrow{\text{slow}} R^2 - \overset{\overset{\displaystyle R^1}{|}}{\underset{\underset{\displaystyle R^3}{|}}{C^+}} + Br^-$$

intermediate
carbocation

$$R^2 - \overset{\overset{\displaystyle R^1}{|}}{\underset{\underset{\displaystyle R^3}{|}}{C^+}} \quad {}^-OH \xrightarrow{\text{fast}} R^2 - \overset{\overset{\displaystyle R^1}{|}}{\underset{\underset{\displaystyle R^3}{|}}{C}} - OH$$

The first step, which is relatively slow, involves breaking the C—Br bond to form the intermediate carbocation. The carbocation is very unstable and reactive, so the second step is fast. The overall rate of the reaction is determined by the slow first step – the **rate determining step** (section 24.11).

Single-step mechanism (S_N2)

Here it is proposed that the reaction occurs in a single step. The ^-OH ion is attracted to the central carbon atom, and as it moves in, it repels the Br atom. At some 'middle' stage in the reaction, Br and OH are both partially bonded to the carbon, the OH on its way in and the Br on its way out. This is the **transition state** of the reaction. It is **not** a reaction intermediate which exists independently, but simply the middle stage in a continuous process during which the ^-OH moves in and the Br^- moves out.

$$\underset{Br}{\overset{R^1}{\underset{\diagdown R^2 \quad R^3}{|}}}{C} \; + \; {}^-OH \longrightarrow \left[Br\text{----}\overset{\overset{\displaystyle R^1}{|}}{\underset{\diagup R^2 \quad R^3 \diagdown}{C}}\text{----}OH \right] \longrightarrow Br^- + \underset{R^2 \quad \underset{R^3}{|} \quad OH}{\overset{R^1}{|}}{C}$$

transition state

Notice that the transition state involves *five* groups around the central carbon atom. Figure 29.3 illustrates the difference between S_N1 and S_N2 in another way.

Which mechanism operates in practice?

In many nucleophilic substitution reactions, both S_N1 and S_N2 mechanisms can and do operate, but in most cases one proceeds much faster than the other. Which of the mechanisms is the dominant one depends on a number of factors. Two of the most important factors are the structure of the halogenoalkane and the nature of the solvent.

Structure of the halogenoalkane

In the two-stage S_N1 mechanism, the rate of the reaction is determined by the ease with which the intermediate carbocation forms. If the substituent groups R^1, R^2, and R^3 are all alkyl groups rather than H atoms, they will tend to donate electrons by the inductive effect (page 424). This stabilises the carbocation and favours its formation.

$$R^2 - \overset{\overset{\displaystyle R^1}{|}}{\underset{\underset{\displaystyle R^3}{|}}{C}} - Br \longrightarrow R^2 - \overset{\overset{\displaystyle R^1}{|}}{\underset{\underset{\displaystyle R^3}{|}}{C^+}} + Br^-$$

If some or all of R^1, R^2 and R^3 are hydrogen atoms, the formation of the carbocation will be less favoured, so the rate of the reaction will be slower. Thus, the two-step S_N1 mechanism is favoured by the presence of substituent alkyl groups. On the other hand, the single step S_N2 mechanism is favoured by the presence of substituent hydrogen atoms. The small hydrogen atoms make it easier to fit five groups around the central carbon in the transition state.

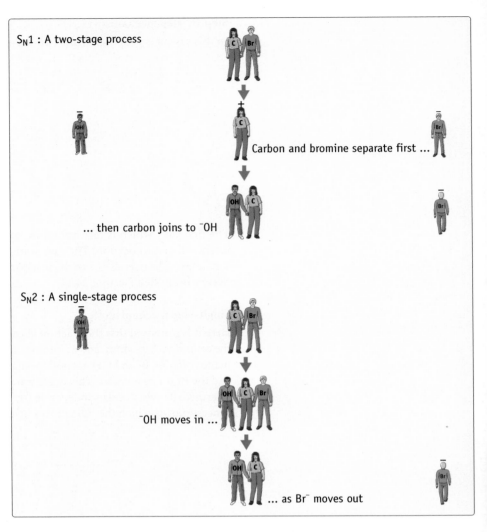

SN1 : A two-stage process

Carbon and bromine separate first ...

... then carbon joins to ⁻OH

SN2 : A single-stage process

⁻OH moves in ...

... as Br⁻ moves out

Figure 29.3
Breaking up is hard to do ...

Nature of the solvent in which the reaction is carried out

Polar solvents, particularly water, favour ion formation and therefore the two-step S_N1 mechanism. Conversely, the single-step mechanism S_N2 is favoured in non-polar solvents.

How do we know which mechanism operates?

Chemists have investigated the mechanism of nucleophilic substitution reactions by studying the rate of reactions of the type

$$RBr + {}^-OH \rightarrow ROH + Br^-.$$

They carried out experiments to find the rate law for different halogenoalkanes RBr, and under different conditions (see chapter 24 for more about reaction rates and rate laws). Chemists have found that two rate laws are possible.

(A) Rate = k[RBr] (i.e. overall first order; rate law does not involve ⁻OH)
(B) Rate = k[RBr] [⁻OH] (i.e. overall second order; rate law includes ⁻OH)

It turns out that the reaction involving 2,2-dimethylbromoethane

$$CH_3 - \underset{\underset{CH_3}{|}}{\overset{\overset{CH_3}{|}}{C}} - Br + {}^-OH \longrightarrow CH_3 - \underset{\underset{CH_3}{|}}{\overset{\overset{CH_3}{|}}{C}} - OH + Br^-$$

is first order, with the type of rate law in **A**. This means that the slowest, rate-determining step of the reaction cannot involve ⁻OH at all, since it does not appear in the rate law equation. Therefore, chemists conclude that for this reaction the mechanism involves the formation of a carbocation as the slow, rate-determining step – it is the two-stage S_N1 mechanism. Once the carbocation has formed, it reacts quickly with ‾OH – this quick second stage has no influence on the overall rate of the reaction. The term S_N1 is used to show that only 1 species, RBr, is involved in the rate-determining step: **S**ubstitution **N**ucleophilic **1**. This mechanism is as we would expect for 2,2-dimethylbromoethane, because the three methyl groups stabilise the intermediate carbocation.

However, chemists found that the corresponding reaction involving bromomethane

$$CH_3Br + ‾OH \rightarrow CH_3OH + Br^-$$

is second order, with the type of rate law in **B**. This means that the rate-determining step must involve both ‾OH and RBr, since both appear in the rate law equation. Therefore, we conclude that for this reaction the mechanism involves a single step in which the two reactants form a transition state: an S_N2 mechanism. The term S_N2 is used to show that **2** species, RBr and ‾OH, are involved in the rate-determining step – **S**ubstitution **N**ucleophilic **2**.

Chemists have used such rate law studies to investigate the reaction mechanisms for substitution reactions involving many different types of compounds.

> **1** Consider the reaction between bromobutane and I‾ ions:
>
> $$CH_3CH_2CH_2CH_2Br + I^- \rightarrow CH_3CH_2CH_2CH_2I + Br^-$$
>
> The reaction is carried out in a propanone solvent. The rate law for this reaction is found to be
>
> $$Rate = k[CH_3CH_2CH_2CH_2Br][I^-]$$
>
> (a) Which mechanism, S_N1 or S_N2, operates in this reaction? How do you know?
> (b) Describe the mechanism of the reaction.
>   You will find the answer on page 462.

Testing for halogenoalkanes

We can use alkaline hydrolysis to find out which halogen is present in a halogenoalkane. The halogenoalkane is boiled with aqueous sodium hydroxide, then cooled and neutralised with dilute nitric acid. Then silver nitrate solution is added. A white precipitate of silver chloride indicates a chloroalkane; a pale yellow precipitate indicates a bromoalkane and a yellow precipitate indicates an iodoalkane. For example:

$$RCl + ‾OH \rightarrow ROH + Cl^-$$

$$Ag^+(aq) + Cl^-(aq) \rightarrow AgCl(s)$$

29.5 Important substitution reactions of halogenoalkanes

Organic halogen compounds have important uses. Several of these have already been mentioned, and their use as insecticides is considered in section 16.10.

Organic halogen compounds are uncommon in nature so they have to be synthesised. They are usually manufactured from alkenes or alkanes. Chloroethane, for example, is made by the addition reaction between ethene and hydrogen chloride. Tetrachloromethane is made by the free radical substitution reaction between chlorine and methane (section 26.6). In the laboratory, halogenoalkanes are usually prepared from alcohols (see section 30.6).

As well as being useful in their own right, halogenoalkanes are important intermediates in synthesis. They can be used to introduce a reactive site into a hydrocarbon molecule. The reactive halogen can then be substituted by another group which could not be introduced directly. Examples of groups that can be introduced in this way are discussed later in this section.

This kind of synthesis is important in small-scale preparations such as those carried out in the laboratory or in the manufacture of pharmaceuticals. Many drugs have complicated organic molecules and their synthesis involves building up a complex molecule from a simple starting compound. Such synthesis often involves many steps. For example, the synthesis of phenobarbitone (a sleep-inducing drug) from methylbenzene involves eight steps (see figure 29.4). Note in figure 29.4 how a chlorine atom is introduced in the first step. Chlorine provides an active site in the relatively unreactive methylbenzene molecule so that other groups can then be introduced.

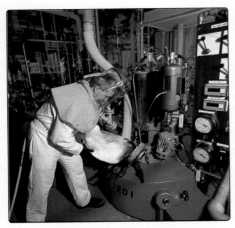

▲ Small-scale synthesis of a pharmaceutical. Laboratory methods are used and the reagents are often expensive.

Figure 29.4
Synthesis of phenobarbitone from methylbenzene. Phenobarbitone is a hypnotic, used in sleeping pills.

The synthetic methods described in this chapter are used for small-scale operations, but they are relatively expensive and therefore not used for large-scale, high-tonnage industrial operations. On this scale high-pressure, high-temperature catalytic processes, very different from those used in the laboratory, tend to be used. These processes involve expensive plant and equipment, but cheap reagents. In the long run they are cheaper than small-scale laboratory processes. The subject of organic synthesis is considered in more detail in chapter 34.

Halogenoalkanes undergo substitution reactions with a wide range of nucleophiles. Species that can act as nucleophiles include not only those carrying a full negative charge (^-OH, ^-CN, CH_3COO^-, $CH_3CH_2O^-$) but also neutral molecules carrying an unshared pair of electrons (H_2O, NH_3).

> 1 Predict the structure of the compound formed by the reaction of bromoethane with each of the nucleophiles mentioned above. The answers are given below, together with the conditions needed for the reaction. Try not to look at these until you have predicted the structures yourself.

Reaction with hydroxide ion
This has already been dealt with (page 456).

Reaction with cyanide ion
When bromoethane is heated under reflux with a solution of potassium cyanide in ethanol, propanenitrile is formed:

$$CH_3CH_2Br + {}^-CN \rightarrow CH_3CH_2CN + Br^-$$
propanenitrile

This reaction is useful in synthesis as a means of increasing the length of a carbon chain (section 34.3).

▲ Chloroalkanes are excellent solvents. Dichloromethane will even dissolve dried paint, so it is used as paint stripper. Unlike trichloromethane and tetrachloromethane, its toxicity is low, so it is safer to use.

Describe the reagents and conditions you would use to prepare:

1 propane-1,2-diol from 1,2-dibromopropane,
2 2-methylpropanenitrile from 2–iodopropane,
3 methoxyethane from chloroethane.

Reaction with ethanoate ion

If bromoethane is warmed with dry silver ethanoate, ethyl ethanoate, an ester, is formed:

$$CH_3COO^-Ag^+(s) + CH_3CH_2Br(l) \rightarrow CH_3COOCH_2CH_3(l) + AgBr(s)$$
silver ethanoate   ethyl ethanoate

The reaction is encouraged by using the silver salt, which removes bromide ion as insoluble silver bromide.

Reaction with ethoxide ion

Bromoethane reacts with an ethanolic solution of sodium ethoxide, $Na^+\ ^-OCH_2CH_3$ (see section 30.5) to form an ether, ethoxyethane:

$$CH_3CH_2Br + CH_3CH_2O^- \rightarrow CH_3CH_2OCH_2CH_3 + Br^-$$
ethoxide ion   ethoxyethane

The ethoxide ion can also act as a base rather than as a nucleophile under these conditions, in which case a different product, ethene, is formed (see next section).

Just as both water and hydroxide ion can act as nucleophiles, so both ethanol and ethoxide ion give nucleophilic substitution reactions with halogenoalkanes. Ethanol reacts considerably more slowly than ethoxide ion, but the product is the same:

$$CH_3CH_2Br + CH_3CH_2OH \rightarrow CH_3CH_2OCH_2CH_3 + HBr$$

Reaction with water

This has already been dealt with (page 456). In the case of bromoethane the product is ethanol, though the reaction is slow even when heated:

$$CH_3CH_2Br + H_2O \rightarrow CH_3CH_2OH + HBr$$
ethanol

Reaction with ammonia

Ammonia is a nucleophile because of the lone pair of electrons on its nitrogen atom. When bromoethane is heated with a concentrated aqueous solution of ammonia in a sealed tube, ethylamine is formed:

$$CH_3CH_2Br\,(l) + NH_3\,(aq) \rightarrow CH_3CH_2NH_2\,(aq) + HBr\,(aq)$$
ethylamine

The nitrogen atom in ethylamine still possesses a lone pair, so it can still act as a nucleophile. With excess bromoethane, further substitution therefore occurs:

$$CH_3CH_2Br + CH_3CH_2NH_2 \rightarrow CH_3CH_2NHCH_2CH_3 + HBr$$
ethylamine   diethylamine

Still further substitution can occur:

$$CH_3CH_2Br + (CH_3CH_2)_2NH \rightarrow (CH_3CH_2)_3N + HBr$$
diethylamine   triethylamine

and finally:

$$CH_3CH_2Br + (CH_3CH_2)_3N \rightarrow (CH_3CH_2)_4N^+Br^-$$
triethylamine   tetraethylammonium bromide

In practice, a mixture of four products is formed: ethylamine, diethylamine, triethylamine and tetraethylammonium bromide.

The reactions mentioned above have all been illustrated by reference to bromoethane, but they are applicable to halogenoalkanes generally. Many different compounds can be synthesised from halogenoalkanes and this is considered further in chapter 34.

Halogenoalkanes can also react with other halide ions. In the example on page 459 the reaction between brombutane and iodide ions takes place by an S_N2 mechanism:

$$CH_3CH_2CH_2 - \overset{\displaystyle \overset{\big|}{C}}{\underset{\overset{\big|}{Br}\ \ H\ \ H}{}} + I^- \longrightarrow \left[CH_3CH_2CH_2 - \underset{H\ \ \ \ \ H}{\overset{\big|}{Br}---- \overset{\big|}{C} ---- I} \right]^- \longrightarrow CH_3CH_2CH_2 - \underset{H\ \ \ H\ \ I}{\overset{\displaystyle \overset{\big|}{C}}{}} + Br^-$$

29.6 Elimination reactions

In the reactions we have considered so far, $^-$OH has acted as a nucleophile in its reactions with halogenoalkanes. For example, when 2-bromopropane is heated with aqueous sodium hydroxide, the hydroxide ion acts as a nucleophile and substitution occurs:

$$\underset{\text{2-bromopropane}}{CH_3 - \overset{\overset{\displaystyle Br}{\big|}}{CH} - CH_3}\ \text{(aq)} + {}^-OH\ \text{(aq)} \rightarrow \underset{\text{propan-2-ol}}{CH_3 - \overset{\overset{\displaystyle OH}{\big|}}{CH} - CH_3}\ \text{(aq)} + Br^-\ \text{(aq)}$$

Under certain conditions, $^-$OH can act as a *base* instead of a nucleophile. When $^-$OH behaves in this way, it removes H^+ from a halogenoalkane. The C—Br bond breaks at the same time as $^-$OH removes H^+ from the neighbouring carbon atom, and an alkene is formed. For example, when 2-bromopropane is heated with a solution of sodium hydroxide in *ethanol* instead of in water, propene is formed:

$$\text{i.e. } CH_3CH_2Br + {}^-OH \longrightarrow CH_2 {=} CH_2 + H_2O + Br^-$$

The overall effect of this reaction is to eliminate HBr from the molecule of 2-bromopropane, so this is an elimination reaction. Notice that the reagent, sodium hydroxide, is the same as for the substitution reaction outlined at the beginning of this section. In the substitution reaction, $^-$OH acted as a nucleophile; in the elimination reaction it acts as a base. By altering the conditions, we can alter the manner in which it acts: in aqueous solution it behaves as a nucleophile, causing substitution; in ethanolic solution it behaves as a base, causing elimination. Under each set of conditions, both substitution and elimination can in fact occur, but by controlling the conditions we can ensure that one particular reaction occurs to a greater extent than the other.

1 Try to predict the effect of an elimination reaction in a halogenoalkane that contains *two* halogen atoms. For example, what would be the product if 1,2-dibromoethane were heated under reflux with a solution of sodium hydroxide in ethanol?

29.7 Acyl halides

So far we have considered molecules in which a halogen atom is attached directly to an alkyl or aryl group. An important group of halogenated compounds is the **acyl halides**,

in which a halogen atom is attached to a $\ \overset{\diagdown}{\underset{\diagup}{C}} {=} O$ group.

An example is ethanoyl chloride, $CH_3 - \overset{\displaystyle C}{\underset{\displaystyle \|}{\underset{\displaystyle O}{}}} - Cl$, usually written CH_3COCl.

Acyl halides have the general formula **RCOHal**, where R— is an alkyl or aryl group. They are named using the suffix **-oyl halide** after a stem indicating the number of carbon atoms in the molecule. Thus CH_3CH_2COBr is called propanoyl bromide and $CH_3CH_2CH_2COCl$ is called butanoyl chloride.

Like halogenoalkanes, acyl chlorides undergo nucleophilic substitution reactions in which the halogen is substituted by a nucleophile. Acyl halides undergo nucleophilic substitution much more readily than the halogenoalkanes. For example, when water is added to ethanoyl chloride (which is a liquid), a violent reaction occurs.

The liquid boils and clouds of hydrogen chloride fumes are evolved:

$$CH_3COCl + H_2O \rightarrow CH_3COOH + HCl$$
$$\text{ethanoic acid}$$

The reaction of acyl halides with nucleophiles involves a different mechanism to that of halogenoalkanes. The first stage of the reaction involves *addition* of the nucleophile across the C=O double bond. (Nucleophilic addition is described further in section 31.3.) The double bond is reformed at a later stage of the reaction. Figure 29.5 shows the mechanism in detail.

Nucleophilic **addition** ...   ... followed by **elimination** of HCl.

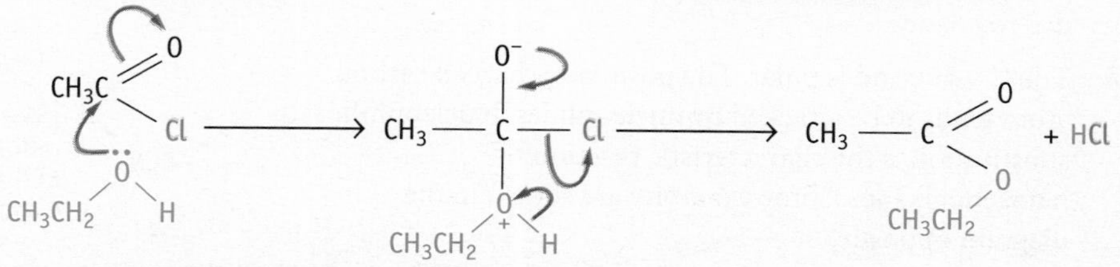

This is a **nucleophilic addition–elimination** reaction.

This different reaction mechanism enables acyl halides to react with nucleophiles much more rapidly than halogenoalkanes do. Acyl chlorides react with all the nucleophiles that will substitute halogenoalkanes.

Like halogenoalkanes, acyl halides are important in synthesis. Just as RHal is used to attach an alkyl group R to a nucleophile, RCOHal is used to attach an acyl group RCO. Because of their greater reactivity, acyl halides will react rapidly with those nucleophiles, such as water and alcohols, that react only slowly with halogenoalkanes.

Some reactions of ethanoyl chloride follow.

- *With ammonia.* Ethanoyl chloride reacts violently with an aqueous solution of ammonia at room temperature. The product is ethanamide, an amide. Amides have the general formula $RCONH_2$ and should not be confused with amines, RNH_2.

$$CH_3COCl + NH_3 \rightarrow CH_3CONH_2 + HCl$$
$$\text{ethanamide}$$

The —C=O group in ethanamide has a strong electron-withdrawing effect. It tends to withdraw electrons from the —NH_2 group, making the nitrogen atom much less nucleophilic than the N in NH_3 or in amines such as $CH_3CH_2NH_2$.
Ethanamide is therefore not a strong nucleophile, despite carrying a lone pair on its nitrogen atom. As a result, further substitution does not occur and ethanamide is the only product. Compare this with the reaction of chloroethane with ammonia, which gives several products.

- *With primary amines.* Primary amines can be thought of as ammonia substituted with an alkyl or aryl group. Primary amines react with acyl chlorides in a similar way to ammonia. The reaction is vigorous at room temperature. For example:

$$CH_3COCl + CH_3CH_2NH_2 \rightarrow CH_3CONHCH_2CH_3 + HCl$$
$$\textit{N}\text{-ethylethanamide}$$

The product is a substituted amide.

- *With alcohols.* Ethanoyl chloride reacts vigorously with alcohols at room temperature to form esters. For example, with ethanol:

$$CH_3CH_2OH + CH_3COCl \rightarrow CH_3COOCH_2CH_3 + HCl$$
$$\text{ethyl ethanoate}$$

As expected, the reaction occurs much more readily than the corresponding reaction of chloroethane with ethanol.

The reactions of ethanoyl chloride described above are quite general and apply to all acyl chlorides.

Figure 29.5
The mechanism of acylation. This example shows the reaction of ethanoyl chloride with ethanol to form ethyl ethanoate.

1 Predict the structure of the product of the reaction between ethanoyl chloride, CH_3COCl, and each of the nucleophiles NH_3 and CH_3CH_2OH. The answers appear below.

Summary

1 Halogenoalkanes are named using a prefix (*chloro-*, etc.) to indicate the nature of the halogen atom and a number to indicate its position on the alkane chain.

2 The C—F bond is very strong. Fluorocarbons are very stable.

3 Chlorofluorocarbons (CFCs) break down in the stratosphere to produce reactive Cl• radicals which destroy ozone.

4 The C—H bond is polar. The positively charged carbon atom tends to be attacked by nucleophiles. Nucleophilic substitution is the characteristic reaction of halogenoalkanes. Some examples are shown in the diagram opposite.

5 There are two mechanisms for nucleophilic substitution. The S_N1 mechanism involves two steps and an intermediate carbonium ion; the S_N2 mechanism involes a single step

6 The Ar—Hal (Ar = aryl group) bond is considerably less reactive than the R—Hal bond.

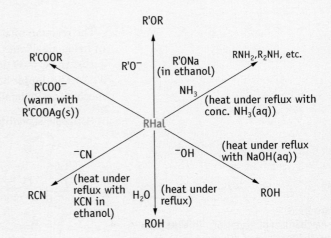

7 In the presence of a strong base, halogenoalkanes can undergo elimination reactions, forming alkenes.

8 Acyl halides contain the COHal group. They undergo nucelophilic substitution considerably more readily than halogenoalkanes.

Review questions

1 The table below gives formulas of some halogenoalkanes and their boiling points.

Compound	Formula	Boiling point/°C
A	CCl_3F	24
B	$F_2C—CF_2$	–6
C	$F_2C—CF_2$ $CCl_2F—CClF_2$	48
D	$CBrF_3$	149

(a) Name each compound.
(b) Which might be used as an aerosol propellant?
(c) Which might be used as a degreasing solvent?
(d) Which might be used as a fire extinguisher?

2 Consider the following compounds.

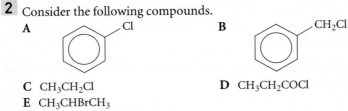

C CH_3CH_2Cl
D CH_3CH_2COCl
E $CH_3CHBrCH_3$

(a) Which are halogenoalkanes?
(b) Which is an acyl halide?
(c) Which would react most readily with cold water?
(d) Which would react least readily with aqueous sodium hydroxide?
(e) Which would form an amide when reacted with aqueous ammonia?
(f) Which would have the lowest boiling point?

3 This question concerns the hydrolysis of three different halogenoalkanes, 1-chlorobutane, 1-bromobutane and 1-iodobutane.
Four drops of each halogenoalkane are added separately to three separate tubes standing in a water-bath at 60°C. Each tube contains 1 cm³ of 0.1 mol dm⁻³ silver nitrate solution. The results are as follows.
1-Chlorobutane – slight cloudiness after three minutes. Still only slightly cloudy after 15 minutes.
1-Bromobutane – slightly cloudy after one minute, opaque after three minutes, coagulation and precipitation after six minutes.
1-Iodobutane – immediately opaque, yellow precipitate within first minute.

(a) What is the precipitate formed in each case?
(b) Why does the precipitate form?
(c) In each case a substitution reaction is occurring between the halogenoalkane and a nucleophile. What is the nucleophile involved?
(d) Which halogenoalkane undergoes substitution most readily and which least readily?
(e) Explain your answer to (d), given the following bond energies in kJ mol⁻¹: C—Cl (in chloroethane) 339; C—Br (in bromoethane) 284; C—I (in iodoethane) 218.

4 Predict the products of reactions between the following pairs of substances.

(a) $CH_3CH_2CH_2Br$ and NH_3

(b) $CH_2ICH_2CH_2I$ and $CH_3CH_2CH_2COO^-Ag^+$

(c) 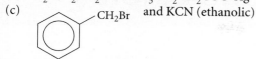 CH_2Br and KCN (ethanolic)

(d) 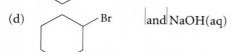 Br and NaOH(aq)

(e) 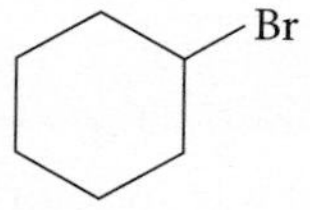 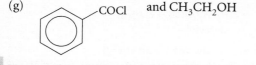 Br and NaOH (ethanolic)

(f) 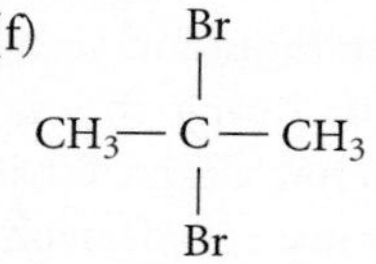
$$CH_3 - \underset{\underset{Br}{|}}{\overset{\overset{Br}{|}}{C}} - CH_3$$
and $CH_3O^-Na^+$

(g) COCl and CH_3CH_2OH

5 Use your knowledge of nucleophilic substitution reactions to predict the structural formulas of the products of reactions between bromoethane and the following.

(a) sodium hydrogensulphide ($Na^{+-}SH$)

(b) potassium nitrite ($K^+NO_2^-$)

(c) potassium chloride

(d) lithium tetrahydridoaluminate, $LiAlH_4$ (nucleophile = H^-)

(e) sodamide ($Na^{+-}NH_2$).

6 Two compounds **X** and **Y** have the same molecular formula, C_3H_5OCl. **X** reacts vigorously with cold water to give a strongly acidic solution, but **Y** has no visible reaction with cold water. Suggest structural formulas for **X** and **Y**.

7 The reaction of 3-chloro-3-ethylpentane with aqueous sodium carbonate at room temperature yields two products, one an alcohol and one an alkene.

(a) Write the structural formula of 3-chloro-3-ethylpentane.

(b) Consider the reaction that produces the alcohol.
   (i) What type of reaction is this?
   (ii) Write the structural formula of the alcohol.
   (iii) Write an equation for the reaction.

(c) Consider the reaction that produces the alkene.
   (i) What type of reaction is this?
   (ii) Suggest a possible structural formula for the alkene.
   (iii) Write an equation for the reaction.

(d) What changes in reaction conditions might favour the formation of the alkene?

8 Give reagents, reaction conditions and equations to show how you would convert

(a) $CH_2{=}CH_2$ to CH_3CH_2CN (2 steps)

(b) 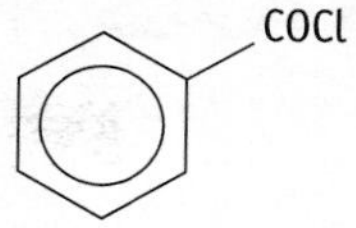 COCl to CONH$_2$ (1 step)

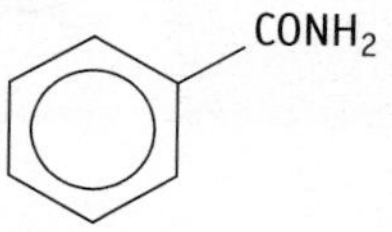

(c) 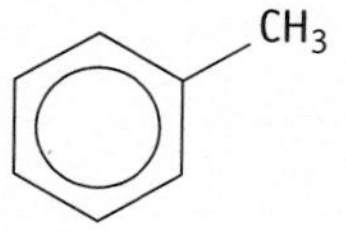 CH$_3$ to CH$_2$OH (2 steps)

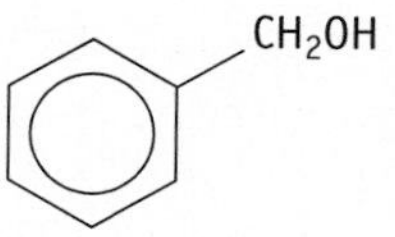

9 Identify the compounds **A** to **D** and the reagents **P** and **Q** in the reaction scheme below.

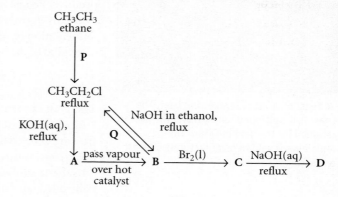

10 Suggest explanations for each of the following observations.

(a) Iodobenzene is more reactive than chlorobenzene towards nucleophiles, but much less reactive than iodoethane.

(b) Tetrachloromethane, CCl_4, is much less reactive towards nucleophiles than chloromethane, CH_3Cl.

(c) Iodo- and bromo-compounds are more useful than chloro-compounds as intermediates in synthesis.

(d) When chloroethanoyl chloride, $ClCH_2COCl$, is added to water, one chlorine atom is quickly substituted to give $ClCH_2COOH$, but the other Cl atom is only substituted slowly, even on boiling.

30 Alcohols, Phenols and Ethers

30.1 Fermentation

When Noah left the ark he promptly planted a vineyard and was soon drinking its produce. People have been fermenting grape juice for at least ten thousand years, and probably fermenting honey for even longer. The reason for doing this is that fermentation of sugar by yeast produces ethanol, the well-known intoxicant. Ethanol is a member of the homologous series of alcohols and is commonly just called 'alcohol'.

$$C_6H_{12}O_6(aq) \rightarrow 2CH_3CH_2OH(aq) + 2CO_2(g)$$
$$\text{a sugar} \qquad \text{ethanol}$$

This is an exothermic reaction that provides the yeast with energy.

Alcoholic fermentation occurs naturally wherever sugar-containing materials such as fruit are allowed to decay. In autumn it is quite common to see drunken (and therefore dangerous) wasps which have been feeding off fermented fruit.

In the manufacture of alcoholic drinks, fermentation is carried out under more controlled conditions. The source of sugar varies: it may be grapes (for wine), honey (for mead), malted grain (for beer), apples (for cider) or indeed any sugar-containing fruit or plant. From the point of view of yeast, the ethanol produced in fermentation is a toxic waste product, which kills the yeast at concentrations greater than about 15% by volume. It is therefore impossible to produce alcoholic drinks containing more than 15% alcohol by fermentation alone.

There are reasons for producing beverages with greater alcoholic content than 15%. For one thing they can be stored longer, because alcohol in high concentration is toxic to bacteria. Another reason is simply that some people like drinks containing alcohol in high concentrations.

To satisfy these requirements, fermented liquids are distilled to increase their alcoholic concentration and produce *spirits*. For example, distillation of wine produces brandy. A typical spirit contains about 40% ethanol.

Alcohols are toxic

All the members of the homologous series of alcohols are toxic to a greater or lesser extent. The first member of the series, methanol, is much more toxic than ethanol. It is added to industrial alcohol (on which no excise tax is charged) to make it undrinkable. It is then called methylated spirit. Unfortunately a few people drink it nevertheless, the eventual result being blindness or death.

The higher alcohols are moderately toxic, unpleasant-tasting compounds. A mixture of these compounds is produced in small amounts during fermentation: the mixture is called *fusel oil* (*fusel* is German for 'bad liquor').

Ethanol itself is intoxicating in small amounts but toxic in large amounts. In small amounts it makes people feel relaxed and for this reason it undoubtedly has its social uses. In large amounts it has a serious effect on mental and physical performance and its use can be very dangerous, particularly to drivers of motor vehicles. It is addictive if regularly taken in large quantities. Ethanol is firmly established as an accepted social drug with many advantages, but there is little doubt that if it were newly introduced today and its properties were known, it would be banned as a dangerous drug.

▲ Yeast on a vat of fermenting beer. The yeast has multiplied ten-fold and the best is retained for the next brewing. The remainder is sold to manufacturers of yeast extract.

▲ Methylated spirit is ethanol with methanol added to make it undrinkable. A dye is often added too.

Manufacturing ethanol

In addition to its use in alcoholic drinks, ethanol is an important industrial chemical. It is used as a solvent for perfumes, lacquers and inks, and as a fuel (see section 30.7). Ethanol can be manufactured by two methods.

- *By fermentation.* Industrial ethanol can be manufactured by fermentation of cheap carbohydrates such as grain or molasses, followed by distillation. This method has the advantage that it uses renewable raw materials, but it is used mainly in tropical countries where the raw materials can be easily and cheaply grown.
- *From ethene.* In the UK, where the raw materials for fermentation are relatively expensive, ethanol is normally manufactured by the catalytic hydration of ethene (see section 27.6).

30.2 Naming alcohols

In this chapter we shall be considering organic **hydroxy compounds,** containing the —OH group. A study of the properties of the —OH group is important to chemists because of the industrial importance of compounds containing this functional group and because of its wide occurrence in biological molecules.

Aliphatic and aromatic hydroxy compounds differ considerably in their properties. They are, therefore, regarded as two distinct groups of compounds: aliphatic hydroxy compounds are called **alcohols** and aromatic ones are called **phenols**.

Alcohols are named using the suffix -ol, preceded if necessary by a number to indicate its position in the carbon skeleton. Thus, CH_3OH is methanol, $CH_3CH_2CH_2OH$ is propan-1-ol, $CH_3CHOHCH_3$ is propan-2-ol and

$$CH_3\text{—}\underset{\underset{OH}{|}}{\overset{\overset{CH_3}{|}}{C}}\text{—}CH_3 \text{ is 2-methylpropan-2-ol.}$$

Alcohols with structures of the form RCH_2OH, where R is an alkyl or aryl group or hydrogen, are called **primary** alcohols. Methanol and propan-1-ol are primary alcohols.

Those with the structure $\overset{R^1}{\underset{R^2}{>}}CHOH$ are called **secondary** alcohols, for example propan-2-ol. Alcohols with the structure $R^2\text{—}\underset{R^3}{\overset{R^1}{C}}OH$ are called **tertiary** alcohols, for example 2-methylpropan-2-ol. Primary, secondary and tertiary alcohols have some important differences in chemical reactivity (see section 30.7).

Some alcohols, particularly biologically occurring ones, contain more than one —OH group in their molecule. They are known as **polyhydric** alcohols. They are named using the suffixes **-diol**, **-triol**, etc., depending on how many —OH groups they contain. Thus,

$HOCH_2CH_2OH$ ethane-1,2-diol,
$HOCH_2CHOHCH_2OH$ propane-1,2,3-triol.

Compounds in which the —OH group is attached to an aromatic ring are called phenols. The simplest and most important is phenol itself:

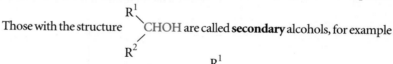

Phenols have the —OH group attached *directly* to the ring.

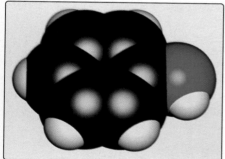

Figure 30.1
Computer-generated models of molecules of ethanol and phenol.

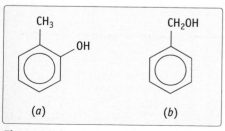

(a)   *(b)*

Figure 30.2
Which of these two compounds would be classed as a phenol?

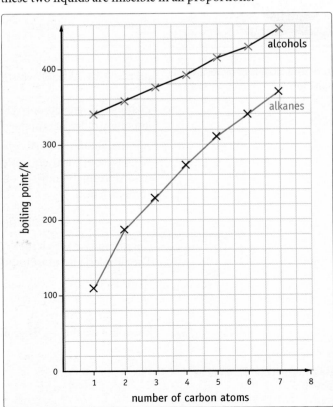
$H — O$
  H
   $O — H$
  H

(*a*) between water molecules

$CH_3CH_2 — O$
   H
    $O — H$
  CH_3CH_2

(*b*) between ethanol molecules

$CH_3CH_2 — O$
   H
    $O — H$
  H

(*c*) between ethanol and water

Figure 30.3
Hydrogen bonding.

Look at the graphs in figure 30.4 and answer the following questions.

1 Why are the boiling points of alcohols higher than those of the corresponding alkanes?

2 Why do the differences in boiling points between corresponding alcohols and alkanes get less as the number of carbon atoms increases?

3 Where would the two graphs intersect, and what is the physical significance of the point of intersection?

Figure 30.4
Boiling points of the first seven straight-chain alkanes and straight-chain primary alcohols.

---

1 One of the two compounds shown in figure 30.2 is a phenol. Which one?

2 Name the following:
   (a) $CH_3CHOHCH_2CH_3$

   (b) 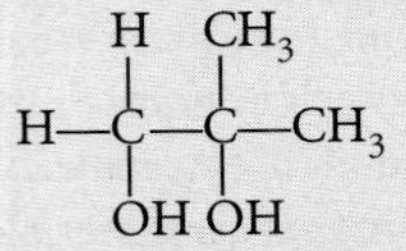

   $H — C — C — CH_3$ with H and CH_3 on top, OH and OH on bottom

3 Write formulas for the following:
   (a) butane-1,2,4-triol

   (b) 2-methylpentan-2-ol

4 From the previous questions, classify those alcohols containing a single —OH group as primary, secondary or tertiary.

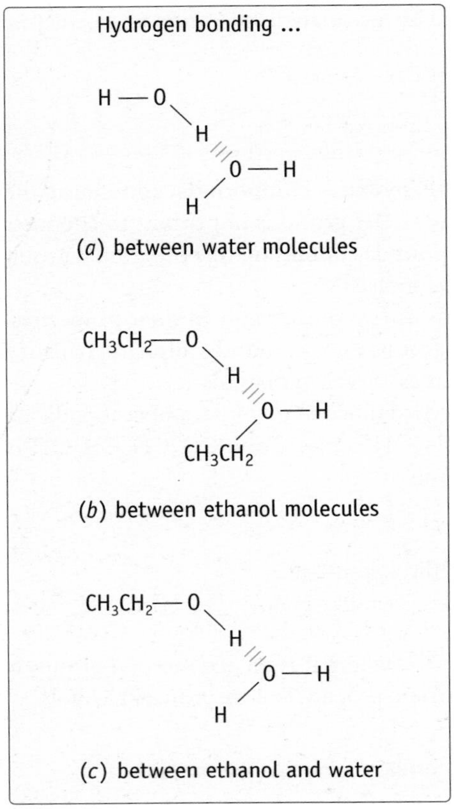

30.3 Alcohols as a homologous series

Industrially, alcohols are manufactured by the hydration of alkenes (see page 425). In the laboratory they can be made by nucleophilic substitution reactions between halogenoalkanes and hydroxide ions, as described in section 29.4.

The alcohols illustrate very well the steady change in physical properties when a homologous series is ascended. The —OH group has a big effect on the physical properties of any molecule of which it is a part. Alcohols show marked physical differences from the organic compounds we have met so far.

Consider three compounds of similar relative molecular mass: propane ($M_r = 44$), chloromethane ($M_r = 50.5$) and ethanol ($M_r = 46$). Ethanol is far less volatile than the other two: it is a liquid at room temperature while the others are gases (boiling points 231 K, 249 K and 351 K, respectively). Furthermore, ethanol is soluble in water in all proportions, while the other two are practically insoluble.

Hydrogen bonding in ethanol

The cause of ethanol's anomalous properties is hydrogen bonding (see section 9.5). Hydrogen bonds form between —OH groups of adjacent ethanol molecules. This gives ethanol relatively high intermolecular forces and therefore relatively low volatility (see figure 30.3). Hydrogen bonding between ethanol molecules and water molecules explains why these two liquids are miscible in all proportions.

Pure ethanol has a strong affinity for water and tends to absorb it from the atmosphere. Furthermore, it is impossible to separate pure water from an ethanol–water mixture by distillation alone. The two liquids form a constant-boiling mixture containing 95.6% ethanol which distils over unchanged.

To produce 100% ethanol (called **absolute alcohol**) the remaining 4.4% water must be removed by a chemical drying agent such as calcium oxide.

Figure 30.4 shows the boiling points of the first seven straight-chain primary alcohols and the first seven straight-chain alkanes.

As the homologous series of alcohols is ascended, the influence of the —OH group becomes less and less important compared with that of the increasingly large hydrocarbon portion. The properties of the higher alcohols therefore tend more and more towards those of the corresponding alkane. This trend can be seen in solubility as well as volatility, as table 30.1 shows.

Table 30.1
Solubility of alcohols in water

Name	Formula	Solubility/g per 100 g of water
methanol	CH_3OH	infinite*
ethanol	CH_3CH_2OH	infinite*
propan-1-ol	$CH_3CH_2CH_2OH$	infinite*
butan-1-ol	$CH_3CH_2CH_2CH_2OH$	8.0
pentan-1-ol	$CH_3CH_2CH_2CH_2CH_2OH$	2.7
hexan-1-ol	$CH_3CH_2CH_2CH_2CH_2CH_2OH$	0.6

* miscible in all proportions

Alcohols as solvents

The lower alcohols such as ethanol tend to be good solvents for polar as well as non-polar solutes. This is because they contain both a highly polar —OH group and a non-polar hydrocarbon portion. For example, both sodium hydroxide (ionic) and hexane (molecular) dissolve well in ethanol. This property makes ethanol, and to a lesser extent methanol and propanol, valuable solvents in the laboratory and in industry. One example is the use of ethanol as a base for perfumes and aftershave lotions: the ethanol dissolves both the water-insoluble oils which provide the aroma and the water which makes up the bulk of the preparation.

Hydrogen bonding between —OH groups also has an effect on the viscosity of alcohols, particularly those with more than one —OH group in their molecule. Thus ethanol has a viscosity of 1.06×10^{-3} N s m^{-2} at 298 K, about the same as water. But propane-1,2,3-triol (commonly called glycerine) is very thick and sticky, with a viscosity of 942×10^{-3} N s m^{-2} at the same temperature. This is because of extensive interaction between its molecules, which carry three —OH groups each.

In phenol, the large, non-polar benzene ring to some extent dominates the OH group. Thus phenol is only partially soluble in water (9.3 g in 100 g water at 20°C). It is a crystalline solid, melting point 43°C.

▲ Windscreen deicing fluid contains propan-2-ol. This alcohol mixes completely with water and has a low freezing point.

30.4 The amphoteric nature of hydroxy compounds

We can think of alcohols as being derived from water, by replacing one hydrogen atom by an alkyl group. If we replace both hydrogens by alkyl groups, we get an **ether** (see section 30.9).

H—O—H	R—O—H	R—O—R
water	an alcohol	an ether

Alcohols might therefore be expected to show some similarity to water, and this is true of their physical properties (see section 30.3).

Water is an amphoteric compound. It can act as an acid, donating a proton, or as a base, accepting a proton:

$$\text{as an acid:} \quad H_2O \rightarrow {}^-OH + H^+$$

$$\text{as a base:} \quad H_2O + H^+ \rightarrow H_3O^+$$

$$\text{overall:} \quad H_2O + H_2O \rightarrow {}^-OH + H_3O^+$$
$$\qquad\qquad\quad \text{acid} \quad \text{base}$$

Alcohols also show amphoteric behaviour:

$$\text{as an acid:} \quad ROH \rightarrow RO^- + H^+$$

$$\text{as a base:} \quad ROH + H^+ \rightarrow ROH_2{}^+$$

$$\text{overall:} \quad ROH + ROH \rightarrow RO^- + ROH_2{}^+$$
$$\qquad\qquad\quad \text{acid} \quad \text{base}$$

When it acts as an acid, the alcohol cleaves at the O—H bond. When it acts as a base, it can subsequently cleave at the R—O bond. Both these forms of bond cleavage are characteristic of hydroxy compounds, and we will consider them separately. In doing so we will take ethanol as a typical aliphatic hydroxy compound and phenol as a typical aromatic one.

30.5 Reactions involving cleavage of the O—H bond

Acidity

Table 30.2 shows how sodium reacts with water, with ethanol and with a solution of phenol in ethanol.

Table 30.2
Reaction of a small piece of sodium with different hydroxy compounds

Water	Ethanol	Solution of phenol in ethanol
sodium floats, melts, rushes about on surface, rapid evolution of hydrogen	sodium sinks, does not melt; steady evolution of hydrogen	sodium sinks, does not melt; rapid evolution of hydrogen

Look at the table and answer these questions.

1 Why does sodium float in water but sink in ethanol?

2 The general reaction in each case is the reduction of H^+ ions by sodium. Write a general ionic equation.

3 Judging from these reactions, which contains a higher concentration of H^+ ions, water or ethanol?

4 Which contains a higher concentration of H^+ ions, ethanol or a solution of phenol in ethanol?

5 Which is the stronger acid, water or ethanol?

6 Which is the stronger acid, phenol or ethanol?

7 Write a full equation for the reaction of ethanol with sodium.

The fact that ethanol reacts with sodium liberating hydrogen suggests that the alcohol contains some hydrogen ions and that these are being reduced by sodium:

$$2Na + 2H^+ \rightarrow 2Na^+ + H_2$$

The fact that the reaction is slower with ethanol than with water suggests that ethanol contains a lower concentration of H^+ than water, i.e. it is a weaker acid. In fact, the K_a of ethanol is 10^{-18} mol dm^{-3} whereas the value for water is 10^{-16} mol dm^{-3}. Remember, the

▲ The antiseptic action of 'Coal Tar Soap' (*above*) depended on the phenol and related compounds it contained. Phenol is still used as an antiseptic, for example in some throat sprays (*below*). ▼

higher the K_a, the stronger the acid (see section 22.9). We can see why ethanol is a weaker acid if we compare the equilibria:

$$CH_3CH_2OH + H_2O \rightleftharpoons CH_3CH_2O^- + H_3O^+$$
$$\text{ethoxide ion}$$

$$H_2O + H_2O \rightleftharpoons HO^- + H_3O^+$$
$$\text{hydroxide ion}$$

The ethoxide ion is more basic than the hydroxide ion, because of the tendency of alkyl groups to donate electrons (the inductive effect). The CH_3CH_2 group donates electrons to the O, increasing the negative charge density on that atom and thus making it more ready to accept protons. Thus the equilibrium above is further to the left in the case of ethanol than with water. Therefore ethanol is the weaker acid. Both water and ethanol are of course extremely weak acids compared with the substances we normally call acids. Correspondingly, the hydroxide and ethoxide ions are strong bases.

In its reaction with sodium, ethanol is converted to ethoxide ion, so the two products are hydrogen and sodium ethoxide:

$$2Na + 2CH_3CH_2OH \rightarrow 2CH_3CH_2O^-Na^+ + H_2$$
$$\text{sodium ethoxide}$$

Phenol as an acid

The reactions with sodium indicate that phenol is a stronger acid than ethanol. In fact it is also a stronger acid than water (K_a for phenol = 10^{-10} mol dm^{-3}; K_a for water = 10^{-16} mol dm^{-3}). Nevertheless, sodium reacts more vigorously with water than with an ethanolic solution of phenol, because the latter only contains a fairly low concentration of phenol.

If we consider the equilibrium:

$$H_2O + \text{(phenol)} \rightleftharpoons \text{(phenoxide ion)} + H_3O^+$$

phenoxide ion

we can see why phenol is a stronger acid than aliphatic alcohols. In the phenoxide ion, the negative charge on the O can to some extent be delocalised round the ring. This reduces the tendency of the phenoxide ion to attract protons, in other words it reduces its strength as a base. Consequently, phenol is a stronger acid than aliphatic alcohols, though it is is still weaker than carboxylic acids such as ethanoic acid.

Because of its acidic nature, phenol is much more soluble in sodium hydroxide solution than in water. Hydroxide ions from sodium hydroxide remove hydrogen ions, displacing the above equilibrium to the right so that the phenol dissolves as sodium phenoxide. However, phenol does not dissolve in sodium carbonate solution, because it is a weaker acid than carbonic acid.

Phenol used to be known as 'carbolic acid' and was used as one of the earliest antiseptics. The Edinburgh doctor Joseph Lister first used it in the 1860s to prevent wounds going septic after surgery. Unwittingly, it had been used as an antiseptic even before this, because the old-fashioned way of treating an amputation wound was to cover it with coal-tar, which contains phenol. Phenol is effective in killing bacteria, but it is also very corrosive to the skin. It has been largely replaced by other antiseptics. Many of these are derivatives of phenol and are better germicides. Two examples are shown in figure 30.5.

Figure 30.5
Derivatives of phenol used as antiseptics.

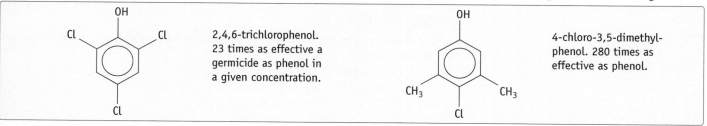

2,4,6-trichlorophenol. 23 times as effective a germicide as phenol in a given concentration.

4-chloro-3,5-dimethyl-phenol. 280 times as effective as phenol.

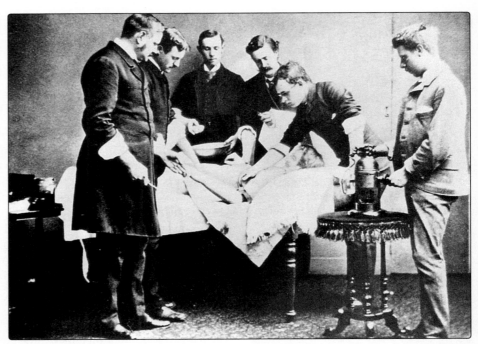

▶ An early example of the use of antiseptics. In this operation a spray of phenol is being directed onto the wound. Notice how primitive the 'aseptic' conditions are compared with a modern operating theatre.

Testing for phenols

Phenols, unlike alcohols, give a violet colour when added to neutral iron(III) chloride solution. The colour is probably due to a complex of Fe^{3+} with phenol as ligand.

$$CH_3-\overset{\underset{||}{O}}{C}-OH + CH_3CH_2-O-H \xrightarrow{H^+} CH_3-\overset{\underset{||}{O}}{C}-O-CH_2CH_3 + H_2O$$

ethanoic acid   ethanol   ethyl ethanoate (an ester)

Esterification

In the presence of an acid catalyst, alcohols react with carboxylic acids to form esters. Water is eliminated. The alcohol cleaves at the O—H bond:
This reaction is considered more fully in chapter 32.

  Phenol is again rather different from aliphatic alcohols: it does not react with acids directly. Phenyl esters can however be made by reacting phenol with an acyl halide (see section 29.7). For example:

$$CH_3-\overset{\underset{||}{O}}{C}-Cl + \underset{\text{phenol}}{\overset{OH}{\bigcirc}} \longrightarrow CH_3-\overset{\underset{||}{O}}{C}-O-\bigcirc + HCl$$

ethanoyl chloride phenol   phenyl ethanoate

> Predict the formulas of the products of reactions between
>
> 1 propan-2-ol and sodium,
>
> 2 propan-1-ol and propanoic acid, in the presence of an acid catalyst.

30.6 Reactions involving cleavage of the C—O bond

Reaction with halide ions

In the presence of a strong acid, the —OH group in an alcohol bonds to H^+ ions, and becomes protonated.

$$\overset{}{\underset{}{>}}C-\ddot{O}\overset{H}{\underset{H^+}{\diagup}} \longrightarrow \overset{}{\underset{}{>}}C-O+\overset{H}{\underset{H}{\diagdown}}$$

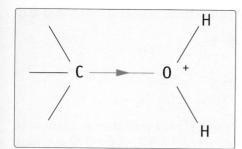

Figure 30.6
Effect of protonation on polarisation of the C—O bond in alcohols. The arrow indicates the direction of displacement of electrons in the bond.

The oxygen atom then carries a positive charge, and tends to attract electrons very strongly from the carbon atom next to it (figure 30.6). This causes a large positive charge on the carbon, making it attractive to nucleophiles. Thus, *in the presence of concentrated sulphuric acid,* ethanol reacts with Br^- to form bromoethane:

$$CH_3CH_2{-}OH + H^+ \longrightarrow CH_3CH_2{-}\overset{+}{O}\!\!\begin{smallmatrix}H\\ \\H\end{smallmatrix} \longrightarrow CH_3CH_2{-}Br + O\begin{smallmatrix}H\\ \\H\end{smallmatrix}$$

Overall, this amounts to the reaction of the alcohol with HBr:

$$CH_3CH_2OH + HBr \rightarrow CH_3CH_2Br + H_2O$$
bromoethane

In some cases an intermediate carbocation may actually be formed before the Br^- attacks. Cleavage of the C—O bond is greatly aided by protonation because it is much easier for the molecule to lose a neutral H_2O molecule than a charged ^-OH ion. We say that H_2O is a better **leaving group** than ^-OH.

This is another example of a nucleophilic substitution reaction. Compare it with the substitution reactions of halogenoalkanes (section 29.4). The reaction is normally carried out by heating ethanol under reflux with potassium bromide and concentrated sulphuric acid. Similar reactions occur between alcohols and Cl^-, Br^- and I^- in acid conditions. Cl^- is the least reactive and I^- the most.

Other halogenation reactions

There are several other reagents that are useful for replacing an —OH group with a halogen atom. Some of the more important are:

phosphorus pentachloride: $ROH + PCl_5 \rightarrow RCl + HCl + POCl_3$

phosphorus tribromide or triiodide (in practice a mixture of red phosphorus and bromine or iodine is used): $3ROH + PBr_3 \rightarrow 3RBr + H_3PO_3$

Phenol is again different in that it cannot be halogenated by normal reagents.

30.7 Reactions involving the carbon skeleton

So far we have considered reactions of alcohols in which part or all of the —OH group is replaced. Alcohols also undergo several reactions which involve *both* the carbon skeleton *and* the —OH group.

Dehydration

(i) Alkene formation

Consider again the protonated form of ethanol that we met in the last section. This ion can readily lose water, forming a carbocation:

$$H{-}\overset{\overset{\displaystyle H}{|}}{\underset{\underset{\displaystyle H}{|}}{C}}{-}\overset{\overset{\displaystyle H}{|}}{\underset{\underset{\displaystyle H}{|}}{C}}{-}\overset{+}{O}\!\!\begin{smallmatrix}H\\ \\H\end{smallmatrix} \longrightarrow H{-}\overset{\overset{\displaystyle H}{|}}{\underset{\underset{\displaystyle H}{|}}{C}}{-}\overset{\overset{\displaystyle H}{|}}{\underset{\underset{\displaystyle H}{|}}{C}}\!{}^+ + H_2O$$

The carbocation is an unstable intermediate. In the reaction with HBr, this ion rapidly reacts with Br^- to form bromoethane. In the absence of any nucleophile like Br^-, however, a different reaction may occur. The ion may lose H^+ and form ethene:

$$H{-}\overset{\overset{\displaystyle H}{|}}{\underset{\underset{\displaystyle H}{|}}{C}}{-}\overset{\overset{\displaystyle H}{|}}{\underset{}{C}}\!{}^+ \longrightarrow \begin{smallmatrix}H\\ \end{smallmatrix}C{=}C\begin{smallmatrix}H\\ \\H\end{smallmatrix} + H^+$$

Thus in the presence of strong acids, ethanol forms ethene in what amounts overall to an elimination reaction:

$$H - \overset{\overset{\displaystyle H}{|}}{\underset{\underset{\displaystyle H}{|}}{C}} - \overset{\overset{\displaystyle H}{|}}{\underset{\underset{\displaystyle OH}{|}}{C}} - H \longrightarrow \overset{\displaystyle H}{\underset{\displaystyle H}{}}C = C\overset{\displaystyle H}{\underset{\displaystyle H}{}} + H_2O$$

Since water is eliminated, this can also be thought of as a dehydration reaction. In practice the intermediate carbocation may never form: the H^+ and the H_2O may leave simultaneously.

The reaction can be carried out in the laboratory by heating ethanol at 170°C with excess concentrated sulphuric acid. Ethene is evolved and can be collected over water. The concentrated sulphuric acid can be thought of as a dehydrating agent, removing water from ethanol, though the actual mechanism as we have seen is more complex than this. Another way of preparing ethene from ethanol is by catalytic dehydration of ethanol vapour. A heated catalyst of aluminium oxide or pumice stone is used, in the apparatus shown in figure 30.7.

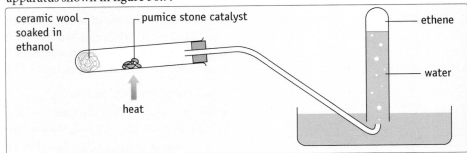

ceramic wool soaked in ethanol — pumice stone catalyst — ethene — water — heat

Figure 30.7
Catalytic dehydration of ethanol.

(ii) Ether formation
The carbocation formed by ethanol in the presence of an acid is open to attack by nucleophiles. Ethanol, with lone pairs on its oxygen atom, is itself a nucleophile, so we might expect it to attack the carbocation:

$$CH_3CH_2{}^+ \quad :\!\overset{|}{\underset{|}{O}} - CH_2CH_3 \longrightarrow CH_3CH_2 - \overset{+}{\underset{|}{O}} - CH_2CH_3$$

ethyl carbocation   H   ethanol   H

This new ion can then lose an H^+ ion:

$$CH_3CH_2 - \overset{+}{\underset{\underset{\displaystyle H}{|}}{O}} - CH_2CH_3 \longrightarrow CH_3CH_2 - O - CH_2CH_3 + H^+$$

ethoxyethane

The product is ethoxyethane, an ether. Overall, this too is effectively a dehydration reaction:

$$\begin{array}{c} CH_3CH_2OH \\ + \\ CH_3CH_2OH \\ \text{ethanol} \end{array} \longrightarrow \begin{array}{c} CH_3CH_2 \\ \diagdown \\ O + H_2O \\ \diagup \\ CH_3CH_2 \\ \text{ethoxyethane} \end{array}$$

The dehydration of ethanol, then, can give two different products. Both this reaction and the one in which ethene is formed occur when ethanol is heated with concentrated sulphuric acid. By adjusting the reaction conditions we can largely determine which product is formed. The formation of a molecule of the ether involves two molecules of ethanol, but the formation of a molecule of the alkene involves only one. Ether formation is therefore favoured by having an excess of ethanol present. Compare this with the excess of sulphuric acid used in the preparation of ethene.

Ethyoxyethane is prepared by heating concentrated sulphuric acid with excess ethanol at 140°C. The ethoxyethane distils off. Ethers are considered further in section 30.9.

The dehydration reactions here are applicable to alcohols in general, but not to phenols.

▯ Predict the formulas of the main products of the following reactions.
(a) Butan-2-ol is heated with excess concentrated sulphuric acid.
(b) Excess propan-1-ol is heated with concentrated sulphuric acid.
(c) Propan-2-ol vapour is passed over heated aluminium oxide.

30 Alcohols, Phenols and Ethers

Oxidation

Table 30.3 shows the effect of warming three different alcohols with acidified potassium dichromate(VI).

Table 30.3
Effect of warming alcohols with acidified potassium dichromate(VI)

Name	Formula	Observation
propan-1-ol	$CH_3CH_2CH_2OH$	orange dichromate(VI) slowly turns green
propan-2-ol	$CH_3CHOHCH_3$	orange dichromate(VI) slowly turns green
2-methylpropan-2-ol	$CH_3\!-\!\overset{\displaystyle CH_3}{\underset{\displaystyle OH}{C}}\!-\!CH_3$	no change

Look at the table and answer these questions.

1. Classify the three alcohols as primary, secondary and tertiary.

2. When potassium dichromate(VI) is reduced in acid solution, green Cr^{3+}(aq) ions are formed. Which of the three alcohols is oxidised by acidified potassium dichromate(VI)?

Primary and secondary alcohols are readily oxidised by a variety of oxidants. Acidified dichromate(VI), acidic or alkaline manganate(VII) or air in the presence of a catalyst are all suitable oxidants. The initial product is a **carbonyl** compound (chapter 31) which in the case of a primary alcohol is an **aldehyde**:

propan-1-ol $\longrightarrow$ propanal, an aldehyde $+ 2H^+ + 2e^-$ electrons accepted by oxidiser

Aldehydes are themselves readily oxidised to acids, thus:

propanal $+ H_2O \longrightarrow$ propanoic acid $+ 2H^+ + 2e^-$

The product of oxidising a primary alcohol is therefore usually an acid, unless the aldehyde is distilled from the reaction mixture as it forms.

In the case of a secondary alcohol the product of oxidation is a **ketone**:

propan-2-ol $\longrightarrow$ propanone, a ketone $+ 2H^+ + 2e^-$

▲ The oxidation of ethanol is used to detect alcohol in motorists' breath. This roadside 'Alcolmeter' contains an electrochemical cell in which alcohol is oxidised at one of the electrodes. The greater the concentration of alcohol, the higher the voltage of the cell and the larger the reading on the meter.

Ketones are not readily oxidised, so the reaction stops at this point.

Tertiary alcohols cannot be readily oxidised since they have no hydrogen atom to be removed from the carbon atom carrying the —OH group. Strong oxidisers cause their molecules to be broken up, giving a mixture of oxidation products.

Phenol, like tertiary alcohols, cannot be oxidised to a carbonyl compound. However, the aromatic ring with its high electron density is susceptible to attack by oxidising agents.

▲ Wine fermenting in a barrel. The curved tube at the top is a trap to keep out bacteria. It allows carbon dioxide to escape from the fermenting wine, but it does not allow entry to bacteria which might turn the wine to vinegar.

1 Predict the products, if any, of oxidising the following alcohols with acidified dichromate(VI):
(a) ethanol (product distilled off immediately),
(b) ethanol (reagents heated together under reflux for some time),
(c) 2-methylbutan-2-ol,
(d) butan-2-ol.

The oxidation of alcohols is important in industry and in living organisms. Methanal (commonly called formaldehyde and used in the production of plastics, see section 30.8) is manufactured by passing a mixture of methanol vapour and air over a silver catalyst at 500°C:

$$2H - \underset{\underset{H}{|}}{\overset{\overset{H}{|}}{C}} - OH + O_2 \xrightarrow[500°C]{Ag} 2H - \underset{\underset{H}{}}{\overset{\overset{O}{\parallel}}{C}} + 2H_2O$$

methanol   methanal

Propanone (acetone) is manufactured from propan-2-ol by a similar process, using air and a copper catalyst at 500°C. In these industrial processes, high-temperature catalytic reactions are used in which air is the oxidiser, rather than aqueous oxidisers like dichromate(VI). This is for economic reasons. Air is much cheaper than dichromate(VI) and although the catalyst is expensive it has a long life before it needs renewing.

Other ways of oxidising ethanol

The bacterial oxidation of ethanol to ethanoic acid (acetic acid) has long been a problem for wine producers. The bacterium, *Acetobacter*, uses air to oxidise ethanol in wine, producing a weak solution of ethanoic acid called vinegar. The bacterium uses this oxidative process as a source of energy. Once a bottle of wine has been opened it will turn to vinegar fairly quickly because of the considerable number of these bacteria in the air. One way of preventing this is to add extra alcohol to the wine so that the concentration of ethanol is too high for the bacteria to tolerate. Wine treated in this way is said to be *fortified*: sherry and port are examples.

Ethanol is disposed of by the body by oxidation: it is oxidised in the liver, first to ethanal and eventually to CO_2 and water. The liver can oxidise about 8 g of ethanol an hour, but very high concentrations are too much for it to cope with. Excessive quantities of ethanol can damage the liver and cause a condition called cirrhosis.

Alcohols as fuels

Ethanol, like all alcohols, undergoes complete oxidation to CO_2 and H_2O when heated in the presence of air. In other words, it burns. Ethanol has a clean, smokeless flame and is sometimes used as a fuel, for example in methylated spirit stoves. Its use as a fuel is limited by its cost in the United Kingdom, but in some tropical countries, where sugar is easily grown, ethanol is cheaply produced by fermentation. In Brazil, for example, ethanol is an important motor fuel because it can be produced at a cost comparable to that of gasoline.

Methanol is also an important fuel. It is quite cheap and can be blended with gasoline to increase its octane number and reduce pollution.

30.8 Substitution reactions of the aromatic ring in phenol

If bromine water is added to a solution of phenol the bromine is immediately decolorised and a white precipitate is formed. This is a substitution reaction, the white precipitate being 2,4,6-tribromophenol:

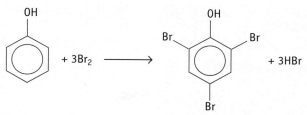

Benzene does not react with bromine except in the presence of a halogen-carrier catalyst (section 28.7). Phenol is more susceptible than benzene to attack by electrophiles. This is because the lone pairs of electrons on the oxygen of the —OH group become partially delocalised round the ring: this increases the electron density there and makes it more susceptible to attack by electrophiles. The 2, 4 and 6 positions are preferentially substituted (see section 28.10).

Phenol is nitrated far more readily than benzene. Dilute nitric acid alone is sufficient, the products being 2- and 4-nitrophenol. A mixture of concentrated nitric acid and concentrated sulphuric acid readily converts it to 2,4,6-trinitrophenol, commonly known as picric acid:

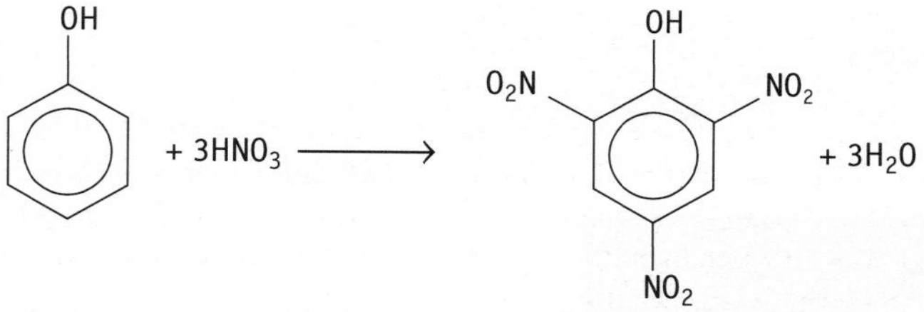

This antiseptic contains halogenated phenols. Can you suggest what 'TCP' stands for? 'Always read the label' is the advice given with this product.

2,4,6-trinitrophenol

This substance is used as a high explosive. It decomposes spontaneously and exothermically, producing large volumes of CO, steam and N_2 which expand and produce an explosive shock.

Plastics from phenol

When phenol is heated with methanal in the presence of an acid or alkali catalyst, a hard, brittle plastic is formed. The initial stage of this reaction is substitution of the phenol in the 2 or 4 position:

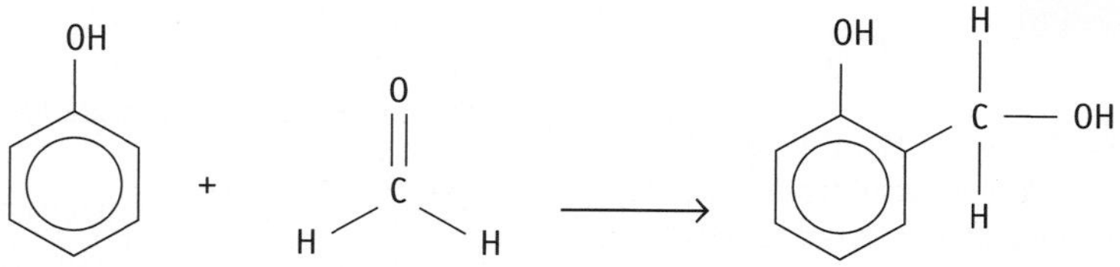

This product then undergoes a condensation reaction in which a molecule of water is eliminated between it and another molecule of phenol.

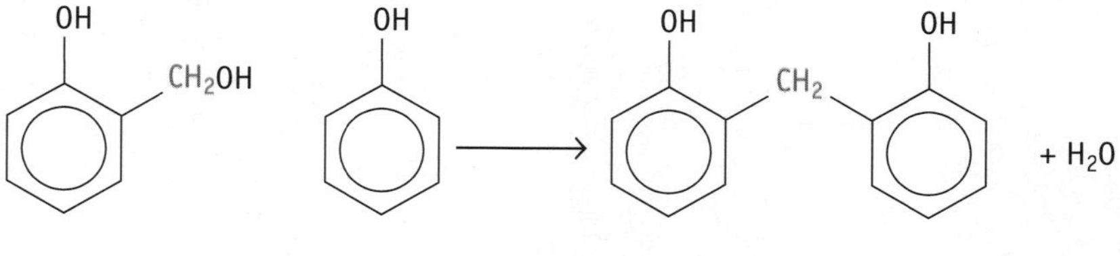

The product may then be substituted by another methanal molecule:

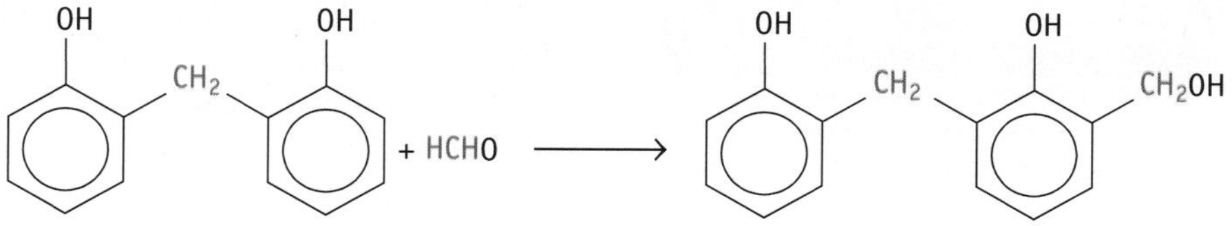

A radio with a 'Bakelite' case. 'Bakelite' was the earliest synthetic plastic.

Condensation then occurs with a further molecule of phenol and the process continues until all the aromatic rings are substituted in the 2, 4 and 6 positions. A molecular network like that shown in figure 30.8 builds up.

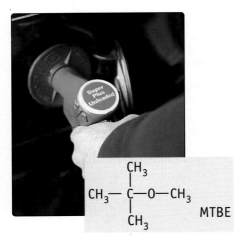

Figure 30.8
Condensation polymer of phenol and methanal.

This polymer is very hard and rather brittle, because of its extensively cross-linked three-dimensional network. It is a dark brown material, known as 'Bakelite'. It has the disadvantage that it sets hard on heating and cannot be remelted, which makes it difficult to mould. Its hardness and cheapness are however a considerable advantage and it was once widely used for such things as electric plugs and motor-car distributor caps. Plastics of this kind, which set hard and cannot be remelted, are called **thermosetting**. Plastics like polythene which can be moulded by melting are called **thermosoftening** or **thermoplastic**.

30.9 Ethers

Ethers are compounds with the general structure R^1OR^2, where R^1 and R^2 are alkyl or aryl groups. They are named by regarding them as alkanes substituted by alkoxy groups. Thus,

$$CH_3OCH_3 \text{ methoxymethane,}$$
$$CH_3CH_2OCH_2CH_3 \text{ ethoxyethane,}$$
$$CH_3OCH_2CH_3 \text{ methoxyethane.}$$

Ethoxyethane is an example of a **symmetrical** ether; methoxyethane is an **unsymmetrical** ether.

Since they have no capacity for hydrogen bonding, ethers are far more volatile than alcohols. Their volatility corresponds closely to alkanes of comparable relative molecular mass. Thus the most commonly encountered ether, ethoxyethane (commonly called simply 'ether'), is a very volatile liquid of boiling point 35°C. This volatility makes ethoxyethane a dangerously flammable substance. Since its relative molecular mass is 74, ethoxyethane vapour is much denser than air. It tends to diffuse slowly and stay at floor or bench level in a laboratory, where it is likely to be ignited by naked flames. Serious explosions and fires have been caused in this way.

The molecule of ethoxyethane is analogous in shape to that of water:
It is therefore slightly polar, but not polar enough to have much effect on the properties

$$CH_3CH_2-\ddot{O}:$$
$$CH_2CH_3$$

of ethoxyethane. Like all ethers, it is only slightly soluble in water and is a good solvent for non-polar substances. It is frequently used as a solvent in practical work because it is volatile and easily distilled off, but its use must be carefully controlled because of the fire risk. The small polarity in the molecule is not enough to make it significantly reactive towards nucleophiles or electrophiles. Ethers are not much more reactive than alkanes.

▲ An ether known as 'MTBE' ('methyl tertiary butyl ether') is added to unleaded petrol to boost its octane rating. MTBE has an octane number of 118 (compared with 95 for premium petrol).

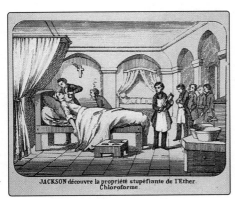

▲ Ethoxyethane (ether) was at one time used as an anaesthetic.

30 Alcohols, Phenols and Ethers

Epoxy compounds: cyclic ethers

Epoxyethane $CH_2\!-\!CH_2$ (with O bridging) is a cyclic ether containing the epoxy group, $-\!CH\!-\!CH\!-$ (with O bridging)

This compound has considerable ring strain (section 26.9) so it is a good deal more reactive than straight-chain ethers. Its chemical reactivity makes it useful in industrial synthesis.

Epoxyethane reacts with water to form ethane-1,2-diol (ethylene glycol):

$$CH_2\!-\!CH_2 \;(O) \; + \; H_2O \; \rightarrow \; HOCH_2CH_2OH$$
ethane-1,2-diol

This is an important organic chemical, used in antifreeze and also for manufacturing polyester fibres.

Epoxyethane can also be made to join to itself, forming a polymer of structure $-OCH_2CH_2OCH_2CH_2O-$. This polymer, when added to the water used in fire hoses in the proportion 30 parts per million, reduces the friction between the water and the hose walls so successfully that the range of the hose is doubled. Epoxyethane is also used to manufacture detergents (section 32.3). Epoxyethane itself is used as a plasticiser to make rigid polymers like PVC more flexible (see section 27.8).

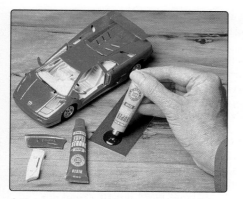

▲ Epoxy resins contain compounds which possess the epoxy group. Under the right conditions, these epoxy compounds polymerise to make the resin set hard.

Summary

1 Alcohols are aliphatic compounds containing the —OH group. They are named using the suffix -ol. Phenols are aromatic compounds with the —OH group attached directly to the ring.

2 Primary alcohols are of the form RCH_2OH; secondary alcohols are of the

form $R^1\atop R^2$ $CHOH$;

tertiary alcohols are of the form $R^2\!-\!\overset{R^1}{\underset{R^3}{C}}\!-\!OH$.

3 The physical properties of alcohols, especially the early members of the series, are strongly influenced by hydrogen bonding.

4 Alcohols can act as proton donors but are weaker acids than water. Phenols are considerably stronger acids.

5 Alcohols can act as bases, becoming protonated on the oxygen atom. This can lead to the formation of a carbocation and subsequent dehydration or substitution reactions.

6 Primary alcohols can be oxidised to aldehydes, then to acids. Secondary alcohols can be oxidised to ketones. Tertiary alcohols cannot be oxidised without fragmenting the molecule.

7 Ethers are compounds of the general form R^1OR^2, where R^1 and R^2 are alkyl or aryl groups. They are rather unreactive.

8 Epoxy compounds contain the $-\!CH\!-\!CH\!-$ (with O bridging) group.

They are more reactive than straight-chain ethers.

9 Some reactions of ethanol are shown in the diagram below.

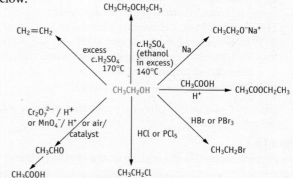

The importance of alcohols in the synthesis of other organic compounds is considered in chapter 34.

10 Some reactions of phenol are shown below.

1 Consider the following compounds.

A

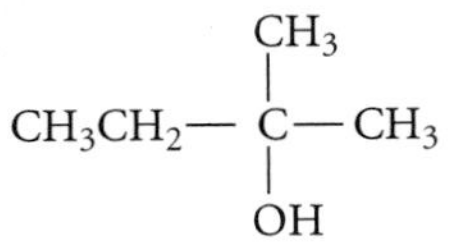

$$CH_3CH_2 \overset{\displaystyle CH_3}{\underset{\displaystyle OH}{\overset{|}{\underset{|}{C}}}} CH_3$$

B $CH_3CH_2CHOHCH_3$

C $CH_3OCH_2CH_2CH_3$

D $CH_3CH_2CH_2CH_2OH$

E

$$\text{(phenol ring with OH at top and } CH_3 \text{ at bottom)}$$

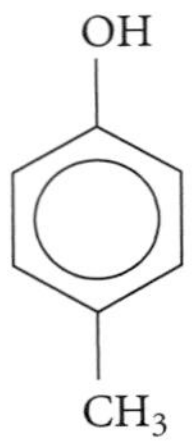

(a) Name each compound.
(b) Which is a primary alcohol?
(c) Which is a tertiary alcohol?
(d) Which is a phenol?
(e) Which is an ether?

2 Refer again to the compounds in question 1.
(a) Which react with sodium metal?
(b) Which could be oxidised to an aldehyde?
(c) Which could be oxidised to a ketone?
(d) Which would be the strongest acid?
(e) Which would form an alkene when heated with excess concentrated sulphuric acid?
(f) Which has the lowest boiling point?

3 Table 30.4 gives some physical properties of water, ethanol and ethoxyethane.

Table 30.4
Some physical properties of water, ethanol and ethoxyethane

Name	water	ethanol	ethoxyethane
Formula	H_2O	CH_3CH_2OH	$CH_3CH_2OCH_2CH_3$
M_r	18	46	74
Boiling point/°C	100	78	35
Density at 273 K/g cm^{-3}	1.00	0.79	0.71
Surface tension at 293 K/N m^{-1}	7.28	2.23	1.69

Suggest explanations in terms of intermolecular forces, why
(a) the boiling point of ethanol is greater than that of ethoxyethane,
(b) the density of water is greater than that of ethanol,
(c) the surface tension of water is greater than that of ethoxyethane.

4 An organic liquid **A** contains carbon, hydrogen and oxygen only. On combustion 0.463 g of **A** gave 1.1 g of carbon dioxide and 0.563 g of water. When vaporised, 0.1 g of **A** occupy 54.5 cm^3 at 208°C and 98.3 kPa (740 mm Hg). (Standard pressure = 101 kPa (760 mm Hg); 1 mole of a gas occupies 22.4 dm^3 at s.t.p.)
(a) What is the percentage composition of **A**?
(b) Find the empirical formula of **A**.

(c) Calculate the relative molecular mass of **A**.
(d) Give the structures of possible non-cyclic isomers of **A**.
(e) Which isomers will react with sodium to give hydrogen?
(f) Which isomers will reduce acidified dichromate(VI) ion to green Cr^{3+}?

5 Suggest explanations for the following observations.
(a) Butan-1-ol is much more soluble in 5 mol dm^{-3} hydrochloric acid than in water.
(b) 2,4,6-Trinitrophenol (picric acid) is a much stronger acid than phenol ($K_a = 10^{-1}$ mol dm^{-3} and 10^{-10} mol dm^{-3}, respectively).
(c) Phenol can be nitrated by dilute nitric acid to give a mixture of 2-nitrophenol and 4-nitrophenol, whereas benzene is only nitrated by a mixture of concentrated nitric and concentrated sulphuric acids.
(d) Ethanol is more acidic than 2-methylpropan-2-ol.
(e) Heating butan-2-ol with excess concentrated sulphuric acid produces a mixture of three isomeric alkenes.

6 Give the reagents and conditions you would use to carry out the following conversions. One or more steps may be involved in the conversions.
(a) Ethanol to ethyl ethanoate (using ethanol as the only organic starting material).
(b) Ethanol to 1,2-dibromoethane.
(c) Ethene to ethanoic acid.
(d) Ethanol to propanenitrile.
(e) Bromoethane to ethoxyethane.

7 Suggest structures for the following compounds **X**, **Y** and **Z**. Explain the reactions involved in each case.
(a) **X**, C_7H_8O, burns with a smoky flame. **X** is only slightly soluble in water, but very soluble in sodium hydroxide solution.
(b) **Y**, $C_6H_{14}O$, is insoluble in water but soluble in concentrated sulphuric acid. **Y** has no reaction with sodium.
(c) **Z**, $C_5H_{12}O$, gives hydrogen when reacted with sodium metal but has no effect on acidified potassium dichromate(VI).

8 Predict the structure of the organic products of the following reactions.
(a) Butan-2-ol is warmed with acidified potassium manganate(VII).
(b) An aqueous solution of chlorine is added to phenol.
(c) Propan-2-ol is warmed with excess concentrated sulphuric acid.
(d) 2-Methylphenol is treated with propanoyl chloride.
(e) Methanol is treated with phosphorus pentachloride.

9 Primary alcohols can be thought of as compounds derived by replacing one hydrogen atom in a water molecule by an alkyl group. Ethers can be thought of as resulting from the replacement of both hydrogen atoms in water by alkyl groups. To what extent do the physical and chemical properties of water, alcohols and ethers fit in with this idea? Illustrate your answer by reference to ethanol and ethoxyethane.

10 For each of the common uses of alcohols given below, explain which property or properties of the alcohols involved make them suitable for that use.

(a) Ethanol is used to clean surfaces before applying adhesive.

(b) Ethane-1,2-diol is used as an antifreeze.

(c) Ethanol is used as the base for many perfumes.

(d) Vinegar is manufactured from a weak solution of ethanol.

(e) Fruits, such as peaches, are sometimes preserved in brandy.

(f) 'Breathalysers' containing potassium dichromate(VI) can be used to test for ethanol in motorists' breath.

11 Five different bottles contain the following liquids

A propan-1-ol $CH_3CH_2CH_2OH$

B butoxybutane $CH_3CH_2CH_2CH_2—O—CH_2CH_2CH_2CH_3$

C cyclopentanol

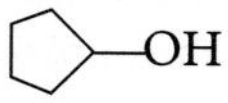

D 2-methylbutan-2-ol

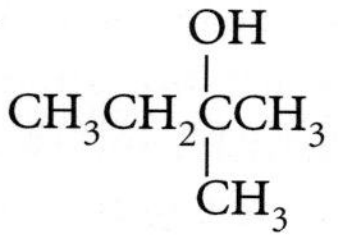

E methanol CH_3OH

If the bottles were unlabelled, what *chemical* tests would you use to find out which liquid was which?

31 Carbonyl Compounds

31.1 The carbonyl group

In chapter 27 we looked at compounds containing C=C double bonds and saw that their typical reactions involves electrophilic addition. In chapter 30 we looked at compounds containing the C—O single bond, which is polar and tends to bring about substitution reactions. We will now turn our attention to the **carbonyl** group, C=O, which is the functional group in aldehydes and ketones. We might expect the reactions of this group to show similarity to the reactions of both

$$\diagdown C = C \diagup \quad \text{and} \quad \diagup C - O -$$

The double bond between C and O in the carbonyl group, like the double bond in alkenes, can be considered to consist of a σ-bond and a π-bond. Unlike the C=C group, however, the carbonyl group does not have an even electron distribution between the two atoms. There is a greater electron density over the more electronegative oxygen atom, as shown in figure 31.1.

▲ Aldehydes and ketones contribute to the flavour of some fruit. The smell of ethanal is reminiscent of apples.

Figure 31.1
Electron distribution in the C=O bond.

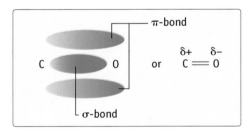

This electron distribution makes the carbon atom attractive towards nucleophiles. Nucleophiles tend to attack and bond to this carbon, breaking the π-bond and resulting eventually in addition. With a general nucleophile $\ddot{X}$—Y, the overall reaction is

$$\diagup^{\delta+}_{C} = \overset{\delta-}{O} + \ddot{X} - Y \longrightarrow \diagup C \diagdown^{OY}_{X}$$

The exact mechanism depends on the nature of X and Y. In most cases, Y is hydrogen. This sort of reaction, which is typical of the carbonyl group, is called **nucleophilic addition**: compare it with the *electrophilic* addition which is typical of alkenes. The mechanism of the nucleophilic addition reaction involving HCN is described in section 31.3.

The reactions of the carbonyl group are important because the group is common in biological molecules, particularly carbohydrates (section 31.7). Carbonyl compounds also have considerable industrial significance, for example as solvents and in the manufacture of thermosetting plastics (section 30.8).

31.2 Aldehydes and ketones – nature and naming

Aldehydes and ketones both contain the carbonyl group, but they differ in its position in the hydrocarbon skeleton. **Aldehydes** have the carbonyl group at the end of a chain. Their general formula is therefore $R—\underset{\underset{O}{\|}}{C}—H$, usually written RCHO, where R is an alkyl or aryl group or hydrogen. **Ketones** have the carbonyl group in a non-terminal position in the chain. The general formula is $R^1—\underset{\underset{O}{\|}}{C}—R^2$, usually written R^1COR^2, where R^1 and R^2 are alkyl or aryl groups.

Thus aldehydes and ketones are structurally quite similar, but their properties differ considerably and they are considered as different homologous series.

Aldehydes are named using the suffix **-al** after a stem indicating the number of carbon atoms (including the one in the carbonyl group). Thus CH_3CHO is called ethanal. Ketones are named using the suffix **-one** after a stem indicating the number of carbon atoms, together with a number, if necessary, to indicate the position of the carbonyl group in the chain. Thus $CH_3COCH_2CH_2CH_3$ is called pentan-2-one. Table 31.1 gives the names and formulas of some important aldehydes and ketones.

> 1 Name the compounds:
> (a) $CH_3CH_2CH_2CHO$,
> (b) $CH_3CH_2COCH_2CH_2CH_3$.
> 2 Give the formula of hexan-2-one.
> 3 Why is there no such compound as ethanone?

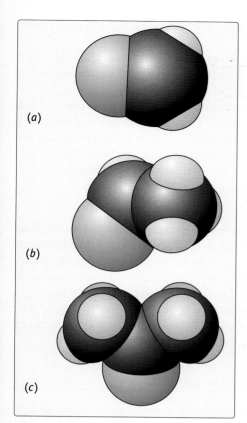

Figure 31.2
Computer-generated models of molecules of (a) methanal, (b) ethanal and (c) propanone.

Table 31.1
Some important aldehydes and ketones

Formula	Systematic name	Other name	State at room temp.	Boiling point/K	Solubility in water
HCHO	methanal	formaldehyde	g	254	soluble
CH_3CHO	ethanal	acetaldehyde	l	294	infinite*
CH_3CH_2CHO	propanal	propionaldehyde	l	321	soluble
CH_3COCH_3	propanone	acetone	l	329	infinite*
$CH_3COCH_2CH_3$	butanone	methylethyl ketone	l	353	very soluble
$CH_3CH_2COCH_2CH_3$	pentan-3-one		l	375	very soluble
⬡—CHO	benzaldehyde		l	451	slightly soluble
⬡—$COCH_3$	phenylethanone	acetophenone	l	475	insoluble

* miscible in all proportions

Physical properties

The polarity of the C=O group has a big influence on the physical properties of aldehydes and ketones. The earlier members of both series are considerably less volatile than alkanes of corresponding relative molecular mass. Thus ethanal (CH_3CHO), with a boiling point of 20°C, is a liquid (though a very volatile one) at room temperature.

▲ Propanone is widely used as a solvent. Nail varnish remover contains propanone.

1 Consider the cases when —R^1 and —R^2 are (a) both —H, (b) —CH_3 and —H, (c) both —CH_3. Place (a), (b) and (c) in order according to the size of positive charge you would expect on the carbonyl carbon in each of the three compounds. (Remember that alkyl groups such as —CH_3 tend to donate electrons to groups to which they are attached, by the inductive effect.)

2 Place (a), (b) and (c) in order of readiness to undergo nucleophilic addition.

3 Name compounds (a), (b) and (c).

4 How would you expect benzaldehyde (C_6H_5—C—H) with O below (double bonded) to compare with ethanal in its readiness to undergo nucleophilic addition?

Propane, $CH_3CH_2CH_3$, has the same relative molecular mass but is a gas at room temperature (boiling point, –42°C).

The polar —C=O group has less effect on intermolecular forces than the —OH group, since the latter is able to participate in hydrogen bonding. Compare the boiling points of ethanal and ethanol, which are 20°C and 78°C, respectively.

The early members of the aldehydes and ketones are soluble in water, and will themselves dissolve both polar and non-polar solutes. Propanone (acetone), for example, is a widely used industrial solvent. As expected, the polar —C=O group has less and less influence on the physical properties of carbonyl compounds as the homologous series are ascended. Table 31.1 gives some physical properties of carbonyl compounds.

Methanal, the simplest aldehyde, is very toxic and possibly carcinogenic.

Making aldehydes and ketones

In the laboratory these compounds can be made by the oxidation of alcohols, as described in section 30.7. In fact, the name *aldehyde* comes from *alcohol dehyd*rogenate. Aldehydes are made by oxidising primary alcohols, ketones by oxidising secondary alcohols. This method is also used industrially for making methanal (from methanol) and propanone (from propan-2-ol). Most propanone, however, is now produced as a by-product from the manufacture of phenol (see review question 7, chapter 28). Some ethanal is manufactured by the oxidation of ethanol, but most is made by direct oxidation of ethene (see review question 3 at the end of this chapter).

31.3 Addition reactions of carbonyl compounds

The characteristic reaction of compounds containing the carbonyl group is nucleophilic addition:

$$R^1 \overset{R^1}{\underset{R^2}{\diagdown}} C \overset{\delta+}{=} \overset{\delta-}{O} + \ddot{X} - Y \longrightarrow \underset{R^2}{\overset{R^1}{\diagdown}} C \overset{OY}{\diagdown} X$$

The readiness with which such an addition reaction occurs is partly determined by the size of the partial positive charge on the carbon atom of the carbonyl group.

Aldehydes tend to be more reactive than ketones in nucleophilic addition reactions. Methanal is the most reactive aldehyde. Benzaldehyde is less reactive than aliphatic aldehydes because of the tendency of the benzene ring to delocalise the positive charge on the carbonyl carbon. On the whole, though, the reactions of aromatic aldehydes and ketones (i.e. those with the carbonyl group attached directly to the benzene ring) are quite similar to those of aliphatic ones.

The important stage of addition to C=O is attack by a nucleophile. Compare this with C=C, where the important stage of addition is attack by an *electrophile*. Carbonyl compounds do react with some of the compounds that undergo addition reactions with alkenes, but many of their reactions are different. Some examples are now given.

Reaction with HCN

In most addition reactions of carbonyl compounds, the molecule adding across the C=O double bond is of the form HX. A good example is HCN.

$$\underset{H}{\overset{CH_3}{\diagdown}} C = O + H - CN \longrightarrow \underset{H}{\overset{CH_3}{\diagdown}} C \overset{OH}{\diagdown} CN$$

31 Carbonyl Compounds

The mechanism of this reaction is as follows. Hydrogen cyanide is a weak acid and dissociates to form $^-$CN ions:

$$HCN(aq) + H_2O(l) \rightleftharpoons H_3O^+(aq) + {}^-CN(aq)$$

The $^-$CN is a nucleophile and attacks the carbon atom of the carbonyl group:

$$\begin{array}{c} CH_3 \\ \\ H \end{array}\!\!\!C \overset{\delta+}{=} \overset{\delta-}{O} \quad \xrightarrow{\ ^-CN\ } \quad \begin{array}{c} CH_3 \quad O^- \\ C \\ H \quad CN \end{array}$$

This intermediate ion then reacts with water to form the product:

$$\begin{array}{c} CH_3 \quad O^- \\ C \\ H \quad CN \end{array} + H{-}O{-}H \quad \longrightarrow \quad \begin{array}{c} CH_3 \quad OH \\ C \\ H \quad CN \end{array} + {}^-OH$$

The reaction occurs with ketones as well as aldehydes:

$$\begin{array}{c} CH_3 \\ CH_3 \end{array}\!\!\!C{=}O + HCN \quad \longrightarrow \quad \begin{array}{c} CH_3 \quad OH \\ C \\ CH_3 \quad CN \end{array}$$

propanone   2-hydroxy-2-methylpropanenitrile

2-Hydroxy-2-methylpropanenitrile is an intermediate in the manufacture of methyl 2-methylpropenoate, the monomer for Perspex.

Reaction with sodium hydrogensulphite

Carbonyl compounds undergo addition reactions when shaken with saturated aqueous sodium hydrogensulphite ($NaHSO_3$). As expected, ethanal reacts more readily than propanone.

$$\begin{array}{c} CH_3 \\ H \end{array}\!\!\!C{=}O + Na^+HSO_3^- \quad \longrightarrow \quad \begin{array}{c} CH_3 \quad OH \\ C \\ H \quad SO_3^- Na^+ \end{array}$$

ethanal   sodium hydrogen-sulphite   Reaction 88% complete after 30 min at 25°C

$$\begin{array}{c} CH_3 \\ CH_3 \end{array}\!\!\!C{=}O + Na^+HSO_3^- \quad \longrightarrow \quad \begin{array}{c} CH_3 \quad OH \\ C \\ CH_3 \quad SO_3^- Na^+ \end{array}$$

propanone   Reaction 47% complete after 30 min at 25°C

The products of these reactions are crystalline ionic compounds, soluble in water.

Reduction

Addition of hydrogen

Like the C=C bond, the C=O bond undergoes addition with hydrogen in the presence of a metal catalyst such as platinum or nickel. Aldehydes give primary alcohols, ketones give secondary alcohols.

$$\begin{array}{c} CH_3 \\ H \end{array}\!\!\!C{=}O + H_2 \quad \xrightarrow{Ni} \quad CH_3{-}\!\!\begin{array}{c} H \\ | \\ C \\ | \\ H \end{array}\!\!{-}OH$$

ethanal   ethanol

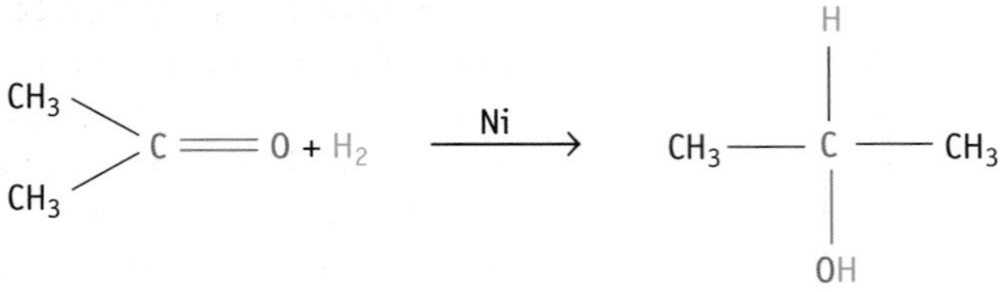

propanone → propan-2-ol

Reduction with metal hydrides

Metal hydrides contain the hydride ion, $^-$H. They are useful reducing agents in organic chemistry. The $^-$H ion acts as a nucleophile, and can attack the carbon atom of the carbonyl group. With ethanal:

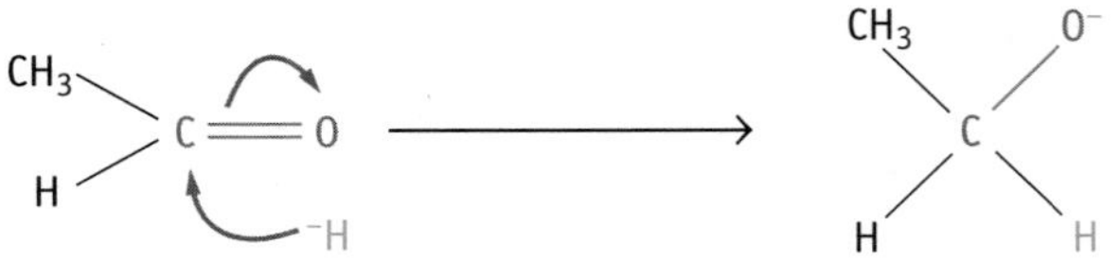

The intermediate ion then reacts with water to give the alcohol:

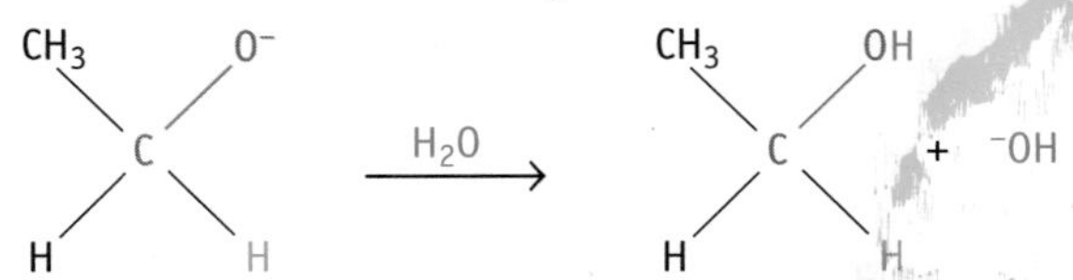

Two hydrides can be used to reduce carbonyl compounds to alcohols. Lithium tetrahydridoaluminate, $LiAlH_4$, also known as lithium aluminium hydride, is a powerful reducing agent: it will reduce carboxylic acids, esters and amides as well as carbonyl compounds. $LiAlH_4$ is easily hydrolysed, so it must be used in a dry ether solvent.

Sodium tetrahydridoborate, $NaBH_4$, also known as sodium borohydride, is a less powerful reducing agent. It reduces carbonyl compounds to alcohols, but does not affect most of the other functional groups that are reduced by $LiAlH_4$. It is therefore useful for the **selective reduction** of carbonyl groups, leaving other functional groups unaffected. $NaBH_4$ is rather more convenient to use than $LiAlH_4$: it is not easily hydrolysed, so it can be used in solution in water or alcohol.

1. Write the formulas of the products you would expect when propanal reacts with (a) $LiAlH_4$ in dry ether, (b) HCN.
2. Write the formula of the product you would expect when (a) butanone, (b) benzaldehyde reacts with $NaHSO_3$.

Table 31.2
Polymers of methanal and ethanal

Method of preparation	Name	Structure	Uses
1 From methanal evaporate aqueous solution of methanal	poly(methanal)	$-O-CH_2-O-CH_2-O-CH_2-$	high-strength plastic (making plastic kettles, gear wheels, etc.)
distil methanal from acidic solution	methanal trimer	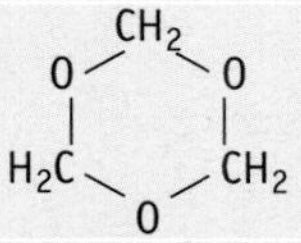	
2 From ethanal add dilute acid to ethanal	ethanal trimer		hypnotic (sleep-inducing) drug
add acid to ethanal below 0°C	ethanal tetramer	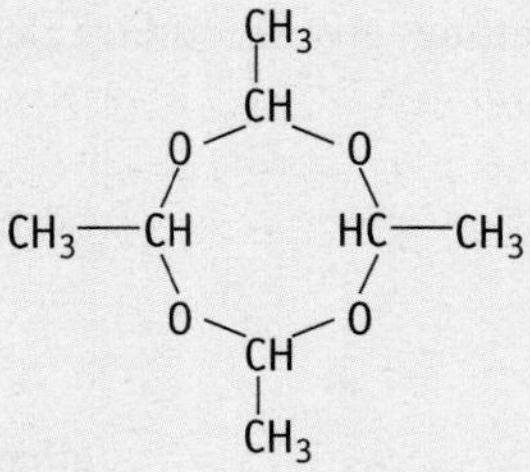	slug poison, solid fuel in portable stoves (commonly called metaldehyde)

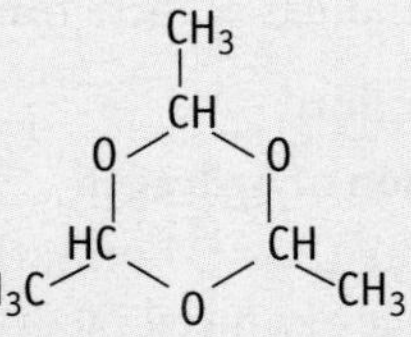

'Self-addition': polymerisation

Just as alkenes can be made to undergo addition polymerisation, carbonyl compounds too can form polymers, by addition across the C=O double bond. Ketones are not reactive enough to polymerise easily, but aldehydes can readily be converted to a variety of addition polymers. As expected, methanal polymerises most readily. Some of the polymers formed by methanal and ethanal are shown in table 31.2 on the previous page.

31.4 Condensation reactions of carbonyl compounds

Sometimes addition reactions are followed by elimination of a molecule of water. Many of these elimination reactions involve derivatives of ammonia of the general form X—NH₂.

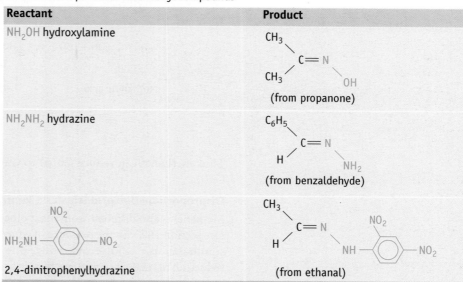

This is a **condensation** or **addition–elimination** reaction. Important examples are shown in table 31.3

Table 31.3
Condensation products of carbonyl compounds

Reactant	Product
NH₂OH hydroxylamine	CH_3, CH_3, C=N, OH (from propanone)
NH₂NH₂ hydrazine	C_6H_5, H, C=N, NH₂ (from benzaldehyde)
NH₂NH—(NO₂)(NO₂) 2,4-dinitrophenylhydrazine	CH_3, H, C=N, NH—(NO₂)(NO₂) (from ethanal)

The products of condensation reactions between 2,4-dinitrophenylhydrazine and carbonyl compounds are all orange crystalline solids with well-defined melting points. They are useful for identifying individual carbonyl compounds. The condensation product is prepared, its melting point is measured accurately and the compound is identified from tables of melting points.

31.5 Oxidation of carbonyl compounds

Table 31.4 on the next page shows the effects of some oxidising agents on different carbonyl compounds.

Aldehydes carry a hydrogen atom next to their carbonyl group. This hydrogen is activated by the carbonyl group and is readily oxidised to —OH. Aldehydes are therefore readily oxidised to carboxylic acids. For example, ethanal is oxidised by Ag^+:

$$CH_3 - \underset{\underset{O}{\|}}{C} - H + H_2O \longrightarrow CH_3\underset{\underset{O}{\|}}{C} - OH + 2H^+ + 2e^-$$

ethanal   ethanoic acid

$$Ag^+ + e^- \longrightarrow Ag$$

▲ Methanal is very toxic, and small quantities of it are released into the air from the plastics used in some kinds of wall insulation. However, some plants absorb methanal from the air – this photo shows spider plants, which are particularly effective at removing methanal vapour.

Look at table 31.4 and answer the questions.

1. Write a half-equation for the reduction of Ag^+ to Ag.
2. Which of the carbonyl compounds in the table are oxidised by Ag^+?
3. Suggest a compound to which ethanal might be oxidised.
4. To what oxidation state is Cu^{2+} reduced by (a) methanal (b) ethanal?
5. Which is the more powerful reducing agent, methanal or ethanal?

Table 31.4
Effect of warming different carbonyl compounds with oxidising agents

Carbonyl compound	Oxidising agent	
	Complexed Ag^+ in alkaline solution (Tollen's reagent)	Complexed Cu^{2+} in alkaline solution (Fehling's or Benedict's solution)
methanal	silver mirror formed on walls of tube	copper metal formed together with copper(I) oxide
ethanal	silver mirror formed on walls of tube	red precipitate of copper(I) oxide formed
propanone	no reaction	no reaction

Overall:

$$CH_3\overset{\displaystyle O}{\underset{\displaystyle \|}{C}}-H + H_2O + 2Ag^+ \longrightarrow CH_3\overset{\displaystyle O}{\underset{\displaystyle \|}{C}}-OH + 2H^+ + 2Ag$$

Methanal has two hydrogen atoms attached to the carbonyl group, both of which are oxidisable. Methanal is very readily oxidised and is consequently a more powerful reducing agent than other aldehydes:

$$H-\overset{\displaystyle O}{\underset{\displaystyle \|}{C}}-H + H_2O \longrightarrow H-\overset{\displaystyle O}{\underset{\displaystyle \|}{C}}-OH + 2H^+ + 2e^-$$

methanal   methanoic acid   accepted by oxidising agent

Then

$$H-\overset{\displaystyle O}{\underset{\displaystyle \|}{C}}-OH + H_2O \longrightarrow \left[HO-\overset{\displaystyle O}{\underset{\displaystyle \|}{C}}-OH\right] + 2H^+ + 2e^-$$

methanoic acid   carbonic acid   accepted by oxidising agent

$$\downarrow$$

$$CO_2 + H_2O$$

Thus methanal can reduce Cu(II) to Cu(0), while other aldehydes reduce it only to Cu(I).

Disproportionation of aldehydes: Cannizzaro's reaction

Methanal is also different from most other aldehydes in being able to **disproportionate** (see section 16.8). In the presence of aqueous sodium hydroxide, molecules of methanal simultaneously oxidise and reduce one another. The products are methanol (from reduction of methanal) and methanoic acid (from oxidation of methanal).

$$HCHO + HCHO \longrightarrow CH_3OH + HCOOH$$
2 molecules of   methanol methanoic
methanal   acid

This is known as Cannizzaro's reaction. Most other aldehydes do not give the reaction because in the presence of sodium hydroxide they polymerise instead (section 31.3). The only aldehydes giving Cannizzaro's reaction are those with no H atom on the carbon atom next to the carbonyl group. Benzaldehyde comes into this category, so it gives Cannizzaro's reaction. When heated with sodium hydroxide solution, benzaldehyde gives phenylmethanol and benzoic acid.

Ketones are not readily oxidised at all. They have no effect on mild oxidising agents such as those in table 31.4. This is because they have no oxidisable hydrogen atom joined to the carbonyl group. Strong oxidising agents such as hot concentrated nitric acid can oxidise ketones, but the effect is to break up the molecule, forming at least two smaller molecules of carboxylic acid.

Tests for the aldehyde group

The oxidising agents in table 31.4 are commonly used to test for the aldehyde group.

Fehling's solution is made by mixing copper(II) sulphate solution with an alkaline solution containing 2,3-dihydroxybutanedioate ions (tartrate ions). These complex the Cu^{2+} ions and prevent precipitation of copper(II) hydroxide from the alkaline solution. *Benedict's solution* is similar.

Tollen's reagent is made by adding excess ammonia solution to a solution of silver(I) ions. It contains the $[Ag(NH_3)_2]^+$ ion in alkaline solution. Complexing of Ag^+ by NH_3 prevents precipitation of silver hydroxide.

Many other oxidising agents can be used to convert aldehydes to carboxylic acids. In the laboratory, acidified dichromate(VI) is commonly used to prepare acids from aldehydes.

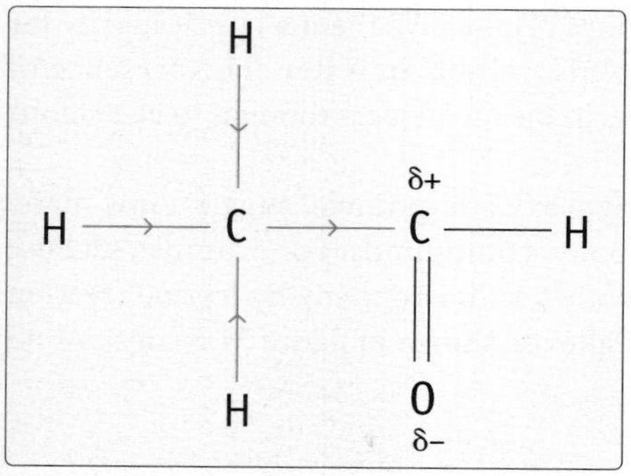

Figure 31.3
Polarising effect of the carbonyl group in ethanal.

31.6 Effect of the carbonyl group on neighbouring atoms

The carbonyl group is polar with a considerable partial positive charge on the carbon atom. This charge has the effect of withdrawing electrons from neighbouring carbon atoms, as shown in figure 31.3.

The result is to make the C—H bond of the neighbouring carbon more polar than normal. As a result, the hydrogen atoms are more readily replaced than those in alkanes. For example, ethanal reacts readily with chlorine even in the dark, forming trichloroethanal:

$$CH_3CHO + 3Cl_2 \longrightarrow CCl_3CHO + 3HCl$$
$$\text{ethanal} \qquad\qquad \text{trichloroethanal}$$

Trichloroethanal is used in the manufacture of the insecticide DDT (see figure 31.4). Section 16.10 gives more information about DDT and particularly the dangers involved in its use.

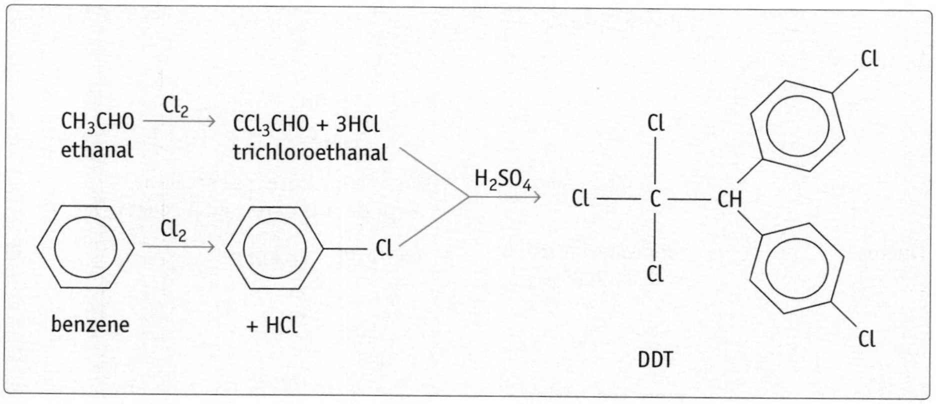

Figure 31.4
Manufacture of DDT.

The iodoform reaction

If chlorine and ethanal are warmed together in the presence of a base, trichloroethanal is initially formed. This then reacts with the base to form trichloromethane, commonly known as chloroform:

$$CCl_3CHO + {}^-OH \longrightarrow CHCl_3 + HCOO^-$$
$$\text{trichloroethanal} \qquad \text{trichloromethane} \quad \text{methanoate ion}$$

A similar reaction occurs when ethanal is warmed with iodine and a base. The product in this case is triiodomethane (iodoform), CHI_3. This kind of reaction is not limited to ethanal. Any compound of the general formula CH_3COR, where R is an alkyl group or hydrogen, will form CHI_3 when warmed with iodine and a base. Triiodomethane is a yellow solid with a characteristic smell. Being insoluble in water, it appears as a yellow crystalline precipitate. This reaction, known as the **iodoform reaction,** is a useful test for compounds of the form CH_3COR. The reaction is also given by compounds of the form CH_3CHOHR, since these are themselves oxidised by the iodine to CH_3COR. For example, ethanol, CH_3CH_2OH, gives the iodoform reaction.

Which of the following compounds would give a yellow precipitate of triiodomethane (iodoform) when heated with a solution of iodine in aqueous sodium carbonate?

1 CH_3CH_2CHO
2 $CH_3CHOHCH_3$
3 $CH_3CH_2CH_2OH$
4 $CH_3CH_2COCH_3$
5 $HCHO$

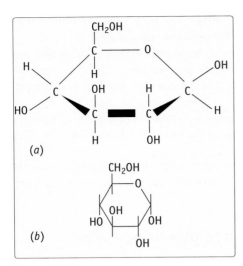

Figure 31.5

Structure of α-glucose: *(a)* full structural formula *(b)* skeletal formula. The hexagonal ring should be thought of as *perpendicular* to the paper, with the —OH groups projecting above or below it. Hydrogen atoms attached to ring carbon atoms are not shown.

31.7 Sugars – naturally occurring carbonyl compounds

Sugars are sweet-tasting soluble carbohydrates. Carbohydrates derive their name from the fact that they are composed of carbon, hydrogen and oxygen with H and O in the ratio of 2 : 1, as in water. **Monosaccharides** such as glucose (figure 31.5) are usually **pentoses** or **hexoses,** i.e. they contain 5 or 6 carbon atoms in their molecules. **Disaccharides** such as sucrose consist of two monosaccharide molecules joined by the elimination of a molecule of water. **Polysaccharides** such as starch are made up of many monosaccharide units joined together. The structure of polysaccharides is considered further in section 9.8. Table 31.5 gives some examples of common carbohydrates. Notice that the monosaccharides all have asymmetric molecules: they therefore exhibit optical isomerism.

The most obvious feature of the structures of the monosaccharides and disaccharides is the presence of large numbers of —OH groups. These give them a large capacity for hydrogen bonding, so they are involatile solids, soluble in water. The presence of —OH groups on several adjacent carbon atoms in the molecule is thought to be responsible for the sweet taste of sugars.

As well as showing the properties of polyhydroxy compounds, sugars show many properties in solution that are typical of carbonyl compounds. For example, glucose gives a crystalline condensation compound with 2,4-dinitrophenylhydrazine (section 31.4). This is surprising since the structure of glucose shown in figure 31.5 contains no carbonyl group.

Table 31.5
Some common carbohydrates. In the ring structures, the C atoms in the ring and the H atoms attached to them have been omitted, for clarity

Name	Type	Structure	Occurrence
glucose	monosaccharide, aldose, hexose	(this is α-glucose: see section 9.8 for details of α- and β-glucose)	occurs abundantly in plants and animals
fructose	monosaccharide, ketose, hexose		in fruit and honey
ribose	monosaccharide, aldose, pentose		component of the molecules of ribonucleic acid (RNA) and vitamin B12
sucrose	disaccharide	glucose — fructose	sugar cane, sugar beet (commonly simply called 'sugar')
maltose	disaccharide	glucose — glucose	malt
lactose	disaccharide	glucose — galactose	milk
starch	polysaccharide	chains of glucose units	plant storage organs, e.g. potato, wheat grain
cellulose	polysaccharide	chains of glucose units (linked differently to those in starch)	structural material of plants

The carbonyl properties possessed by glucose arise from the fact that in addition to its normal ring form it can exist as an 'open-chain' form.

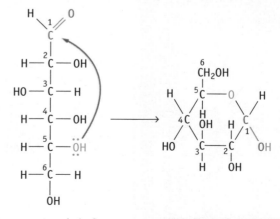

ring form   open-chain form

The two forms are readily interconverted and in aqueous solution about 1% of glucose molecules exist in the open-chain form. This form carries an aldehyde group, so glucose has several properties typical of an aldehyde. It is sometimes called an **aldose**. Thus, in addition to the condensation reaction already mentioned, glucose shows the reducing properties typical of an aldehyde. The reduction of Fehling's solution (or Benedict's solution) is a standard test for glucose and other reducing sugars.

The open-chain form of fructose is

Fructose is therefore a **ketose.**

Why does the open-chain form of glucose and other sugars change to the ring form? It is a result of the tendency of the carbonyl group to undergo nucleophilic addition. The nucleophile involved is the oxygen atom of one of the —OH groups of the same molecule. An internal nucleophilic addition reaction occurs, forming a ring.

▲ Honey is a concentrated solution of sugars including glucose and fructose. Hydrogen bonding between the sugars and water makes the honey viscous.

glucose: open-chain form   glucose: ring form
                                             (α-glucose)

▲ Sucrose dissolves because its —OH groups form hydrogen bonds to water.

This reaction occurs spontaneously. Under normal conditions in aqueous solution the two forms exist in equilibrium, with the ring form predominating.

An understanding of the chemistry of carbohydrates is vital to an understanding of biology. These molecules occur in all living organisms, as structural materials (e.g. cellulose), energy storage compounds (e.g. starch) and primary energy sources (e.g. glucose).

Consider the following sugars. In each case the normal ring form is shown, with the open-chain form below it.

A   B   C

1 Which is a heptose?
2 Which would reduce Tollen's reagent to silver?
3 Which is a ketose?
4 Which would undergo a condensation reaction with 2,4-dinitrophenylhydrazine?

31.8 Starch

Starch is polysaccharide. It is a natural condensation polymer, formed by joining α-glucose units together by eliminating water molecules between them.

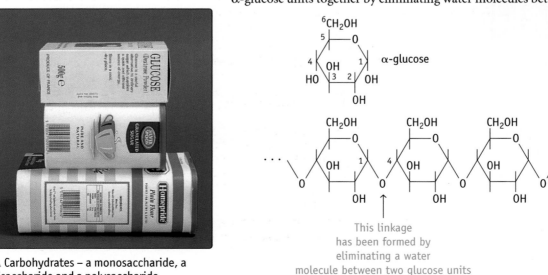

This linkage has been formed by eliminating a water molecule between two glucose units

▲ Carbohydrates – a monosaccharide, a disaccharide and a polysaccharide.

Compare this with the structure of cellulose, another polysaccharide. Cellulose is made by joining β-glucose units together (see section 9.8 for details of α- and β-glucose).

cellulose

Although their structures are similar, starch and cellulose have very different properties. Starch is powdery and forms a hydrated gel when you heat it with water. Cellulose is fibrous and insoluble, and gives plant cell walls their strength. There is more about the structure of cellulose in section 9.8.

The hydrolysis of starch

We can split up starch by putting the water molecules back – this is hydrolysis. As in most hydrolysis reactions, a catalyst is needed. There are two ways to hydrolyse starch: by acid catalysis and by using enzymes.

(i) Acid-catalysed hydrolysis

Starch can be hydrolysed by boiling with dilute acid. The product is glucose:

more simply :

| glucose | glucose | glucose | glucose | acid hydrolysis | 4 | glucose |

(ii) Enzyme-catalysed hydrolysis

In biological systems, starch is hydrolysed by enzymes called amylases. Saliva contains an amylase: when you eat starch, digestion begins in your mouth. In this case, the product of hydrolysis is not glucose but maltose.

| glucose | glucose | glucose | glucose | enzyme hydrolysis | 2 | glucose | glucose |

starch   maltose

Investigation of starch hydrolysis using paper chromatography

You can use paper chromatography to investigate the products of hydrolysis of starch. Figure 31.6 on the next page shows the method. Different samples of starch are hydrolysed by acid and by enzyme catalysis. The resulting solutions are concentrated and spotted onto the baseline of the chromatogram. Spots of glucose and maltose solution are added to the baseline, and the chromatogram is run using a suitable solvent. The spots can be be shown up using a **locating agent** which makes them coloured. You can analyse the results by comparing the positions of the spots.

1 Using a ruler, measure the R_f values for each of the four spots in figure 31.6(b). (See figure 31.7 for an explanation of R_f values).
2 What does figure 31.6(b) tell you about the starch hydrolysis products?

▲ Starch forms a hydrated gel when you heat it with water – very useful for thickening sauces and gravy.

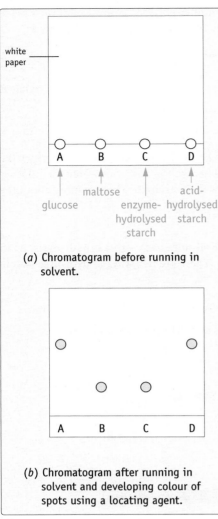

(a) Chromatogram before running in solvent.

(b) Chromatogram after running in solvent and developing colour of spots using a locating agent.

Figure 31.6
Using chromatography to investigate the products of hydrolysis of starch.

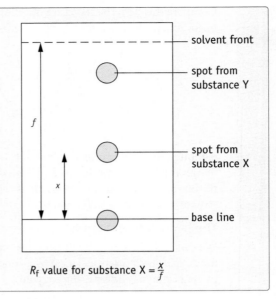

R_f value for substance X $= \dfrac{x}{f}$

Figure 31.7
R_f values are used to identify the different spots on a chromatogram. R_f stands for 'relative to the front', i.e. the distance a substance travels relative to the solvent front. The R_f value of a substance is constant for a given set of conditions: the temperature, the solvent and the type of chromatography paper.

Summary

1   The carbonyl group, —C=O, is present in aldehydes and ketones. In aldehydes it is in a terminal position in the carbon chain. In ketones it is in a non-terminal position.

2   Aldehydes and ketones are named using the suffixes -al and -one, respectively.

3   Aldehydes are prepared by oxidising primary alcohols, ketones by oxidising secondary alcohols.

4   The carbonyl group readily undergoes nucleophilic addition. This is sometimes followed by the elimination of a molecule of water, resulting in a condensation reaction.

5   Aldehydes are generally more reactive than ketones.

6   The tendency of aldehydes to undergo nucleophilic addition makes them polymerise readily.

7   Aldehydes can be oxidised to carboxylic acids by a variety of reagents. Ketones are not readily oxidised.

8   The carbonyl group activates the hydrogen atoms on neighbouring carbon atoms, making them more readily substituted than those in alkanes.

9   Monosaccharides such as glucose or fructose are polyhydroxy compounds, usually containing five or six carbon atoms. They show some properties typical of carbonyl compounds.

10 Starch and cellulose are polysaccharides made by joining glucose units together by eliminating water molecules between them.

11 Table 31.6 gives some of the important reactions of ethanal and propanone.

Table 31.6
Important reactions of ethanal and propanone

Reagent	Reaction type	Product from ethanal	Product from propanone
HCN	addition	CH$_3$CH with OH and CN groups	CH$_3$CCH$_3$ with OH and CN groups
NaHSO$_3$	addition	CH$_3$CH with OH and SO$_3^-$Na$^+$ groups	CH$_3$CCH$_3$ with OH and SO$_3^-$Na$^+$ groups
LiAlH$_4$, NaBH$_4$ or H$_2$/Ni	addition/reduction	CH$_3$CH$_2$OH	CH$_3$CHOHCH$_3$
dilute acid	addition/ polymerisation	(CH$_3$CHO)$_3$ above 0°C (CH$_3$CHO)$_4$ below 0°C	does not polymerise
NH$_2$OH	condensation	CH$_3$CH=NOH	CH$_3$CCH$_3$=NOH
NH$_2$NH$_2$	condensation	CH$_3$CH=NNH$_2$	CH$_3$CCH$_3$=NNH$_2$
Ag$^+$(complexed) (Tollen's reagent)	oxidation	CH$_3$COOH Ag$^+$ reduced to Ag	no reaction
Cu^{2+} (complexed) (Fehling's or Benedict's solution)	oxidation	CH$_3$COOH Cu^{2+} reduced to Cu$_2$O	no reaction
I$_2$/base	iodoform reaction	yellow ppte of CHI$_3$	yellow ppte of CHI$_3$

1 Consider the following compounds:

A $CH_3COCHCH_2CH_3$
           |
          CH_3

B $CH_3CH_2CH_2CH_2CHO$

C

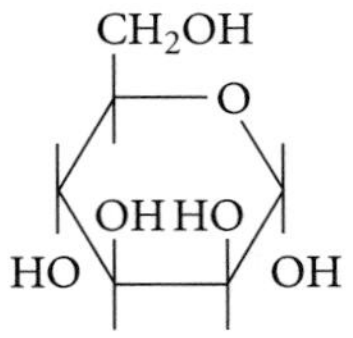

D

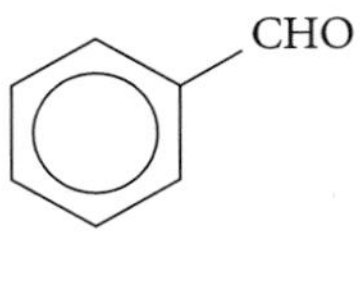

E

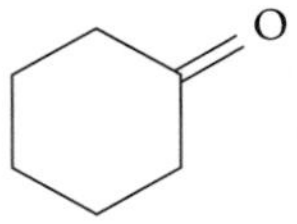

(a) Which are aldehydes?

(b) Which are ketones?

(c) Which is a hexose?

(d) Name compounds **A**, **B**, **D** and **E**.

(e) Which would produce a red precipitate of copper(I) oxide when boiled with Fehling's solution?

(f) Which would give a yellow crystalline precipitate when reacted with iodine in basic solution?

(g) Which would be reduced to a secondary alcohol by hydrogen in the presence of a nickel catalyst?

2 'The —C=C— and —C=O groups might be expected to show chemical similarity, but in fact they show very little.' Is this true? Illustrate and discuss the statement, referring to a wide range of examples and to the underlying chemical principles.

3 This question is about the manufacture of ethanal. Read the passage below and then answer the questions on it.

Ethanal used to be made from ethyne, CH≡CH, manufactured from coke. This process is now being replaced by a method based on ethene. The overall process amounts to oxidation. Ethene and oxygen are bubbled together through an aqueous solution containing $CuCl_2$ and $PdCl_2$ catalysts:

$$CH_2{=}CH_2 + \tfrac{1}{2}O_2 \xrightarrow[CuCl_2]{PdCl_2} CH_3CHO$$

The mechanism of the reaction involves initial attack on the ethene molecule by water. Since H_2O is a nucleophile, it is necessary to render the ethene, which is normally subject only to electrophilic attack, attractive to nucleophiles. This is done by the Pd^{2+} catalyst, which forms a complex with the ethene, decreasing the electron density in the double bond and making the molecule prone to nucleophilic attack.

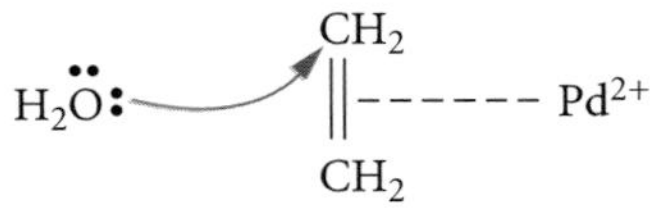

The complex thus formed breaks down by a number of stages to form ethanal and palladium metal. The palladium metal is oxidised back to Pd^{2+} by oxygen. The $CuCl_2$ is a catalyst for this oxidation.

(a) Why is ethanal an important industrial chemical?

(b) Why do you think ethene has replaced ethyne as the starting material for ethanal manufacture?

(c) The suggested mechanism involves the formation of a complex between ethene and Pd^{2+}. How do you think ethene is bonded to Pd^{2+} in this complex?

(d) Suggest a reason why the formation of this complex renders ethene attractive to nucleophiles rather than electrophiles.

(e) Could this method be used to manufacture *propanone*? If so, what starting material would be used instead of ethene?

4 Predict the formulas of the products of the following reactions.

(a) $CH_3COCH_2CH_3 + H_2 \xrightarrow{Ni}$

(b) $C_6H_5COCH_3 + NH_2OH \longrightarrow$

(c) $CH_3CH_2COCH_2CH_3 + HCN \longrightarrow$

(d) $C_6H_5CHO + KMnO_4 \longrightarrow$

(e) $CH_3CH_2CHO + Cl_2 \longrightarrow$

5 Write structural formulas for all compounds of molecular formula C_4H_8O containing a carbonyl group. How would you distinguish between the different compounds, using simple chemical tests?

6 (a) A compound **A** has molecular formula $C_4H_6O_2$. **A** reacts with HCN to form compound **B**, $C_6H_8O_2N_2$. **A** is readily oxidised by acidified potassium dichromate(VI) to an acidic compound **C**, $C_4H_6O_4$. When 1.0 g of **C** is dissolved in water and titrated with 1.0 mol dm^{-3} sodium hydroxide, 16.9 cm^3 of sodium hydroxide are required for neutralisation.
Suggest structural formulas for **A**, **B** and **C** and explain the above reactions.

(b) A compound **X** contains 64.3% C, 7.1% H, and 28.6% O by mass. Its relative molecular mass is 56. **X** reduces Fehling's solution to copper(I) oxide. **X** reacts with hydrogen in the presence of a nickel catalyst: 0.1 g of **X** was found to react with 80 cm^3 of hydrogen (measured at s.t.p.). (1 mole of gas occupies 22 400 cm^3 at s.t.p.)
Suggest a structural formula for **X** and explain the above reactions.

7 Suggest explanations for the following.
  (a) Five different oxidation products of ethane-1,2-diol are known (excluding carbon dioxide and water).
  (b) When ethanal is added to heavy water (D_2O) containing a small amount of base, trideuteroethanal (CD_3CHO) is formed.
  (c) Trichloroethanal undergoes addition reactions far more readily than ethanal.

8 This question is about Grignard reagents, compounds that are of great use in organic synthesis for forming carbon–carbon bonds.

Grignard reagents are compounds of general formula $RMgX$ where $X = Br$ or I. They are very reactive, giving rise to the highly nucleophilic ion R^-. When a solution of a Grignard reagent in dry ether is added to a carbonyl compound, the R^- attacks the carbonyl group:

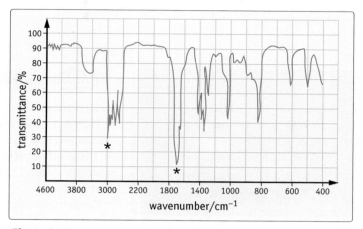

If water is now added to this product, an alcohol is formed:

Overall:

Predict the formulas of the compounds formed when the following are treated with the Grignard reagent methyl magnesium bromide, followed by water.
  (a) Methanal
  (b) Ethanal
  (c) Propanone
  (d) Carbon dioxide.

9 X is a carbonyl compound with three carbon atoms in its molecule.
Figure 31.8 shows the infrared spectrum of **X**.
Figure 31.9 shows the NMR spectrum of **X**.
Use table 25.5 (page 386), giving the characteristic IR absorptions of some common bonds, and table 25.6 (page 389), giving the chemical shifts for some types of protons, when you answer this question.

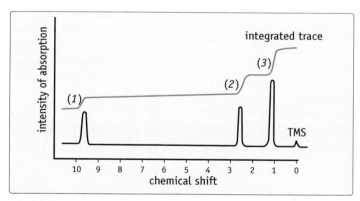

Figure 31.8
The infrared spectrum of **X**.

Figure 31.9
The NMR spectrum of **X**.

  (a) Use the spectra to identify **X**. Give its name and formula.
  (b) Identify the bonds which are responsible for the peaks marked * on the IR spectrum (figure 31.8).
  (c) Identify the protons responsible for each of the peaks in the NMR spectrum (figure 31.9).

32 Carboxylic Acids and their Derivatives

32.1 Carboxylic acids

Why is an ant like a pickled onion?

They both owe their powerful effect to carboxylic acids: in the case of the ant to methanoic acid injected into the skin when the ant bites; in the case of the pickled onion to the ethanoic acid present in vinegar. Carboxylic acids and their derivatives occur widely in nature. They are also present in many manufactured products, such as soaps and polyesters.

Carboxylic acids contain the **carboxyl** group, $-\overset{\underset{\|}{O}}{C}-OH$. Their general formula is usually written RCOOH, where R is an alkyl or aryl group or hydrogen. They are named using the suffix **-oic acid** after a stem indicating the number of carbon atoms (including the one in the carboxyl group). The first two members of the series, systematically named methanoic acid and ethanoic acid, are often called by their older names, formic acid and acetic acid.

Where two carboxyl groups are present, the acid is dibasic and the suffix **-dioic acid** is used. Table 32.1 further illustrates the naming of carboxylic acids.

▲ An ant bites a termite. The ant makes a wound with its jaws, then sprays on methanoic acid.

Table 32.1
Some important carboxylic acids

Formula	Systematic name	Other name	Occurrence and uses
HCOOH	methanoic acid	formic acid (from Latin *formica*, an ant)	used in textile processing and as a grain preservative; ants use it as a poison, injecting it when they bite their victims
CH_3COOH	ethanoic acid	acetic acid (from Latin *acetum*, vinegar)	in vinegar; used in making artificial textiles
CH_3CH_2COOH	propanoic acid	propionic acid	calcium propanoate is used as an additive in bread manufacture
COOH (benzene ring)	benzoic acid		food preservative
COOH / COOH	ethanedioic acid	oxalic acid	in rhubarb leaves

The —C═O and —O—H are so close together on the carboxyl group that they modify one another's properties a great deal. Carboxylic acids therefore have many reactions that are different from those of both alcohols and carbonyl compounds.

The properties of the —COOH group are modified only slightly when it is attached to a benzene ring. As a result, aromatic carboxylic acids have many properties in common with aliphatic ones.

Consider the following acids:

CH₃CH₂CH₂COOH

A (diacid with COOH-CH₂-CH₂-COOH)

B

C (methylbenzene with COOH)

CH₃(CH₂)₁₆COOH

D

1. Which is octadecanoic acid?
2. Which could be prepared by oxidation of butane-1,4-diol?
3. Which is a diabasic acid?
4. Which is butanoic acid?
5. Which is an aromatic acid?

▲ Both these bottles contain pure ethanoic acid (acetic acid). Pure ethanoic acid freezes at 17°C. The bottle on the left has been kept in the fridge.

Table 32.2
Some properties of carboxylic acids

Making carboxylic acids

In the laboratory, carboxylic acids are normally prepared by oxidising primary alcohols (section 30.7) or aldehydes (section 31.5). Carboxylic acids can also be prepared by hydrolysis of nitriles (section 33.2). The hydrolysis is carried out by heating with strong acid or base, e.g.

$$RCN + 2H_2O + HCl \rightarrow RCOOH + NH_4Cl$$

Nitriles are themselves prepared from halogenoalkanes (section 29.5).

The oxidation of alcohols and aldehydes was at one time the major industrial method of preparing carboxylic acids. Nowadays the most important aliphatic acids are produced by direct oxidation of alkanes. A volatile petroleum fraction is oxidised in the liquid phase by passing air through it under pressure in the presence of a catalyst. Ethanoic acid is the major and most important product. By-products include methanoic and propanoic acids.

Benzoic acid is made industrially and in the laboratory by the oxidation of methylbenzene (section 28.9).

Physical properties

Ethanoic acid melts at 17°C and boils at 118°C. This means that although it is normally a liquid in British laboratories, it freezes in cold weather. In fact it is a fairly good rule that when the ethanoic acid freezes it is time to switch the heating on. Because of the readiness with which it freezes, and the similarity of solid ethanoic acid to ice, pure ethanoic acid is often described as 'glacial'. Dilute solutions of the acid, of course, freeze at about the same temperature as water.

The boiling point of ethanoic acid is higher than that of either ethanol (78°C), which has the same number of carbon atoms, or propan-1-ol (97°C), which has the same relative molecular mass. The relatively high boiling point of ethanoic acid and other carboxylic acids reflects a considerable degree of hydrogen bonding. Carboxylic acids form stronger hydrogen bonds than alcohols. This is because their —OH group is more polarised due to the presence of the electron-withdrawing —C=O group:

Figure 32.1
Carboxylic acids can form hydrogen-bonded dimers.

Furthermore, carboxylic acids have the possibility of forming doubly hydrogen-bonded dimers (figure 32.1):

Acid	Formula	State at room temperature	Boiling point/°C	K_a at 25°C/mol dm^{-3}	Solubility in water
methanoic	HCOOH	l	101	1.6×10^{-4}	infinite*
ethanoic	CH₃COOH	l	118	1.7×10^{-5}	infinite*
propanoic	CH₃CH₂COOH	l	141	1.3×10^{-5}	infinite*
butanoic	CH₃CH₂CH₂COOH	l	164	1.5×10^{-5}	infinite*
octanoic	CH₃(CH₂)₆COOH	l	237	1.4×10^{-5}	slightly soluble
chloroethanoic	ClCH₂COOH	s	189	1.3×10^{-3}	very soluble
dichloroethanoic	Cl₂CHCOOH	l	194	5.0×10^{-2}	infinite*
trichloroethanoic	Cl₃CCOOH	s	196	2.3×10^{-1}	very soluble
benzoic	C₆H₅COOH	s	249	6.4×10^{-5}	slightly soluble
ethanedioic	(COOH)₂	s	—	3.5×10^{-2} (first dissociation) 4.0×10^{-5} (second dissociation)	soluble

*miscible in all proportions

Carboxylic acids in the liquid and solid states exist mostly in this form (see section 9.8).

Because of their capacity for hydrogen bonding, the early members of the series are miscible with water in all proportions. As with all homologous series, solubility decreases with increasing molecular size. Table 32.2 gives physical properties of some carboxylic acids.

Methanoic, ethanoic and propanoic acids have strong, sharp vinegary odours. The C_4 to C_8 acids (butanoic to octanoic) have very strong, unpleasant odours. Butanoic acid is responsible for the smell of rancid butter and is present in human sweat. Its smell can be detected at concentrations of 10^{-11} mol dm^{-3} by humans and at concentrations of 10^{-17} mol dm^{-3} by dogs.

32.2 The carboxyl group and acidity

In chapter 30 we discussed the acidity of the —OH group in alcohols, and found that alcohols are even less acidic than water. Carboxylic acids also contain the —OH group, but they are much stronger acids than alcohols and water. Most of the acids we taste in our food, such as ethanoic acid in vinegar and citric acid in lemons, are carboxylic acids. They appear to have a very strong acid taste; nevertheless, they are only present in low concentrations (vinegar is about 7% ethanoic acid) and they are weak acids compared with inorganic ones such as hydrochloric acid.

Carboxylic acids react with bases to form salts. For example, ethanoic acid reacts with sodium hydroxide to form sodium ethanoate:

$$CH_3COOH + NaOH \rightarrow CH_3COO^-Na^+ + H_2O$$
ethanoic acid $\qquad$ sodium ethanoate

Carboxylic acids also form salts when they react with carbonates and hydrogencarbonates. In this case, carbon dioxide is evolved. For example:

$$2CH_3COOH + Na_2CO_3 \rightarrow 2CH_3COO^-Na^+ + CO_2 + H_2O$$
sodium carbonate

$$CH_3COOH + NaHCO_3 \rightarrow CH_3COO^-Na^+ + CO_2 + H_2O$$
sodium
hydrogencarbonate

The reaction with sodium hydrogencarbonate is used as a test for carboxylic acids. All carboxylic acids react with sodium hydrogencarbonate to release carbon dioxide; this distinguishes them from weaker acids such as phenols.

When a carboxylic acid dissociates in water, it forms a carboxylate anion RCOO$^-$:

$$H_2O + R-C\!\!\stackrel{\displaystyle O}{\diagdown}_{OH} \longrightarrow H_3O^+ + R-C\!\!\stackrel{\displaystyle O}{\diagdown}_{O^-}$$

carboxylic acid $\qquad\qquad\qquad$ carboxylate anion

X-ray diffraction studies of the anion show that the two carbon–oxygen bonds are of equal length. This implies that the two bonds are identical. It suggests that the negative charge and double-bond character are distributed evenly over the whole carboxylate group, delocalising the negative charge:

$$R-C\!\!\stackrel{\displaystyle O}{\diagdown}_{O}$$

Thus each oxygen effectively carries half a negative charge. This delocalisation of charge on the anion makes it less likely to join up with H$^+$ again. As a result, the equilibrium

$$RCOOH + H_2O \rightleftharpoons RCOO^- + H_3O^+$$

is much further to the right than in the case of alcohols, so carboxylic acids are stronger acids. Nevertheless, the majority of the acid is in the un-ionised form and a solution of ethanoic acid of concentration 0.1 mol dm^{-3} is only 0.3% ionised.

▲ The ability of a dog to track a person is due to its ability to detect carboxylic acids in the sweat from the person's feet.
Each person's sweat glands produce a characteristic blend of carboxylic acids which can be detected and recognised by the sensitive nose of the dog.

Consider the following acids:

CH$_3$COOH $\qquad$ CH$_2$ClCOOH
CCl$_3$COOH $\qquad$ HCOOH

Which would be the stronger acid:

1 CH$_3$COOH or CH$_2$ClCOOH?
2 CCl$_3$COOH or CH$_2$ClCOOH?
3 CH$_3$COOH or HCOOH?

methanoate ion

ethanoate ion

chloroethanoate ion

Figure 32.2
Anions formed from methanoic, ethanoic and chloroethanoic acids.

In general, the higher the density of negative charge on the RCOO⁻ ion, the more it will attract H^+ and reform the un-ionised RCOOH. So we can say that the more negative charge there is on RCOO⁻, the weaker the acid will be. The nature of the group to which —COOH is attached can have a considerable effect on the strength of a carboxylic acid. In general, an electron-withdrawing group reduces the density of negative charge on RCOO⁻, and so increases the strength of the acid. An electron-donating group does the reverse.

Table 32.2 gives the K_a values for different carboxylic acids (remember, the higher the K_a, the stronger the acid). The electron-donating —CH_3 group tends to increase the negative charge density on the carboxylate group in the ethanoate ion. Ethanoic acid is therefore weaker than methanoic, which carries no electron-donating alkyl group. On the other hand, the electron-withdrawing chlorine atom reduces the charge density on the chloroethanoate ion so chloroethanoic acid is stronger than ethanoic (figure 32.2).

The electron-donating effect of all alkyl groups is roughly equal, so aliphatic carboxylic acids with more than one carbon atom (ethanoic, propanoic, butanoic, etc.) are roughly equal in strength.

32.3 Salts of carboxylic acids

Like all acids, carboxylic acids react with bases to form salts. Table 32.3 gives some experimental data about benzoic acid. Benzoic acid is not very soluble in cold water, but is soluble in hot water.

Table 32.3
Behaviour of benzoic acid in basic and acidic solution

Experiment	Result
1 solid benzoic acid is shaken with water	most of the benzoic acid is undissolved
2 sodium hydroxide is added to the aqueous suspension from experiment (1)	benzoic acid dissolves
3 hydrochloric acid is added to the solution from experiment (2)	precipitate of benzoic acid appears

Look at table 32.3, then answer these questions.

1 Why is benzoic acid not very soluble in cold water?

2 Why does it dissolve in sodium hydroxide solution?

3 Why does hydrochloric acid precipitate benzoic acid from its solution in sodium hydroxide?

Carboxylic acids are weak acids and so only slightly dissociated in aqueous solution.

$$RCOOH + H_2O \rightleftharpoons RCOO^- + H_3O^+$$
carboxylic acid   carboxylate ion

Addition of a base removes H_3O^+, shifting this equilibrium to the right. This converts the acid to the carboxylate ion and forms the salt. Benzoic acid is much more soluble in basic solution than in water because in the presence of a base it is converted almost entirely to soluble benzoate ion. The addition of a strong acid to a solution of the carboxylate salt reverses the equilibrium, re-forming free acid. The anions of carboxylic acids are themselves basic: a 1.0 mol dm⁻³ solution of sodium ethanoate has a pH of about 9.5.

Salts of carboxylic acids show marked differences in properties from the free acids, because they are ionic rather than molecular substances. Thus all carboxylate salts are solids, and most are soluble in water. Not surprisingly, as the hydrocarbon chain increases in length, the properties of these salts tend more and more to those of molecular hydrocarbons rather than ionic salts.

Salts of carboxylic acids with medium-length chains have the characteristics of both hydrocarbons and salts. This makes them valuable as detergents.

Detergents

Detergents are substances that improve the cleaning properties of water. The most common 'dirt' is that on human clothes and bodies. Typically this consists of a mixture of natural fats, long-chain carboxylic acids, skin debris, silicon(IV) oxide and carbon. The main problem is caused by greasy fats and carboxylic acids, which tend to bind the other materials to the skin or fabric surface. This prevents them being washed away by water. A detergent works by enabling water to mix with and remove these greasy materials.

▲ Rhubarb leaves are poisonous because they contain ethanedioic acid (oxalic acid). This acid is poisonous, because of the toxic nature of its anion, the ethanedioate ion. The correct way to treat this kind of poisoning is to give an alkali which can form an insoluble salt with ethanedioic acid. For example, magnesium hydroxide can be used, because magnesium ethanedioate is insoluble.

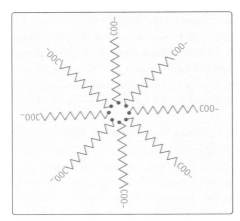

Figure 32.3
Arrangement of soap chains in a micelle, shown in cross-section. The charged ends are on the outside, since they are attracted by water. (Each micelle contains far more chains than shown here.)

Soap

Soap was the first detergent, dating from Roman times, and it is still one of the commonest. Soap is a mixture of the salts of medium- and long-chain carboxylic acids. Its manufacture is described on page 510.

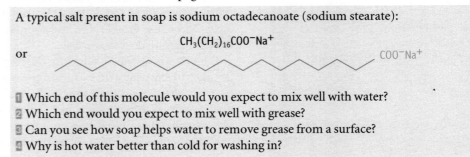

A typical salt present in soap is sodium octadecanoate (sodium stearate):

$$CH_3(CH_2)_{16}COO^-Na^+$$

1. Which end of this molecule would you expect to mix well with water?
2. Which end would you expect to mix well with grease?
3. Can you see how soap helps water to remove grease from a surface?
4. Why is hot water better than cold for washing in?

The charged —COO⁻ group at the end of a soap chain enables it to dissolve in water. When soap dissolves, it does so in the form of spherical clusters called **micelles** (figure 32.3). These micelles are large enough to cause the solution to scatter light: this is why soapy water is cloudy. The non-polar hydrocarbon chain mixes well with greasy substances. When soapy water comes into contact with grease, the non-polar end of the molecule mixes with the non-polar grease (figure 32.4). This leaves the grease surrounded by an outer sheath of polar —COO⁻ groups which are attracted to water. The forces between water and grease are thus much increased, so that the grease is lifted off the surface in the form of small globules. These can then be rinsed away. The sheath of negatively charged groups causes the globules to repel one another, so they do not coagulate and redeposit on the surface.

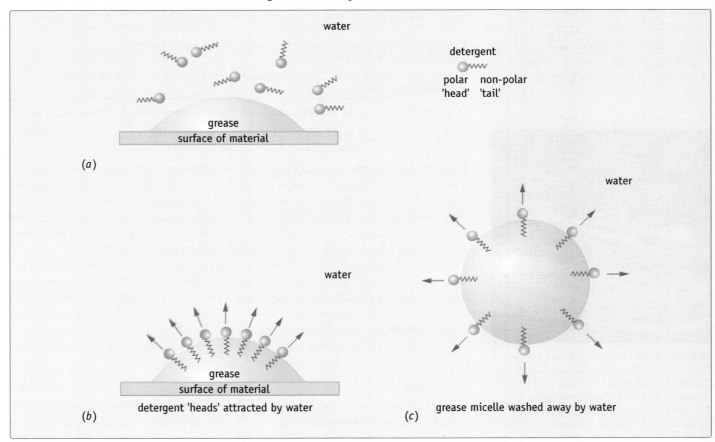

Figure 32.4
Effect of detergent on grease.

Soapless detergents

Soap is not the only detergent. Any molecule with a fairly large non-polar 'tail' and a charged or polar 'head' will have detergent properties. The head may carry a negative charge (**anionic** detergents are the most common), a positive charge (**cationic**), or may consist of a polar, uncharged group (**non-ionic**).

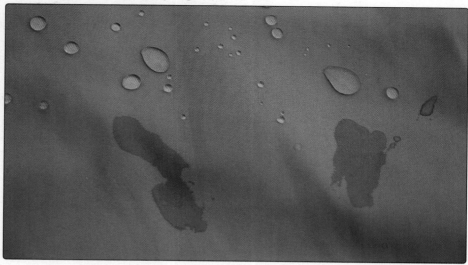

◀ The effect of soap on the wetting ability of water. The droplets at the top are pure water; those at the bottom are a soap and water solution.

Soaps have the disadvantage that they form a 'scum' with hard water. Hard water contains calcium ions. These ions react with the soap to form the insoluble calcium salt of the carboxylic acid. This is the scum. For example:

$$2CH_3(CH_2)_{16}COO^-(aq) \ + \ Ca^{2+}(aq) \ \rightarrow \ (CH_3(CH_2)_{16}COO)_2Ca(s)$$

octadecanoate ions from soap — scum

The scum is itself undesirable and its formation leads to the use of excessive quantities of soap.

Soapless detergents have the advantage that they form no scum with hard water. In addition, they are made from by-products of oil refining. Soap, on the other hand, is made from vegetable or animal fats which are often expensive.

A typical and very common soapless detergent has the structure:

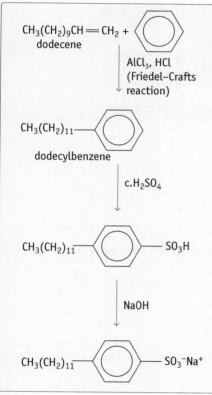

Figure 32.5
Manufacture of a soapless detergent.

Which are the water-attracting and water-repelling parts of this compound?

This detergent is manufactured by the sequence shown in figure 32.5. Look back to section 28.7 to remind yourself of the reactions involved. This kind of soapless detergent is present in many well-known washing powders and liquids.

Earlier synthetic detergents included branched-chain compounds with the structure:

These are cheap to make. Unfortunately because of their side chains they cannot be degraded by bacteria, unlike the synthetic detergent shown earlier. They therefore tended to persist in water after sewage treatment and caused foaming in rivers and streams. They have now been replaced by detergents with unbranched side chains.

Synthetic detergents have replaced soap for many uses. Many different compounds with detergent properties are now known and manufactured. They are used not only for cleaning purposes but also as wetting agents, foam stabilisers and emulsifying agents. All these uses depend on their ability to mix both with water and with water-repelling materials. Some more examples are given in table 32.4.

▶ Households use large quantities of detergent every day, and it all ends up in the sewage. It is important that detergents are biodegradable so they can be quickly broken down by bacteria – otherwise the detergents end up in the rivers.

Table 32.4
Some synthetic detergents

Type	Non-polar portion	Polar portion	Examples of uses
Non-ionic	C_9H_{19}—⬡—	$O(CH_2CH_2O)_xCH_2CH_2OH$ ($x = 5$ to 10)	liquid detergents emulsifying agents wetting agents
Cationic	$C_{15}H_{31}$————	$CH_2\overset{+}{N}(CH_3)_3Br^-$	hair conditioner
Anionic	$C_{12}H_{23}$————	$COOCH_2CHOHCH_2OSO_3^-Na^+$	toothpastes

32.4 Some important reactions of carboxylic acids

Reaction with bases
This has already been considered (section 32.3).

Esterification
This important reaction is considered in the next section.

Reaction with phosphorus pentachloride
Carboxylic acids react with phosphorus pentachloride to form acyl chlorides (section 29.7). For example

$$CH_3COH + PCl_5 \longrightarrow CH_3CCl + HCl + POCl_3$$

with the $\overset{||}{O}$ (ethanoic acid) and $\overset{||}{O}$ (ethanoyl chloride)

Other halogenating agents such as $SOCl_2$ and PBr_3 (which introduces a Br atom) react similarly.

Oxidation
As a rule, carboxylic acids are not readily oxidised. This is because they are themselves the end products of the oxidation sequence

$$\text{primary alcohol} \rightarrow \text{aldehyde} \rightarrow \text{carboxylic acid.}$$

There are however two exceptions to this general rule.

(i) Methanoic acid and methanoates are readily oxidised, for example by potassium manganate(VII), to carbon dioxide:

$$HCOOH \longrightarrow CO_2 + 2H^+ + 2e^-$$
methanoic acid — accepted by oxidising agent

(ii) Ethanedioic acid and its salts are oxidised by warm potassium manganate(VII), to carbon dioxide:

$$(COOH)_2 \longrightarrow 2CO_2 + 2H^+ + 2e^-$$
ethanedioic acid — accepted by oxidising agent

Phosphorus pentachloride reacts with the —OH group in alcohols (section 30.6).

1 What is the product of such a reaction?
2 Given that carboxylic acids contain the —OH group, what product would you expect from the reaction of, say, ethanoic acid with phosphorus pentachloride?

Reduction

Carboxylic acids are rather hard to reduce. Powerful reducing agents such as lithium tetrahydridoaluminate, $LiAlH_4$, will convert them to the corresponding primary alcohol:

$$R-\underset{\underset{O}{\|}}{C}-OH \xrightarrow[\substack{\text{in dry} \\ \text{ether}}]{LiAlH_4} RCH_2OH$$

Esters are somewhat easier to reduce than acids: see section 32.5.

Dehydration

Methanoic and ethanedioic acids are exceptional as the only carboxylic acids that can be readily dehydrated.

Methanoic acid is dehydrated to carbon monoxide when warmed with concentrated sulphuric acid:

$$HCOOH \xrightarrow{c.H_2SO_4} CO + H_2O$$

Ethanedioic acid is dehydrated by the same method to carbon monoxide and carbon dioxide:

$$(COOH)_2 \xrightarrow{c.H_2SO_4} CO + CO_2 + H_2O$$

Use your knowledge of the reaction of carboxylic acids to write the structural formulas of the products formed when

1. benzoic acid is treated with phosphorus pentachloride,
2. butanoic acid is treated with lithium tetrahydridoaluminate,
3. ethanedioic acid is treated with excess sodium hydroxide.

32.5 Esters

Figure 32.6 illustrates an experiment in which ethanol and ethanoic acid are heated in the presence of concentrated sulphuric acid. Study it, then answer the questions below.

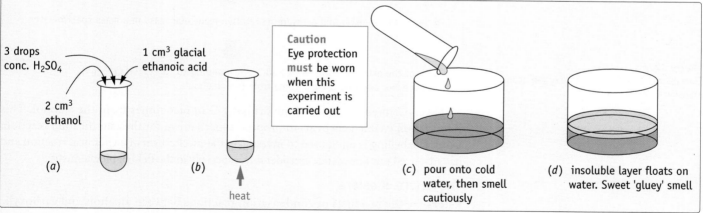

Figure 32.6
Reaction between ethanol and ethanoic acid in the presence of concentrated sulphuric acid.

Ethanol and ethanoic acid react together to form an ester, ethyl ethanoate:

$$\underset{\text{ethanoic acid}}{CH_3\underset{\underset{O}{\|}}{C}-OH} + \underset{\text{ethanol}}{CH_3CH_2OH} \longrightarrow \underset{\text{ethyl ethanoate}}{CH_3\underset{\underset{O}{\|}}{C}-OCH_2CH_3} + H_2O$$

1. Does (a) ethanol, (b) ethanoic acid, (c) sulphuric acid, mix with water?
2. Do any of the substances above have a sweet, 'gluey' smell?
3. What can you say about the product of this reaction between ethanol and ethanoic acid?
4. Why do you think the reaction mixture is poured onto cold water before smelling?

This reaction proceeds extremely slowly under ordinary conditions. However, it goes at an appreciable rate in the presence of a strong acid catalyst such as sulphuric or hydrochloric acid. This is a general method of preparing esters, and can be applied to any combination of acid and aliphatic alcohol:

$$R^1—\underset{\underset{O}{\|}}{C}—OH + R^2OH \ \rightarrow \ R^1—\underset{\underset{O}{\|}}{C}—OR^2 + H_2O$$

Notice that the bridging oxygen atom between the R^1CO and R^2 groups in the ester could have come either from the alcohol or from the acid. As the equation stands, there is no way of telling from which molecule it originated. To put the problem another way, does the oxygen atom in the alcohol molecule end up in the ester or in the water?

The answer to this question was found by two American chemists, Roberts and Urey, using a technique known as **isotopic labelling**. The method is outlined in figure 32.7.

1 Prepare methanol 'labelled' with the oxygen isotope ^{18}O

$$CH_3{}^{18}OH$$

2 React this with benzoic acid in the presence of an acid catalyst

$$\bigcirc\!\!\!\!\bigcirc—COOH + CH_3{}^{18}OH$$

3 Two sets of products are possible according to which bond breaks in $CH_3{}^{18}OH$

$$\bigcirc\!\!\!\!\bigcirc—CO^{18}OCH_3 + H_2O \qquad \text{or} \qquad \bigcirc\!\!\!\!\bigcirc—COOCH_3 + H_2{}^{18}O$$

$$(M_r = 138) \qquad\qquad\qquad\qquad (M_r = 136)$$

4 Separate the ester and measure its relative molecular mass in a mass spectrometer

5 M_r of the ester is found to be 138, corresponding to
Hence the bridging O must come from the alcohol.

This experiment shows that the bridging O in fact comes from the alcohol. This information enabled chemists to propose a mechanism for the esterification reaction. Isotopic labelling is often used to investigate the mechanism of a chemical reaction and has advanced our knowledge considerably, especially in the field of biochemistry.

The nature of esters

Esters are the products of condensation reactions between alcohols and carboxylic acids. They can also be prepared by reacting alcohols with acyl chlorides (section 29.7) or acid anhydrides (section 32.7). Esters contain the $—\underset{\underset{O}{\|}}{C}—O$ functional group.

Esters are named by regarding them as alkyl (or aryl) derivatives of carboxylic acids. Thus the name is obtained from a stem indicating the alcohol from which the ester is derived, with the suffix **-yl**, followed by a stem indicating the acid, with the suffix **-oate**. So the ester derived from methanol and propanoic acid is called methyl propanoate. The name of an ester is most easily worked out from its formula by these stages:

1 Divide the formula into two portions by mentally drawing a line after the bridging O of the —COO group, e.g.

$$CH_3CH_2\underset{\underset{O}{\|}}{C}—O\ \vert\ CH_2CH_2CH_2CH_3$$

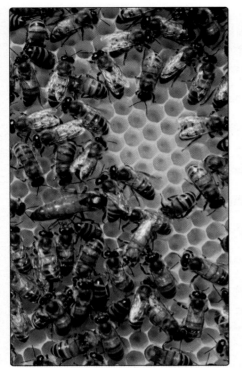

▲ Beeswax contains esters made from long-chain alcohols and long-chain acids. A typical ester in beeswax is $C_{15}H_{31}COOC_{30}H_{61}$.

Figure 32.7
Investigation of the fate of oxygen atoms in an esterification reaction.

32 Carboxylic Acids and their Derivatives

Try these questions to test your understanding of how esters are named and prepared.

1 Give the name of
   (a) $C_6H_5COOCH_3$,
   (b) $HCOOCH_2CH_3$
2 Write the formula of
   (a) heptyl decanoate,
   (b) phenyl benzoate.
3 How would you prepare propyl propanoate?
4 Write the formula of the product formed when ethanoic acid and butan-2-ol are warmed together in the presence of an acid catalyst.

▲ Audio tape need to be strong, so it is made from polyester.

2 Name the portion that does *not* carry the —COO group. This portion is an alkyl group – *butyl* in this example.

3 Name the portion that carries the —COO group. This portion is a carboxylate group – *propanoate* in this example.

4 Combine **2** and **3** to give the name of the ester – *butyl propanoate*.

Unlike the acids and alcohols from which they are derived, esters have no free —OH groups so they cannot form hydrogen bonds. They are therefore volatile compared with acids and alcohols of similar molecular mass and they are not very soluble in water. Ethyl ethanoate, for example, is a liquid, boiling point 77°C. At 25°C its solubility is 8.5 g per 100 g of water.

Volatile esters have characteristically pleasant, fruity smells. The flavour and fragrance of many fruits and flowers are due to mixtures of compounds, many of them esters. Artificial fruit flavourings are made by mixing synthetic esters to give the approximate flavour (raspberry, pear, cherry, etc.) required. Organic acids are usually added to give the sharp taste characteristic of fruit. Artificial fruit flavours can only approximate to the real thing, because it would be too costly to include all the components of the complex mixture of compounds present in the real fruit.

The smell of some adhesives is reminiscent of ethyl ethanoate. Polystyrene cement, for example, consists of polystyrene dissolved in ethyl ethanoate. When the cement is applied, the ethyl ethanoate evaporates, leaving behind the solid plastic which binds together the surfaces being joined.

Polyesters

Polyesters are synthetic fibres used as substitutes for cotton and wool. They are marketed under such trade names as 'Terylene' and 'Dacron'. As the name 'polyester' suggests, they are polymers joined by an ester linkage. By far the commonest polyester is 'Terylene' poly(ethane-1,2-diyl benzene-1,4-dicarboxylate). This is made by esterifying ethane-1,2-diol with benzene-1,4-dicarboxylic acid (terephthalic acid) (figure 32.8).

Having two —OH groups on the alcohol and two —COOH groups on the acid makes possible the formation of the polymer. This is an example of **condensation polymerisation**, because the polymer is formed by a condensation reaction.

Monomers
$HOCH_2CH_2OH$
ethane-1,2-diol

$HOOC$ —⬡— $COOH$

benzene-1,4-dicarboxylic acid

Polymerisation

$\ldots + HOCH_2CH_2OH + HOOC$—⬡—$COOH + HOCH_2CH_2OH + HOOC$—⬡—$COOH + \ldots$

$\ldots — OCH_2CH_2OOC$—⬡—$COOCH_2CH_2OOC$—⬡—$CO — \ldots$

+   +   +   +
H_2O H_2O H_2O H_2O

Figure 32.8
The formation of a polyester.

Polyesters have great tensile strength. They are used as the bonding resin in glass fibre plastics as well as being widely used to make fabrics.

Esterification as an equilibrium reaction: hydrolysis

The reaction of a carboxylic acid with an alcohol to form an ester is fairly slow, even in the presence of an acid catalyst. As well as being kinetically quite slow, esterification is also an equilibrium reaction that does not normally reach completion.

If known quantities of ethanol, ethanoic acid and hydrochloric acid catalyst are sealed together and left for two or three weeks, an equilibrium is reached.

Ethanol, ethanoic acid, ethyl ethanoate and water are all present, as well as unchanged acid catalyst.

$$CH_3CH_2OH + CH_3COOH \overset{H^+}{\rightleftharpoons} CH_3COOCH_2CH_3 + H_2O$$
$$\text{ethanol} \qquad \text{ethanoic acid} \qquad \text{ethyl ethanoate}$$

If the reaction mixture is now titrated with standard sodium hydroxide, the amount of ethanoic acid present at equilibrium can be found.

From this the amounts of all the other components of the equilibrium mixture can be worked out. Using these results the equilibrium constant K_c for the esterification reaction can easily be found: its value is about 4 at 25°C.

1 Write an expression for the equilibrium constant for the esterification of ethanol and ethanoic acid as in the last equation.
2 If 1 mole each of ethanoic acid and ethanol are allowed to reach equilibrium at 25°C, what will be the number of moles of ethyl ethanoate present at equilibrium? (Assume $K_c = 4$ at 25°C.)
3 What would you expect to happen if ethyl ethanoate were refluxed with an excess of water for a long time?

A mixture of 1 mole of ethanol and 1 mole of ethanoic acid brought to equilibrium at 25°C contains 2/3 mole each of ethyl ethanoate and water and 1/3 mole each of ethanol and ethanoic acid. The esterification reaction is thus far from complete, and it is readily reversed.

Hydrolysis of esters

The reverse of esterification is **ester hydrolysis**, in which an ester reacts with water to form an alcohol and a carboxylic acid. Like esterification, ester hydrolysis is a slow reaction, but it is speeded up by an acid catalyst. It is also catalysed by alkali, but in this case the carboxylic acid formed by hydrolysis reacts with excess alkali to form the carboxylate salt. This removes the carboxylic acid from the equilibrium mixture as it is formed, which means that ester hydrolysis can proceed to completion in the presence of alkali. This cannot happen when an acid catalyst is used.

For example:

$$CH_3COOCH_2CH_3 + {}^-OH \longrightarrow CH_3COO^- + CH_3CH_2OH$$
$$\text{ethyl ethanoate} \qquad\qquad \text{ethanoate ion} \qquad \text{ethanol}$$

or

$$CH_3COOCH_2CH_3 + NaOH \longrightarrow CH_3COO^-Na^+ + CH_3CH_2OH$$

Esters are therefore hydrolysed more effectively in alkaline than in acidic solution. In both cases the reaction is still quite slow and the mixture must be boiled under reflux. Strictly speaking, the alkali is not acting catalytically in this reaction because, as the equation shows, the $^-$OH is used up. The hydrolysis of an ester by alkali is sometimes called **saponification**, a word derived from Latin words meaning 'soap-making' (see below).

Ester hydrolysis is important in biological systems. The photograph in figure 32.9 (pyrethrum insecticides) is an example.

Reduction of esters

Esters can be reduced to alcohols using lithium tetrahydridoaluminate, LiAlH$_4$. Molecules of two alcohols are produced:

$$R^1\underset{\underset{O}{\|}}{C}{-}OR^2 \xrightarrow[\text{dry ether}]{LiAlH_4} R^1CH_2OH + R^2OH$$

This reduction can also be carried out using hydrogen under high pressure with a suitable catalyst.

Figure 32.9
Pyrethrum flowers contain a natural insecticide, called pyrethrin. It is an ester, which is hydrolysed quickly when eaten by mammals, making it harmless to them, but is toxic to insects. Pyrethroid insecticides can now be manufactured which work in the same way but are safe in the environment.

32.6 Fats

Fats and oils are naturally occurring esters which are used as energy storage compounds by plants and animals. They are members of a larger class of naturally occurring compounds called **lipids**: compounds which are insoluble in water but soluble in organic solvents.

Fats are derived from propane-1,2,3-triol, $HOCH_2$—CHOH—CH_2OH (commonly known as glycerol or glycerine). This molecule has the capacity to combine with one, two or three molecules of carboxylic acid. In practice, most fats are *triesters* derived from propane-1,2,3-triol and a variety of long-chain carboxylic acids. These long-chain acids are sometimes called **fatty acids**. For example, a simple fat molecule is derived from a molecule of propane-1,2,3-triol and three molecules of octadecanoic acid:

$$
\begin{array}{c}
CH_2OH \\
| \\
CHOH \\
| \\
CH_2OH
\end{array}
\quad + \quad
\begin{array}{c}
HOOC(CH_2)_{16}CH_3 \\
HOOC(CH_2)_{16}CH_3 \\
HOOC(CH_2)_{16}CH_3
\end{array}
\quad \longrightarrow \quad
\begin{array}{c}
CH_2OOC(CH_2)_{16}CH_3 \\
| \\
CHOOC(CH_2)_{16}CH_3 \\
| \\
CH_2OOC(CH_2)_{16}CH_3
\end{array}
\quad + \quad 3H_2O
$$

1 mole of propane-1,2,3-triol   3 moles of octadecanoic acid   1 mole of propane-1,2,3-triyl trioctadecanoate

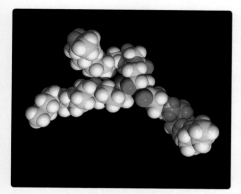

▲ A typical molecule found in fats. This is the triester formed from glycerol and hexadecanoic acid.

Table 32.5 gives some common fatty acids. Fifty or so are found in nature, the vast majority of them having an even number of carbon atoms in their molecule.

Table 32.5
Some common fatty acids

Structure	Systematic name	Common name	Occurrence
$CH_3(CH_2)_{16}COOH$	octadecanoic acid	stearic acid	mainly in animal fats
$CH_3(CH_2)_{10}COOH$	dodecanoic acid	lauric acid	coconut oil, palm-kernel oil
$CH_3(CH_2)_{14}COOH$	hexadecanoic acid	palmitic acid	most fats, especially palm oil
$CH_3(CH_2)_7CH{=}CH(CH_2)_7COOH$	octadec-9-enoic acid	oleic acid	most fats, especially olive oil

Fats containing a large proportion of unsaturated acids tend to have low melting points. Many are liquid at room temperature and these are called **oils**. They can be converted to solid fats by hydrogenation. There is evidence that unsaturated fats are healthier than saturated ones (see section 27.6).

The chart in figure 32.10 gives the proportion of saturated and unsaturated acids in fats and oils from different sources.

Figure 32.10
Components of fats and oils, classified according to the position and degree of saturation of their fatty acids.
U = unsaturated
S = saturated. For example, UUU indicates a triester in which all three fatty acids are unsaturated.

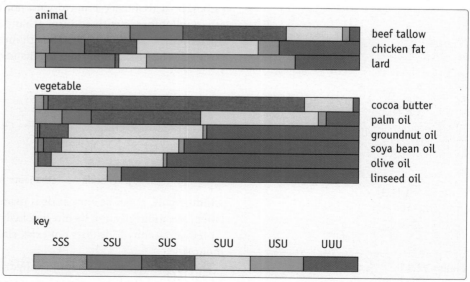

Like all esters, fats can be hydrolysed. The products are propane-1,2,3-triol and a mixture of carboxylic acids. If the hydrolysis is carried out by boiling the fat with an

alkali such as sodium hydroxide, the products are propane-1,2,3-triol and a mixture of the sodium salts of fatty acids. For example:

$$CH_2OOC(CH_2)_{16}CH_3$$
$$|$$
$$CHOOC(CH_2)_{16}CH_3 \ + \ 3NaOH \ \longrightarrow \ 3CH_3(CH_2)_{16}COO^-Na^+ \ + \ \begin{array}{l} CH_2OH \\ | \\ CHOH \\ | \\ CH_2OH \end{array}$$
$$|$$
$$CH_2OOC(CH_2)_{16}CH_3$$

propan-1,2,3,-triyl   sodium octadecanoate   propane-1,2,3-triol
trioctadecanoate

Sodium octadecanoate is a soap.

This reaction is the basis of **soap-making**, one of the oldest chemical manufacturing processes. Soap is manufactured by boiling fat with sodium or potassium hydroxide. The fat is usually tallow (storage fat of cattle and sheep) or vegetable oil such as coconut oil. Because fats contain a mixture of fatty acids, soaps are in practice a mixture of the salts of fatty acids. Potassium soaps are softer than sodium ones, so potassium hydroxide is the alkali used in the manufacture of toilet soaps. Propane-1,2,3-triol (glycerol) is a useful by-product of soap manufacture. It is used to make paints and the explosive 'nitroglycerine'.

32.7 Acid anhydrides

Acid anhydrides can be regarded as the products of a dehydration reaction between two molecules of carboxylic acid. For example, ethanoic anhydride:

ethanoic acid   ethanoic anhydride

Despite their name, acid anhydrides are not readily formed by direct dehydration of carboxylic acids. Ethanoic anhydride is normally made in the laboratory by the reaction of ethanoyl chloride with sodium ethanoate. Ethanoyl chloride reacts readily with nucleophiles (section 29.7). The ethanoate ion is a nucleophile, and it attacks the positively charged carbon atom in ethanoyl chloride.

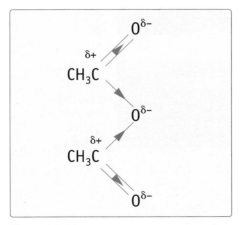

ethanoyl chloride   ethanoate ion

Industrially, ethanoic anhydride is made from ethanal or from ethanoic acid (in the latter case, indirectly, not by direct dehydration).

Methanoic anhydride does not exist, since methanoic acid dehydrates to give carbon monoxide.

The oxygen atoms in ethanoic anhydride withdraw electrons from the carbons to which they are attached, giving them a large partial positive charge (figure 32.11).

▲ Sunflower seeds are a good source of unsaturated oils.

▲ What oils do you think this soap was made from?

Figure 32.11
Polarisation in ethanoic anhydride.

32 Carboxylic Acids and their Derivatives

Ethanoic anhydride, like ethanoyl chloride, is readily attacked by nucleophiles such as H_2O, NH_3, ROH, RNH_2, etc. Representing a general nucleophile as $\ddot{X}$—Y, we can show the reaction as

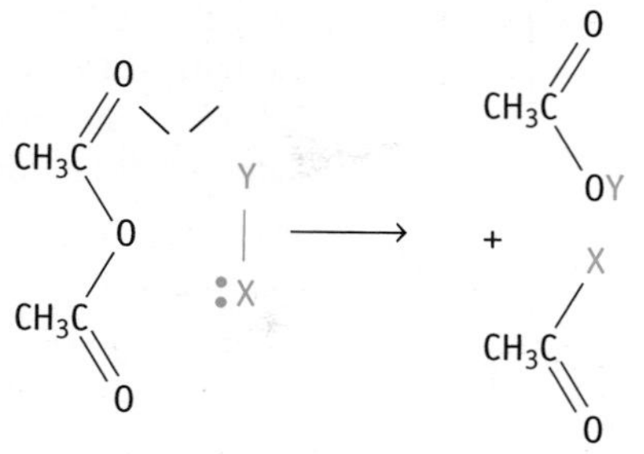

In most cases, Y is hydrogen.

In general, acid anhydrides, $(RCO)_2O$, act as **acylating agents**, joining the acyl group RCO onto a nucleophile. In this respect they behave like acyl chlorides which are also acylating agents, though acyl chlorides react more vigorously than acid anhydrides. Compare the general acylating reactions of acid anhydrides and acyl chlorides:

Acid anhydrides react with all those nucleophiles which react with acyl chlorides:

with **water** to give the carboxylic acid:

e.g.
$$(CH_3CO)_2O + H_2O \rightarrow 2CH_3COOH$$

with **alcohols** to give esters:

e.g.
$$(CH_3CO)_2O + CH_3CH_2OH \rightarrow CH_3COOCH_2CH_3 + CH_3COOH$$
ethyl ethanoate

with **ammonia** to give amides:

e.g.
$$(CH_3CO)_2O + NH_3 \rightarrow CH_3CONH_2 + CH_3COOH$$
ethanamide

with **amines** to give substituted amides:

e.g.
$$(CH_3CO)_2O + CH_3CH_2NH_2 \rightarrow 2CH_3CONHCH_2CH_3 + CH_3COOH$$

In the above reactions, ethanoic anhydride is used as an example, though the reactions apply quite generally to all acid anhydrides.

In industry, ethanoic anhydride is used in preference to ethanoyl chloride when an **ethanoylating agent** is required. This is because it is cheaper to make and less vigorous in its reactions. Aspirin, for example, is manufactured by ethanoylating 2-hydroxybenzoic acid:

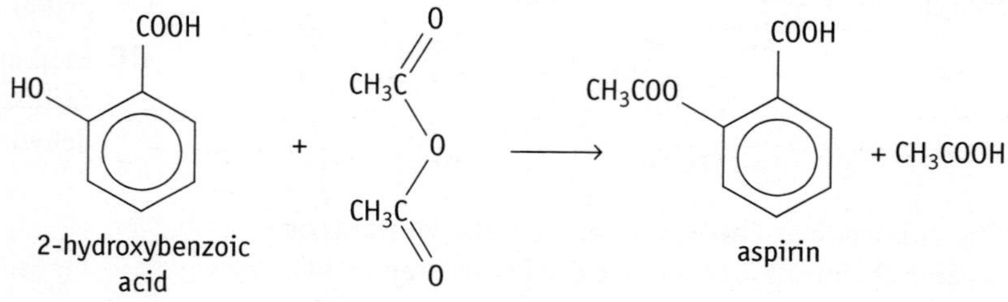

2-hydroxybenzoic acid

aspirin

> ❶ Predict the formulas of the products of reactions between the following pairs of substances:
> (a) ethanoic anhydride and phenol,
> (b) propanoic anhydride and methylamine,
> (c) butanoic anhydride and ammonia.

The artificial fibre cellulose triethanoate (known as 'triacetate' and sold under the trade name 'Tricel') is made by ethanoylating cellulose with ethanoic anhydride. Three of the —OH groups on each glucose molecule of the cellulose chain are ethanoylated. This gives a fibre with good 'wash and wear' properties.

Certain dicarboxylic acids can form **internal anhydrides** by eliminating a molecule of water from one rather than two molecules of the acid. For example, *cis*-butenedioic acid (maleic acid) readily eliminates water when heated, forming its anhydride:

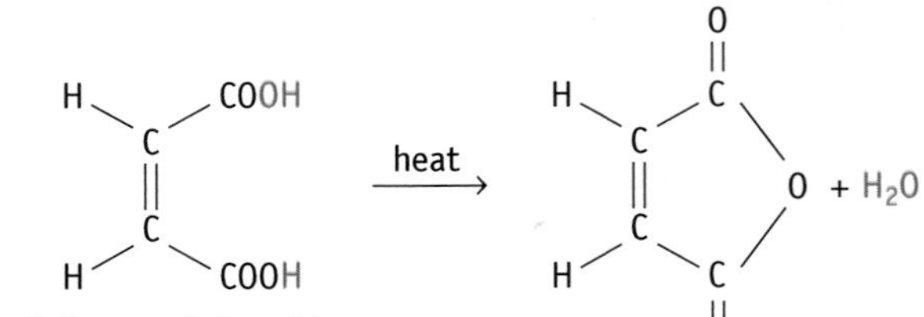

cis-butenedioic acid
(maleic acid)

However, the *trans* isomer of this compound, commonly called fumaric acid,

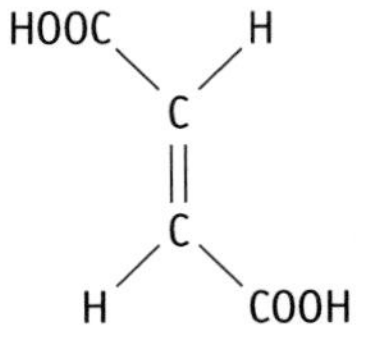

does not readily dehydrate on heating. With the *trans* isomer it is impossible to form a cyclic internal anhydride.

Summary

1   Carboxylic acids contain the carboxyl group, —COOH. They are named using the suffix *-oic acid.*

2   Carboxylic acids are prepared by oxidising primary alcohols or aldehydes or by the hydrolysis of nitriles.

3   The —COOH group is strongly acidic compared with the —OH group, but nevertheless carboxylic acids are weak compared with mineral acids such as hydrochloric acid. The strength of a carboxylic acid is increased by the presence of electron-withdrawing substituents.

4   Some important reactions of ethanoic acid are shown in the following diagram:

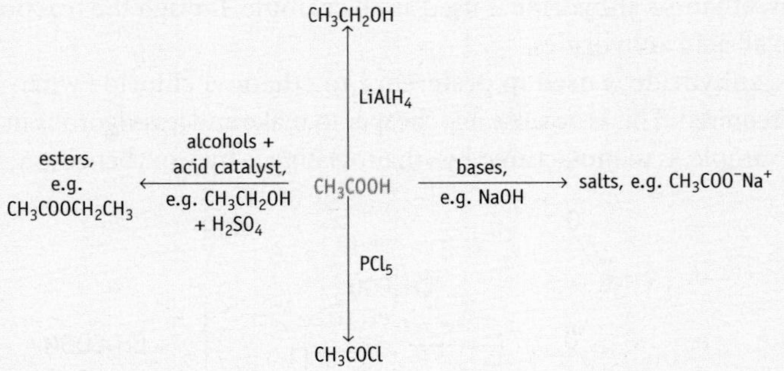

The importance of carboxylic acids in the synthesis of organic chemicals is developed further in chapter 34.

5   Esters have the general formula R^1COOR^2. They are formed by a condensation reaction between an alcohol and a carboxylic acid, usually in the presence of an acid catalyst.

6   Esterification is a reversible reaction. Esters can be hydrolysed by boiling with dilute acid or alkali.

7   Acid anhydrides have the general formula $(RCO)_2O$. they are acylating agents, with the ability to join the RCO group to nucleophiles.

8   Detergents are substances that improve the cleaning effect of water. They have both a polar and a non-polar portion in their molecules. Soap is a detergent: soaps are the sodium or potassium salts of long-chain carboxylic acids.

9   Fats are esters derived from propane-1,2,3-triol and long-chain carboxylic acids. Alkaline hydrolysis of fats produces soaps.

10 Methanoic acid and ethanedioic acid differ from other carboxylic acids in being readily oxidised and dehydrated.

Review questions

1 Consider the following compounds:

A $CH_3CHCOOH$
      |
     CH_3

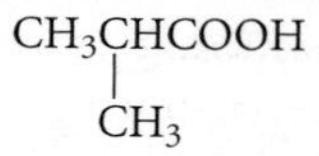

B benzene ring with COOH and CH_3 substituents

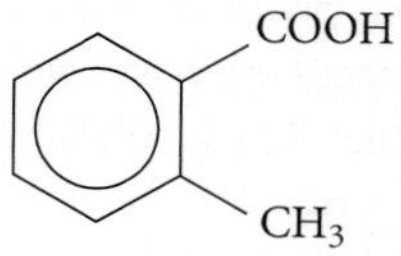

C $CH_3CH_2COOCH_3$

D $HOOCCH_2CH_2COOH$

E CH_3CH_2CO
           O
  CH_3CH_2CO

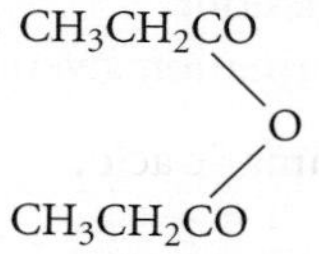

(a) Which is an ester?
(b) Which is a dibasic acid?
(c) Which is an acid anhydride?
(d) Name each compound.
(e) Which would be almost insoluble in water, but would slowly dissolve when boiled with sodium hydroxide solution?
(f) Which would be almost insoluble in water, but soluble in cold sodium hydroxide solution?
(g) Which would form a pleasant-smelling liquid when warmed with ethanol and concentrated sulphuric acid?
(h) Which would react with ammonia to form a mixture of propanamide and propanoic acid?

2 Predict the formulas of the products of the following reactions.
(a) $CH_3CH_2COOH + PBr_3 \longrightarrow$
(b) $C_6H_5COOH + LiAlH_4 \longrightarrow$
(c) $CH_3COO(CH_2)_4CH_3 + NaOH(aq) \xrightarrow{boil}$
(d) $CH_3COOH + Ca(OH)_2(aq) \longrightarrow$
(e) 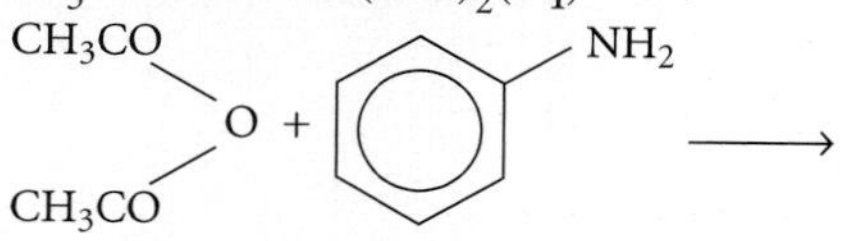 CH_3CO
          $O +$ benzene ring with NH_2 $\longrightarrow$
  CH_3CO

3 (a) When a carboxylate salt (e.g. $RCOO^-Na^+$) is heated with 'soda lime' (a mixture of sodium and calcium hydroxides) a *decarboxylation* reaction occurs. In this reaction, CO_2 is effectively eliminated from the molecule, forming an alkane:

    $RCOO^-Na^+ + NaOH \rightarrow RH + Na_2CO_3$
                       an alkane

What do you think would be formed when *soap* is heated with soda lime?
(b) Suggest explanations for the following observations.
(i) Butane, propan-1-ol, propanal and ethanoic acid all have approximately the same relative molecular mass, but their boiling points are 273 K, 371 K, 322 K and 391 K, respectively.

(ii) When dilute hydrochloric acid is added to an aqueous solution of soap, a white insoluble substance is formed.

4 Three isomeric acids **A**, **B** and **C** have molecular formula $C_8H_6O_4$ and all contain a benzene ring. In each case, one mole of the acid will react with 2 moles of sodium hydroxide. Suggest structures for the acids.
When the three acids are separately heated, **A** and **B** melt without decomposing, but **C** loses a molecule of water at about 250°C to form **D**, $C_8H_4O_3$. Suggest structures for **C** and **D**.

5 How would you carry out the following conversions in the laboratory? One or more steps may be involved in each case.
(a) CH_3CHO to $CH_3COOCH_2CH_3$
(b) $CH_2{=}CH_2$ to CH_3COOH
(c) CH_3CH_2COOH to $(CH_3CH_2CO)_2O$
(d) 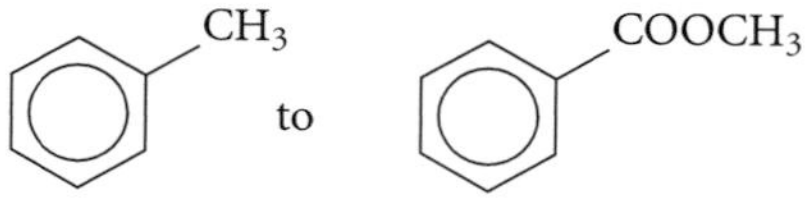 benzene with CH_3 to   benzene with $COOCH_3$

(e) CH_3COOH to CH_3CONH_2

6 What simple chemical tests would you use to distinguish one compound from the other in the following pairs:
(a) CH_3CH_2CHO and CH_3CH_2COOH,
(b) $HCOOH$ and CH_3COOH,
(c) $CH_3COCH_2CH_3$ and $CH_3COOCH_2CH_3$,
(d) CH_3COOCH_3 and $HCOOCH_2CH_3$,
(e) $CH_2FCOO^-Na^+$ and $CH_3COO^-Na^+$.

7 Write structural formulas for all acids and esters of molecular formula $C_4H_8O_2$. What simple chemical tests would you use to distinguish between the different esters?

8 Consider the following compounds:
A $CH_2OOC(CH_2)_{14}CH_3$
    |
   $CHOOC(CH_2)_{16}CH_3$
    |
   $CH_2OOC(CH_2)_{14}CH_3$
B $HOCH_2CH_2OH$
C $CH_3(CH_2)_{15}COOH$
D $CH_3(CH_2)_{15}SO_3^-Na^+$
E $HOOC(CH_2)_4COOH$
(a) Which is a fatty acid?
(b) Which is an anionic detergent?
(c) Which would form soap on boiling with sodium hydroxide solution?
(d) Which pair of compounds could together be used to make a polyester?
(e) Which is a fat?
(f) Which would be oxidised to carbon dioxide and water on boiling with acidified potassium manganate(VII)?
(g) Which might form a cyclic internal anhydride?

9 This question is about an investigation into the mechanism of hydrolysis of ethyl ethanoate:

$$CH_3COOCH_2CH_3 + H_2O \xrightarrow[\text{catalyst}]{\text{HCl}} CH_3COOH + CH_3CH_2OH$$
$$\text{ethyl ethanoate} \qquad\qquad \text{ethanoic acid} \quad \text{ethanol}$$

Two experiments, **A** and **B**, were carried out. Read the accounts of the experiments, then answer the questions following.

Experiment A Ethyl ethanoate was refluxed with deuterated water, D_2O, containing deuterium chloride, DCl ($D = {}^2_1H$). The two hydrolysis products were separated and purified and their relative molecular masses were measured using a mass spectrometer.

The alcohol formed had $M_r = 47$, and the acid had $M_r = 61$.

(a) Compare M_r for the alcohol formed in this experiment with the value expected from the equation above. Account for any difference.

(b) Compare M_r for the acid formed in this experiment with the value expected from the equation above. Account for any difference.

(c) What information if any, does experiment **A** give about the mechanism of the hydrolysis reaction?

Experiment B Ethyl ethanoate was refluxed with $H_2{}^{18}O$ containing HCl. The two hydrolysis products were separated and purified and their relative molecular masses were measured using a mass spectrometer. The alcohol formed had $M_r = 46$, and the acid had $M_r = 62$.

(d) and (e) Repeat steps (a) and (b) above, this time using the results from experiment **B**.

(f) What information, if any, does experiment **B** give about the mechanism of the hydrolysis reaction?

(g) Compare your answer with the information given on page 506.

Organic Nitrogen Compounds

33

▲ Sir William Perkin, discoverer of mauve, the first synthetic dye.

33.1 Dyestuffs and the development of the organic chemical industry

The structure of the organic chemical industry is very different from its inorganic counterpart. The inorganic chemical industry is rather static. It tends to stick for long periods to established processes producing a small number of essential chemicals in very large quantities – for example, sulphuric acid, sodium hydroxide and steel. In organic chemistry, on the other hand, the number of possible compounds and the number of routes by which they can be made is far greater. As a result, the industry is more dynamic, constantly discovering new compounds to do a particular job, and new ways of making them. Pharmaceuticals and polymers are just two areas of the organic chemical industry that show this rapid evolution. With new compounds and new processes to be investigated, the research chemist is more closely involved.

How it all began

The evolution of the organic chemicals industry started with the discovery of synthetic dyestuffs. Until the 1850s dyes were all produced from living materials, usually plants. They were expensive, limited in colour range and tended to fade.

In 1856 in London, William Perkin, an 18-year-old student of the German chemist Hofmann, was attempting to prepare the drug quinine from aniline. Although he failed, he managed, quite by accident, to produce a brilliant purple dye that was named *mauve*. This was the first synthetic dyestuff. It was the forerunner of many other dyes produced from aniline, or phenylamine as it is now called. The presence of the —NH_2 group attached to a benzene ring makes phenylamine especially suited to the preparation of coloured materials, particularly **azo dyes**.

Germany pulls ahead

The synthetic dyestuffs industry was dominated by Britain, its country of origin, until the 1870s. However, many German chemists had gained experience of synthetic dyestuffs in Britain, and Germany slowly began to draw ahead and to develop a dynamic dyestuffs industry. This was partly due to the German education system, which placed great importance on the training of scientists in universities. Germany thus had the large reserve of trained scientists needed to carry out the research necessary to develop the new industry. Then, as now, the key to the development of science and technology was education.

By the First World War, Britain was importing most of its dyes from Germany, and producing only 20% of its own needs. Indeed, even after the outbreak of war, dyes had to be imported secretly from Germany to dye British soldiers' uniforms.

The experience gained by the German chemical industry in the development of dyestuffs helped it diversify into pharmaceuticals (aspirin was first used in Germany in 1898), synthetic polymers, photographic chemicals and many other fields. By the eve of the First World War the German chemical industry was superior to that of any other country.

The organic chemical industry has evolved a long way since its start with dyestuffs.

These materials now account for only a very small proportion of the total tonnage of organic chemicals produced. Dyestuffs are manufactured on a fairly small scale by relatively lengthy and complex processes. Their production tends to resemble scaled-up laboratory methods rather than the giant processes used in the manufacture of petrochemicals. Nevertheless, dyestuffs are essential materials and their high price makes them commercially important. Section 33.7 gives further details of the production of azo dyes.

33.2 Important organic nitrogen compounds

In this chapter we shall be mainly concerned with compounds containing the $-NH_2$ group, but we will briefly meet other nitrogen compounds. The range of nitrogen compounds covered and the methods used to name them are dealt with in this section.

Amines

Amines can be considered as compounds formed from ammonia by substituting hydrogen atoms with alkyl or aryl groups. If one of the hydrogen atoms in NH_3 is substituted, we get a compound of the form RNH_2, called a **primary amine**. Such compounds are named using the suffix **-ylamine** after a stem indicating the number of carbon atoms in the molecule. Thus CH_3NH_2 is methylamine. Substitution of two of the hydrogens of NH_3 gives compounds of the form R^1R^2NH, called **secondary amines**, named in a similar way to primary amines. Thus $(CH_3)_2NH$ is dimethylamine. You should be able to work out the general structure and nomenclature of **tertiary amines**. **Quaternary ammonium salts** have four alkyl groups substituted on the nitrogen atom. This gives the nitrogen a positive charge, which is balanced by a negative ion such as Cl^-. Thus, $(CH_3)_4 N^+Br^-$ is tetramethylammonium bromide. Table 33.1 gives some examples of amines and other nitrogen compounds together with their systematic and other names.

The prefix **amino-** can be used to indicate the presence of an $-NH_2$ group in molecules containing more than one functional group, e.g. aminoethanoic acid.

Amides

Not to be confused with amines, **amides** have the general structure

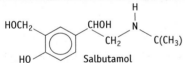

▲ Salbutamol is a very effective drug for treating asthma attacks. Look at its structure: is the amine group primary, secondary or tertiary?

Table 33.1
Some important organic nitrogen compounds

Formula	Type of compound	Name	Other name
$CH_3CH_2NH_2$	primary amine	ethylamine	aminoethane
⬡NH_2	primary amine	phenylamine	aniline
$(CH_3)_3N$	tertiary amine	trimethylamine	—
CH_3CONH_2	amide	ethanamide	acetamide
⬡NO_2	nitro-compound	nitrobenzene	—
CH_3CN	nitrile	ethanenitrile	acetonitrile

$CH_3CHCOOH$
     |
     NH_2
	amino acid	2-aminopropanoic acid	alanine

The presence of the $-C=O$ group 'next door' has a big effect on the properties of the $-NH_2$ group, so amides behave very differently from amines. Amides are named using the suffix **-amide** after a stem indicating the number of carbon atoms in the molecule, including that in the $-C=O$ group. Thus CH_3CONH_2 is ethanamide.

▲ The peculiar smell of fish is mainly due to amines.

Nitro-compounds

Compounds containing the —NO_2 group are called **nitro-compounds**. Only aromatic ones need concern us here. They are named using the prefix **nitro-**. Thus $C_6H_5NO_2$ is called nitrobenzene.

Nitriles (cyano-compounds)

Compounds containing the cyano group —C≡N are called **nitriles** or **cyano-compounds**. They are named using the suffix **-nitrile** after the stem indicating the number of carbon atoms, including that in the —CN group. Thus CH_3CN is ethanenitrile.

Amino acids

As their name implies, **amino acids** contain both the —NH_2 and —COOH groups. By far the most important are those in which the —NH_2 and the —COOH are both attached to the same carbon, thus:

$$R—CH—COOH$$
$$\underset{NH_2}{|}$$

Most of the amino acids found in nature are of this form: they are called α-amino acids. Amino acids are named as amino derivatives of carboxylic acids. Thus

$$CH_3CHCOOH$$
$$\underset{NH_2}{|}$$

is called 2-aminopropanoic acid. Since many natural amino acids have quite complex structures, they are often referred to by trivial names, such as alanine in this case. There is more about amino acids in section 33.8.

1 Name the following compounds:
   (a) $CH_3CH_2CH_2CONH_2$
   (b) CH_3CH_2CN
   (c) $(CH_3CH_2)_2NH$
2 Give the formulas of the following compounds:
   (a) 2-nitrophenylamine
   (b) aminoethanoic acid
   (c) propylamine
   (d) propanamide

33.3 The nature and occurrence of amines

The —NH_2 group is very widespread in biological molecules, especially proteins. Normally the group is associated with other functional groups. Free amines are relatively rare in nature, though they do occur in decomposing protein materials. Amines are formed by the action of bacteria on amino acids. For example, $NH_2(CH_2)_4NH_2$ and $NH_2(CH_2)_5NH_2$ are found in decaying animal flesh, as their common names, putrescine and cadaverine, suggest. Dimethylamine and trimethylamine are found in rotting fish and are partly responsible for its peculiar smell.

The physical and chemical properties of the early members of the amine series resemble those of ammonia. Their smell is very similar to that of ammonia, though with a slightly fishy character. Like ammonia, the early amines are gaseous and very soluble in water. Their high solubility is due to hydrogen bonding between the NH_2 group and water molecules.

Table 33.2 on the next page gives some physical properties of amines. Can you suggest a reason why trimethylamine has a lower boiling point than dimethylamine even though its relative molecular mass is higher?

Phenylamine, with its large hydrocarbon portion, is only sparingly soluble in water, but it dissolves well in organic solvents. Its ability to dissolve in fats means that it is readily absorbed through the skin. Since it is toxic, great care must be taken when using phenylamine.

▲ Caffeine is a stimulant found in coffee; theobromine is the stimulant in chocolate. Note the similarity of the two molecules – can you spot the difference?

caffeine

theobromine

Table 33.2
Physical properties of ammonia and some amines

Name	Formula	State at 25°C	Boiling point/°C
ammonia	NH_3	g	−33
methylamine	CH_3NH_2	g	−6
ethylamine	$CH_3CH_2NH_2$	g	17
propylamine	$CH_3CH_2CH_2NH_2$	l	49
butylamine	$CH_3CH_2CH_2CH_2NH_2$	l	78
phenylamine	$C_6H_5NH_2$	l	184
dimethylamine	$(CH_3)_2NH$	g	7
trimethylamine	$(CH_3)_3N$	g	3
triethylamine	$(CH_3CH_2)_3N$	l	90

33.4 Making amines

There are many ways of making amines. Four of the most important methods are listed here.

From ammonia and halogenoalkanes

Halogenoalkanes undergo substitution reactions with ammonia to form a mixture of primary, secondary and tertiary amines and a quaternary ammonium salt (section 29.5):

$$RCl + NH_3 \longrightarrow RNH_2 + HCl$$

$$RCl + RNH_2 \longrightarrow R_2NH + HCl$$

$$RCl + R_2NH \longrightarrow R_3N + HCl$$

$$RCl + R_3N \longrightarrow R_4N^+Cl^-$$

In practice the product is the salt of the amine, formed by combination of the HCl with the free amine (section 33.5). The problem with this method is that it produces a mixture of products which must be separated by distillation. Nevertheless, the method is widely used both in the laboratory and industrially, because of the ready availability of the starting materials.

By reduction of nitro-compounds

This method is normally used only for aromatic amines.

Aromatic nitro-compounds are readily prepared by the nitration of aromatic hydrocarbons (see section 28.6). Reduction of nitro-compounds in acid solution produces amines:

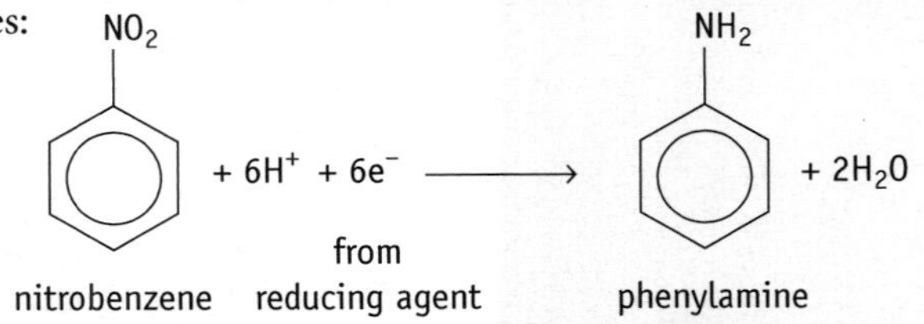

nitrobenzene   from reducing agent   phenylamine

In the laboratory the reducing agent used can be lithium tetrahydridoaluminate, $LiAlH_4$. Another reducing agent is tin in concentrated hydrochloric acid:

$$Sn \longrightarrow Sn^{4+} + 4e^-$$

Overall:

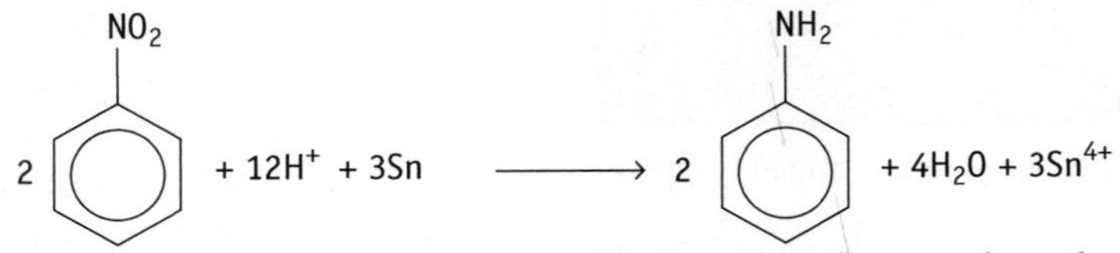

This method is also used industrially to manufacture phenylamine, the reducing agent in this case being iron in concentrated hydrochloric acid.

1 How would you prepare:
  (a) propylamine from chloroethane
     (2 stages),
  (b) ethylamine from chloroethane,
  (c) 2-methylphenylamine from
     methylbenzene (2 stages)?

By reduction of nitriles

Primary amines can be produced by the reduction of nitriles using powerful reducing agents such as lithium tetrahydridoaluminate ($LiAlH_4$):

$$R-C\equiv N \xrightarrow[\text{dry ether}]{LiAlH_4} RCH_2NH_2$$

This method is useful in the laboratory but because of the relatively high cost of the reducing agent it is not used industrially.

By reduction of amides

Amides, like nitriles, are reduced by $LiAlH_4$ to primary amines:

$$RCONH_2 \xrightarrow[\text{dry ether}]{LiAlH_4} RCH_2NH_2$$

33.5 Amines as bases

Figure 33.1 illustrates an experiment in which acid is added to a solution of ethylamine.

Answer these questions.

1 What evidence is there for a chemical reaction between ethylamine and hydrochloric acid?
2 Why does the smell of ethylamine disappear when hydrochloric acid is added?
3 Why does the smell reappear when sodium hydroxide is added?

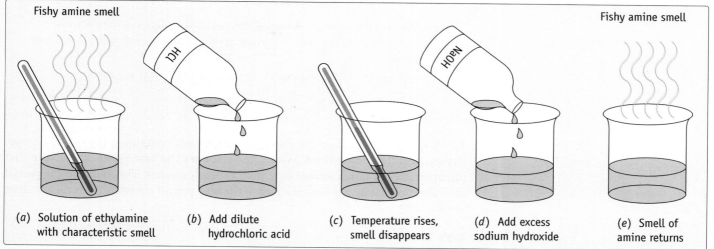

Fishy amine smell   Fishy amine smell

(a) Solution of ethylamine with characteristic smell   (b) Add dilute hydrochloric acid   (c) Temperature rises, smell disappears   (d) Add excess sodium hydroxide   (e) Smell of amine returns

Figure 33.1
Effect of acid on aqueous ethylamine.

Like ammonia and all primary amines, ethylamine carries a lone pair of electrons on its nitrogen atom. This enables it to bond to a hydrogen ion:

Ethylamine is therefore a base like ammonia.

When an acid is added to a solution of ethylamine, a salt is formed – in this case **ethylammonium chloride**, $CH_3CH_2NH_3^+Cl^-$. Like all salts it is involatile and therefore has no smell. When a strong base such as sodium hydroxide is added to this salt, protons are removed from it. This reforms the free amine.

$$CH_3CH_2NH_3^+(aq) + {}^-OH(aq) \rightarrow CH_3CH_2NH_2(aq) + H_2O(l)$$
ethylammonium ion   ethylamine

Compare this series of reactions with the corresponding reactions of ammonia:

$$NH_3(aq) + H^+(aq) \longrightarrow NH_4^+(aq)$$
$$NH_4^+(aq) + {}^-OH(aq) \longrightarrow NH_3(aq) + H_2O(l)$$

Amine salts are soluble white crystalline solids, similar to ammonium compounds.

Figure 33.2 shows another experiment that can be carried out with ethylamine and HCl.

1 What would you expect to happen?

2 Why?

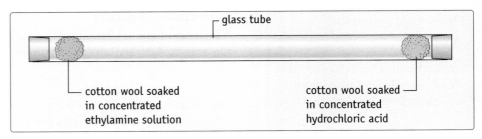

Figure 33.2
An experiment with ethylamine and HCl. The apparatus is set up as shown. What happens?

Is ethylamine a stronger or weaker base than ammonia?

To answer this question, look at the following equilibria:

$$CH_3CH_2NH_2(aq) + H^+(aq) \rightleftharpoons CH_3CH_2NH_3^+(aq)$$

$$NH_3(aq) + H^+(aq) \rightleftharpoons NH_4^+(aq)$$

The stronger the base, the further the equilibrium is to the right. The position of equilibrium depends on the stability of the cation on the right-hand side. The more stable the cation, the further the equilibrium will be to the right – and the stronger the base. Remember that alkyl groups have a positive inductive effect (page 424) – the $CH_3CH_2—$ group in $CH_3CH_2—NH_3^+$ pushes electrons onto the $—NH_3^+$ group. This stabilises the ion, so $CH_3CH_2NH_2$ is a stronger base than NH_3. We can use similar ideas to predict the basic strength of other compounds containing the $—NH_2$ group.

> 1 By considering the electron-donating effect of alkyl groups and the electron-withdrawing effect of the $—C{=}O$ group, try to arrange the following in order of basic strength:
>
> (a) NH_3 (b) $CH_3\underset{\underset{O}{\|}}{C}NH_2$ (c) $(CH_3CH_2)_2NH$ (d) $CH_3CH_2NH_2$

Table 33.3 gives K_b values for different compounds containing the $—NH_2$ group. Remember, the higher the K_b value, the stronger the base (page 330). Note that phenylamine is a much weaker base than aliphatic amines. This is due to the partial delocalisation round the benzene ring of the lone pair of electrons from the nitrogen atom.

Table 33.3
K_b values of ammonia, amines and ethanamide

Formula	Name	K_b at 25°C/mol dm^{-3}
NH_3	ammonia	1.8×10^{-5}
$CH_3CH_2NH_2$	ethylamine	5.4×10^{-4}
$(CH_3CH_2)_2NH$	diethylamine	1.3×10^{-3}
CH_3CONH_2	ethanamide	10^{-15}
$C_6H_5NH_2$	phenylamine	5×10^{-10}

33.6 Other reactions of amines

With acyl chlorides

The lone pair of electrons on the nitrogen atom gives amines nucleophilic character. Thus they tend to react readily with electrophiles such as ethanoyl chloride (section 29.7).

For example:

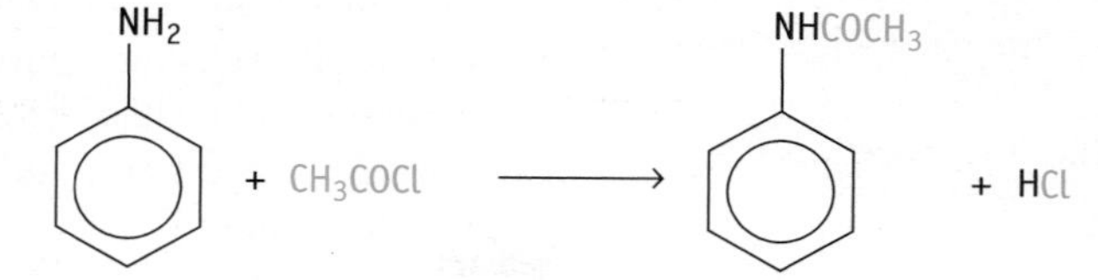

phenylamine   ethanoyl chloride   *N*-phenylethanamide (m.p. 114°C)

The products of this type of reaction are called **acyl derivatives**. Unlike the amines from which they are made, they are crystalline solids with sharp melting points. They are therefore useful for identifying (**characterising**) unknown amines. The acyl derivative is prepared and its melting point is taken. This is then checked against the melting points of known amides in published tables.

With nitrous acid

Nitrous acid (HNO_2) is rather unstable, so when it is used as a chemical reagent it is usually generated on the spot from sodium nitrite and hydrochloric acid.

When ethylamine is reacted with nitrous acid, nitrogen is rapidly evolved and a mixture of ethanol, ethene and other products is formed. However, when phenylamine is reacted with nitrous acid at low temperatures (below 10°C), a clear solution is formed, but no nitrogen appears. When this solution is warmed, nitrogen is evolved and phenol is formed.

These observations are explained as follows. When amines react with nitrous acid, **diazonium compounds** are formed:

$$RNH_2 + HNO_2 \longrightarrow \underset{\text{diazonium ion}}{R{-}\overset{+}{N}{\equiv}N} + {}^{-}OH + H_2O$$

The diazonium group, $-\overset{+}{N}{\equiv}N$, group is rather unstable. In the case of the ethyldiazonium ion, it decomposes at once:

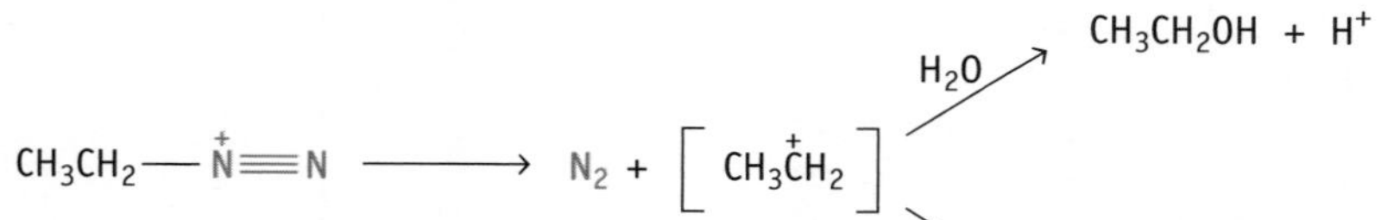

ethyldiazonium ion

When the $-\overset{+}{N}{\equiv}N$ group is attached to a *benzene ring*, though, the ion is stabilised to some extent by the delocalised electrons of the ring. The benzenediazonium ion is therefore much more stable than its aliphatic counterparts. Nevertheless it decomposes readily above 10°C. Aromatic diazonium compounds are considered further in section 33.7.

Reactions of the aromatic ring in phenylamine

The lone pair of electrons on the nitrogen atom in phenylamine tends to get partly delocalised round the ring. We have already seen how this reduces the basic strength of phenylamine relative to aliphatic amines. Another result is that the electron density round the ring in phenylamine is considerably increased. This makes it undergo electrophilic substitution much more readily than benzene. For example, phenylamine reacts with bromine water even in the absence of a halogen carrier catalyst:

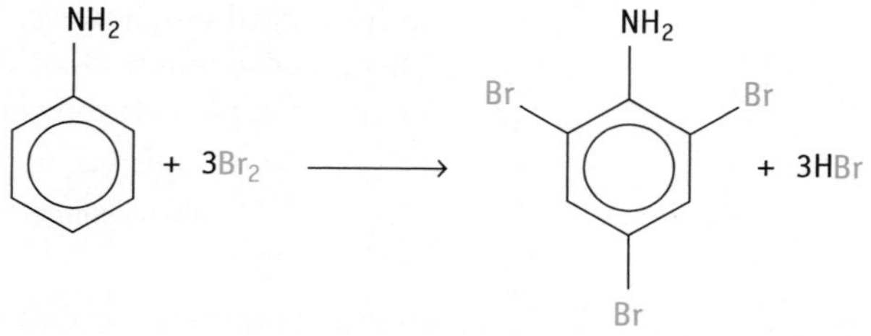

2,4,6-tribromophenylamine

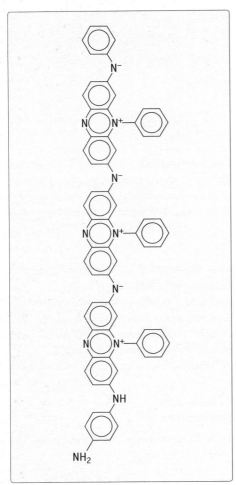

Figure 33.3
The pigment Aniline Black is a mixture of several substances. The structure of one of them is believed to be that shown here (the + and – charges are effectively delocalised over the molecule, which is electrically neutral overall).

Figure 33.4
Maternal death rate from childbed fever. The figures show the number of mothers who died per 100 000 total births, 1850–1964, in England and Wales (logarithmic scale). Sulphonamides were introduced in 1935.

The high electron density in the phenylamine molecule makes it very easy to oxidise. Pure phenylamine is colourless, but it quickly darkens owing to atmospheric oxidation. Chemical oxidising agents can convert it to a host of different products, including the pigment Aniline Black (figure 33.3).

An important ring-substituted derivative of phenylamine is 4-aminobenzene-sulphonamide:

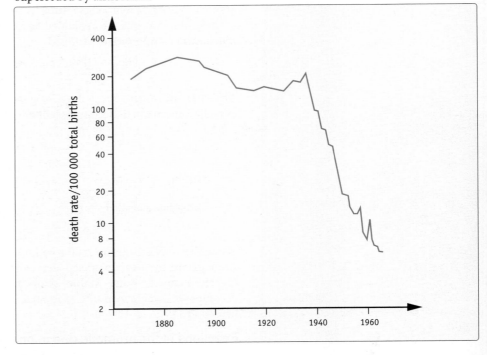

This compound and other compounds derived from it are the active agents in the **sulphonamides**, one of the earliest groups of antibacterial drugs. The sulphonamide drugs played an important part in the decline of puerperal or childbed fever, once a major cause of death in childbirth. They were first used in 1935 (see figure 33.4). They also helped to achieve the vast reduction in deaths from pneumonia and tuberculosis since the 1940s. They are still being used today though they have been largely superseded by antibiotics.

33.7 Diazonium salts

The only stable diazonium salts are aromatic ones (section 33.6), and even these are not particularly stable. In aqueous solution benzenediazonium chloride decomposes above about 5–10°C, and the compound is explosive when solid. Its reactive nature makes it useful in synthesis, however.

Benzenediazonium chloride is prepared by adding a cold solution of sodium nitrite to a solution of phenylamine in concentrated hydrochloric acid below 5°C:

$$C_6H_5NH_2 + HNO_2 + HCl \longrightarrow C_6H_5N_2^+Cl^- + 2H_2O$$

phenylamine · · · · · · · · · · · · benzenediazonium chloride

Owing to the explosive nature of the solid, the compound is always used in solution.

33 Organic Nitrogen Compounds

Consider the structure of the benzenediazonium ion, C_6H_5 —$\overset{+}{N}\equiv N$.

1. What stable molecule can be eliminated from the ion?
2. What would be the product left when this molecule is eliminated?
3. What kind of reaction would this product tend to undergo?

The benzenediazonium ion reacts with nucleophiles:

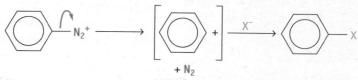

+ N$_2$

The introduction of the diazonium group is therefore a way of making the aromatic ring susceptible to *nucleophilic* substitution. (Remember the characteristic mode of reaction of an aromatic ring is normally *electrophilic* substitution.) This makes diazonium compounds useful in synthesis. For example:

$$C_6H_5N_2^+ + I^- \longrightarrow \underset{\text{iodobenzene}}{C_6H_5I} + N_2 \qquad \text{(warm benzenediazonium chloride with KI solution)}$$

$$C_6H_6N_2^+ + H_2O \longrightarrow \underset{\text{phenol}}{C_6H_5OH} + N_2 + H^+ \qquad \text{(warm the aqueous solution)}$$

$$C_6H_5N_2^+ + Cl^- \longrightarrow \underset{\text{chlorobenzene}}{C_6H_5Cl} + N_2 \qquad \text{(warm benzenediazonium chloride with CuCl catalyst)}$$

Coupling reactions

The positive charge on the —$\overset{+}{N_2}$ group of the benzenediazonium ion means that this groups is itself a strong electrophile. Thus we might expect it to attack another benzene ring, particularly one that has an electron-donating group such as —OH attached to it:

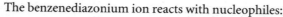

(4-hydroxyphenyl)azobenzene

This is an example of a **coupling reaction**. If a cold solution of benzenediazonium chloride is added to a cold solution of phenol in sodium hydroxide, a bright orange precipitate is immediately formed. This is (4-hydroxyphenyl)azobenzene. It is an example of an **azo compound**. Many different azo compounds can be formed by coupling reactions between diazonium compounds and activated aromatic rings. They are all brightly coloured. Their colour arises from the extensive delocalised electron systems they possess, delocalisation extending from one ring through the —N=N— group to the next ring.

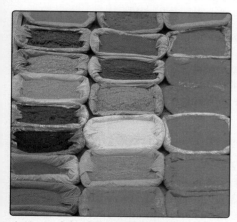

▲ Many of the dyes used for clothes and fabrics are azo dyes.

Figure 33.5
Structures of some azo dyes.

1. Write the structures of the products you would expect to be formed at each stage when:
   (a) phenylamine is dissolved in excess concentrated hydrochloric acid,
   (b) sodium nitrite solution is added to the cooled solution from the first stage,
   (c) the product from the second stage is added to a fresh solution of phenylamine.

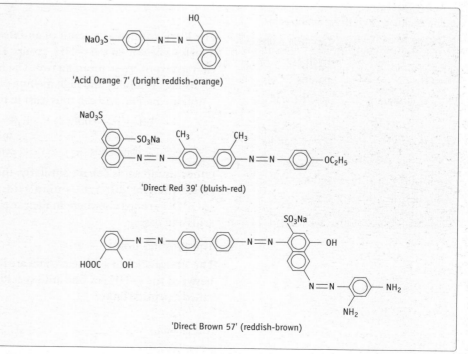

'Acid Orange 7' (bright reddish-orange)

'Direct Red 39' (bluish-red)

'Direct Brown 57' (reddish-brown)

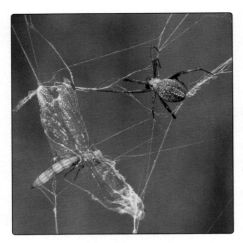

Spider webs and silkworm silk are made from a strong protein called fibroin. It is stronger than steel with the same diameter. Spiders recycle the protein in the web by eating it, along with the prey trapped inside.

Figure 33.6
The asymmetry of the amino acid molecule.

Table 33.4
Some naturally occurring amino acids. In this table X stands for the

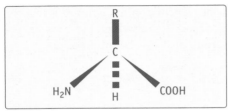

group common to all these amino acids.

Formula	Common name
X—H	glycine
X—CH$_3$	alanine
X—CH$_2$COOH	aspartic acid
X—CH$_2$⬡	phenylalanine
X—CH$_2$OH	serine
X—(CH$_2$)$_4$NH$_2$	lysine
X—CH$_2$SH	cysteine

Coupling reactions are important in the dyestuffs industry for making azo dyes. Many different colours can be obtained by adjusting the structure of the azo compound. These can be quite complex as figure 33.5 shows.

Unlike diazonium compounds, azo compounds are quite stable. Azo dyes do not fade or lose their colour.

33.8 Amino acids and proteins

All proteins are chemically similar, but they perform many different functions in living things. They act as structural materials (e.g. skin and fingernails). They make up the muscle fibres that enable animals to move. They are the basic material of the enzymes that catalyse the many chemical reactions that are the driving force of all organisms. Protein molecules are chemically very similar, yet their diverse functions mean there must be many different ways of putting such molecules together.

Proteins are long-chain molecules made by linking together relatively small molecules called amino acids. There are 22 different amino acids found widely in nature and many proteins contain several thousand amino acid units. For a protein containing 5000 such units the number of possible arrangements using 22 different amino acids is 22^{5000}, i.e. about 10^{6700}, so it is not surprising that proteins come in many different forms. Indeed, the surprising thing is that a given organism can control precisely the form of protein it produces.

Amino acids were introduced in section 33.2. Table 33.4 gives some examples of naturally occurring amino acids. All amino acids except aminoethanoic acid have four different groups attached to the central carbon atom (see figure 33.6). Because of this, they exhibit optical isomerism (section 25.8).

Amino acids carry at least two functional groups, —NH$_2$ and —COOH. They therefore show the properties of both amines and carboxylic acids.

Consider the simplest amino acid, aminoethanoic acid (glycine), H$_2$NCH$_2$COOH.

1 Write the structure of the ion formed by aminoethanoic acid when it reacts with a base.
2 Write the structure of the ion formed by aminoethanoic acid when it reacts with an acid.
3 Could aminoethanoic acid form an ion under neutral conditions?

In neutral solution and in the solid state, aminoethanoic acid exists as a dipolar ion:

$$H_3\overset{+}{N}—CH_2—COO^-$$

The ion is formed as a result of an *internal* acid–base reaction. The —COOH group donates a proton to the —NH$_2$ group. This kind of ion is called a **zwitterion**, from the German word 'zwei' meaning two. The ionic character of aminoethanoic acid accounts for its high solubility and high melting point (234°C).

Aminoethanoic acid can thus exist in three forms, depending on the pH:

$H_3\overset{+}{N}—CH_2—COOH$	$H_3\overset{+}{N}—CH_2—COO^-$	$H_2N—CH_2—COO^-$
in acid	in neutral	in basic
conditions	conditions	conditions

Other amino acids behave similarly, though the pH ranges in which the three forms exist differ for different amino acids. Some amino acids have extra —NH$_2$ and —COOH groups elsewhere in their molecule, and this affects their behaviour towards acids and bases.

The peptide link

The amino acids in a protein chain are linked by the elimination of a molecule of water between the —NH$_2$ of one amino acid and the —COOH of the next. Such a link is called a **peptide link**:

$$H_2N - \underset{\underset{H}{|}}{\overset{\overset{R^1}{|}}{C}} - \overset{O}{\underset{OH}{C}} \quad + \quad \underset{\underset{H}{|}}{\overset{\overset{H}{|}}{N}} - \underset{\underset{H}{|}}{\overset{\overset{R^2}{|}}{C}} - \overset{O}{\underset{OH}{C}}$$

$$H_2N - \underset{\underset{H}{|}}{\overset{\overset{R^1}{|}}{C}} - \overset{\overset{O}{\|}}{C} - \underset{\underset{H}{|}}{\overset{\overset{H}{|}}{N}} - \underset{\underset{H}{|}}{\overset{\overset{R^2}{|}}{C}} - \overset{O}{\underset{OH}{C}}$$

peptide link
$+H_2O$

▲ A computer graphic of a model of insulin, which maintains the balance of glucose metabolism in the body. Insulin is a protein made up of 51 amino acids.

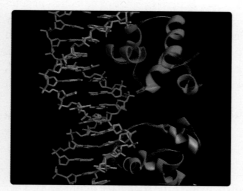

▲ A protein (blue and orange) binding to a section of DNA. Both protein and DNA are organic, nitrogen-containing polymers.

This kind of linkage is not easily formed under laboratory conditions, but it forms readily in living systems under the influence of enzymes. Long chains of amino acids, called **polypeptides**, can be formed in this way. Once formed, these polypeptide chains can take up very precise three-dimensional arrangements, leading to the formation of proteins. The structure of proteins is considered further in section 9.8.

The formation of a peptide link between amino acids is reversible, so proteins can be hydrolysed to re-form amino acids. Hydrolysis can be achieved by boiling with aqueous acid, but it is carried out more effectively by digestive enzymes like pepsin and trypsin.

Hydrolysis is used to identify the amino acids present in a protein. The protein is hydrolysed, then subjected to paper chromatography. The individual amino acids separate on the chromatogram and can be identified.

33.9 Synthetic fibres and nylon

Synthetic fibres

Naturally occurring fibres such as wool, cotton and silk are gradually being replaced by synthetic fibres for many uses. Synthetic fibres are cheaper and often have better wearing and washing characteristics. However, consumers tend to prefer synthetic fibres that feel and look like natural ones. It is, therefore, not surprising that several synthetic fibres are structurally similar to natural ones.

Wool and silk are protein fibres. Cotton is cellulose. Many of the artificial fibres used in Britain are in fact only semi-synthetic, being based on natural cellulose. These are the fibres known as rayon (viscose) and cellulose triacetate. They are made by using chemical reactions to convert cellulose (wood pulp, straw, etc.) into a form suitable for use as a textile. (See section 9.8 for the structure of cellulose.)

Nylon

Nylon is one of the major wholly synthetic fibres. Nylon has some structural similarity to protein fibres such as wool and silk. Like proteins, nylon is a condensation polymer. The monomer units are joined by a link similar to the peptide link. There are several forms of nylon. One of the commonest is nylon 66, which is made by a condensation reaction between 1,6-diaminohexane and hexanedioic acid. The product is named nylon 66 because both monomers contain 6 carbon atoms.

$$\underset{H}{\overset{H}{\diagdown}}N - (CH_2)_6 - N\underset{H}{\overset{H}{\diagup}} \quad \underset{HO}{\overset{O}{\diagdown}}C - (CH_2)_4 - C\underset{OH}{\overset{O}{\diagup}} \quad \underset{H}{\overset{H}{\diagdown}}N - (CH_2)_6 - N\underset{H}{\overset{H}{\diagup}}$$

1,6-diaminohexane · · · · · · · · · · · hexanedioic acid · · · · · · · · · · · 1,6-diaminohexane

$$- HN - (CH_2)_6 - NH - \overset{\overset{}{|}}{\underset{\underset{O}{\|}}{C}} - (CH_2)_4 - \overset{}{\underset{\underset{O}{\|}}{C}} - NH - (CH_2)_6 - NH -$$

$+ H_2O$ · · · · · · · · · · · $+ H_2O$

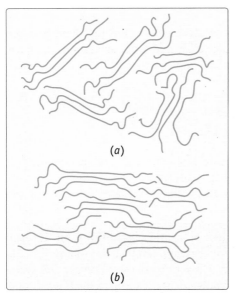

Figure 33.7
Arrangement of polymer chains, (a) before, and (b) after drawing.

▲ Extruding nylon through a spinnaret.

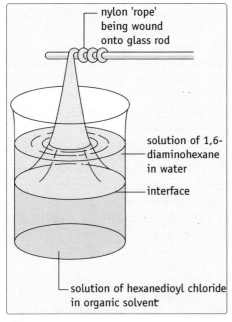

nylon 'rope' being wound onto glass rod

solution of 1,6-diaminohexane in water

interface

solution of hexanedioyl chloride in organic solvent

Figure 33.8
The nylon rope trick. The hexanedioyl chloride and 1,6-diaminohexane react at the interface between the aqueous and organic layers, and a nylon 'rope' can be continuously pulled out.

Notice the structural similarity of this polymer to a polypeptide. The CONH grouping is also present in amides (see section 33.2), so nylons are sometimes given the general name **polyamides**. The 1,6-diaminohexane and hexanedioic acid used in this process are manufactured from phenol or cyclohexane.

The polymer produced in this way is melted and squeezed (extruded) through small holes in a plate called a spinneret. As the jets emerge they are cooled and they solidify. The fibres produced in this way contain molecular chains that are tangled and disarrayed. The fibres are now stretched or **drawn**, with the result that the molecules become more aligned and the fibre is strengthened (see figure 33.7).

Nylon 66 can be more readily produced on a laboratory scale from hexanedioyl chloride instead of hexanedioic acid, because acyl chlorides are more reactive than carboxylic acids. Hexanedioyl chloride reacts readily with 1,6-diaminohexane at room temperature to form nylon 66, with the elimination of HCl.

$$H_2N-(CH_2)_6-NH_2 \quad ClOC-(CH_2)_4-COCl \quad H_2N-(CH_2)_6-NH_2$$

1,6-diaminohexane   hexanedioyl chloride   1,6-diaminohexane

$$-HN-(CH_2)_6-NH-\underset{O}{\overset{}{C}}-(CH_2)_4-\underset{O}{\overset{}{C}}-NH-(CH_2)_6-NH-$$

$+ HCl \qquad\qquad + HCl$

This reaction forms the basis of the well-known 'nylon rope trick' (see figure 33.8). The reaction is not used industrially because of the high cost of hexanedioyl chloride.

Despite its structural similarity to wool, nylon lacks the softness and moisture-absorbing properties of the natural fibre. It is, however, harder wearing and has good 'wash and wear' characteristics. One of the earliest uses of nylon was as a substitute for silk in the manufacture of stockings. The elasticity and strength of nylon make it ideal for making stockings and tights. About 75% of UK nylon consumption goes into clothing. Other uses include tufted carpets, tyre cords and machine gear wheels and bearings.

▲ In the 1940s nylon was newly discovered and nylon stockings were a luxurious novelty.

33 Organic Nitrogen Compounds

Aramids

'Aramids' are aromatic polyamides. They are similar to nylon, but they have benzene rings between the amide groups, instead of alkane chains. *Kevlar* is an aramid, and it has the structure

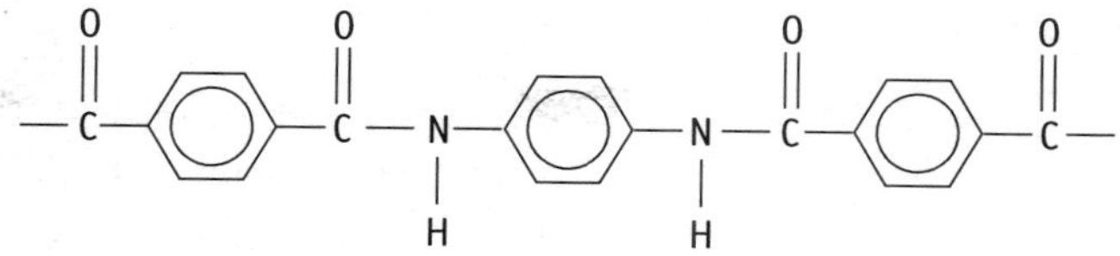

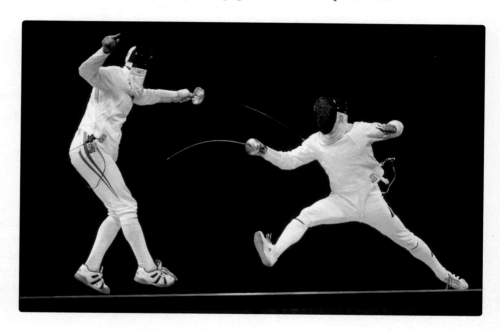

The flat, rigid benzene rings make these molecular chains stiff and rigid. Because of this, Kevlar is a very strong material: weight for weight, it is five times stronger than steel. Kevlar is used in ropes, tyres, sports equipment and bullet-proof vests.

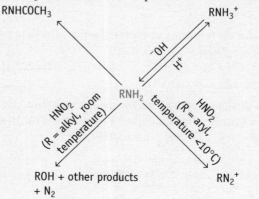

▶ Fencers' protective jackets are often made from Kevlar, an aramid.

Summary

1   Primary amines have the structure RNH_2. They are named using the suffix *-ylamine*.

2   Four ways of making amines are summarised below.

   (a) $RCl \xrightarrow{NH_3} RNH_2$

   (b) $RNO_2 \xrightarrow{reduce} RNH_2$ (R = aryl)

   (c) $RCN \xrightarrow{reduce} RCH_2NH_2$

   (d) $RCONH_2 \xrightarrow{reduce} RCH_2NH_2$

3   Amines, particularly the early members of the series, show considerable physical and chemical similarity to ammonia.

4   Amines are basic. Alkyl amines are stronger bases than ammonia, aryl amines are weaker than ammonia.

5   This diagram shows some important reactions of amines.

$RNHCOCH_3$ $RNH_3{}^+$

^-OH H^+

RNH_2

HNO_2 (R = alkyl, room temperature)

HNO_2 (R = aryl, temperature <10°C)

ROH + other products + N_2

$RN_2{}^+$

6 This diagram shows some important reactions of benzenediazonium chloride.

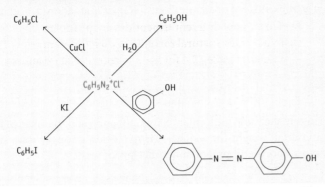

7 Amino acids contain both —NH$_2$ and —COOH groups. They can behave as bases and as acids.

8 Proteins contain chains of amino acids joined by a condensation linkage between the —COOH of one amino acid and the —NH$_2$ of the next. This is called a peptide link.

9 Nylons are condensation polymers containing the —CONH group. Nylon 66 is made by a condensation reaction between H$_2$N(CH$_2$)$_6$NH$_2$ and HOOC(CH$_2$)$_4$COOH.

Review questions

1 Consider the following compounds:

A CH$_3$CHCH$_3$
       |
     CN

B CH$_3$CH$_2$CH$_2$NH$_2$

C
  O$_2$N   NO$_2$

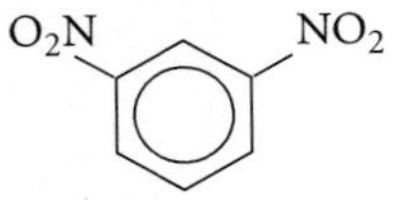

D 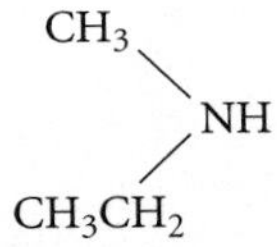
    CH$_3$
        NH
  CH$_3$CH$_2$

E CH$_3$CH$_2$CONH$_2$

F (CH$_3$CH$_2$CH$_2$)$_3$N

(a) Which is a primary amine?
(b) Which is a nitrile?
(c) Which is an amide?
(d) Which is a tertiary amine?
(e) Name each compound.

2 (a) Place the following in expected order of basic strength. Give reasons for your answer.

(i)   (ii)   (iii)

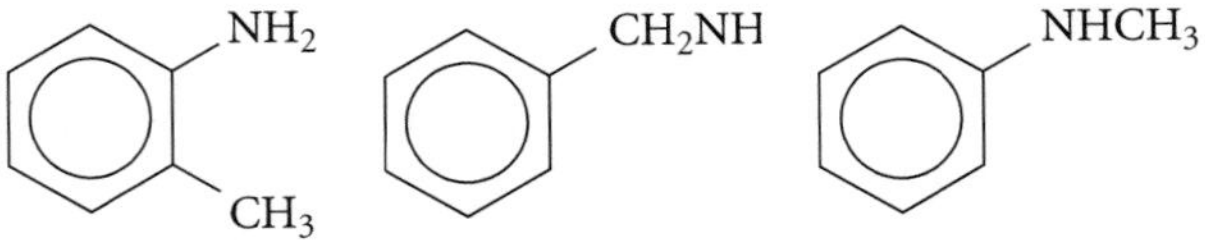

NH$_2$ CH$_2$NH   NHCH$_3$

CH$_3$

(b) Place the following in expected order of boiling point. Give reasons for your answer.
(i) CH$_3$CH$_2$CH$_2$NH$_2$ (ii) CH$_3$CH$_2$CH$_2$CH$_3$
(iii) CH$_3$CH$_2$CH$_2$OH

3 Draw up a table comparing the main physical and chemical properties of methylamine and ammonia. Do the two compounds in general behave similarly?

4 Consider the following compounds:

A
       CH$_3$
       NH$_2$

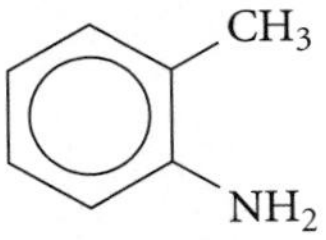

B CH$_3$CH$_2$NH$_2$
C (CH$_3$CH$_2$)$_2$NH
D H$_2$N(CH$_2$)$_5$NH$_2$

(a) Which would be among the products of the reaction of chloroethane with ammonia?
(b) Which would be converted to a diazonium compound by the action of nitrous acid below 10°C?
(c) Which could be made by reduction of ethanenitrile?
(d) Which would react with hydrochloric acid in the ratio one mole of the compound to two moles of hydrochloric acid?
(e) Which is the weakest base?
(f) Which could be one of the reagents in the manufacture of a nylon (polyamide)?

5 Suggest explanations for the following.
(a) Ethylamine can be readily made from chloroethane and ammonia, but it is difficult to make phenylamine from chlorobenzene and ammonia.
(b) If electrodes are placed in a solution of aminoethanoic acid at pH 0 the aminoethanoic acid migrates to the cathode. If the same experiment is repeated at pH 14 the aminoethanoic acid migrates to the anode.
(c) Phenylamine is much more soluble in dilute hydrochloric acid than in water.

6 In the following, explain how you would convert the first compound to the second. Each conversion may involve one or more steps.

(a) $CH_2{=}CH_2$ to $H_2NCH_2CH_2NH_2$

(b)

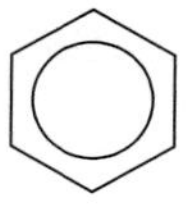

to

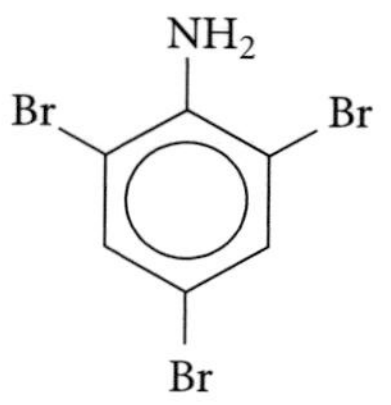

(c) $CH_2{=}CH_2$ to $CH_3CH_2CH_2NH_2$

(d) $CH_3CH_2NH_2$ to $CH_3CH_2NH{-}C(O)C_6H_5$

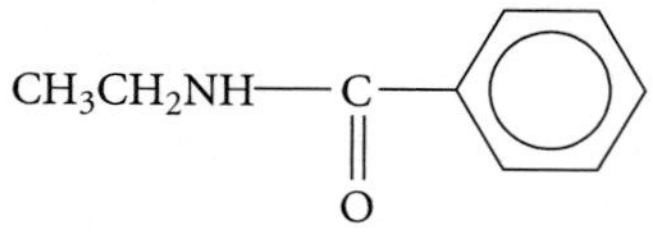

(e)

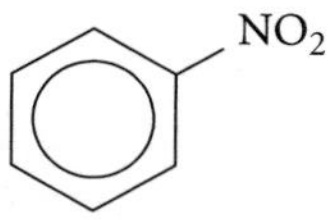

to

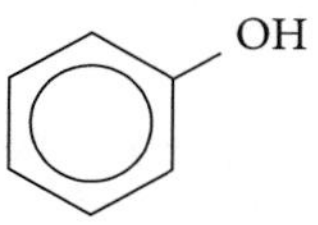

7 A compound **X** containing carbon, hydrogen and nitrogen only has relative molecular mass 88. When reacted with nitrous acid, 0.1 g of **X** released 50.9 cm^3 of nitrogen gas, measured at s.t.p. (1 mole of a gas occupies 22 400 cm^3 at s.t.p.)

(a) How many moles of nitrogen gas are produced by 1 mole of **X** when it reacts with nitrous acid?

(b) How many —NH_2 groups are there in one molecule of **X**?

(c) Write three possible structural formulas for **X**.

(d) What volume of 0.1 mol dm^{-3} hydrochloric acid would be needed to neutralise 50 cm^3 of a 0.2 mol dm^{-3} solution of **X**?

8 Copy out the following reaction sequences, inserting the formulas of the products formed in the blank spaces.

(a) $CH_3COCl \xrightarrow{NH_3} \quad \xrightarrow{LiAlH_4} \quad \xrightarrow{HCl}$

(b) $CH_3CH_2CH_2CH_2Br \xrightarrow{NH_3}$

$CH_3CH_2CH_2CH_2NH_2 \xrightarrow{CH_3COCl}$

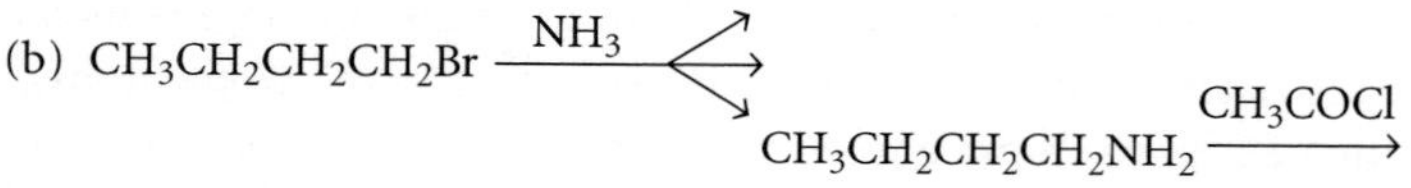

(c)

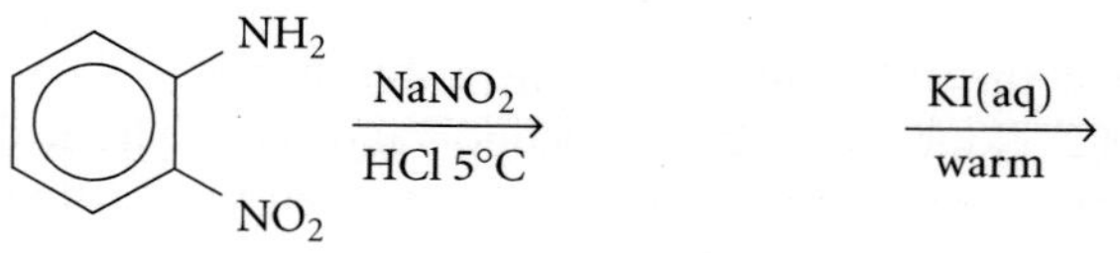

$\xrightarrow[\text{HCl 5°C}]{NaNO_2}$ $\xrightarrow[\text{warm}]{KI(aq)}$

9 Suppose a molecule of each of the amino acids serine, lysine and alanine are joined together by peptide links to form a tripeptide.

(a) How many different tripeptides are possible?

(b) Write the structural formula of each different tripeptide. (Refer to table 33.4 for the structures of these amino acids.)

10 A compound **A** of molecular formula C_7H_8 was treated with a mixture of concentrated nitric acid and concentrated sulphuric acid to form two isomeric compounds. One of these isomers was **B**, $C_7H_7O_2N$. **B** could be converted to **C**, C_7H_9N by reduction. **C** reacted with dilute hydrochloric acid to form **D**, $C_7H_{10}NCl$. When cold sodium nitrite solution was added to a cold solution of **D** in hydrochloric acid, a solution of **E**, $C_7H_7N_2Cl$, was produced. When a cold solution of **E** was added to a cold solution of phenol, a brightly coloured substance **F** was produced. When the solution **E** was warmed alone an unreactive gas was evolved and **G**, C_7H_8O, was formed. Give possible structural formulas for **A** to **G**.

34 Synthetic Routes for Organic Chemicals

34.1 Introduction

One of the most important applications of organic chemistry is the synthesis of new compounds. These new compounds have known structures, and properties that can be accurately predicted. This approach is particularly important in the synthesis of drugs, dyes, anaesthetics and pesticides.

Once an active compound has been discovered, a wide range of similar compounds can be synthesised. These are then tested to find the most effective compound with the required properties.

For example, one of the earliest pain-killers was 2-hydroxybenzoic acid (salicylic acid) (figure 34.1). Unfortunately this acid irritated the patient's mouth and stomach. In severe cases, it caused internal bleeding. Between 1850 and 1900, various derivatives of salicylic acid were prepared in order to find an equally therapeutic but less toxic form of the drug. In 1899, the German chemist Hofmann synthesised the ethanoyl (acetyl) derivative of salicylic acid. This compound was found to have the beneficial properties of the original drug without causing the same irritation to the mouth and stomach. It is still used today as **aspirin** (figure 34.1).

34.2 Starting materials for synthesis

Many of the organic chemicals synthesised in industry rely on crude oil, vegetable oils or fats for their starting materials. The cracking of alkanes in crude oil produces large quantities of short-chain alkenes. These alkenes have anything from two to eight carbon atoms. They include ethene ($CH_2{=}CH_2$) and propene ($CH_3CH{=}CH_2$) and can be converted into alcohols, ketones, acids and other derivatives by appropriate reactions.

The hydrolysis of naturally occurring fats and oils provides an important source of carboxylic acids. These carboxylic acids contain between eight and eighteen carbon atoms but they are restricted to compounds with an even number of carbon atoms.

Figure 34.2 on the next page shows some important synthetic routes based on the availability of alkenes from petrochemicals and the availability of carboxylic acids from naturally occurring fats and oils. This overall scheme contains many of the reactions included in the summaries at the end of the organic chemistry chapters.

Look closely at figure 34.2. Notice that alkenes, alcohols and carboxylic acids are the starting points for several reactions. These compounds are often the starting point of many organic syntheses. The cracking of crude oil fractions provides a ready supply of alkenes for industry while fats and oils can be used to manufacture alcohols and carboxylic acids. Primary alcohols and carboxylic acids can be interconverted via aldehydes using suitable oxidants and reductants.

34.3 Ascending and descending a homologous series

The other key reagents in preparing organic aliphatic chemicals are halogenoalkanes and nitriles. Indeed, the conversion of halogenoalkane to nitrile provides an important

▲ Pharmaceutical companies synthesise thousands of compounds to test in the search for new drugs.

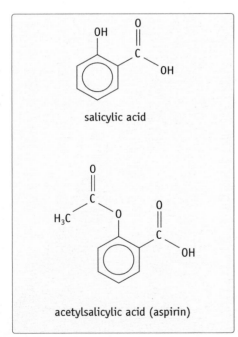

salicylic acid

acetylsalicylic acid (aspirin)

Figure 34.1 The structures of 2-hydroxybenzoic acid (salicylic acid) and 2-ethanoyloxybenzoic acid (acetylsalicylic acid, aspirin).

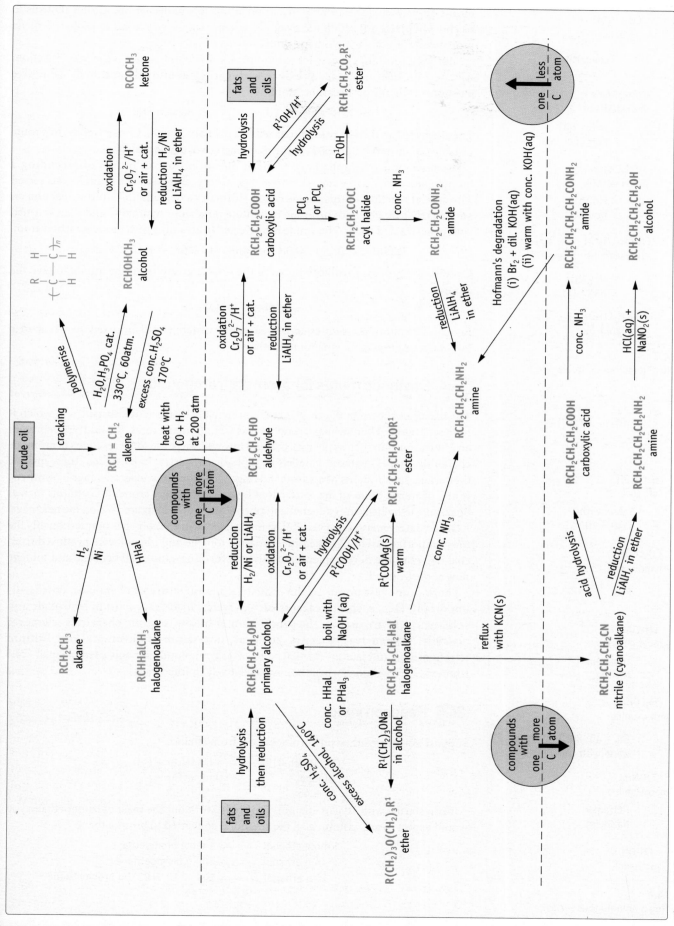

Figure 34.2
Important synthetic routes in organic chemistry.

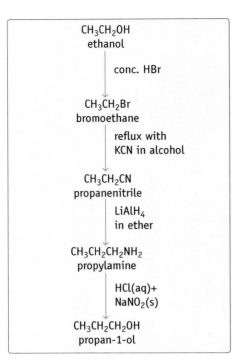

CH₃CH₂OH
ethanol

↓ conc. HBr

CH₃CH₂Br
bromoethane

↓ reflux with
KCN in alcohol

CH₃CH₂CN
propanenitrile

↓ LiAlH₄
in ether

CH₃CH₂CH₂NH₂
propylamine

↓ HCl(aq)+
NaNO₂(s)

CH₃CH₂CH₂OH
propan-1-ol

Figure 34.3
Ascending a homologous series.

CH₃CH₂OH
ethanol

↓ reflux with
Cr₂O₇²⁻/H⁺

CH₃COOH
ethanoic acid

↓ PCl₅

CH₃COCl
ethanoyl chloride

↓ cold conc.
NH₃(aq)

CH₃CONH₂
ethanamide

↓ Br₂ + dil. KOH(aq) then
warm with conc. KOH(aq)

CH₃NH₂
methylamine

↓ HCl(aq) +
NaNO₂(s)

CH₃OH
methanol

Figure 34.4
Descending a homologous series.

means of increasing the number of carbon atoms in a molecule. So, nitriles allow us to ascend a homologous series. For example, ethanol can be converted to propan-1-ol via propanenitrile along the synthetic route shown in figure 34.3.

Another important process for ascending a homologous series is the 'oxo' reaction. Alkenes, on heating under pressure with carbon monoxide and hydrogen, give aldehydes with one more carbon atom.

$$RCH=CH_2 + CO + H_2 \rightarrow RCH_2CH_2CHO$$

The aldehyde can then be readily reduced to a primary alcohol. Consequently, this route provides an important process for the manufacture of primary alcohols.

The last few paragraphs have focused our attention on means of ascending a homologous series. How can we reverse this process and *descend* a homologous series? The crucial reaction in this case is called Hofmann's degradation. In this reaction an amide is first treated with bromine and dilute potassium hydroxide and then warmed with concentrated alkali. The amide is converted to an amine with one less carbon atom.

$$RCONH_2 + Br_2 + 4KOH \rightarrow RNH_2 + 2KBr + K_2CO_3 + 2H_2O$$

Overall, the process involves removing the $C=O$ group from the amide to give the amine

$$RCONH_2 \rightarrow RNH_2$$

Figure 34.4 shows a sequence of reactions which could be used to descend a homologous series from ethanol to methanol.

34.4 Synthetic routes for aromatic compounds

The reactions shown in figure 34.2 relate principally to aliphatic substances in which R represents either H or an alkyl group such as CH_3. In many cases, R could also represent an aromatic group such as phenyl (C_6H_5). However, there are some reactions of aromatic chemicals (arenes) without parallels in aliphatic chemistry. In particular, these include the various substitutions of an aromatic ring. Thus, there is a separate range of synthetic routes of arenes. Some of the conversions linking the simpler aromatic chemicals such as benzene, methylbenzene (toluene) and phenol are shown in figure 34.5 on the next page.

Drugs, detergents and dyes can all be made from benzene, which is unquestionably the most important arene. At one time, benzene, toluene and phenol were produced from coal tar. With the development of catalytic reforming, almost all benzene and toluene now come from oil.

There is one last point. In the industrial manufacture of chemicals, cost is all-important. Hence, cheap reagents such as water, hydrogen, sodium hydroxide and sulphuric acid are used in the production of 'heavy' organic chemicals whenever possible. Expensive reagents such as potassium dichromate(VI), lithium tetrahydridoaluminate(III) (LiAlH₄) and silver compounds are only used in small-scale laboratory preparations when the cost is relatively unimportant.

34.5 Reaction yield

Suppose you are synthesising bromoethane from ethanol.

$$CH_3CH_2OH + HBr \xrightarrow{c.H_2SO_4} CH_3CH_2Br + H_2O$$
ethanol   bromoethane
$M_r = 46$ $M_r = 109$

If you started with 10 g of ethanol, you could work out the mass of bromoethane you would expect to get, assuming all the ethanol is converted to bromoethane.

so
1 mole ethanol ⟶ 1 mole bromoethane
46 g ethanol ⟶ 109 g bromoethane
10 g ethanol ⟶ $\frac{109 \times 10}{46}$ = 23.7 g bromoethane.

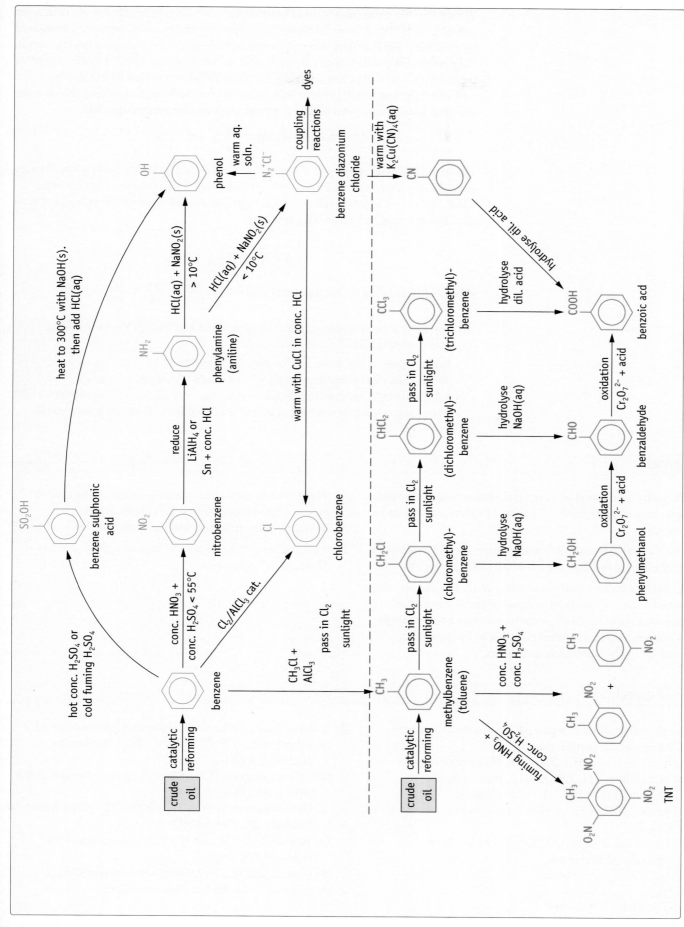

Figure 34.5
Important synthetic routes for arenes.

We say that the **maximum yield** of bromoethane is 23.7 g. In reality you would not get as much as 23.7 g of bromoethane. This is because the reaction produces unwanted **by-products**. For example, some of the ethanol will undergo an elimination reaction in the presence of concentrated sulphuric acid, to form unwanted ethene. Also, some of the bromoethane will be lost when the reaction mixture is purified by distillation.

Realistically, you might get as little as 15.0 g of bromoethane. Comparing this with the theoretical maximum yield of 23.7 g, we can work out the **percentage yield**.

$$\text{Percentage yield} = \frac{15.0}{23.7} \times 100 = 63\%$$

Organic syntheses never give a 100% yield, and a yield as high as 90% is usually very good. In many synthetic reactions, you are doing well if you get a yield of over 50%.

If a synthetic reaction involves several steps, the overall yield will be reduced at each step. For example, in a three-stage synthesis

$$A \xrightarrow{60\%} B \xrightarrow{60\%} C \xrightarrow{60\%} D$$

if the yield of each stage is 60%, the overall yield will be $\dfrac{60}{100} \times \dfrac{60}{100} \times \dfrac{60}{100} = 21.6\%$

> Suppose you are synthesising butanoic acid, $CH_3CH_2CH_2COOH$, by oxidising butan-1-ol, $CH_3CH_2CH_2CH_2OH$
>
> - What reagent could you use to oxidise the butan-1-ol?
> - What unwanted by-products might you get in addition to butanoic acid?
> - What is the maximum yield of butanoic acid you could get from 10.0 g of butan-1-ol?
> - Suppose you actually got 5.5 g of butanoic acid. What is your percentage yield?

Summary

1 Chemists have found ways of synthesising new compounds with known structures and with properties that can be predicted.

2 The natural resources for many of the organic chemicals synthesised in industry are crude oil, fats and vegetable oils.

3 Alkenes, alcohols and carboxylic acids are the starting materials for the synthesis of many organic compounds. The cracking of crude oil fractions provides a ready supply of alkenes. Fats and vegetable oils can be hydrolysed to produce alcohols and carboxylic acids.

4 There are two important reactions which can be used to ascend a homologous series: the conversion of halogenoalkane to nitrile, and the 'oxo' reaction.

5 The crucial reaction in descending a homologous series is Hofmann's Degradation.

Review questions

1 Use the information in figure 34.2 to construct synthetic routes for the manufacture of
(a) butanoic acid ($CH_3(CH_2)_2COOH$) from crude oil,
(b) hexylamine ($CH_3(CH_2)_5NH_2$) from crude oil,
(c) propanone (CH_3COCH_3) from crude oil,
(d) decanal ($CH_3(CH_2)_8CHO$) from naturally occurring fats,
(e) 1-bromododecane ($CH_3(CH_2)_{10}CH_2Br$) from naturally occurring fats.

2 Write synthetic routes for the laboratory preparation of
(a) propylamine ($CH_3(CH_2)_2NH_2$) from butylamine ($CH_3(CH_2)_3NH_2$),
(b) 1-bromopentane ($CH_3(CH_2)_4Br$) from 1-bromobutane ($CH_3(CH_2)_3Br$),
(c) butyl butanoate ($CH_3(CH_2)_2COO(CH_2)_3CH_3$) from butanal ($CH_3(CH_2)_2CHO$),
(d) propan-2-ol ($CH_3CHOHCH_3$) from propan-1-ol ($CH_3CH_2CH_2OH$),
(e) benzoic acid (C_6H_5COOH) from benzene (C_6H_6),
(f) phenol (C_6H_5OH) from benzene (C_6H_6).

34 Synthetic Routes for Organic Chemicals

Table of relative atomic masses

The table gives the relative atomic masses of elements correct to one decimal place.

Element	Symbol	A_r	Element	Symbol	A_r
Aluminium	Al	27.0	Molybdenum	Mo	95.9
Antimony	Sb	121.8	Neodymium	Nd	144.2
Argon	Ar	39.9	Neon	Ne	20.2
Arsenic	As	74.9	Nickel	Ni	58.7
Barium	Ba	137.3	Niobium	Nb	92.9
Beryllium	Be	9.0	Nitrogen	N	14.0
Bismuth	Bi	209.0	Osmium	Os	190.2
Boron	B	10.8	Oxygen	O	16.0
Bromine	Br	79.9	Palladium	Pd	106.4
Cadmium	Cd	112.4	Phosphorus	P	31.0
Caesium	Cs	132.9	Platinum	Pt	195.1
Calcium	Ca	40.1	Potassium	K	39.1
Carbon	C	12.0	Praseodymium	Pr	140.9
Cerium	Ce	140.1	Rhenium	Re	186.2
Chlorine	Cl	35.5	Rhodium	Rh	102.9
Chromium	Cr	52.0	Rubidium	Rb	85.5
Cobalt	Co	58.9	Ruthenium	Ru	101.1
Copper	Cu	63.5	Samarium	Sm	150.4
Dysprosium	Dy	162.5	Scandium	Sc	45.0
Erbium	Er	167.3	Selenium	Se	79.0
Europium	Eu	152.0	Silicon	Si	28.1
Fluorine	F	19.0	Silver	Ag	107.9
Gadolinium	Gd	157.3	Sodium	Na	23.0
Gallium	Ga	69.7	Strontium	Sr	87.6
Germanium	Ge	72.6	Sulphur	S	32.1
Gold	Au	197.0	Tantalum	Ta	180.9
Hafnium	Hf	178.5	Tellurium	Te	127.6
Helium	He	4.0	Terbium	Tb	158.9
Holmium	Ho	164.9	Thallium	Tl	204.4
Hydrogen	H	1.0	Thorium	Th	232.0
Indium	In	114.8	Thulium	Tm	169.9
Iodine	I	126.9	Tin	Sn	118.7
Iridium	Ir	192.2	Titanium	Ti	47.9
Iron	Fe	55.8	Tungsten	W	183.9
Krypton	Kr	83.8	Uranium	U	238.0
Lanthanum	La	138.9	Vanadium	V	50.9
Lead	Pb	207.2	Xenon	Xe	131.3
Lithium	Li	6.9	Ytterbium	Yb	173.0
Lutetium	Lu	175.0	Yttrium	Y	88.9
Magnesium	Mg	24.3	Zinc	Zn	65.4
Manganese	Mn	54.9	Zirconium	Zr	91.2
Mercury	Hg	200.6			

The modern periodic table (wide form)

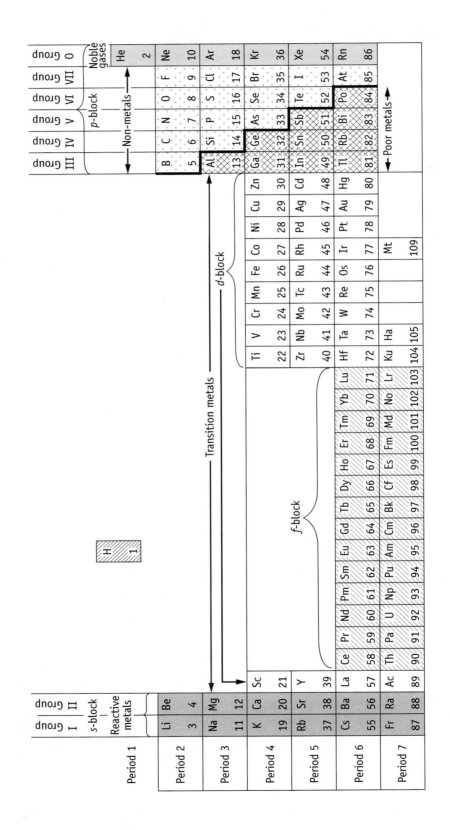

Assessment Questions

These assessment questions are taken from Advanced GCE past papers and are, in general, of A2 standard. Those associated with AS content can be used as whole questions to develop A2 skills, and appropriate subsections of these past papers can be used for AS practice and test preparation.

1 (a) A proton, a neutron and an electron all travelling at the same velocity enter a magnetic field. State which particle is deflected the most and explain your answer. (2)

(b) Give two reasons why particles must be ionised before being analysed in a mass spectrometer. (2)

(c) A sample of boron with a relative atomic mass of 10.8 gives a mass spectrum with two peaks, one at $m/z = 10$ and one at $m/z = 11$. Calculate the ratio of the heights of the two peaks. (2)

(d) Compound **X** contains only boron and hydrogen. The percentage by mass of boron in **X** is 81.2%. In the mass spectrum of **X** the peak at the largest value of m/z occurs at 54.

  (i) Use the percentage by mass data to calculate the empirical formula of **X**.

  (ii) Deduce the molecular formula of **X**. (4)

(Total 10 marks)

NEAB 1998

2 The table below shows some accurate relative atomic masses.

Atom	^1H	^{12}C	^6Li
Relative atomic mass	1.0078	12.0000	6.0149

(a) Why is ^{12}C the only atom with a relative atomic mass which is an exact whole number? (1)

(b) Calculate the mass of 1 mol of $^1H^+$ ions. The mass of a single electron is 9.1091×10^{-28} g. (Avogadro's number, L, is 6.0225×10^{23}) (2)

(c) (i) Explain briefly the process by which a sample is ionised in a mass spectrometer.

  (ii) Give **one** reason why it is important to use the minimum possible energy to ionise a sample in a mass spectrometer.

  (iii) After ionisation and before deflection, what happens to the ions in a mass spectrometer; how is this achieved? (5)

(d) Why is it a good approximation to consider that the relative atomic mass of the $^6Li^+$ ion, determined in a mass spectrometer, is the same as that of ^6Li? (1)

(Total 9 marks)

NEAB 1997

3 Magnesium oxide, MgO, is a white solid with a high melting temperature which is used as a furnace lining.

  (i) State **two** ways in which magnesium oxide can be obtained giving a balanced equation in each case. (4)

  (ii) Using **outer** electrons only, draw a dot and cross diagram showing the bonding in magnesium oxide. (2)

  (iii) State why magnesium oxide is described as a basic oxide. (1)

  (iv) One industrial method for obtaining magnesium from magnesium oxide involves heating magnesium oxide with silicon at a high temperature in the absence of air.

$$2MgO + Si \rightarrow 2Mg + SiO_2$$

  I. State why the process must be carried out in the absence of air. (1)

  II. Calculate the mass of silicon required to convert completely 500 kg of magnesium oxide into magnesium. (3)

(Total 11 marks)

WJEC 1998

4 Malachite is a green, naturally-occurring mineral. When powdered malachite is heated in air it decomposes to give black copper(II) oxide, carbon dioxide and water as the only products. When reacted with dilute hydrochloric acid the malachite completely dissolves and carbon dioxide is released.

A student suggested that malachite is hydrated copper(II) carbonate, $CuCO_3.xH_2O$, and carried out the following two experiments to determine the formula of the mineral.

Experiment A

0.240 g of malachite produced 24.0 cm³ of carbon dioxide when reacted with excess dilute hydrochloric acid.

[Take the volume of one mole of gas as 24.0 dm³ under the conditions used. $A_r(H)$ 1.0; $A_r(C)$ 12.0; $A_r(Cu)$ 63.5; $A_r(O)$ 16.0]

Experiment B

On heating to *constant mass* 0.240 g of malachite decomposed to give 0.160 g of copper(II) oxide. On further reduction in a stream of hydrogen gas the copper oxide produced 0.127 g of metallic copper.

(a) (i) Calculate the amount (in moles) of carbon dioxide formed in experiment A. (1)
   (ii) What mass of carbon dioxide was produced in this reaction? (1)
(b) (i) In experiment B why was the malachite heated to constant mass? (1)
   (ii) Calculate the mass of water lost when the malachite was heated in experiment B. (2)
   (iii) How many moles of water were produced in this reaction? (1)
   (iv) How many moles of copper were produced on reduction of the copper oxide? (1)
(c) (i) From the results of both experiments deduce the mole ratio $CO_2 : H_2O : Cu$. (1)
   (ii) Explain why the suggested formula for malachite is not consistent with the experimental observations. (2)
(d) If malachite is composed of two copper-containing compounds, one of which is copper(II) carbonate, suggest a possible formula for the mineral and explain your reasoning. (2)

(Total: 12 marks)
OCSEB 1997

5 (a) Define the terms
   (i) atomic number; (1)
   (ii) mass number; (1)
   (iii) relative atomic mass. (1)
(b) In 1919 F. W. Aston, using an early form of the mass spectrometer, showed that neon exists as a mixture of isotopes.
The mass spectrum is shown below; determine the relative atomic mass of neon.

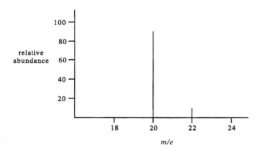

(2)

(c) (i) Copy and complete the following table:

Radiation	Nature of the Radiation	Relative Penetrating Power
α		
β		
γ		

(3)

(ii) Define the term **half-life**. (1)

(iii) Radioactive sources are used in metal rolling mills to monitor the thickness of sheet metal. Suggest how this might be done. (2)
(iv) Suggest why the isotope used in this application should have a long half-life. (2)
(d) (i) Define the second ionisation energy of fluorine. (2)
   (ii) Copy the axes below and sketch a graph to show the successive ionisation energies of fluorine. Give reasons for the shape of the line you draw.

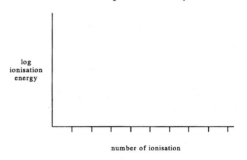

(4)
(Total 19 marks)
London 1997

6 Below are several nuclides of elements

$$^{24}_{12}\text{Mg} \qquad ^{84}_{36}\text{Kr} \qquad ^{90}_{38}\text{Sr} \qquad ^{131}_{53}\text{I} \qquad ^{137}_{55}\text{Cs}$$

From the above choose
   (i) **one** element whose atom contains 52 neutrons, (1)
   (ii) **two** elements whose atoms each have the same number of electrons in the outer shell, (2)
   (iii) **one** element which exists as a molecular crystalline lattice at room temperature. (1)
(Total 4 marks)
WJEC 1998

7 (a) What is meant by the terms
   (i) electron; (3)
   (ii) energy level? (2)

(b) Using a Data Book identify the species **X**, **Y** and **Z**:

species	atomic number	electronic configuration
X	23	$1s^2 2s^2 2p^6 3s^2 3p^6 3d^2$
Y	11	$1s^2 2s^2 2p^6 3s^1$
Z	17	$1s^2 2s^2 2p^6 3s^2 3p^6$

(3)

(c) Which of the above species, **X**, **Y** or **Z**, would be most likely to
   (i) react violently with water;

(ii) form an octahedral cyano complex;
(iii) be part of the raw material used in the industrial production of sodium hydroxide;
(iv) be found in an array of positive ions in a cloud of delocalised electrons? (4)

(Total 12 marks)

OCSEB 1997

8 (a) (i) Give the electronic configuration of an atom of the isotope of calcium, $^{45}_{20}Ca$. (1)
(ii) Give the names and numbers of each type of particle present in a nucleus of this isotope. (2)
(iii) State one reason why the information in (a)(i) is usually more useful to chemists than that in (a)(ii). (1)
(b) The isotope $^{45}_{20}Ca$ can be made by bombarding $^{45}_{21}Sc$ with neutrons.
(i) Write a nuclear equation for this reaction. (2)
(ii) $^{45}_{20}Ca$ is radioactive, decaying by beta (β) emission. A sample of calcium, containing 6.00×10^{-3} g of $^{45}_{20}Ca$ initially, was found to contain only 7.50×10^{-4} g of this isotope after 540 days. Calculate the half-life of $^{45}_{20}Ca$ showing how you arrive at your answer. (2)
(iii) State and explain what would happen to the radiation from this source if it was passed through an electrical field. (2)

(Total 10 marks)

London 1997

9 The graph shows the first ionisation energies for the atoms from hydrogen to argon.

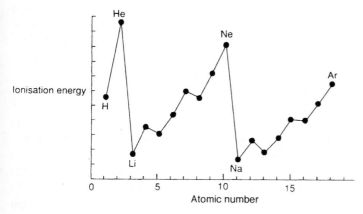

(a) (i) Give the equation which represents the first ionisation energy of oxygen atoms. (1)
(ii) Why is the first ionisation energy of helium the largest of all the atoms? (2)
(iii) Why is the first ionisation energy of oxygen atoms less than that of nitrogen atoms? (2)
(iv) Why do the first ionisation energies of the atoms in Group I decrease as the atomic number increases? (2)

(b) (i) Write equations which represent the first and second electron affinities of oxygen atoms. (2)
(ii) The first electron affinity of oxygen atoms is $-141\,kJ\,mol^{-1}$ and the second is $+798\,kJ\,mol^{-1}$. Suggest why the first is exothermic and the second endothermic. (2)
(c) Magnesium burns brightly in oxygen to give magnesium oxide, which contains the ions Mg^{2+} and O^{2-}. The formation of both these ions from their elements is strongly endothermic. Why, therefore, should magnesium combine with oxygen? (2)
(d) The table below gives the successive ionisation energies of sodium.

No. of ionisation	1	2	3	4	5	6
Energy/kJ mol⁻¹	496	4563	6913	9544	13 352	16 611

No. of ionisation	7	8	9	10	11
Energy/kJ mol⁻¹	20 115	25 491	28 934	141 367	159 079

What information about the electronic structure of sodium is provided by this data? (2)

(Total 15 marks)

London 1998

10 The following table contains ionisation energy data.

Element	N	O	F	Ne	Na
First ionisation energy /kJmol⁻¹	1400	1310	1680	2080	494

(a) Explain the meaning of the term *first ionisation energy* of an element. (2)
(b) Explain why neon has a higher first ionisation energy than fluorine. (2)
(c) Explain why oxygen has a lower first ionisation energy than nitrogen. (2)
(d) Explain why sodium has a lower first ionisation energy than neon. (2)
(e) Predict an approximate value for the first ionisation energy of carbon and explain your answer. (3)

(Total 11 marks)

NEAB 1997

11 The table below gives data on chlorides of elements in Period 3 of the Periodic Table.

chloride	NaCl	MgCl₂	AlCl₃	SiCl₄	PCl₃
addition of water	dissolves	dissolves	vigorous reaction	vigorous reaction	vigorous reaction

(a) (i) What is the trend in the pH of the solutions obtained after the addition of water to each of the chlorides?
(ii) Write an equation for the reaction of SiCl₄ with water. (3)

(b) Copy and complete the diagram below to show the structure of sodium chloride using the given key to represent sodium and chloride ions.

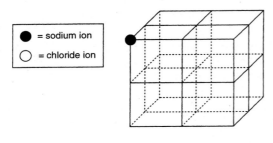

(2)

(c) Showing outer electron shells only, draw a 'dot-and-cross' diagram of
(i) $MgCl_2$,
(ii) PCl_3. (3)

(d) Draw a diagram to show the expected shape and bond angles in a molecule of PCl_5. (2)

(e) Phosphorus also forms a pentachloride, PCl_5, which is thought to exist in the solid form as $[PCl_4]^+[PCl_6]^-$. Suggest the shapes of these two ions. (2)

(Total 12 marks)
UCLES 1997

12 Nitrogen and carbon monoxide have similarities. The gases have diatomic molecules and they have simple molecular structures. Both gases are colourless, odourless and tasteless, but nitrogen is harmless whereas carbon monoxide is extremely toxic.

(a) Nitrogen atoms exist in two natural isotopic forms, ^{14}N and ^{15}N. Copy and complete the table below to show the atomic structures of these isotopes of nitrogen.

Isotope	Number of		
	protons	neutrons	electrons
^{14}N			
^{15}N			

(2)

(b) State **two** typical properties of a compound with a *simple molecular structure*. (2)

(c) A 'dot-and cross' diagram, showing outer shell electrons only, of a carbon monoxide molecule is shown below.

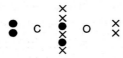

Copy the following diagrams and in each case show, by means of a circle, the electrons associated with

(i) a lone pair,

(ii) a covalent bond,

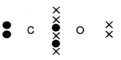

(iii) a co-ordinate bond (dative covalent bond).

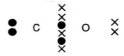

(3)

(d) 'Dot-and-cross' diagrams, showing all electrons, of nitrogen and carbon monoxide molecules are shown below.

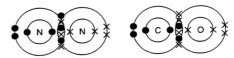

(i) These diagrams show that a nitrogen molecule is isoelectronic with a carbon monoxide molecule. Suggest what is meant by the word *isoelectronic*.

(ii) Hydrogen cyanide, HCN, is also isoelectronic with N_2 and with CO.
Construct a 'dot-and-cross' diagram for the HCN molecule showing outer shell electrons only.

(iii) Suggest why carbon monoxide and hydrogen cyanide are both more reactive than nitrogen. (4)

(Total 11 marks)
UCLES 1997

13 Fluorine is the most electronegative element, forming both ionic and covalent compounds.

(a) Explain in terms of electronic structures, how calcium atoms combine with fluorine atoms to form calcium fluoride. (4)

(b) (i) Boron trifluoride, BF_3, is a covalent molecule which does not obey the octet rule. Draw a "dot and cross" diagram to show the bonding in BF_3. (2)

(ii) Draw and explain the shape of the BF_3 molecule. (2)

(iii) Suggest why BF_3 is a non-polar molecule even though it contains polar bonds. (2)

(iv) Boron trifluoride can react with ammonia to form the compound $BF_3.NH_3$ which is a solid. What type of bond is formed between the nitrogen and the boron atoms? (1)

(Total 11 marks)
NICCEA 1998

14 The table below shows some boiling temperatures (T_b) at a pressure of 100 kPa.

Substance	H_2	CH_4	HCl
T_b /K	21	112	188

(a) In liquid hydrogen, the atoms are held together by covalent bonds.

   (i) What is a covalent bond?

   (ii) How are the hydrogen molecules held together in liquid hydrogen? (2)

(b) Explain why methane has a higher boiling temperature than hydrogen. (2)

(c) (i) Give the meaning of the term *electronegativity*.

   (ii) The electronegativities of hydrogen, carbon and chlorine are 2.1, 2.5 and 3.0, respectively. Use these values to explain why the boiling temperature of hydrogen chloride is greater than that of methane. (6)

(Total 10 marks)

NEAB 1997

15 Some data relating to propane, ethanol and methanoic acid are given in the table.

Compound	Relative molecular mass	Boiling point/ °C
Propane	44	−42.2
Ethanol	46	78.5
Methanoic acid	46	101.0

(a) (i) State what is meant by the term *polar bond* and explain how one can arise within a molecule. (2)

   (ii) Draw the graphical formula for each of the compounds in the table and clearly show the polarity of any polar bonds present. (4)

(b) (i) State the type of intermolecular force present in propane and explain why it has a low boiling point. (2)

   (ii) State the main type of intermolecular force present in both pure ethanol and pure methanoic acid. Draw diagrams to show clearly this force in each of the compounds. Suggest a reason for the higher boiling point of methanoic acid. (4)

(Total 12 marks)

AEB 1998

16 (a) Explain the meaning of the term *electronegativity*. (2)

(b) The electronegativities of some atoms are given below.

H	Li	B	C	N	O	F
2.1	1.0	2.0	2.5	3.0	3.5	4.0

Arrange the following bonds in order of expected increasing ionic character, putting the least ionic first.

N—H, F—H, B—H, C—H.

(1)

Explain briefly how you have used the electronegativities to arrive at this order. (2)

(c) (i) With the aid of a diagram describe the nature of hydrogen bonding. (4)

   (ii) Which of the following molecules is/are likely to form hydrogen bonds? (D represents deuterium, ^2_1H)

     CH_4 NH_3 CH_3OH D_2O

(2)

(Total 11 marks)

OCSEB 1997

17 (a) Define the term *electronegativity*. (2)

(b) Figure 1 shows the trend in the boiling points of the hydrides of the Group 6 elements, oxygen to tellurium.

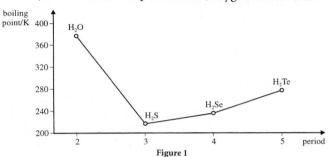

Figure 1

   (i) Explain the rising trend in the boiling points of the compounds H_2S to H_2Te. (3)

   (ii) Hydrogen bonding is said to account for the anomalously high boiling point of H_2O. With reference to the nature of the atoms involved, explain why intermolecular forces are so strong in H_2O. Draw a diagram, containing at least three molecules, to show hydrogen bonding in H_2O. (5)

(c) Protein molecules are composed of sequences of amino acid molecules joined together in long chains. The sequence of amino acids in a protein chain is illustrated in Figure 2.

Figure 2

The protein chains are organised into complex three-dimensional structures. Briefly describe how the three-dimensional structure of a protein is held in place. (3)

(Total 13 marks)

AEB 1998

18 (a) (i) State **two** bonding factors which influence the solubility of a solid compound in water, at a given temperature. (2)

   (ii) Explain why ethanol is soluble in water in all proportions but a hydrocarbon such as hexane is not. (3)

(b) Solid sodium chloride has a **face-centred cubic** crystal structure.

   (i) Sketch a diagram to illustrate the structure of the sodium chloride crystal and state the crystal coordination number of the sodium ion. (2)

(ii) State why the coordination number of the caesium ion in caesium chloride is different from that of the sodium ion in sodium chloride. (1)

(iii) Explain why solid iodine, which also has a face-centred cubic lattice, is readily volatile on warming but sodium chloride has a melting temperature of 1074 K. (3)

(Total 11 marks)

WJEC 1997

19 (a) White phosphorus melts at 317 K. In a mass spectrometer, white phosphorus gives only one peak at $m/z = 124$ in the region of the molecular ion.

(i) What can you deduce from the fact that there is only one peak in the region of the molecular ion? Give the formula of the species responsible for this peak.

(ii) Explain in terms of its structure and bonding why white phosphorus has a low melting point.

(iii) Explain why the melting point of silicon (1683 K) is higher than that of white phosphorus. (10)

(b) When phosphorus is burned in a limited supply of air, it forms two oxides **A** and **B**. Oxide **A** contains 56.4% by mass of phosphorus.

(i) Calculate the empirical formula of **A**. (3)

(ii) What additional information is required in order to calculate the molecular formula of **A**? (1)

(iii) Oxide **B** has the molecular formula P_4O_{10}. Write an equation for the formation of P_4O_{10} from phosphorus and oxygen. Calculate the mass of P_4O_{10} which would be formed by complete oxidation of 3.20 g of phosphorus. (2)

(iv) With the aid of an equation explain why P_4O_{10} is classified as an acidic oxide. (2)

(c) State the type of structure possessed by each of the following oxides. In each case, predict the resulting pH when the oxide is added to water and write an equation for any reaction which occurs.

Na_2O MgO SiO_2 SO_2 (12)

(Total 30 marks)

NEAB 1997

$[A_r(P) = 31.]$

20 In an experiment to find the enthalpy change of neutralisation, aqueous sodium hydroxide was added to 25 cm^3 of 0.60 mol dm^{-3} sulphuric acid. The following readings were obtained.

Total volume of sodium hydroxide added/cm^3	Temperature/ °C
0	20.0
10	21.7
20	23.5
30	25.3
40	26.4
50	25.7
60	25.0

(a) Define the term *standard enthalpy change of neutralisation*. (2)

(b) Draw a labelled diagram of the apparatus that could be used in this experiment. (2)

(c) Transfer the axes below to the graph paper provided and plot a graph of temperature against total volume of NaOH(aq) added. (2)

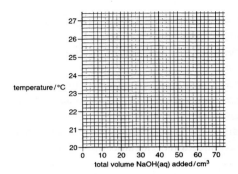

(d) (i) Estimate the volume of aqueous sodium hydroxide needed to neutralise 25 cm^3 of the sulphuric acid.

(ii) Estimate the temperature **rise** corresponding to the volume of NaOH(aq) in (d)(i). (2)

(e) (i) Calculate the enthalpy change in this experiment. (Assume the specific heat capacity of all solutions is 4.2 J cm^{-3} K^{-1}.)

(ii) Calculate the corresponding standard enthalpy change of neutralisation. (6)

(Total 14 marks)

UCLES 1997

21 The tables below contain data which are needed to answer the questions.

Name	hydrazine	ethane
Formula of compound	N_2H_4	C_2H_6
Boiling temperature /K	387	184

Formula and state of compound	$C_2H_6(g)$	$CO_2(g)$	$H_2O(l)$
Standard enthalpy of formation (at 298 K) /kJ mol^{-1}	−85	−394	−286

(a) Suggest why hydrazine has a much higher boiling temperature than ethane. (2)

(b) When liquid hydrazine burns in oxygen it forms nitrogen and water. The standard enthalpy change for this reaction when one mole of hydrazine forms water in the liquid state is −624 kJ mol^{-1}.

(i) Write a balanced equation for the combustion of hydrazine in oxygen.

(ii) Calculate the standard enthalpy of formation of liquid hydrazine. (4)

(c) (i) Write an equation for the complete combustion of ethane.

(ii) Use the appropriate standard enthalpies of formation to calculate the standard enthalpy of combustion of ethane. (4)

(d) Suggest one reason why hydrazine is more suitable than ethane for use as a rocket fuel. (1)

(Total 11 marks)

NEAB 1998

22 Hydrocarbons, such as heptane, C_7H_{16}, are used as fuels making use of their combustion reaction with oxygen.

(a) (i) Define the term *standard enthalpy change of combustion*.

(ii) State the temperature and pressure that are conventionally chosen for quoting standard enthalpy changes. (3)

(b) Use the data below to calculate the standard enthalpy change of combustion of heptane.

Compound	$\Delta H_f^{\ominus}$ /kJ mol^{-1}
$C_7H_{16}(l)$	−224.4
$CO_2(g)$	−393.5
$H_2O(l)$	−285.9

$$C_7H_{16}(l) + 11O_2(g) \rightarrow 7CO_2(g) + 8H_2O(l)$$

(3)

(c) Oil companies use cracking to modify the structures of hydrocarbon fractions in order to produce compounds that are in greater demand.

Construct a possible equation for the cracking of heptane. (2)

(Total 8 marks)

UCLES 1997

23 (a) Define the term *standard molar enthalpy change of formation*. (3)

(b) State *Hess's law*. (1)

(c) The equation below shows the reaction between ammonia and fluorine.

$$NH_3(g) + 3F_2(g) \rightarrow 3HF(g) + NF_3(g)$$

(i) Use the standard molar enthalpy change of formation ($\Delta H_f^{\ominus}$) data in **Figure 1** to calculate the molar enthalpy change for this reaction.

Compound	NH_3	HF	NF_3
$\Delta H_f^{\ominus}$ kJ mol^{-1}	−46	−269	−114

Figure 1

(4)

(ii) Use the average bond enthalpy data in **Figure 2** to calculate a value for the molar enthalpy change for the same reaction between ammonia and fluorine.

$$NH_3(g) + 3F_2(g) \rightarrow 3HF(g) + NF_3(g)$$

Bond	N—H	F—F	H—F	N—F
Average bond enthalpy/kJmol^{-1}	388	158	562	272

Figure 2

(3)

(d) The answer you have calculated in (c)(i) is regarded as being the more reliable value. Suggest why this is so. (3)

(Total 14 marks)

AEB 1997

24 This question concerns the manufacture of ethanol by the direct hydration of ethene. The reaction is represented by the following equation.

$$H_2O(g) + C_2H_4(g) \rightleftharpoons C_2H_5OH(g) \quad \Delta H = -46.0 \text{ kJ mol}^{-1}.$$

The catalyst used is phosphoric(V) acid on a silica support at 570 K and the mechanism is thought to involve protonation of the ethene molecule.

(a) (i) Draw the structure of a protonated ethene molecule. (1)

(ii) The protonated ethene molecule then reacts with a molecule of water. State what feature of the water molecule allows it to behave as a nucleophile in this reaction. (1)

(iii) The operating temperature is said to be the optimum temperature for the process. State **two** opposing economic effects which arise when the operating temperature is raised for the direct hydration of ethene. (2)

(iv) State what effect an increase in pressure will have on the equilibrium yield of ethanol and give a reason for your answer. (2)

(v) At pressures higher than 70 atmospheres polymerisation of one of the compounds in the reaction mixture may occur. Give the name of the polymer produced and the structure of the repeating unit. (2)

(b) State **one** important economic consideration when deciding on the operating pressure of a chemical plant. (1)

(c) The enthalpy changes of formation of ethene and steam are +52.3 kJ mol^{-1} and −242 kJ mol^{-1} respectively. Calculate the enthalpy change of formation of gaseous ethanol.

$$H_2O(g) + C_2H_4(g) \rightarrow C_2H_5OH(g) \quad \Delta H = -46.0 \text{ kJ mol}^{-1}$$

(2)

(Total 11 marks)

WJEC 1998

25 (a) (i) Using the data provided, construct a Born-Haber cycle for magnesium chloride, $MgCl_2$, and from it determine the electron affinity of chlorine.

	ΔH/kJ mol^{-1}
Enthalpy of atomisation of chlorine	+122
Enthalpy of atomisation of magnesium	+148
First ionisation energy of magnesium	+738
Second ionisation energy of magnesium	+1451
Lattice enthalpy of magnesium chloride	−2526
Enthalpy of formation of magnesium chloride	−641

(5)

(ii) The theoretically calculated value for the lattice enthalpy of magnesium chloride is −2326 kJ mol^{-1}. Explain the difference between the theoretically calculated value and the experimental value given in the data in (a)(i), in terms of the bonding of magnesium chloride. (3)

(b) The table below gives some information about the sulphates of elements in Group 2.

Sulphate	Solubility /mol dm^{-3}	Lattice enthalpy /kJ mol^{-1}	Hydration enthalpy of M^{2+}/kJ mol^{-1}
$CaSO_4$	4.6×10^{-2}	−2480	−1650
$SrSO_4$	7.1×10^{-4}	−2484	−1480
$BaSO_4$	9.4×10^{-6}	−2374	−1360

(i) Suggest an explanation for the trend shown in the hydration enthalpies of the cations. (2)

(ii) Comment on the trend in the solubilities of these sulphates in relation to the lattice and hydration enthalpies given in the table. (4)

(iii) Barium sulphate, which is opaque to X-rays, is used for the "barium meal" to enable X-ray pictures to be taken of the gut. Barium ions are very toxic; why is this not a problem here? (1)

(iv) Give the equation for the reaction of barium with cold water. (2)

(v) Suggest the practical procedure by which you might convert the solution of the product in reaction (iv) into a reasonably pure sample of barium sulphate. (3)

(Total 20 marks)
London 1998

26 Potassium chlorate(v) decomposes on heating according to the equation

$$4KClO_3(s) \rightarrow 3KClO_4(s) + KCl(s) \quad \Delta H^{\ominus} = +16.8 \text{ kJ mol}^{-1}$$

The standard molar entropy, $S^{\ominus}$, for each species involved in the reaction is shown below.

	$S^{\ominus}$/J K^{-1} mol^{-1}
$KClO_3(s)$	112
$KClO_4(s)$	134
$KCl(s)$	83

(a) Calculate the standard entropy change for the above reaction. (2)

(b) Calculate the standard free energy change at 298 K for the above reaction. (3)

(c) (i) In terms of free energy, state the necessary condition for spontaneous change.

(ii) Assuming that ΔH and ΔS do not vary with temperature, determine the lowest temperature at which the above reaction will become feasible. (4)

(Total 9 marks)
NEAB 1997

27 The combustion of fossil fuels produces sulphur dioxide, a major contributor to acid rain.

(a) Write **two** equations to show the formation of acid rain from sulphur dioxide. (2)

(b) Acid rain is responsible for extensive damage to trees.

(i) State **two** damaging effects that acid rain can have on trees, either directly or indirectly through the soil.

(ii) Why would chalky soil help to limit this damage? (3)

(c) A 1 tonne sample of coal, containing 1.92% of sulphur by mass, was used as a fuel.

(i) Calculate the mass of sulphur dioxide that was formed. (1 tonne = 10^6 g)

(ii) Calculate the volume of sulphur dioxide that was produced at room temperature and pressure (r.t.p).

(iii) List **two** methods that could be used to reduce the sulphur dioxide emissions from the combustion of fossil fuels. (6)

(Total 11 marks)
UCLES 1997

[1 mole of gas occupies 24 dm^3 at r.t.p.]

28 (a) A rate equation shows how the rate of a chemical reaction depends upon the concentration of the reactants. For a reaction between two substances A and B, the rate equation is of the form:

$$\text{Rate} = k[A]^m[B]^n$$

Use this rate equation to explain the terms:

(i) rate of reaction. (1)

(ii) overall order of reaction. (3)

(b) The data below were obtained when substances A and B reacted in solution at constant temperature.

$\dfrac{[A]}{\text{mol dm}^{-3}}$	$\dfrac{[B]}{\text{mol dm}^{-3}}$	initial rate of formation of product $\text{mol dm}^{-3}\,\text{s}^{-1}$
0.10	0.01	1.23×10^{-3}
0.10	0.03	1.11×10^{-2}
0.05	0.03	5.55×10^{-3}

(i) Determine the order of reaction with respect to

    A   (1)

    B   (1)

(ii) Calculate the rate constant, k, giving appropriate units. (2)

(c) State and explain the effect on the rate of a chemical reaction of:

(i) an increase in temperature; (4)

(ii) the addition of a catalyst. (4)

(Total 16 marks)

OCSEB 1998

29 (a) The rate of reaction between compounds **C** and **D** was studied at a fixed temperature and some results obtained are shown in the table below.

Experiment	Initial concentration of **C**/mol dm^{-3}	Initial concentration of **D**/mol dm^{-3}	Initial rate r/mol dm^{-3} s^{-1}
1	0.010	0.010	1.0×10^{-6}
2	0.020	0.010	4.0×10^{-6}
3	0.030	0.020	9.0×10^{-6}
4	0.040	0.020	to be calculated

Use the data in the table to deduce the order of reaction with respect to compound **C** and the order of reaction with respect to compound **D**. Hence calculate the initial rate of reaction in experiment 4. (3)

(b) Methyl ethanoate is hydrolysed by sodium hydroxide. In a series of experiments at a given temperature, the reaction was found to be first order with respect to this ester and also first order with respect to hydroxide ions. In one of these experiments, when the initial concentration of both reagents was 0.020 mol dm^{-3}, the initial rate of reaction was found to be 8.8×10^{-5} mol dm^{-3} s^{-1}

(i) Write an equation for the reaction between methyl ethanoate and hydroxide ions.

(ii) Write the rate equation for the reaction and calculate the value of the rate constant, with its units, at this temperature.

(iii) In a further experiment at the same temperature, the initial concentration of ester was 0.020 mol dm^{-3} and the initial concentration of hydroxide ions was 2.0 mol dm^{-3}. Under these conditions the reaction

appears to be zero order with respect to hydroxide ions. Suggest why this is so. (8)

(Total 11 marks)

NEAB 1998

30 The rate law for the oxidation of iodide ions by acidified hydrogen peroxide solution may be determined by measuring the time taken for the iodine concentration to reach 1×10^{-5} mol dm^{-3} using different initial concentrations of the reactants.

$$H_2O_2 + 2I^- + 2H^+ \rightarrow 2H_2O + I_2$$

(a) Suggest a convenient method for determining when the iodine concentration reaches 1×10^{-5} mol dm^{-3}. (1)

(b) In a series of experiments the time taken for the iodine concentration to reach 1×10^{-5} mol dm^{-3} was recorded as shown below.

Experiment	Initial concentration/mol dm^{-3}			Time/s
	H_2O_2	I^-	H^+	
A	0.02	0.02	2.0	12
B	0.01	0.02	2.0	24
C	0.02	0.06	2.0	4
D	0.02	0.02	0.5	48

Calculate the order with respect to each of the reactants, write a rate equation for this reaction and state the units of the rate constant, k. (4)

(Total 5 marks)

NICCEA 1998

31 (a) In an experiment to investigate reaction rates, an excess of magnesium powder was added to 20 cm^3 of aqueous hydrochloric acid, of concentration 0.20 mol dm^{-3}, at 20°C.

$$Mg(s) + 2HCl(aq) \rightarrow MgCl_2(aq) + H_2(g)$$

The results of the experiment are shown in **Figure 1**.

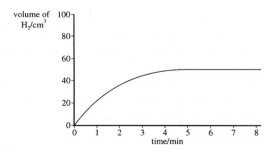

Figure 1

Copy **Figure 1** on a sheet of graph paper, and on the same axes draw lines to show the results you would expect to obtain if the experiment were repeated with the changes described below. Label the lines **C**, **T** and **P** respectively.

(i) **Line C**   An excess of magnesium powder is added to 20 cm³ of aqueous hydrochloric acid of concentration 0.40 mol dm⁻³ at a temperature of 20°C. (1)

(ii) **Line T**   An excess of magnesium powder is added to 20 cm³ of aqueous hydrochloric acid of concentration 0.20 mol dm⁻³ at a temperature of 50°C. (1)

(iii) **Line P**   A single piece of magnesium ribbon (an excess) is added to 20 cm³ of aqueous hydrochloric acid of concentration 0.20 mol dm⁻³ at a temperature of 20°C. (1)

(b) The reaction in which hydrogen iodide is thermally decomposed was investigated.

The results of the investigation led to the following rate equation.

$$\text{rate} = k[\text{HI}]^2$$

(i) What does the symbol k represent in this equation? (1)

(ii) With an initial concentration of hydrogen iodide of 0.20 mol dm⁻³, the initial rate of the reaction was found to be 2.0×10^{-4} mol dm⁻³ s⁻¹. Calculate the value of k, stating its units. (3)

(iii) Deduce the value of the initial rate of reaction when the experiment is repeated, at the same temperature, but with an initial concentration of hydrogen iodide of 0.60 mol dm⁻³. (2)

(c) (i) Copy the axes provided in **Figure 2**, and draw the distribution of the energies of gas molecules at a temperature T_1. Label this distribution T_1. On the same axes show the distribution for the same sample of gas at a **higher** temperature T_2. Label this distribution T_2.

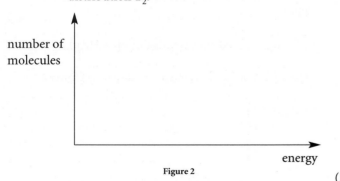

number of molecules

energy

Figure 2

(3)

(ii) Explain, in terms of the collision theory, why rates of reactions are increased by an increase in temperature. (3)

(Total 15 marks)
AEB 1998

32 (a) Define the term *activation energy*. (1)

(b) The endothermic reaction between substances **P** and **Q** can be represented by the following equation.

$$\textbf{P}(g) + \textbf{Q}(g) \rightarrow \textbf{R}(g) + \textbf{S}(g)$$

(i) Copy and complete the reaction profile for this reaction, shown in **Figure 1**. Indicate and label clearly the activation energy and the enthalpy change for the reaction. (3)

(ii) Also on your copy of **Figure 1**, draw and label the reaction profile for the reaction when it is catalysed. (1)

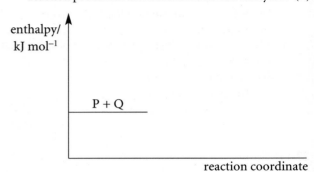

enthalpy/ kJ mol⁻¹

P + Q

reaction coordinate

Figure 1

(iii) Explain why a catalyst increases the rate of a chemical reaction. (2)

(c) Catalytic converters are fitted to many motor cars in order to reduce the levels of some of the pollutants in the exhaust gases. The poisonous pollutants are converted into non-toxic products by the catalytic action of a metal catalyst.

(i) Give the name of one such pollutant and state the non-toxic product(s) to which it is converted. (2)

(ii) Explain why some commonly available types of petrol should never be used in a motor car fitted with a catalytic converter. (2)

(d) (i) **Figure 2** shows the distribution of molecular energies for the mixture of reactants in (b) at a temperature T_1. Copy and label the axes and, on the same axes, draw the distribution of molecular energies for the same gaseous mixture at a higher temperature T_2.

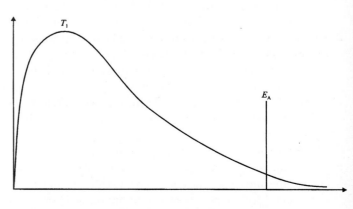

T_1

E_A

Figure 2

(4)

(ii) In **Figure 2**, the energy corresponding to the activation energy of the reaction (E_A) is shown. On your drawing, indicate a possible energy corresponding to the activation energy of the catalysed reaction and label it E_A *catalysed*. (1)

(Total 16 marks)

AEB 1997

33 Consider the equilibrium·

$$N_2O_4(g) \rightleftharpoons 2NO_2(g)$$

(a) (i) Write an expression for K_c, indicating the units. (2)

(ii) 1 mol of dinitrogen tetroxide, N_2O_4, was introduced into a vessel of volume 10.0 dm³ at a temperature of 70°C. At equilibrium 50% had dissociated. Calculate K_c. (4)

(iii) Using the following data calculate the enthalpy change for the forward reaction.

	$\Delta H_f^{\ominus}$ /kJ mol⁻¹
N_2O_4	+9.70
NO_2	+33.9

(2)

(iv) If the same experiment is carried out at 100°C, state qualitatively, giving your reasons, how the equilibrium composition will change. (2)

(b) Explain what you would do to increase the degree of dissociation of N_2O_4(g) at constant temperature. (2)

(c) What is the effect of a catalyst on the following:
(i) the value of K_c; (1)
(ii) the equilibrium position; (1)
(iii) the rate of attainment of equilibrium? (1)

(d) Suggest why the reaction

$$N_2 + 2O_2 \rightarrow 2NO_2$$

is not a very useful method of making NO_2. (2)

(Total 17 marks)

London 1998

34 The equilibrium yield of product in a gas-phase reaction varies with changes in temperature and pressure as shown below.

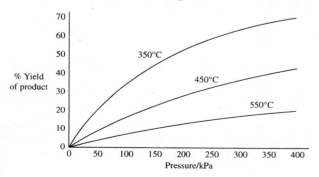

(a) Use the information given above to deduce whether the forward reaction involves an increase, a decrease, or no change in the number of moles present. Explain your deduction. (4)

(b) Use the information given above to deduce whether the forward reaction is exothermic or endothermic. Explain your answer. (3)

(c) (i) Estimate the percentage yield of product which would be obtained at 350°C and a pressure of 250 kPa.
(ii) State what effect, if any, a catalyst has on the position of the equilibrium. Explain your answer. (4)

(d) A 70% equilibrium yield of product is obtained at a temperature of 350°C and a pressure of 400 kPa. Explain why an industrialist may choose to operate the plant at
(i) a temperature higher than 350°C;
(ii) a pressure lower than 400 kPa. (2)

(Total 13 marks)

NEAB 1997

35 Part of the Contact process for the manufacture of sulphuric acid involves the reversible reaction:

$$2SO_2(g) + O_2(g) \rightleftharpoons 2SO_3(g) \quad \Delta H = -198 \text{ kJ mol}^{-1}$$

(a) State *Le Chatelier's principle*. (2)

(b) For the above equilibrium, state and explain the effect on the equilibrium position of
(i) increasing the pressure, at constant temperature (2)
(ii) increasing the temperature, at constant pressure. (2)

(c) Write an expression for the equilibrium constant, K_c, for the above equilibrium. (1)

(d) State and explain the effect on K_c of
(i) increasing the pressure, at constant temperature (1)
(ii) increasing the temperature, at constant pressure. (2)

(e) In the commercial production of sulphuric acid, this reaction is usually carried out at about 150 kPa, about 500°C and using a suitable catalyst.
(i) Name a suitable catalyst for this process. (1)
(ii) Use the information and your earlier answers to comment on why these conditions of pressure, temperature and use of catalyst are chosen. (3)

(f) Any unreacted sulphur dioxide and oxygen are recycled. Why is it important, economically and environmentally, to recycle these gases? (2)

(Total 16 marks)

UCLES 1997

36 Phosphorus(V) chloride dissociates at high temperatures according to the equation

$$PCl_5(g) \rightleftharpoons PCl_3(g) + Cl_2(g)$$

83.4 g of phosphorus(V) chloride are placed in a vessel of volume 9.23 dm³. At equilibrium at a certain temperature, 11.1 g of chlorine are produced at a total pressure of 250 kPa. Use these data, where relevant, to answer the questions that follow.

(a) Calculate the number of moles of each of the gases (Cl_2, PCl_3, PCl_5) in the vessel at equilibrium. (3)

(b) (i) Write an expression for the equilibrium constant, K_c, for the above equilibrium.

(ii) Calculate the value of the equilibrium constant, K_c, and state its units. (4)

(c) (i) Write an expression for the equilibrium constant, K_p, for the above equilibrium.

(ii) Calculate the mole fraction of chlorine present in the equilibrium mixture.

(iii) Calculate the partial pressure of PCl_5 present in the equilibrium mixture.

(iv) Calculate the value of the equilibrium constant, K_p, and state its units. (7)

(Total 14 marks)
NEAB 1998

$$[A_r(P) = 31, A_r(Cl) = 35.5.]$$

37 (a) Write an expression for the ionic product of water, K_w (1)

(b) K_w varies with temperature as indicated by the data below.

Temperature /K	298	373
K_w/mol^2 dm^{-6}	1.00×10^{-14}	51.3×10^{-14}

(i) Calculate the hydrogen ion concentration in water at 373 K and hence its pH.

(ii) Explain why water at 373 K is neutral.

(iii) Write an equation for the dissociation of water and explain why the values of K_w given above indicate that this dissociation is an endothermic process. (6)

(c) Use the value of K_w at 298 K to calculate the pH of a solution formed when 5.0 cm^3 of 1.0 M NaOH are added to 995 cm^3 of water. (3)

(Total 10 marks)
NEAB 1997

38 (a) The dissociation of ethanoic acid in aqueous solution is represented by the equation:

$$CH_3COOH(aq) \rightleftharpoons H^+(aq) + CH_3COO^-(aq)$$

(i) Write a K_a expression for this dissociation. (1)

(ii) Given that $K_a(CH_3COOH) = 1.70 \times 10^{-5}$ mol dm^{-3} at 298 K, calculate the pK_a value. (2)

(iii) Calculate the pH of an aqueous solution of ethanoic acid of concentration 0.10 mol dm^{-3} at 298 K. (4)

(iv) How is pH measured accurately in the laboratory? (1)

(b) When a titration is carried out between aqueous solutions of ethanoic acid and sodium hydroxide, the reaction that occurs is represented by the equation:

$$CH_3COOH(aq) + NaOH(aq) \rightarrow CH_3COONa(aq) + H_2O(l)$$

(i) State how ethanoic acid acts as a Brønsted-Lowry acid in this reaction. (2)

(ii) Suggest the name of a suitable indicator for use in the titration. (1)

(c) A mixture of aqueous solutions of ethanoic acid and sodium ethanoate acts as a buffer solution.

(i) State and explain what happens to the pH of this buffer solution when a small amount of hydrochloric acid is added to it. (4)

(ii) Give **one** example of the biological importance of buffer solutions. (1)

(Total 16 marks)
AEB 1997

39 Propanoic acid occurs naturally in Swiss cheese in a concentration that can be as high as 1%. Its sodium and calcium salts are food additives used in processed cheeses to retard the formation of moulds.

(a) Propanoic acid is a weak acid with an acid dissociation constant of 1.22×10^{-5} mol dm^{-3}.

(i) Write an equation for the ionisation of propanoic acid. (2)

(ii) Calculate pK_a for propanoic acid. (2)

(b) Sodium propanoate is made by the reaction of sodium hydroxide with propanoic acid.

(i) Write an equation for the reaction. (1)

(ii) If the reaction were carried out by titration using 0.1 M solutions of the hydroxide and the acid, state the name of a suitable indicator. (1)

(c) A mixture of sodium propanoate and propanoic acid acts as a buffer solution.

(i) What is meant by the term **buffer solution**? (2)

(ii) Explain how the mixture of sodium propanoate and propanoic acid acts as a buffer solution. (Up to 2 marks may be obtained for the quality of language in this part) (5)

(iii) Calculate the pH of the solution formed by adding 15.0 cm^3 of 0.1 M sodium hydroxide to 30.0 cm^3 of 0.1 M propanoic acid. (3)

(iv) Name a buffer solution found in a biological system and explain its importance. (2)

(Total 18 marks)
NICCEA 1998

40 Carbon dioxide dissolves in water to form carbonic acid, H_2CO_3, and the ions, HCO_3^- and CO_3^{2-}. The diagram shows the relative percentages of these species in water at different pH values.

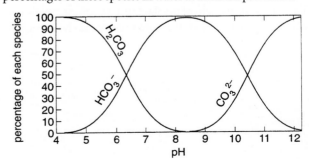

(a) Using the diagram, comment on the relative concentrations of H_2CO_3, HCO_3^- and CO_3^{2-} over this pH range. (3)

(b) Most natural waters have a pH value between 6 and 8. Which of the carbon-containing species in the diagram are dominant in natural water? (1)

(c) Write ionic equations linking
(i) H_2CO_3 and HCO_3^-,
(ii) HCO_3^- and CO_3^{2-}. (2)

(d) How is the solubility of carbon dioxide in water affected by
(i) an increase in temperature,
(ii) an increase in pressure? (2)

(e) Rain water of pH 5 ran over some calcium carbonate rock.
(i) Using the diagram, describe, with an equation, what you would expect to happen.
(ii) The product in (e)(i) was heated. Describe, with an equation, what you would expect to happen. (3)

(Total 11 marks)
UCLES 1997

41 The table gives data on some solubilities and solubility products. Some of the data entries are missing.

(a) Copy the table and fill in the missing data.
(A_r Ag, 107.9; A_r C, 12.0; A_r Ca, 40.0; A_r Cl, 35.5; A_r O, 16.0) (6)

Compound	Solubility product /$mol^2\ dm^{-6}$	Solubility /$mol\ dm^{-3}$	Solubility /$g\ dm^{-3}$
AgI	8.30×10^{-17}	9.11×10^{-9}	2.14×10^{-6}
AgCl	1.80×10^{-10}		
$CaCO_3$			0.0300

(b) Copy the table below, then state and explain what will happen to the solubility of the silver chloride if the salts mentioned are dissolved in a saturated solution of silver chloride. (Give the effect on the solubility as *increase*, *decrease*, or *no change*.)
$2 \times (2)$

Salt	Sodium nitrate	Sodium chloride
Effect		
Explanation		

(c) A student analysed the chloride ion content of a sample of seawater by adding an excess of silver nitrate, filtering off the precipitate of silver chloride, rinsing the precipitate with water, and finally drying and weighing the silver chloride. In an experiment the student obtained 7.253 g of silver chloride from 100 cm^3 of seawater.
(i) What was the molar concentration of chloride ions in the seawater sample? (3)
(ii) What is the maximum percentage error in the measured mass of the silver chloride precipitate

introduced by rinsing the precipitate with 100cm^3 of water? (Assume that the water is saturated with silver chloride after rinsing.) (3)

(d) Carbon dioxide is one of the main gases associated with global warming.
Suggest how the calcium compounds dissolved in the world's oceans can act to reduce the impact of rising levels of carbon dioxide in the atmosphere. (3)

(Total 19 marks)
OCSEB 1997

42 (a) Copy and complete the boxes below to show the electronic configuration of a V^{3+} ion using the convention shown for the V atom.

		3d					4s
V atom (Ar)		↑	↑	↑			↑↓
V^{3+} ion (Ar)							

(1)

(b) Vanadium is a transition element. Give two characteristic properties of such a transition element other than the ability to form coloured ions. (2)

(c) Ammonium vanadate(V) dissolves in sulphuric acid to give a yellow solution, the colour being due to the VO_2^+ ion.
(i) What is the oxidation number of vanadium in the VO_2^+ ion? (1)
(ii) Give the systematic name of the VO_2^+ ion. (1)

(d) Treatment of the yellow solution from (c) with zinc causes the colour to change to green then to blue, followed by green again and finally violet. Give the formulae of the ions responsible for each of these colours:
The first green colour
The second green colour
The violet colour (3)

(e) In the sequence of changes in (d), zinc acts as a reducing agent.
(i) State the meaning of the term **reducing agent**. (1)
(ii) Write a half equation showing how zinc acts as a reducing agent. (1)

(f) (i) Write the half equation for the conversion of VO_2^+ to VO^{2+} in acid solution. (1)
(ii) Hence write the equation for the reduction of VO_2^+ to VO^{2+} by zinc. (1)

(Total 12 marks)
London 1998

43 (a) Define the term *standard electrode potential* for a metal/metal ion system. (2)

(b) The apparatus shown below was used to measure the potential difference between copper and zinc electrodes:

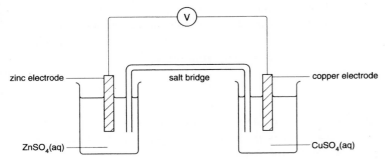

(i) Give the half equation for the reaction taking place at the zinc electrode. (1)

(ii) Describe the direction of the flow of electrons between the two metals. (1)

(iii) The e.m.f. of the cell was measured as 1.10 volts. Explain how this value can also be deduced from the data in a Data Book. (2)

(c) The e.m.f. of the cell was measured at various concentrations of the aqueous solutions:

Concentration of Zn^{2+}/mol dm^{-3}	Concentration of Cu^{2+}/mol dm^{-3}	e.m.f./volts
1.0	1.0	1.10
1.0	0.5	1.09
1.0	0.1	1.07
0.5	1.0	1.11
0.1	1.0	1.13

(i) What is the effect on the e.m.f. of diluting the copper(II) sulfate solution? (1)

(ii) Suggest why diluting the copper(II) sulfate solution has this effect. (3)

(iii) Suggest a value for the e.m.f. if the concentration of the copper(II) sulfate was 0.001 mol dm^{-3} and the concentration of the zinc(II) sulfate was 1.0 mol dm^{-3}. (1)

(iv) By referring to a Data Book suggest what the effect on the value of the e.m.f. will be of adding aqueous ammonia to the beaker containing the copper(II) sulfate solution. Give your reasoning. (3)

(d) (i) Outline the electrochemical processes involved in the rusting of iron. (2)

(ii) Explain how coating iron with zinc protects the iron from rusting. (2)

(Total 18 marks)

OCSEB 1998

44 (a) Explain what is meant by the term *standard electrode potential*. (2)

(b) The emf of a number of copper/copper(II) sulphate cells with different aqueous copper(II) ion concentrations was measured against the standard hydrogen electrode at 298 K and the results shown on the graph below.

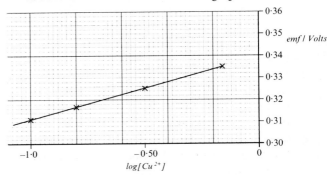

(i) From the graph above determine the value of the emf when log (Cu^{2+}) is zero. (1)

(ii) State the significance of the value when log (Cu^{2+}) is zero. (1)

(iii) State what can be deduced from the shape of the graph. (1)

(c) Some standard redox potentials are shown below.

System	Standard redox potential/V
$Cr^{2+} + 2e^- \rightleftharpoons Cr$	−0.91
$Cr^{3+} + 3e^- \rightleftharpoons Cr$	−0.74
$Cr_2O_7^{2-} + 14H^+ + 6e^- \rightleftharpoons 2Cr^{3+} + 7H_2O$	+1.33
$2H^+(aq) + 2e^- \rightleftharpoons H_2(g)$	0.00

(i) State which ion would be the *weakest oxidising agent*. (1)

(ii) State, giving a reason, whether metallic chromium would be expected to react with dilute hydrochloric acid. (2)

(iii) Given the equation,

$$Cr_2O_7^{2-} + 2OH^- \rightleftharpoons 2CrO_4^{2-} + H_2O$$

state the colour change when aqueous sodium hydroxide is added to potassium dichromate(VI). (1)

(d) Interpret the reaction scheme below.

(i) Identify **A, B, C, D, E** and **R** and write a balanced equation for each of the reactions **I** to **V**. (5)

(ii) Give one reason why such a scheme would not be applicable to sodium carbonate. (1)

(Total 15 marks)

WJEC 1997

45 The following is a list of some standard electrode potentials.

$$\textbf{A} \quad I_2 + 2e^- \rightleftharpoons 2I^- \quad +0.54V$$
$$\textbf{B} \quad Cu^+ + e^- \rightleftharpoons Cu \quad +0.52V$$
$$\textbf{C} \quad Fe^{2+} + 2e^- \rightleftharpoons Fe \quad -0.44V$$

(a) Define the term *standard electrode (redox) potential*. (2)

(b) Using the standard electrode potentials **A**, **B** and **C** above, suggest
  (i) one equation for a reaction which would go to completion.
  (ii) one equation for a reaction which might proceed but only to an equilibrium position. (4)

(c) (i) Draw a labelled diagram of the cell produced by connecting half-cells corresponding to **A** and **C**.
  (ii) Draw an arrow on the diagram to show the direction of electron flow in the external circuit.
  (iii) Calculate the standard cell potential (e.m.f.) of this cell.
  (iv) What would be the effect on the potential of this cell of decreasing the concentration of $Fe^{2+}(aq)$? Explain your answer. (9)

(Total 15 marks)

UCLES 1997

46 This question refers to the elements in the **third period** of the Periodic Table.

(a) Name the element which forms an amphoteric oxide. (1)

(b) Write the formula of the oxide which is covalent, insoluble in water and has a high melting point. (1)

(c) Name the metal which forms a basic oxide which is only slightly soluble in water. (1)

(d) Write an equation for the reaction of one of these elements with cold water in which a flammable gas is rapidly evolved. (2)

(Total 5 marks)

NICCEA 1998

47 (a) In the Periodic Table, there is a trend in the atomic radii of the elements in Period 3 and in Group 1.
  (i) Describe and explain the trend for Period 3.
  (ii) Predict, with reasons, the trend for Group 1. (8)

(b) Describe and explain the variations in the boiling points of the elements in Period 3 of the Periodic Table. Your answer should be in terms of the structure and bonding of the elements. (9)

(c) The table below provides data on the successive ionisation energies of carbon.

ionisation number	1st	2nd	3rd	4th	5th	6th
ionisation energy /kJ mol^{-1}	1090	2350	4610	6220	37800	47300

(i) Explain why each successive ionisation energy increases in value.
(ii) Write an equation to represent the 5th ionisation energy of carbon.
(iii) Explain how these data can be used to provide evidence for the electronic configuration of carbon. (8)

(Total 25 marks)

UCLES 1997

48 (a) The following table shows some physical properties of two *s*-block metals.

Metal	Hardness	Melting temperature /°C	Density /g cm^{-3}
caesium	very soft	28.7	1.9
barium	quite hard	714	3.51

(i) Suggest reasons for the differences in the physical properties of caesium and barium as shown in the table. The metals have the same crystal structure. (3)
(ii) Caesium gets its name from the blue colour it or its salts impart to a Bunsen flame. What process within the atom is responsible for the emission of this colour? (1)
(iii) If the light emitted from excited caesium atoms is passed through a spectrometer, what would you expect to see? (1)

(b) Sodium burns in excess oxygen to give a yellow solid, **Y**.
  (i) **Y** contains 58.97% sodium. Find its empirical formula. (2)
  (ii) The relative molecular mass of **Y** is 78. What is its molecular formula? (1)
  (iii) If **Y** is reacted with ice-cold dilute sulphuric acid, a solution of **Z** is obtained which will react with potassium manganate(VII) solution. Describe the experimental procedure you would use to determine the mole ratio in which **Z** and potassium manganate(VII) react together. (3)

(Total 11 marks)

London 1997

$$[A_r(Na) = 23.0, A_r(O) = 16.0.]$$

49 This question concerns some compounds of the Group II elements magnesium to barium.

(a) Copy and complete the following equations:
  (i) $CaCO_3(s) \xrightarrow{\text{heat}}$
  (ii) $Mg(NO_3)_2(s) \xrightarrow{\text{heat}}$ (2)

(b) (i) Which Group II nitrate decomposes at the lowest temperature?
  (ii) Explain your answer in (i). (3)

(c) Magnesium sulphate is more soluble in water than barium sulphate. Suggest an explanation for this. (2)

(d) Metallic calcium can be obtained by passing an electric current through molten calcium chloride.
  (i) Write ionic equations for reactions that occur at the cathode and at the anode.
  (ii) Calcium chloride also conducts electricity if it is dissolved in water. State the likely products at the cathode and at the anode. (4)

(Total 11 marks)

UCLES 1997

50 (a) Magnesium occurs naturally as the mineral *carnallite*, $KCl.MgCl_2.6H_2O$.
  (i) State what is **observed** and give a balanced equation for the reaction which occurs when a solution of carnallite is treated with sodium hydroxide solution. (2)
  (ii) State how to test for the presence of chloride ions in the carnallite solution, giving details of the reagents added, **observation** and an **ionic** equation for any precipitation reaction which may occur. (3)
(b) Both magnesium sulphate and barium sulphate occur naturally as minerals but only magnesium sulphate is soluble in water.
  (i) Explain, in terms of hydration and lattice enthalpies, the reason why only one of the compounds is soluble in water. (2)
  (ii) Name **two** features of magnesium sulphate which identify it as a **typical ionic compound** (1)
(c) State why magnesium is an essential element for plant growth. (1)

(Total 9 marks)

WJEC 1998

51 (a) State and explain the trend in the boiling points of the halogens chlorine, bromine and iodine. (3)
(b) State what is meant by the term *electronegativity*. Explain the trend in electronegativity for the halogens chlorine, bromine and iodine. (5)
(c) State and explain the trend in the reducing properties of the halide ions Cl^-, Br^- and I^-. Describe what you would observe when an aqueous solution containing chlorine is added to separate aqueous solutions containing bromide and iodide ions. Write equations for any reactions which occur. (8)

(Total 16 marks)

NEAB 1997

52 (a) Sodium hydroxide is manufactured using the Gibbs Diaphragm Cell with an electrolyte of purified saturated brine.

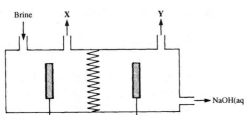

  (i) Copy the diagram and label the anode and the cathode. (1)
  (ii) Write equations for the reactions at the anode and the cathode. (2)
  (iii) What is the main impurity in the sodium hydroxide product? (1)
  (iv) The main impurities in the brine which must be removed before electrolysis are magnesium and calcium ions. Suggest a reason why they must be removed. (2)
(b) Electrolysis of brine under different conditions gives sodium chlorate(I); in this case the sodium hydroxide and chlorine are allowed to mix at room temperature.
  (i) Give the ionic equation for the reaction of chlorine with cold dilute aqueous sodium hydroxide. (1)
  (ii) If the solution of sodium chlorate(I) is heated, the chlorate(I) ion disproportionates. Write the ionic equation for the reaction, and use it to explain the meaning of disproportionation. (3)
(c) Give one use for
  (i) sodium hydroxide (1)
  (ii) chlorine (1)
  (iii) substance **Y** on the diagram. (1)
(d) Suggest how using a solution of sodium chlorate(I), or otherwise, you could distinguish between separate aqueous solutions of potassium bromide and potassium iodide. (4)

(Total 17 marks)

London 1997

53 (a) Concentrated sulphuric acid is added to separate unlabelled solid samples of sodium chloride, sodium bromide and sodium iodide.
  (i) Write an equation to represent the reaction between sulphuric acid and sodium chloride. (2)
  (ii) Describe **one** different observation you could make in each case that would enable you to identify the halide ion present in that sample. (3)
  (iii) Account for the different behaviours of the halide ions in their reactions with concentrated sulphuric acid. (3)
(b) Describe, giving details of the observations you could make in each case, how you would use aqueous solutions of ammonia and silver nitrate to confirm the identities of the halide ions present in aqueous solutions of the unlabelled samples in (a). (5)
(c) Aqueous silver nitrate is added to an aqueous solution of sodium fluoride. State and explain how the observation(s) you could make here would differ from those described in (b). (2)

(Total 15 marks)

AEB 1998

54 Water is treated with various chemicals before it is fit to drink. Typical processing uses aluminium sulphate, $Al_2(SO_4)_3$, and chlorine, Cl_2.

(a) What is the purpose of using
   (i)   aluminium sulphate,
   (ii) chlorine? (2)

(b) The amount of aluminium sulphate that is added must be carefully controlled. The European Union recommends a maximum level of dissolved aluminium in drinking water of 6×10^{-7} mol dm^{-3}.
   (i)   Why is it important that not too much aluminium sulphate is added to the water?
   (ii) Calculate the maximum mass of aluminium sulphate that is permitted in 1000 tonnes of water. (Assume that 1000 dm^3 of water has a mass of 1 tonne.) (4)

(c) One problem with the use of chlorine is that it may react with traces of methane in the water. Suggest an equation for this reaction. (1)

(Total 7 marks)

UCLES 1997

$$[A_r(Al) = 27, A_r(S) = 32, A_r(O) = 16.]$$

55 This question is about the elements of Group IV of the Periodic Table.

(a) (i)   Give the oxidation numbers of carbon in carbon monoxide and in carbon dioxide. (2)
   (ii) State the type of reaction involved in the conversion of carbon monoxide to carbon dioxide. (1)
   (iii) How would you carry out this conversion in the laboratory? (1)

(b) (i)   Write the electronic structure of a carbon atom. (1)
   (ii) Draw the shape of a molecule of carbon dioxide. (1)
   (iii) Draw the shape of a molecule of water. (1)
   (iv) Explain why the shapes of these two molecules are different even though both compounds have a formula of the type AB_2. (3)

(c) Give one way in which carbon dioxide differs from silicon dioxide, SiO_2, either in its chemical reactions or its physical properties. (2)

(d) Give one way in which lead dioxide, PbO_2, behaves differently from both CO_2 and SiO_2. (2)

(e) (i)   Outline the general trend in chemical character of the elements on descending Group IV from carbon to lead. (2)
   (ii) Explain, with the help of diagrams of their structures, why carbon and lead have such different characters. (4)

(Total 20 marks)

OCSEB 1998

56 (a) Explain how a bond is formed between a metal ion and a ligand in a complex ion. (2)

(b) Consider the complex compound $[Co(NH_2CH_2CH_2NH_2)_3]Cl_3$
   (i)   Name the ligand bonded to cobalt.
   (ii) What name is given to this type of ligand?

   (iii) What is the oxidation state of cobalt in this compound?
   (iv) What is the co-ordination number of cobalt in this compound?
   (v) Deduce the likely shape around cobalt in this complex. (5)

(c) From your knowledge of the reaction of cobalt(II) ions with ammonia, outline how you would prepare a solution of the complex $[Co(NH_2CH_2CH_2NH_2)_3]Cl_3$ starting from solid cobalt(II) chloride. (3)

(Total 10 marks)

NEAB 1998

57 Examine the reaction scheme below.

(a) (i)   Give the formula of each of the species in the products lettered **A** to **D**.
   (ii) Describe what you would see and explain what would happen if solution **B** were diluted with an excess of water. (6)

(b) When aqueous sodium hydroxide is added to a solution of copper(II) sulphate, a blue precipitate is formed. If, however, an excess of the ligand EDTA is first added to a solution of copper(II) sulphate and then aqueous sodium hydroxide added, no precipitate results.
   (i)   Give the formula of the blue precipitate.
   (ii) What type of ligand is EDTA?
   (iii) Suggest why no precipitate appears if EDTA is added before the aqueous sodium hydroxide. (4)

(Total 10 marks)

NEAB 1997

58 (a) Give the electron structure of the vanadium atom and the V^{2+} ion:

(2)

(b) (i)   Suggest why the hydrated ion $[V(H_2O)_6]^{2+}$ is coloured.
   (ii) Name the types of bonding within ions of this type. (3)

(c) Ammonium vanadate, NH_4VO_3, dissolves in aqueous sodium hydroxide with the evolution of a colourless gas. The solution becomes yellow after acidification.

The gas has a pungent odour and produces a pale blue precipitate with copper(II) sulphate solution. The precipitate dissolves as more gas is passed in, to give a deep blue solution.

   (i) Write an ionic equation for the reaction of the cation in NH_4VO_3 with alkali.
   (ii) Name the pale blue precipitate.
   (iii) Give the formula of the ion responsible for the colour of the deep blue solution.
   (iv) Ammonium vanadate, on treatment with sulphuric acid, gives a yellow colour due to the $[VO_2]^+$ ion. Addition of zinc to the solution causes the solution colour to change to blue, then green, then violet. Give the oxidation number of vanadium in the vanadium-containing ions in each coloured solution. (5)

(d) The industrial production of sulphur trioxide from sulphur dioxide and oxygen is catalysed by vanadium(V) oxide. It has been proposed that the first stage of the reaction is

$$SO_2 + V_2O_5 \rightarrow SO_3 + 2VO_2$$

Write an equation for the second stage, thus showing the behaviour of vanadium(V) oxide as a catalyst. (1)

(e) Give the systematic name for each of these ions:
   (i) $[VO_2]^+$
   (ii) $[Cr(NH_3)_4Cl_2]^+$ (2)
(f) Draw and describe the shape of the ion in (e)(ii). (2)

(Total 15 marks)
London 1996

59 Keto-ethers contain both the 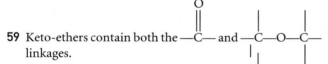 linkages.

The proton NMR spectrum of keto-ether **A**, $C_6H_{12}O_3$, is shown below.

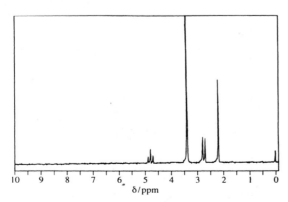

The measured integration trace gives the ratio 0.6 to 3.6 to 1.2 to 1.8 for the peaks at δ 4.8, 3.4, 2.7 and 2.2, respectively.

Proton chemical shift data	
Type of proton	δ/ppm
RCH_3	0.7–1.2
R_2CH_2	1.2–1.4
R_3CH	1.4–1.6
$RCOCH_3$	2.1–2.6
$RCOCH_3$	3.3–3.9
$RCH(OR)_2$	4.4–5.0

Refer to the spectrum and to the information given in the table in order to answer the following questions.
(a) How many different types of proton are present in compound **A**? (1)
(b) What is the actual ratio of the numbers of each type of proton? (1)
(c) What type of proton is responsible for the peak at δ 2.2? (1)
(d) What can be deduced from the splitting of the peaks at δ 4.8 and δ 2.7? (2)
(e) Suggest the structure of compound **A**. (2)

(Total 7 marks)
NEAB 1997

60 A mass spectrum of CH_4 is shown below. Naturally-occurring carbon contains 1% of ^{13}C. The deuterium (^2H) content of hydrogen is negligible.

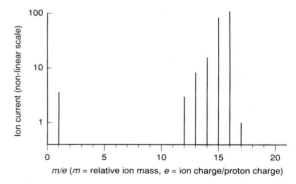

(a) What part of the mass spectrometer causes the particles being analysed to move in a curve? (1)
(b) Make a copy of the mass spectrum. Mark on it with a cross the feature which demonstrates that the relative molecular mass of $^{12}CH_4$ is 16. (1)
(c) State which feature of the mass spectrum demonstrates that carbon has a naturally occurring isotope ^{13}C. (1)
(d) Give the formula of **one** ion which contributes to the line at $m/e = 16$. (1)
(e) Suggest why other lines also appear in the mass spectrum, but none between 1 and 12. (2)

(Total 6 marks)
OCSEB 1997

61 Two types of reaction in organic chemistry are electrophilic addition and nucleophilic substitution.

(a) Define the terms:
  (i)   Nucleophile; (1)
  (ii) Electrophile; (1)
  (iii) Substitution; (1)
  (iv) Addition. (1)

(b) Give the mechanism of a reaction of your choice which proceeds by electrophilic addition. (3)

(c) (i)   Describe, briefly, a reaction you could carry out in the laboratory which occurs by nucleophilic substitution. (1)
  (ii) Write a mechanism for this reaction. (3)
  (iii) What type of nucleophilic substitution is occurring in this reaction? Justify your answer. (1)

(d) When a solution of bromine is shaken with cyclohexene, ⬡, the bromine is decolourised. However, when bromine is added to benzene there is no decolourisation.
  (i)   Write an equation for the reaction of cyclohexene with bromine. (1)
  (ii) Explain, in terms of the bonding why no reaction occurs when a solution of bromine is shaken with benzene. (2)

(Total 15 marks)
London 1998

62 Acid-catalysed dehydration of an optically active compound **A**, $C_4H_{10}O$, yields two isomeric products **B** and **C**, C_4H_8. Compound **B** exists in stereoisomeric forms. The reaction between **C** and hydrogen bromide produces compound **D**, C_4H_9Br, which yields **A** on hydrolysis. Oxidation of **A** gives compound **E**, C_4H_8O.

Compound **A** has a broad absorption at 3350 cm^{-1} in the infra-red, compound **E** has a strong absorption band at 1715 cm^{-1} and compounds **B** and **C** each have significant absorption bands close to 1650 cm^{-1}.

Use the data in the table and the information provided in the question to deduce structures for compounds **A, B, C, D** and **E**. Name each of these compounds.

Infra-red absorption data

Bond	Wavenumber/cm^{-1}
C—H	2850 – 3300
C—C	750 – 1100
C=C	1620 – 1680
C=O	1680 – 1750
C—O	1000 – 1300
O—H (alcohols)	3230 – 3550
O—H (acids)	2500 – 3000

(Total 15 marks)
NEAB 1997

63 Spices are often responsible for the flavour in many foods. Two types of compounds present in the spice ginger are gingerols and shogoals. An example of each is shown below.

Substance **E** (a gingerol)

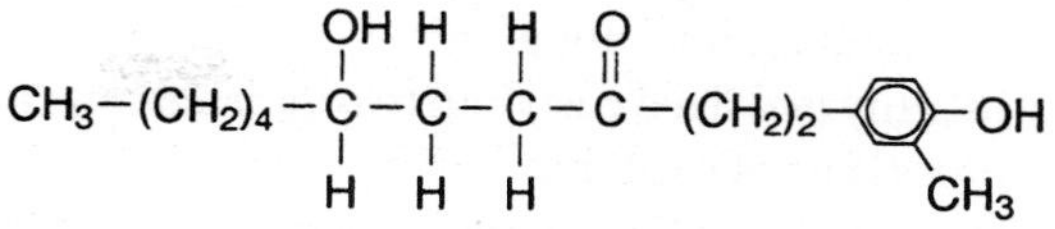

$CH_3-(CH_2)_4-\overset{OH}{\underset{H}{C}}-\overset{H}{\underset{H}{C}}-\overset{H}{\underset{H}{C}}-\overset{O}{C}-(CH_2)_2-⬡-OH$ (with CH_3)

Substance **F** (a shogoal)

$CH_3-(CH_2)_4-\overset{H}{\underset{H}{C}}=\overset{H}{\underset{H}{C}}-\overset{H}{\underset{H}{C}}-\overset{O}{C}-(CH_2)_2-⬡-OH$ (with CH_3)

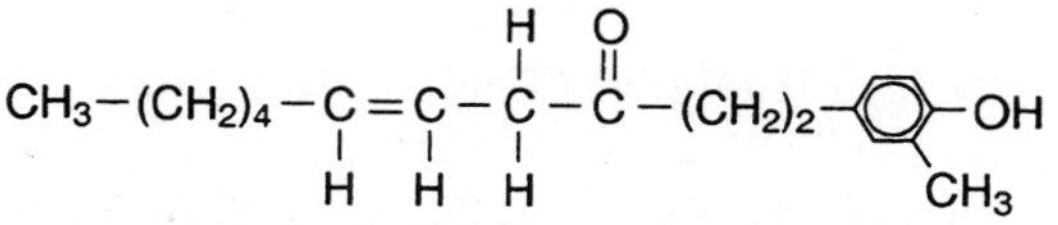

(a) Substances **E** and **F** exist as stereoisomers.
  (i)   Copy the two structures and on each draw a circle around the feature that gives rise to the stereoisomerism.
  (ii) State the type of stereoisomerism associated with **E** and with **F**. (4)

(b) (i)   Name **two** functional groups common to both **E** and **F**.
  (ii) Name **one** functional group present in **E** but not in **F**.
  (iii) Name **one** functional group present in **F** but not in **E**. (4)

(c) Choose a suitable reagent for each of the following situations and state what you would expect to see.
  (i)   A reagent that reacts **both** with **E** and with **F**.
  (ii) A reagent that reacts with **E** but **not** with **F**.
  (iii) A reagent that reacts with **F** but **not** with **E**. (6)

(d) Suggest how **E** could be converted into **F** in the laboratory. (2)

(Total 16 marks)
UCLES 1997

64 (a) Naphtha is one of the fractions obtained by the primary distillation of crude oil. Much of the naphtha produced undergoes further processing, including cracking.
  (i)   Explain the term *fractions* and outline the principles of the primary distillation process of crude oil. State why the process is necessary. (5)
  (ii) Give a brief outline of the principles of cracking and explain why such processes are of economic importance. Illustrate your answer by writing an equation to represent the process, using the alkane C_6H_{14} as an example. (6)
  (iii) Cracking involves free radical chain reactions. What is a free radical? Write an equation for a possible initiation step in the cracking of C_6H_{14}. (3)

(b) Another process using naphtha as a feedstock is catalytic reforming. Many reactions occur including the dehydrogenation of cycloalkanes and the simultaneous cyclisation and dehydrogenation (dehydrocyclisation) of alkanes to give aromatic hydrocarbons (arenes) with higher octane numbers than naphtha.

(i) Write the equation to show the dehydrogenation of cyclohexane, C_6H_{12}. (1)

(ii) Write the equation to show the dehydrocyclisation of hexane, C_6H_{14}. (1)

(iii) Suggest a reason for the economic importance of catalytic reforming. (1)

(c) Identify **two** fixed costs and **two** variable costs in the processing of naphtha. (4)

(d) Steam reforming is another process associated with the petrochemical industry; naphtha and natural gas are used as the major feedstocks. The key reaction using natural gas is:

$$CH_4(g) + H_2O(g) \rightleftharpoons CO(g) + 3H_2(g) \quad \Delta H = +210 kJ \, mol^{-1}$$

The gaseous mixture produced is called synthesis gas.

(i) Suggest, with reasons, the qualitative conditions of temperature and pressure that might be used to ensure a good yield of synthesis gas. (4)

(ii) Comment on any additional factors which may have to be considered when deciding upon the actual conditions used. (3)

(Total 28 marks)

AEB 1997

65 (a) The enthalpy of hydrogenation of cyclohexene, C_6H_{10} is -120 kJ mol^{-1} and that of benzene, C_6H_6, is -208 kJ mol^{-1}. Explain why the value for benzene is not three times that of cyclohexene and account for the difference. Hence discuss why bromine in aqueous solution will react with cyclohexene to form 1,2-dibromocyclohexane but bromine does not react with benzene under these conditions. (8)

(b) Substitution reactions in organic chemistry can involve three different mechanisms. Name these three types of mechanism and, using the compounds methane, bromomethane and benzene, write equations to illustrate each type of mechanism. For your examples with methane and bromomethane, state the reaction conditions and outline the mechanisms involved. (17)

(c) Give the reagent and conditions needed to produce hex-1-ene from hexan-1-ol. The infra-red spectrum of one of these two compounds is shown below. Use the spectrum and the table of data to identify this compound. State **two** regions, other than the fingerprint region, where the infra-red spectrum of the other compound would be different. (5)

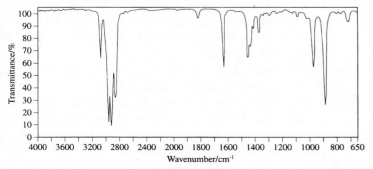

Table of infra-red absorption data

Bond	Wavenumber/cm^{-1}
C—H	2850 – 3300
C—C	750 – 1100
C=C	1620 – 1680
C=O	1680 – 1750
C—O	1000 – 1300
O—H (alcohols)	3230 – 3550
O—H (acids)	2500 – 3000

(Total 30 marks)

NEAB 1997

66 (a) Alkane **C** has a relative molecular mass of 170 and occurs in the kerosene fraction obtained by the fractional distillation of petroleum.

(i) Write the general formula for the homologous series of alkanes.

(ii) Deduce the molecular formula of alkane **C**.

(iii) Give one use for the kerosene fraction.

(iv) Name one fraction which is obtained higher up the fractionating column than kerosene and explain why it is obtained higher up. (6)

(b) Three hydrocarbons, **D**, **E** and **F**, all have the molecular formula C_6H_{12}.

D decolourises an aqueous solution of bromine and shows geometric isomerism.

E also decolourises an aqueous solution of bromine but does not show geometric isomerism .

F does not decolourise an aqueous solution of bromine. Draw one possible structure each for **D**, **E** and **F**. (3)

(Total 9 marks)

NEAB 1997

67 (a) Using mechanisms where appropriate, describe the reactions of alkanes and alkenes with bromine. (12)

(b) Reduction of a branched alkene, **A**, with hydrogen formed a compound **B**. On analysis, **B** had the composition by mass of C, 82.8%; H, 17.2% ($M_r = 58$). Reaction of **A** with steam in the presence of a catalyst produced a mixture of two isomers **C** and **D**. Identify possible structures for compounds **A**–**D**. Your answer should include displayed formulae for each structure and equations for each reaction. (9)

(c) Compound **E**, C_4H_6, is a **diene**. Predict a likely structure for **E** and the saturated products that could be formed by reaction of **E** with steam in the presence of a catalyst. (4)

(Total 25 marks)

UCLES 1997

68 (a) Using appropriate examples from carbon chemistry, describe and explain what is meant by structural isomerism and *cis–trans* isomerism. (6)

(b) The chemical properties of but-2-ene are similar to those of ethene.

(i) Use this information to predict the organic products in, and the equations for, the reactions of but-2-ene with bromine, hydrogen bromide and steam.

(ii) How, and under what conditions, would you expect but-2-ene to react with potassium manganate(VII)?

(iii) Draw a section of the polymer formed from but-2-ene by showing **two** repeat units. (12)

(c) Explain the following.

(i) Methane reacts with chlorine only in ultraviolet light.

(ii) Trichloromethane is polar whereas tetrachloromethane is non-polar. (7)

(Total 25 marks)

UCLES 1997

69 (a) (i) Give **one** example of an addition polymer, and **one** example of a condensation polymer. (2)

(ii) Give the structure of an example of each of these two types of polymer.

addition polymer (2)

condensation polymer (2)

(iii) Explain why the word *addition* is used in describing a polymer as an addition polymer. (1)

(b) (i) Give two uses for each of these two types of polymer.

addition polymer: (2)

condensation polymer: (2)

(ii) Protective clothing is often required for use in an environment involving exposure to aqueous acids. Which of these two types of polymer would you select as the basis of material for the protective clothing? Give your reasons. (2)

(Total 13 marks)

OCSEB 1997

70 Write an essay on delocalisation in organic chemistry. You should cover the concept of delocalisation as well as its chemical/structural consequences. You should refer, in particular, to benzene, carboxylate anions (i.e. the anions derived from carboxylic acids) and amides.

(Total 20 marks)

OCSEB 1997

71 A hydrocarbon is known to contain a benzene ring. It has a relative molecular mass of 106 and has the following composition by mass: C, 90.56%; H, 9.44%.

(a) (i) Use the data above to show that the empirical formula is C_4H_5.

(ii) Deduce the molecular formula.

(iii) Draw structures for all possible isomers of this hydrocarbon that contain a benzene ring. (7)

(b) In the presence of a catalyst (such as aluminium chloride), one of these isomers **A** reacts with chlorine to give only **one** monochloro-product **B**.

(i) Deduce which of the isomers in (a)(iii) is **A**.

(ii) Draw the structure of **B**. (2)

(c) In the presence of ultraviolet light, **A** reacted differently with chlorine to give another monochloro-product **C** as well as several polychloro-products.

(i) State the type of reaction involved in the conversion of **A** into **C**.

(ii) Suggest why several other polychloro-products were formed in this reaction. (2)

(Total 11 marks)

UCLES 1997

72 (a) Draw structures of the organic product(s) when

(i) ethene reacts with bromine at room temperature, (1)

(ii) benzene reacts with bromine at room temperature, in the absence of light, but in the presence of iron filings. (1)

(b) (i) The benzene molecule is said to have π-**electron delocalisation**. Explain what is meant by this term. (2)

(ii) Show the mechanism of the nitration of benzene. (2)

(c) Five derivatives of benzene, compounds **A** to **E**, are all liquids above 35°C and have the following structures:

(i) Describe a simple test tube experiment to distinguish between **A** and **B** stating the reagents, conditions and the observation which enables the distinction to be made. (2)

(ii) State which **one** of the five compounds, on mild oxidation, yields compound **A**. (1)

(iii) State which **one** of the five compounds, on oxidation, yields compound **B**. (1)

(iv) Choose **one** of the five compounds which exhibits optical isomerism. (1)

(v) Choose, from the five compounds, a pair of structural isomers. (1)

(Total 12 marks)

WJEC 1998

73 (a) Methylbenzene (*toluene*), $C_6H_5CH_3$, can be nitrated in a similar way to benzene. Give the reagents for the nitration of methylbenzene and name the type of mechanism. (3)

(b) Under extreme conditions, however, further nitration occurs, forming 2-methyl-1,3,5-trinitrobenzene (*TNT*).

Write the molecular formula of TNT and, given that it has a relative molecular mass of 227, calculate the percentage by mass of hydrogen in the compound. (3)

(Total 6 marks)

NEAB 1997

74 (a) (i) Explain why 2-bromopropane undergoes nucleophilic substitution whereas bromobenzene does not.

(ii) Explain why 1-bromopropane and 2-bromopropane undergo nucleophilic substitution at different rates. (8)

(b) 2-Bromopropane can react with sodium hydroxide to form two different organic products **W** and **X** depending upon the reaction conditions used.

W rapidly decolourises bromine and has a much lower boiling point than **X**.

(i) Draw the displayed formulae of **W** and **X**. Using equations where appropriate, show clearly your reasoning in the identification of each compound.

(ii) Name the types of reaction involved in the formation of **W** and **X**. (8)

(c) Suggest how 1-bromopropane could be converted into

(i) 1-aminopropane

(ii) butanoic acid. (9)

(Total 25 marks)

UCLES 1997

75 (a) (i) One product of the reaction of ethene with aqueous bromine in the presence of sodium chloride is 1,2-dibromoethane. Give the mechanism for the formation of 1,2-dibromoethane .

(ii) Give the structural formula of ONE other compound formed.

(iii) Explain mechanistically why the following compound is not formed. (6)

(b) (i) Give a reagent, and the conditions under which it might be used, to convert 1,2-dibromoethane to ethane-1,2-diol.

(ii) How would the rate of this reaction differ if 1,2-diiodoethane was used instead of 1,2-dibromoethane at the same temperature? (4)

(Total 10 marks)

London 1996

76 (a) (i) 1-Bromobutane reacts with aqueous sodium hydroxide according to the equation

$$CH_3CH_2CH_2CH_2Br + OH^- \rightarrow CH_3CH_2CH_2CH_2OH + Br^-$$

Give the mechanism for the reaction. (2)

(ii) Compare the reactivities of chlorobenzene and 1-bromobutane when refluxed with hot aqueous sodium hydroxide. Explain your answer. (1)

(b) $CFCl_3$ is a useful non-toxic compound.

(i) Give **two** uses to which $CFCl_3$ and related compounds have been put. (1)

(ii) Explain briefly why the use of compounds such as $CFCl_3$ has been restricted in recent years. (2)

(c) A number of organo-chlorine compounds have been used as pesticides. One example is dieldrin.

(i) In the molecule of dieldrin there are six chlorine

Dieldrin

atoms. State how many of the six are attached to unsaturated carbon atoms. (1)

(ii) Give **two** properties of organo-chlorine pesticides which have been responsible for the restriction of their use in recent years. (2)

(Total 9 marks)

WJEC 1998

77 Write an essay on nucleophilic substitution reactions at saturated carbon (i.e. tetrahedral carbon), illustrating your answer by reference to the hydrolysis of bromoethane with water and with aqueous sodium hydroxide.

Make clear the meanings of the terms *nucleophile*, *electrophile*, *leaving group*, as well as the roles of these species in the reaction, and explain the use of the terms *strong* and *weak* as applied to nucleophiles and electrophiles. You should cover the structural requirements for a species to be a nucleophile or a leaving group, and you should also consider a halogenated alkane which is resistant to nucleophilic substitution.

(Total 20 marks)

OCSEB 1997

78 Butan-1-ol can be oxidised by acidified potassium dichromate(VI) using two different methods.

(a) In the first method, butan-1-ol is added dropwise to acidified potassium dichromate(VI) and the product is distilled off immediately.

   (i) Using the symbol [O] for the oxidising agent, write an equation for this oxidation of butan-1-ol, showing clearly the structure of the product. State what colour change you would observe.

   (ii) Butan-1-ol and butan-2-ol give different products on oxidation by this first method. By stating a reagent and the observation with each compound, give a simple test to distinguish between these two oxidation products. (6)

(b) In a second method, the mixture of butan-1-ol and acidified potassium dichromate(VI) is heated under reflux. Identify the product which is obtained by this reaction. (1)

(c) Give the structures and names of two branched chain alcohols which are both isomers of butan-1-ol. Only isomer 1 is oxidised when warmed with acidified potassium dichromate(VI). (4)

(Total 11 marks)
NEAB 1997

79 Alcohols and ethers have the same general formula, $C_nH_{2n+2}O$. Ethers contain a C—O—C linkage as, for example, in methoxypropane, H_3C—O—$CH_2CH_2CH_3$.

(a) (i) Give the number of peaks in the low resolution proton n.m.r. spectrum of methoxypropane and the ratio of the areas under the peaks in the spectrum.

   (ii) Draw the structure of an ether which is an isomer of methoxypropane and which produces only 2 peaks in its low resolution proton n.m.r. spectrum. (3)

(b) (i) Alcohols can be prepared from haloalkanes by reaction with hydroxide ions. Name and outline the mechanism for the preparation of propan-1-ol from bromopropane .

   (ii) Ethers can be prepared from haloalkanes by a similar reaction. Copy and complete the following equation which shows the formation of methoxypropane.

..................+ $BrCH_2CH_2CH_3 \rightarrow CH_3OCH_2CH_2CH_3 + Br^-$
(5)

(c) Ethers are not oxidised by acidified potassium dichromate(VI). Name the type of alcohol which is also not oxidised by acidified potassium dichromate(VI) and draw the structure of the alcohol of this type which is an isomer of methoxypropane. (2)

(d) Write an equation for the complete combustion of methoxypropane in an excess of oxygen. (1)

(e) Ethers and alcohols can be distinguished by studying their infra-red spectra. Using the table of data given below, state where, other than in the fingerprint region, their infra-red spectra will be different and explain what causes this difference.

Table of infra-red absorption data

Bond	Wavenumber/cm^{-1}
C—H	$2850 - 3300$
C—C	$750 - 1100$
C=C	$1620 - 1680$
C=O	$1680 - 1750$
C—O	$1000 - 1300$
O—H in alcohols	$3230 - 3550$
O—H in acids	$2500 - 3000$

(2)

(Total 13 marks)
NEAB 1997

80 The terms *primary*, *secondary* and *tertiary* are used in classifying both alcohols and amines, although they are used in different ways for each of these two classes of organic compound.

(a) (i) Outline concisely the meanings of these terms as applied to alcohols. (2)

   (ii) Explain how the meanings of these terms differ when they are applied to amines. (2)

(b) For each of the following types of alcohol state clearly **one** chemical property which distinguishes it from the other two types.

   (i) primary   (ii) secondary   (iii) tertiary (4)

(Total 8 marks)
OCSEB 1997

81 An alcohol has a relative molecular mass of 74.0 and has the following composition by mass: C, 64.9%; H, 13.5%; O, 21.6%.

(a) (i) Calculate the empirical formula of the alcohol and show that its molecular formula is the same as the empirical formula.

   (ii) Draw the displayed formula of the **four** possible isomers of this alcohol. (7)

(b) Compound **A**, one of these isomers, can be oxidised to form a neutral compound **B** which does **not** react with Fehling's (Benedict's) solution.

   (i) Identify compound **B**.

   (ii) Deduce which of the four alcohols in **(a)(ii)** is compound **A**. (2)

(Total 9 marks)
UCLES 1997

82 Propenal, CH_2=CHCHO, is one of the materials that gives crispy bacon its sharp odour. In the following question assume that the carbon-carbon double bond and the aldehyde group in propenal behave independently.

(a) Give the structural formulae of the compounds formed when propenal reacts with:

   (i) hydrogen bromide (2)

(ii) hydrogen cyanide (1)
(iii) 2,4-dinitrophenylhydrazine. (2)
(b) (i) Give the mechanism for the reaction between hydrogen cyanide and the aldehyde group. You may represent the aldehyde group as

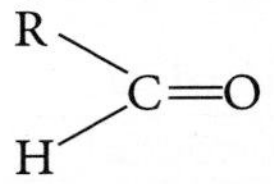

(3)
(ii) The reaction in (i) occurs best in slightly acidic conditions. It is slower if the pH is high or low. Suggest reasons why this is so. (3)
(c) Explain why lithium tetrahydridoaluminate(III) (lithium aluminium hydride), $LiAlH_4$, reacts only with the $C=O$ bond and not with the $C=C$ bond, even though these bonds have the same electronic structure. (2)
(d) Suggest reactions, giving equations and conditions, which would convert propenal into a compound which would react with iodine in the presence of sodium hydroxide solution. (4)

(Total 17 marks)
London 1998

83 Carvone,

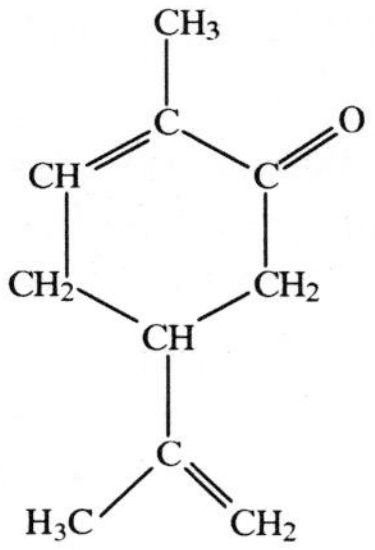

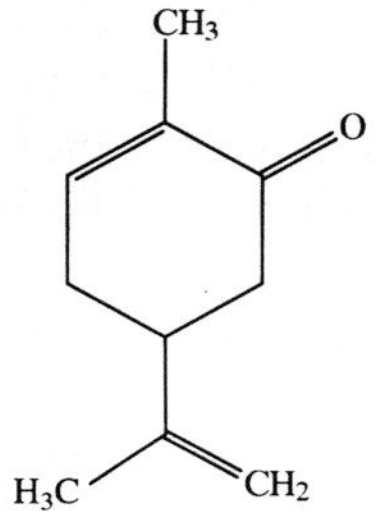

is the main flavouring material in spearmint oil.
(a) Draw the structure of the product obtained when carvone reacts with
(i) 2,4-dinitrophenylhydrazine (2)
(ii) bromine. (2)
(b) Explain why carvone does not react with ammoniacal silver nitrate. (1)
(c) Carvone is chiral and shows optical activity.
(i) Explain the meaning of **chiral**. (1)
(ii) Copy the structure of carvone and circle the feature of the molecule which makes it chiral. (1)
(iii) What is optical activity? (2)
(d) Suggest a synthetic method, including reference to reagents and conditions, by which carvone could be converted to,

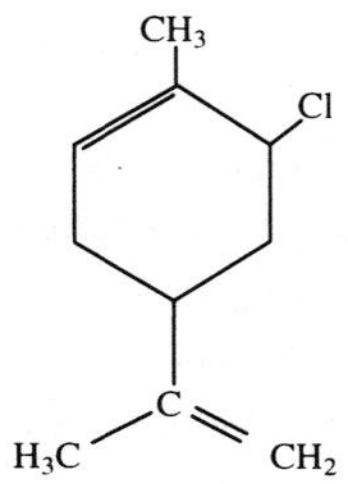

(5)
(Total 14 marks)
London 1997

84 (a) Outline the reaction of propanone with the following reagents. Give the equation for the reaction, the conditions, and the name of the organic product.
(i) Hydrogen cyanide (3)
(ii) Sodium tetrahydridoborate(III) (sodium borohydride). (You may represent $NaBH_4$ as [H].)(3)
(b) (i) Give the mechanism for the reaction in (a)(i). (3)
(ii) What type of mechanism is this? (1)
(iii) What feature of the carbonyl group makes this type of mechanism possible? Explain how this feature arises. (2)
(iv) Explain briefly, by reference to its structure, why ethene would not react with HCN in a similar way. (1)

(Total 13 marks)
London 1998

85 Benzaldehyde is known as "oil of bitter almonds" since it is found in the glucoside, amygdalin, which occurs in bitter almonds.

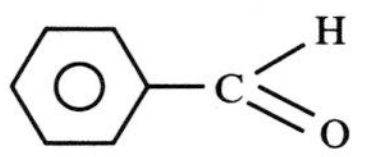

benzaldehyde

Amygdalin may be hydrolysed by dilute acids, or by the enzyme emulsin, to benzaldehyde, glucose and hydrogen cyanide.

$$C_{20}H_{27}O_{11}N + 2H_2O \rightarrow C_6H_5CHO + HCN + 2C_6H_{12}O_6$$

(a) Explain the term **enzyme**. (2)
(b) Glucose has the following structure:

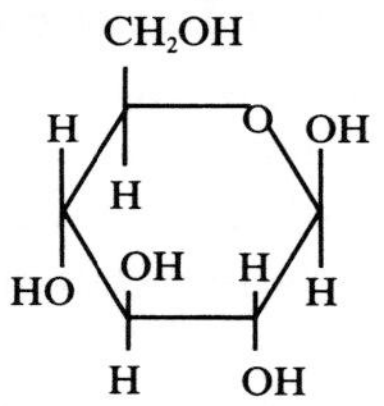

(i) How many primary, secondary and tertiary alcohol groups are there in glucose? (2)
(ii) Suggest why glucose is extremely soluble in water. (2)
(c) Benzaldehyde reacts with hydrogen cyanide to form a cyanohydrin called mandelonitrile.

$$C_6H_5CHO + HCN \rightarrow C_6H_5CHOHCN$$

The reaction is speeded up by the presence of sodium hydroxide which forms cyanide ions.
(i) What type of reagent is the cyanide ion? (1)
(ii) Write an equation for the reaction of hydrogen cyanide with sodium hydroxide. (2)
(iii) In practice hydrogen cyanide is generated by the reaction of sodium cyanide with hydrochloric acid. Addition of hydrogen cyanide does not occur in the presence of excess hydrochloric acid. Suggest an explanation. (2)

(d) The mechanism for the formation of the cyanohydrin is shown below.

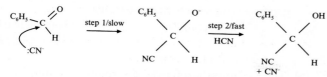

(i) Write a rate equation for the reaction. (2)
(ii) What is the overall order of the reaction? (1)
(iii) Deduce the units of the rate constant. (2)

(e) Mandelonitrile is a chiral molecule but the product from the reaction of benzaldehyde with hydrogen cyanide does not rotate the plane of polarised light.
(i) Explain why mandelonitrile is a chiral molecule. (2)
(ii) Suggest why a sample of mandelonitrile prepared in the laboratory does not rotate the plane of polarised light. (2)

(f) Cyanohydrins are useful intermediates in synthetic organic chemistry. Name, or give the formulae of, the reagents A to D in the following flow scheme.

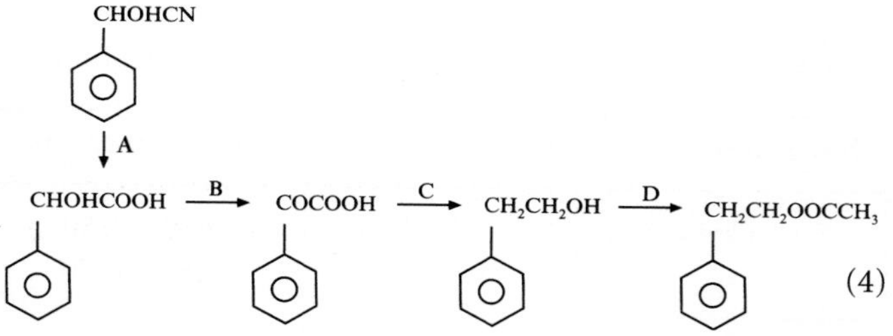

CHOHCN

$\downarrow$ A

CHOHCOOH $\xrightarrow{\text{B}}$ COCOOH $\xrightarrow{\text{C}}$ CH₂CH₂OH $\xrightarrow{\text{D}}$ CH₂CH₂OOCCH₃ (4)

(Total 24 marks)
NICCEA 1998

86 (a) (i) Give the observation which would be made if aqueous propanone is added to 2,4-dinitrophenylhydrazine reagent. (1)
(ii) Give the observation which would be made if propanone is warmed with an aqueous mixture of potassium iodide and sodium chlorate(I) **OR** aqueous sodium hydroxide and iodine. (1)
(iii) Give the structure of the observed product in (a)(ii) and a use for the reaction. (2)

(b) A compound, **X**, has a molar mass of 58 08 g mol⁻¹ and the following composition by mass.

C 62.04%;   H 10.41%;   O 27.55%.

(i) Calculate the molecular formula of **X**. (2)
(ii) **X** gives a silver mirror when warmed with Tollen's reagent. Draw the structure of **X**. (1)
(iii) Draw the structure of an isomer of **X**. (1)
(iv) On oxidation with acidified potassium dichromate(VI), **X** gives propanoic acid. In a reaction 11.6g of pure **X** gave 10 .6g of purified propanoic acid. Calculate the percentage yield in the reaction. (2)

(Total 10 marks)
WJEC 1997

87 (a) Salicylic acid is the trivial name of the compound with the formula:

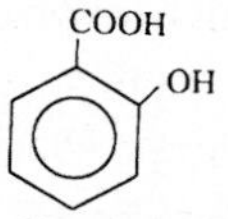

$$\text{C}_6\text{H}_4(\text{COOH})(\text{OH})$$

It has the reactions of both a carboxylic acid and a phenol.
(i) Suggest the systematic name of salicylic acid. (1)
(ii) Write an equation for the reaction between salicylic acid and aqueous sodium hydroxide. (2)
(iii) Draw a structure to represent the organic product of the reaction between salicylic acid and methanol. State the conditions used for the reaction. (3)

(b) The salicylate ion, HO.C₆H₄.COO⁻, can act as a bidentate ligand in the formation of complex ions. Explain the terms *ligand* and *complex ions*, and from your knowledge of other ligands, state which atoms in the salicylate ion could be used when acting as a ligand. (5)

(c) The pain reliever Aspirin is made by the reaction of ethanoic anhydride with salicylic acid. Before it can be used in the preparation of medicines, a sample of Aspirin must be carefully analysed.
(i) Given that ethanoic anhydride does not react with carboxylic acids, write an equation for its reaction with salicylic acid. (1)
(ii) In a student experiment, 0.500 g of a sample of the solid product of the reaction of ethanoic anhydride with salicylic acid was dissolved in ethanol. In the titration of this solution, 27.60 cm³ of 0.100 mol dm⁻³ sodium hydroxide solution were required for neutralisation. Calculate the percentage by mass of Aspirin in the sample. (4)
(iii) Suggest an explanation for the use of ethanol to dissolve the sample. (1)
(iv) Suggest a suitable indicator for the titration, giving a reason for your choice. (2)

(d) Outline how the technique of thin layer chromatography could be used to show the presence of Aspirin in a commercially available tablet. (6)

(Total 25 marks)
AEB 1996

88 (a) Write an equation for the formation of ethyl ethanoate from ethanoyl chloride and ethanol. Name and outline the mechanism for the reaction taking place. (6)

(b) Suggest why the above reaction is a more efficient way of preparing ethyl ethanoate than the reaction between ethanoic acid and ethanol. (2)

(c) The naturally-occurring polymer cellulose contains three alcohol groups per repeating unit and can be represented by the formula [C₆H₇O₂(OH)₃]ₙ. The acylation of cellulose to form cellulose 'triacetate' is normally carried out using ethanoic (*acetic*) anhydride. Give the formula of the reaction product and state **two** reasons why ethanoic anhydride, rather than ethanoyl chloride, is used to manufacture cellulose 'triacetate'. (3)

(Total 11 marks)
NEAB 1997

89 The polymer poly(methyl 2-methylpropenoate) (*Perspex*) can be made by a process which involves the following reactions.

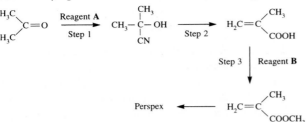

(a) (i) Identify reagent **A**.
   (ii) Name and outline the mechanism for the reaction in Step 1. (6)
(b) (i) Name reagent **B**.
   (ii) Name the type of reaction occurring and give a substance which would act as a catalyst for the reaction in Step 3. (3)
(c) Draw the repeating unit of Perspex. (1)
(d) Write an equation for the reaction between the product of Step 2 with sodium hydroxide, showing clearly the structure of the new product. (2)

(Total 12 marks)

NEAB 1997

90 Consider the effect of concentrated aqueous hydrochloric acid upon the three compounds, **B**, **C** and **D**, below.

$$CH_3CONH_2 \qquad CH_3CO_2CH_3 \qquad CH_3CH_2NH_2$$
$$\textbf{B} \qquad\qquad \textbf{C} \qquad\qquad \textbf{D}$$

Copy and complete the table below by inserting the appropriate product(s), if any. You should make an entry in each box, writing 'no change', 'no further change', or the structural formula or the name of any products.

	Cold	Boiling for half an hour
B		
C		
D		

(Total 8 marks)

OCSEB 1997

91 (a) Explain why ethylamine is a Brønsted-Lowry base. (2)
(b) Why is phenylamine a weaker base than ethylamine? (2)
(c) Ethylamine can be prepared from the reaction between bromoethane and ammonia.
   (i) Name the type of reaction taking place and outline a mechanism.
   (ii) Give the structures of **three** other organic substitution products which can be obtained from the reaction between bromoethane and ammonia. (8)

(d) Write an equation for the conversion of ethanenitrile into ethylamine and give one reason why this method of synthesis is superior to that in part (c). (2)

(Total 14 marks)

NEAB 1997

92 (a) Figure 1 shows the infra-red absorption spectrum for "nylon-6,6".

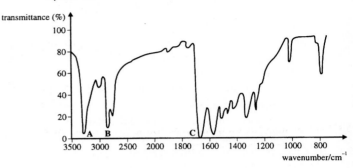

Figure 1

Figure 2 shows some characteristic infra-red absorptions.

Molecule or group	Bond absorbing	Wavenumber/cm^{-1}
alkyl	C—H C—H	2960 – 2850 1460 – 1370
aldehyde	C—H C=O	3250 – 3200 1740 – 1650
alkene	C—H C—H	3095 – 3075 990 – 890
amide	N—H	3500 – 3300
alcohol	C—O O—H	1200 – 1050 3650 – 3590

Figure 2

(i) Using the data in **Figure 2**, identify the cause of the absorption peaks **A**, **B** and **C** in **Figure 1**. (3)
(ii) Draw graphical formulae of **two** monomers that can be used to produce "nylon-6,6". (4)

(b) In an experiment involving enzymes, a student used urease, which is found in plants. Urease converts urea to ammonia. The student took some urea solution and added 3 drops of litmus solution, followed by drops of dilute hydrochloric acid until the solution just changed to a red colour. A small quantity of 1% urease solution was then added and the solution quickly changed colour to blue.

(i) Explain why the colour of the solution becomes blue. (2)

(ii) Copy, complete and balance the following equation for the reaction in which urea is hydrolysed.

$$\underset{H_2N}{\overset{H_2N}{>}}C=O + H_2O \rightarrow$$
(2)

(iii) The student repeated the experiment, using ethanamide (CH_3CONH_2) in place of urea. There was no change in the colour of the solution when the urease was added. Explain this observation in terms of the activity of the enzyme. (3)

(iv) Suggest how the enzyme urease could be denatured in this reaction. (1)

(Total 15 marks)

AEB 1997

93 (a) Consider the following reaction sequence, then answer the questions which follow.

$$C_2H_5Br \overset{\text{step}}{\underset{1}{\rightarrow}} C_2H_5CN \overset{\text{step}}{\underset{2}{\rightarrow}} C_2H_5CO_2H \overset{\text{step}}{\underset{3}{\underset{PCl_5}{\rightarrow}}} A \overset{\text{step}}{\underset{4}{\underset{NH_3}{\rightarrow}}} C_2H_5CONH_2 \overset{\text{step}}{\underset{5}{\rightarrow}} C_2H_5NH_2$$

(i) Give the reagents and conditions necessary for:
Step 1
Step 2
Step 5 (7)

(ii) Identify **A**. (1)

(b) Nylon 6:6, a polyamide, has a structure containing the following repeat unit:

$$-[CO-(CH_2)_4-CONH-(CH_2)_6-NH]-$$

(i) Give the structures of the monomers from which this polymer could be made. (2)

(ii) What type of polymer is this? (1)

(c) (i) Show the structure of poly(tetrafluoroethene) (2)

(ii) State one use of this polymer. (1)

(d) Which of the two polymers in (b) and (c) is the easier to break down and hence constitutes the smaller environmental hazard? Give a reason for your answer. (2)

(Total 16 marks)

London 1998

94 3-Aminophenylethanone can be obtained from benzene in three steps:

(a) For **each** step, name the type of reaction taking place and suggest a suitable reagent or a combination of reagents. (7)

(b) Write an equation showing how the electrophile is formed from the reagent(s) in Step 1. Outline a mechanism for the subsequent reaction between benzene and this electrophile. (5)

(Total 12 marks)

NEAB 1997

95 (a) Explain why sodium and magnesium are both good conductors of electricity, but sodium has a much lower melting point and density than magnesium. (9)

(b) The standard electrode potentials for sodium, magnesium and water are given in the table.

Electrode reaction	$E^{\ominus}$/V
$Na^+(aq) + e^- \rightleftharpoons Na(s)$	−2.71
$Mg^{2+}(aq) + 2e^- \rightleftharpoons Mg(s)$	−2.38
$2H_2O(l) + 2e^- \rightleftharpoons H_2(g) + 2OH^-(aq)$	−0.83

(i) Write a balanced ionic equation for the reaction between sodium and water, and explain why it can be described as a redox reaction. Calculate the e.m.f. that would be expected if the reaction were set up as a cell under standard conditions. (4)

(ii) Use the data above to explain why a reaction might be expected to occur between magnesium and water. Your answer should include the calculation of the e.m.f. of a standard magnesium-water cell. (3)

(iii) Write down the cell diagram relevant to the reaction between magnesium and water. (2)

(iv) When magnesium is added to water, little apparent reaction takes place. Suggest **two** reasons to account for this. (2)

(c) (i) State and explain the trend in the thermal stability of Group 2 carbonates. (3)

(ii) Suggest, with a reason, how the thermal stability of sodium carbonate compares with that of magnesium carbonate. (2)

(Total 25 marks)

AEB 1996

96 One of the major ammonia-producing plants in this country is located at Billingham on the North Sea coast. The source of the hydrogen needed is methane, CH_4, although water has been used in the past and could become important again in the future.

The synthesis gas from which ammonia is produced has a typical composition of

74.2% hydrogen, 24.7% nitrogen, 0.8% methane, 0.3% argon.

The gas is compressed to about 200 atmospheres pressure and passed over a catalyst at a temperature of 450°C. Under these conditions about 15% of the synthesis gas is converted into ammonia.

$$N_2(g) + 3H_2(g) \rightleftharpoons 2NH_3(g) \qquad \Delta H = -92.0 \text{ kJ mol}^{-1}$$

On leaving the converter, ammonia is removed from the unreacted gases.

Some ammonia is used to produce nitric acid. A further reaction between ammonia and nitric acid gives ammonium nitrate, most of which is used as fertiliser. Since ammonium nitrate is potentially explosive, great care has to be taken in its manufacture.

(a) (i) Suggest **one** reason why an ammonia plant is located at Billingham. (1)
   (ii) Under what circumstances could water *become important again in the future* as a source of hydrogen? (1)
   (iii) Explain the presence of argon in the synthesis gas. (1)

(b) Explain **one** advantage and **two** disadvantages of using high pressure for the production of ammonia. (3)

(c)

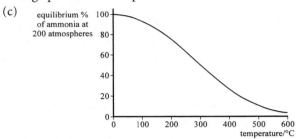

Figure 1

   (i) By referring to **Figure 1**, discuss the choice of 450°C as a temperature and the use of a catalyst for the conversion. (4)
   (ii) Give the name of the catalyst used and the type of catalysis occurring. (2)

(d) Describe:
   (i) how ammonia is removed from the unreacted gases; (2)
   (ii) what happens to the unreacted gases. (1)

(e) Write equations for the reactions in which:
   (i) ammonia is oxidised to give nitrogen monoxide (NO) and water as the first stage of nitric acid production; (2)
   (ii) ammonia reacts with nitric acid. (1)

(f) (i) Explain why an ammonium nitrate plant is located on the same site as the ammonia production unit in Billingham. (3)
   (ii) State a safety precaution which would be taken in and around an ammonium nitrate plant. (1)

(g) Ammonia is used to manufacture 1,6-diaminohexane, which reacts with hexanedioic acid to form a polymer.

$$H_2N(CH_2)_6NH_2 \qquad HOOC(CH_2)_4COOH$$
1,6-diaminohexane   hexanedioic acid

Give the name of the polymer formed, draw the graphical formula of the repeating unit and classify the type of polymerisation reaction.
Give **two** uses of the polymer. (6)

(Total 28 marks)
AEB 1998

97 This question is about sulfur dioxide, SO_2. Sulfur dioxide is used in the production of sulfuric acid, H_2SO_4, by the Contact Process. In this process sulfur dioxide is catalytically oxidised to sulfur trioxide, SO_3, according to the gaseous equilibrium:

$$2SO_2(g) + O_2(g) \rightleftharpoons 2SO_3(g)$$

Some quantitative information about this reaction is given below. The standard enthalpies of formation are:

	$O_2(g)$	$SO_2(g)$	$SO_3(g)$
$\Delta H_f^{\ominus}$(298 K)/kJ mol^{-1}	0	−297	−395

(a) (i) Calculate the standard enthalpy change for the reaction at 298 K. (2)
   (ii) Why is the value of $\Delta H_f^{\ominus}$ for oxygen zero? (1)

(b) (i) Write down the expression for the equilibrium constant, K_p, for this equilibrium. (1)
   (ii) What determines the way in which the equilibrium constant changes with temperature? (1)
   (iii) Equilibrium constants and rates of reaction are both affected by changes of temperature. For the equilibrium, show the effect on the rate and the equilibrium constant of increasing the temperature, by copying and completing the table.

$$2SO_2(g) + O_2(g) \rightleftharpoons 2SO_3(g)$$

	Effect of increasing the temperature
The value of the equilibrium constant	
The rate of the forward reaction	
The rate of the reverse reaction	

(3)

(c) Sulfur dioxide is a major pollutant in the atmosphere and must be closely monitored. The amount of sulfur dioxide in a sample of air can be determined by reaction with hydrogen peroxide, H_2O_2, and titrating the sulfuric acid formed with standard alkali.

   (i) Write the equation for the reaction between sulfur dioxide and hydrogen peroxide. (1)
   (ii) In a typical experiment 200 m^3 of polluted air was bubbled slowly through 25.0 cm^3 of a solution of hydrogen peroxide. The resulting solution was neutralised by 15.3 cm^3 of sodium hydroxide solution of concentration 0.100 mol dm^{-3}.
Calculate the sulfur dioxide concentration in g m^{-3}, in the sample of air.
(A_r (S) 32.0; A_r (O) 16.0) (4)

(Total 13 marks)
OCSEB 1998

98 (a) Explain why light is essential for the reaction between methane and chlorine to form chloromethane at room temperature. Explain also why this method is not regarded as a good way of synthesising pure chloromethane. (7)

(b) Explain why the first ionisation energy of phosphorus is greater than that of both silicon and sulphur. (5)

(c) An acidic solution of potassium manganate(VII) will liberate chlorine from dilute sodium chloride solution but an acidic solution of potassium dichromate(VI) will not. Solid potassium dichromate(VI) will liberate chlorine gas from concentrated hydrochloric acid.

Explain these observations with reference to the data below.

$$
\begin{array}{lr}
 & E^{\ominus}/V \\
MnO_4^-(aq) + 8H^+(aq) + 5e^- \rightleftharpoons Mn^{2+}(aq) + 4H_2O & +1.52 \\
Cl_2(g) + 2e^- \rightleftharpoons 2Cl^-(aq) & +1.36 \\
Cr_2O_7^{2-}(aq) + 14H^+(aq) + 6e^- \rightleftharpoons 2Cr3+(aq) + 7H_2O & +1.33
\end{array}
$$
(7)

(d) By application of the concept of bond energy, predict a value for ΔH for the reaction:

$$C_2H_5OH + CH_3CO_2H \rightleftharpoons CH_3CO_2C_2H_5 + H_2O$$

Hence predict the effect of an increase in temperature on the equilibrium constant for this reaction. (6)

(Total 25 marks)
London 1996

99 A student determined the % of iron in an iron(II) salt **X** as follows:

1.20g of the salt was placed in a beaker and 50.0cm^3 of water was added to dissolve the salt. The solution was heated to boiling point with a small quantity of concentrated nitric acid and eventually became a yellow colour.

One drop of this yellow solution was then extracted and tested by adding one drop of a freshly prepared dilute solution of potassium hexacyanoferrate(III). This latter solution will react with iron(II) ions to give a blue colouration; no such colouration was obtained.

The further treatment of the solution included the addition of ammonia solution, a little at a time, with constant stirring, until a slight smell of ammonia persisted. The brown precipitate which had formed was allowed to settle. This precipitate was separated and washed and heated strongly to give iron(III) oxide.

A further 1.20 g sample of **X** was treated in the same way and in both experiments the mass of iron(III) oxide obtained was 0.245 g.

(a) Suggest a colour for the iron(II) salt solution. (1)

(b) Why was concentrated nitric acid added? (1)

(c) What ion gave rise to the yellow colour in the solution after boiling with concentrated nitric acid? (1)

(d) (i) The formula of potassium hexacyanoferrate(III) is $K_3Fe(CN)_6$. Write the formula of the complex anion in this compound.

(ii) Draw a diagram showing the shape of the anion in (i). Describe the shape.

(iii) Why was the hexacyanoferrate(III) test carried out?

(iv) Had a blue colouration appeared in the test, what would the student then have to do? (6)

(e) Explain why ammonia solution was added until a slight smell of ammonia persisted. (2)

(f) (i) Name the brown precipitate formed on the addition of the ammonia solution.

(ii) Write an ionic equation showing the formation of this precipitate. (2)

(g) Write the equation for the action of heat on the precipitate. (1)

(h) Calculate the mass of iron present in 0.245 g of iron(III) oxide. (2)

(i) Calculate the % of iron present in **X**. (2)

(Total 18 marks)
London 1996

100 The elements in Group 4 of the periodic table show a trend from non-metallic to metallic behaviour as atomic number increases. At the same time oxidation state + 2 becomes more stable compared with oxidation state + 4.

(a) Explain why metallic character increases with increasing atomic number in Group 4. (4)

(b) Lead forms oxides in the +2 and +4 oxidation state.

(i) State the acid-base character of lead(II) oxide, illustrating your answer with equations. (3)

(ii) Write an equation for the reaction of lead(IV) oxide with concentrated HCl at room temperature, and explain the difference between this reaction and that in part (b)(i). (4)

(c) Tetrachloromethane does not react with cold water, whereas silicon tetrachloride hydrolyses rapidly. In contrast, in Group 3 both boron trichloride and aluminium trichloride hydrolyse rapidly.

(i) State the shape of a molecule of boron trichloride and justify it in terms of electron structures. (2)

(ii) Write the equation for the reaction of boron trichloride with water. Explain why boron trichloride reacts with water whereas tetrachloromethane, CCl_4, does not. (3)

(iii) Explain how you would use solutions of silver nitrate and ammonia to show that the reaction product from boron trichloride and water contains chloride ions. (3)

(d) Silver nitrate can also be used to show the presence of iodide ions in solution. Silver iodide is sparingly soluble in water.

(i) Given the thermochemical data below, suggest why silver iodide is so insoluble.

lattice enthalpy of AgI	-876 kJ mol^{-1}
hydration enthalpies: Ag$^+$	-464kJ mol^{-1}
I$^-$	-293 kJ mol^{-1} (2)

(ii) Explain why your result from (i) only *suggests* why AgI is so insoluble. (1)

(iii) The following equilibrium is established when excess solid silver iodide is in contact with a saturated solution of silver iodide:

$$AgI(s) \rightleftharpoons Ag^+(aq) + I^-(aq)$$

Silver nitrate is used to precipitate halides in quantitative analysis, where the precipitate is weighed to find the amount of halide. Why, in such experiments, is an excess of silver nitrate added to the test solution? (3)

(Total 25 marks)

London 1997

101 A sequence of reactions by which nitric acid could be made is given below.

I $\quad N_2(g) + O_2(g) \rightleftharpoons 2NO(g) \quad \Delta H = +181 \text{ kJ mol}^{-1}$

II $\quad 2NO(g) + O_2(g) \rightarrow 2NO_2(g) \quad \Delta H = -113 \text{ kJ mol}^{-1}$

III $\quad 2NO_2(g) + H_2O(l) + \frac{1}{2}O_2(g) \rightarrow 2HNO_3(aq)$
$$\Delta H = -128 \text{ kJ mol}^{-1}$$

(a) Calculate ΔH for the reaction:

$$N_2(g) + H_2O(l) + 2\frac{1}{2}O_2(g) \rightarrow 2HNO_3(aq)$$

Given this value, explain why large amounts of nitric acid are NOT found in the earth's atmosphere, given that all necessary reagents are present. (6)

(b) Reaction I is the rate determining step in this sequence of reactions. Deduce qualitatively the conditions of temperature and pressure necessary in order to increase the amount of HNO_3 formed in a given time and thus make the process more economic. (9)

(c) One use of nitric acid is to make inorganic nitrate fertilisers such as ammonium nitrate. Organic compounds, e.g. urea, are also used. Discuss the relative merits of these compounds as nitrogenous fertilisers. (4)

(d) The structure $-\overset{\displaystyle O}{\overset{\displaystyle \diagup \, \diagdown}{CH\!-\!CH_2}}$ is found as an end group in epoxy resins. In the final 'curing' of such polymers, compounds which may contain three or more $-\ddot{N}H_2$ or $-\ddot{N}H-$ groups are used.

Describe how these interact in co-polymer formation, and the property imparted to the final polymer. (6)

(Total 25 marks)

London 1997

102 (a) A buffer solution of pH = 3.87 contains 7.40 g dm^{-3} of propanoic acid together with a quantity of sodium propanoate. K_a for propanoic acid = 1.35×10^{-5} mol dm^{-3} at 298 K.

(i) Explain what a buffer solution is and how this particular solution achieves its buffer function. (8)

(ii) Calculate the concentration in g dm^{-3} of sodium propanoate, $C_2H_5CO_2Na$, in the solution, stating any assumptions made. (6)

(iii) If the sodium propanoate were to be replaced by anhydrous magnesium propanoate, calculate the concentration of magnesium propanoate in g dm^{-3}, required to give a buffer of the same pH. (2)

(b) Suggest a sequence of reactions by which propanoic acid, $C_2H_5CO_2H$, could be converted into 2-aminopropanoic acid, stating the important conditions. (5)

(c) Pure 2-aminopropanoic acid has a relatively high melting point and is insoluble in hydrocarbon solvents.

(i) Sketch a structure for this compound which would account for these observations. (1)

(ii) Write ionic equations to show the action of an aqueous solution of 2-aminopropanoic acid with:
A. an excess of acid;
B. an excess of alkali. (2)

(iii) Hence explain how 2-aminopropanoic acid might behave as a buffer in solution. (1)

(Total 25 marks)

London 1997

103 (a) Explain how an aqueous solution containing ethanoic acid and sodium ethanoate resists changes in pH when contaminated with small amounts of acid or alkali. (2)

(b) It can be shown that for solutions described in (a)

$$pH = pK_a + \log \frac{[\text{sodium ethanoate}]}{[\text{ethanoic acid}]}$$

where $pK_a = -\log K_a$.

(i) State the significance of the value of the pH when [sodium ethanoate] = [ethanoic acid]. (1)

(ii) Calculate the mass of sodium ethanoate which must be dissolved in 1 dm^3 of ethanoic acid of concentration 0.10 mol dm^{-3} to produce a solution with a pH value of 5.5 at 298 K. It can be assumed that there is no volume change on dissolving the salt. (At 298 K the value of K_a for ethanoic acid is 1.8×10^{-5} mol dm^{-3}.) (3)

(c) (i) Name an indicator which would be suitable for the titration of aqueous ethanoic acid with aqueous sodium hydroxide, giving a reason for your choice. (1)

(ii) State whether the pH of an aqueous solution of pure ammonium chloride would be greater or less than 7 at 298 K, giving a reason for your answer. (1)

(d) The pH of human blood plasma is maintained at a value between 7.39 and 7.41 by the buffering action of dissolved carbon dioxide, hydrogencarbonate ions and carbonic acid, H_2CO_3. Write equations to show what reactions may occur when the blood absorbs small amounts of acid or small amounts of alkali. (2)

(Total 10 marks)

WJEC 1997

104 (a) Concentrated sulphuric acid reacts with sodium chloride as follows:
$$H_2SO_4 + Cl^- \rightleftharpoons HCl + HSO_4^-$$
   (i) Identify the conjugate acid/base pairs in this reaction by copying the equation and writing suitable identifying symbols under it. (2)
   (ii) What would be the observable result of this reaction? (1)
   (iii) Explain why this reaction goes almost completely to the right despite the fact that both hydrochloric and sulphuric acids are strong. (3)

(b) When concentrated sulphuric acid reacts with solid sodium iodide, hydrogen sulphide, sodium hydrogensulphate and water are formed together with one other product.
   (i) Identify this product and state how you would recognise it. (2)
   (ii) Write an ionic half equation to show the conversion of sulphuric acid to hydrogen sulphide. (2)
   (iii) Hence write the full ionic equation for the reaction between concentrated sulphuric acid and sodium iodide. (2)
   (iv) What is the function of the sulphuric acid in this reaction? (1)

(c) Explain concisely why the type of reaction occurring in (b) does not occur with sodium chloride. (2)

(Total 15 marks)

London 1997

105 This question concerns the reaction between aqueous potassium manganate(VII), $KMnO_4$, and an aqueous solution containing both ethanedioic acid, $H_2C_2O_4$, and sulphuric acid.

In the reaction the purple manganate(VII) ion is reduced slowly, in the presence of hydrogen ions, to manganese(II) ions. The resulting solution is colourless.

(a) Design an experiment to determine how the rate of this reaction varies with temperature. (5)

(b) During this reaction the concentration of the Mn^{2+} ions varies with time as shown on the sketch below. The temperature of the mixture remains constant throughout.

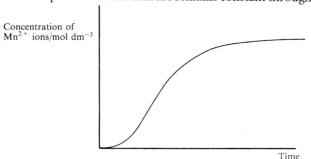

Concentration of Mn^{2+} ions/mol dm^{-3}

Time

Comment on and explain the shape of the curve. (4)

(c) In the reaction the ethanedioate ion from ethanedioic acid is oxidised to carbon dioxide. Construct ionic half equations and use them to write an overall ionic equation for the reaction of manganate(VII) ions and ethanedioate ions in the presence of hydrogen ions. (5)

(d) 12.5 cm^3 of $0.0800 \text{ mol dm}^{-3}$ sodium hydroxide solution was required to completely neutralise both the sulphuric acid and ethanedioic acid in 10.0 cm^3 of the acid solution. Calculate the total acid concentration. (4)

(Total 18 marks)

London 1997

106 Some modern British coins are made from an alloy, nickel-brass, which consists essentially of the metals copper, nickel and zinc. A one pound coin weighing 9.50 g was completely dissolved in concentrated nitric acid, in which all three metals dissolve, to give solution **A**.

Dilute sodium hydroxide solution was then added carefully with stirring, until present in excess. Zinc hydroxide is amphoteric. The precipitate formed, **B**, was filtered off from the supernatant liquid, **C**. The precipitate, **B**, was quantitatively transferred to a graduated flask of 500 cm^3 capacity. Dilute sulphuric acid was then added to dissolve the whole of the precipitate, **B**, and the solution made up to 500 cm^3 with distilled water.

25.0 cm^3 portions of this solution was pipetted into a titration flask and an excess of potassium iodide solution added. The liberated iodine was then titrated against sodium thiosulphate solution of concentration $0.100 \text{ mol dm}^{-3}$. 18.7 cm^3 of the sodium thiosulphate solution was required for complete reaction.

$$
\begin{array}{lr}
 & E^{\ominus}/V \\
Ni^{2+}(aq) + 2e^- \rightleftharpoons Ni(s) & -2.71 \\
Zn^{2+}(aq) + 2e^- \rightleftharpoons Zn(s) & -0.76 \\
\tfrac{1}{2}S_4O_6^{2-} + e^- \rightleftharpoons S_2O_3^{2-} & +009 \\
Cu^{2+}(aq) + e^- \rightleftharpoons Cu^+(aq) & +0.15 \\
Cu^{2+}(aq) + 2e^- \rightleftharpoons Cu(s) & +034 \\
\tfrac{1}{2}I_2(aq) + e^- \rightleftharpoons I^-(aq) & +0\ 54 \\
NO_3^-(aq) + 2H^+(aq) + e^- \rightleftharpoons NO_2(g) + H_2O(l) & +0.81 \\
\end{array}
$$

(a) (i) Using the appropriate half equations, write an equation for the reaction of any one of the metals in nickel-brass with concentrated nitric acid. (2)
   (ii) What type of reaction is taking place? (1)

(b) Identify by giving full formulae:
   (i) the complex cations present in **A**; (2)
   (ii) the precipitates in **B**; (2)
   (iii) any metal-containing anion in **C**. (1)

(c) (i) Write an equation for the precipitation of any one of the metal ions in **A** with sodium hydroxide. (2)
   (ii) What type of reaction is occurring in (c)(i)? (1)

(d) Suggest an explanation why it is necessary to add sodium hydroxide, followed by dilute sulphuric acid, before performing the titration. (3)

(e) On addition of the potassium iodide solution, the **only** reaction which occurs is:
$$2Cu^{2+}(aq) + 4I^-(aq) \rightarrow 2CuI(s) + I_2(aq)$$
   (i) Write an equation for the reaction between sodium thiosulphate and the liberated iodine. What indicator would you use in this titration? At what stage would you add it? Give a reason for your answer. (3)

(ii) Calculate the percentage of copper in the alloy. (5)

(iii) Suggest why this reaction occurs in the light of the $E^{\ominus}$ values given. (3)

(Total 25 marks)

London 1998

107 For the reaction:

$$2NO(g) + O_2(g) \rightleftharpoons 2NO_2(g)$$

the rate equation for the forward reaction is:

$$\text{Rate} = k[NO]^2[O_2]$$

(a) (i) Deduce the units for the rate constant k. (1)

(ii) State and explain how the rate of reaction would change if the concentration of O_2 was doubled, all other factors remaining constant. (2)

(iii) State and explain how the rate of reaction would change if the concentration of NO was halved, all other factors remaining constant. (2)

(b) (i) What is the overall order of the reaction? (1)

(ii) Explain on the basis of collision theory why this reaction is unlikely to occur in a single step. (2)

(c) The first step in a possible mechanism for the reaction above is:

$$2NO \rightleftharpoons N_2O_2$$

(i) Draw dot and cross diagrams to show the electronic structures of the two molecules NO and N_2O_2. What feature of the electronic structure of NO would suggest that this is a likely first step in the reaction? (4)

(ii) Explain why the enthalpy change for this step is -163 kJ mol^{-1}, given that the average bond energy for the N—N bond in compounds of nitrogen is $+163 \text{ kJ mol}^{-1}$. (2)

(iii) Explain why this step does not control the rate of the reaction. Assuming there is one further step in the reaction write an equation for this. (2)

(d) (i) Deduce the effect of increasing the temperature on the position of equilibrium in the first step of the mechanism in (c). (2)

(ii) Discuss the economic implications of increasing the temperature on the overall process, if NO_2 was made industrially by this process. (3)

(e) NO_2 reacts with aqueous sodium hydroxide according to the following equation:

$$2OH^- + 2NO_2 \rightarrow NO_2^- + NO_3^- + H_2O$$

(i) What type of reaction is this? Justify your answer. (2)

(ii) Deduce the ionic half equations for this reaction. (2)

(Total 25 marks)

London 1998

108 A carboxylic acid **A** contains 40.0% carbon, 6.70% hydrogen and 53.3% oxygen by mass. When 10.0 cm^3 of an aqueous solution of **A**, containing 7.20 g dm^{-3}, was titrated against 0.050 mol dm^{-3} sodium hydroxide, the following pH readings were obtained.

Volume NaOH/cm^3	0 00	2.5	5.0	7.5	10.0	14.0	15.0	16.0	17.5	20.0	22.5
pH		2.5	3.2	3.5	3.8	4.1	4.71	5.2	9.1	11.5	11.8	12.0

(a) (i) Plot a graph of pH (on the y axis) against volume of NaOH (on the x axis). Use the graph to determine the end point of the titration. Hence calculate the relative molecular mass of **A**. (8)

(ii) Calculate the value of K_a for **A** and state its units. (4)

(b) Calculate the molecular formula of **A**. Given that **A** contains one asymmetric carbon atom, deduce its structure. Briefly indicate your reasoning. (4)

(c) The mass spectrum of **A** shows major peaks at m/e values of 15, 30, 45 and 75. Suggest the formula for the species responsible for each of these four peaks. (4)

(d) Describe a series of tests you would perform in order to confirm the structure obtained in (b), given that you already know that it is an acid. (5)

(Total 25 marks)

London 1998

109 Citric acid is used in foodstuffs as an antioxidant and, together with its sodium salt, as an acidity regulator. It occurs naturally in fruit juices.

$$\text{A formula of citric acid is } HO-\underset{\underset{CH_2CO_2H.}{|}}{\overset{\overset{CH_2CO_2H}{|}}{C}}-CO_2H$$

(a) (i) Assuming citric acid behaves in aqueous solution as a monoprotic acid:

$$RCO_2H + H_2O \rightleftharpoons RCO_2^- + H_3O^+$$

write an expression for K_a for this acid. (1)

(ii) Calculate the pH of lemon juice which contains citric acid at a concentration of 0.200 mol dm^{-3}. (K_a for citric acid = 7.4×10^{-4} mol dm^{-3}). (3)

(b) The use of citric acid together with its salt, sodium citrate, as an acidity regulator depends on the ability of this mixture to act as a buffer.

(i) What is the function of a buffer solution? (2)

(ii) Describe how the mixture of citric acid and sodium citrate achieves this buffering action. Give equations for the TWO reactions you describe. (3)

(iii) Calculate the pH of a buffer solution containing 0.200 mol dm^{-3} of citric acid and 0.400 mol dm^{-3} of sodium citrate. (2)

(c) Citric acid forms a liquid ester which has the structural formula

$$CH_2COOC_2H_5$$
$$HO—C—COOC_2H_5$$
$$CH_2COOC_2H_5$$

(i) Describe a test you could use to show that the ester contains an —OH group. (2)

(ii) What reagent would you use to hydrolyse the ester? (1)

(iii) Treatment of the products of the reaction in (c)(ii) leads to the production of a pure sample of citric acid. How would you show the presence of the —CO_2H group in the citric acid other than by the use of an indicator? (2)

(Total 16 marks)
London 1998

110 (a) (i) Write the electron configurations for Cu and for Cu^+. (1)

(ii) In what way is the electronic structure of copper unusual in terms of the general trend across the first transition series? (1)

(iii) Explain why the copper(I) ion is not coloured. (2)

(b) Use the electrode potential data concerning copper to answer the questions that follow.

$$Cu^{2+}(aq) + e^- \rightleftharpoons Cu^+(aq) \quad E^\ominus = +0.15\,V$$
$$Cu+(aq) + e^- \rightleftharpoons Cu(s) \quad E^\ominus = +0.52\,V$$

Suggest what would happen if a sample of copper(I) sulphate, Cu_2SO_4, were added to water. State in general terms the nature of the process which is occurring and state what you would see if copper(I) sulphate were to be added to water. (7)

(c) Copper(II) sulphate solution contains the complex ion $[Cu(H_2O)_6]^{2+}$.
State what you would see if a solution of aqueous ammonia was added dropwise to a solution of this ion. Show by means of equations how this reaction proceeds and state the type of reaction occurring at each stage. (6)

(d) If copper(II) sulphate is dissolved in liquid ammonia, a solution containing the ion $[Cu(NH_3)_6]^{2+}$ is formed.

(i) Draw the shape of this ion, and explain why it is this shape. (2)

(ii) Draw the 'electrons in boxes' diagram for this complex ion, distinguishing clearly the copper electrons from the ligand electrons. (1)

(e) The principal oxidation states of copper are +1 and +2, but a few compounds of copper(III) are known.
Explain why transition metals are able to show a variety of oxidation states but compounds containing copper(III) are rare. Which property might you expect copper(III) compounds to have and why? (5)

(Total 25 marks)
London 1998

111 (a) The free radical reaction between methane and chlorine in ultraviolet light has a mechanism whose initiation step is

$$Cl_2 \rightarrow 2Cl^\bullet$$

Followed by one of two possible propagation steps

$$CH_4 + Cl^\bullet \rightarrow {}^\bullet CH_3 + HCl \quad I$$
$$CH_4 + Cl^\bullet \rightarrow {}^\bullet CH_3Cl + H^\bullet \quad II$$

(i) Use the data given below to predict which is the more likely propagation step.
bond enthalpies/kJ mol^{-1}: Cl—Cl 243; C—H 435; H—Cl 432; C—Cl 346. (4)

(ii) Identify one product you might expect from a termination step following each reaction that would not be found in the alternative reaction scheme. (2)

(b) The reaction in part (a) is an important industrial source of hydrochloric acid, HCl(aq). The hydrogen halides HX(aq), sometimes called the hydrohalic acids, are all Bronsted-Lowry acids.

(i) What is a Bronsted-Lowry acid? (1)

(ii) Explain why HF(aq) is the weakest of these hydrohalic acids. (2)

(c) Boron trichloride is a gaseous compound which reacts readily with water.

(i) Write an equation for this reaction. (2)

(ii) Explain why boron trichloride reacts readily with water. (3)

(Total 14 marks)
London 1998

112 In diffused light methane combines readily with chlorine in a photochemical substitution reaction forming chloromethane as the initial organic product. Further chlorination produces a mixture of chlorinated hydrocarbons.

(a) Explain the term **photochemical substitution**. (2)

(b) Write equations to represent the following steps in the reaction:

(i) The initiation step (1)

(ii) Two propagation steps (2)

(iii) A termination step. (1)

(c) Including an equation, define **standard enthalpy change of formation** of chloromethane. (4)

(d) Given the standard enthalpy changes of formation below, calculate the enthalpy change for the overall reaction.

$$CH_4(g) + Cl_2(g) \rightarrow CH_3Cl(g) + HCl(g) \quad (3)$$

Compound	$\Delta H_f^\ominus$ /kJ mol^{-1}
HCl	−72
CH_4	−75
CH_3Cl	−82

(e) An organic compound has the composition by mass shown in the table.

Element	Percentage
Hydrogen	4.1
Carbon	24.2
Chlorine	71.7

(i) Calculate the empirical formula of the compound. (3)

(ii) The mass spectrum of the compound showed the molecular ion peak at a mass/charge ratio of 98. Suggest a structural formula for the compound. (1)

(iii) Suggest why smaller peaks may be found at 100 and 102. (2)

(Total 19 marks)
NICCEA 1998

113 (a) When equimolar amounts of ethanol and ethanoic acid are mixed at 298 K and allowed to reach equilibrium, the equilibrium mixture contains twice the number of moles of ethyl ethanoate as moles of ethanoic acid.

$$CH_3CH_2OH(l) + CH_3COOH(l) \rightleftharpoons CH_3COOCH_2CH_3(l) + H_2O(l)$$

(i) Write the expression for the equilibrium constant, K_c, for this reaction. (1)

(ii) Calculate the value of K_c for the reaction at 298 K. (2)

(b) Ethyl ethanoate can be prepared by the action of ethanoic anhydride on ethanol according to the equation

$$CH_3CH_2OH + (CH_3CO)_2O \rightarrow CH_3COOCH_2CH_3 + CH_3COOH.$$

Given the standard molar enthalpy changes of combustion in the table below, calculate the standard enthalpy change for the above reaction at 298 K. (3)

Formula of compound	Standard enthalpy change of combustion at 298 K/kJ mol^{-1}
CH_3CH_2OH	-1367
$(CH_3CO)_2O$	-1794
$CH_3COOCH_2CH_3$	-2238
CH_3COOH	-874.1

(c) In Brazil ethanol is produced by fermentation of plants such as sugar cane and then blended with petrol to produce a vehicle fuel.

(i) State **one** environmental **or** economic advantage of using ethanol in this way. (1)

(ii) The standard enthalpy change of combustion of octane is -5470 kJ mol^{-1} and that of ethanol is -1367 kJ mol^{-1}. Calculate the heat produced when 1 g of octane is burned completely **and** when 1 g of ethanol is burned completely. (2)

(iii) State, **briefly**, how ethanol is produced from ethene on a large scale. (1)

(Total 10 marks)
WJEC 1998

114 Consider the following reaction scheme:

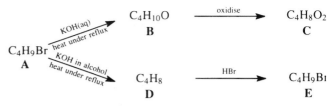

The compound **A** is not chiral, but **E** is. **C** will liberate carbon dioxide from sodium hydrogencarbonate solution. **B** and **C** will react with one another under suitable conditions.

(a) Draw the structural formulae of the following:
A, B, C, E. (4)

(b) Give a qualitative test for the functional group in **D**. (2)

(c) Write the equation for the reaction of **C** with sodium hydrogencarbonate. (2)

(d) Give the structure of the organic compound formed by reaction between **B** and **C**. (1)

(e) Suggest suitable reagents and conditions for the conversion of **B** to **C**. (2)

(f) Substance **C** has K_a value of 1.51×10^{-5} mol dm^{-3}.

(i) Find the pH of a 0.100 mol dm^{-3} solution of **C**. (3)

(ii) What property is shown by a solution which contains a mixture of **C** and its sodium salt? (1)

(iii) Calculate the pH of a solution formed by adding 5.5 g of the sodium salt of **C** to 500 cm^3 of a solution of **C** of concentration 0.100 mol dm^{-3}. (4)

(Total 19 marks)
London 1998

115 Phenylamine, $C_6H_5NH_2$, can be made in several ways from benzene. The flowchart shows two of these in outline.

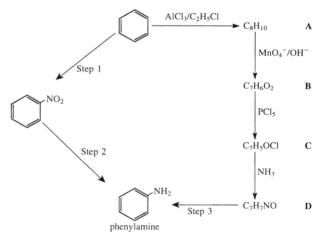

(a) (i) Give the reagents and conditions needed for step 1. (2)

(ii) Write the mechanism for this reaction. (4)

(iii) Give the reagents and conditions for step 2. (3)

(b) (i) In the alternative pathway identify the compounds **A** to **D** inclusive. (4)

(ii) State the reagents and conditions needed for step 3. (3)

(iii) Why is the first pathway commercially preferable for the manufacture of phenylamine? (1)

(c) Phenylamine reacts with nitrous acid in the presence of concentrated hydrochloric acid at 0°–5°C to produce benzenediazonium chloride.

If the temperature of the solution rises, this compound reacts with the solvent water to give phenol, C_6H_5OH, and nitrogen gas. The rate of the reaction can be followed by measuring the volume of nitrogen produced at various times, pressure and temperature remaining constant. The amount of benzenediazonium chloride remaining in the solution is proportional to $V_\infty - V_t$ where V_t is the volume of gas at time t and V_∞ is that at the end of the reaction.

(i) Give the equation for the reaction between benzenediazonium chloride and water. (2)

(ii) Some results of this experiment are shown below.

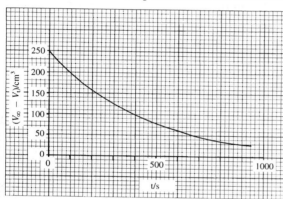

Deduce the order of reaction with respect to benzenediazonium chloride from the graph. (4)

(iii) Is the reaction necessarily of this order overall? Explain your answer. (2)

(Total 25 marks)
London 1997

Answers to Assessment Questions

Selected answers only are provided here. Detailed marking schemes are available on the associated website at www.chemistryincontext.co.uk

1 (c) 1 : 4
   (d) (i) B_2H_5
        (ii) B_4H_{10}
2 (b) 1.0073 g
3   (iv) (II) 174.3 kg
4 (a) (i) 0.001 mole
        (ii) 0.012 g
   (b) (ii) 0.068 g
        (iii) 0.0038
        (iv) 0.0020
   (c) (i) 1 : 6.8 : 2
   (d) $CuCO_3.Cu(OH)_2.xH_2O$.
5 (b) 20.2
8 (b) (ii) 180 days
10 (e) 1200 kJ mol^{-1}
19 (b) (i) P_2O_3
        (iii) 7.33 g
20 (d) (i) 37 ± 1 cm^3
        (ii) 6.6°C
   (e) (i) 1719 ± 28 J
        (ii) −115 kJ
21 (b) (ii) +52 kJ mol^{-1}
   (c) (ii) −1561 kJ mol^{-1}
22 (b) −4817.3 kJ mol^{-1}
23 (c) (i) −875 kJ mol^{-1}
        (ii) −864 kJ mol^{-1}
24 (c) −235.7 kJ
25 (a) (i) −348 kJ mol^{-1}
26 (a) +37 JK^{-1} mol^{-1}
   (b) +5.8 kJ mol^{-1}
   (c) (ii) 454 K
27 (c) (i) 38 400 g (38.4 kg)
        (ii) 14 400 dm^3
28 (b) (ii) 1.23×10^2 (mol dm^{-3})$^{-2}$ s^{-1} or mol^{-2} dm^6 s^{-1}
29 (a) Order w.r.t. **C** = 2; order w.r.t. **D** = 0; initial rate in expt 4 = 1.6×10^{-5} mol dm^{-3} s^{-1}
   (b) (ii) 2.2×10^{-1} mol^{-1} dm^3 s^{-1}
30 (b) order w.r.t. H_2O_2 = 1, order w.r.t. I^- = 1, order w.r.t. H^+ = 1

(mol dm^{-3})$^{-2}$ s^{-1} or mol^{-2} dm^6 s^{-1}
33 (a) (ii) 0.2 mol dm^{-3} (iii) +58.1 kJ mol^{-1}
34 (c) (i) $60 \pm 3\%$
36 (a) Cl_2 = 0.156, PCl_3 = 0.156, PCl_5 = 0.244
   (b) (ii) 0.0108 mol dm^{-3}
   (c) (ii) 0.281
        (iii) 109.7 kPa
        (iv) 45.0 kPa
37 (b) (i) $[H^+] = 7.16 \times 10^{-7}$ mol dm^{-3}; pH = 6.14
   (c) 11.7
38 (a) (ii) 4.77   (iii) 2.89
39 (a) (ii) 4.91
   (c) (iii) 4.91
41 (a) AgCl: solubility 1.34×10^{-5} mol dm^{-3}; 1.92×10^{-3} g dm^{-3}
        $CaCO_3$: solubility product 9×10^{-8} mol^2 dm^{-6}; solubility 3×10^{-4} mol dm^{-3}
   (c) (i) 5.06×10^{-2} mol dm^{-3}
        (ii) 0.0026%
43 (c) (iii) 1.10 V
44 (b) (i) +0.34 V
45 (c) (iii) +0.98 V
48 (b) (i) NaO   (ii) Na_2O_2
54 (b) (ii) 102.6 g
57 (a) (i) **A** $[Cu(H_2O)_4]^{2+}$(aq), SO_4^{2-}(aq); **B** $[CuCl_4]^{2-}$(aq); **C** $Cu(OH)_2$(s); **D** $[Cu(NH_3)_4]^{2+}$(aq)
   (b) (i) $Cu(OH)_2$(s)
59 (b) 1 : 6 : 2 : 3
60 (d) CH_4^+ or $^{13}CH_3^+$
62 A butan-2-ol, $CH_3CH_2CH(OH)CH_3$;
   B but-2-ene, $CH_3CH=CHCH_3$;
   C but-1-ene, $CH_3CH_2CH=CH_2$;
   D 2-bromobutane, $CH_3CH_2CHBrCH_3$;
   E butanone, $CH_3CH_2COCH_3$
66 (a) (ii) $C_{12}H_{26}$
   (b) D $CH_3CH_2CH=CHCH_2CH_3$ or $CH_3CH=CHCH_2CH_2CH_3$
        E $CH_2=CHCH_2CH_2CH_2CH_3$
        F

67 (b)
   A methylpropene

   B methylpropane

   C and **D** methyl propan-2-ol and methyl propan-1-ol

   (c) **E**

   $HOCH_2CH_2CH_2CH_2OH$,
   $CH_3CHOHCHOHCH_3$,
   $CH_3CHOHCH_2CH_2OH$
71 (a) (ii) C_8H_{10}
        (iii)

   (b) (i) 1,4-dimethylbenzene
        (ii)

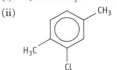
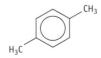

73 (b) 2.2%

74 (b) (i) **W** CH_3—CH=CH_2;
  X CH_3—$CHOH$—CH_3

79 (a) (i) 4 peaks, ratio 3 : 2 : 2 : 3
  (ii) CH_3CH_2–O–CH_2CH_3
  (b) (ii) CH_3O^-
  (c)

$$H_3C - \underset{\underset{CH_3}{|}}{\overset{\overset{CH_3}{|}}{C}} - OH$$

83 (a) (i)

[structure: 2-methyl-5-(prop-1-en-2-yl)cyclohex-2-en-1-one 2,4-dinitrophenylhydrazone]

  (ii)

[structure: 2,3-dibromo-2-methyl-... cyclohexadienone with H_3C, CH_2Br, Br, CH_3, Br substituents]

86 (b) (i) C_3H_6O
  (ii)

$$H - \underset{\underset{H}{|}}{\overset{\overset{H}{|}}{C}} - \underset{\underset{H}{|}}{\overset{\overset{H}{|}}{C}} - \overset{\overset{O}{\|}}{C} - H$$

  (iii)

$$H - \underset{\underset{H}{|}}{\overset{\overset{H}{|}}{C}} - \underset{\underset{O}{\|}}{C} - \underset{\underset{H}{|}}{\overset{\overset{H}{|}}{C}} - H$$

  (iv) 71.5%

87 (c) (ii) 99.4%

88 (c) $[C_6H_7O_2(OCOCH_3)_3]_n$

89 (a) (i) HCN
  (b) (i) methanol

90

Cold	Boiling for half an hour
B No change	Ethanoic acid and ammonium chloride
C No change	Ethanoic acid and methanol
D $CH_3CH_2NH_3^+Cl^-$ Ethylammonium chloride	No further change

93 (a) (ii) C_2H_5COCl, propanoyl chloride

95 (b) (i) +1.88V
  (ii) +1.55V

97 (a) (i) –196 kJ mol^{-1}
  (c) (ii) 2.4×10^{-4} g m^{-3}

98 (d) $\Delta H \approx 0$

99 (d) (i) $[(Fe(CN)_6]^{3-}$
  (h) 0.1715 g
  (i) 14.3%

100 (d) (i) ΔH_{soln} = +119 kJ mol^{-1}

101 (a) –60 kJ mol^{-1}

102 (a) (ii) 0.96 g dm^{-3}
  (iii) 0.85 g dm^{-3}
  (c) (i)

$$H - \underset{\underset{H}{|}}{\overset{\overset{H}{|}}{C}} - \underset{\underset{\underset{+}{NH_3}}{|}}{\overset{\overset{H}{|}}{C}} - \overset{\overset{O}{\|}}{C} - O^-$$

103 (b) (ii) 47.15 g

105 (d) 0.1 mol dm^{-3}

106 (b) (i) $[Cu(H_2O)_6]^{2+}$, $[Ni(H_2O)_6]^{2+}$, $[Zn(H_2O)_6]^{2+}$
  (ii) $Ni(OH)_2$, $Cu(OH)_2$
  (iii) $[Zn(OH)_4]^{2-}$ or $[Zn(OH)_3]^-$ or ZnO_2^{2-}
  (e) (ii) 25%

107 (a) (i) mol^{-2} dm^6 s^{-1}
  (b) (i) 3

108 (a) (i) 90 g mol^{-1}
  (ii) 1.25×10^{-4} mol dm^{-3}
  (b) $C_3H_6O_3$; CH_3—$CH(OH)$—$COOH$ [1]
  (c) 15 $^+CH_3$
  30 ^+CHOH
  45 $^+COOH/^+CH_3CHOH$
  75 $^+CH(OH)COOH$

109 (a) (ii) 1.91
  (b) (iii) 3.43

112 (d) –79 kJ mol^{-1}
  (e) (i) CH_2Cl
  (ii)

$$H - \underset{\underset{Cl}{|}}{\overset{\overset{H}{|}}{C}} - \underset{\underset{Cl}{|}}{\overset{\overset{H}{|}}{C}} - H$$

113 (a) (ii) 4
  (b) –48.9 kJ mol^{-1}
  (c) (ii) octane –47.89 kJ g^{-1}; ethanol –29.67kJ g^{-1}

114 (a) **A**

$$H - \underset{\underset{H}{|}}{\overset{\overset{H}{|}}{C}} - \underset{\underset{H}{|}}{\overset{\overset{H}{|}}{C}} - \underset{\underset{H}{|}}{\overset{\overset{H}{|}}{C}} - \underset{\underset{H}{|}}{\overset{\overset{H}{|}}{C}} - Br$$

  B

$$H - \underset{\underset{H}{|}}{\overset{\overset{H}{|}}{C}} - \underset{\underset{H}{|}}{\overset{\overset{H}{|}}{C}} - \underset{\underset{H}{|}}{\overset{\overset{H}{|}}{C}} - \underset{\underset{H}{|}}{\overset{\overset{H}{|}}{C}} - OH$$

  C

$$H - \underset{\underset{H}{|}}{\overset{\overset{H}{|}}{C}} - \underset{\underset{H}{|}}{\overset{\overset{H}{|}}{C}} - \underset{\underset{H}{|}}{\overset{\overset{H}{|}}{C}} - \overset{\overset{O}{\|}}{C} - OH$$

  E

$$H - \underset{\underset{H}{|}}{\overset{\overset{H}{|}}{C}} - \underset{\underset{H}{|}}{\overset{\overset{H}{|}}{C}} - \underset{\underset{Br}{|}}{\overset{\overset{H}{|}}{C}} - \underset{\underset{H}{|}}{\overset{\overset{H}{|}}{C}} - H$$

  (d)

$$CH_3CH_2CH_2 - \overset{\overset{O}{\|}}{C} - O - CH_2CH_2CH_2CH_3$$

  (f) (i) 2.91
  (iii) 4.82

115 (b) (i) **A**

[structure: benzene ring with CH_2CH_3]

  B

[structure: benzene ring with $\overset{\overset{O}{\|}}{C}$—OH]

  C

[structure: benzene ring with $\overset{\overset{O}{\|}}{C}$—Cl]

  D

[structure: benzene ring with $\overset{\overset{O}{\|}}{C}$—$NH_2$]

  (c) (ii) first order

Answers to Review Questions

1. (a) 16 g (b) 31 g (c) 32 g (d) 124 g (e) 44 g (f) 215.8 g (g) 12.82 g
   (h) 120 g
2. (a) 0.1 (b) 0.1 (c) 0.1 (d) 0.2 (e) 1.0 (f) 0.125 (g) 3.0 (h) 0.002
3. (a) 1.2×10^{24} (b) 6×10^{22} (c) 1.8×10^{24} (d) 1.8×10^{22}
4. (a) To prevent air molecules impeding the passage of particles through the mass spectrometer.
   (b) The accelerating force on X^{2+} will be twice that on X^+.
   (c) The deflecting field on X^{2+} will be twice that on X^+.
   (d) Relative masses, i.e. just numbers
   (e) None
5. (a) $\frac{15.3}{27}$ (0.57) moles of X combine with $\frac{13.6}{16}$ (0.85) moles of O
   (b) 1 mole of X combines with $\frac{0.85}{0.57}$ (1.5) moles of O
   (c) X_2O_3
6. (a) $Ba^{2+}(aq) + SO_4^{2-}(aq) \rightarrow BaSO_4(s)$
   Moles of $BaSO_4$ = Moles of $SO_4^{2-} = \frac{4.66}{233} = 0.02$
   (b) Moles of M_2SO_4 = moles of $SO_4^{2-} = 0.02$
   (c) 0.02 moles of $M_2SO_4 = 5.34$ g ∴ 1 mole $M_2SO_4 = \frac{5.34}{0.02} = 267$ g
   (d) $2A_r(M) + 32 + 4 \times 16 = 267$ ∴ $A_r(M) = 85.5$
   (e) Rubidium
7. (a) Mass of carbon $= \frac{3.52 \times 12}{44} = 0.96$ g
   Mass of hydrogen $= \frac{1.44 \times 2}{18} = 0.16$ g
   ∴ Mass of oxygen $= 2.40 - 0.96 - 0.16 = 1.28$ g
   (b) Ratio of masses C : H : O = 0.96 : 0.16 : 1.28
   Ratio of moles C : H : O = 0.08 : 0.16 : 0.08 = 1 : 2 : 1
   ∴ Empirical formula = CH_2O
   $M_r(CH_2O) = 30$ ∴ Molecular formula = $C_2H_4O_2$
8. (a) $H^{37}Cl^+$ (b) $^{35}Cl^+$ and $^{37}Cl^+$
   (c) There is 3 times as much ^{35}Cl to ^{37}Cl and therefore 3 times as much $H^{35}Cl$ to $H^{37}Cl$.
   (d) $^1H^+$, $^2H^+$, $^{12}C^+$, $^{12}C^1H^+$, $^{12}C^1H_2^+$, $^{12}C^1H_3^+$, $^{12}C^1H_3{}^2H^+$

Chapter 2

1. (a) 0.01 (b) 0.03 (c) $AlBr_3$ (d) 266.7 (e) Al_2Br_6
   (f) $2Al + 3Br_2 \rightarrow Al_2Br_6$
2. C and D
3. (a) One molecule of hydrogen peroxide contains two atoms of hydrogen and two atoms of oxygen.
   (b) $2H_2O_2 \rightarrow 2H_2O + O_2$
   $\quad\quad$ 68 g $\quad\quad$ 36 g $\quad$ 32 g
   ∴ 17 g $H_2O_2 \rightarrow$ 8 g O_2
   (c) 8 g
4. (a) Oxygen, O_2
   (b) Lithium chloride, LiCl
   (c) $LiClO_4 \rightarrow LiCl + 2O_2$
   (d) $LiCl(aq) + AgNO_3(aq) \rightarrow LiNO_3(aq) + AgCl(s)$
   (e) 106.4 g $LiClO_4 \rightarrow$ 42.4 g LiCl
   ∴ 2.13 g $LiClO_4 \rightarrow \frac{42.4}{106.4} \times 2.13$ g (0.85 g) LiCl

5. $M_r(Na_2CO_3.10H_2O) = 286$
   $M_r(Na_2CO_3) = 106$
   ∴ 1000 g (1kg) of $Na_2CO_3.10H_2O = \frac{1000}{286} = 3.5$ moles for 4p
   1000 g (1kg) of $Na_2CO_3 = \frac{1000}{106} = 9.4$ moles for 8p
   It is better to buy the anhydrous salt at 8p per kg.
6. (a) $NaCl(aq) + AgNO_3(aq) \rightarrow AgCl(s) + NaNO_3(aq)$
   (b) $2NaCl(l) \rightarrow 2Na(l) + Cl_2(g)$
   (c) $2Fe(s) + 3Cl_2(g) \rightarrow 2FeCl_3(s)$
   (d) $CH_4(g) + 2O_2(g) \rightarrow CO_2(g) + 2H_2O(g)$
   (e) $CuO(s) + H_2SO_4(aq) \rightarrow CuSO_4(aq) + H_2O(l)$

   (a) $\cancel{Na^+}(aq) + Cl^-(aq) + Ag^+(aq) + \cancel{NO_3^-}(aq) \rightarrow Ag^+Cl^-(s) + \cancel{Na^+}(aq) + \cancel{NO_3^-}(aq)$
   i.e. $Cl^-(aq) + Ag^+(aq) \rightarrow Ag^+Cl^-(s)$
   (b) $2Na^+(l) + 2Cl^-(l) \rightarrow 2Na(l) + Cl_2(g)$
   (c) $2Fe(s) + 3Cl_2(g) \rightarrow 2Fe^{3+}(Cl^-)_3(s)$
7. A. T B. T C. F D. T E. F
8. (b) As more acid is added, more 'thio' can react and more ppte. forms. When all the 'thio' has reacted, extra amounts of acid cannot produce any more ppte.
   (c) 10 cm^3 of 0.1 mol dm^{-3} 'thio' react with 5 cm^3 of 0.2 mol dm^{-3} acid.
   i.e. $\frac{10}{1000} \times 0.1$ mol 'thio' react with $\frac{5}{1000} \times 0.2$ mol acid
   i.e. 1 : 1 molar proportions.
9. (a) 7.93 g (b) 2.07 g (c) $\frac{7.93}{138.2} = 0.057$ (d) $\frac{2.07}{18} = 0.115$
   (e) $\frac{0.115}{0.057} = 2$ (f) $K_2CO_3.2H_2O$
10. (a) $2NaOH(aq) + H_2SO_4(aq) \rightarrow Na_2SO_4(aq) + 2H_2O(l)$
    (b) $\frac{28}{1000} \times 1 = 0.028$ (c) 0.056 (d) 0.056
    (e) $\frac{0.056}{25} \times 1000 = 2.24$

Chapter 3

1. (a) $-1, +1, +5, -1, -1, -1$
   (b) $+1, +2, +4, +5, -2, -3$
2. (a) e.g. H_2S, H_2S_2, S, S_2Cl_2, SCl_2, SO_2, SO_3
   (b) (i) SO_3^{2-} oxidised, MnO_4^- reduced
   (ii) I^- oxidised, H_2SO_4 reduced
   (iii) $S_2O_3^{2-}$ oxidised to SO_2 and reduced to S
   (iv) SO_3^{2-} oxidised, Ce^{3+} reduced
   (v) $S_2O_3^{2-}$ oxidised, I_2 reduced
3. (b) Cl_2 oxidised to ClO^- and reduced to Cl^-
   (d) Ca oxidised to Ca^{2+}, F_2 reduced to F^-
4. (a) $Cu \rightarrow Cu^{2+} + 2e^-$, $(NO_3^- + 2H^+ + e^- \rightarrow NO_2 + H_2O) \times 2$
   (b) $2I^- \rightarrow I_2 + 2e^-$, $H_2O_2 + 2H^+ + 2e^- \rightarrow 2H_2O$
   (c) $(SO_3^{2-} + H_2O \rightarrow SO_4^{2-} + 2H^+ + 2e^-) \times 3$, $Cr_2O_7^{2-} + 14H^+ + 6e^- \rightarrow 2Cr^{3+} + 7H_2O$
   (d) $2Cl^- \rightarrow Cl_2 + 2e^-$, $MnO_2 + 4H^+ + 2e^- \rightarrow Mn^{2+} + 2H_2O$
   (e) $Zn \rightarrow Zn^{2+} + 2e^-$, $(Ag^+ + e^- \rightarrow Ag) \times 2$
5. (a) $(Fe^{3+} + e^- \rightarrow Fe^{2+}) \times 2$, $2I^- \rightarrow I_2 + 2e^-$
   (b) From right to left through the voltmeter

(c) The salt bridge allows the movement of ions and completes the circuit for flow of charge. K^+ ions move into salt bridge on the right and out of it on the left. Negative ions move into salt bridge on the left and NO_3^- out of it on the right.

(d) If $[KI]$ increases, I^- is more likely to give up electrons, so the voltage will increase.

6 Oxidation is loss of electrons or increase in oxidation number. Reduction is gain of electrons or decrease in oxidation number.

(a) $2Fe^{2+} \rightarrow 2Fe^{3+} + 2e^-$ oxidation, $Cl_2 + 2e^- \rightarrow 2Cl^-$ reduction

(b) $O_2^- + H_2 \rightarrow H_2O + 2e^-$ oxidation, $Cu^{2+} + 2e^- \rightarrow Cu$ reduction

(c) $Cu \rightarrow Cu^{2+} + 2e^-$ oxidation,
$NO_3^- + 4H^+ + 3e^- \rightarrow NO + 2H_2O$ reduction

(d) $2Na \rightarrow 2Na^+ + 2e^-$ oxidation, $H_2 + 2e^- \rightarrow 2H^-$ reduction

(e) This is not a redox reaction.

7 (a) A +6, B +6, C +3, D +6, E +6, F +3

(b) $B \rightarrow C$, $C \rightarrow D$ and $E \rightarrow F$

(c) $CrO_3 + H_2O + 2NH_3 \rightarrow (NH_4)_2CrO_4$
$2(NH_4)_2CrO_4 \rightarrow Cr_2O_3 + 5H_2O + 2NH_3 + N_2$
$Cr_2O_3 + 4NaOH + \frac{3}{2}O_2 \rightarrow 2Na_2CrO_4 + 2H_2O$
$2CrO_4^{2-} + 2H^+ \rightarrow Cr_2O_7^{2-} + H_2O$
$Cr_2O_7^{2-} + 14H^+ + 6Fe^{2+} \rightarrow 2Cr^{3+} + 7H_2O + 6Fe^{3+}$

8 (a) A +5, B +4, C +4, D +3, E +3, F +2, G +5, H +5, I +4

(b) F_2 is the strongest oxidising agent, oxidising V to +5 state, then Cl_2 which oxidises V to +4 state, then Br_2 which oxidises V to +3 state.

(c) Cl_2 is a stronger oxidising agent than HCl. Cl_2 oxidises V to +4 state, HCl oxidises V to +2 state.

9 0.5g $FeSO_4.7H_2O = \frac{0.5}{277.8}$ moles (0.00234)
$(Fe^{2+} \rightarrow Fe^{3+} + e^-) \times 5$
$MnO_4^- + 8H^+ + 5e^- \rightarrow Mn^{2+} + 4H_2O$
$\therefore$ 0.00234 moles Fe^{2+} reacts with $\frac{0.0018}{5}$ moles MnO_4^-
If volume of 0.1 mol dm^{-3} $KMnO_4 = x$
$\frac{x}{1000} \times 0.1 = \frac{0.0018}{5}$; $x = 3.6$ cm^3
Assumptions made : All substances are pure. The reaction goes to completion.
5 moles $Fe^{2+} \equiv 1$ mole MnO_4^-

10 (a) $MnO_4^- + 8H^+ + 5e^- \rightarrow Mn^{2+} + 4H_2O$
$(Fe^{2+} \rightarrow Fe^{3+} + e^-) \times 5$
(b) 5
(c) $\frac{25}{1000} \times 0.04 \times 5 = 5 \times 10^{-3}$
(d) $5 \times 10^{-3} \times 10 \times 56 = 2.8$ g
(e) 28%

Chapter 4

1 (a) (i) Li and Be (ii) B and N (iii) LiF, BeO, BN, CO, NO (iv) N, O, F

(b) (i) Increase from Li to C and then fall to very low values
(ii) Increase from Li(+1) to C (+4), then fall N (+3), O(−2) and F (−1)
(iii) Li and Be chloride – ionic, high boiling points
BCl$_3$ to ClF – simple molecular, low boiling points

(c) Li and Be – lose outer shell electrons – +1 and +2 oxidation numbers. The other chlorides are simple molecular – B (3 bonds), C (4 bonds), N (3 bonds), O (2 bonds), F (1 bond).

(d) (i) Li and Be – metals $\rightarrow$ B, C metalloids $\rightarrow$ N_2, O_2, F_2, Ne simple molecules
(ii) LiCl, BeCl$_2$ – ionic, BCl$_3$, CCl$_4$, NCl$_3$, OCl$_2$, FCl – simple molecular

2 (a) (i) C and Si (ii) C and Si
(b) (i) He, Ne, Ar (ii) He, Ne, Ar
(c) (i) Giant molecular (ii) Simple molecular (monatomic)
(d) (i) Li, Be, B, C, Al, Si (ii) H, He, N, O, F, Ne, Ar

(e) (i) Relatively strong forces remain even after melting.
(ii) The energy needed to melt the substance is almost sufficient to overcome the forces between liquid particles and cause it to vaporise.

4 (a) (i) Electrons (ii) Ions, A^{2+} and B^-.

(b) AB_2 is ionic. In the molten state, A^{2+} and B^- ions can move to the electrodes and allow conduction. In solid AB_2, A^{2+} and B^- ions are held in a lattice structure by their oppositely charged neighbours. The ions can vibrate but cannot move from their positions in a solid lattice, so solid AB_2 cannot conduct electricity.

(c) A is a metal from its conductivity as a solid. A is below H in the activity series with an oxidation number of +2 in its compounds, e.g. copper.
B_2 has a simple molecular structure (melting point very low). B_2 forms B^- ions with metals. B is probably a halogen. Melting point of B is <-50 °C so it is probably F_2 or Cl_2.

5 (a) In MgO(l), Mg^{2+} and O_2^- ions can move towards the electrodes and conduct. In MgO(s), Mg^{2+} and O_2^- ions are held in a regular lattice by the attraction of oppositely charged ions. In the solid, the Mg^{2+} and O^{2-} ions can only vibrate about fixed positions.

(b) MgO(s) is held by 2+ charges on Mg^{2+} ions and 2− charges on O^{2-} ions. In NaCl, the ionic bonds are due to single 1+ and 1− charges only.

(c) In both solid and liquid Mg, the outer electrons are very mobile and can move from one atom to the next under the influence of charge.

(d) SiO_2 forms a giant molecular structure with Si and O atoms joined by covalent bonds in a regular lattice structure. CO_2 forms simple molecules of CO_2 with very weak Van der Waals forces between separate molecules.

6 (a) From Na to Mg to Al there is an increase in the number of outer shell electrons and therefore increasing forces between the mobile electrons and the metal ions in the lattice. This causes the melting point to increase. Si has a giant molecular structure in which each Si is joined to four other Si atoms in its structure giving it a very high melting point.

(b) The element P to Ar have simple molecular structures. Within the molecules of P_4, S_8 and Cl_2, there are strong covalent bonds, but only weak Van der Waals forces between the separate molecules of P_4, S_8, Cl_2 and Ar. Therefore the boiling points of these 4 elements are much lower.

(c) The boiling points relate to very different molecules – P_4, S_8, Cl_2 and Ar.

7 1 g X reacts with 141 cm^3 H_2. $\therefore$ 1 mole H_2 (22 400 cm^3) reacts with 158.8 g X.
Now formula could be XH, XH$_2$, XH$_3$ etc. $\therefore$ A_r(X) = 79.4 or 158.8 or 238.2
37.3 g Cl_2 reacts with 62.7 g X $\therefore$ 1 mole Cl_2 (71g) reacts with 119.3 g X
Now formula could be XCl, XCl$_2$, XCl$_3$ etc. $\therefore$ A_r(X) = 59.7 or 119.3 or 179.0 or 238.7
(a) Probably 238
(b) Uranium
(c) Y is UH$_3$ and Z is UCl$_4$

Chapter 5

1 (a) Te – 52, I – 53
(b) Te – 127.6, I – 126.9
(c) ^{128}Te has 52p, 76n, 52e; ^{127}I has 53p, 74n, 53e
(d) Because it has one proton fewer
(e) Te has 1 less proton but its common isotope has 2 more neutrons than the common isotope of I.
(f) Ar/K, Co/Ni, Th/Pa

2 (a) 1 – is untrue. Atoms are not indestructible as indicated by radioactive isotopes and atom smashing experiments.

2 – is untrue. Atoms of elements beyond uranium in the periodic table have been created (synthesised). Atoms are destroyed in radio isotopes.

3 – is untrue. Isotopes of the same element are not alike.

4 – is true.

5 – is true but atoms can combine in very large numbers, e.g. in proteins, carbohydrates and nucleic acids.

(b) It provides a good summary for non-nuclear reactions.

3 (a) (i) 9 (ii) 19

(b) $\dfrac{19}{6 \times 10^{23}}$ g $= 3.17 \times 10^{-23}$ g

(c) $d = \dfrac{m}{v} = \dfrac{3.17 \times 10^{-23}}{\frac{4}{3}\pi\left(\frac{10^{-14}}{2}\right)^3} = 6 \times 10^{19}$ g m^{-3}

(d) They are extremely strong

(e) $\dfrac{\frac{4}{3}\pi\left(\frac{10^{-10}}{2}\right)^3}{\frac{4}{3}\pi\left(\frac{10^{-14}}{2}\right)^3} = 10^{12}$

4 (a) 14

(b) 28, 29 and 30

(c) $\dfrac{92}{100} \times 28 + \dfrac{5}{100} \times 29 + \dfrac{3}{100} \times 30 = 25.76 + 1.45 + 0.9 = 28.11$

(d) They contain different proportions of the 3 isotopes.

5 (a) $_1^1H_2$, $_1^1H_1^2H$, $_1^2H_2$

(b) 2, 3 or 4

(c) 1 dm^3 of hydrogen has a mass of 0.10 g

22.4 dm^3 of hydrogen has a mass of 2.24 g (i.e. 1 mole of H$_2$)

$\Rightarrow$ relative atomic mass of hydrogen = 1.12

Suppose the hydrogen has $x\%$ H, then $\dfrac{x}{100} \times 1 + \dfrac{(100-x)}{100} \times 2 = 1.12$

$\therefore x + 200 - 2x = 112$ $\therefore x = \% \, _1^1H = 88, \% \, _1^2H = 12$

6 (a) (i) 28.0062 (ii) 28.0172 (iii) 27.9949 (iv) 28.0312 (v) 28.0282

(b) DCN

7 Relative molecular mass = 60

(a) 17 could be —OH, 29 could be 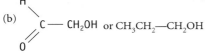, CH$_3$CH$_2$ or

31 could be —CH$_2$OH

(b) 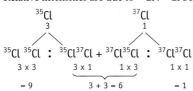 or CH$_3$CH$_2$—CH$_2$OH

8 (a) CH$_2^{35}$Cl$_2$, CH$_2^{35}$Cl^{37}Cl and CH$_2^{37}$Cl$_2$

(b) Relative intensities are due to ^{35}Cl : ^{37}Cl being 3 : 1

^{35}Cl ^{35}Cl : ^{35}Cl^{37}Cl + ^{37}Cl^{35}Cl : ^{37}Cl^{37}Cl

3 x 3   3 x 1   1 x 3   1 x 1

= 9   3 + 3 = 6   = 1

9 (a) Carry out measurements in a mass spectrometer to find the relative atomic mass of the isotopes.

Find the areas below the relative abundance versus atomic mass graphs to obtain the percentage of each isotope. Hence calculate the relative atomic mass.

(b) Bombard a sample of the element with high speed electrons and measure the wavelength of the X-rays produced. Calculate the frequency of the X-rays and then use the graph of $\sqrt{(\text{X-ray frequency})}$ against atomic number to find the atomic number.

Chapter 6

1 (a)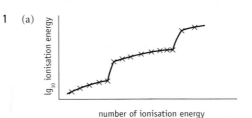

(b) From the nucleus the electronic shell structure is 2, 8, 5.

(c) $1s^2 2s^2 2p^6 3s^2 3p^3$

2 (a) **A** – First ionisation energy is relatively small, i.e. electron easily removed. Second ionisation energy is much larger (×9), i.e. second electron very difficult to remove due to breakage into a complete shell.

(b) **B** and **D** – Group II

(c) Group III – The first three ionisation energies are relatively low, i.e. three electrons are relatively easy to remove. The fourth ionisation energy, corresponding to the removal of a fourth electron is much larger due to the breakage into a new (full) shell.

(d) Element **C**, 630 + 1600 = 2230 kJ mol^{-1}

3 (a)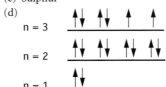

(b) 16

(c) Sulphur

(d)

n = 3   ↑↓ ↑↓ ↑ ↑

n = 2   ↑↓ ↑↓ ↑↓ ↑↓

n = 1   ↑↓

4 (a)
- The number of protons (+ charge) in the nucleus
- The distance of the outermost electron from the nucleus
- The shielding of inner shells of electrons
- The number of electrons in the outer shell (any three)

(b) From Li to Cs, the outermost electron is further from the nucleus and is shielded more effectively by inner shells of electrons. These two factors outweigh the increasing positive charge on the nucleus from Li to Cs and therefore the first ionisation energy falls from Li to Cs.

(c) See sections 6.3 and 6.4.

5 (a) Frequency or wavelength.

(b) Frequency – decreasing, wavelength – increasing, L to R.

(c) When sufficient energy is supplied, electrons can be promoted (excited) to higher energy levels in an atom. The electrons are unstable in higher levels and emit excess energy as radiation and fall back into lower energy levels.

As the energy levels are fixed, the energy lost between any higher level and a lower level is also of a certain fixed value so the radiation emitted will only have certain fixed frequencies (i.e. specific colours). This means that the atomic spectrum of an element will consist of a series of lines of different colours.

(d) Level 2 i.e. $n = 2$

(e) Each element has its own characteristic line spectrum. So, an element can be identified by its line spectrum just as a criminal can be identified from a fingerprint.

6 (a) Neon and argon have filled outer shells of electrons. It is difficult to break into these stable electron structures, so their first ionisation energies are relatively large.

The elements sodium and potassium, immediately after Ne and Ar, have only one electron in their outer shells. This is easily removed, so their first ionisation energies are relatively low.

(b) In the outer shell, Mg has a filled $3s^2$ sub-shell. This confers some partial stability on an atom. Aluminium, immediately after Mg, has the outer shell structure, $3s^2 3p^1$.

The single electron in the $3p$ sub-shell is removed relatively easily, so the first ionisation energy of Mg is greater than that of Al.

(c) The first ionisation energy involves removing an electron from an atom which is initially uncharged. The second ionisation energy involves removing an electron from an ion which is oppositely charged.

(d) The maxima of the graphs correspond to the removal of an electron from atoms or ions with filled and therefore more stable outer shells.

Chapter 7

1 (a) α-particles 4.0 and +2, β-particles $\frac{1}{1800}$ and -1

(b) Electromagnetic waves

(c) See section 7.2

2 (a) Energy gained by absorption of α-particles = Energy lost due to loss of heat.

(b) Let temperature of material = $x\,°C$ above its surroundings

∴ Heat loss from material = $8.4 \times \frac{x}{10}$ J min^{-1}

Heat gained from α-particle absorption = $8 \times 10^{-12} \times 4.2 \times 10^{10} \times 60$ J min^{-1}

When the steady state is achieved:

$8.4 \times \frac{x}{10} = 8 \times 10^{-12} \times 4.2 \times 10^{10} \times 60$

$x = \frac{8 \times 4.2 \times 6}{8.4} = 24$

∴ Temperature of material is 24 °C above its surroundings.

3 (a) Change in Mass No.   Change in At. No.

(i) -4 -2
(ii) 0 $+1$
(iii) 0 0
(iv) -1 0

(b) $^{232}_{90}$Th loses $6 \times {}^4_2$He^{2+} and $4 \times {}^0_{-1}$e Loss of mass = 24,

Loss of protons = $(6 \times 2) - 4$ Loss of protons = 8 ∴ Final product is $^{208}_{82}$Pb

(c) (i) $^7_3$Li + $^1_1$p → $2\,^4_2$He (ii) $^{232}_{90}$Th → $^{228}_{88}$Ra + $^4_2$He

(d) 1 mg Po emits 3×10^{18} α-particles

Assume that each Po atom emits only 1 α-particle and that 1 mole of Po contains 6×10^{23} atoms

For 6×10^{23} Po atoms (A_r for Po), mass = $\frac{0.001}{3 \times 10^{18}} \times 6 \times 10^{23}$ g = 200 g

∴ A_r(Po) = 200

4 (a) $9 \pm 1\%$

(b) (i) $^{14}_7$N + $^4_2$He → $^{17}_8$O + $^1_1$H

(ii) $^2_1$H + $^{56}_{26}$Fe → $^{57}_{27}$Co + $^1_0$n

5 $^{60}_{27}$Co is used to treat inaccessible cancer tumours because it emits very penetrating gamma-rays.

$^{60}_{27}$Co → $^{60}_{28}$Ni + $^0_{-1}$e + γ

$^{32}_{15}$P is used to treat superficial cancers such as skin cancer and as part of phosphate fertilisers to study the metabolism of phosphate and phosphorus by plants. It emits less penetrating beta-particles.

$^{32}_{15}$P → $^{32}_{16}$S + $^0_{-1}$e

$^{131}_{53}$I is used in the diagnosis and treatment of thyroid diseases and in research into iodine metabolism.

6 (a) By collision of neutrons with nitrogen atoms in the upper atmosphere.

$^{14}_7$N + $^1_0$n → $^{14}_6$C + $^1_1$H

(b) $^{14}_6$C → $^{14}_7$N + $^0_{-1}$e

(c) When a specimen containing carbon and hence a fixed concentration (say $x\%$) of ^{14}C dies, the ^{14}C decays and is not replaced. Knowing the half-life of ^{14}C and the percentage of ^{14}C in the archaeological remains, it is possible to 'date' the remains.

(d) See the last part of section 7.5.

Chapter 8

1 (a)

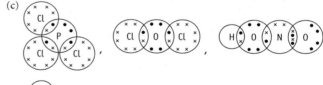

(b)

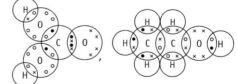

(c)

(d)

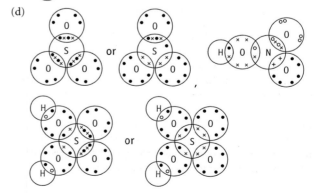

2 (a) (i) Moving from left to right across the period, the bonding changes from ionic in $NaCl$ and $MgCl_2$ to covalent in $AlCl_3$, $SiCl_4$, PCl_3 and SCl_2.

  (ii) Elements with one or two electrons in the outer shell lose these electrons to form ions (e.g. Na^+ and Mg^{2+}) in ionic compounds. Elements with 3 or more electrons in the outer shell form covalent bonds and gain further electrons in their outer shell.

(b) An ionic bond involves the transfer of one or more electrons from one atom to another forming oppositely charged ions. The electrostatic attraction between the oppositely charged ions constitutes the ionic bond.

e.g. $Na^x + {}^\bullet H \rightarrow [Na]^+ [{}^x_\bullet H]^-$

A covalent bond is formed by the sharing of a pair of electrons by two atoms. One electron is contributed by each atom to the shared pair.

e.g.

A co-ordinate bond is formed by the sharing of a pair of electrons by two atoms. Both the electrons in the shared pair are contributed by only one of the atoms.

e.g. The ammonium ion in $NH_4^+Cl^-$

3 Using — to represent a covalent bond

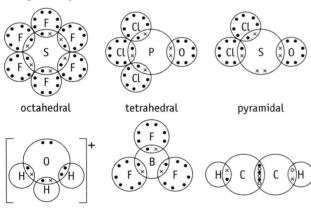

octahedral   tetrahedral   pyramidal

pyramidal   trigonal planar   linear

4 (a) C 2, 4; N 2, 5; O 2, 6; F 2, 7

(b)

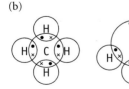

(c) CH_4 – tetrahedral with respect to atoms and tetrahedral with respect to negative centres around the C atom

NH_3 – pyramidal with respect to atoms, but tetrahedral with respect to negative centres (3 shared pairs and 1 lone pair of electrons) around the N atom

H_2O – bent (V-shaped) with respect to atoms, but tetrahedral with respect to negative centres (2 shared pairs and 2 lone pairs of electrons) around the O atom

HF – linear with respect to atoms, but tetrahedral with respect to negative centres (one shared pair and 3 lone pairs of electrons) around the F atom

(d) NH_4^+ – tetrahedral with respect to atoms around N atom

NH_3 – pyramidal with respect to atoms around N atom

NH_2 – bent (V-shaped) with respect to atoms around N atom

5 (a) X is 2, 7; Y is 2, 8, 8, 1 and Z is 2, 8, 18, 6

(b) (i) Ionic (ii) Covalent (iii) Ionic

(c)

  (i)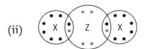

  (ii)

  (iii)

(d) (i) XY will have low volatility – ionic bonds between ions in lattice structure.

   X_2Z will have high volatility – weak intermolecular forces between separate molecules.

  (ii) XY will conduct in the liquid phase but not as a solid.

   $X^-(l)$ and $Y^+(l)$ ions can move to electrodes of opposite charge and conduct when liquid. In the solid phase, the ions are held in the lattice by oppositely charged neighbours.

   X_2Z cannot conduct electricity in solid or liquid state as it has no ions or mobile electrons.

  (iii) XY is soluble in water (a polar solvent) owing to the attraction of charged ions for polar water molecules.

   X_2Z is insoluble in water. It will be polar but is unable to form strong enough attractions with water to dissolve in it.

6 (a) It is relatively easier to form Sn^{4+} and Pb^{4+} ions than to form C^{4+} and Si^{4+} ions. In any group of the periodic table, electrons are lost more readily as A_r increases.

(b) The carbon–oxygen links in methoxymethane comprise one single covalent bond, those in CO_2 comprise two covalent bonds and those in carbon monoxide comprise one double covalent bond plus one co-ordinate bond.

i.e.

As the bonding increases, the bond length gets shorter.

(c)

Delocalisation allows 3 possible structures and the actual structure is somewhere in between all 3 with three equivalent carbon–oxygen bonds.

(d) Aluminium fluoride is an ionic compound in which Al^{3+} and F^- ions are strongly attracted, giving a high melting point.

Aluminium chloride forms dimers of Al_2Cl_6 in the solid state.

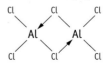

There are strong bonds within the Al_2Cl_6 dimers, but only weak inter-molecular forces between the dimers so the solid melts at relatively low temperatures.

7 (a) From Na to Cl, electron affinities get more exothermic.

(b) Atoms gain an electron more readily on moving across the periodic table. Atoms on the left prefer to lose electrons.

(c) Mg has an outer shell with two electrons in a filled *s* sub-shell. This confers partial stability on the Mg atom and it is less ready to accept an electron than might be expected. This makes its electron affinity more positive (endothermic) than expected.

(d) Si has an outer shell with two electrons in a *p* sub-shell. By gaining one more electron, it has a half-filled *p* sub-shell which confers extra stability. Thus Si gains an electron more readily and more exothermically than expected.

Chapter 9

1 (a) ICl and CH_2Cl_2

(b) (i) Trigonal planar   (ii) Linear   (iii) Linear   (iv) Tetrahedral

(c) $C_2H_5NH_2$, CH_3OH, H_2SO_3

(d) CO_2, BF_3

2 (a) (i) Non-polar   (ii) Non-polar   (iii) Polar

(b) $I_2 - CCl_4$ attractions are roughly similar in strength to $I_2 - I_2$ and $CCl_4 - CCl_4$ attractions, so I_2 mixes readily with (and is very soluble in) CCl_4.

(c) $I_2 - H_2O$ attractions are much weaker than $H_2O - H_2O$ polar attractions, so I_2 does not mix readily with water and is only slightly soluble in water.

(d) No; non-polar CCl_4 molecules will attract polar water molecules much less than the $H_2O - H_2O$ attractions.

(e) The water and CCl_4 do not mix but as I_2 is more soluble in CCl_4 than in H_2O, the yellow I_2 in water becomes much paler as most of the I_2 dissolves in CCl_4 to give a pink/purple solution.

3 (a)

cis   *trans*

Cis-dichloroethene is polar whereas the *trans* isomer is non-polar. The polarity leads to stronger intermolecular forces and therefore a higher boiling point for the *cis*-form.

(b) (i) Pentane is a linear molecule whilst 2,2-dimethylpropane is more spherical (globular). The linear molecules have greater possible surface areas of contact and therefore greater Van der Waals attractions making the boiling point higher.

(ii) Greater surface area (Van der Waals) attractions in pentane, pull the molecules closer together 'on average', so the density is greater for pentane.

(iii) 2-Methylbutane has a structure in between the linear of pentane and the spherical of 2,2-dimethylpropane;

$$CH_3 \text{---} CH \text{---} CH_2 \text{---} CH_3$$
$$|$$
$$CH_3$$

Its boiling point and density are therefore probably in between these two.

4 (a) Ensure you draw and label axes including units, plot points accurately and then draw the line/curve of best fit.

(b) NH_3 has significant H-bonding unlike PH_3 and AsH_3.

(c) $+13.3 \pm 0.3$ kJ mol^{-1}

(d) $+10$ kJ mol^{-1}

(e) The NH_3 molecule has only one lone pair of electrons and 3 N—H bonds, so it is limited to 'an average' of only two H-bonds per molecule – one involving a lone pair and one involving an H atom. The H_2O

molecule has two lone pairs of electrons and 2 O—H bonds. Thus it can form 'an average' of 4 H-bonds per molecule.

5 (a) Each water molecule in ice can form 4 H-bonds – 2 involving its 2 lone pairs and 2 involving its H atoms.

∴ energy of 1 mole of H-bonds in ice $= \frac{22}{4} = +5.5$ kJ mol^{-1}

(b) % H-bonds broken $= \frac{6}{22} \times 100 = 27\%$

(c) H_2O molecule is V-shaped (bent). The two H—O bonds are 'polarised' owing to the high electronegativity of oxygen. This makes the region of the O atom slightly negative and the regions around the H atoms slightly positive.

i.e.

(d) Each of the H—X bonds will be polar (possibly only slightly) as no two elements have exactly the same electronegativity. If the molecule is non-polar, the effect of one H—X bond must counteract that of the other, so the molecule must be linear.

6 (a) **B**

(c)

(d)

(e)

(b) (i) approx. 67 (ii) approx. 67 (iii) approx. 134

7 (a)

(b) 1000

(c)

(d) (i) Proteins have one —N̈H group and one C=Ö: group with lone pairs on the O atom per amino acid unit. These can H-bond with water molecules allowing certain proteins to mix with and dissolve readily in water.

(ii) Hydrogen bonds within a protein between —N̈—H groups and between —N̈—H and C=Ö: groups can be aligned precisely between specific amino acid units. This will give the protein a precise configuration.

(iii) H-bonding between fibrous protein molecules such as hair, wool and silk causes kinks in the generally linear structure. When the material is pulled, some hydrogen bonds are broken and the structure becomes straighter as the material stretches. When the material is released, hydrogen bonds reform again and the material contracts.

8 (a) All three molecules have weak Van der Waals forces of attraction. Water molecules also have 2 O—H groups for H-bonding. Ethanol molecules (CH_3CH_2OH) also have one O—H group for H-bonding. Ethoxyethane has no O—H groups and cannot H-bond, so there are only weak Van der Waals forces between the molecules. Consequently, water molecules have the strongest attractions between them and ethoxyethane molecules have the weakest attractions between them, so the order of boiling points is $H_2O > CH_3CH_2OH > CH_3CH_2OCH_2CH_3$. The analogous S compounds cannot H-bond, but they do have Van der Waals forces between molecules which are greater as M_r increases.

(b) BF_3 has a trigonal planar molecule with 3 identical B—F bonds and no lone pair on the B atom. It is therefore non-polar. NF_3 has a pyramidal molecule with 3 identical N—F bonds plus a lone pair on the N atom. It is therefore polar.

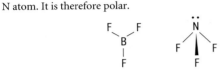

Chapter 10

1 (a) An atom is the smallest part of an element which can ever exist. A molecule is the smallest part of an element or a compound which can exist alone under ordinary conditions. An ion is a charged particle formed from one or more atoms by loss or gain of one or more electrons.

(b) (i) Cu atoms (ii) CO_2 molecules (iii) a giant molecule of C atoms

(c) See sections 10.5 and 10.6

2 (a) A
(b) D
(c) C
(d) B
(e) E

3 (a) D
(b) C
(c) B
(d) E
(e) D
(f) E

4 (a) A material which will scratch or mark another when the two are rubbed together
(b) $2C + SiO_2 \rightarrow CSi + CO_2$
(c) It is cheaper.
(d) Diamonds are beautiful and rare. Carborundum is neither.
(e) It provided a cheap abrasive, for the manufacture of machines.

5 I_2 is lustrous. I_2 has a low conductivity. These 2 facts are unexpected as it is a non-metal. ICl liberates I_2 at the cathode. This is unexpected as ICl would be expected to be simple molecular. If I_2 is liberated at the cathode, it suggests the presence of I^+ ions.

6 (a) Differences in the electronegativity of constituent elements result in $MgCl_2$ being ionic, while $AlCl_3$ and $SiCl_4$ are simple molecules.
(b) SiO_2 forms a giant (3-D) molecule of covalently bonded Si and O atoms with a very high melting point. CO_2 forms separate small CO_2 molecules in which the carbon atom is strongly bonded to two oxygen atoms, but there are only very weak Van der Waals forces between separate CO_2 molecules. Therefore CO_2 has a very low melting point.
(c) $Ca^{2+}O^{2-}(s)$ has double charges on the constituent ions, whereas $Na^+Cl^-(s)$ has single charges.
(d) Water is polar and can H-bond. Benzene is non-polar. Glucose is polar and can H-bond. Cyclohexane is non-polar. Glucose can form Glu–H_2O attractions similar in strength to H_2O–H_2O attractions and therefore mixes readily with and dissolves in water. The Glu–Glu bonds

are much stronger than possible Glu–benzene attractions and so glucose does not dissolve in benzene.

7 (a) $n\lambda = 2d \sin\theta$. Assuming $n = 1$
$0.1537 = 2d \sin 14°$, $d = 0.318$ nm $= 3.18 \times 10^{-10}$ m
(b) Assuming $n = 2$ and $d = 0.318$ nm, $\theta = 29°$
(c) (i) 3.22×10^{-29} m^3 (ii) 0.5 (iii) 6.44×10^{29} m^3
(d) density $= \frac{m}{v}$ volume of one mole $= \frac{74.6}{2} = 37.3$ cm$^3 = 3.73 \times 10^{-5}$ m^3
(e) $L = \dfrac{\text{volume of 1 mole}}{\text{volume of 1 'ion pair'}} = \dfrac{3.73 \times 10^{-5}}{6.44 \times 10^{-29}} = 5.8 \times 10^{23}$

8 (a) Co-ordination number = number of nearest neighbours
(b) c.p. – most metals except alkali metals, Fe and Mn
b.c.c. – alkali metals, Fe and Mn
(c) (i) 12 (ii) 8
(d) (i) Higher density in c.p. structure – atoms closer 'on average'
(ii) Higher melting point in c.p. structure – atoms closer and therefore more strongly bonded (iii) Greater malleability in b.c.c. structure – atoms not so strongly bonded and therefore able to move over each other more easily

9 (a) XCl_x – ionic, YCl_y – covalent bonds between Y and Cl atoms within YCl_y molecules, but Van der Waals bonds between the separate YCl_y molecules.
(b) Positive X ions are strongly bonded to Cl^- ions in lattice, which leads to high melting and boiling points. Separate YCl_y mols are only held together by weak Van der Waals forces so have low melting and boiling points.
Ions in XCl_x form attractions to polar water molecules so they dissolve in and mix with water easily. The attractions of the oppositely charged ions for each other means that XCl_x will not dissolve in non-polar benzene.
Molecules of YCl_y can form similar attractions to benzene as for each other so they mix with and dissolve in benzene easily. But H-bonded water molecules will not mix with YCl_y because the water molecules are too strongly attracted to each other.

10 (a)

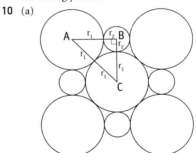

(b) ΔABC is a 45°/45°/90° triangle
$\therefore \dfrac{r_1 + r_2}{2r_1} = \dfrac{1}{\sqrt{2}}$, $r_1 + r_2 = \sqrt{2}r_1$
$\Rightarrow \dfrac{r_2}{r_1} = \sqrt{2} - 1$
(c) NaI

1 (a) (i) [graph: p vs V, decreasing curve]
(ii) [graph: p vs $\frac{1}{V}$, straight line increasing]
(iii) [graph: pV vs p, horizontal line]
(iv) [graph: pV vs V, horizontal line]

(b)

(i)
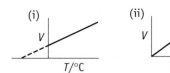

(ii)

(c)

(i)

(ii)

(d)
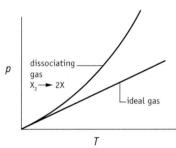

2   $\dfrac{p_1 V_1}{T_1} = \dfrac{p_2 V_2}{T_2}$, $p_1 = p_2$, $\dfrac{975}{278} = \dfrac{V_2}{298}$, $V_2 = 1045$ cm^3

∴ The balloon will burst.

3   (i) An increase in pressure will increase the rate of diffusion, as particles will move away more rapidly from a point of higher pressure.
(ii) Assuming that an increase in volume results in a decrease in pressure, the rate of diffusion will decrease.
(iii) An increase in temperature will cause particles to move faster, so the rate of diffusion will increase.

4   (a) $p_A + p_B = 800$, $800 \times 300 = p_A \times 400$, ∴ $p_A = 600$ mm Hg
⇒ $p_B = 200$ mm Hg
(b) There are 2 isotopes of argon with different relative atomic masses.

5   $pV = nRT$, $pV = \dfrac{m}{M} RT$, $0.25 \times 1 = \dfrac{0.5}{M} \times 0.082 \times 364$
$M_r = 60$

6   Let volume of ethyne $= x$ cm^3, ∴ volume of ethene $= (10 - x)$ cm^3
$C_2H_2(g) + 2\frac{1}{2}O_2(g) \rightarrow 2CO_2(g) + H_2O(l)$
   x $2\frac{1}{2}x$ $2x$ 0
$C_2H_4(g) + 3O_2(g) \rightarrow 2CO_2(g) + 2H_2O(l)$
$(10 - x)$ $3(10 - x)$ $2(10 - x)$
volume of O_2 used $= 30 - 2 = 28$
∴ $28 = 2\frac{1}{2}x + 3(10 - x)$, ∴ $-2 = -\frac{1}{2}x$, $x = 4$ cm^3
∴ volume of ethyne $= 4$ cm^3, so volume of ethene $= 6$ cm^3

7   $V_{CO} + V_{CO_2} + V_{N_2} = 60$ cm^3 volume O_2 taken $= 20$ cm^3
On explosion $2CO(g) + O_2(g) \rightarrow 2CO_2(g)$
           2 volumes   1 volume   2 volumes
Decrease in volume $= 1$ volume $= 10$ cm^3,
∴ volume of CO $= 2$ volumes $= 20$ cm^3
Volume of CO_2 after explosion $= 35$ cm^3
∴ volume of CO_2 initially $= 15$ cm^3
∴ volume of N_2 $= 25$ cm^3
⇒ % CO_2 in exhaust gas $= \dfrac{15}{60} \times 100 = 25\%$

8   (a) 2 volumes phosphorus hydride → solid P + 3 volumes H_2
1 volume   phosphorus hydride   → solid P + $\frac{3}{2}$ volumes H_2
∴ n molecules phosphorus hydride → solid P + $\frac{3}{2}n$ molecules H_2
1 molecule phosphorus hydride   → solid P + $\frac{3}{2}$ molecules H_2
1 molecule phosphorus hydride   → solid P + 3 atoms H
(b) ∴ Empirical formula of phosphorus hydride $= P_xH_3$
(c) PH_3
(d) Determine relative molecular mass

(e) PH_3 will probably resemble NH_3:
physical properties – gaseous, slightly denser than air, pungent smell, very soluble in water (any 2)
chemical properties – basic nature, reaction with HCl(g), decomposition at high temperature (any 2)

9   (a) $2H_2(g) + O_2(g) \rightarrow 2H_2O(l)$
$2CO(g) + O_2(g) \rightarrow 2CO_2(g)$
(b) 2 volumes H_2 + 1 volume O_2 → zero volume
2 volumes CO + 1 volume O_2 → 2 volumes CO_2
(c) CO_2 is absorbed by concentrated KOH. ∴ volume CO_2 = 10 cm^3
(d) Volume of CO = 10 cm^3
(e) Total decrease in volume as a result of explosion $= 80 - 51 = 29$ cm^3
(f) Decrease due to CO = 5 cm^3
Decrease due to H_2 = 24 cm^3
(g) Volume of H_2 in original 40 cm^3 = 16 cm^3
∴ volume of N_2 in original 40 cm^3 $= 40 - 10 - 16 = 14$ cm^3

10 (a) $C_xH_y(g) + (x + \frac{y}{4})O_2(g) \rightarrow xCO_2(g) + \frac{y}{2}H_2O(l)$
1 volume $(x + \frac{y}{4})$ volume x volume   0 volume
20 cm^3 $(150 - 30)$ cm^3 80 cm^3
1 cm^3 6 cm^3 4 cm^3
⇒ $x = 4$ and $x + \frac{y}{4} = 6$ ∴ $y = 8$
∴ Molecular formula of A $= C_4H_8$
(b)

CH$_3$ — CH$_2$ — CH ═ CH$_2$
but-1-ene

cis-but-2-ene

trans-but-2-ene

2-methylpropene

cyclobutane

methylcyclopropane

<hr>

Chapter 12

1   (a) $Q = m \times c \times \Delta T = 400 \times 4.2 \times 10 = 16\,800$ J $= 16.8$ kJ
(b) Number of moles ethanol used = 0.02.
Energy produced per mole = 840 kJ.
(c) The experiment is not carried out under standard conditions (i.e. 298 K, 1 atm.)
(d)
• Heat loss to surroundings
• Heat loss to calorimeter and thermometer
• Incomplete combustion
• Evaporation of ethanol when not burning
• Water not stirred
• Loss of heat from water   (Any 4)

2   (a) $\Delta H_f^{\ominus}$ is the enthalpy change when 1 mole of a substance is formed from its elements under standard conditions.
$\Delta H_c^{\ominus}$ is the enthalpy change when 1 mole of a substance is completely burnt in oxygen under standard conditions.
(b) $2C$ (graphite) $+ 3H_2(g) + \frac{1}{2}O_2 \rightarrow C_2H_5OH$
(c) $2C$ (graphite) $+ 3H_2(g) + \frac{1}{2}O_2(g) + 3O_2(g) \xrightarrow{\Delta H_f^{\ominus}} C_2H_5OH(g) + 3O_2(g)$

$2\Delta H_c^{\ominus}$ (graphite) $\Delta H_c^{\ominus}(C_2H_5OH(l))$
$+3\Delta H_c^{\ominus}(H_2(g))$

$2CO_2 + 3H_2O(l)$

(d) $\Delta H_f^{\ominus}$ $(C_2H_5OH(l)) = (2 \times -393) + (3 \times -286) - (-1368)$
$= -276$ kJ mol^{-1}

3 (a) Number of moles of butane $= \dfrac{1.12}{22.4} = 0.05$

Heat liberated by burning butane $= 0.05 \times 3000 \times \dfrac{3}{4} = 112.5$ kJ

∴ If volume of water which they could boil is x cm^3,

$x \times 4.2 \times 80 = 112\,500$

$\Rightarrow x = 335$ cm^3

(b) Other assumptions:
- 1 mole of butane at 0 °C and 1 atm occupies 22.4 dm^3.
- The butane is completely burnt.
- The mass of 1 cm^3 of water is 1.0 g.
- The specific heat capacity of water = 4.2 J g^{-1} K^{-1}.
- The water boils at 100 °C.

4 (a) $C_6H_{12}O_6(s) + 6O_2(g) \rightarrow 6CO_2(g) + 6H_2O(l)$
$C_2H_5OH(l) + 3O_2(g) \rightarrow 2CO_2(g) + 3H_2O(l)$

(b)

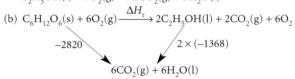

$\Delta H_r = -2820 - (2 \times -1368) = -84$ kJ

(c) The process is exothermic so the temperature of the water may rise. If the temperature becomes too high, it would kill the yeasts.

(d) The home reaction vessel is much smaller than those used in breweries. This means that a higher proportion of the heat produced is lost. So, home brewers put their fermentation vessel in a warm place to reduce heat loss.

5 (a) ΔH (step 1) $= +413 + 158 - 565 = +6$ kJ; ΔH (step 2) $= -495$ kJ.

(b) Yes, step 1 with a small endothermic value will be relatively slow compared to step 2 with a large exothermic value.

(c) $CH_4 + Cl_2 \rightarrow CH_3^{\bullet} + HCl + Cl^{\bullet}$ $\Delta H = +413 + 242 - 431 = +224$ kJ
$CH_3^{\bullet} + Cl^{\bullet} \rightarrow CH_3Cl$ $\Delta H = -339$ kJ

(d) Yes

(e) The activation energy for step 1 is much lower for CH_4/F_2 than for CH_4/Cl_2. The energy in sunlight is required to provide part of the activation energy for the CH_4/Cl_2 reaction.

6 (a) A solute dissolves in a solvent provided the relative sizes of solute–solvent attractions, solute–solute attractions and solvent–solvent attractions are roughly the same.

A solute is unlikely to dissolve in a solvent if the solute–solute attractions are very different to the solvent–solvent attractions.

(b) The lattice energy of an ionic compound, ΔH_{latt}, is the enthalpy change of formation for one mole of the substance from gaseous ions under standard conditions.

i.e. $X^+(g) + Y^-(g) \rightarrow X^+Y^-(s)$

The hydration energy of an ionic compound, ΔH_{hyd}, is the enthalpy change when gaseous ions making up one mole of the substance are solvated by water molecules to form an infinitely dilute solution.

$X^+(g) + Y^-(g) + (aq) \rightarrow X^+(aq) + Y^-(aq)$

The enthalpy change of solution, ΔH_{soln}, of an ionic compound is the enthalpy change when one mole of the substance dissolves in water to form an infinitely dilute solution.

$X^+Y^-(s) + (aq) \rightarrow X^+(aq) + Y^-(aq)$

(c)

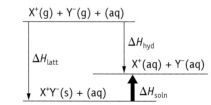

(d) $\Delta H_{\text{latt}} = \Delta H_{\text{hyd}} - \Delta H_{\text{soln}}$
$\Rightarrow -642 = \Delta H_{\text{hyd}} - 21$
$\Delta H_{\text{hyd}} = -621$ kJ mol^{-1}

7 (a) (i) The (first) ionisation energy of an element is the energy required to remove one electron from each atom in a mole of gaseous atoms under standard conditions.

(ii) The atomisation energy of an element is the enthalpy change when one mole of gaseous atoms is formed from the element under standard conditions.

(iii) The lattice enthalpy of an ionic compound is the enthalpy change of formation for one mole of the substance from gaseous ions under standard conditions.

(b) See section 12.12 for full labelling.
(All values on the cycle are in kJ mol^{-1})

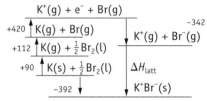

(c) $\Delta H_{\text{latt}}(KBr(s)) = +342 - 420 - 112 - 90 - 392 = -672$ kJ mol^{-1}

(d) From KF to KI, the lattice energy gets less exothermic. As the ionic radius of the halide ion increases, its charge density decreases. Its attraction to the positive K^+ ion also decreases, so the lattice energy gets less exothermic.

8 (a) $\Delta H_{\text{latt}}(CaO(s)) = -790 + 141 - 249 - 1100 - 590 - 177 - 636$
$= -3401$ kJ mol^{-1}

(b) ΔH_{i_1} (Ca) involves the removal of one electron from neutral atoms. ΔH_{i_2} (Ca) involves the removal of a second electron from positive Ca^+ ions. The attraction of negative electrons for Ca^+ ions make ΔH_{i_2} more endothermic.

(c) O atoms readily gain electrons. The first electron is being added to neutral O atoms and this process, $\Delta H_e(O)$, is exothermic. However, the second electron must be added to a negative O^- ion. The repulsion between the O^- ion and an approaching electron means that this process, $\Delta H_e(O^-)$, will be endothermic.

(d) The first ionisation energy of Mg will be more endothermic than that for Ca. The atomic radius of Mg is smaller than that of Ca and the screening effect of inner shells of electrons will be less for Mg than Ca. Both these factors will increase the attraction of the nucleus for the outermost electron. However, the nuclear charge in Mg is less than Ca which would cause the attraction of the outermost electron to be less. Overall, the distance and shielding factors outweigh the nuclear charge factor so the outermost electron is held more strongly by Mg than Ca atoms, so the first ionisation energy is more endothermic for Mg.

Chapter 13

1 (a) From Na to Cl, electrons are being added to the third shell at about the same distance from the nucleus. At the same time, protons are being added to the nucleus and as the positive nuclear charge increases the outer electrons are drawn in closer.

(b) Na^+, Mg^{2+} and Al^{3+} have lost all the outer shell electrons from their respective atoms.

Consequently, ionic radii are less than atomic radii.

Cl^- and S^{2-} have gained electrons and filled their outermost shells. Accordingly, ionic radii are greater than atomic radii.

(c) From Na to Cl, the atomic radius decreases and the nuclear charge increases. Both of these factors mean that the outermost electrons are held more strongly by the nuclear charge from Na to Cl. This means that the first ionisation energy will show a general increase from Na to Cl.

(d) The outer shell electron structure of Mg is $3s^2$ and that of Al is $3s^23p^1$. Mg has a filled $3s$ sub-shell whereas Al has a filled $3s$ sub-shell and a single electron in the $3p$ sub-shell. It is easier to remove the single $3p$ electron in Al than to disturb the pair of electrons filling the $3s$ sub-shell in Mg.

2 (a) LiCl and $BeCl_2$ have giant structures with strong bonds between the constituent ions. The strong bonds result in high boiling points and these chlorides are solids at 20°C. Chlorides of the other elements have simple molecular structures. They consist of small molecules containing a few atoms with only weak forces between the separate molecules. This means that they are easily separated so these chlorides have low boiling points and are either gases or liquids at 20°C.

(b) LiCl is ionic. In the liquid state, the ions Li^+ and Cl^- can move to the electrodes and conduct electricity.

(c) LiCl will dissolve in the water to form a neutral solution of aqueous ions, $Li^+(aq)$ and $Cl^-(aq)$. $BeCl_3$ will react with water to form a solution containing boric acid, $H_3BO_3(aq)$ and hydrochloric acid, $HCl(aq)$.

(d) As the elements change from low electronegativity (Li) to high electronegativity (F), the bonding in their chlorides changes from ionic (LiCl) to covalent (BCl_3 to ClF).

3 (a) $NaH + H_2O \rightarrow NaOH + H_2$
$MgH_2 + 2H_2O \rightarrow Mg(OH)_2 + H_2$

(b) SiH_4 is a simple molecular substance which is non–polar with little attraction to water molecules.

(c) $PH_3 + H_2O \rightleftharpoons PH_4^+ + OH^-$

(d) $HCl + H_2O \rightleftharpoons H_3O^+ + Cl^-$

(e) Metal hydrides contain H^- ions which react with water to form H_2 and OH^- ions. The OH^- ions make the final solution alkaline.
SiH_4 is a non–polar molecule which has no reaction with water.
PH_3 is a simple molecule in which the P atom has a lone pair of electrons to attract H^+ ions from water molecules. This leaves OH^- ions to make the solution alkaline.
H_2S and HCl contain electronegative S and Cl atoms which form negative ions S^{2-} and Cl^- and release H^+ ions. The H^+ ions make the solution acidic.

4 (a) N, F   (b) B   (c) B, N   (d) Ne   (e) Be   (f) Li, Be, N

5 (a)
(i)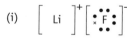

(ii)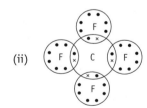

(iii)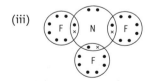

(b) (i) trigonal planar (ii) tetrahedral (iii) pyramidal (iv) V–shaped

(c) LiF is ionic. It contains Li^+ and F^- ions in a regular lattice arrangement in which all the Li^+ ions are attracted to all the F^- ions.

6 (a) Na, Mg and Al form oxides in which the bonding is ionic.
S forms SO_3 in which the bonding is covalent/dative covalent.

(b) (i) $Na_2O + H_2O \rightarrow 2NaOH$ sodium hydroxide – alkaline
$MgO + H_2O \rightarrow Mg(OH)_2$ magnesium hydroxide – alkaline
Al_2O_3 – insoluble in water
$SO_3 + H_2O \rightarrow H_2SO_4$ sulphuric acid – acidic

(ii) $Na_2O + 2H^+ \rightarrow 2Na^+ + H_2O$ neutralisation
       salt   water
$MgO + 2H^+ \rightarrow Mg^{2+} + H_2O$ neutralisation
       salt   water
$Al_2O_3 + 6H^+ \rightarrow 2Al^{3+} + 3H_2O$ neutralisation
           salt   water
SO_3 – no reaction with H^+ ions (acid), but it reacts with the water present.

(iii) Na_2O ⎱ no reaction with alkali, but react
      MgO ⎰ with water present
$Al_2O_3 + 2OH^- + 3H_2O \rightarrow 2Al(OH)_4^-$
                              tetrahydroxoaluminate(III)
$SO_3 + 2OH^- \rightarrow SO_4^{2-} + H_2O$ neutralisation
         salt   water

(c) $MgCl_2$ is ionic ($Mg^{2+}(Cl^-)_2$) with strong ionic bonds between Mg^{2+} and Cl^- ions in a regular lattice – hence high melting point. Aluminium chloride forms simple molecular dimers of Al_2Cl_6 On heating, the Al_2Cl_6 sublimes as simple molecules.

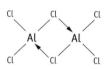

$SiCl_4$ is a simple molecule with covalent bonds between each Si and 4Cl atoms. There are only weak bonds between the separate $SiCl_4$ molecules so the liquid is volatile and vaporises easily.

7 (a) Na Mg Al Si P S Cl Ar

(b) (i) Na, Mg   (ii) none

(c) (i) $AlCl_3$

(ii) Pass dry Cl_2 over heated Al and allow the aluminium chloride to sublime into a collection vessel whilst the excess Cl_2 passes out. The preparation must be carried out in a fume cupboard.

8 (a) P is Na or K, Q is Cr

(b) P – s-block, Q – d-block

Chapter 14

1 (a) Acids – substances which dissociate in water to produce H^+ ions. Bases – substances which react with H^+ ions to form water.

(b) An acid is a proton donor; a base is a proton acceptor.

(c) (i) $HSO_4^-(aq) \rightleftharpoons H^+(aq) + SO_4^{2-}(aq)$

(ii) $HSO_4^- + H^+ \rightarrow H_2SO_4$

(d) According to the Arrhenius definition, bases were restricted to the oxides and hydroxides of metals which could react with H^+ ions to form water.
The Brønsted–Lowry definition for bases covered a much wider range of substances which simply accepted a proton (H^+ ion).
e.g. almost all oxyanions, $CO_3^{2-} + H^+ \rightarrow HCO_3^-$,
almost all anions, $Cl^- + H^+ \rightarrow HCl$,
molecules with lone pairs of electrons, $NH_3 + H^+ \rightarrow NH_4^+$

2 (a) 0.5   (b) 1.75×10^{-3} (c) 0.875×10^{-3} (d) 8.75×10^{-3}

(e) $\dfrac{2.5 - (8.75 \times 10^{-3} \times 106)}{18} = 0.087$ (f) 10

3 (a) Oxidation is the loss of electrons. Reduction is the gain of electrons.

(b) (i) $Mg \rightarrow Mg^{2+} + 2e^-$ oxidation
$Cl_2 + 2e^- \rightarrow 2Cl^-$ reduction

(ii) $I_2 + 2e^- \rightarrow 2I^-$ reduction

      $2S_2O_3^{2-} \rightarrow S_4O_6^{2-} + 2e^-$ oxidation

   (iii) $MnO_2 + 4H^+ + 2e^- \rightarrow Mn^{2+} + 2H_2O$ reduction

      $2Cl^- \rightarrow Cl_2 + 2e^-$ oxidation

   (iv) $H_2O_2 + 2H^+ + 2e^- \rightarrow 2H_2O$ reduction

      $2I^- \rightarrow I_2 + 2e^-$ oxidation

   (v) $2Fe^{3+} + 2e^- \rightarrow 2Fe^{2+}$ reduction

      $2I^- \rightarrow I_2 + 2e^-$ oxidation

(c) (i) Zn metal, 1.0 M $ZnSO_4$(aq), 2 beakers, salt bridge (filter paper soaked in KNO_3(aq)), H_2(g) at 1 atm. and 25°C, platinised Pt electrode, 1.0 M HCl, glass 'bell' with holes near base, voltmeter.

   (ii)

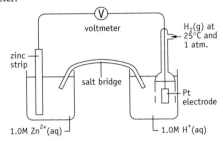

zinc strip   salt bridge   voltmeter H_2(g) at 25°C and 1 atm.   Pt electrode   1.0M Zn^{2+}(aq)   1.0M H^+(aq)

4   (a) Cr is oxidised from +3 to +6

   (b) no redox

   (c) S is oxidised from +2 to +2.5

   (d) C is oxidised from +3 to +4

   (e) I is reduced from +5 to 0

5   (a) See figure 14.6. Electrons flow in the external circuit from Zn rod to Cu rod.

   (b) and (c) $Zn \rightarrow Zn^{2+} + 2e^-$ +0.76 V

             $Cu^{2+} + 2e^- \rightarrow Cu$ +0.34 V   } e.m.f. = +0.76 + 0.34 = +1.10 V

   (d) (i) E.m.f. will rise since the process $Cu^{2+} + 2e^- \rightarrow Cu$ will be promoted.

     (ii) E.m.f. will fall since the process $Zn \rightarrow Zn^{2+} + 2e^-$ will be reduced (i.e. the reverse reaction will be promoted).

6   (a) $Fe \rightarrow Fe^{2+} + 2e^-$ Iron forms Fe^{2+} ions.

      $H_2O + \frac{1}{2}O_2 + 2e^- \rightarrow 2OH^-$ Dissolved O_2 is reduced to OH^- ions.

      Then $Fe^{2+} + 2OH^- \rightarrow Fe(OH)_2$ $Fe(OH)_2$ is precipitated.

      $2Fe(OH)_2 + \frac{1}{2}O_2 \rightarrow Fe_2O_3 + 2H_2O$ $Fe(OH)_2$ is oxidised by dissolved O_2 to form rust.

   (b) If the Zn coating is pierced, the Zn and Fe metals form a cell in aqueous solution and the Zn will be more likely to go into solution as Zn^{2+} ions than Fe as Fe^{2+}.

   If the Sn coating is pierced, the Sn and Fe metals also form a cell in aqueous solution, but in this case the Fe will go into solution as Fe^{2+} ions, so making rusting more likely.

   (c) Sn is much less reactive than Zn, so is less likely to contaminate food. Zn is much cheaper than Sn, so is used to galvanise buckets.

7   (a) $2H^+ + 2e^- \rightarrow H_2$ (b) (i) Ce^{4+} (ii) Zn

   (c) Fe^{3+} and Ce^{4+}

   $2I^- + 2Fe^{3+} \rightarrow I_2 + 2Fe^{2+}$

   $2I^- + 2Ce^{4+} \rightarrow I_2 + 2Ce^{3+}$

   (d) (i) +0.77 + 0.44 = 1.21 V

     (ii) $Fe^{3+} + e^- \rightarrow Fe^{2+}$ and $Fe \rightarrow Fe^{2+} + 2e^-$

     (iii) From the Fe/Fe^{2+} half cell to the Fe^{3+}/Fe^{2+} half cell

   (e) $2Fe^{3+} + Fe \rightarrow 3Fe^{2+}$

8   (a) $\frac{2.75}{1000} \times 0.0015 \times 100 \times 40 \times 1000$ mg Ca per 100 cm^3 serum

   = 16.5 mg per 100 cm^3

   (b) All the Ca^{2+} ions in blood serum react with edta. No other substances in the blood serum react with edta.

   (c) The level is abnormal.

Chapter 15

1   (a) and (b) (i) Atomic radius increases as electrons are occupying an additional shell.

   (ii) Ionisation energy decreases. The distance of the outermost electrons and the shielding of these electrons increases. These factors make it easier to remove the outermost electrons despite the increasing nuclear charge.

   (iii) Strength as reducing agents increases; $E^{\ominus}$ for the process $M(s) \rightarrow M^{2+}(aq) + 2e^-$ is more positive with increasing atomic number.

   The process $M(s) \rightarrow M^{2+}(aq)$ involves atomisation, ionisations and then hydration of ions. Overall this is more likely to happen (i.e. the elements react more readily as reducing agents) as atomic number increases.

   (iv) The vigour of the reaction increases with atomic number. The reaction involved is

   $M(s) + Cl_2(g) \rightarrow M^{2+}(Cl^-)_2(s)$ and the metal is acting as a reducing agent (see part (iii)).

   (v) Electropositivity is a measure of the tendency of the metal to lose electrons and form positive ions, (i.e. to act as a reducing agent). This increases with atomic number.

2   (a) Metals – they form positive ions and are good conductors of electricity.

   (b) Ca^{2+} ions (c) See table 15.3 and also question 1.

   (d) Reactivity for Group II involves the process $M(s) \rightarrow M^{2+}(s) + 2e^-$ or $M(s) \rightarrow M^{2+}(aq) + 2e^-$.

   Part of each of these processes involves ionisation energies,

   i.e. $M(g) \rightarrow M^+(g) + e^-$ and $M^+(g) \rightarrow M^{2+}(g) + e^-$

   whilst the redox (electrode) potential of a Group II metal involves the process

   $M^{2+}(aq) + 2e^- \rightleftharpoons M(s)$.

   (e) The process $Mg(s) \rightarrow Mg^+(g) + e^-$ is easier (less endothermic) than the process $Mg(s) \rightarrow Mg^{2+}(g) + 2e^-$, but the energy evolved when $Mg^{2+}(g)$ ions are hydrated in water or surrounded by negative ions in a crystal is much greater than the corresponding processes for $Mg^+(g)$. So, magnesium compounds contain Mg^{2+} ions and not Mg^+ ions.

   (f) Group II compounds contain M^{2+} ions whereas Group I compounds contain M^+ ions. M^{2+} ions will have a greater charge density and therefore have a greater attraction for polar water molecules.

3   (a) Be, Mg, Ca, Sr, Ba, Ra

   (b) Mg is $1s^2\,2s^2\,2p^6\,3s^2$, Ca is $1s^2\,2s^2\,2p^6\,3s^2\,3p^6\,4s^2$

   (c)

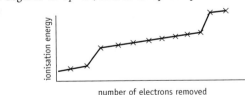

ionisation energy   number of electrons removed

   (d) Their most stable ions are those of the form M^{2+}.

   (e) The process $M(s) \rightarrow M^{2+}(aq) + 2e^-$ or $M(s) \rightarrow M^{2+}(s) + 2e^-$ is more exothermic than the corresponding processes for M^+ or M^{3+}.

   (f) The first ionisation energy will decrease (become less endothermic, see Q.1 (a)(ii)).

4   (a) $LiNO_3$ decomposes on heating forming Li_2O; other Group I nitrates form nitrites.

   Li_2CO_3 decomposes on heating at much lower temperatures than other Group I carbonates.

   LiOH decomposes on heating forming Li_2O; other Group I hydroxides are stable.

(b) See table 15.2 and Question 1.

(c) The first ionisation energy is the energy required to convert one mole of gaseous atoms to one mole of gaseous ions each with one positive charge (under standard conditions).

(d) The first ionisation energy is relatively low. The next 8 ionisation energies rise gradually through intermediate values. The last 2 ionisation energies are very large.

(e) The Na^+ ion is more stable than the Na^{2+} ion.

(f) The reactivity of Group I involves $M(s) \rightarrow M^+(aq) + e^-$ or $M(s) \rightarrow M^+(s) + e^-$.

These processes can be seen as atomisation, followed by ionisation and then finally hydration by water or combination with anions to form a solid crystal. Ionisation is an important part of these processes.

5 (a)

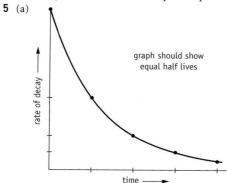

graph should show equal half lives

rate of decay

time

(b) (i) $^{42}_{19}K \rightarrow ^{42}_{20}Ca + ^{0}_{-1}e$

(ii) $^{42}_{20}Ca$

(iii) 0.005 g

6 (a) React dolomite with dilute HCl until no more will dissolve.
$CaCO_3(s) + 2HCl(aq) \rightarrow CaCl_2(aq) + CO_2(g) + H_2O(l)$
$MgCO_3(s) + 2HCl(aq) \rightarrow MgCl_2(aq) + CO_2(g) + H_2O(l)$
Filter off any excess insoluble dolomite.
Precipitate $CaSO_4$ by adding dilute $H_2SO_4(aq)$
$Ca^{2+}(aq) + SO_4^{2-}(aq) \rightarrow CaSO_4(s)$
Filter off the $CaSO_4(s)$, wash with water and allow to dry at room temperature.
Add $Na_2CO_3(aq)$ to the filtrate to precipitate $MgCO_3(s)$
$Mg^{2+}(aq) + CO_3^{2-}(aq) \rightarrow MgCO_3(s)$
Filter off the precipitate of $MgCO_3(s)$, wash it with water and then react it with the minimum amount of dilute H_2SO_4.
$MgCO_3(s) + H_2SO_4(aq) \rightarrow MgSO_4(aq) + CO_2(g) + H_2O(l)$
Now evaporate the solution of $MgSO_4$ until crystals start to form around the edges of the evaporating basin.
Finally, set the solution aside to cool and form crystals of $MgSO_4.7H_2O(s)$.

(b) Heat the crystals of $MgSO_4.7H_2O(s)$ until they reach a constant minimum mass.
$MgSO_4.7H_2O(s) \xrightarrow{\text{heat}} MgSO_4(s) + 7H_2O(g)$

7 (a) 1 mole $SO_2 \equiv$ 1 mole I_2
$\therefore$ number of moles of $SO_2 = \frac{12}{1000} \times 0.025 = 3 \times 10^{-4}$

(b) 1 mole $Na_2SO_3 \rightarrow$ 1 mole SO_2
$\therefore$ Mass of Na_2SO_4 present $= 3 \times 10^{-4} \times 126 = 3.78 \times 10^{-2}$ g

(c) 0.038 g per 100 g meat $\Rightarrow 0.0378 \times 10^4$ g per 10^6 g meat = 378 p.p.m.

(d) $Ba^{2+}(aq) + SO_4^{2-}(aq) \rightarrow BaSO_4(s)$

(e) 1 mole $SO_2 \rightarrow$ 1 mole SO_4^{2-} $\therefore 3 \times 10^{-4}$ moles SO_4^{2-} are formed.
Mass of $BaSO_4(s)$ precipitate $= 3 \times 10^{-4} \times 233$ g = 0.0699 g.

8 (a) (i) Without water   (ii) Containing water of crystallisation
(iii) The water present in hydrated crystals

(b) (i) $CaSO_4(s) + C(s) \rightarrow CaO(s) + SO_2(g) + CO(g)$

(ii) $(CaSO_4)_2.H_2O(s) + 3H_2O(l) \rightarrow 2CaSO_4.2H_2O(s)$

(iii) $CaSO_4.2H_2O(s) \rightarrow CaSO_4(s) + 2H_2O(g)$

(c) (i) When mixed with water it hardens and sets quickly to form a firm solid. Broken limbs can therefore be set quickly.

(ii) When mixed with water it expands and sets firmly. This means that the mould is filled and an accurate model is produced.

(d) Anhydrite sets only very slowly when mixed with water.

Chapter 16

1 (a) (i) Anode(+) $2Cl^-(aq) \rightarrow Cl_2(g) + 2e^-$ OH^- ions attracted to anode but not discharged.

Cathode(−) $2H^+(aq) + 2e^- \rightarrow H_2(g)$ Na^+ ions attracted to cathode but not discharged.

H^+ and Cl^- ions are discharged, leaving behind Na^+ and OH^- ions which constitute sodium hydroxide.

(ii) $Cl_2(g) + 2NaOH(aq) \rightarrow NaCl(aq) + NaClO(aq) + H_2O(l)$

(b) $IO_3^- + 6H^+ + 5e^- \rightarrow \frac{1}{2}I_2 + 3H_2O$
$HSO_3^- + H_2O \rightarrow SO_4^{2-} + 3H^+ + 2e^-$

(c) KF(s) in HF(l) produces F_2 at the anode on electrolysis.
KF(s) in H_2O produces O_2 at the anode on electrolysis because of the discharge of OH− ions from the water.
Anode(+) $F^-(aq) \not\longrightarrow$ not discharged
$4OH^-(aq) \longrightarrow 4e^- + O_2(g) + 2H_2O(l)$

(d) (i) $^{207}_{83}Bi + ^{4}_{2}He \rightarrow ^{211}_{85}At$

(ii) By loss of α–particles $^{211}_{85}At \rightarrow ^{207}_{83}Bi + ^{4}_{2}He$

2 (a) Slow reaction to form an equilibrium mixture containing H_2, At_2 and hydrogen astatide. $H_2(g) + At_2(g) \rightarrow 2HAt(g)$

(b) Solid At_2 will react with NaOH(aq) to form sodium astatide and sodium astatate(v).
$3At_2(s) + 6NaOH(aq) \rightarrow NaAtO_3(aq) + 5NaAt(aq) + 3H_2O(l)$

(c) Products are hydrogen astatide (steamy vapour) plus astatine (dark vapour and dark, probably black) solid. Products from H_2SO_4 include $NaHSO_4(s)$ and $SO_2(g)$.
$NaAt(s) + H_2SO_4(l) \rightarrow NaHSO_4(s) + HAt(g)$
$2At^-(s) \rightarrow At_2(g) + 2e^-$
$H_2SO_4 + 2H^+ + 2e^- \rightarrow SO_2 + 2H_2O$

(d) A precipitate of silver astatide (predictably orange) will form
$AgNO_3(aq) + NaAt(aq) \rightarrow NaNO_3(aq) + AgAt(s)$

(e) Solid At_2 will react with $Na_2S_2O_3$ to form sodium astatide and sodium tetrathionate.
$At_2 + 2e^- \rightarrow 2At^-$
$2S_2O_3^{2-} \rightarrow S_4O_6^{2-} + 2e^-$

3 (a) (i) An additional filled shell from one element to the next

(ii) Decreasing strength as oxidising agents from F_2 to I_2 (section 16.6)

(iii) Colour, state, melting point, boiling point, odour – all change gradually from F_2 to I_2 (section 16.4).

(iv) Usual oxidation state in compounds is −1 owing to their readiness to gain one electron to obtain a stable outer shell.

(b) See part (a) and sections 16.4 to 16.8.

(c) The electron affinity of an element is the enthalpy change when 1 mole of gaseous atoms of the element each gain one electron to form gaseous ions with one negative charge.
i.e. $X(g) + e^- \rightarrow X^-(g)$
This process is most exothermic for F becoming gradually less exothermic as atomic number increases.
When halogens react to form negative ions, the process can be viewed in 3 stages: atomisation e.g. $\frac{1}{2}Hal_2 \rightarrow Hal(g)$,
electron affinity e.g. $Hal(g) + e^- \rightarrow Hal^-(g)$

and finally hydration e.g. $Hal^-(g) + (aq) \rightarrow Hal(aq)$ or crystal formation with positive ions. Electron affinity is obviously one important part of the overall reaction.

(d) Any three of the following six points.

• F^- cannot be oxidised to F_2 by common oxidising agents because F_2 is such a powerful oxidising agent.

• F has only one oxidation state of -1 and never has positive oxidation states like other halogens because it is the most electronegative element.

• F_2 reacts with all non–metals except N_2, He, Ne and Ar, with SiO_2 and with hydrocarbons. (Even Cl_2 is much less reactive – section 16.5.)

• F_2 reacts rapidly with water to form HF and O_2 unlike other halogens (section 16.7).

• F_2 reacts with alkalis to give OF_2 and F^-. Other halogens give halates and halides.

• F^- gives no precipitate with $AgNO_3(aq)$ as F^- ions have a strong affinity for polar water molecules and AgF is soluble in water.

4 (a) The mass spectrograph will identify $^{35}_{17}Cl^+$, $^{37}_{17}Cl^+$, $^{35}_{17}Cl_2^+$, $^{35}_{17}Cl\,^{37}_{17}Cl^+$ and $^{37}_{17}Cl_2^+$ at relative masses 35, 37, 70, 72 and 74 respectively.

(b) As the group is descended, the attraction of halogens for electrons (i.e. their oxidising power) gets less. This is because the electrons are further from the attraction of the nucleus and are shielded by inner shells. This reduces their affinity for electrons even though there is an increasing positive charge on the nucleus.

(c) F_2 is the most electronegative element, therefore it must always have a negative oxidation number in its compounds. In addition, F can only gain one electron to form a very stable filled outer shell so it has only the one oxidation state of -1 in its compounds.

(d) HF contains highly electronegative F atoms, which make the HF molecule very polar. This results in H-bonding between HF molecules which does not occur to this degree in other hydrogen halides. The H-bonding creates extra strong intermolecular forces which results in a boiling point for HF higher that that of HI which has much heavier molecules.

5 (a) $I_2 + 2S_2O_3^{2-} \rightarrow 2I^- + S_4O_6^{2-}$

(b) $\frac{25}{1000} \times 0.1 \times \frac{1}{2} = 1.25 \times 10^{-3}$ mol

(c) $1.25 \times 10^{-3} \times 2 \times 10 = 2.5 \times 10^{-2}$ mol

(d) $\frac{2.5 \times 10^{-2} \times 64 \times 100}{2.0} = 80\%$ Cu

6 (a) 1,2,3,4,5,6-hexachlorocyclohexane

(b) They are either non–polar or weakly polar or contain large non-polar parts of their molecules which have intermolecular forces of similar strength to those in fats but which are much weaker than the intermolecular forces in water.

Therefore, they are attracted to, mix with and dissolve in fats but cannot do so with water in which the molecules retain stronger attractions to each other.

(c) Seed-eating birds often migrate at the end of the summer and if present in the autumn they are not feeding chicks. This means that pesticides are less likely to be ingested by these species. Furthermore, autumn-sown crops are likely to have pesticides washed off them more rapidly and more thoroughly during the autumn/winter period.

(d) DDT is absorbed/ingested by insects. The DDT becomes associated with/reacts with chemicals (e.g. enzymes) critical to the insects' metabolism. As a result of this association/reaction the critical chemical is ineffective. The insect can no longer metabolise effectively and dies.

(e) Small doses may reduce the effectiveness of a critical chemical (e.g. enzyme) without removing it completely.

(f) Specific action on targeted insects; no action on other organisms (animals or plants); effective in small quantities/low concentrations; easily stored and transported; biodegradable; easily and cheaply manufactured.

Chapter 17

1 (a)

• Increasing metallic character and decreasing non-metallic character as atomic number increases

• Giant molecular lattices for C and Si becoming giant metallic structures in tin and lead; bonding becomes weaker from covalent to metallic causing changes in reactivity.

• Only tin and lead are reactive enough (as metals/reducing agents) to liberate H_2 from dilute acids.

$$Pb(s) + 2H^+(aq) \rightarrow Pb^{2+}(aq) + H_2(g)$$

• Elements react with oxygen to form acidic, non–metallic oxides with carbon,

$$C(s) + O_2(g) \rightarrow CO_2(g)$$

to basic, metallic oxides with Pb.

$$2Pb(s) + O_2(g) \rightarrow 2PbO(s)$$

(b) See table 17.5

(c) See table 17.6.

(d) See table 17.7.

(e)

• With increase in atomic number there is a decrease in the stability of the +4 oxidation state and an increase in the stability of the +2 oxidation state.

• In C and Si compounds, the +4 state is very stable and any +2 compound is easily oxidised to +4.

e.g. $CO(g) + \frac{1}{2}O_2(g) \rightarrow CO_2(g)$

• In Sn compounds, the +4 state is only slightly more stable than +2 so $Sn^{2+}(aq)$ is converted to $Sn^{4+}(aq)$ under mild oxidising conditions.

$\begin{cases} Sn^{2+}(aq) \rightarrow Sn^{4+}(aq) + 2e^- \\ I_2(aq) + 2e^- \rightarrow 2I^-(aq) \end{cases}$

• In Pb compounds, the +2 state has become the more stable. PbO is stable, whereas PbO_2 is a good oxidising agent, being converted to the +2 state.

e.g. $PbO_2(s) + 4HCl(l) \rightarrow PbCl_2(s) + Cl_2(g) + 2H_2O(l)$

2 (a) Tinstone (cassiterite) – impure tin(IV) oxide, SnO_2

(b) Reduction of SnO_2 with carbon (coke)

$SnO_2(s) + 2C(s) \rightarrow Sn(l) + 2CO(g)$

(c) Tin's main uses are as a relatively inert metal and in alloys. It is used to tin-plate steel for canning meats, soups, vegetables, etc. It is used in several alloys, e.g. solder (Sn/Pb) and pewter (Sn/Pb).

(d) Tin is less reactive than zinc, so tin plate forms a thin layer of tin oxide much more slowly than zinc plating (galvanising) forms zinc oxide. However, once the surface plating is scratched, and the steel (iron) becomes exposed, a cell will be set up. In zinc plating, the zinc (more reactive than iron) will form Zn^{2+} ions and prevent the iron rusting, but in tin plating the tin is less reactive than iron, so Fe^{2+}/Fe^{3+} ions form and rusting occurs rapidly.

3 (a) (i) Pb forms a layer of lead(II) oxide on heating in air (oxygen).

$2Pb(s) + O_2(g) \rightarrow 2PbO(s)$

(ii) Pb reacts very slowly with soft water containing dissolved oxygen to form lead(II) hydroxide, $Pb(OH)_2$.

$2Pb(s) + O_2(g) + 2H_2O(l) \rightarrow 2Pb(OH)_2(s)$

In hard water, the lead forms a protective coating of insoluble lead carbonate or sulphate.

(iii) Lead reacts slowly with dilute nitric acid to form hydrogen and lead(II) nitrate.

$Pb(s) + 2HNO_3(aq) \rightarrow Pb(NO_3)_2(aq) + H_2(g)$

(b) SnO   React tin with dilute acid to form $Sn^{2+}(aq)$

$Sn(s) + 2H^+(aq) \rightarrow Sn^{2+}(aq) + H_2(g)$

Add NaOH(aq) to precipitate $Sn(OH)_2(s)$

$Sn^{2+}(aq) + 2OH^-(aq) \rightarrow Sn(OH)_2(s)$

Filter off the $Sn(OH)_2$ and heat to decompose to $SnO(s)$
$$Sn(OH)_2(s) \rightarrow SnO(s) + H_2O$$
SnO_2 Heat powdered tin strongly in oxygen.
$$Sn(s) + O_2(g) \rightarrow SnO_2(s)$$
(c) (i) $SnO(s)$ reacts with oxygen on heating to form $SnO_2(s)$.
$$SnO(s) + \tfrac{1}{2}O_2(g) \rightarrow SnO_2(s)$$
SnO_2 does not react with oxygen.

(ii) SnO reacts with dilute HCl to form $SnCl_2(aq)$ and water.
$$SnO(s) + 2HCl(aq) \rightarrow SnCl_2(aq) + H_2O(l)$$
SnO_2 reacts to form $SnCl_4(aq)$ and water
$$SnO_2(s) + 4HCl(aq) \rightarrow SnCl_4(aq) + 2H_2O$$

(iii) SnO reacts with NaOH(aq) to form sodium trihydroxostannate(II).
$$SnO + OH^- + H_2O \rightarrow \quad Sn(OH)_3^-$$
$$\text{trihydroxostannate(II)}$$
SnO_2 reacts with NaOH(aq) to form sodium stannate(IV).
$$SnO_2 + 2OH^- \rightarrow SnO_3^{2-} + H_2O$$
$$\text{stannate(IV)}$$

4 (a) $Sn + 2I_2 \rightarrow SnI_4$

(b) $I_2(s)$ is soluble in $CCl_4(l)$; SnI_4 is soluble in hot $CCl_4(l)$ but insoluble in cold $CCl_4(l)$.

(c) Number of moles Sn taken $= \dfrac{4}{119} = 0.034$

Number of moles I_2 taken $= \dfrac{12.7}{254} = 0.05$

$\therefore$ Sn is in excess.

Yield of $SNI_4 = 0.025$ mol $= 0.025 \times 627 = 15.68$ g

(d) When all the I_2 is used up and there is no purple coloration in the CCl_4

(e) Because SnI_4 will begin to crystallise at room temperature and clog up the funnel

(f) To wash any SNI_4 off the excess tin

(g) (i) Tetrahedral (ii) Decomposes to SnI_2 and I_2 on heating

5 (a)

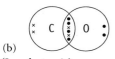

(b)
'Iso–electronic' means same number of electrons. N_2 is

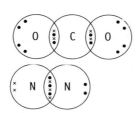

(c)
$$CO_2(g) + C(s) \xrightarrow{\quad} 2CO(g) \quad \Delta H = +172\,kJ$$
$$\searrow^{-394} \qquad \nearrow^{2 \times -111}$$
$$2C(s) + O_2(g)$$

$$CO(g) + \tfrac{1}{2}O_2(g) \xrightarrow{\quad} CO_2(g) \quad \Delta H = -283\,kJ$$
$$\searrow^{-111} \qquad \nearrow^{-394}$$
$$C(s) + O_2(g)$$

CO_2 is much more stable than CO. CO_2 is stable with respect to C and CO. CO is stable with respect to C, but unstable with respect to CO_2.

6 (a)
• Tin and lead are relatively inert metals.
• Tin and lead form important alloys.
• Both form compounds in +2 and +4 oxidation states.
• Both react slowly with dilute acids to give +2 salts and H_2, etc.
(b) With increasing atomic number, the compounds of Group IV elements change from being molecular with covalent bonding to ionic. With increase in atomic number, it is easier for elements to lose electrons and form ions.

In tin compounds, the +4 state is more stable, whilst in lead the +2 state is more stable.
Lead(IV) chloride which does exist as simple molecules with four covalent bonds to the four Cl atoms. Lead(II) chloride is, however, ionic, $Pb^{2+}(Cl^-)_2$.
(c) Tin and lead do have variable oxidation state and form some complex ions, but they do not have coloured ions nor do they have the catalytic properties that are typical of transition metals.
More importantly, Sn and Pb do not have any ions with a partially filled d sub-shell which defines a transition element.
(d) (i) $Pb^{2+}(s) + H_2S(g) \rightarrow PbS(s) + 2H^+(aq)$
$$\text{black}$$
(ii) $H_2O_2(aq)$ reacts with black PbS to form white lead(II) sulphate and the colour of the pigment is therefore clear again.
$$PbS(s) + 4H_2O_2(aq) \rightarrow PbSO_4(s) + 4H_2O(l).$$

Chapter 18

1 (a)
• Coloured compounds – Cu^{2+} compounds are blue.
• Variable oxidation state – Cu has stable +1 and +2 compounds.
• Form complex ions – $[Cu(H_2O)_4]^{2+}$, $[Cu(NH_3)_4]^{2+}$ and $[CuCl_4]^{2-}$ for Cu^{2+} and $[Cu(NH_3)_2]^+$ and $[CuCl_3]^{2-}$ for Cu^+.
• Catalytic properties – Cu or CuO act as catalysts for the oxidation of primary alcohols to aldehydes using $O_2(g)$. (Any 3 of the above 4 properties)

(b) (i) When Cu^{2+} salts are added to water, Cu^{2+} ions are attracted to the negative end of polar water molecules to form tetraaquacopper(II) ions. The electric field associated with the Cu^{2+} ion is sufficiently intense to draw electrons in the O—H bonds of the water molecules towards itself, causing these water molecules to become proton donors. Hence the solution is acidic, turning blue litmus red.
$$[Cu(H_2O)_4]^{2+}(aq) + H_2O(l) \rightleftharpoons [Cu(H_2O)_3OH]^+(aq) + H_3O^+(aq)$$

(ii) Ammonia solution contains $NH_3(aq)$ and $OH^-(aq)$ ions. Initially, the OH^- ions react with $Cu^{2+}(aq)$ to form a pale blue ppte. of copper(II) hydroxide.
$$Cu^{2+}(aq) + 2OH^-(aq) \rightarrow Cu(OH)_2(s)$$
When more ammonia solution is added, the $NH_3(aq)$ displace the OH^- ions to form a deep blue solution containing tetraamminecopper(II) ions.
$$Cu(OH)_2(s) + 4NH_3(aq) \rightarrow [Cu(NH_3)_4]^{2+}(aq) + 2OH^-(aq)$$

2 (a) $[Cu(NH_3)_4]^{2+}$ and $[CuCl_4]^{2-}$ are suitable examples.
(b) Both these ions are square planar.
(c) The Cu^{2+} ion in each case will have the electron structure $(Ar)3d^9$. The NH_3 and Cl^- donate electrons into the fourth shell for copper.

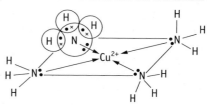

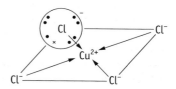

The donation from all four NH_3 molecules and all four Cl^- ions are similar.

3 (a) Transition elements are filling up a *d* sub-shell of electrons. They each have at least one ion with a partially filled *d* sub-shell.

(b) V is (Ar) $3d^3 4s^2$
V forms stable oxidation states of 2+ (loss of the two 4s electrons), 3+, 4+ and 5+ (loss of the two 4s and the three *3d* electrons).

(c)
• Coloured compounds – Vanadate(V) in acid solution is yellow, oxovanadium(IV) ions are blue, vanadium(III) ions are green and vanadium(II) ions are violet.
• Catalytic properties – V_2O_5 acts as catalyst for the Contact Process.
• Complex ions – oxovanadium(IV) ions $[V^{4+}{\leftarrow}\ddot{\underset{\cdot\cdot}{O}}{:}^{2-}]^{2+}$, hexachlorovanadate(IV) $[VCl_6]^{2-}$

4 (a) (i) $[Co(NH_3)_5Br]^{2+}SO_4^{2-}$ will give $[Co(NH_3)_5Br]^{2+}(aq)$ and $SO_4^{2-}(aq)$
$[Co(NH_3)_5SO_4]^+Br^-$ will give $[Co(NH_3)_5SO_4]^+(aq)$ and $Br^-(aq)$

(ii) Dissolve a little of each compound in water. One aqueous solution will give a white precipitate with $Ba(NO_3)_2(aq)$ showing that it contains free $SO_4^{2-}(aq)$.
$Ba^{2+}(aq) + SO_4^{2-}(aq) \rightarrow BaSO_4(s)$
white ppte.
The other aqueous solution will give a pale yellow precipitate with $AgNO_3(aq)$ showing that it contains free $Br^-(aq)$.
$Ag^+(aq) + Br^-(aq) \rightarrow AgBr(s)$
pale yellow ppte.

(iii) In each complex, the oxidation state of cobalt is 3+ and its co-ordination number is 6.

(iv)

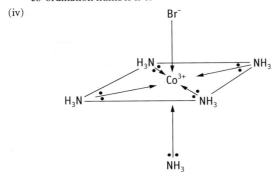

Shape – octahedral; 6 co-ordinate bonds from the 5 NH_3 molecules and the Br^- ion.

(b) (i) Square planar – this structure can have *cis–trans* isomers. Tetrahedral could not have isomers.

(ii)

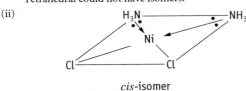

cis-isomer

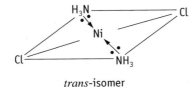

trans-isomer

5 (a) $1s^2\ 2s^2\ 2p^6 3s^2 3p^6\ 3d^9$

(b) Cu^{2+} ions probably act as catalysts by being able to change oxidation state. Cu^{2+} ions are reduced to Cu^+ ions in the first part of the reaction and then oxidised back to Cu^{2+}.

(c) (i) Without Cu^{2+} ions, the catalysis process is impossible so the enzyme-protein alone has no catalytic activity.

(ii) The enzyme-protein probably allows the reactant(s) to orient itself (themselves) in the ideal position for the reaction to occur. Without this steric help from the enzyme-protein, the reaction is much less efficient, but Cu^{2+} ions alone can catalyse the reaction.

(iii) Egg albumen, being a protein, is able to form interactions with the reactant(s) similar to the enzyme-protein and this helps the Cu^{2+} ions to catalyse the reaction, though not as efficiently as the specific enzyme-protein.

6 (a) 4+ in all the compounds.

(b) **A** is $[Pt(NH_3)_6]^{4+}(Cl^-)_4$; **B** is $[Pt(NH_3)_5Cl]^{3+}(Cl^-)_3$; **C** is $[Pt(NH_3)_4Cl_2]^{2+}(Cl^-)_2$; **D** is $[Pt(NH_3)_3Cl_3]^+Cl^-$; **E** is $[Pt(NH_3)_2Cl_4]$.

(c), (d), (e)

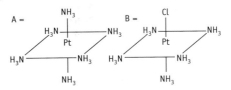

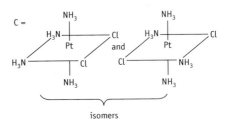

isomers

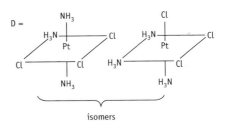

isomers

1 (a) A transition metal is an element with at least one ion with a partially filled *d* sub-shell.

(b) (i) Sc is not a transition metal. It only forms one stable oxidation state (+3), and Sc^{3+} does not have a partially filled *d* sub-shell.

(ii) Fe is a transition metal, forming 2 stable oxidation states of +2 and +3, both of which have ions with partially filled *d* sub-shells.

(iii) Zn is not a transition metal. It only forms one stable oxidation state (+2), and Zn^{2+} ions do not have a partially filled *d* sub-shell.

(c) Copper(I) compounds contain Cu^+ ions with the electron structure (Ar) $3d^{10}$. The Cu^+ ion has a fully filled 3d sub-shell. Electron transitions are therefore impossible within the 3d levels of Cu^+ and so copper(I) compounds are white.

(d) In crossing a period of transition metals, electrons are being added to the same *d* shell whilst the nuclear charge and nuclear mass are increasing. As the positive nuclear charge increases, it pulls the outer electrons closer. Thus, the mass per unit volume (i.e. the density) increases.

2 (a) An iron pipe joined to magnesium in wet earth sets up a simple cell. The magnesium, being the more reactive metal, will give up electrons which pass to the iron, and Mg^{2+} ions form.

$Mg(s) \rightarrow Mg^{2+}(aq) + 2e^-$

At the iron, water molecules and oxygen will take the electrons to form hydroxide ions.

$H_2O(l) + \frac{1}{2}O_2(g) + 2e^- \rightarrow 2OH^-(aq)$

(b) The two isomers will be a 'cis' form (with the 2 Cl^- ions forming a 90° angle with the central M^{3+} ion) and a 'trans' form (in which the $Cl—M^{3+}—Cl^-$ angle is 180°).

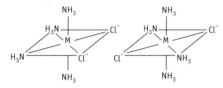

(c) Cu has virtually no reaction with dilute HCl and reacts very slowly with concentrated HCl. When strong complexing agents are added, they react with any Cu^{2+} ions which form and this causes further reaction of copper with the HCl(aq)

$Cu(s) + 2HCl(aq) \rightarrow CuCl_2(aq) + H_2(g)$

(d) Iron is more reactive than copper. Therefore, the Fe goes into solution as Fe^{2+} or Fe^{3+} whilst Cu^{2+} ions are converted to Cu which coats the iron.

$Fe(s) \rightarrow Fe^{2+}(aq) + 2e^-, \quad Cu^{2+}(aq) + 2e^- \rightarrow Cu(s)$

When the iron is dipped in concentrated HNO_3, a thin, coherent, unreactive layer of iron(III) oxide (Fe_2O_3) forms. This protects the iron and prevents reaction with the $CuSO_4(aq)$.

3 (a) Heat ZnS in air to form ZnO. React the ZnO with dilute H_2SO_4 to form $ZnSO_4(aq)$ and then electrolyse this. Zn is deposited at the cathode.

$Zn^{2+}(aq) + 2e^- \rightarrow Zn(s)$

(b) Electrolyse molten $MgCl_2$ as the usual chemical reducing agents will not reduce Mg^{2+} and electrolysis of an aqueous solution will produce H_2 rather than Mg at the cathode. Use a steel cathode and ensure the products Mg and Cl_2 do not come into contact and react again. Mg(l) is produced at the cathode.

$Mg^{2+}(l) + 2e^- \rightarrow Mg(l)$

(c) Iron is obtained by reduction in a blast furnace with CO. A mixture containing coke and Fe_2O_3 is added to the furnace. Hot air is blasted into the furnace to react with coke forming CO.

$2C(s) + O_2(g) \rightarrow 2CO(g)$

The CO reduces the Fe_2O_3 to iron.

$Fe_2O_3(s) + 3CO(g) \rightarrow 2Fe(l) + 3CO_2(g)$

The molten iron runs to the bottom of the furnace and is tapped off. This provides a relatively cheap way of making huge quantities of iron.

(d) React insoluble Ag_2S with a solution of cyanide ions.

$Ag_2S(s) + 4CN^-(aq) \rightarrow 2[Ag(CN)_2]^-(aq) + S^{2-}(aq)$

Then precipitate silver from the solution by adding zinc dust.

$2[Ag(CN)_2]^-(aq) + Zn(s) \rightarrow 2Ag(s) + [Zn(CN)_4]^{2-}(aq)$

4 (a) $MnO_4^- + 8H^+ + 5e^- \rightarrow Mn^{2+} + 4H_2O$
   $(Fe^{2+} \rightarrow Fe^{3+} + e^-) \times 5$

(b) 5

(c) $\frac{25}{1000}$ $0.03 \times 5 = 3.75 \times 10^{-3}$

(d) $3.75 \times 10^{-3} \times 10 \times 56 = 2.1$ g

(e) $\frac{2.1}{10} \times 100 = 21\%$

5 (a) (i) Any Group I or Group II metal or Al or Zn (any two)
   (ii) Fe, Pb, Cu, Ag (any two)
   (iii) Au, Pt

(b) Magnetic separation, froth flotation, water jets on a sloping agitated table (any two) acting on the crushed ore

(c) Because metal ores are often mixed with earthy materials such as clay and sand

(d) Blister copper is purified by electrolysis of copper sulphate solution. The blister copper is the anode and a thin sheet of pure copper is the cathode. Cu dissolves into solution at the anode and pure copper is deposited at the cathode.

Anode (+) $Cu(s) \rightarrow Cu^{2+}(aq) + 2e^-$
Cathode (−) $Cu^{2+}(aq) + 2e^- \rightarrow Cu(s)$

Impure pig iron from the blast furnace contains impurities – C, S, Si, P and Mn. These impurities are removed by oxidation. C and S are oxidised to CO_2 and SO_2 which escape as gases. Si and P are oxidised to less volatile oxides, SiO_2 and P_2O_5 which combine with added lime to form slag.

6 (a) The ore used should be:
• abundant – especially for a metal required in large quantity like Fe;
• cheap;
• of high quality – Sn ores in Cornwall are so poor in quality that it is uneconomic to extract tin there;
• accessible – open cast mining is cheaper than deep mining;
• easily concentrated/purified.

(b) The reducing agent should be:
• cheap – Fe is needed in such large quantity that only coke could be used; if a metal ore is reduced by electrolysis, that will require cheap electricity. Al is therefore often manufactured near cheap hydroelectric plants;
• accessible – most blast furnaces are situated near coalfields to provide an accessible supply of coal for coke.

(c) The location of the extraction plant should give:
• ready access to the ore and reducing agent,
• ease of transport of ore, reducing agent and metal,
• closeness to the market for the metal and any by-products.

7 See table 19.3, section 19.13 and section 19.14.

Chapter 20

1 (a) When a solute X distributes itself between two solvents Y and Z, it is found, at equilibrium, that

$\dfrac{\text{concentration of X in solvent Y}}{\text{concentration of X in solvent Z}}$ is a constant.

That is provided that the temperature is constant, the solvents are immiscible and none of the solvents or X react with each other, associate or dissociate.

(b) Solvent extraction can be used to obtain and purify some solutes. The solute (usually in aqueous solution) is shaken with a second solvent in which it is more soluble. Most of the solute then dissolves in the second solvent which is separated off. The second solvent is usually much more volatile than the solute, so the two can be separated easily by distillation.

(c) $\dfrac{[M_{(petrol)}]_{eq}}{[M_{(s)}]_{eq}} = 0.01$ $\therefore \dfrac{\left(\dfrac{5-x}{1000}\right)}{\left(\dfrac{x}{100}\right)} = 0.01$ where x is the mass of M removed

$\Rightarrow \dfrac{5-x}{1000} = \dfrac{0.01 \times x}{100}$ so $x = 4.5$ g

2 (a) $CH_3CH_2OH + CH_3COOH \rightleftharpoons CH_3CH_2OOCCH_3 + H_2O$

(b) $K_c = \dfrac{[CH_3CH_2OOCCH_3]eq[H_2O]eq}{[CH_3CH_2OH]eq[CH_3COOH]eq}$

(c)

	ethanol	+ ethanoic acid	⇌ ethyl ethanoate	+ water
initial moles	5	6	6	4
equilibrium moles	3	4	8	6

(d) $K_c = \dfrac{\frac{8}{v} \times \frac{6}{v}}{\frac{3}{v} \times \frac{4}{v}} = 4$ where v is the volume of the mixture

(e) Suppose x moles of ethanoic acid form. $\therefore$ x moles of ethanol form and x moles of both ethyl ethanoate and water react.

	ethanol	+ ethanoic acid	$\rightleftharpoons$ ethyl ethanoate	+ water
initial moles	1	1	3	3
equilibrium moles	$1 + x$	$1 + x$	$3 - x$	$3 - x$

$\therefore K_c = 4 = \dfrac{(3-x)(3-x)}{(1+x)(1+x)}$

Taking square roots on both sides of this equation, $2 = \dfrac{3-x}{1+x}$

$\therefore 2(1+x) = 3 - x$, so $2 + 2x = 3 - x$, $x = \frac{1}{3}$ mole

At equilibrium, there are $1\frac{1}{3}$ mol of both ethanol and ethanoic acid, and $2\frac{2}{3}$ mol of both ethyl ethanoate and water.

3 (a) $\qquad I_2(g) \rightleftharpoons 2I(g)$

At equilibrium 0.6×10^5 Pa 0.4×10^5 Pa $K_p = \dfrac{(0.4 \times 10^5)^2}{(0.6 \times 10^5)} = 2.67 \times 10^4$ Pa

(b) $\qquad I_2(g) \rightleftharpoons 2I(g)$ where P is the new final pressure

At equilibrium $\quad 0.8 \times P \qquad 0.2 \times P$

So, $K_p = 2.67 \times 10^4 = \dfrac{(0.2 \times P)^2}{0.8 \times P}$

$2.67 \times 10^4 = \dfrac{0.04P^2}{0.8P}$, $P = 5.34 \times 10^5$ Pa

4 (a)

(i) $K_p = \dfrac{(P_{NO_2})^2}{(P_{NO})^2(P_{O_2})}$, $K_c = \dfrac{[NO_2(g)]^2}{[NO(g)]^2[O_2(g)]}$

but $P_{NO_2} = [NO_2(g)]RT$, $P_{NO} = [NO(g)]RT$ and $P_{O_2} = [O_2(g)]RT$

$\Rightarrow K_p = \dfrac{[NO_2(g)]^2(RT)^2}{[NO(g)]^2(RT)^2[O_2(g)]RT} = K_c(RT)^{-1}$

(ii) $K_p = K_c(RT)^{-\frac{1}{2}}$

(b) For (a)(i), units of K_p are $(Pa)^{-1}$ and those for K_c are $(mol\ dm^{-3})^{-1}$ or $mol^{-1}\ dm^3$.

For (a)(ii), units of K_p are $(Pa)^{-\frac{1}{2}}$ and those for K_c are $(mol\ dm^{-3})^{-\frac{1}{2}}$ or $mol^{-\frac{1}{2}}\ dm^{3/2}$.

5 (a) $K_p = \dfrac{(P_{HI})^2}{(P_{H_2})(P_{I_2})}$

(b) $\dfrac{(4.0 \times 10^4)^2}{(2.5 \times 10^4)(1.6 \times 10^4)} = 4.0$, no units

(c)

	$H_2(g)$	+ $I_2(g)$	$\rightleftharpoons$	$2HI(g)$
initial pressures	0	3×10^4		3×10^4
equilibrium pressures	x	$3 \times 10^4 + x$		$3 \times 10^4 - 2x$

So, $K_p = 4 = \dfrac{(3 \times 10^4 - 2x)^2}{x(3 \times 10^4 + x)}$

$\Rightarrow 4(3 \times 10^4 x + x^2) = (3 \times 10^4 - 2x)^2$

$12 \times 10^4 x + 4x^2 = 9 \times 10^8 - 12 \times 10^4 x + 4x^2$

$x = 3.75 \times 10^3$ Pa

$\therefore$ Final pressures, $P_{H_2} = 3.75 \times 10^3$ Pa, $P_{I_2} = 3.375 \times 10^4$ Pa and $P_{HI} = 2.25 \times 10^4$ Pa

(d)

	$H_2(g)$	+ $I_2(g)$	$\rightleftharpoons$	$2HI(g)$
initial pressures	0	0		6×10^4
equilibrium pressures	x	x		$6 \times 10^4 - 2x$

So, $K_p = 4 = \dfrac{(6 \times 10^4 - 2x)^2}{x \times x}$, $\therefore 4x^2 = 3.6 \times 10^9 - 2.4 \times 10^5 x + 4x^2$

$\Rightarrow x = 1.5 \times 10^4$ Pa

At equilibrium, $P_{H_2} = P_{I_2} = 1.5 \times 10^4$ Pa, $P_{HI} = 3 \times 10^4$ Pa

Chapter 21

1 (a) $K_p = \dfrac{(P_{CO})(P_{H_2})^3}{(P_{CH_4})(P_{H_2O})}$

(b) (i) K_p will not change.

(ii) K_p will increase.

(iii) K_p will not change.

(c) (i) Concentration of products will decrease and concentration of reactants will increase.

(ii) Concentration of products will increase and concentration of reactants will decrease.

(iii) Composition of equilibrium mixture will not change.

2 (a) $4NH_3(g) + 5O_2(g) \rightarrow 4NO(g) + 6H_2O(g)$

(b) Maximum yield of NO requires low temperature and low pressure.

(c) Low temperature and low pressure give a high yield at equilibrium, but a slow reaction rate. Industry requires a relatively fast reaction rate for economic reasons. This is achieved at high temperature and a pressure of 7 atm.

(d) The NO(g) is allowed to react with oxygen in the air to form nitrogen dioxide.

$2NO(g) + O_2(g) \rightarrow 2NO_2(g)$

The $NO_2(g)$ is then passed up a tower partially filled with inert material and down which water flows. The NO_2 reacts with water to form nitric acid and NO which can be cycled through the process again.

$3NO_2(g) + H_2O(l) \rightarrow 2HNO_3(aq) + NO(g)$

3 (a) $K_c = \dfrac{[Sn^{4+}(aq)][Fe^{2+}(aq)]^2}{[Sn^{2+}(aq)][Fe^{3+}(aq)]^2}$

(b) The concentration terms cancel in the expression for K_c

(c) (i) $K_c = \dfrac{[Sn^{2+}(aq)][Fe^{3+}(aq)]^2}{[Sn^{4+}(aq)][Fe^{2+}(aq)]^2} = \dfrac{1}{10^{10}} = 10^{-10}$

(ii) $K_c = \dfrac{[Sn^{4+}(aq)]^2[Fe^{2+}(aq)]^4}{[Sn^{2+}(aq)]^2[Fe^{3+}(aq)]^4} = (10^{10})^2 = 10^{20}$

4 (a) $K_c = \dfrac{[PCl_3(g)][Cl_2(g)]}{[PCl_5(g)]}$

(b) $mol\ dm^{-3}$

(c) $\dfrac{1}{8 \times 10^{-3}} = 125\ mol^{-1}\ dm^3$

(d) (i) After the addition, $[PCl_5]$ will fall, $[PCl_3]$ and $[Cl_2]$ will rise.

(ii) $[PCl_5]$ will rise, $[PCl_3]$ and $[Cl_2]$ will fall.

(iii) $[PCl_5]$ will fall, $[PCl_3]$ and $[Cl_2]$ will rise.

(e) (i) K_c is unaffected. (ii) K_c is unaffected. (iii) K_c increases.

(f)

	$PCl_5(g)$	$\rightleftharpoons$	$PCl_3(g)$	+	$Cl_2(g)$
initial concentrations	x		0		0
final concentrations	0.5×10^{-1}		$(x - 0.5 \times 10^{-1})$		$(x - 0.5 \times 10^{-1})$

$K_c = 8 \times 10^{-3} = \dfrac{(x - 0.5 \times 10^{-1})^2}{0.5 \times 10^{-1}}$

$\Rightarrow (x - 0.5 \times 10^{-1})^2 = 4 \times 10^{-4}$

$\therefore$ concentration of PCl_3 = concentration of $Cl_2 = (x - 0.5 \times 10^{-1})$

$= 2 \times 10^{-2}\ mol\ dm^{-3}$

5 (a)

	$COCl_2(g)$	$\rightleftharpoons$	$CO(g)$	+	$Cl_2(g)$
initial moles	1		0		0
final moles	$1 - \alpha$		α		α
final pressure	$\dfrac{1-\alpha}{1+\alpha}P$		$\dfrac{\alpha}{1+\alpha}P$		$\dfrac{\alpha}{1+\alpha}P$

$K_p = \dfrac{(P_{CO})(P_{Cl_2})}{(P_{COCl_2})} = \dfrac{\frac{\alpha}{1+\alpha}P \times \frac{\alpha}{1+\alpha}P}{\frac{1-\alpha}{1+\alpha}P} = \dfrac{\alpha^2 P}{1-\alpha^2}$

(b) (i) $K_p = \dfrac{\alpha^2 P}{1 - \alpha^2}$ $\therefore 0.2 = \dfrac{\alpha^2}{1-\alpha^2} \times 10^5 \Rightarrow \alpha = 1.4 \times 10^{-3}$

(ii) $K_p = \dfrac{\alpha^2 P}{1 - \alpha^2}$ $\therefore 0.2 = \dfrac{\alpha^2}{1-\alpha^2} \times 2 \times 10^5 \Rightarrow \alpha = 10^{-3}$

6 (a) Using Le Chatelier's principle; the pressure suddenly increases so $[CO(g)]$ and $[H_2(g)]$ decrease and $[CH_3OH(g)]$ increases.

(b) Using Le Chatelier's principle; $[CO(g)]$ and $[H_2(g)]$ increase and $[CH_3OH(g)]$ decreases.

(c) Using Le Chatelier's principle; after the addition of more H_2, $[CO(g)]$ and $[H_2(g)]$ will decrease and $[CH_3OH(g)]$ will increase.

(d) The catalyst has no effect on the position of equilibrium.

(e) The inert gas has no effect on the equilibrium.

7 (a) By Le Chatelier's principle; if the temperature is increased, a reaction will tend to move in the endothermic direction. Thus, the formation of diamond is favoured by high temperature.

(b) By Le Chatelier's principle, increasing pressure will cause a reaction to respond by trying to reduce the pressure (i.e. solids occupy lower volume and become more dense). So, the formation of diamond will be favoured by high pressure.

8 For sulphuric acid – see section 21.5.

Sulphuric acid can act as an acid (when dilute), as an oxidising agent (when concentrated) and as a dehydrating agent (when concentrated).

For nitric acid – see sections 21.6 and 21.7.

Nitric acid can act as an acid (when dilute) and as an oxidising agent.

For ammonia – see section 21.5.

Ammonia can act as a base and as a complexing agent.

Chapter 22

1 $AgBr(s) \rightleftharpoons Ag^+(aq) + Br^-(aq)$, Let solubility $= s$ mol dm^{-3}

$K_{s.p.}(AgBr) = [Ag^+][Br^-] = s \times s = 7 \times 10^{-7} \times 7 \times 10^{-7}$

$K_{s.p.}(AgBr) = 4.9 \times 10^{-13}$ mol^2 dm^{-6}

2 (a) $Ag_2C_2O_4(s) \rightleftharpoons 2Ag^+(aq) + C_2O_4^{2-}(aq)$, Let solubility $= s$ mol dm^{-3}

$[Ag^+] = 2s$ and $[C_2O_4^{2-}] = s$

$K_{s.p.}(Ag_2C_2O_4) = [Ag^+]^2[C_2O_4^{2-}] = (2s)^2 \times s$

$\Rightarrow 5 \times 10^{-12} = 4s^3$

$\therefore$ solubility $= s = 1.08 \times 10^{-4}$ mol dm^{-3}

(b) Solubility normally increases with temperature, so this will result in an increase in the value of $K_{s.p.}$ as temperature rises.

3 (a) (i) Suppose the solubility of $PbSO_4$ in water $= s$ mol dm^{-3},

$PbSO_4(s) \rightleftharpoons Pb^{2+}(aq) + SO_4^{2-}(aq)$

$K_{s.p.}(PbSO_4) = [Pb^{2+}][SO_4^{2-}]$

$\Rightarrow 1.6 \times 10^{-8} = s \times s$

$\therefore$ solubility $PbSO_4$ in water $= 1.26 \times 10^{-4}$ mol dm^{-3}

(ii) Suppose the solubility of $PbSO_4$ in 0.1 mol dm^{-3} $Pb(NO_3)_2 = s^1$

In this case $[Pb^{2+}] = (s^1 + 0.1)$ mol dm^{-3}, $[SO_4^{2-}] = s^1$ mol dm^{-3}

$\therefore K_{s.p.}(PbSO_4) = (s^1 + 0.1) \times s^1 = 1.6 \times 10^{-8}$

Now, as $s^1 \ll 0.1$, $0.1 \times s^1 = 1.6 \times 10^{-8}$

$\therefore$ solubility of $PbSO_4$ in 0.1 mol dm^{-3} $Pb(NO_3)_2 = 1.6 \times 10^{-7}$ mol dm^{-3}

(iii) Suppose the solubility of $PbSO_4$ in 0.01 mol dm^{-3} $Na_2SO_4 = s^1$

In this case, $[Pb^{2+}] = s^1$ mol dm^{-3}, $[SO_4^{2-}] = (0.01 + s^1)$ mol dm^{-3}

$\therefore K_{s.p.}(PbSO_4) = s^1 \times (0.01 + s^1) = 1.6 \times 10^{-8}$

Now, as $s^1 \ll 0.01$, $s \times 0.01 = 1.6 \times 10^{-8}$

$\therefore$ Solubility of $PbSO_4$ in 0.01 mol dm^{-3} $Na_2SO_4 = 1.6 \times 10^{-6}$ mol dm^{-3}

(b) If a solution contains Pb^{2+} or SO_4^{2-} ions, they will add to the concentration of either Pb^{2+} or SO_4^{2-} and suppress the amount of $PbSO_4$ which can dissolve.

(c) The generalisation known as the common ion effect says that the solubility of a salt A^+B^- is reduced by the presence of either A^+ or B^- ions from a second source.

This is very well illustrated by the calculations in parts (ii) and (iii) of section (a) in this question. i.e. the solubility of $PbSO_4$ has been reduced by the presence of Pb^{2+} ions from $Pb(NO_3)_2$ in part (ii) and by the presence of SO_4^{2-} ions from Na_2SO_4 in part (iii).

4 (a) $HX(aq) + H_2O(l) \rightleftharpoons H_3O^+(aq) + X^-(aq)$

(i) If water is added, the equilibrium will move more to the right and the degree of dissociation of HX increases.

(ii) If gaseous HCl is added, it reacts with water to form $H_3O^+(aq)$. This pushes the equilibrium to the left and reduces the degree of dissociation.

(iii) If NaX(s) is added, the concentration of $X^-(aq)$ rises. This pushes the equilibrium to the left, reducing the degree of dissociation.

(b) (i) HA

(ii) $K_a(HA) = \dfrac{[H^+][A^-]}{[HA]} = 10^{-6}$

$[H^+] \simeq [A^-]$ and $[HA] = 1 - [H^+] \simeq 1$ as $[H^+]$ is very small.

$\therefore [H^+] = 10^{-3}$ mol dm^{-3}

5 (a) $HClO \rightleftharpoons H^+ + ClO^-$ $K_a = \dfrac{[H^+][ClO^-]}{[HClO]} = 3.2 \times 10^{-8}$

$[H^+] \simeq [ClO^-]$

and $[HClO] = 1.25 \times 10^{-2} - [H^+] \simeq 1.25 \times 10^{-2}$ as $[H^+]$ is very small.

$\Rightarrow 3.2 \times 10^{-8} = \dfrac{[H^+]^2}{1.25 \times 10^{-2}}$

$\therefore [H^+] = 2 \times 10^{-5}$ mol dm^{-3}

Now, as $[H^+][OH^-] = 10^{-14}$ at 298 K

$[OH^-] = 5 \times 10^{-10}$ mol dm^{-3}

(b) pH $= -\lg[H^+]$

$= -\lg[2 \times 10^{-5}] = -(0.30 - 5.0) = 4.7$

6 (a) Assuming the 1.0 mol dm^{-3} HCl is fully dissociated, $[H^+] = 1.0$, so the pH $= -\lg[H^+] = 0$. As water is added, its concentration will fall to say 10^{-1} mol dm^{-3} (pH = 1.0), then 10^{-2} mol dm^{-3} (pH = 2.0), etc. Eventually, the HCl will be so dilute that the pH will be virtually the same as that of water. i.e. pH = 7 (at 298 K) and the pH will not change on further dilution.

(b) $FeCl_3(aq)$ contains $Fe^{3+}(aq)$ and $Cl^-(aq)$ ions.

The $Fe^{3+}(aq)$ contains Fe^{3+} ions surrounded by water molecules. The charge density on Fe^{3+} ions is great enough to pull electrons in the O—H bonds of the water molecules towards them, causing the release of H^+ ions and making the solution acidic to litmus.

$[Fe(H_2O)_6]^{3+}(aq) \rightleftharpoons [Fe(H_2O)_5(OH)]^{2+}(aq) + H^+(aq)$

(c) When ethanoic acid is titrated with sodium hydroxide, the pH changes rapidly from about 7.0 to 10.0. This requires an indicator which changes colour in this range. Phenolphthalein changes colour at pH 8 to 10, but methyl orange changes colour between 3 and 4.5 which is useless.

(d) The $[H^+]$ from pure water is 10^{-7}. If the solution also contains 10^{-8} mol dm^{-3} HCl, then the pH will be slightly less than 7.

7 (a) The buffer contains a reservoir of C_6H_5COOH and $C_6H_5COO^-$. When HCl(aq) (i.e. $H^+(aq)$) is added, the H^+ ions are trapped by $C_6H_5COO^-$ ions forming more C_6H_5COOH, so the $[H^+]$ does not change very much and the pH stays constant. When NaOH(aq) (i.e. $OH^-(aq)$) is added, the OH^- ions react with C_6H_5COOH to form H_2O and $C_6H_5COO^-$ so the $[H^+]$ does not change very much again.

(b) $C_6H_5COOH \rightleftharpoons C_6H_5COO^- + H^+$ $K_a = \dfrac{[H^+][C_6H_5COO^-]}{[C_6H_5COOH]}$

$[H^+] \simeq [C_6H_5COO^-]$

and $[C_6H_5COOH] = 0.02 - [H^+] \simeq 0.02$ as $[H^+]$ will be very small.

So, $K_a = 6.4 \times 10^{-5} = \dfrac{[H^+]^2}{0.02}$, $\therefore [H^+] = 1.1 \times 10^{-3}$ mol dm^{-3}

(c) pH $= -\lg[H^+] = -\lg[1.1 \times 10^{-3}] = -(0.04 - 3.00) = 2.96$

(d) $[C_6H_5COO^-]$ from sodium benzoate $= \dfrac{7.2}{144} = 0.05$ mol dm^{-3}

In this solution $[C_6H_5COOH] = 0.02 - [H^+] \simeq 0.02$

and $[C_6H_5COO^-] = 0.05 + [H^+] \simeq 0.05$

$\therefore K_a = 6.4 \times 10^{-5} = \dfrac{[H^+] \times 0.05}{0.02}$, so $[H^+] = 2.56 \times 10^{-5}$ mol dm^{-3}

$\Rightarrow$ pH $= -lg[H^+] = -lg[2.56 \times 10^{-5}] = -(0.41 - 5.00) = 4.59$

(e) Moles of NaOH added $= 10^{-3} =$ moles $[OH^-]$ added.

When OH^- is added, $C_6H_5COOH + OH^- \rightarrow C_6H_5COO^- + H_2O$

In this solution, $[C_6H_5COOH] \simeq 0.02 - 0.001 = 0.019$

$[C_6H_5COO^-] \simeq 0.05 + 0.001 = 0.051$

$\therefore K_a = 6.4 \times 10^{-5} = \dfrac{[H^+] \times 0.051}{0.019}$, so $[H^+] = 2.38 \times 10^{-5}$ mol dm^{-3}

$\Rightarrow$ pH $= -lg[H^+] = -lg[2.38 \times 10^{-5}] = -(0.38 - 5.00) = 4.62$

$\therefore$ pH changes by 0.03

8 $H_2PO_4^- \rightleftharpoons H^+ + HPO_4^{2-}$ $K_a = \dfrac{[H^+][HPO_4^{2-}]}{[H_2PO_4^-]}$

pH $= -lg[H^+]$, if pH $= 7.4$, $[H^+] = 3.98 \times 10^{-8}$

So, $6.4 \times 10^{-8} = 3.98 \times 10^{-8} \dfrac{[HPO_4^{2-}]}{[H_2PO_4^-]}$; $\therefore \dfrac{[H_2PO_4^-]}{[HPO_4^{2-}]} = 0.62$

9 (a) 10^{-4} mol dm^{-3} HCl; $[H^+] = 10^{-4}$ $\therefore$ pH $= 4.0$

(b) 10^{-4} mol dm^{-3} Ba(OH)$_2$; $[OH^-] = 2 \times 10^{-4}$

Now, $[H^+][OH^-] = 10^{-14}$ at 298 K, $\therefore [H^+] = 5 \times 10^{-11}$, $\therefore$ pH $= 10.3$

(c) 1.0 mol dm^{-3} H$_2$X; $[H^+]$ would be 2.0 mol dm^{-3} if 100% dissociated. In 50% dissociated solution, $[H^+] = 1.0$ mol dm^{-3}.

$\therefore$ pH $= -lg[H^+] = -lg[1.0] = 0$

(d) $CH_3CH_2COOH \rightleftharpoons CH_3CH_2COO^- + H^+$

$K_a = \dfrac{[CH_3CH_2COO^-][H^+]}{[CH_3CH_2COOH]}$

Now, $[H^+] \simeq [CH_3CH_2COO^-]$

and $[CH_3CH_2COOH] = 0.01 - [H^+] \simeq 0.01$ as $[H^+]$ is very small.

$\therefore K_a = 1.45 \times 10^{-5} = \dfrac{[H^+]^2}{0.01}$, so $[H^+] = 3.81 \times 10^{-4}$

$\Rightarrow$ pH $= -lg[H^+] = -lg(3.81 \times 10^{-4}) = -(0.58 - 4.00) = 3.42$

(e) $NH_4OH \rightleftharpoons NH_4^+ + OH^-$ $K_b = \dfrac{[NH_4^+][OH^-]}{[NH_4OH]}$

$[OH^-] \simeq [NH_4^+]$

and $[NH_4OH] = 1.0 - [OH^-] \simeq 1.0$ as $[OH^-]$ is very small.

$\therefore K_b = 1.7 \times 10^{-5} = \dfrac{[OH^-]^2}{1.0}$, so $[OH^-] = 4.12 \times 10^{-3}$ mol dm^{-3}

Now, $[H^+][OH^-] = 10^{-14}$ at 298 K, $\therefore [H^+] \times 4.12 \times 10^{-3} = 10^{-14}$,

So, $[H^+] = 2.43 \times 10^{-12}$, pH $= 11.61$

10 (a) HCl(aq) is fully dissociated. So, in 0.1 mol dm^{-3} HCl, $[H^+] = 10^{-1}$ mol dm^{-3}, so pH $= 1.0$.

CH_3COOH is only partially dissociated, and in 0.1 mol dm^{-3} CH_3COOH, $[H^+]$ is approx. 10^{-3}, so pH $\simeq 3$.

(b) No. The pH changes rapidly from about 7.0 to 10.5 during the titration. Methyl orange does not change colour in this pH range, but between 3.0 and 4.5.

(c) Yes. Phenolphthalein changes colour between pH 8 and 10 which corresponds with the rapid pH change in the titration.

(d) There is no sharp pH change at any point, so no indicator will change colour with the addition of a small excess of acid or alkali.

(e) Using a pH meter.

Chapter 23

1 (a) One iodine molecule could displace each of the 10 CCl$_4$ molecules. Therefore one iodine molecule could arrange itself in 10 different ways.

(b) One

(c) Among the CCl$_4$ molecules

(d) 6.02×10^{23} (Avogadro's number)

(e) One – this is not likely.

(f) Iodine dissolves as this produces the most likely arrangement of particles.

(g) Maximum disorder has been reached and iodine molecules return to the solid lattice at the same rate as they leave it.

2 (a) $2 \times 32.7 = 65.4$; $65.4 + 204.9 = 270.3$ J K^{-1} mol^{-1}

(b) $2 \times 26.8 = 53.6$ J K^{-1} mol^{-1}

(c) $\Delta S = 53.6 - 270.3 = -216.7$ J K^{-1} mol^{-1}

(d) No

(e) $\Delta S_{surr} = -\Delta H/T$; $\Delta S_{surr} = -(-1\,204\,000)/298 = 4040.3$ J K^{-1} mol^{-1}

(f) $\Delta S_{total} = \Delta S_{surr} + \Delta S_{syst}$; $\Delta S_{total} = 4040.3 + (-216.7)$
$= 3823.6$ J K^{-1} mol^{-1}

(g) Yes

3 (a) A

(b) C

(c) A

(d) C

(e) A

(f) B

(g) B

(h) A

4 (a) $S^{\ominus}$ increases down the group. As the number of electrons increases so the number of electronic energy levels increases, so there are more ways of distributing the quanta of energy.

(b) Hydrogen is a diatomic molecule. Therefore it has electronic and vibrational energy levels. Helium only has electronic energy levels. $S^{\ominus}$ is higher for the hydrogen molecules as there are more ways of distributing the quanta of energy.

(c) Solids have the lowest standard entropies because they are the most ordered. From solid to liquid to gas, the standard entropy increases.

5 (a) $\Delta G^{\ominus} = +128.8$ kJ mol^{-1}. The reaction is not spontaneous at 298 K. Raising the temperature will make ΔG more negative. $\Delta G = 0$ when $T = 1079$ K. This is the minimum temperature to which CaCO$_3$ must be heated in order to make it decompose.

(b) $\Delta G^{\ominus} = +48.2$ kJ mol^{-1}; $\Delta G = 0$ when $T = 574$ K

(c) MgCO$_3$ is less thermally stable than CaCO$_3$. It decomposes at a lower temperature.

(d) Mg^{2+} has a higher charge density than Ca^{2+}, therefore it polarises CO$_3^{2-}$ more than Ca^{2+} does, and MgCO$_3$ decomposes more easily.

Chapter 24

1 (a) (i)   Any explosion or precipitation
       (ii) Slow burning, cooking, etc.
       (iii) Iron burning in oxygen

(b) Discussion should involve consideration of activation energies.

2 (a) $\times 8$ (b) $\times 4$ (c) $\times \frac{1}{8}$ (d) no change   (e) approximately $\times 8$

3 (a)

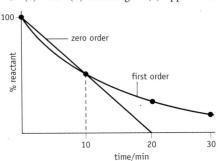

(b) Assuming uranium disintegrates directly to lead and nothing else and assuming there is no lead present initially;

then, initially, molar proportions   U : Pb = 1 : 0

after one half-life, molar proportions   U : Pb = 1 : 1

after two half-lives, molar proportions   U : Pb = 1 : 3

$\therefore$ age of rock $= 9 \times 10^9$ years

4 (a) (i) Order of reaction with respect to a given reactant is the power of that reactant's concentration in the rate equation.

  (ii) The rate constant is the constant of proportionality in the rate equation. The rate constant is numerically equal to the reaction rate, when all reactant concentrations are 1.0 mol dm^{-3}.

  (iii) Half-life is the time taken for the concentration of a reactant to fall to half its original value.

  (iv) Activation energy is the minimum energy needed by the reactants for a reaction to occur.

  (v) As a reaction proceeds, bonds must stretch and break for a reaction to occur. The 'activated state' describes the state of the reacting molecules at their position of maximum energy during the course of the reaction.

 (b) (i) The gradient of the graph = -2.24×10^4

  (ii) The gradient of the graph = $-2.24 \times 10^4 = \dfrac{-E_A}{R}$

   $\therefore E_A = 2.24 \times 10^4 \times 8.3 = 186\,000$ J $mol^{-1} = 186$ kJ mol^{-1}

5 (a) reaction rate =
$\dfrac{\text{change in } [I_2]}{\text{time}} = \dfrac{10^{-5}}{6}$ mol dm^{-3} $s^{-1} = 1.67 \times 10^{-6}$ mol dm^{-3} s^{-1}

 (b) reaction rate = $\dfrac{10^{-5}}{12}$ mol dm^{-3} $s^{-1} = 8.33 \times 10^{-7}$ mol dm^{-3} s^{-1}

 (c) When $[H_2O_2]$ is halved, the reaction rate is halved, i.e. rate is proportional to $[H_2O_2]$.
 $\therefore$ order with respect to $H_2O_2 = 1$

 (d) Reaction rate = $k[H_2O_2][H^+][I^-]$, for the reaction in part (a).
 $\dfrac{10^{-5}}{6} = k \times 0.01 \times 0.01 \times 0.1 \therefore k = \dfrac{1}{6} = 1.67 \times 10^{-1}$

 (e) mol^{-2} dm^6 s^{-1}

 (f) Rate = $k[H_2O_2][H^+][I^-] = 1.67 \times 10^{-1} \times 0.05 \times 0.10 \times 0.02$
 $\therefore$ rate = 1.67×10^{-5} mol dm^{-3} s^{-1}

6 (a) To neutralise the acid catalysing the reaction and allow the titration to be carried out with care and without further reaction occurring.

 (b) From the graph, the volume of $Na_2S_2O_3(aq)$ required for titration falls by 3 cm^3 in 30 minutes.
 $\therefore$ reaction rate = $\dfrac{-3}{30} = -0.1$ cm^3 $Na_2S_2O_3(aq)$ min^{-1}

 (c) It does not change.

 (d) No

 (e) 0

 (f) $2S_2O_3^{2-} + I_2 \rightarrow S_4O_6^{2-} + 2I^-$

 (g) At time = 0 min, $[I_2] = 0.01$ mol dm^{-3}

 (h) 20 cm^3

 (i) Number of mol I_2 at $t = 0$ min = $\dfrac{10}{1000} \times 0.01 = 10^{-4}$
 10^{-4} mol I_2 react with 2×10^{-4} mol $S_2O_3^{2-}$
 2×10^{-4} mol are present in 20 cm^3 solution
 $\therefore$ molarity of $Na_2S_2O_3 = \dfrac{2 \times 10^{-4} \times 1000}{20}$ mol dm^{-3}
 = 0.01 mol dm^{-3}

 (j) -0.2 cm^3 $Na_2S_2O_3(aq)$ min^{-1}

7 (a) (i) 0 (ii) 2

 (b) Rate = $k[X]^0[Y]^2 = k[Y]^2$

 (c) $Y + Y \xrightarrow{\text{slow}} Y_2$
 $X + Y_2 \xrightarrow{\text{fast}} XY_2$

 (d) Rate = $k[Y]^2$, so $0.0001 = k(0.1)^2$; $k = 0.01$

 (e) mol^{-1} dm^3 s^{-1}

 (f) Experiments to obtain the value of k at different temperatures

 (g) To explore and predict reaction mechanisms and understand how reactions occur.

8 (a) By following the volume of CO_2 produced with time.

 (b) By following the change in pressure with time as 4 moles of gas form 3 moles of gas

 (c) By following the change in the colour of the solution with time. $Br_2(aq)$ is a yellow/orange colour whereas all the other substances are colourless.

9 (a)

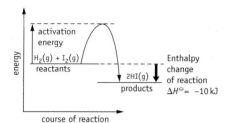

 (b)
 (i) Order with respect to $I_2 = 1$ (ii) overall order = 2

 (c) (i) Rate = $k[H_2][I_2]$, $\therefore 2.3 \times 10^{-5} = k[0.1][0.2]$
   $\therefore k = \dfrac{2.3 \times 10^{-5}}{0.1 \times 0.2} = 1.15 \times 10^{-3}$ mol^{-1} dm^3 s^{-1}

  (ii) mol^{-1} dm^3 s^{-1}

10 (a)

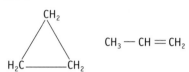

$CH_3 - CH = CH_2$

 (b) Plot a graph of % cyclopropane remaining (vertical) against time in hours (horizontal). Show that the half-life is constant (15 h). As the half-life is constant, the reaction is first order with respect to cyclopropane.

Chapter 25

1 (a) C = $(12/44 \times 1.37) \times 100 = 37.36\%$
  H = $(2/18 \times 1.12) \times 100 = 12.44\%$

 (b) Oxygen

 (c) % O = 50.2 % ; ratio C : H : O is 1 : 4 : 1 so empirical formula is CH_4O

 (d) (i) $M_r = 32$, molecular formula is CH_4O

  (ii) CH_3OH

  (iii) $32 = CH_4O^+$
   $31 = CH_3O^+$
   $30 = CH_2O^+$
   $29 = CHO^+$
   $17 = OH^+$
   $15 = CH_3^+$

2 (a) **A** and **C**
 (b)

H — C — C — C — OH (with H, H, H above and H, H, H below)

H — C — C — C — H (with H, H, H above and H, OH, H below)

H — C — C — O — C — H (with H, H, H above and H, H, H below)

 (c) **A** and **C**: alcohols
  B: ethers

 (d) **A** is propan-1-ol; **B** is methoxyethane; **C** is propan-2-ol
 The alcohols have higher boiling points than the ether. The primary alcohol has a slightly higher boiling point than the secondary.

3 (a) Alcohols
(b) Ketones
(c) Halogenoalkanes
(d) Carboxylic acids
(e) Alkenes
(f) Amino acids
(g) Ethers

4 (a)

```
    H   H   H
    |   |   |
H — C — C — C — Cl
    |   |   |
    H   H   H

    H   H   H
    |   |   |
H — C — C — C — H
    |   |   |
    H   Cl  H
```

(c)

```
    H   Cl
    |   |
H — C — C — Cl
    |   |
    H   Br

    H   H
    |   |
H — C — C — Cl
    |   |
    Br  Cl

    H   H
    |   |
H — C — C — Br
    |   |
    Cl  Cl
```

This exists as two optical isomers:

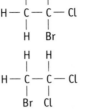

(b)

```
    H   H   H   H   H   H
    |   |   |   |   |   |
H — C — C — C — C — C — C — H
    |   |   |   |   |   |
    H   H   H   H   H   H
```

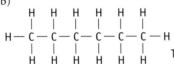

```
    H   H   H   H   H
    |   |   |   |   |
H — C — C — C — C — C — H
    |   |   |   |   |
    H   H   H H-C-H H
             |
             H
```

```
    H   H   H   H   H
    |   |   |   |   |
H — C — C — C — C — C — H
    |   |   |   |   |
    H   H H-C-H H   H
           |
           H
```

```
           H
         H-C-H
    H   H |   H   H
    |   | |   |   |
H — C — C — C — C — H
    |   | |   |   |
    H   H H-C-H H   H
           |
           H
```

```
         H
       H-C-H
    H   |   H   H
    |   |   |   |
H — C — C — C — C — H
    |   |   |   |
    H   H H-C-H H
           |
           H
```

5 Si—O = 464; Si—H = 318. Si—O bonds are strong, therefore the silicones are unreactive. Si—H bonds are easily broken, so the silanes are very reactive.

6 C—F bond is difficult to break and forming the F—O bond does not release much energy. C—H bond is easier to break, and forming the H—O bond releases more energy than is required for this process.

7 (a) C: $\dfrac{12/44 \times 0.2280}{12} = 5.166 \times 10^{-3}$ mol

H: $\dfrac{2/18 \times 0.124}{1} = 1.377 \times 10^{-2}$ mol

N: $\dfrac{17.2}{1000} \times 0.1 = 1.72 \times 10^{-3}$ mol

(b) C_3H_8N

(c) $C_6H_{16}N_2$

(d)

```
      H   H   H   H   H   H
      |   |   |   |   |   |
H₂N — C — C — C — C — C — C — NH₂
      |   |   |   |   |   |
      H   H   H   H   H   H
```

8 D ≡ G ≡ E; B ≡ H ≡ F; A ≡ C

9 The ability to form bonds with itself.
The ability to form four covalent bonds which leads to a large variety of compounds.
The ability to form multiple bonds.

10 (a) (i) Phenyl hydrogens

(ii) $-\overset{\displaystyle\;}{\underset{\displaystyle\parallel}{C}}-CH_2$ with $\parallel\ O$

(iii) $-CH_3$

(b)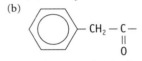

11 (a) **A** is butan–1–ol; **B** is ethoxyethane.

(b)
1 : O—H in butan–1–ol
2 : C—H in butan–1–ol
3 : C—O in butan–1–ol
4 : C—O in ethoxyethane

12 (a) At pH = 6 phenolphthalein absorbs in the UV region of the spectrum.
(b) Colourless
(c) Blue/green
(d) Red/violet (pink)
(e) When the pH changes from 6 to 10, phenolphthalein would change from colourless to pink.

Chapter 26

1 (a) Hexane
(b) 4–Ethyl–2–methylhexane
(c) Cycloheptane
(d) Hexane
(e) Methylcyclopropane
(f) Ethylcyclopentane

2 (a) **A** diesel oil
B gasoline
C gasoline
D gasoline
E refinery gas
(b) **C** and **D**
(c) **C** and **E**
(d) **E B D C A**
(e) **A**
(f) **B**

3 (i)

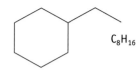

C_8H_{16}

(ii)

C_7H_{14}

(iii)

$$CH_3—\underset{\underset{CH_3}{|}}{\overset{\overset{CH_3}{|}}{C}}—\underset{}{\overset{\overset{CH_3}{|}}{CH}}—CH_3 \quad C_7H_{16}$$

(iv)

$$CH_3—\underset{\underset{CH_3}{|}}{\overset{\overset{CH_3}{|}}{C}}—\overset{\overset{CH_2CH_3}{|}}{CH}—CH—CH_2—CH_2—CH_3 \quad C_{13}H_{28}$$

with CH_2CH_3 below.

4 (b) Yes

(c) 661 kJ mol^{-1}

(d) —CH$_2$—

(e) Molecules contain the same numbers of C and H atoms – therefore the same numbers of bonds are broken and made.

(f) butane: $^{-2877}/_{58} = -49.6$ kJ g^{-1}

pentane: $^{-3509}/_{72} = -48.7$ kJ g^{-1}

hexane: $^{-4195}/_{86} = -48.8$ kJ g^{-1}

The values are very similar.

5 (a) To help the petrol burn more smoothly

(b) $C_8H_{18} + 12\frac{1}{2}O_2 \rightarrow 8CO_2 + 9H_2O$

(c) 1 dm^3 = 750g; number of moles = $^{750}/_{114} = 6.58$

(d) volume of oxygen = $12.5 \times 6.58 \times 24 = 1974$ dm^3

volume of air = $1974 \times 5 = 9870$ dm^3

(e) volume of nitrogen = 7896 dm^3

(f) number of moles = $^{7896}/_{24} = 329$

mass of nitrogen = $329 \times 28 = 9212$ g

(g) 6.58 moles of petrol produces 8×6.58 moles of carbon dioxide;

$6.58 \times 8 = 52.64$ moles

mass of 52.64 moles is $52.64 \times 44 = 2316$ g

(h) mass of 0.5 cm^3 of tetraethyllead = $1.6 \times 0.5 = 0.8$ g

mass of lead = $(^{207}/_{323}) \times 0.8 = 0.5$

6 (a) $CH_3CH_2CH_2CH_2CH_2CH_2CH_2CH_3$ (b)

(c) Octane:

Cyclooctane:

7 (a) $(CH_2)_3$ ΔH_c per $CH_2 = -697$ kJ mol^{-1}

$(CH_2)_4$ ΔH_c per $CH_2 = -685$ kJ mol^{-1}

$(CH_2)_5$ ΔH_c per $CH_2 = -664$ kJ mol^{-1}

$(CH_2)_6$ ΔH_c per $CH_2 = -658$ kJ mol^{-1}

$(CH_2)_7$ ΔH_c per $CH_2 = -662$ kJ mol^{-1}

$(CH_2)_8$ ΔH_c per $CH_2 = -664$ kJ mol^{-1}

$(CH_2)_9$ ΔH_c per $CH_2 = -664$ kJ mol^{-1}

$(CH_2)_{10}$ ΔH_c per $CH_2 = -663$ kJ mol^{-1}

(b) (i) Yes

(ii) Yes

(iii) No

(c) Strain will affect the bond energies and hence the value of ΔH_c. The larger cycloalkanes are unstrained, so the values of ΔH_c per CH_2 are very similar.

8 (a) HBr

(b) $C_6H_{14}+Br_2 \rightarrow C_6H_{13}Br + HBr$

(c) Sunlight is required to initiate the reaction by breaking the Br—Br bond.

(d) $Br_2 \rightarrow Br\cdot + Br\cdot$ Initiation

$\left. \begin{array}{l} C_6H_{14} + Br\cdot \rightarrow C_6H_{13}\overset{.}{.} + HBr \\ C_6H_{13}\overset{.}{.} + Br_2 \rightarrow C_6H_{13}Br + Br\cdot \end{array} \right\}$ Propagation

$\left. \begin{array}{l} C_6H_{13}\overset{.}{.} + C_6H_{13}\overset{.}{.} \rightarrow C_{12}H_{26} \\ Br\cdot + Br\cdot \rightarrow Br_2 \\ C_6H\cdot_{13} + Br\cdot \rightarrow C_6H_{13}Br \end{array} \right\}$ Termination

9

10 (a) $Cl_2 + C_6H_{14} \rightarrow C_6H_{13}Cl + HCl$

(b) No reaction

(c) No reaction

(d) No reaction

(e) No reaction

(f) Cracking e.g. $C_6H_{14} \rightarrow C_2H_4 + C_4H_{10}$

11 Modifications: larger, pressurised storage tank; modified system for mixing fuel with air

Advantages: burns more efficiently – less C produced in the engine, less CO in the exhaust.

12 (a) Reaction with an alkali – sodium hydroxide, or calcium hydroxide

(b) Cooling the mixture – ethane liquifies at $-100\,°C$

(c) Fuel

(d) Nitrogen is unreactive.

(e) It is the basis of many polymers and for producing many other organic chemicals.

(f) $C_2H_6 \rightarrow C_2H_4 + H_2$

High temperature, catalyst

(g) Naphtha is more available than natural gas in Europe.

Chapter 27

1 (a) (i) Pent-1-ene

(ii) *trans*-hex-2-ene

(iii) *cis*-hex-3-ene

(iv) But-2-yne

(v) Cyclohexene

(b) (i) CH₃CH=CH₂

(ii)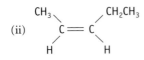

(iii)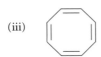

(iv)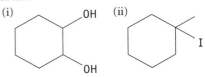

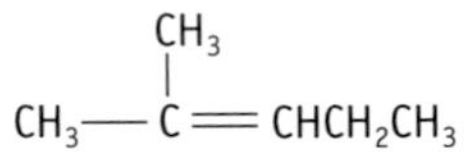

2 (a) poly(1,1-dichlororethene)

(b)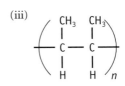

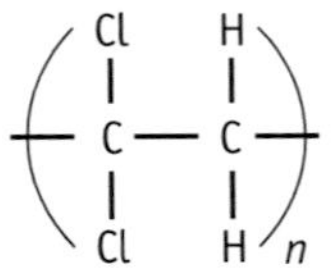

(c) Higher, because the intermolecular forces will be greater.

3 (a) (i) No   (ii)

(iii) No   (iv)

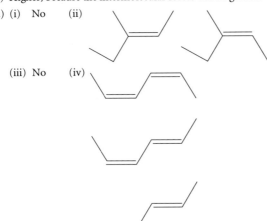

(b) The 3-membered ring acts like a double bond. The two Cl atoms can both be above the ring, or one above and one below. There is no free rotation about the C—C bond.

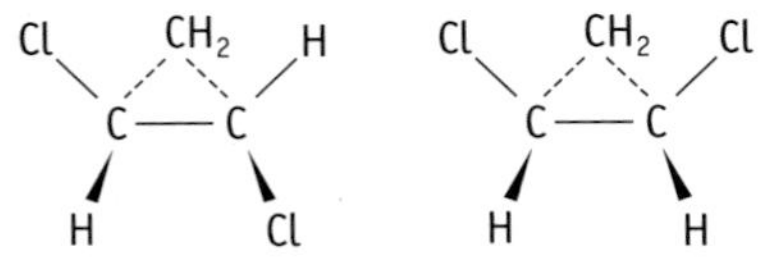

4 (a)   Proportion by moles   C $\frac{87.8}{12}$: H $\frac{12.2}{1}$

= 1 : 1.67

Empirical formula is C₃H₅

(b) C₆H₁₀

(c) Moles H₂: $^{27.3}/_{22\,400} = 1.22 \times 10^{-3}$

moles A: $^{0.1}/_{82} = 1.22 \times 10^{-3}$

so 1 mole of H₂ reacts with 1 mole of A

(d) One

(e) C₆H₁₂

(f) A   B

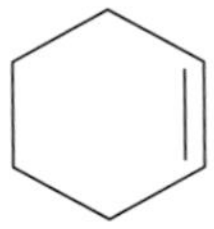

5 (a) Cracking of petroleum fractions.
Ethene is also produced in cracking reactions, which are difficult to control.

(b) (i) $CH_3—\overset{+}{C}H—CH_3$ (ii) $CH_3—CH_2—\overset{+}{C}H_2$

(c) (i) is more stable

(d) 1-Propyl hydrogensulphate would be formed from carbocation (ii), which is less stable.

(e) $CH_3—\overset{+}{C}H_2$

(f) Less stable

(g) Propene will be attacked preferentially as a more stable carbocation is formed.

(h) It means that a high yield of propan-2-ol is formed.

6 (a) Shake with bromine water. Cyclohexene would be decolorised, cyclohexane would not.

(b)   (i)   (ii)

(iii)

7 (a) 140 °C and Ni catalyst

(b) 252 kJ mol⁻¹

(c) 436 kJ mol⁻¹

(d) 826 kJ mol⁻¹

(e) −138 kJ mol⁻¹

8 The π-bond is easily broken, so ethene and propene are relatively reactive. They react with various reagents to form useful products:
e.g. with hydrogen – manufacture of margarine
with sulphuric acid – manufacture of alcohols
with oxygen/silver – manufacture of epoxyethane.
They react with themselves to form polymers (e.g. polyethene and polypropene).
The alkanes have no π-bonds, only σ-bonds which are difficult to break. Therefore the alkanes are unreactive and less useful.

9 (a) A, B

(b) C, D

(c) A, B

(d) A

(e) A, B

10 (a) To break the π-bond and generate free radicals.

(b) (i) Both consist of (CH₂)ₙ; both are saturated.

(ii) Saturated alkane molecules are generally shorter. They have a CH₃— group at each end.

(c) 'Back biting' introduces a radical site into the middle of a chain. From this site a branch can grow.

(d) The growing chain could react with, for example, an RO· radical, or another growing chain.

Chapter 28

1 (a) 1,3-Dibromobenzene

(b) 1-Ethyl-2-methylbenzene

(c) 1,2,3-Trimethylbenzene

(d) 1-Bromo-3-chloro-5-nitrobenzene

(e) 4-Nitrobenzoic acid

(f) 3-Chlorophenol

2

(a)

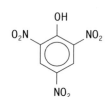

O_2N — OH — NO_2 ring with NO_2

(d)

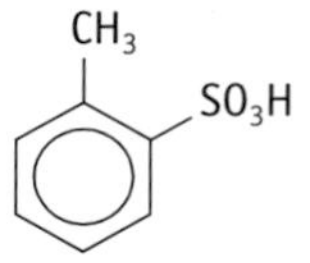

CH_3 ring with SO_3H

(b)

Cl — ring — Cl

(e)

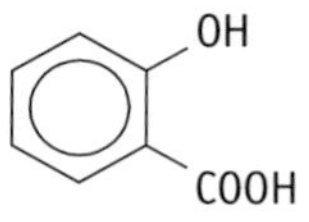

ring with OH and COOH

(c)

NH_2 ring with NO_2

(f)

NH_2 ring with Cl

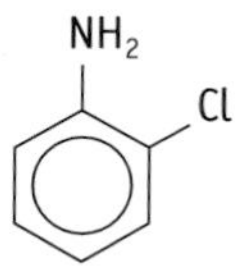

3 (a) Nickel

(b) $3 \times (-120) = -360$ kJ mol^{-1}

(c) The experimental value is much less negative.

(d) The difference between the two values represents the delocalisation energy of benzene.

4 (a) D—Cl →AlCl$_3$

(b) AlCl$_3$ increases the polarisation of the DCl. Ionisation occurs, forming AlCl$_4^-$ and D$^+$

(c) AlCl$_3$ is electron deficient. It attracts the lone pairs on the Cl in DCl, making the D electrophilic.

(d) The C—H bonds in the side chain cannot be broken. There are no free radicals present. The π-electrons in the benzene ring are attracted to the D$^{\delta+}$ and this produces substitution of the ring hydrogen atoms.

5 (a) Shake with bromine water – cyclohexadiene causes bromine to be decolorised, benzene does not.

(b) Shake with alkaline potassium manganate(VII) – benzene has no reaction, methylbenzene changes the purple manganate(VII) to green (or brown).

(c) Add bromine in the dark, followed by iron filings. Cyclohexane has no reaction. Benzene decolorises the bromine and forms HBr.

(d) Add bromine water. Hexatriene decolorises it, benzene does not.

6 (a) (b) 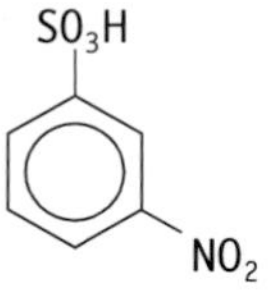 (c)  (d)

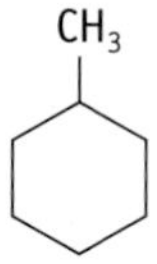

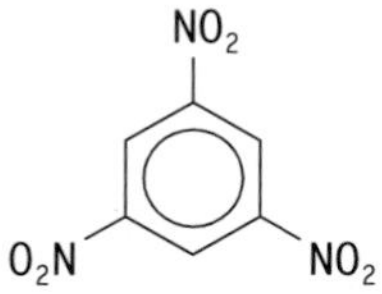

(a) CH_3 cyclohexane

(b) SO_3H ring with NO_2

(c) COOH ring with COOH

(d) NO_2 ring with O_2N and NO_2

(e) CH_3—CH—CH_3 on ring

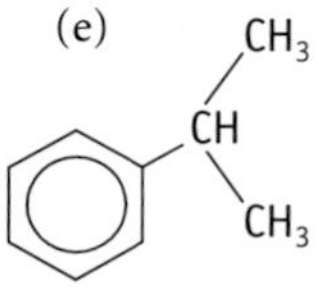

(f) CH_3 ring with Br and CH_3 ring with Br

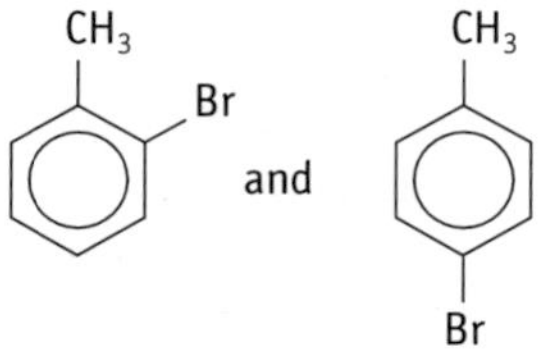

7 (a) $CH_3—\overset{+}{C}H—CH_3$ and $CH_3—CH_2—\overset{+}{C}H_2$

(i)   (ii)

(b) (i)

(c)

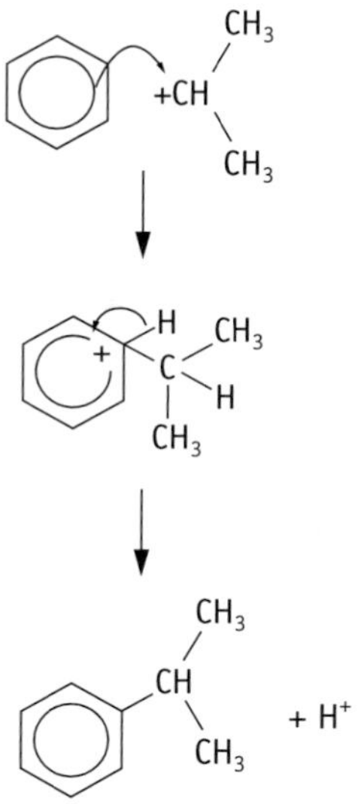

(reaction scheme with benzene and carbocation steps) + H$^+$

The concentration of carbocation (i) will be much greater than that of carbocation (ii), because it is more stable.

(d) Benzene is readily available from oil fractions via catalytic reforming. Propene is produced during the cracking of larger alkanes, so is cheaply available. Both products are valuable.

8 (a) (i) (ii)

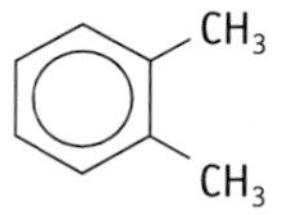

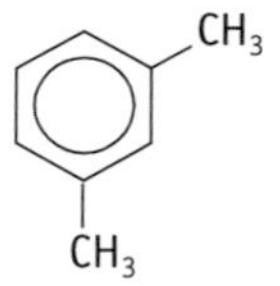

(i) ring with CH_3, CH_3

(ii) ring with CH_3, CH_3

(iii) 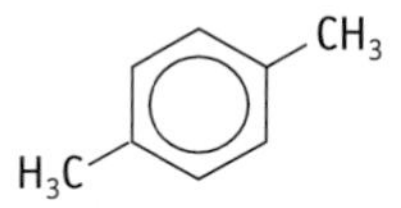 (iv)

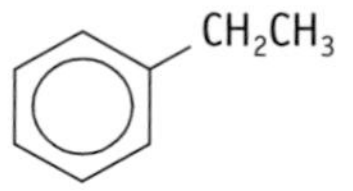

(iii) H_3C ring CH_3

(iv) ring CH_2CH_3

(b) (i) **A** 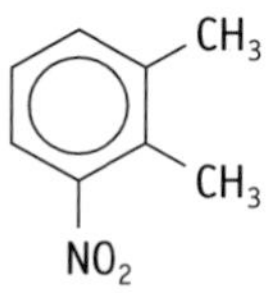   **B**

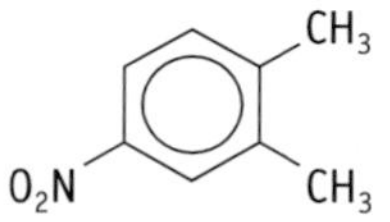

A: ring with CH_3, CH_3, NO_2

B: O_2N ring with CH_3, CH_3

(ii) **A** 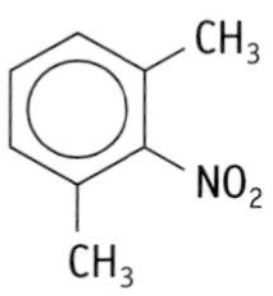   **B**

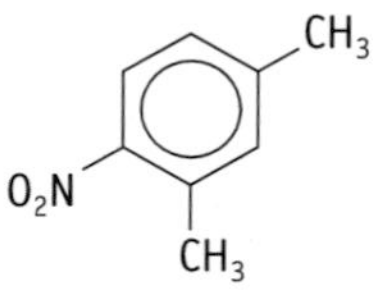

A: ring with CH_3, NO_2, CH_3

B: O_2N ring with CH_3, CH_3

C

O_2N ring with CH_3, CH_3

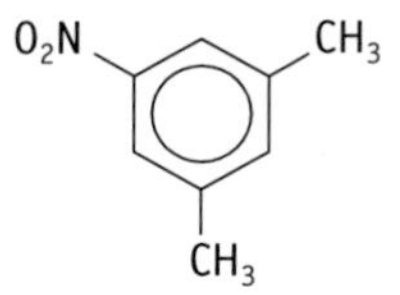

(iii)

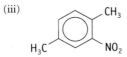

(iv) **A**

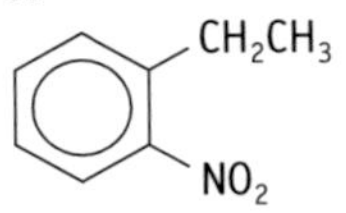

CH$_2$CH$_3$

NO$_2$

B

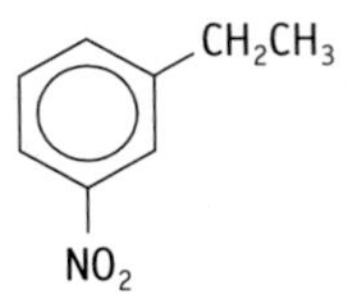

CH$_2$CH$_3$

NO$_2$

C

CH$_2$CH$_3$

O$_2$N

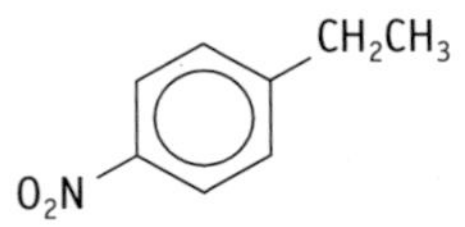

(c) (i) **B**   (ii) **B**   (iv) **C**

9 (a) (i) and (ii)
   (b) (i), (ii), (iii) and (iv)
   (c) (i), (ii), (iv) and (v)
   (d) (i) and (ii)
   (e) (iv) and (v)
   (f) (i) and (ii)
   (g) (i), (iv) and (v)

10 (a) substitution – **A**
      addition – **B**
   (b) (i)   484 kJ
      (ii)   860 kJ
      (iii) 678 kJ
      (iv)   862 kJ
      (v)   (484 + 860) − (678 + 862) = −196 kJ
   (c) (i)   242 kJ
      (ii)   678 kJ
      (iii) (242+434) − 678 = −2 kJ
   (d) Reaction **A** is more likely to occur as more energy is released.

Chapter 29

1 (a) A Trichlorofluoromethane
      B Octafluorocyclobutane
      C 1,1,2-Trichloro-1,2,2-trifluoroethane
      D Bromotrifluoromethane
   (b) B
   (c) C
   (d) D

2 (a) B, C and E
   (b) D
   (c) D
   (d) A
   (e) D
   (f) C

3 (a) The silver halide: AgCl, AgBr, AgI
   (b) Halide ions are formed by nucleophilic substitution. The halide ions then react with Ag$^+$(aq) to form a precipitate.
   (c) Water
   (d) Iodobutane – most readily; chlorobutane – least readily
   (e) The C—Cl bond is stronger than C—Br which is stronger than C—I. Nucleophilic substitution involves breaking the C—Hal bond, so it is slowest for the chloro compound.

4 (a) CH$_3$CH$_2$CH$_2$NH$_2$ + HBr
   (b) CH$_3$CH$_2$CH$_2$COO—CH$_2$CH$_2$CH$_2$—OOCCH$_2$CH$_2$CH$_3$ + AgI

(c)

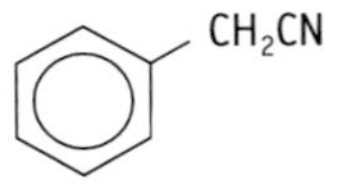

CH$_2$CN

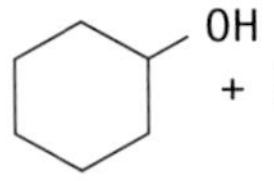

(d)

OH

+ HBr

(e)

+ H$_2$O + Br$^-$

(f)

OCH$_3$

CH$_3$—C—CH$_3$ + NaBr

OCH$_3$

(g)

COOCH$_2$CH$_3$ + HCl

5 (a) CH$_3$CH$_2$—SH
   (b) CH$_3$CH$_2$NO$_2$
   (c) CH$_3$CH$_2$Cl
   (d) CH$_3$CH$_3$
   (e) CH$_3$CH$_2$NH$_2$

6

X CH$_3$CH$_2$C

O

Cl

Y CH$_3$CHClCHO or CH$_2$ClCH$_2$CHO

7 (a)

CH$_2$CH$_3$

CH$_3$CH$_2$CCH$_2$CH$_3$

Cl

   (b) (i) Nucleophilic substitution
      (ii)

CH$_2$CH$_3$

CH$_3$CH$_2$CCH$_2$CH$_3$

OH

      (iii)

CH$_2$CH$_3$ CH$_2$CH$_3$

CH$_3$CH$_2$CCH$_2$CH$_3$ + $^-$OH ⟶ CH$_3$CH$_2$CCH$_2$CH$_3$ + Br$^-$

Cl   OH

   (c) (i) Elimination
      (ii)

CH$_2$CH$_3$

CH$_3$CH=C—CH$_2$CH$_3$

      (iii)

CH$_2$CH$_3$

CH$_3$CH$_2$–C–CH$_2$CH$_3$ + $^-$OH

Cl

CH$_2$CH$_3$

CH$_3$–CH=C–CH$_2$CH$_3$ + H$_2$O + Cl$^-$

   (d) Elimination is more likely to occur when the solvent is non-aqueous.

8

(a)   CH$_2$=CH$_2$ $\xrightarrow[\text{KBr}]{\text{cH}_2\text{SO}_4}$ CH$_3$=CH$_2$Br $\xrightarrow[\text{ethanol}]{\text{Reflux} \atop \text{KCN in}}$ CH$_3$CH$_3$CN

Equations:
    CH$_2$=CH$_2$+HBr ⟶ CH$_3$CH$_2$Br
    CH$_3$CH$_2$Br + KCN ⟶ CH$_3$CH$_2$CN + KBr

(b)

Equation:

$$C_6H_5COCl + NH_3 \longrightarrow C_6H_5CONH_2 + HCl$$

(c)

Equations:

9   $P = Cl_2$/UV radiation
    $Q = HCl$(aq)

    A = ethanol
    B = ethene
    C = 1,2-dibromoethane
    D = ethane-1,2-diol

10 (a) Iodobenzene is more reactive than chlorobenzene as the C—I bond is weaker than C—Cl and is therefore more easily broken. It is much less reactive than iodoethane as the delocalised electrons on the benzene ring reduce the likelihood of attack by nucleophiles.
    (b) CCl_4 is tetrahedral and has no overall dipole. It is effectively non-polar – the nucleophile cannot attack the $C^{\delta+}$ as it is in the centre of the molecule.
    (c) Iodo and bromo compounds are more useful as they are more reactive. The C—Cl bond is relatively strong, so chloro compounds are less reactive.
    (d) The chlorine atom next to the carbonyl group is easily substituted. This C atom is very attractive to nucleophiles as it has two electronegative groups attached to it. The other Cl atom is less readily substituted.

Chapter 30

1 (a) A 2-Methylbutan-2-ol
      B Butan-2-ol
      C Methoxypropane
      D Butan-1-ol
      E 4-Methylphenol
   (b) D
   (c) A
   (d) E
   (e) C
2 (a) A, B, D and E
   (b) D
   (c) B
   (d) E
   (e) A, B and D
   (f) C

3 (a) Ethanol has hydrogen bonds between molecules, ethoxyethane does not.
   (b) Both water and ethanol form hydrogen bonds. In water, the bonds are stronger, and the molecules are smaller. Therefore, the molecules are closer together.
   (c) The surface tension in water arises as a result of hydrogen bonds between molecules. The intermolecular forces in ethoxyethane are much weaker so the surface tension is correspondingly lower.
4 (a) Mass C $= \frac{12}{44} \times 1.1 = 0.3$ g
     mass H $= \frac{2}{18} \times 0.563 = 0.0626$ g
     % C $= \frac{0.3}{0.463} \times 100 = 64.8$
     % H $= \frac{0.0626}{0.463} \times 100 = 13.5$
     % O $= 100 - (64.8 + 13.5) = 21.7$
   (b) C : H : O $= 64.8/_{12} : 13.5/_1 : 21.7/_{16}$
             $= 5.4 : 13.5 : 1.356$
             $= 4 : 10 : 1$
     Empirical formula $= C_4H_{10}O$
   (c) Vol. of 0.1 g at s.t.p $= 54.4 \times \frac{273}{481} \times \frac{98.3}{101} = 30.1$ cm^3
     $M_r = \frac{22\,400}{30.1} \times 0.1 = 74.4$

   (d) Molecular formula $= C_4H_{10}O$
     A $CH_3CH_2CH_2CH_2OH$ **B** $CH_3CH_2CHCH_3$
                                            |
                                           OH

                                            OH
     C CH_3CHCH_2OH **D** $CH_3-\overset{\,}{C}-CH_3$
             |
             CH_3 CH_3

     E $CH_3CH_2CH_2-O-CH_3$ **F** $CH_3CH_2-O-CH_2CH_3$

   (e) **A** to **D** all react with sodium to give hydrogen.
   (f) **A** to **C** all reduce $Cr_2O_7{}^{2-}$ to Cr^{3+}.
5 (a) Butan-1-ol acts as a base. The lone pairs of electrons on the oxygen atom accept a proton from the HCl.
     $CH_3CH_2CH_2CH_2OH + H^+ \rightarrow CH_3CH_2CH_2CH_2OH_2{}^+$
     The charged ion is more soluble than the neutral molecule.
   (b) The base formed by picric acid is stabilised relative to that formed by phenol, by the electron-withdrawing effect of the nitro groups.

   (c) The —OH group activates the benzene ring. The lone pairs on the oxygen interact with the delocalised electron system and enhance it. This makes the ring more attractive to electrophiles such as $NO_2{}^+$.

   (d) In 2-methylpropan-2-ol, the methyl groups surrounding the carbon atom containing the —OH are electron releasing. This strengthens the O—H bond, relative to that in ethanol.
   (e) The —OH group is lost along with an H either from carbon atom 1 or from carbon atom 3. This produces but-1-ene or but-2-ene. The latter has *cis* and *trans* isomers.
6 (a) (i) Reflux with acidifed potassium chromate(VI) (or acidified potassium manganate(VII)), to make ethanoic acid.

(ii) Warm ethanoic acid and ethanol in a water bath with a few drops of concentrated sulphuric acid.

(b) (i) Heat with excess concentrated sulphuric acid to produce ethene.

(ii) Shake ethene with bromine.

(c) (i) Bubble ethene into concentrated sulphuric acid. Add water and warm gently to form ethanol.

(ii) Reflux the ethanol with acidified potassium chromate(VI) to form ethanoic acid.

(d) (i) Reflux ethanol with potassium bromide and concentrated sulphuric acid to make bromoethane.

(ii) Heat bromoethane under reflux in ethanol with potassium cyanide.

(e) (i) Heat under reflux with aqueous sodium hydroxide to make ethanol.

(ii) Heat excess ethanol with concentrated sulphuric acid at 140 °C.

7 (a) **X** is methylphenol. It burns to produce carbon, carbon dioxide and water. It reacts with sodium hydroxide to form sodium methylphenoxide.

(or isomers)

(b) **Y** is an ether. It dissolves in concentrated sulphuric acid, because the oxygen becomes protonated.
$CH_3CH_2CH_2$—O—$CH_2CH_2CH_3$ (or isomers)

(c) **Z** is a tertiary alcohol. It reacts with sodium metal owing to the —OH group. It does not react with oxidising agents.

8 (a) $CH_3CH_2COCH_3$

(b)

(c) CH_3—CH$=$CH$_2$

(d)

(e) CH_3—Cl

9 Alcohols are miscible with water and have significant hydrogen bonding. They are polar and therefore react with nucleophiles (e.g. HBr). Like water, they act as acids in reaction with sodium – but are much weaker. Alcohols can also act as nucleophiles, but are weaker nucleophiles than water.

Ethers do not show the above properties, as both the H atoms attached to the oxygen have been replaced.

10 (a) Good solvent – removes grease

(b) Miscible with water and has low freezing point

(c) Good solvent for a variety of organic compounds and mixes with water; volatile therefore evaporates easily

(d) Ethanol can be oxidised to ethanoic acid.

(e) Ethanol kills bacteria.

(f) Ethanol is oxidised by potassium dichromate(VI), which changes colour from orange to green.

11 (i) Add metallic sodium. All except **B** would react.

(ii) Heat the others with acidified potassium dichromate(VI). Only **D** would not be oxidised.

(iii) Collect the products of oxidation by distillation. Test with damp indicator paper. **C** would not form an acid (it would be a ketone).

(iv) Dehydrate the remaining two by heating with concentrated sulphuric acid. **A** would form an alkene which could be tested with bromine water.

Chapter 31

1 (a) **B, C and D**

(b) **A and E**

(c) **C**

(d) **A** = 3-methylpentan-2-one; **B** = pentanal; **D** = benzaldehyde; **E** = cyclohexanone

(e) **B, C and D**

(f) **A**

(g) **A and E**

2 Both groups are unsaturated and undergo addition reactions, e.g. with H_2 in the presence of a nickel catalyst. Alkenes, however, react with electrophiles, while the carbonyl group is attacked by nucleophiles.

In the case of the carbonyl group, some of the addition products undergo further reaction, in condensation reactions.

The carbonyls also undergo a greater variety of reactions: oxidation to acids (in the case of aldehydes) and the iodoform reaction. These arise as a result of the effect of the carbonyl group on adjacent atoms.

3 (a) Ethanal is used in the manufacture of other chemicals, including medicines. It is also used to make slug poison and solid fuel.

(b) Ethene is more readily available from the petrochemical industry.

(c) The π-electrons occupy the vacant d orbitals in Pd^{2+}.

(d) The π-electrons are no longer available, and their removal makes the ethene positively charged.

(e) Yes. Propene.

4 (a) $CH_3CH(OH)CH_2CH_3$

(b)

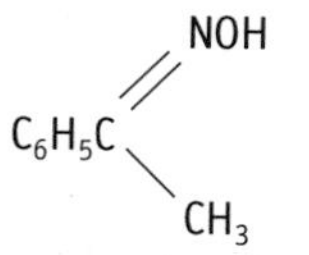

(c) $CH_3CH_2C(OH)(CN)CH_2CH_3$
(d) C_6H_5COOH
(e) CH_3CCl_2CHO

5   A $CH_3CH_2CH_2CHO$

   B   CH₃CH — CHO
         |
         CH₃

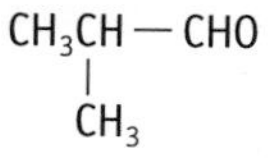

   C $CH_3CH_2COCH_3$

   A and B react with Tollen's reagent, Fehling's solution or other oxidising agents. To distinguish between them, make derivatives (with 2,4-dinitrophenylhydrazine) and measure the melting points.
   C does not react with oxidising agents.

6   (a)   **A** = $OHCCH_2CH_2CHO$

   B = HO   OH
            \   /
             CH — CH₂CH₂ — CH
            /   \
          NC   CN

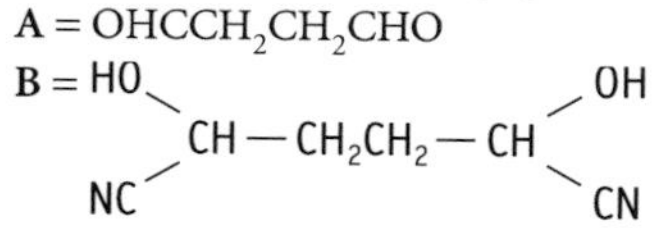

   C = $HOOCCH_2CH_2COOH$

   A to B Nucleophilic addition: ⁻CN attacks carbonyl groups to give addition product
   A to C Oxidation of aldehyde to acid
   Number of moles NaOH added = $\frac{16.9}{1000} \times 1.0 = 0.0169$
   Number of moles C that this reacts with = $\frac{1}{118} = 0.008474$
   ∴ they react in ratio 2 moles NaOH to 1 mole C so C has 2 —COOH groups.

   (b)   C : H : O
           5.36 : 7.11 : 1.7875
           3 : 4 : 1
   so empirical formula = C_3H_4O
   $M_r = 56$ ∴ molecular formula = C_3H_4O
   number of moles X = $\frac{0.1}{56} = 0.0178$
   number of moles $H_2 = \frac{80}{22\,400} = 0.00357$

   ∴ 1 mole **X** reacts with 2 moles H_2; **X** has 2 double bonds.
   X is

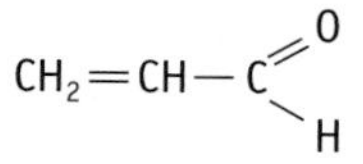

   CH₂=CH — C⟍O
                   ⟍H

   with Fehling's solution   CH₂=CH — C⟍O   is formed (oxidation)
                                           ⟍OH

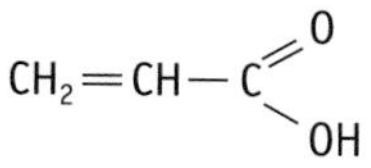

   with H_2 $CH_3CH_2CH_2OH$ is formed (reduction)

7   (a)   The —OH groups can both be oxidised to an aldehyde or an acid:

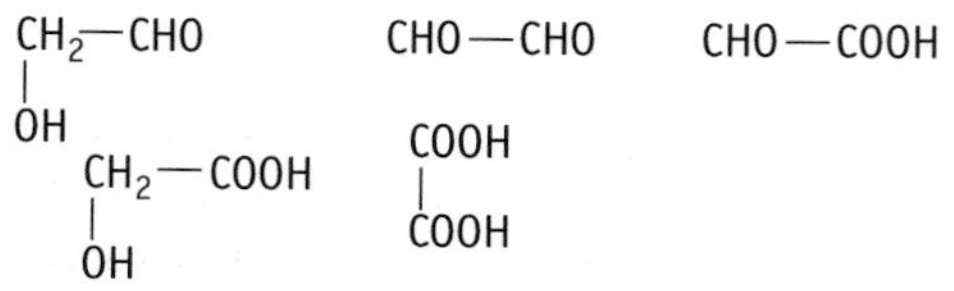

   CH₂—CHO   CHO—CHO   CHO—COOH
   |
   OH
          CH₂—COOH   COOH
          | |
          OH   COOH

   (b)   The C=O group makes the H atoms on the adjacent C atom slightly acidic:
         $CH_3CHO + {}^-OH \rightleftharpoons {}^-CH_2CHO + H_2O$
   This ion removes D^+ from D_2O
         ${}^-CH_2CHO + D_2O \rightleftharpoons CH_2DCHO + {}^-OD$
   This continues until all three H atoms have been replaced
   (c)   The Cl atoms are electron-withdrawing, and make the C atom more susceptible to attack by nucleophiles.

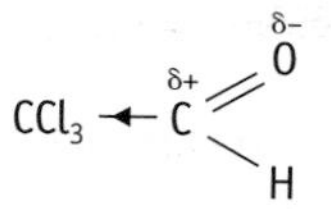

CCl₃ ←$\overset{\delta+}{C}$=$\overset{\delta-}{O}$
                   \H

8   (a) CH_3CH_2OH
       (b) $CH_3CH(OH)CH_3$
       (c) $CH_3C(OH)(CH_3)CH_3$
       (d) CH_3COOH
9   (a) CH_3CH_2CHO propanal
       (b)   3000 C—H   1700 C=O
       (c)   CH₃CH₂CHO
                ↑   ↑   ↑
               1.2 2.5 9.8

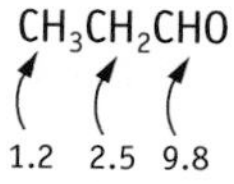

Chapter 32

1   (a)   C
       (b)   D
       (c)   E
       (d)   A Methylpropanoic acid
             B 2-Methylbenzoic acid
             C Methyl propanoate
             D Butanedioic acid
             E Propanoic anhydride
       (e)   C
       (f)   B
       (g)   A, B, D and E
       (h)   E

2   (a) CH_3CH_2COBr
       (b) $C_6H_5CH_2OH$
       (c) $CH_3COONa + CH_3(CH_2)_3CH_2OH$
       (d) $Ca(CH_3COO)_2 + H_2O$
       (e)
             ⬡—NHCOCH₃

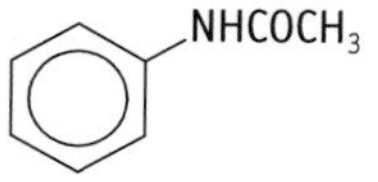

3   (a)   An alkane; for example $C_{17}H_{35}COO^-Na^+$ would form $C_{17}H_{36}$.
       (b)   (i)   The boiling point depends on the intermolecular forces. Butane has the lowest as it has only weak Van der Waals forces between the molecules. Propanal is next as it has dipole–dipole interactions between the molecules. Propan-1-ol and ethanoic acid both have hydrogen bonds between the molecules, but they are stronger in ethanoic acid.
            (ii)   Soap is a salt of an acid. If a strong acid is added, the fatty acid from which the soap was made is displaced. The fatty acid is insoluble so appears as a precipitate.
   $CH_3(CH_2)_{16}COO^-(aq) + HCl(aq) \rightarrow CH_3(CH_2)_{16}COOH(s) + Cl^-(aq)$

4

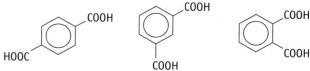

           A and B   **C**

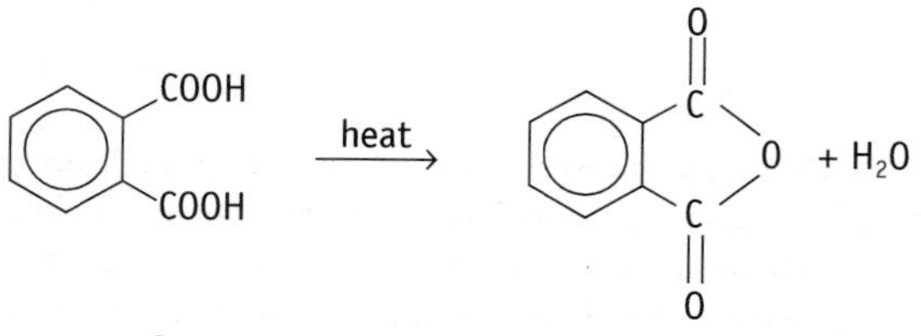

   C   D

5 (a) (i) Reflux ethanal with acidified potassium dichromate(VI) to make ethanoic acid.
   (ii) Reduce ethanal with LiAlH$_4$ in dry ether to form ethanol.
   (iii) Heat ethanoic acid and ethanol, with concentrated sulphuric acid, in a water bath.
  (b) (i) Bubble ethene into concentrated sulphuric acid. Add water and warm to produce ethanol.
   (ii) Oxidise as in (a) to ethanoic acid.
  (c) (i) Add phosphorus pentachloride to make propanoyl chloride.
   (ii) React propanoic acid with sodium hydroxide to make sodium propanoate.
   (iii) React the propanoyl chloride with the sodium propanoate to make propanoic anhydride.
  (d) (i) Heat methylbenzene with alkaline potassium manganate(VII) to make benzoic acid.
   (ii) Gently warm a mixture of benzoic acid, methanol and concentrated sulphuric acid to form the ester.
  (e) (i) React ethanoic acid and phosphorus pentachloride to make ethanoyl chloride.
   (ii) React the ethanoyl chloride with ammonia to produce ethanamide.

6 (a) Add damp universal indicator paper, or solid sodium carbonate. The acid would turn the indicator orange and fizz with sodium carbonate. The aldehyde would do neither.
  (b) Heat each one separately with concentrated sulphuric acid. Methanoic acid would produce carbon monoxide gas. Ethanoic acid would have no reaction.
  (c) The ketone would form a precipitate with 2,4-dinitrophenylhydrazine; the ester would not. Or reflux each with aqueous sodium hydroxide. The ester would slowly hydrolyse and neutralise the sodium hydroxide; the ketone would not.
  (d) Heat both esters with aqueous sodium hydroxide. Methyl ethanoate would give a mixture of methanol and sodium ethanoate; ethyl methanoate would give a mixture of ethanol and sodium methanoate. Distil off the alcohols and carry out an iodoform test. Ethanol would give a positive reaction, methanol would not. Oxidise each alcohol product with acidified potassium dichromate(VI), then test with Tollen's reagent. The product from ethyl methanoate will give a silver mirror.
  (e) Compare the pH values of solutions of the two salts. The fluoro substituted ethanoate will have a lower pH, because fluoroethanoic acid is the stronger.

7 There are six possible compounds.

A CH$_3$CH$_2$CH$_2$COOH B HC$\diagdown\!\!\!\diagup$$^O_{OCH_2CH_2CH_3}$

C CH$_3$C$\diagup\!\!\!\diagdown$$^O_{OCH_2CH_3}$ D CH$_3$CH$_2$C$\diagup\!\!\!\diagdown$$^O_{OCH_3}$

E H–C$\diagup\!\!\!\diagdown$$^O_{O–CH(CH_3)_2}$ F CH$_3$CH(CH$_3$)C$\diagup\!\!\!\diagdown$$^O_{OH}$

B, C, D and E are esters. They could be distinguished by hydroysing each by refluxing with aqueous sodium hydroxide and distilling off the alcohol product. B would give propan-1-ol, C would give ethanol, D would give methanol and E would give propan-2-ol. The alcohols from B, C and E would be dehydrated by heating with concentrated sulphuric acid to form an alkene. The alcohol from D would not. The alcohols from C and E would give positive iodoform reactions. The alcohol from B would not. E is oxidised to a ketone, which does not give a reaction with Tollen's reagent. B, C and D are oxidised to aldehydes, which react with Tollen's reagent.

8 (a) C
  (b) D
  (c) A (and C)
  (d) B and E
  (e) A
  (f) B
  (g) E

9 (a) M_r is greater than expected by 1. One D atom must have replaced an H atom. C$_2$H$_5$OD is formed.
  (b) M_r is greater than expected by 1. One D atom must have replaced an H atom. CH$_3$COOD is formed.
  (c) No useful information
  (d) C$_2$H$_5$OH is formed.
  (e) M_r is greater than expected by 2. CH$_3$CO^{18}OH is formed.
  (f) and (g) Experiment B indicates that the C—O bond joined to the carbonyl group in the ester is the one that breaks. This is consistent with the information given.

Chapter 33

1 (a) B
  (b) A
  (c) E
  (d) F
  (e) A: 2-methylpropanenitrile
   B: propylamine
   C: 1,3-dinitrobenzene
   D: ethylmethylamine
   E: propanamide
   F: tripropylamine

2 (a) (ii) strongest (i) weakest
  Phenylmethylamine (ii) is strongest as the nitrogen is not next to the ring so the lone pairs do not interact with the π electrons and are available for bonding to H$^+$. Methylphenylamine (iii) is weaker than (ii) because the lone pair will interact with the π electron system, but stronger than (i) as the methyl group is electron releasing, and will counteract this to some extent. 1-Methylphenylamine (i) is the weakest as the lone pair will interact with the π electron system and be less available for bonding to H$^+$.
  (b) highest: (iii) strong hydrogen bonding; middle (i): hydrogen bonding, but not as strong as (iii) because N is less electronegative than O; lowest: (ii) no hydrogen bonding; only Van der Waals forces between molecules.

3

Property	Ammonia	Methylamine
Melting point/°C	–77	–92.5
Boiling point/°C	–33	–6.3
pK_a	9.25	10.64
Solubility	Soluble in water	Soluble in water
Basic character	Reacts with HCl to form a salt	Reacts with HCl to form a salt
Smell	Pungent	Pungent and fishy

The properties of methylamine and ammonia are very similar.

4 (a) B, C
  (b) A
  (c) B
  (d) D
  (e) A
  (f) D

5 (a) Chloroethane undergoes nucleophilic substitution. Ammonia is a good nucleophile. Chlorobenzene is not attacked by nucleophiles because the ring of delocalised electrons strengthens the C—Cl bond.

(b) In acidic solution, aminoethanoic acid is protonated and forms a positive ion. In alkaline solution, it exists as the negative ion:

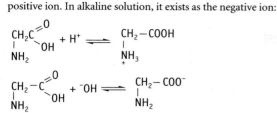

(c) Phenylamine acts as a base in the presence of hydrochloric acid; the nitrogen is protonated so phenylamine exists as the phenylammonium ion which is soluble.

6 (a) (i) React ethene with bromine to make 1,2–dibromoethane.
   (ii) Heat with excess concentrated aqueous ammonia solution in a sealed tube.

(b) (i) Heat benzene with a mixture of concentrated nitric and concentrated sulphuric acids at 50°C to make nitrobenzene.
   (ii) Reduce the nitrobenzene to phenylamine by heating with tin and concentrated hydrochloric acid.
   (iii) Shake the phenylamine with bromine water.

(c) (i) Bubble ethene into a mixture of KBr and concentrated sulphuric acid to make bromoethane.
   (ii) Reflux with KCN in ethanol to make propanenitrile.
   (iii) Reduce with H_2 and Ni catalyst or with $LiAlH_4$ in ether.

(d) React ethylamine with benzoyl chloride. A condensation reaction takes place.

(e) (i) Reduce the nitrobenzene to phenylamine by heating with tin and concentrated hydrochloric acid.
   (ii) React with sodium nitrite and dilute hydrochloric acid above 10°C.

7 (a) moles of X $= \dfrac{0.1}{88} = 0.001136$

moles of $N_2 = \dfrac{50.9}{22\,400} = 0.00227$

∴ 1 mole of X produces 2 moles of N_2

(b) 2

(c) (i) $CH_2(NH_2)CH_2CH_2CH_2NH_2$ (ii) $CH_3-CH-CH_2CH_2-NH_2$
$\qquad\qquad\qquad\qquad\qquad\qquad\qquad\quad\;\; |$
$\qquad\qquad\qquad\qquad\qquad\qquad\qquad\quad NH_2$

(iii) $CH_3-CH_2-CH-CH_2NH_2$
$\qquad\qquad\qquad\quad |$
$\qquad\qquad\qquad\; NH_2$

(d) $50 \times 2 \times \dfrac{0.2}{0.1} = 200\ cm^3$

8

(a)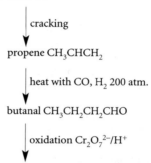

(b)
$CH_3CH_2CH_2CH_2Br \longrightarrow$ $(CH_3CH_2CH_2CH_2)_2NH$
$\qquad\qquad\qquad\qquad\quad\;\; (CH_3CH_2CH_2CH_2)_3N$
$\qquad\qquad\qquad\qquad\quad\;\; CH_3CH_2CH_2CH_2NH_2 \xrightarrow{CH_3COCl} CH_3CH_2CH_2CH_2NH-C-CH_3$
$\qquad\qquad\qquad\qquad\qquad\qquad\qquad\qquad\qquad\qquad\qquad\qquad\qquad\;\; \| $
$\qquad\qquad\qquad\qquad\qquad\qquad\qquad\qquad\qquad\qquad\qquad\qquad\qquad\; O$

(c)
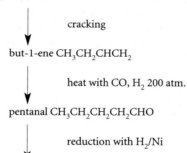

9 (a) 6

(b) The tripeptides are
ser–lys–ala
ser–ala–lys
ala–ser–lys
ala–lys–ser
lys–ala–ser
lys–ser–ala

For example, the structural formula of ser–lys–ala is

$H_2N-\;CH-CO-NH-C-CO-NH-CH-COOH$
$\qquad\quad |\qquad\qquad\qquad |\qquad\qquad\qquad\quad |$
$\qquad\quad CH_2OH\qquad\quad (CH_2)_4NH_2\qquad CH_3$

10

A — benzene ring with CH_3

B — benzene ring with CH_3 and NO_2

C — benzene ring with CH_3 and NH_2

D — benzene ring with CH_3 and NH_3Cl

E — benzene ring with CH_3 and N_2Cl

F — H_3C- ring $-N=N-$ ring $-OH$

G — benzene ring with CH_3 and OH

Chapter 34

1 (a) crude oil

↓ cracking

propene CH_3CHCH_2

↓ heat with CO, H_2 200 atm.

butanal $CH_3CH_2CH_2CHO$

↓ oxidation $Cr_2O_7^{2-}/H^+$

butanoic acid $CH_3CH_2CH_2COOH$

(b) crude oil

↓ cracking

but-1-ene $CH_3CH_2CHCH_2$

↓ heat with CO, H_2 200 atm.

pentanal $CH_3CH_2CH_2CH_2CHO$

↓ reduction with H_2/Ni

pentan-1-ol $CH_3CH_2CH_2CH_2CH_2OH$

↓ PBr_3

1-bromopentane $CH_3(CH_2)_4Br$

↓ reflux KCN/ethanol

hexanenitrile $CH_3(CH_2)_4CN$

↓ reduction with H_2/Ni

hexylamine $CH_3(CH_2)_4CH_2NH_2$

(c) crude oil

    ↓ cracking

propene CH_3CHCH_2

    ↓ $H_2O/H_3PO_4/330°C/60$ atm.

propan-2-ol $CH_3CH(OH)CH_2$

    ↓ oxidation $Cr_2O_7^{2-}/H^+$

propanone CH_3COCH_3

(d) fats

    ↓ hydrolysis

decanoic acid $CH_3(CH_2)_8COOH$

    ↓ reduction $LiAlH_4$/ether

decanal $CH_3(CH_2)_8CHO$

(e) fats

    ↓ hydrolysis

dodecanoic acid $CH_3(CH_2)_{10}COOH$

    ↓ reduction $LiAlH_4$/ether

dodecanol $CH_3(CH_2)_{10}CH_2OH$

    ↓ PBr_3

1-bromododecane $CH_3(CH_2)_{11}Br$

2   (a) butylamine $CH_3(CH_2)_3NH_2$

    ↓ dilute $HCl/NaNO_2$

butan-1-ol $CH_3(CH_2)_3OH$

    ↓ oxidation $Cr_2O_7^{2-}/H^+$

butanoic acid $CH_3(CH_2)_3COOH$

    ↓ concentrated NH_3

butanamide $CH_3(CH_2)_3CONH_2$

    ↓ Hofmann degradation
    (i) Br_2 + dilute KOH(aq)
    (ii) warm with concentrated KOH(aq)

propylamine $CH_3(CH_2)_2NH_2$

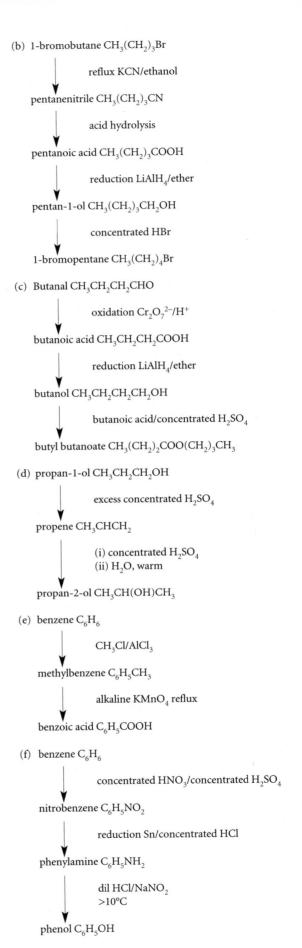

(b) 1-bromobutane $CH_3(CH_2)_3Br$

    ↓ reflux KCN/ethanol

pentanenitrile $CH_3(CH_2)_3CN$

    ↓ acid hydrolysis

pentanoic acid $CH_3(CH_2)_3COOH$

    ↓ reduction $LiAlH_4$/ether

pentan-1-ol $CH_3(CH_2)_3CH_2OH$

    ↓ concentrated HBr

1-bromopentane $CH_3(CH_2)_4Br$

(c) Butanal $CH_3CH_2CH_2CHO$

    ↓ oxidation $Cr_2O_7^{2-}/H^+$

butanoic acid $CH_3CH_2CH_2COOH$

    ↓ reduction $LiAlH_4$/ether

butanol $CH_3CH_2CH_2CH_2OH$

    ↓ butanoic acid/concentrated H_2SO_4

butyl butanoate $CH_3(CH_2)_2COO(CH_2)_3CH_3$

(d) propan-1-ol $CH_3CH_2CH_2OH$

    ↓ excess concentrated H_2SO_4

propene CH_3CHCH_2

    ↓ (i) concentrated H_2SO_4
    (ii) H_2O, warm

propan-2-ol $CH_3CH(OH)CH_3$

(e) benzene C_6H_6

    ↓ $CH_3Cl/AlCl_3$

methylbenzene $C_6H_5CH_3$

    ↓ alkaline $KMnO_4$ reflux

benzoic acid C_6H_5COOH

(f) benzene C_6H_6

    ↓ concentrated HNO_3/concentrated H_2SO_4

nitrobenzene $C_6H_5NO_2$

    ↓ reduction Sn/concentrated HCl

phenylamine $C_6H_5NH_2$

    ↓ dil $HCl/NaNO_2$
    >10°C

phenol C_6H_5OH

Index

Acknowledgements

Photographs

The authors and publisher are grateful to the following for permission to reproduce copyright material. Every effort has been made to trace all relevant copyright holders, but if any have been inadvertently overlooked, the publishers will be pleased to make necessary arrangements at the first opportunity.

Photo research by Zooid Pictures Ltd.

p1 left Manchester City Council
p1 right Bettmann/Corbis UK Ltd.
p2 Philippe Plially/Science Photo Library
p3 Science Museum//Science & Society Picture Library
p5 Cath Gibson/Andy Clarke
p10 Chinch Gryniewicz/Corbis UK Ltd.
p14 Dean Conger/Corbis UK Ltd.
p16 left James L. Amos/Corbis UK Ltd.
p16 centre Adam Woolfitt/Corbis UK Ltd.
p16 right Charles E. Rotkin/Corbis UK Ltd.
p20 Science Photo Library
p21 Science Photo Library
p21 Science Photo Library
p23 top Robert Aberman
p23 centre Robert Aberman
p23 bottom Jeffrey L. Rotman/Corbis UK Ltd.
p26 David Parker/Science Photo Library
p27 Martyn Austin/Corbis UK Ltd.
p28 top Roger Wood/Corbis UK Ltd.
p28 bottom Corbis UK Ltd.
p29 top Peticolas/Megna/Science Photo Library
p29 bottom Catherine Pouedras/Science Photo Library
p32 Dean Conger/Corbis UK Ltd.
p35 Science Photo Library
p38 Bettmann/Corbis UK Ltd.
p40 top Science Photo Library
p40 bottom University of Leicester
p46 Adam Hart-Davis/Science Photo Library
p52 Science Museum/Science & Society Picture Library
p54 Bettmann/Corbis UK Ltd.
p55 Robert Aberman
p56 Science Photo Library
p57 London Aerial Photo Library/Corbis UK Ltd.
p60 Kevin Fleming/Corbis UK Ltd.
p64 top Jonathan Blair/Corbis UK Ltd.
p64 bottom Lester V. Bergman/Corbis UK Ltd.
p65 top left Imperial College/Science Photo Library

p65 top centre Physics Department, Imperial College/Science Photo Library
p65 top right Physics Department, Imperial College/Science Photo Library
p65 bottom Jan Butchofsky-Houser/Corbis UK Ltd.
p67 Michael St. Maur Sheil/Corbis UK Ltd.
p73 University of Leicester
p74 top Vince Streano/Corbis UK Ltd.
p74 bottom Stephen Frink/Corbis UK Ltd.
p76 Bettmann/Corbis UK Ltd.
p80 Martin Dohrn/Science Photo Library
p81 John Reader/Science Photo Library
p85 top Patrick Ward/Corbis UK Ltd.
p85 bottom Philip Harris Education
p87 Sinclair Stammers/Science Photo Library
p89 Shelia Terry/Science Photo Library
p103 Shelia Terry/Science Photo Library
p106 Clause Nuridsany & Maria Perennou/Science Photo Library
p107 Simon Fraser/Science Photo Library
p111 Science Source/Science Photo Library
p115 left David Parker/Science Photo Library
p115 right James L. Amos/Corbis UK Ltd.
p116 left Hulton-Deutsch Collection/Corbis UK Ltd.
p116 right Science Museum/Science & Society Picture Library
p117 Stoe & Cie GmbH
p118 right Professor Asprey/University of Leeds
p119 Kevin R. Morris/Corbis UK Ltd.
p122 top Manfred Kage/Science Photo Library
p122 bottom Paul A. Souders/Corbis UK Ltd.
p123 left De Beers
p123 right De Beers
p124 De Beers
p125 Dr. Jeremy Burgess/Science Photo Library
p126 Dave G. Houser/Corbis UK Ltd.
p135 top Mary Evans Picture Library
p135 bottom Robert Pickett/Corbis UK Ltd.
p137 Graham Ewens/Science Photo Library
p139 Michael Nicholson/Corbis UK Ltd.
p146 Scott Camazine/Science Photo Library
p150 Philip Harris Education
p152 Corbis UK Ltd.
p153 Science Photo Library
p155 Andy Clarke
p164 top AKG - London
p164 bottom Michael St. Maur Sheil/Corbis UK Ltd.
p165 top Bettmann/Corbis UK Ltd.
p165 bottom Paul Almasy/Corbis UK Ltd.
p166 Johnathan Smith/Corbis UK Ltd.

p173 University of California, Berkley
p186 Charles E. Rotkin/Corbis UK Ltd.
p196 Will & Deni McIntyre/Science Photo Library
p197 W & E Electric Vehicles
p198 top Adam Hart-Davis/Science Photo Library
p198 bottom CBS News/Associated Press
p205 Dr. Patricia J. Shulz/Peter Arnold Inc./Science Photo Library
p207 Andy Clarke
p211 Galen Rowell/Corbis UK Ltd.
p214 Martin Bond/Science Photo Library
p217 top left Janez Skok/Corbis UK Ltd.
p217 top right Staffan Widstrand/Corbis UK Ltd.
p217 bottom Yann Arthus-Bertrand/Corbis UK Ltd.
p218 Hulton-Deutsch Collection/Corbis UK Ltd.
p220 top Martin Bond/Science Photo Library
p220 bottom Frank Leather Collection/Corbis UK Ltd.
p221 left British Gypsum
p221 Dean Conger/Corbis UK Ltd.
p224 right Treak Cliff Cavern
p225 Richard T. Nowitz/Corbis UK Ltd.
p235 top Bob Winsett/Corbis UK Ltd.
p235 bottom Andy Clarke
p236 top Galen Rowell/Corbis UK Ltd.
p236 bottom Ian Beames/Ecoscene
p239 left Charles O'Rear/Corbis UK Ltd.
p239 right De Beers
p241 top Lowell Georgia/Corbis UK Ltd.
p241 bottom Tim Hazael/Science Photo Library
p242 John Heseltine/Corbis UK Ltd.
p243 Ferranti Electronics/A. Sternberg/Science Photo Library
p247 bottom left Victoria & Albert Museum
p247 bottom right Victoria & Albert Museum
p247 top Robbie Jack/Corbis UK Ltd.
p251 Andy Clarke
p255 Charles D. Winters/Science Photo Library
p261 J. Allan Cash Photo Library
p265 top Martin Land/Science Photo Library
p265 bottom Science Museum/Science & Society Picture Library
p266 top Charles E. Rotkin/Corbis UK Ltd.
p266 bottom Heine Schneebeli/Science Photo Library
p267 left Philippa Lewis/Corbis UK Ltd.
p267 right Harvey Pincis/Science Photo Library